McGraw-Hill

Dictionary of
Materials
Science

McGraw-Hill

New York   Chicago   San Francisco   Lisbon   London   Madrid
Mexico City   Milan   New Delhi   San Juan   Seoul   Singapore
Sydney   Toronto

Materials in this dictionary are derived from the McGRAW-HILL DICTIONARY OF SCIENTIFIC AND TECHNICAL TERMS, Sixth Edition, copyright © 2003 by the McGraw-Hill Companies, Inc., and the McGRAW-HILL ENCYCLOPEDIA OF SCIENCE & TECHNOLOGY, Ninth Edition, copyright © 2002 by The McGraw-Hill Companies, Inc. All rights reserved.

McGRAW-HILL DICTIONARY OF MATERIALS SCIENCE, copyright © 2003 by The McGraw-Hill Companies, Inc. All rights reserved. Printed in the United States of America. Except as permitted under the United States Copyright Act of 1976, no part of this publication may be reproduced or distributed in any form or by any means, or stored in a database or retrieval system, without the prior written permission of the publisher.

2 3 4 5 6 7 8 9 0   DOC/DOC   0 9 8 7 6 5 4

ISBN 0-07-142176-9

 This book is printed on recycled, acid-free paper containing a minimum of 50% recycled, de-inked fiber.

This book was set in Helvetica Bold and Novarese Book by TechBooks, Fairfax, Virginia. It was printed and bound by RR Donnelley, The Lakeside Press.

McGraw-Hill books are available at special quantity discounts to use as premiums and sales promotions, or for use in corporate training programs. For more information, please write to the Director of Special Sales, Professional Publishing, McGraw-Hill, Two Penn Plaza, New York, NY 10121-2298. Or contact your local bookstore.

Library of Congress Cataloging-in-Publication Data

McGraw-Hill dictionary of materials science.
     p.   cm.
  ISBN 0-07-142176-9
  1. Materials science—Dictionaries.

  TA402.D57   2003
  620.1′1′03—dc21   2003051203

On the cover: Simulation of methane dissociation on nickel surface. (Accelrys http://www.accelrys.com)

Contents

Preface

The *McGraw-Hill Dictionary of Materials Science* provides a compendium of more than 11,000 terms that are central to materials science and the numerous related branches of engineering and science. The coverage includes terminology in the fields of biomaterials, chemical engineering and other branches of engineering, cryogenics, crystallography, electricity, electromagnetism, electronics, fluid mechanics, graphic arts, inorganic and organic chemistry, metallurgy, optics, physical chemistry, physics, polymer chemistry, textiles, and thermodynamics.

The definitions are derived from the *McGraw-Hill Dictionary of Scientific and Technical Terms*, 6th edition (2003). Each one is classified according to the field with which it is primarily associated. The pronunciation of each term is provided along with synonyms, acronyms, and abbreviations where appropriate. A guide to the use of the *Dictionary* is included, explaining the alphabetical organization of terms, the format of the book, cross referencing, and how synonyms, variant spellings, abbreviations, and similar information are handled. A pronunciation key is also provided to assist the reader. An appendix provides conversion tables for commonly used scientific units as well as charts and listings of useful mathematical, engineering, and scientific data.

Many of the terms used in materials science are found in specialized dictionaries and glossaries; the *McGraw-Hill Dictionary of Materials Science*, however, aims to provide the user with the convenience of a single, comprehensive reference. It is the editors' hope that it will serve the needs of scientists, engineers, students, teachers, librarians, and writers for high-quality information, and that it will contribute to scientific literacy and communication.

Mark D. Licker
Publisher

Staff

Mark D. Licker, Publisher—Science

Elizabeth Geller, Managing Editor
Jonathan Weil, Senior Staff Editor
David Blumel, Staff Editor
Alyssa Rappaport, Staff Editor
Charles Wagner, Digital Content Manager
Renee Taylor, Editorial Assistant

Roger Kasunic, Vice President—Editing, Design, and Production

Joe Faulk, Editing Manager
Frank Kotowski, Jr., Senior Editing Supervisor

Ron Lane, Art Director

Thomas G. Kowalczyk, Production Manager
Pamela A. Pelton, Senior Production Supervisor

Henry F. Beechhold, Pronunciation Editor
Professor Emeritus of English
Former Chairman, Linguistics Program
The College of New Jersey
Trenton, New Jersey

How to Use the Dictionary

ALPHABETIZATION. The terms in the *McGraw-Hill Dictionary of Materials Science* are alphabetized on a letter-by-letter basis; word spacing, hyphen, comma, solidus, and apostrophe in a term are ignored in the sequencing. Also ignored in the sequencing of terms (usually chemical compounds) are italic elements, numbers, small capitals, and Greek letters. For example, the following terms would be alphabetized as:

abalyn	high-alloy steel
A stage	high polymer
Beer's law	mho
beeswax	micelle

FORMAT. The basic format for a defining entry provides the term in boldface, the field is small capitals, and the single definition in lightface:

> **term** [FIELD] Definition.

A term may be followed by multiple definitions, each introduced by a boldface number:

> **term** [FIELD] **1.** Definition. **2.** Definition. **3.** Definition.

A term may have difinitions in two or more fields:

> **term** [ENG] Definition. [MATER] Definition.

A simple cross-reference entry appears as:

> **term** *See* another term.

A cross reference may also appear in combination with definitions:

> **term** [ENG] *See* another term. [MATER] Definition.

CROSS REFERENCING. A cross-reference entry directs the user to the defining entry. For example, the user looking up "branched chain" finds:

> **branched chain** *See* side chain.

The user then turns to the "S" terms for the definition. Cross references are also made from variant spellings, acronyms, abbreviations, and symbols.

> **ABS** *See* acrylonitrile butadiene styrene resin.
> **aluminium** *See* aluminum.
> **at. wt** *See* atomic weight.
> **Au** *See* gold.

CHEMICAL FORMULAS. Chemistry definitions may include either an empirical formula (say, for abietic acid, $C_{20}H_{30}O_2$) or a line formula (for acrylonitrile, CH_2CHCN), whichever is appropriate.

ALSO KNOWN AS ..., etc. A definition may conclude with a mention of a synonym of the term, a variant spelling, an abbreviation for the term, or other such information, introduced by "Also known as ...," "Also spelled ...," "Abbreviated ...," "Symbolized ...," "Derived from ...." When a term has more than one definition, the positioning of any of these phrases conveys the extent of applicability. For example:

> **term** [MATER] **1.** Definition. Also known as synonym. **2.** Definition. Symbolized T.

In the above arrangement, "Also known as ..." applies only to the first definition; "Symbolized ..." applies only to the second definition.

> **term** [ENG] **1.** Definition. **2.** Definition. [MATER] Definition. Also known as synonym.

In the above arrangement, "Also known as ..." applies only to the second field.

> **term** [ENG] Also known as synonym. **1.** Definition. **2.** Definition. [MATER] Definition.

In the above arrangement, "Also known as ..." applies only to both definitions in the first field.

> **term** Also known as synonym. [ENG] **1.** Definition. **2.** Definition. [MATER] Definition.

In the above arrangement, "Also known as ..." applies to all definitions in both fields.

Fields and Their Scope

[BIOMATER] **biomaterials**—The technological area involving natural or synthetic nondrug materials that are compatible with living tissue and used to enhance, treat, or replace organs, tissues, and functions in living organisms.

[CHEM] **chemistry**—The scientific study of the properties, composition, and structure of matter, the changes in structure and composition of matter, and accompanying energy changes.

[CHEM ENG] **chemical engineering**—A branch of engineering that involves the design of chemical products and processes for a wide range of engineering fields, including petroleum, materials science, agricultural, energy, environmental, pharmaceutical, and biomedical.

[CIV ENG] **civil engineering**—The planning, design, construction, and maintenance of fixed structures and ground facilities for industry, for transportation, for use and control of water, for occupancy, and for harbor facilities.

[CRYO] **cryogenics**—The science of producing and maintaining very low temperatures, of phenomena at those temperatures, and of technical operations performed at very low temperatures.

[CRYSTAL] **crystallography**—The branch of science that deals with the geometric description of crystals, their internal arrangement, and their properties.

[ELEC] **electricity**—The science of physical phenomena involving electric charges and their effects when at rest and when in motion.

[ELECTR] **electronics**—The technological area involving the manipulation of voltages and electric currents through the use of various devices for the purpose of performing some useful action with the currents and voltages; this field is generally divided into analog electronics, in which the signals to be manipulated take the form of continuous currents or voltages, and digital electronics, in which signals are represented by a finite set of states.

[ELECTROMAG] **electromagnetism**—The branch of physics dealing with the observations and laws relating electricity to magnetism, and with magnetism produced by an electric current.

[ENG] **engineering**—The science by which the properties of matter and the sources of power in nature are made useful to humans in structures, machines, and products.

[FL MECH] **fluid mechanics**—The science concerned with fluids, either at rest or in motion, and dealing with pressures, velocities, and accelerations in the fluid, including fluid deformation and compression or expansion.

[GRAPHICS] **graphic arts**—The fine and applied arts of representation, decoration, and writing or printing on flat surfaces together with the techniques and crafts associated with each; includes painting, drawing, engraving, etching, lithography, photography, and printing arts.

[INORG CHEM] **inorganic chemistry**—The branch of chemistry that deals with the reactions and properties of all chemical elements and their compounds, excluding hydrocarbons but usually including carbides and other simple carbon compounds (such as CO_2, CO, and HCN).

[MATER] **materials**—A multidisciplinary field concerned with the properties and uses of materials in terms of composition, structure, and processing.

[MECH] **mechanics**—The branch of physics which seeks to formulate general rules for predicting the behavior of a physical system under the influence of any type of interaction with its environment.

[MET] **metallurgy**—The branch of engineering concerned with the production of metals and alloys, their adaptation to use, and their performance in service; and the study of chemical reactions involved in the processes by which metals are produced, and of the laws governing the physical, chemical, and mechanical behavior of metallic materials.

[OPTICS] **optics**—The study of phenomena associated with the generation, transmission, and detection of electromagnetic radiation in the spectral range extending from the long-wave edge of the x-ray region to the short-wave edge of the radio region; and the science of light.

[ORG CHEM] **organic chemistry**—The study of the structure, preparation, properties, and reactions of carbon compounds.

[PHYS] **physics**—The science concerned with those aspects of nature which can be understood in terms of elementary principles and laws.

[PHYS CHEM] **physical chemistry**—The branch of chemistry that deals with the interpretation of chemical phenomena and properties in terms of the underlying physical processes, and with the development of techniques for their investigation.

[POLYM CHEM] **polymer chemistry**—The study of the composition, structure, and properties of high-molecular-weight materials made up of chains of repeating units; these materials may be organic, inorganic, or organometallic, and may be synthetic or natural in origin.

[SOLID STATE] **solid-state physics**—The branch of physics centering on the physical properties of solid materials; it is usually concerned with the properties of crystalline materials only, but it is sometimes extended to include the properties of glasses or polymers.

[TEXT] **textiles**—The area of industry involving the production of fibers, filaments, or yarn, and the cloth made from these materials.

[THERMO] **thermodynamics**—The branch of physics which seeks to derive, from a few basic postulates, relations between properties of substances, especially those which are affected by changes in temperature, and a description of the conversion of energy from one form to another.

Pronunciation Key

Vowels

a	as in b**a**t, th**a**t
ā	as in b**ai**t, cr**a**te
ä	as in b**o**ther, f**a**ther
e	as in b**e**t, n**e**t
ē	as in b**ee**t, tr**ea**t
i	as in b**i**t, sk**i**t
ī	as in b**i**te, l**igh**t
ō	as in b**oa**t, n**o**te
ȯ	as in b**ough**t, t**au**t
u̇	as in b**oo**k, p**u**ll
ü	as in b**oo**t, p**oo**l
ə	as in b**u**t, sof**a**
au̇	as in cr**ow**d, p**ow**er
ȯi	as in b**oi**l, sp**oi**l
yə	as in form**u**la, spectac**u**lar
yü	as in f**ue**l, m**u**le

Semivowels/Semiconsonants

w	as in **w**ind, t**w**in
y	as in **y**et, on**i**on

Stress (Accent)

'	precedes syllable with primary stress
ˌ	precedes syllable with secondary stress
¦	precedes syllable with variable or indeterminate primary/ secondary stress

Consonants

b	as in **b**i**b**, dri**bb**le
ch	as in **ch**arge, stre**tch**
d	as in **d**og, ba**d**
f	as in **f**ix, sa**f**e
g	as in **g**ood, si**g**nal
h	as in **h**and, be**h**ind
j	as in **j**oint, di**g**it
k	as in **c**ast, bri**ck**
k̲	as in Ba**ch** (used rarely)
l	as in **l**oud, be**ll**
m	as in **m**ild, su**mm**er
n	as in **n**ew, de**n**t
n̲	indicates nasalization of preceding vowel
ŋ	as in ri**ng**, si**ng**le
p	as in **p**ier, sli**p**
r	as in **r**ed, sca**r**
s	as in **s**ign, po**s**t
sh	as in **s**ugar, **sh**oe
t	as in **t**imid, ca**t**
th	as in **th**in, brea**th**
th̲	as in **th**en, brea**the**
v	as in **v**eil, wea**v**e
z	as in **z**oo, crui**s**e
zh	as in bei**g**e, trea**s**ure

Syllabication

·	indicates syllable boundary when following syllable is unstressed

a See ampere.

A See ampere.

a axis [CRYSTAL] One of the crystallographic axes used as reference in crystal description, usually oriented horizontally, front to back. { 'ā 'ak₁sis }

abalyn [ORG CHEM] A liquid rosin that is a methyl ester of abietic acid; prepared by treating rosin with methyl alcohol; used as a plasticizer. { 'ab·ə₁lin }

Abelian quantum Hall state [CRYO] A quantum Hall state that contains two or more components of incompressible fluid, has a filling factor equal to p/q, where q is not divisible by p, and has a topological order that can be described as a pattern of dancing steps of the electrons and can be characterized by a symmetric matrix and a charge vector, both with integer entries. { ə'bē·lē·ən ¦kwän·təm ¦hól ₁stāt }

abherent [MATER] A substance that inhibits a material from adhering to itself or another material. Also known as abhesive. { ab'hēr·ənt }

abhesive See abherent. { ab'hē·ziv }

abietic acid [ORG CHEM] $C_{20}H_{30}O_2$ A tricyclic, crystalline acid obtained from rosin; used in making esters for plasticizers. { ₁a·bē'et·ik 'as·əd }

ablating material See ablative agent. { ə'blād·iŋ mə₁tir·ē·əl }

ablative agent [MATER] A material from which the surface layer is to be removed, often for the purpose of dissipating extreme heat energy, as in space vehicles reentering the earth's atmosphere. Also known as ablating material; ablative material; ablator. { ə·blā·div 'ā·jənt }

ablative material See ablative agent. { 'a·blə·div mə'tir·ē·əl }

ablator See ablative agent. { 'ab₁lād·ər }

aboundikro See Sapele mahogany. { ə'baún·dē ₁krō }

about-sledge [MET] A large hammer that is utilized in blacksmithing. { ə'baút ₁slej }

abradant See abrasive. { ə'brād·ənt }

abrasion [ENG] **1.** The removal of surface material from any solid through the frictional action of another solid, a liquid, or a gas or combination thereof. **2.** A surface discontinuity brought about by roughening or scratching. { ə'brā·zhən }

abrasion resistance [MATER] The ability of a surface to resist wearing due to contact with another surface moving with respect to it. { ə'brā·zhən rə'zis·təns }

abrasion-resistance index [MATER] In vulcanized material or synthetic rubber compounds, a measure of abrasion resistance relative to a standard rubber compound under defined conditions. { ə'brā·zhən rə'zis·təns 'in·deks }

abrasive [MATER] Also known as abradant. **1.** A material used, usually as a grit sieved by a specified mesh but also as a solid shape or as a paste or slurry or air suspension, for grinding, honing, lapping, superfinishing, polishing, pressure blasting, or barrel tumbling. **2.** A material sintered or formed into a solid mass such as a hone or a wheel disk, cone, or burr for grinding or polishing other materials. **3.** Having qualities conducive to or derived from abrasion. { ə'brā·siv }

abrasiveness [MATER] **1.** The property of a material causing wear of a surface. **2.** The quality or characteristic of being able to scratch, abrade, or wear away another material. { ə'brās·əv·nəs }

abrasive paper [MATER] Tough paper to whose surface an abrasive, such as sand or emery, has been bonded for use in grinding or polishing. { ə'brās·əv 'pā·pər }

abrasive sand [MATER] Grit used as abrasive, usually graded as to which sieve mesh it will pass through. { ə'brās·əv 'sand }

ABS See acrylonitrile butadiene styrene resin.

absolute index of refraction See index of refraction. { 'ab·sə₁lüt 'in₁deks əv ri'frak·shən }

absolute refractive constant See index of refraction. { 'ab·sə₁lüt ri'frak·tiv 'kän·stənt }

absolute scale See absolute temperature scale. { 'ab·sə₁lüt ₁skāl }

absolute temperature [THERMO] **1.** The temperature measurable in theory on the thermodynamic temperature scale. **2.** The temperature in Celsius degrees relative to the absolute zero at $-273.16°C$ (the Kelvin scale) or in Fahrenheit degrees relative to the absolute zero at $-459.69°F$ (the Rankine scale). { 'ab·sə₁lüt 'tem·prə·chür }

absolute temperature scale [THERMO] A scale with which temperatures are measured relative to absolute zero. Also known as absolute scale. { 'ab·sə₁lüt 'tem·prə·chür ₁skāl }

absolute viscosity [FL MECH] The tangential force per unit area of two parallel planes at unit distance apart when the space between them is filled with a fluid and one plane moves with unit velocity in its own plane relative to the other. Also known as coefficient of viscosity. { 'ab·sə‚lüt vis'käs·ə·dē }

absolute zero [THERMO] The temperature of −273.16°C, or −459.69°F, or 0 K, thought to be the temperature at which molecular motion vanishes and a body would have no heat energy. { 'ab·sə‚lüt 'zir·ō }

absorb [CHEM] To take up a substance. [ELECTROMAG] To take up energy from radiation. [PHYS] To take up matter or radiation. { əb'sȯrb }

absorbance [PHYS CHEM] The common logarithm of the fraction of incident light that is transmitted. { əb'sȯr·bəns }

absorbed charge [ELEC] Charge on a capacitor which arises only gradually when the potential difference across the capacitor is maintained, due to gradual orientation of permanent dipolar molecules. { əb'sȯrbd 'chärj }

absorbency [CHEM] Penetration of one substance into another. { əb'sȯr·bən·sē }

absorbent [MATER] A material which, in contact with a liquid or gas, extracts one or more substances for which it has an affinity, and is altered physically or chemically during the process. { əb'sȯr·bənt }

absorbent cotton [MATER] A cotton fiber that absorbs water because its natural waxes have been removed. { əb'sȯr·bənt 'kät·ən }

absorbent paper [MATER] Paper capable of absorbing and holding liquids by the capillarity of the pores between or within the closely matted cellulosic fibers. { əb'sȯr·bənt 'pā·pər }

absorber [ELECTR] A material or device that takes up and dissipates radiated energy; may be used to shield an object from the energy, prevent reflection of the energy, determine the nature or the radiation, or selectively transmit one or more components of the radiation. { əb'sȯr·bər }

absorptance [PHYS] The ratio of the total unabsorbed radiation to the total incident radiation; equal to one (unity) minus the transmittance. { əb'sȯrp·təns }

absorption [CHEM] The taking up of matter in bulk by other matter, as in dissolving of a gas by a liquid. [ELEC] The property of a dielectric in a capacitor which causes a small charging current to flow after the plates have been brought up to the final potential, and a small discharging current to flow after the plates have been short-circuited, allowed to stand for a few minutes, and short-circuited again. Also known as dielectric soak. [ELECTROMAG] Taking up of energy from radiation by the medium through which the radiation is passing. { əb'sȯrp·shən }

absorption band [PHYS] A range of wavelengths or frequencies in the electromagnetic spectrum within which radiant energy is absorbed by a substance. { əb'sȯrp·shən ‚band }

absorption coefficient [PHYS] If a flux through a material decreases with distance x in proportion to e^{-ax}, then a is called the absorption coefficient.

Also known as absorption factor; absorption ratio; coefficient of absorption. { əb'sȯrp·shən ‚kō·ə'fish·ənt }

absorption current [ELEC] The component of a dielectric current that is proportional to the rate of accumulation of electric charges within the dielectric. { əb'sȯrp·shən 'kər·ənt }

absorption curve [PHYS] A graph showing the curvilinear relationship of the variation in absorbed radiation as a function of wavelength. { əb'sȯrp·shən ‚kərv }

absorption factor See absorption coefficient. { əb'sȯrp·shən ‚fak·tər }

absorption ratio See absorption coefficient. { əb'sȯrp·shən ‚rā·shō }

absorptivity [THERMO] The ratio of the radiation absorbed by a surface to the total radiation incident on the surface. { əb‚sȯrp'tiv·əd·ē }

abtesla See gauss. { ab'tes·lə }

ac See alternating current.

Ac₀ [MET] The temperature at which a magnetic change occurs in cementite; the Curie point of cementite.

Ac₁ [MET] The temperature at which austenite begins to be formed upon heating a steel.

Ac₂ [MET] The Curie point of ferrite.

Ac₃ [MET] The temperature at which the transformation of ferrite to austenite is completed upon heating a steel.

Ac₄ [MET] The temperature at which delta iron is formed from gamma iron upon heating a steel.

Ac_cm [MET] The temperature at which the solution of cementite in austenite is completed upon heating a hypereutectoid steel.

acacia gum See gum arabic. { ə'kā·shyə ‚gəm }

acapau [MATER] A type of ebony obtained from *Voucapapoua americana* in the Amazon valley; used for fine inlay work. Also known as partridge wood. Properly spelled acapu. { ‚äk·ə'pau̇ }

acapu See acapau. { ‚ak·ə'pü }

acaroid resin [ORG CHEM] A gum resin from aloelike trees of the genus *Xanthorrhoea* in Australia and Tasmania; used in varnishes and inks. Also known as gum accroides; yacca gum. { 'a·kə‚rȯid 'rez·ən }

accelerated aging [ENG] Hastening the deterioration of a product by a laboratory procedure in order to determine long-range storage and use characteristics. { ak'sel·ə‚rād·əd 'āj·iŋ }

accelerated life test [ENG] Operation of a device, circuit, or system above maximum ratings to produce premature failure; used to estimate normal operating life. { ak'sel·ər‚ā·dəd 'līf ‚test }

accelerated weathering [ENG] A laboratory test used to determine, in a short period of time, the resistance of a paint film or other exposed surface to weathering. { ak'sel·ər‚ā·dəd 'weth·ər·iŋ }

accelerating agent [MATER] **1.** A substance which increases the speed of a chemical reaction. **2.** A compound which hastens and improves the curing of rubber. { ak'sel·ər‚ād·iŋ 'ā'jənt }

accelerator [MATER] **1.** Any substance added to stucco, plaster, mortar, concrete, cement, and so on to hasten the set. **2.** In the vulcanization process, a substance, added with a curing agent, to speed processing and enhance physical characteristics of a vulcanized material. { ak'sel·ə,rād·ər }

accentuator [MATER] A material that acts to increase the selectivity or intensity of a stain. { ak'sen·chə,wād·ər }

acceptor [SOLID STATE] An impurity element that increases the number of holes in a semiconductor crystal such as germanium or silicon; aluminum, gallium, and indium are examples. Also known as acceptor impurity; acceptor material. { ak'sep·tər }

acceptor atom [SOLID STATE] An atom of a substance added to a semiconductor crystal to increase the number of holes in the conduction band. { ak'sep·tər 'ad·əm }

acceptor impurity See acceptor. { ak'sep·tər im 'pyür·ə·dē }

acceptor level [SOLID STATE] An energy level in a semiconductor that results from the presence of acceptor atoms. { ak'sep·tər ,lev·əl }

acceptor material See acceptor. { ak'sep·tər mə'tir·ē·əl }

accumulator See storage battery. { ə'kyü·myə ,lād·ər }

accumulator battery See storage battery. { ə'kyü·myə,lād·ər 'bad·ə·rē }

acetal resins [POLYM CHEM] Linear, synthetic resins produced by the polymerization of a monomeric aldehyde, formaldehyde (acetal homopolymers), or of one or more aldehydes with cyclic trimers of aldehydes (acetal copolymers); hard, tough plastics. Also known as polyacetals. { 'as·ə,tal 'rez·ənz }

acetate [TEXT] The official name for the textile fiber produced from partially hydrolyzed cellulose acetate. Formerly known as acetate rayon. { 'as·ə,tāt }

acetate dye [CHEM] 1. Any of a group of water-insoluble azo or anthroquinone dyes used for dyeing acetate fibers. 2. Any of a group of water-insoluble amino azo dyes that are treated with formaldehyde and bisulfate to make them water-soluble. { 'as·ə,tāt ,dī }

acetate film [MATER] A transparent cellulose acetate resin sheet that is resistant to grease, oil, and dust; used for photographic film, magnetic tapes, and packaging. May contain dyes or pigments if used for other applications such as stage lighting. { 'as·ə,tāt ,film }

acetate rayon See acetate. { 'as·ə,tāt 'rā,än }

acetylated cotton [MATER] Mildewproof cotton made by the chemical conversion of part of the raw cotton fiber to cellulose acetate. { ə'sed·əl ,āt·əd 'kät·ən }

acetylene welding See oxyacetylene welding. { ə'sed·əl,ēn 'weld·iŋ }

a-c fracture [CRYSTAL] A type of tension fracture lying parallel to the a-c fabric plane and normal to plane b in a crystal. { 'a·sē 'frak·chər }

achiral molecules [ORG CHEM] Molecules which are superimposable to their mirror images. { ¦ā,kī·rəl 'mäl·ə,kyülz }

achromatic [OPTICS] Capable of transmitting light without decomposing it into its constituent colors. { ¦a·krə¦mad·ik }

achromatic color [OPTICS] A color that has no hue or saturation but only brightness, such as white, black, and various shades of gray. { ¦a·krə ¦mad·ik 'kəl·ər }

acicular powder [MET] A metal powder whose grains are ellipsoids. { ə'sik·yə·lər 'paúd·ər }

acid [CHEM] 1. Any of a class of chemical compounds whose aqueous solutions turn blue litmus paper red, react with and dissolve certain metals to form salts, and react with bases to form salts. 2. A compound capable of transferring a hydrogen ion in solution. 3. A substance that ionizes in solution to yield the positive ion of the solvent. 4. A molecule or ion that combines with another molecule or ion by forming a covalent bond with two electrons from the other species. { 'as·əd }

acid acceptor [ORG CHEM] A stabilizer compound added to plastic and resin polymers to combine with trace amounts of acids formed by decomposition of the polymers. { 'as·əd ək 'sep·tər }

acid-base catalysis [CHEM] The increase in speed of certain chemical reactions due to the presence of acids or bases. { 'as·əd 'bās kə 'tal·ə·sis }

acid-base equilibrium [CHEM] The condition where the measurable ratio of acidic and basic ions in a solution is temporarily stable. { 'as·əd 'bās ,ik·wə'lib·rē·əm }

acid-base pair [CHEM] A concept in the Brønsted theory of acids and bases; the pair consists of the source of the proton (acid) and the base generated by the transfer of the proton. { 'as·əd 'bās 'pār }

acid bottom and lining [MET] A melting furnace's inner bottom and lining composed of materials that at operating temperatures of the furnace react with the melt and slag to give an acid reaction; examples of materials are sand, siliceous rock, and silica brick. { 'as·əd 'bät·əm an 'līn·iŋ }

acid brittleness [MET] Low ductility of a metal due to its absorption of hydrogen gas, which may occur during an electrolytic process or during cleaning. Also known as hydrogen embrittlement. { 'as·əd 'brit·əl·nəs }

acid bronze [MET] A copper-tin alloy containing lead and nickel; used in pumping equipment. { 'as·əd 'bränz }

acid calcium phosphate See calcium phosphate. { 'as·əd 'kal·sē·əm 'fäs,fāt }

acid cell [PHYS CHEM] An electrolytic cell whose electrolyte is an acid. { 'as·əd ,sel }

acid cure [MET] The removal of some gangue carbonates from uranium ore by agitation with sulfuric acid prior to the leaching process. { 'as·əd ,kyür }

acid electrolyte [INORG CHEM] A compound, such as sulfuric acid, that dissociates into ions when dissolved, forming an acidic solution that conducts an electric current. { 'as·əd ə'lek· trə,līt }

acid halide |ORG CHEM| A compound of the type RCOX, where R is an alkyl or aryl radical and X is a halogen. { 'as·əd 'hā,līd }

acidic group |ORG CHEM| The radical COOH, present in organic acids. { ə'sid·ik 'grüp }

acidic oxide |INORG CHEM| An oxygen compound of a nonmetal, for example, SO_2 or P_2O_5, which yields an oxyacid with water. { ə'sid·ik 'äk ,sīd }

acidity |CHEM| The state of being acid. { ə'sid· ə·tē }

acid lead |MET| A 99.9% pure commercial lead made by adding copper to fully refined lead. { 'as·əd 'led }

acid number See acid value. { 'as·əd ¦nəm·bər }

acid open-hearth process |MET| A steelmaking process employing an open-hearth furnace lined with siliceous-type refractories. { 'as·əd ,ō·pən 'härth ,prä·səs }

acid phosphate |INORG CHEM| A mono- or dihydric phosphate, for example, M_2HPO_4 or MH_2PO_4, where M represents a metal atom. { 'as·əd 'fäs,fāt }

acid pickle |MATER| Industrial waste water that is the spent liquor from a chemical process used to clean metal surfaces. { 'as·əd 'pik·əl }

acid potassium sulfate See potassium bisulfate. { 'as·əd pə'tas·ē·əm 'səl,fāt }

acid process |CHEM ENG| In paper manufacture, a pulp digestion process that uses an acidic reagent, for example, a bisulfite solution containing free sulfur dioxide. |MET| A melting process carried out in a furnace lined with acidic materials which combine readily with the oxides in the ore. { 'as·əd ,prä·səs }

acid-proof coating |MATER| Material in liquid form suitable for application, by spraying, to the wall of projectile or bomb cavities to protect the metal from attack by explosives or other shell fillers. { 'as·əd ¦prüf 'kōt·iŋ }

acid refractory |MATER| A refractory that is composed principally of silica and reacts at high temperatures with bases such as lime, alkalies, and basic oxides. { 'as·əd rə'frak·tə·rē }

acid-resistant |MATER| Able to withstand chemical reaction with, or degeneration by, acids. { 'as·əd rə¦zis·tənt }

acid salt |CHEM| A compound derived from an acid and base in which only a part of the hydrogen is replaced by a basic radical; for example, the acid sulfate $NaHSO_4$. { 'as·əd ,sòlt }

acid slag |MET| Furnace slag in which there is more silica and silicates than lime and magnesia. { 'as·əd ,slag }

acid solution |CHEM| An aqueous solution containing more hydrogen ions than hydroxyl ions. { 'as·əd sə'lü·shən }

acid steel |MET| Steel produced in a melting furnace employing siliceous-type refractories. { 'as·əd 'stēl }

acid value |ORG CHEM| A number indicating the amount of nonesterified carboxylic acid present in a sample as determined by alkaline titration. Also known as acid number. { 'as·əd 'val·yü }

acid wash |MATER| A solution of phosphoric acid applied to steel parts that removes and neutralizes the alkaline solutions used for grease removal after machining; also leaves a metallic phosphate coating which accepts paint well and, of itself, provides a degree of protection against rust. { 'as·əd ,wash }

acieration |MET| Electrolytic coating of a thin metal plate with iron; the iron hardens to steellike strength. { ,a·sī·ə'rā·shən }

aclastic |OPTICS| Having the property of not refracting light. { ¦ā'klas·tik }

acoustical insulation board |MATER| A porous board designed or used for acoustical applications or for sound-insulating construction. { ə'küs·tə·kəl in·sə'lā·shən ,bòrd }

acoustical material |MATER| Any natural or synthetic material that absorbs sound; acoustic tile is an example. { ə'küs·tə·kəl mə'tir·ē·əl }

acoustic branch |SOLID STATE| One of the parts of the dispersion relation, frequency as a function of wave number, for crystal lattice vibrations, representing vibration at low (acoustic) frequencies. { ə'küs·tik ,branch }

acoustic insulation |MATER| A material used to diminish sound energy that passes through it or strikes its surface. { ə'küs·tik in·sə'lā·shən }

acoustic lens |MATER| Selected materials shaped to refract sound waves in accordance with the principles of geometrical optics, as is done for light. Also known as lens. { ə'küs·tik 'lenz }

acoustic mode |SOLID STATE| The type of crystal lattice vibrations which for long wavelengths act like an acoustic wave in a continuous medium, but which for shorter wavelengths approach the Debye frequency, showing a dispersive decrease in phase velocity. { ə'küs·tik 'mōd }

acoustic phonon |SOLID STATE| A quantum of excitation of an acoustic mode of vibration. { ə'küs·tik 'fōn,än }

acoustic plaster |MATER| Plaster having good acoustic absorbing properties; it contains metal which, upon contact with water, evolves gas to aerate the mass. { ə'küs·tik 'plas·tər }

acoustic tile |MATER| A thin, often decorative tile with sound-absorbing properties, used to cover ceilings and walls. { ə'küs·tik 'tīl }

acoustooptical cell |ELEC| An electric-to-optical transducer in which an acoustic or ultrasonic electric input signal modulates or otherwise acts on a beam of light. { ə¦küs·tō¦äp·tə·kəl 'sel }

acoustooptical material |MATER| A material in which the refractive index or some other optical property can be changed by an acoustic wave. { ə¦küs·tō¦äp·tə·kəl mə'tir·ē·əl }

a-c plane |CRYSTAL| A plane at right angles to the surface of movement in a crystal. { 'ā'sē 'plān }

acrolein dimer |ORG CHEM| A flammable, water-soluble liquid used as an intermediate for resins, dyestuffs, and pharmaceuticals. { ə'krōl·ē·ən 'dī·mər }

acrylamide |ORG CHEM| $CH_2CHCONH_2$ Colorless, odorless crystals with a melting point of 84.5°C; soluble in water, alcohol, and acetone; used in organic synthesis, polymerization, sewage treatment, ore processing, and permanent press fabrics. { ə'kril·ə‚mīd }

acrylamide copolymer [POLYM CHEM] A thermosetting resin formed of acrylamide with other resins, such as the acrylic resins. { ə'kril·ə‚mīd kō'päl·ə·mər }

acrylate [ORG CHEM] A salt or ester of acrylic acid. [POLYM CHEM] See acrylate resin. { 'ak·rə‚lāt }

acrylate resin [POLYM CHEM] Acrylic acid or ester polymer with a —CH₂–CH(COOR)– structure; used in paints, sizings and finishes for paper and textiles, adhesives, and plastics. Also known as acrylate. { 'ak·rə‚lāt 'rez·ən }

acrylate rubber [MATER] A member of a class of elastomers based on acrylate esters. { 'ak·rə‚lāt 'rəb·ər }

acrylic acid [ORG CHEM] CH₂CHCOOH An easily polymerized, colorless, corrosive liquid used as a monomer for acrylate resins. { ə'kril·ik 'as·əd }

acrylic ester [ORG CHEM] An ester of acrylic acid. { ə'kril·ik 'es·tər }

acrylic fiber [TEXT] Any of numerous synthetic textile fibers made by polymerization of acrylonitrile. { ə'kril·ik 'fī·bər }

acrylic resin [ORG CHEM] A thermoplastic synthetic organic polymer made by the polymerization of acrylic derivatives such as acrylic acid, methacrylic acid, ethyl acrylate, and methyl acrylate; used for adhesives, protective coatings, and finishes. { ə'kril·ik 'rez·ən }

acrylic rubber [POLYM CHEM] Synthetic rubber containing acrylonitrile; for example, nitrile rubber. { ə'kril·ik 'rəb·ər }

acrylic syrup [MATER] Lucite in a liquid form; used as a low-pressure laminating resin; produces stiff, strong, tough laminates that can be adapted to bright or translucent colors. { ə'kril·ik 'sir·əp }

acrylonitrile [ORG CHEM] CH₂CHCN A colorless liquid compound used in the manufacture of acrylic rubber and fibers. Also known as vinyl-cyanide. { ‚ak·rə‚lō'nī·trəl }

acrylonitrile-butadiene rubber See nitrile rubber. { ‚ak·rə‚lō'nī·trəl ‚byüd·ə'dī‚ēn 'rəb·ər }

acrylonitrile butadiene styrene resin [POLYM CHEM] A polymer made by blending acrylonitrile-styrene copolymer with a butadiene-acrylonitrile rubber or by interpolymerizing polybutadiene with styrene and acrylonitrile; combines the advantages of hardness and strength of the vinyl resin component with the toughness and impact resistance of the rubbery component. Abbreviated ABS. { ‚ak·rə‚lō'nī·trəl ‚byüd·ə'dī‚ēn 'stī·rēn 'rez·ən }

acrylonitrile copolymer [POLYM CHEM] Oil-resistant synthetic rubber made by polymerization of acrylonitrile with compounds such as butadiene or acrylic acid. { ‚ak·rə‚lō'nī·trəl kō'päl·ə·mər }

acrylonitrile rubber See nitrile rubber. { ‚ak·rə‚lō'nī·trəl 'rəb·ər }

actinic [PHYS] Pertaining to electromagnetic radiation capable of initiating photochemical reactions, as in photography or the fading of pigments. { ‚ak'tin·ik }

actinic achromatism [OPTICS] **1.** The design of a photographic lens system so that light sources at the wavelength of the Fraunhofer D line near 589 nanometers and the G line at 430.8 nanometers are focused at the same point and produce images of the same size. **2.** The design of an astronomical lens system so that light sources at the wavelength of the Fraunhofer F line at 486.1 nanometers and the G line at 430.8 nanometers are focused at the same point and produce images of the same size. Also known as FG achromatism. { ‚ak'tin·ik ‚ā'krōm·ə‚tiz·əm }

actinic glass [OPTICS] Glass that transmits more of the visible components of incident radiation and less of the infrared and ultraviolet components. { ‚ak'tin·ik 'glas }

actinide series [CHEM] The group of elements of atomic number 89 through 103. Also known as actinoid elements. { 'ak·tə‚nīd 'sir‚ēz }

actinium [CHEM] A radioactive element, symbol Ac, atomic number 89; its longest-lived isotope is ²²⁷Ac with a half-life of 21.7 years; the element is trivalent; chief use is, in equilibrium with its decay products, as a source of alpha rays. { ak'tin·ē·əm }

actinodielectric [ELEC] Of a substance, exhibiting an increase in electrical conductivity when electromagnetic radiation is incident upon it. { ‚ak·tə‚nō‚dī·ə'lek·trik }

actinoelectricity [ELEC] The electromotive force produced in a substance by electromagnetic radiation incident upon it. { ‚ak·tə‚nō·i‚lek'tris·ə·dē }

actinoid elements See actinide series. { 'ak·tə‚nóid 'el·ə·məns }

actinometer [ENG] Any instrument used to measure the intensity of radiant energy, particularly that of the sun. { ‚ak·tə'näm·əd·ər }

activate [ELEC] To make a cell or battery operative by addition of a liquid. { 'ak·tə‚vāt }

activated alumina [MATER] Highly porous, granular aluminum oxide that preferentially absorbs liquids from gases and vapors, and moisture from some liquids; also used as a catalyst or catalyst carrier, as an absorbent to remove fluorides from drinking water, and in chromatography. { 'ak·tə‚vād·əd ə'lüm·ə·nə }

activated carbon [MATER] A powdered, granular, or pelleted form of amorphous carbon characterized by very large surface area per unit volume because of an enormous number of fine pores. Also known as activated charcoal. { 'ak·tə‚vād·əd 'kär·bən }

activated charcoal See activated carbon. { 'ak·tə‚vād·əd 'char‚kōl }

activated clay [MATER] Bentonite, or other clay, treated with acid to enhance its ability to absorb or bleach. { 'ak·tə‚vād·əd klā }

activated diffusion [SOLID STATE] Movement of atoms, ions, or lattice defects across a potential barrier in a solid. { 'ak·də‚vād·əd di'fyü·zhən }

activated rosin flux [MATER] Soldering flux containing activating agents which promote wetting by the solder. { 'ak·tə‚vād·əd 'räz·ən 'fläks }

activated sintering |MET| Sintering of a metal powder compact in contact with a gaseous atmosphere which reacts with the metal surfaces and enhances the joining of metal particles. { 'ak·tə‚vād·əd 'sin·tər·iŋ }

activating reagent |MATER| Material added to another material or mixture so that a physical or chemical change will take place more rapidly or completely. { 'ak·tə‚vād·iŋ ‚rē'ā·jənt }

activation |CHEM| Treatment of a substance by heat, radiation, or activating reagent to produce a more complete or rapid chemical or physical change. |MET| **1.** A process of facilitating the separation and collection of ore powders by the use of substances which change the response of the particle surfaces to a flotation fluid. **2.** A process that increases the rate of pressing and heating a metal powder into cohesion. { ‚ak·tə'vā·shən }

activation energy |PHYS CHEM| The energy, in excess over the ground state, which must be added to an atomic or molecular system to allow a particular process to take place. { ‚ak·tə'vā·shən 'en·ər·jē }

activator |CHEM| **1.** A substance that increases the effectiveness of a rubber vulcanization accelerator; for example, zinc oxide or litharge. **2.** A trace quantity of a substance that imparts luminescence to crystals; for example, silver or copper in zinc sulfide or cadmium sulfide pigments. { 'ak·tə‚vād·ər }

active center |CHEM| **1.** Any one of the points on the surface of a catalyst at which the chemical reaction is initiated or takes place. **2.** See active site. { 'ak·tiv 'sen·tər }

active material |ELEC| **1.** A fluorescent material used in screens for cathode-ray tubes. **2.** An energy-storing material, such as lead oxide, used in the plates of a storage battery. **3.** A material, such as the iron of a core or the copper of a winding, that is involved in energy conversion in a circuit. **4.** In a battery, the chemically reactive material in either of the electrodes that participates in the charge and discharge reactions. |ELECTR| The material of the cathode of an electron tube that emits electrons when heated. { 'ak·tiv mə'tir·ē·əl }

active site |CHEM| The effective site at which a given heterogeneous catalytic reaction can take place. Also known as active center. { 'ak·tiv 'sīt }

active solid |CHEM| A porous solid possessing adsorptive properties and used for chromatographic separations. { 'ak·tiv 'säl·əd }

active substrate |SOLID STATE| A semiconductor or ferrite material in which active elements are formed; also a mechanical support for the other elements of a semiconductor device or integrated circuit. { 'ak·tiv 'səb‚strāt }

activity series |CHEM| A series of elements that have similar properties—for example, metals—arranged in descending order of chemical activity. { ak'tiv·əd·ē ‚sir·ēz }

adapter |MET| A connecting piece, usually made of fireclay, between a horizontal zinc retort and the condenser in which the molten zinc collects. { ə'dap·tər }

addition polymer |POLYM CHEM| A polymer formed by the chain addition of unsaturated monomer molecules, such as olefins, with one another without the formation of a by-product, as water; examples are polyethylene, polypropylene, and polystyrene. Also known as addition resin. { ə'di·shən 'päl·ə·mər }

addition polymerization |POLYM CHEM| A reaction initiated by an anion, cation, or radical in which a large number of monomer units are added rapidly (a chain reaction) until terminated by some mechanism, forming a high-molecular-weight polymer in a very short time; an example is the free-radical polymerization of propylene to polypropylene. { ə‚dish·ən pə‚lim·ə·rə'zā·shən }

addition reaction |ORG CHEM| A type of reaction of unsaturated hydrocarbons with hydrogen, halogens, halogen acids, and other reagents, so that no change in valency is observed and the organic compound forms a more complex one. { ə'di·shən rē'ak·shən }

addition resin See addition polymer. { ə'di·shən 'rez·ən }

addition solid solution |CRYSTAL| Random addition of atoms or ions in the interstices within a crystal structure. { ə'di·shən 'säl·əd sə'lü·shən }

additive |MATER| **1.** A substance added to another to strengthen or otherwise alter it for the purpose of improving the performance of the finished product. **2.** See admixture. { 'ad·əd·iv }

additive level |MATER| The total percentage of all the additives in an oil sample. { 'ad·əd·iv 'lev·əl }

additive primary colors |OPTICS| The three colors, usually red, green, and blue, which are mixed together in an additive process. { 'ad·əd·iv 'prīm·ə·rē 'kəl·ərz }

additive process |OPTICS| The process of producing colors by mixing lights of additive primary colors in various proportions. { 'ad·əd·iv ‚prä·səs }

adduct |CHEM| **1.** A chemical compound that forms from chemical addition of two species; for example, reaction of butadiene with styrene forms an adduct, 4-phenyl-1-cyclohexene. **2.** The complex compound formed by association of an inclusion complex. { 'a‚dəkt }

adherend |MATER| **1.** A body attached to another by means of an adhesive substance. **2.** The surface to which an adhesive adheres. { ‚ad'hir·ənd }

adhesion |ELECTROMAG| Any mutually attractive force holding together two magnetic bodies, or two oppositely charged nonconducting bodies. |ENG| Intimate sticking together of metal surfaces under compressive stresses by formation of metallic bonds. |MECH| The force of static friction between two bodies, or the effects of this force. |PHYS| The tendency, due to intermolecular forces, for matter to cling to other matter. { ad'hē·zhən }

adhesional work |THERMO| The work required to separate a unit area of a surface at which two substances are in contact. Also known as work of adhesion. { ad'hē·zhən·əl ,wərk }

adhesive |MATER| A substance used to bond two or more solids so that they act or can be used as a single piece; examples are resins, formaldehydes, glue, paste, cement, putty, and polyvinyl resin emulsions. { ad'hēz·iv }

adhesive bond |MECH| The forces such as dipole bonds which attract adhesives and base materials to each other. { ad'hēz·iv 'bänd }

adhesive bonding |ENG| The fastening together of two or more solids by the use of glue, cement, or other adhesive. { ad'hēz·iv 'bänd·iŋ }

adhesive strength |ENG| The strength of an adhesive bond, usually measured as a force required to separate two objects of standard bonded area, by either shear or tensile stress. { ad'hēz·iv 'streŋkth }

adhesive tape |MATER| Tape coated with a substance that binds or sticks to a surface. { ad'hēz·iv ,tāp }

adiabatic |THERMO| Referring to any change in which there is no gain or loss of heat. { ¦ad·ē·ə ¦bad·ik }

adiabatic compression |THERMO| A reduction in volume of a substance without heat flow, in or out. { ¦ad·ē·ə¦bad·ik kəm'presh·ən }

adiabatic cooling |THERMO| A process in which the temperature of a system is reduced without any heat being exchanged between the system and its surroundings. { ¦ad·ē·ə¦bad·ik 'kül·iŋ }

adiabatic demagnetization |CRYO| A method of cooling paramagnetic salts to temperatures of 10^{-3} K; the sample is cooled to the boiling point of helium in a strong magnetic field, thermally isolated, and then removed from the field to demagnetize it. Also known as Giaque-Debye method; magnetic cooling; paramagnetic cooling. { ¦ad·ē·ə¦bad·ik ,de,mag·nəd·ə'zā·shən }

adiabatic envelope |THERMO| A surface enclosing a thermodynamic system in an equilibrium which can be disturbed only by long-range forces or by motion of part of the envelope; intuitively, this means that no heat can flow through the surface. { ¦ad·ē·ə¦bad·ik 'en·və,lōp }

adiabatic expansion |THERMO| Increase in volume without heat flow, in or out. { ¦ad·ē·ə ¦bad·ik ik'span·chən }

adiabatic process |THERMO| Any thermodynamic procedure which takes place in a system without the exchange of heat with the surroundings. { ¦ad·ē·ə¦bad·ik prä·səs }

adiabatic vaporization |THERMO| Vaporization of a liquid with virtually no heat exchange between it and its surroundings. { ¦ad·ē·ə¦bad·ik ,vā·pər·ə'zā·shən }

adipic acid |ORG CHEM| HOOC(CH$_2$)$_4$COOH A colorless crystalline dicarboxylic acid, sparingly soluble in water; used in nylon manufacture. { ə'dip·ik 'as·əd }

admiralty brass |MET| An alloy containing 71% copper, 28% zinc, and 0.75–1% tin for additional corrosion resistance. { 'ad·mrəl·tē 'bras }

admittance |ELEC| A measure of how readily alternating current will flow in a circuit; the reciprocal of impedance, it is expressed in siemens. { əd'mit·əns }

admixture |MATER| A material (other than aggregate, cement, or water) added in small quantities to concrete to produce some desired change in properties. Also known as additive. { ¦ad ¦miks·chər }

adobe brick |MATER| A brick made from clay mixed with straw, of large but varying dimensions, roughly molded and sun-dried. { ə'dō·bē 'brik }

adsorbate |CHEM| A solid, liquid, or gas which is adsorbed as molecules, atoms, or ions by such substances as charcoal, silica, metals, water, and mercury. { ad'sȯr,bāt }

adsorbent |CHEM| A solid or liquid that adsorbs other substances; for example, charcoal, silica, metals, water, and mercury. { ad'sȯr·bənt }

adsorption |CHEM| The surface retention of solid, liquid, or gas molecules, atoms, or ions by a solid or liquid, as opposed to absorbtion, the penetration of substances into the bulk of the solid or liquid. { ad'sȯrp·shən }

Ae |MET| The temperature of equilibrium for corresponding phase change Ac; Ae$_{cm}$, Ae$_3$, and Ae$_4$ are similarly related to Ac$_{cm}$, Ac$_3$, and Ac$_4$.

Ae$_1$ |MET| The temperature, attained without thermal lag, at which cementite-austenite conversion takes place in a hypoeutectoid steel or ferrite-austenite conversion takes place in a hypereutectoid steel.

aeolotropic See anisotropic. { ¦ē·ə·lō¦träp·ik }

aeolotropy See anisotropy. { ,ē·ə'lä·trə·pē }

aerated concrete |MATER| Concrete made by adding substances which will liberate gases by chemical reaction; entrapped gases reduce its density and enhance insulation properties. { 'e,rād·əd ,kän'krēt }

aerator |MET| A device which decreases the density of sand by mixing it with air, thus facilitating the movement of sand particles in packing. { 'e,rād·ər }

aerobic adhesive |MATER| A two-compartment structural adhesive that contains an acrylic resin and retains its adhesiveness in the presence of oxygen. { e¦rō,bik ad'hēz·iv }

aerogel |CHEM| A porous solid formed from a gel by replacing the liquid with a gas with little change in volume so that the solid is highly porous. { 'e·rō,jel }

aerosol |CHEM| A suspension of small particles in a gas; the particles may be solid or liquid or a mixture of both; aerosols are formed by the conversion of gases to particles, the disintegration of liquids or solids, or the suspension of powdered material. { 'e·rə,sȯl }

affinity |CHEM| The extent to which a substance or functional group can enter into a chemical reaction with a given agent. Also known as chemical affinity. { ə'fin·əd·ē }

A15 compound See A15 phase. { ¦ā'fif¦tēn 'käm ¦paund }

A15 phase |SOLID STATE| An intermetallic compound having the chemical formula A$_3$B, where

A represents a transition element, and a crystal structure in which the B atoms are located at the corners and in the center of a cubic unit cell, while the A atoms are arranged in pairs on the cube faces. Also known as A15 compound. { ¦ā-fif'tēn ‚fāz }

African greenheart [MATER] The yellowish-brown wood of *Piptadena africana*; used for shipbuilding and dock timbers. Also known as dahoma. { 'af·ri·kən 'grēn‚härt }

afterblow [MET] A blow in a Bessemer process, occurring after the flame for the removal of carbon has dropped, to remove phosphorus. { 'af·tər‚blō }

afterchromed dye [MATER] A dye that is improved in color quality or fastness after the textile is dyed by treatment with sodium dichromate, copper sulfate, or similar materials. { 'af·tər ‚krōmd 'dī }

aftertack [MATER] Of a paint film, tackiness or stickiness which remains over a long period of time. Also known as residual tack. { 'af·tər‚tak }

Ag See silver.

agar [MATER] A gelatinous product extracted from certain red algae and used chiefly as a gelling agent in culture media. { 'äg·ər }

agar attar See oriental linaloe. { 'äg·ər 'at·ər }

agate glass [MATER] Multicolor glass made by blending glasses of two or more colors or by rolling transparent glass into powdered glass of various colors. { 'ag·ət ‚glas }

age hardening [MET] Increasing the hardness of an alloy by a relatively low-temperature heat treatment that causes precipitation of ‚components or phases of the alloy from the supersaturated solid solution. Also known as precipitation hardening. { ‚āj 'härd·ən·iŋ }

agel fiber [MATER] Fiber from the leaves and stem of the gebang palm; used for rope, sailcloth, and fishnets. { 'ā·jəl 'fī·bər }

age softening [MET] The loss of strength and hardness which takes place at room temperature in some alloys because of the spontaneous decrease of residual stresses in the strain-hardened structure. { 'āj 'sof·niŋ }

agglomerate [MET] A rigid mass of metallic particles that have been joined together by a powder metallurgical technique, such as sintering. { ə'gläm·ə·rət }

agglomeration [MET] Conversion of small pieces of low-grade iron ore into larger lumps by application of heat. { ə‚gläm·ə'rā·shən }

aggregate [CHEM] A group of atoms or molecules that are held together in any way, for example, a micelle. [MATER] The natural sands, gravels, and crushed stone used for mixing with cementing material in making mortars and concretes. { 'ag·rə·gət }

aggregation [CHEM] A process that results in the formation of aggregates. { ‚ag·rə'gā·shən }

aging [CHEM] All irreversible structural changes that occur in a precipitate after it has formed. [ELEC] Allowing a permanent magnet, capacitor, meter, or other device to remain in storage for a period of time, sometimes with a voltage applied, until the characteristics of the device become essentially constant. [ELECTROMAG] Change in the magnetic properties of iron with passage of time, for example, increase in the hysteresis. [ENG] **1.** The changing of the characteristics of a device due to its use. **2.** Operation of a product before shipment to stabilize characteristics or detect early failures. [MATER] **1.** Change in the properties of any substance with time. **2.** Change occurring in powders or slips with the passage of time. **3.** Curing of ceramic materials, such as clays and glazes, by a definite period of time under controlled storage conditions. [MET] **1.** Change in properties of an alloy or metal which generally proceeds slowly at room temperatures and faster at elevated temperatures. **2.** Strain relief, occurring through long storage outdoors under varying temperatures, of iron castings intended for use as toolroom plates or lathe-bed supports. **3.** A second heat treatment of an alloy at a lower temperature, causing precipitation of the unstable phase and increasing hardness, strength, and electrical conductivity. { 'āj·iŋ }

agreement residual [SOLID STATE] The sum of the differences between the observed and calculated structure amplitudes of a crystal, for all observed reflections, divided by the sum of the observed amplitudes. { ə¦grē·mənt rə'zij·ə·wəl }

air [CHEM] A predominantly mechanical mixture of a variety of individual gases forming the earth's enveloping atmosphere. { er }

air-acetylene welding [MET] A gas-welding process in which the heat is obtained from the combustion of acetylene and air. { er ə'sed·əl·ən 'weld·iŋ }

air-bend die [MET] A device for forming metals in which only the two edges of the lower section are in contact with the metal. { 'er ‚bend 'dī }

air-blown asphalt [MATER] A bituminous product made by reacting the residual oil of petroleum distillation with air at 400–600°F (204–316°C). { 'er ¦blōn 'as‚fölt }

air brick [MATER] A brick or brick-sized metal box that is hollow or perforated and used for ventilation. { 'er ‚brik }

air carbon arc cutting [MET] A carbon arc cutting process in which the molten metal in the cut is removed by an airblast. { 'er 'kär·bən ‚ärk 'kəd·iŋ }

air content [MATER] The volume of air voids in cement paste, mortar, or concrete, exclusive of pore space in aggregate particles; usually expressed as a percentage of total volume of the mixture. { 'er 'kän‚tent }

air-cooled blast-furnace slag [MET] The material resulting from solidification of molten blast-furnace slag under atmospheric conditions. { 'er ‚küld ¦blast ¦fər·nəs ‚slag }

air-cure [CHEM ENG] To vulcanize at ordinary room temperatures, or without the aid of heat. { 'er ‚kyür }

air-dried lumber [MATER] Wood dried by exposure to air under natural conditions; usually has a moisture content not greater than 24. Also known as air-seasoned lumber; natural-seasoned lumber. { ¦er ¦drīd 'ləm·bər }

air-dried strength [MET] Tenacity of a sand mixture after a core or mold has been dried in air without application of heat. { ¦er ¦drīd 'streŋkth }

air drying [ENG] Removing moisture from a material by exposure to air to the extent that no further moisture is released on contact with air; important in lumber manufacture. { ¦er 'drī·iŋ }

air-entrained cement [MATER] Cement with improved qualities due to the introduction of air bubbles in its preparation. { ¦er in'trānd sə'ment }

air-entraining agent [MATER] An admixture, usually a resin, soap, or grease, for portland cement or concrete to effect air entrainment and, thus, superior properties. { ¦er in¦trān·iŋ 'ā·jənt }

air feed [MET] In thermal spraying, transmittal of powdered material by air pressure through the gun into the heat source. { 'er ¸fēd }

air furnace [MET] **1.** A furnace using a natural air draft. **2.** A furnace in which the metal is melted by a flame originating from fuel burned at one end, passing over the hearth in the middle, and exiting at the other end. { 'er ¸fər·nəs }

air gas [MATER] A gaseous fuel made by blowing air through a coal or coke bed so that CO_2 is reduced to CO. { 'er ¸gas }

air-hardening steel [MET] A steel whose content of carbon and other alloying elements is sufficient for the steel to harden fully by cooling in air or any other atmosphere from a temperature above its transformation range. Also known as self-hardening steel. { 'er ¦härd·niŋ 'stēl }

air knife [ENG] A device that uses a thin, flat jet of air to remove the excess coating from freshly coated paper. [MET] A device that employs a high-velocity airstream to separate metallic particles with different densities by shifting the trajectory of lighter particles more than that of heavier particles. { 'er ¸nīf }

air-knife coating [ENG] An even film of coating left on paper after treatment with an air knife. { 'er ¦nīf ¸kōd·iŋ }

airless spraying [ENG] The spraying of paint by means of high fluid pressure and special equipment. Also known as hydraulic spraying. { 'er·ləs 'sprā·iŋ }

air quench [MET] In a heat-treatment process, the rapid cooling of a metal by a blast of·cold air. { 'er ¸kwench }

air-seasoned lumber *See* air-dried lumber. { 'er ¦sēz·ənd 'ləm·bər }

air-sensitive crystal [CHEM] A crystal that decomposes when exposed to air. { 'er¦sen·səd·iv 'krist·əl }

air-setting mortar [MATER] Mortar that sets in air at atmospheric pressure and ordinary temperatures. { 'er ¦sed·iŋ ¦mord·ər }

air-standard cycle [THERMO] A thermodynamic cycle in which the working fluid is considered to be a perfect gas with such properties of air as a volume of 12.4 cubic feet per pound at 14.7

pounds per square inch (approximately 0.7756 cubic meter per kilogram at 101.36 kilopascals) and 492°R and a ratio of specific heats of 1:4. { ¦er ¦stan·dərd 'sī·kəl }

air-suspension encapsulation [CHEM ENG] A technique for microencapsulation of various types of solid particles; the particles undergo a series of cycles in which they are first suspended by a vertical current of air while they are sprayed with a solution of coating material, and are then moved by the airstream into a region where they undergo a drying treatment. Also known as Wurster process. { 'ər sə¦spen·shən in¸kap·sə'lā·shən }

air void [MATER] A space which is filled with air in cement paste, mortar, or concrete. { 'er ¸void }

akund [MATER] A silky cotton fiber from the shrub *Calotropis gigantea*; used in combination with kapok fiber for insulation. { 'ä¸künd }

Al *See* aluminum.

alabaster glass [MATER] A glass that contains small inclusions of different diffractive indexes and shows no color reaction to light. { 'al·ə ¸bas·tər ¸glas }

Alaska cedar [MATER] Wood from *Chamaecyparis nootkaensis* or *Cupressus sitkaensis*; has a fine, uniform, straight grain and is light and moderately hard; used for furniture and boat building. Also known as Sitka cypress; yellow cedar; yellow cypress. { ə'las·kə 'sēd·ər }

albarium [MATER] A white lime used for stucco; made by burning marble. { al'bar·ē·əm }

albite law [CRYSTAL] A rule specifying the orientation of alternating lamellae in multiple twin feldspar crystals; the twinning plane is brachypinacoid and is common in albite. { 'al¸bīt ¸lo }

albolite [MATER] A plastic cement composed principally of magnesia and silica. { 'al·bə¸līt }

albumin glue [MATER] A bonding agent composed of soluble dried blood with minor additives and giving strong, durable bonds when coagulated in plywood joints at temperatures of 160–180°F (71–82°C). { ¸al'byü·mən 'glü }

alcian blue [MATER] A copper phthalocyanin derivative used as a stain for connective tissue mucins and a number of epithelial mucins and as a gelling agent for lubricating fluids. { 'al·shən 'blü }

alcogel [CHEM] A gel formed by an alcosol. { 'al·kə¸jel }

alcohol [ORG CHEM] Any member of a class of organic compounds in which a hydrogen atom of a hydrocarbon has been replaced by a hydroxy (—OH) group. { 'al·kə¸hol }

alcosol [CHEM] Mixture of an alcohol and a colloid. { 'al·kə¸sol }

aldehyde [ORG CHEM] One of a class of organic compounds containing the CHO radical. { 'al·də¸hīd }

aldehyde polymer [POLYM CHEM] Any of the plastics based on aldehydes, such as formaldehyde, acetaldehyde, butyraldehyde, or acrylic aldehyde (acrolein). { 'al·də¸hīd 'päl·ə·mər }

alfenol [MET] A permeability alloy that has 16% aluminum and 84% iron; it is brittle and at 572°F

(300°C) can be rolled into thin sheets; used for transformer cores and tape recorder heads. { 'al·fə,nōl }

algicide [MATER] A chemical used to kill algae. { 'al·jə,sīd }

algin [MATER] A hydrophilic polysaccharide extracted from brown algae, such as giant kelp. [ORG CHEM] *See* sodium alginate. { 'al·jən }

alginic acid sodium salt *See* sodium alginate. { al'jin·ik 'as·əd 'sōd·ē·əm 'sȯlt }

alicyclic [ORG CHEM] **1.** Having the properties of both aliphatic and cyclic substances. **2.** Referring to a class of saturated hydrocarbon compounds whose structures contain one ring. Also known as cycloaliphatic; cycloalkane. **3.** Any one of the compounds of the alicyclic class. Also known as cyclane. { ¦al·ə¦sī·klik }

aliphatic [ORG CHEM] Of or pertaining to any organic compound of hydrogen and carbon characterized by a straight chain of the carbon atoms; three subgroups of such compounds are alkanes, alkenes, and alkynes. { ¦al·ə¦fad·ik }

alite [MATER] A constituent of portland cement clinker consisting mostly of calcium silicate. { 'ā,līt }

alizarin yellow [MATER] A dye useful as an acid-base indicator; solutions change color from yellow (acid) to purple (basic) in the pH range 10.1 to 12.0. { ə'liz·ə·rən 'yel·ō }

alkali [CHEM] Any compound having highly basic qualities. { 'al·kə,lī }

alkali cellulose [MATER] Product of wood pulp steeped with sodium hydroxide; first step in manufacture of viscose rayon and other cellulosics. { 'al·kə,lī 'sel·yə,lōs }

alkalide [INORG CHEM] A member of a class of crystalline salts with an alkali metal atom. { 'al·kə,līd }

alkali lead [MET] An alloy of lead hardened with small quantities of alkali metals; used as bearing metals. { 'al·kə,lī 'led }

alkali lignin [MATER] A type of lignin produced by treating the black liquor from the soda process with acid; used as an extender in the negative plates of storage batteries, in asphalt, and in paperboard products. { 'al·kə,lī 'lig·nən }

alkali metal [CHEM] Any of the elements of group I in the periodic table: lithium, sodium, potassium, rubidium, cesium, and francium. { 'al·kə,lī ,med·əl }

alkaline [CHEM] **1.** Having properties of an alkali. **2.** Having a pH greater than 7. { 'al·kə,līn }

alkaline cleaner [MET] An aqueous solution of an alkali used for metal cleaning. { 'al·kə,līn 'klēn·ər }

alkaline-earth oxide [INORG CHEM] The stable reaction of an alkaline-earth element with oxygen. { 'al·kə,līn 'ərth 'äk,sīd }

alkalinity [CHEM] The property of having excess hydroxide ions in solution. { ,al·kə'lin·ə·dē }

alkali reactivity [MATER] Susceptibility of a concrete aggregate to alkali-aggregate reaction. { 'al·kə,lī ,rē·ak'tiv·əd·ē }

alkali-resisting paint [MATER] A paint, such as one made with a synthetic resin, that does not undergo saponification when used in such places as bathrooms or on such materials as new concretes. { ¦al·kə'li rə,zist·iŋ 'pānt }

alkane [ORG CHEM] A member of a series of saturated aliphatic hydrocarbons having the empirical formula C_nH_{2n+2}. Also known as paraffin; paraffinic hydrocarbon. { 'al,kān }

alkene [ORG CHEM] One of a class of unsaturated aliphatic hydrocarbons containing one or more carbon-to-carbon double bonds. { 'al ,kēn }

alkoxy [ORG CHEM] An alkyl radical attached to a molecule by oxygen, such as the ethoxy radical. { al'käk,sē }

alkyd paint [MATER] A paint using an alkyd resin as the vehicle for the pigment. { 'al·kəd ,pānt }

alkyd resin [ORG CHEM] A class of adhesive resins made from unsaturated acids and glycerol. { 'al·kəd 'rez·ən }

alkyl [ORG CHEM] An organic group that results from removal of a hydrogen atom from a saturated hydrocarbon; may be represented in a chemical formula by R–. { 'al,kil }

alkylene [ORG CHEM] An organic radical formed from an unsaturated aliphatic hydrocarbon; for example, the ethylene radical C_2H_3–. { 'al·kə ,lēn }

alkyl halide [ORG CHEM] A compound consisting of an alkyl group and a halogen; an example is ethylbromide. { 'al·kəl 'hāl,īd }

alkyne [ORG CHEM] One of a group of organic compounds containing a carbon-to-carbon triple bond. { 'al,kīn }

allene [ORG CHEM] C_3H_4 An unsaturated aliphatic hydrocarbon with two double bonds. Also known as propadiene. { ə'lēn }

Allen red metal [MET] An alloy of copper and lead containing 50% lead and a small quantity of sulfur to hold the lead in solution. { ¦al·ən ¦red ,med·əl }

alligator effect *See* orange peel. { 'al·ə,gād·ər ə'fekt }

alligatoring [MATER] Cracking of a film of paint or varnish, with broad, deep cracks through one or more coats. Also known as crocodiling. [MET] **1.** A splitting of an end of a rolled steel slab in which the plane of the split is parallel to the rolled surface. Also known as fishmouthing. **2.** The roughening of a sheet-metal surface during forming due to the coarse grain of the metal used. { 'al·ə,gād·ər·iŋ }

alligator squeezer [MET] A tool with a fixed upper jaw and a movable lower jaw used to squeeze a ball of iron produced by the paddling process into a bloom or billet. { 'al·ə,gād·ər ,skwēz·ər }

allochromatic crystal [CRYSTAL] A crystal having photoconductive properties due to the presence of small particles within it. { ,a·lə·krə'mad·ik 'kris·təl }

allochromy [PHYS] Emission of electromagnetic radiation that results from incident radiation at a different wavelength, as occurs in fluorescence or the Raman effect. { ¦a·lə¦krōm·ē }

allomerism [CRYSTAL] A constancy in crystal

form in spite of a variation in chemical composition. { ə'läm·ə,riz·əm }

allowable load [MECH] The maximum force that may be safely applied to a solid, or is permitted by applicable regulators. { ə'laủ·ə·bəl 'lōd }

allowable stress [MECH] The maximum force per unit area that may be safely applied to a solid. { ə'laủ·ə·bəl 'stres }

allowed energy bands [SOLID STATE] The restricted regions of possible electron energy levels in a solid. { ə'laủd 'en·ər·jē ,banz }

alloy [MATER] A composite plastic produced by blending and melting together different polymers. [MET] A metal product containing two or more elements as a solid solution, as an intermetallic compound, or as a mixture of metallic phases. { 'a,lȯi }

alloy adhesive [MATER] An adhesive compounded from resins of two or more differential chemical classes, such as thermosetting and elastomeric, in order to combine performance characteristics. { 'a,lȯi ad,hē·ziv }

alloying [MET] The addition of a metal or alloy to another metal or alloy. { ə'lȯi·iŋ }

alloy junction [ELECTR] A junction produced by alloying one or more impurity metals to a semiconductor to form a *p* or *n* region, depending on the impurity used. Also known as fused junction. { 'a,lȯi ,jəŋk·shən }

alloy-junction diode [ELECTR] A junction diode made by placing a pill of doped alloying material on a semiconductor material and heating until the molten alloy melts a portion of the semiconductor, resulting in a *pn* junction when the dissolved semiconductor recrystallizes. Also known as fused-junction diode. { 'a,lȯi ¦jəŋk·shən 'dī,ōd }

alloy plating [MET] The codeposition of two or more metals on an electrode by electrolysis. { 'a,lȯi ,plāt·iŋ }

Alloy 750 [MET] A bearing alloy, containing 6.5% tin, 2.5% silicon, 1% copper, 0.5% nickel, and the remainder aluminum; used for automobile engine bearings. { 'a,lȯi ,sev·ən 'fif·tē }

alloy steel [MET] A steel whose distinctive properties are due to the presence of one or more elements other than carbon. { ¦a,lȯi ¦stēl }

all-weld-metal test specimen [MET] A specimen composed entirely of weld metal used in a weld tension test wherein the axis of the weld from which it is derived is parallel to the axis of the test bar. { ¦ȯl ¦weld ¦med·əl 'test ,spes·ə·mən }

allyl- [ORG CHEM] A prefix used in names of compounds whose structure contains an allyl cation. { 'al·əl }

allylic hydrogen [ORG CHEM] In an organic molecule, a hydrogen attached to a carbon atom that is adjacent to a double bond. { ə'lil·ik 'hī·drə·jən }

allyl plastic See allyl resin. { 'al·əl 'plas·tik }

allyl resin [POLYM CHEM] Any of a class of thermosetting synthetic resins derived from esters of allyl alcohol or allyl chloride; used in making cast and laminated products. Also known as allyl plastic. { 'al·əl 'rez·ən }

almendro [MATER] The wood of tonka bean tree species in Panama (*Coumarouna panamensis*); resistant to marine borers and used as a substitute for greenheart in marine construction. { äl'men·drō }

almon [MATER] A type of lauan from the tree *Shorea eximia*; the wood is fairly strong and hard with a coarse texture. { 'al,mōn }

alnico [MET] One of a series of ferrous alloys containing aluminum, nickel, and cobalt, valued because of their highly retentive magnetic properties; usually designated with a roman-numeral number, such as alnico VII. Also known as aluminum-nickel-cobalt alloy. { 'al·ni,kō }

alnico magnet [ELECTROMAG] A permanent magnet made of alnico. { 'al·ni,kō 'mag·nət }

aloe wood oil See oriental linaloe. { 'a·lō ¦wůd ,ȯil }

alpha brass [MET] An alloy of copper and zinc containing up to 36% zinc dissolved, rather than chemically combined, with the copper; ductile, easily cold-worked, and corrosion resistant; used for hot-water pipes. { 'al·fə ,bras }

alpha bronze [MET] A copper-tin alloy; commercial forms containing 4–5% tin are used in coinage, springs, and turbine blades. { 'al·fə ,bränz }

alpha gypsum [MATER] A specially processed gypsum having low consistency and high compressive strength, often exceeding 5000 pounds per square inch (34.5 megapascals). { 'al·fə 'jip·səm }

alpha iron [MET] Iron with a body-centered cubic structure which is stable below 1670°F (910°C). { 'al·fə ,ī·ərn }

alpha olefin [ORG CHEM] An olefin where the unsaturation (double bond) is at the alpha position, that is, between the two end carbons of the carbon chain. { 'al·fə 'ō·lə,fən }

alsifilm [MATER] A bentonite gel in the form of sheets used primarily for electrical insulation, because of its properties of heat and oil resistance. { 'alz·ə,film }

alternate immersion test [MET] A corrosion test in which a specimen is repeatedly carried through a cycle of immersion in and removal from a corrosive medium over definite time intervals. { 'ȯl·tər·nət im'ər·zhən ,test }

alternate polarity operation [MET] A resistance welding method which utilizes alternating polarity for the progression of weld pulses. { 'ȯl·tər·nət pə'lar·əd·ē ,äp·ə'rā·shən }

alternating copolymer [POLYM CHEM] A polymer formed of two different monomer molecules that alternate in sequence in the polymer chain. { 'ȯl·tər,näd·iŋ ,kō'päl·ə·mər }

alternating current [ELEC] Electric current that reverses direction periodically, usually many times per second. Abbreviated ac. { ¦ȯl·tər ,näd·iŋ ¦kər·ənt }

alternating-current Josephson effect [CRYO] The oscillating current flow resulting from the tunneling of electron pairs through a thin insulating barrier between two superconductors when a steady voltage, V, is maintained across the barrier, with a frequency equal to $2eV/h$, where e is the magnitude of the charge of the electron and h is Planck's constant. { ¦ȯl·tər·nād·iŋ ¦kər·ənt 'jō·səf·sən i‚fekt }

alternating-current resistance See high-frequency resistance. { ¦ȯl·tər‚nād·iŋ ¦kər·ənt ri'zis·təns }

alternating stress [MECH] A stress produced in a material by forces which are such that each force alternately acts in opposite directions. { 'ȯl·tər·nād·iŋ 'stres }

alum [INORG CHEM] **1.** Any of a group of double sulfates of trivalent metals such as aluminum, chromium, or iron and a univalent metal such as potassium or sodium. **2.** See aluminum sulfate; ammonium aluminum sulfate; potassium aluminum sulfate. { 'al·əm }

alum cake [MATER] A material composed of silica and aluminum sulfate produced by the action of sulfuric acid on clay. { 'al·əm ‚kāk }

alumel [MET] An alloy containing nickel, aluminum, chromium, and silicon; used to form the chromel-alumel thermocouple. { 'al·yü·mel }

alumetized steel See aluminized steel. { ə'lüm·ə‚tīzd 'stēl }

alumina [INORG CHEM] Al_2O_3 The native form of aluminum oxide occurring as corundum or in hydrated forms, as a powder or crystalline substance. { ə'lüm·ə·nə }

alumina balls [MATER] Alumina in the form of balls $\frac{1}{4}$ to $\frac{3}{4}$ inch (6.4 to 19 millimeters) in diameter; usually composed of 99% alumina and having high resistance to chemicals and heat; used in reactor and catalytic beds. { ə'lüm·ə·nə ‚bȯlz }

alumina brick [MATER] A group of fireclay bricks containing 50, 60, or 70% alumina, used in high temperature applications. { ə'lüm·ə·nə ‚brik }

alumina bubble brick [MATER] A lightweight refractory brick used to line kiln walls; manufactured by passing an air jet over molten alumina to produce small hollow bubbles which are pressed into bricks and other shapes. { ə'lüm·ə·nə ¦bəb·əl ‚brik }

alumina cement [MATER] A cement made with bauxite and containing a high percentage of aluminate, having the property of setting to high strength in 24 hours. { ə'lüm·ə·nə si 'ment }

alumina fibers [MATER] Short, linear crystals of alumina which have a strength of up to 200,000 pounds per square inch (1.38 gigapascals); used in plastics as a filler to improve heat resistance and dielectric properties. Also known as sapphire whiskers. { ə'lüm·ə·nə 'fī·bərz }

alumina porcelain [MATER] Porcelain composed principally of alumina; used to make spark plugs. { ə'lüm·ə·nə 'pȯr·slən }

aluminate [INORG CHEM] A negative ion usually assigned the formula AlO_2^- and derived from aluminum hydroxide. { ə'lüm·ə‚nāt }

aluminate cement [MATER] Cement made with bauxite, with high percentage of alumina; sets to a high strength in 24 hours and is used for constructing bank walls and laying roads. Also

known as aluminous cement; high-alumina cement; high-speed cement. { ə'lüm·ə‚nāt si 'ment }

alumina trihydrate [INORG CHEM] $Al_2O_3 \cdot 3H_2O$, or $Al(OH)_3$ A white powder; insoluble in water, soluble in hydrochloric or sulfuric acid or sodium hydroxide; used in the manufacture of ceramic glasses and in paper coating. Also known as aluminum hydrate; aluminum hydroxide; hydrated alumina; hydrated aluminum oxide. { ə'lüm·ə·nə ‚trī'hī‚drāt }

aluminium See aluminum. { ‚al·yü'min· ē·əm }

aluminize [ENG] To apply a film of aluminum to a material, such as glass. [MET] To form a protective surface alloy on a metal by treatment at elevated temperature with aluminum or an aluminum compound. { ə'lüm·ə‚nīz }

aluminized steel [MET] A steel coated with an aluminum-iron alloy coating; prepared by dip-coating and diffusing aluminum into steel at 1600°F (870°C); resists scaling and oxidation up to 1650°F (900°C). Also known as alumetized steel; calorized steel. { ə'lüm·ə‚nīzd 'stēl }

aluminosilicate [INORG CHEM] $3Al_2O_3 \cdot 2SiO_2$ A colorless, crystalline combination of silicate and aluminate in the form of rhombic crystals. { ə¦lüm·ə‚nō¦sil·ə‚kāt }

aluminothermy [MET] The process of reducing a metallic oxide to the metal and producing great heat by mixing finely divided aluminum with the oxide, which is reduced as the aluminum is oxidized. { ə'lüm·ə·nō‚thər·mē }

aluminous cement See aluminate cement. { ə'lüm·ə·nəs si'ment }

aluminum A chemical element, symbol Al, atomic number 13, and atomic weight 26.9815. Also spelled aluminium. { ə'lüm·ə·nəm }

aluminum alloy [MET] An alloy of aluminum and relatively small amounts of other metals, such as copper, magnesium, or manganese. { ə'lüm·ə·nəm 'a‚lȯi }

aluminum ammonium sulfate See ammonium aluminum sulfate. { ə'lüm·ə· nəm ə'mon·ē·əm 'səl‚fāt }

aluminum-base grease See aluminum-soap grease. { ə'lüm·ə·nəm ‚bās 'grēs }

aluminum brass [MET] **1.** A casting brass to which aluminum has been added as a flux to improve the casting qualities and, with the addition of lead, the machining qualities. **2.** A wrought brass to which aluminum has been added to improve the extruding and forging qualities and the oxidation resistance. { ə'lüm·ə·nəm 'bras }

aluminum bronze [MET] A copper-aluminum alloy which may also contain iron, manganese, nickel, or zinc. { ə'lüm·ə·nəm 'bränz }

aluminum chloride [INORG CHEM] $AlCl_3$ or Al_2Cl_6 A deliquescent compound in the form of white to colorless hexagonal crystals; fumes in air and reacts explosively with water; used as a catalyst. { ə'lüm·ə·nəm 'klȯr‚īd }

aluminum coating [MET] A film of aluminum applied to a metallic surface by, for example, spraying, electrolysis, or hot dipping. { ə'lüm·ə·nəm 'kōd·iŋ }

aluminum conductor [ELEC] Any of several aluminum alloys employed for conducting electric current; because its weight is one-half that of copper for the same conductance, it is used in high-voltage transmission lines. { ə'lüm·ə·nəm kən'dək·tər }

aluminum fluoride [INORG CHEM] $AlF_3 \cdot 3\frac{1}{2}H_2O$ A white, crystalline powder, insoluble in cold water. { ə'lum·ə·nəm 'flür,īd }

aluminum fluosilicate [INORG CHEM] $Al_2(SiF_6)_3$ A white powder that is soluble in hot water; used for artificial gems, enamels, and glass. Also known as aluminum silicofluoride. { ə'lüm·ə·nəm ,flü·ə'sil·i·kət }

aluminum foil [MET] Aluminum in the form of a sheet of thickness not exceeding 0.005 inch (0.127 millimeter). { ə'lüm·ə·nəm 'föil }

aluminum grease See aluminum-soap grease. { ə'lüm·ə·nəm 'grēs }

aluminum halide [INORG CHEM] A compound of aluminum with a halogen element, such as aluminum chloride. { ə'lüm·ə·nəm 'ha,līd }

aluminum hydrate See alumina trihydrate. { ə'lüm·ə·nəm 'hī·drāt }

aluminum hydroxide See alumina trihydrate. { ə'lüm·ə·nəm hī'dräk,sīd }

aluminum-magnesium alloy [MET] An alloy of aluminum, 5–10% magnesium, and sometimes small amounts of other metals, characterized by high resistance to corrosion and high machinability. { ə'lüm·ə·nəm mag'nēz·ē·əm 'a,lói }

aluminum-nickel-cobalt alloy See alnico. { ə'lüm·ə·nəm ¦nik·əl ¦kō,bólt 'a,lói }

aluminum nitrate [INORG CHEM] $Al(NO_3)_3 \cdot 9H_2O$ White, deliquescent crystals with a melting point of 73°C; soluble in alcohol and acetone; used as a mordant for textiles, in leather tanning, and as a catalyst in petroleum refining. { ə'lüm·ə·nəm 'nī,trāt }

aluminum orthophosphate [INORG CHEM] $AlPO_4$ White crystals, melting above 1500°C; insoluble in water, soluble in acids and bases; useful in ceramics, paints, pulp, and paper. Also known as aluminum phosphate. { ə'lüm·ə·nəm ,ór·thō 'fäs,fāt }

aluminum oxide [INORG CHEM] Al_2O_3 A compound in the form of a white powder or colorless hexagonal crystals; melts at 2020°C; insoluble in water; used in aluminum production, paper, spark plugs, absorbing gases, light bulbs, artificial gems, and manufacture of abrasives, refractories, ceramics, and electrical insulators. { ə'lüm·ə·nəm 'äk,sīd }

aluminum paint [MATER] A mixture of oil varnish and aluminum pigment in the form of thin flakes which overlap in the paint film; reflects the sun's radiation well and retains the heat in hot-air or hot-water pipes or tanks. { ə'lüm·ə·nəm 'pānt }

aluminum paste [MATER] Aluminum powder finely ground in oil; used in aluminum paints. { ə'lüm·ə·nəm 'pāst }

aluminum phosphate See aluminum orthophosphate. { ə'lüm·ə·nəm 'fäs,fāt }

aluminum potassium sulfate See potassium aluminum sulfate. { ə'lüm·ə·nəm pə'tas·ē·əm 'səl,fāt }

aluminum powder [MATER] Small flakes of aluminum metal obtained by stamping or ball-milling foil in the presence of a fatty lubricant, such as stearic acid, which causes the flakes to orient in a pattern to give high brilliance. { ə'lüm·ə·nəm 'paüd·ər }

aluminum silicate [INORG CHEM] $Al_2(SiO_3)_3$ A white solid that is insoluble in water; used as a refractory in glassmaking. { ə'lüm·ə·nəm 'sil·ə ,kāt }

aluminum silicofluoride See aluminum fluosilicate. { ə'lüm·ə·nəm ¦sil·ə·kō¦flür,īd }

aluminum-silicon alloy [MET] An alloy of aluminum, 5–22% silicon, and sometimes small amounts of other metals, characterized by ease of casting and welding, light weight, and high resistance to corrosion. { ə'lüm·ə·nəm 'sil·i·kən 'a,lói }

aluminum-silicon bronze [MET] An alloy consisting chiefly of copper, with aluminum and silicon added to give greater strength and hardness. { ə'lüm·ə·nəm 'sil·i·kən 'bränz }

aluminum-soap grease [MATER] A lubricating grease consisting of a petroleum oil thickened with aluminum soap. Also known as aluminum-base grease; aluminum grease. { ə'lüm·ə·nəm 'sōp 'grēs }

aluminum sodium sulfate [INORG CHEM] $AlNa(SO_4)_2 \cdot 12H_2O$ Colorless crystals with an astringent taste and a melting point of 61°C; soluble in water; used as a mordant and for waterproofing textiles, as a food additive, and for matches, tanning, ceramics, engraving, and water purification. Abbreviated SAS. Also known as porous alum; soda alum; sodium aluminum sulfate. { ə'lüm·ə·nəm 'sōd·ē·əm 'səl,fāt }

aluminum solder [MET] A solder containing up to 15% aluminum, having a melting point above that of the tin-lead solders, and applied with a brazing torch. { ə'lüm·ə·nəm 'säd·ər }

aluminum sulfate [INORG CHEM] $Al_2(SO_4)_3 \cdot 18H_2O$ A colorless salt in the form of monoclinic crystals that decompose in heat and are soluble in water; used in papermaking, water purification, and tanning, and as a mordant in dyeing. Also known as alum. { ə'lüm·ə·nəm 'səl ,fāt }

alysoid See catenary. { 'al·ə,sóid }

Am See americium.

A/m See ampere per meter.

Am² See ampere meter squared.

amalgam [MET] An alloy of mercury. { ə'mal·gəm }

amalgamate [MET] **1.** To unite a metal in an alloy with mercury. **2.** To unite two dissimilar metals. **3.** To cover the zinc elements of a galvanic battery with mercury. { ə'mal·gə,māt }

amalgamating table [MET] A sloping wooden table covered with a copper plate on which mercury is spread to amalgamate with precious-metal particles. { ə'mal·gə,mād·iŋ ,tā·bəl }

amalgamation [MET] Also known as amalgam

treatment. **1.** The process of separating metal from ore by alloying the metal with mercury; formerly used for gold and silver recovery, where it has been superseded by the cyanide process. **2.** The formation of an alloy of a metal with mercury. { ə,mal·gə'mā·shən }

amalgamation pan [MET] A circular cast-iron pan in which gold or silver ore is ground and the precious metal particles are amalgamated with mercury added to the pan. { ə,mal·gə'mā·shən ,pan }

amalgamator [MET] A device for bringing pulverized ore into contact with mercury to form an amalgam from which the metal is subsequently recovered. { ə'mal·gə,mād·ər }

amalgam barrel [MET] A small batching mill used to grind auriferous concentrates with mercury. { ə'mal·gəm ,bar·əl }

amalgam retort [MET] A retort in which mercury is distilled off from gold, or silver amalgam is obtained in amalgamation. { ə'mal·gəm ri,tòrt }

amalgam treatment See amalgamation. { ə'mal·gəm ,trēt·mənt }

amber glass [MATER] A tinted glass made by using different mixtures of sulfur and iron oxide; the color can vary from pale yellow to ruby amber. { 'am·bər 'glas }

ambident [ORG CHEM] A chemical species or molecule that possesses two alternative reactive sites, either of which can bond in a reaction; examples include cyanate ions, thiosulfate ions, oxime anions, and enolate ions. Also known as ambidentate. { 'am·bə·dənt }

ambidentate See ambident. { ,am· bə'den,tāt }

ambient temperature [PHYS] The temperature of the surrounding medium, such as gas or liquid, which comes into contact with the apparatus. { 'am·bē·ənt 'tem·prə·chər }

American wire gage [MET] A particular series of specified diameters and thicknesses established as a standard in the United States and used for nonferrous sheets, rods, and wires. Abbreviated AWG. Also known as Brown and Sharp gage (B and S gage). { ə'mer·ə·kən 'wīr ,gāj }

americium [CHEM] A chemical element, symbol Am, atomic number 95; the mass number of the isotope with the longest half-life is 243. { ,am·ə'ris·ē·əm }

americyl ion [INORG CHEM] A dioxo monocation of americium, with the formula $(AmO_2)^-$. { ə'mer·ə·səl 'ī,än }

A-metal [MET] A type of permeability alloy containing 44% nickel and a small amount of copper; used to give nondistortion characteristics upon magnetization in transformers and loudspeakers. { 'ā ,med·əl }

amide [ORG CHEM] One of a class of organic compounds containing the $CONR_2$ radical, where R may represent H or an organic (usually hydrocarbon) radical. { 'am,īd }

amidine [ORG CHEM] A compound which contains the radical $CNHNH_2$. { 'am·ə,dēn }

amido [ORG CHEM] Indicating the NR_2 functional group, where R represents either hydrogen or an organic (usually hydrogen) radical, when it is present in a molecule and bonded to the C atom of a functional group. { ə'mē,dō }

amidol [ORG CHEM] $C_6H_3(NH_2)_2OH·HCl$ A grayish-white crystalline salt; soluble in water, slightly soluble in alcohol; used as a developer in photography and as an analytical reagent. { 'am·i,dòl }

amine [ORG CHEM] One of a class of organic compounds which can be considered to be derived from ammonia by replacement of one or more hydrogens by functional groups. { ə'mēn }

amino-, amin- [CHEM] Having the property of a compound in which the group NH_2 is attached to a radical other than an acid radical. { ə'mē,nō }

amino acid [BIOMATER] Any of the organic compounds that contain one or more basic amino groups and one or more acidic carboxyl groups and that are polymerized to form peptides and proteins; only 20 of the more than 80 amino acids found in nature serve as building blocks for proteins; examples are tyrosine and lysine. { ə'mē,nō 'as·əd }

aminodiborane [INORG CHEM] Any compound derived from diborane (B_2H_6) in which one H of the bridge has been replaced by NH_2. { ə¦mē·nō ,dī¦bòr,än }

amino group [ORG CHEM] A functional group $(-NH_2)$ formed by the loss of a hydrogen atom from ammonia. { ə'mē·nō ,grüp }

amino plastic [MATER] Any plastic made of compounds derived from ammonia. { ə'mē·nō 'plas·tik }

amino resin [POLYM CHEM] A type of resin prepared by condensation polymerization, with an aldehyde, of a compound containing an amino group. { ə'mē·nō 'rez·ən }

Am²/Js See ampere square meter per joule second.

ammine [INORG CHEM] One of a group of complex compounds formed by coordination of ammonia molecules with metal ions. { 'a,mēn }

ammonation [INORG CHEM] A reaction in which ammonia is added to other molecules or ions by covalent bond formation utilizing the unshared pair of electrons on the nitrogen atom, or through ion-dipole electrostatic interactions. { ,a·mə'nā·shən }

ammonia [INORG CHEM] NH_3 A colorless gaseous alkaline compound that is very soluble in water, has a characteristic pungent odor, is lighter than air, and is formed as a result of the decomposition of most nitrogenous organic material; used as a fertilizer and as a chemical intermediate. { ə'mōn·yə }

ammonia alum See ammonium aluminum sulfate. { ə'mōn·yə 'al·əm }

ammoniac [INORG CHEM] See ammoniacal. [MATER] A gum resin obtained from the stems of the ammoniac plant; used in medicine, perfume, plaster, concrete, and adhesive. { ə'mōn·ē,ak }

ammoniacal [INORG CHEM] Pertaining to ammonia or its properties. Also known as ammoniac. { ¦a·mə¦nī·ə·kəl }

ammoniated mercuric chloride See ammoniated mercury. { ə¦mōn·ē·ād·əd mər¦kyür·ik ¦klòr,īd }

ammoniated mercury [INORG CHEM] $HgNH_2Cl$ A white powder that darkens on light exposure; insoluble in water and alcohol, soluble in ammonium carbonate solutions and in warm acids; used in pharmaceuticals and as a local anti-infective in medicine. { ə'mōn·ē·ād·əd 'mər·kyə·rē }

ammoniated superphosphate [INORG CHEM] A fertilizer containing 5 parts of ammonia to 100 parts of superphosphate. { ə'mōn·ē·ād·əd ˌsü·pər'fäs,fāt }

ammonium alum See ammonium aluminum sulfate. { ə'mōn·yəm 'al·əm }

ammonium aluminum sulfate [INORG CHEM] $NH_4Al(SO_4)_2 \cdot 12H_2O$ Colorless, odorless crystals that are soluble in water; used in manufacturing medicines and baking powder, dyeing, papermaking, and tanning. Also known as alum; aluminum ammonium sulfate; ammonia alum; ammonium alum. { ə'mōn·yəm əˌlü·mə·nəm 'səl,fāt }

ammonium bicarbonate [INORG CHEM] NH_4- HCO_3 White, crystalline, water-soluble salt; used in baking powders and in fire-extinguishing mixtures. Also known as ammonium hydrogen carbonate. { ə'mōn·yəm bī'kär·bə,nāt }

ammonium bichromate See ammonium dichromate. { ə'mōn·yəm bī'krō,māt }

ammonium bifluoride [INORG CHEM] $NH_4F \cdot HF$ A salt that crystallizes in the orthorhombic system and is soluble in water; prepared in the form of white flakes from ammonia treated with hydrogen fluoride; used in solution as a fungicide and wood preservative. Also known as ammonium acid fluoride; ammonium hydrogen fluoride. { ə'mōn·yəm bī'flúr,īd }

ammonium borate [INORG CHEM] NH_4BO_3 A white, crystalline, water-soluble salt which decomposes at $198°C$; used as a fire retardant on fabrics. { ə'mōn·yəm 'bòr,āt }

ammonium bromide [INORG CHEM] NH_4Br An ammonium halide that crystallizes in the cubic system; made by the reaction of ammonia with hydrobromic acid or bromine; used in photography and for pharmaceutical preparations (sedatives). { ə'mōn·yəm 'brō,mīd }

ammonium carbamate [INORG CHEM] NH_4- NH_2CO_2 A salt that forms colorless, rhombic crystals, which are very soluble in cold water; an important, unstable intermediate in the manufacture of urea; found in commercial ammonium carbonate. { ə'mōn·yəm 'kär·bə ˌmāt }

ammonium carbonate [INORG CHEM] **1.** $(NH_4)_2$- CO_3 The normal ammonium salt of carbonic acid, prepared by passing gaseous carbon dioxide into an aqueous solution of ammonia and allowing the vapors (ammonia, carbon dioxide, water) to crystallize. **2.** $NH_4HCO_3 \cdot NH_2COONH_4$ A white, crystalline double salt of ammonium bicarbonate and ammonium carbamate obtained commercially; the principal ingredient of smelling salts. { ə'mōn·yəm 'kär·bə,nāt }

ammonium chloride [INORG CHEM] NH_4Cl A white crystalline salt that occurs naturally as a sublimation product of volcanic action or is manufactured; used as an electrolyte in dry cells, as a flux for soldering, tinning, and galvanizing, and as an expectorant. { ə'mōn·yəm 'klór ˌīd }

ammonium chromate [INORG CHEM] $(NH_4)_2$- CrO_4 A salt that forms yellow, monoclinic crystals; made from ammonium hydroxide and ammonium dichromate; used in photography as a sensitizer for gelatin coatings. { ə'mōn·yəm 'krō,māt }

ammonium dichromate [INORG CHEM] $(NH_4)_2$- Cr_2O_7 A salt that forms orange, monoclinic crystals; made from ammonium sulfate and sodium dichromate; soluble in water and alcohol; ignites readily; used in photography, lithography, pyrotechnics, and dyeing. Also known as ammonium bichromate. { ə'mōn· yəm dī'krō,māt }

ammonium fluoride [INORG CHEM] NH_4F A white, unstable, crystalline salt with a strong odor of ammonia; soluble in cold water; used in analytical chemistry, glass etching, and wood preservation, and as a textile mordant. { ə'mōn·yəm 'flúr,īd }

ammonium fluosilicate [INORG CHEM] $(NH_4)_2$- SiF_6 A toxic, white, crystalline powder; soluble in alcohol and water; used for mothproofing, glass etching, and electroplating. Also known as ammonium silicofluoride. { ə'mōn·yəm ˌflü·ə'sil·ə ˌkāt }

ammonium halide [INORG CHEM] A compound with the ammonium ion bonded to an ion formed from one of the halogen elements. { ə'mōn·yəm 'hal,īd }

ammonium hydrogen carbonate See ammonium bicarbonate. { ə'mōn·yəm 'hi·drə·jən 'kär·bə,nāt }

ammonium hydrogen fluoride See ammonium bifluoride. { ə'mōn·yəm 'hi·drə·jən 'flúr,īd }

ammonium hydroxide [INORG CHEM] NH_4OH A hydrate of ammonia, crystalline below $-79°C$; it is a weak base known only in solution as ammonia water. Also known as aqua ammonia. { ə'mōn·yəm ˌhī'dräk,sīd }

ammonium iodide [INORG CHEM] NH_4I A salt prepared from ammonia and hydrogen iodide or iodine; it forms colorless, regular crystals which sublime when heated; used in photography and for pharmaceutical preparations. { ə'mōn·yəm 'ī·ə,dīd }

ammonium molybdate [INORG CHEM] $(NH_4)_2$- MoO_4 White, crystalline salt used as an analytic reagent, as a precipitant of phosphoric acid, and in pigments. { ə'mōn·yəm mə'lib,dāt }

ammonium nickel sulfate See nickel ammonium sulfate. { ə'mōn·yəm ˌnik·əl 'səl,fāt }

ammonium nitrate [INORG CHEM] NH_4NO_3 A colorless crystalline salt; very insensitive and stable high explosive; also used as a fertilizer. { ə'mōn·yəm 'nī,trāt }

ammonium perchlorate [INORG CHEM] NH_4- ClO_4 A salt that forms colorless or white rhombic and regular crystals, which are soluble in water; it decomposes at $150°C$, and the reaction is explosive at higher temperatures. { ə'mōn·yəm pər'klor,āt }

ammonium persulfate [INORG CHEM] $(NH_4)_2S_2O_8$ White crystals which decompose on melting; soluble in water; used as an oxidizing agent and bleaching agent, and in etching, electroplating, food preservation, and aniline dyes. { ə'mōn·yəm pər'səl,fāt }

ammonium phosphate [INORG CHEM] $(NH_4)_2HPO_4$ A salt of ammonia and phosphoric acid that forms white monoclinic crystals, which are soluble in water; used as a fertilizer and fire retardant. { ə'mōn·yəm 'fäs,fāt }

ammonium salt [INORG CHEM] A product of a reaction between ammonia and various acids; examples are ammonium chloride and ammonium nitrate. { ə'mōn·yəm 'solt }

ammonium silicofluoride *See* ammonium fluosilicate. { ə'mōn·yəm ¦sil·ə·kō'flur,īd }

ammonium sulfamate [INORG CHEM] $NH_4OSO_2NH_2$ White crystals with a melting point of 130°C; soluble in water; used for flameproofing textiles, in electroplating, and as an herbicide to control woody plant species. { ə'mōn·yəm 'səl·fə,māt }

ammonium sulfate [INORG CHEM] $(NH_4)_2SO_4$ Colorless, rhombic crystals which melt at 140°C and are soluble in water. { ə'mōn·yəm 'səl,fāt }

ammonium sulfide [INORG CHEM] $(NH_4)_2S$ Yellow crystals, stable only when dry and below 0°C; decomposes on melting; soluble in water and alcohol; used in photographic developers and for coloring brasses and bronzes. { ə'mōn·yəm 'səl,fīd }

ammonium vanadate [INORG CHEM] NH_4VO_3 A white to yellow, water-soluble, crystalline powder; used in inks and as a paint drier and textile mordant. { ə'mōn·yəm 'van·ə,dāt }

amorphous [PHYS] Pertaining to a solid which is noncrystalline, having neither definite form nor structure. { ə'mor·fəs }

amorphous film [MATER] A magnetically ordered metallic film that can be deposited on a semiconductor chip or on almost any other material without need for a crystal substrate, for use in magnetic bubble memories. { ə'mor·fəs 'film }

amorphous ribbon [MET] A metallic alloy that has an amorphous structure and is formed into a strip 25 to 50 micrometers thick and 1 to 150 millimeters (0.04 to 6 inches) wide by a process in which a melt of the required composition is ejected through an orifice onto a copper drum where it is instantly quenched and formed into a ribbon by rotation of the drum. { ə'mor·fəs 'rib·ən }

amorphous semiconductor [SOLID STATE] A semiconductor material which is not entirely crystalline, having only short-range order in its structure. { ə'mor·fəs ¦sem·ē·kən¦dək·tər }

amorphous solid [SOLID STATE] A rigid material whose structure lacks crystalline periodicity; that is, the pattern of its constituent atoms or molecules does not repeat periodically in three dimensions. { ə'mor·fəs 'säl·əd }

amount of substance [CHEM] A measure of the number of elementary entities present in a substance or system; usually measured in moles. { ə'maunt əv 'səb·stəns }

amp *See* amperage;ampere. { amp }

ampacity [ELEC] Current-carrying capacity in amperes; used as a rating for power cables. { am'pas·əd·ē }

amperage [ELEC] The amount of electric current in amperes. Abbreviated amp. { 'am·prij }

ampere [ELEC] The unit of electric current in the rationalized meter-kilogram-second system of units; defined in terms of the force of attraction between two parallel current-carrying conductors. Abbreviated a; A; amp. { 'am,pir }

Ampère currents [ELECTROMAG] Postulated "molecular-ring" currents to explain the phenomena of magnetism as well as the apparent nonexistence of isolated magnetic poles. { 'äm,per 'kər·əns }

ampere-hour [ELEC] A unit for the quantity of electricity, obtained by integrating current flow in amperes over the time in hours for its flow; used as a measure of battery capacity. Abbreviated Ah; amp-hr. { 'am,pir ¦au·ər }

ampere-hour capacity [ELEC] The charge, measured in ampere-hours, that can be delivered by a storage battery up to the limit to which the battery may be safely discharged. { 'am,pir ¦au·ər kə'pas·əd·ē }

Ampère law [ELECTROMAG] **1.** A law giving the magnetic induction at a point due to given currents in terms of the current elements and their positions relative to the point. Also known as Laplace law. **2.** A law giving the line integral over a closed path of the magnetic induction due to given currents in terms of the total current linking the path. { 'äm,per ,lo }

ampere meter squared [ELECTROMAG] The SI unit of electromagnetic moment. Abbreviated Am^2. { ¦am,pir ¦mēd·ər 'skwerd }

ampere-minute [ELEC] A unit of electrical charge, equal to the charge transported in 1 minute by a current of 1 ampere, or to 60 coulombs. Abbreviated A min. { ¦am,pir ¦min·ət }

ampere per meter [ELECTROMAG] The SI unit of magnetic field strength and magnetization. Abbreviated A/m. { 'am,pir pər 'mēd·ər }

ampere per square inch [ELEC] A unit of current density, equal to the uniform current density of a current of 1 ampere flowing through an area of 1 square inch. Abbreviated A/in². { 'am,pir pər ,skwer 'inch }

ampere per square meter [ELEC] The SI unit of current density. Abbreviated A/m^2. { 'am,pir pər ,skwer 'mēd·ər }

Ampère rule [ELECTROMAG] The rule which states that the direction of the magnetic field surrounding a conductor will be clockwise when viewed from the conductor if the direction of current flow is away from the observer. { 'äm ,per ,rül }

ampere square meter per joule second [ELECTROMAG] The SI unit of gyromagnetic ratio.

Abbreviated Am^2/Js. { ¦am,pir ¦skwer ¦mēd·ər pər ¦jül 'sek·ənd }

Ampère theorem [ELECTROMAG] The theorem which states that an electric current flowing in a circuit produces a magnetic field at external points equivalent to that due to a magnetic shell whose bounding edge is the conductor and whose strength is equal to the strength of the current. { 'äm,per ,thir·əm }

ampere-turn [ELECTROMAG] A unit of magnetomotive force in the meter-kilogram-second system defined as the force of a closed loop of one turn when there is a current of 1 ampere flowing in the loop. Abbreviated amp-turn. { 'am,pir ,tərn }

amphiprotic See amphoteric. { ¦am· fə¦präd·ik }

amphoteric [CHEM] Having both acidic and basic characteristics. Also known as amphiprotic. { ¦am·fə¦ter·ik }

amp-hr See ampere-hour.

amu See atomic mass unit.

amyl [ORG CHEM] Any of the eight isomeric arrangements of the radical C_5H_{11} or a mixture of them. Also known as pentyl. { 'am·əl }

anacardium gum See cashew gum. { ,an·ə 'kärd·ē·əm 'gəm }

anaerobic adhesive [MATER] A single-component adhesive that hardens rapidly to form a strong bond between surfaces from which air is excluded. { ¦an·ə¦rōb·ik əd'hēz·iv }

analog [CHEM] A compound whose structure is similar to that of another compound but whose composition differs by one element. { 'an·əl ,äg }

analogous pole [SOLID STATE] The pole of a crystal that acquires a positive charge when the crystal is heated. { ə'nal·ə·gəs ,pōl }

analytic mechanics [MECH] The application of differential and integral calculus to classical (nonquantum) mechanics. { ,an·əl'id·ik mi 'kan·iks }

anchor [ENG] A device, such as a metal rod, wire,or strap, for fixing one object to another, such as specially formed metal connectors used to fasten together timbers, masonry, or trusses. [MET] A device that prevents the movement of sand cores in molds. { 'aŋ·kər }

anchored catalyst See immobilized catalyst. { 'aŋ·kərd 'kad·ə,list }

anchored-type ceramic veneer [MATER] Ceramic veneer which is attached to a backing by grout and nonferrous metal anchors. { 'aŋ·kərd ,tīp sə'ram·ik və'nēr }

anchor pattern [MET] The pattern of minute projections formed on a metal surface by sandblasting, shot blasting, or chemical etching; used to enhance the adhesiveness of a surface coating. { 'aŋ·kər ,pad·ərn }

Anderson-Dayem bridge [CRYO] A Josephson junction in a superconducting film,formed by a constriction with length and width on the order of a few micrometers or less. { ¦an·dər·sən ¦dā·əm ,brij }

Andrade's creep law [MECH] A law which states that creep exhibits a transient state in which strain is proportional to the cube root of time and then a steady state in which strain is proportional to time. { 'an,drādz 'krēp ,lò }

Andrews's curves [THERMO] A series of isotherms for carbon dioxide, showing the dependence of pressure on volume at various temperatures. { 'an,drüz ,kərvz }

Andronikashvili experiment [CRYO] An experiment to determine the fractional densities of the superfluid and normal fluid components of liquid helium by measuring the period and decrement of a torsional pendulum immersed in the helium. { ,an·drə¦ni·kəsh¦vil·ē ik'sper·ə·mənt }

anelastic creep [MATER] Time-dependent elastic (nonpermanent) creep. . { ,an·ə¦las·tik 'krēp }

anelasticity [MECH] Deviation from a proportional relationship between stress and strain. { ¦an·ə·las¦tis·əd·ē }

anelectric [PHYS] Not becoming charged by friction. { ¦an·ə¦lek·trik }

angle of contact [FL MECH] The angle between the surface of a liquid and the surface of a partially submerged object or of the container at the line of contact. Also known as contact angle. { 'aŋ·gəl əv 'kän,takt }

angle of reflection [PHYS] The angle between the direction of propagation of a wave reflected by a surface and the line perpendicular to the surface at the point of reflection. Also known as reflection angle. { 'aŋ·gəl əv ri'flek·shən }

angle of refraction [PHYS] The angle between the direction of propagation of a wave that is refracted by a surface and the line that is perpendicular to the surface at the point of refraction. { 'aŋ·gəl əv ri'frak·shən }

angular aggregate [MATER] An aggregate whose particles possess well-defined edges formed at the intersection of roughly planar faces. { 'aŋ·gyə·lər 'ag·ri·gət }

anhydride [CHEM] A compound formed from an acid by removal of water. { an'hī,drīd }

anhydrous [CHEM] Being without water, especially water of hydration. { an'hī·drəs }

anhydrous ammonia [INORG CHEM] Liquid ammonia, a colorless liquid boiling at −33.3°C. { an'hī·drəs ə'mōn·yə }

anhydrous calcium sulfate [MATER] Gypsum from which all the water of crystallization has been removed. Also known as dead-burnt gypsum. { an'hī·drəs ,kal·sē·əm 'səl,fāt }

anhydrous ferric chloride See ferric chloride. { an'hī·drəs ,fer·ik 'klòr,īd }

anhydrous gypsum plaster [MATER] Plaster which has a greater percentage of the water of crystallization removed than normal gypsum plasters; used as a finish plaster. { an'hī·drəs ¦jip·səm ¦plas·tər }

anhydrous hydrogen chloride [INORG CHEM] HCl Hazardous, toxic, colorless gas used in polymerization, isomerization, alkylation, nitration, and chlorination reactions; becomes hydrochloric acid in aqueous solutions. { an'hī·drəs ,hī·drə·jən 'klòr,īd }

anhydrous phosphoric acid *See* phosphoric anhydride. { an'hī·dres fä'sfórik 'as·əd }

anhydrous plumbic acid *See* lead dioxide. { an'hī·dres 'pləmb·ik 'as·əd }

anhydrous sodium carbonate *See* soda ash. { an'hi·dres ˌsōd·ē·əm 'kärb·ə,nāt }

anhydrous sodium sulfate [INORG CHEM] Na_2SO_4 Water-soluble, white crystals with bitter, salty taste; melts at $888°C$; used in the manufacture of glass, paper, pharmaceuticals, and textiles, and as an analytical reagent. { an'hī·dres ˌsōd·ē·əm 'səl,fāt }

aniline-formaldehyde resin [CHEM] A thermoplastic resin made by polymerizing aniline and formaldehyde. { 'an·əl·ən ˌfór'mal·də,hīd ˌrez·ən }

animal black [MATER] Finely divided carbon made by calcination of animal bones or ivory; used for pigments, decolorizers, and purifying agents; varieties include bone black and ivory black. { 'an·ə·məl ˌblak }

animal glue [MATER] A glue made from the bones, hide, horns, and connective tissues of animals. { 'an·ə·məl ˌglü }

anion [CHEM] An ion that is negatively charged. { 'a,nī·ən }

anion exchange [CHEM] A type of ion exchange in which the immobilized functional groups on the solid resin are positive. { 'a,nī·ən iks'chānj }

anionic detergent [MATER] A class of detergents having a negatively charged surface-active ion, such as sodium alkylbenzene sulfonate. { ¦a,nī ¦än·ik di'tər·jənt }

anionic polymerization [POLYM CHEM] A type of polymerization in which Lewis bases, such as alkali metals and metallic alkyls, act as imitators. { ¦a,nī¦än·ik pə,lim·ə·rə'zā·shən }

anisotropic [PHYS] Showing different properties as to velocity of light transmission, conductivity of heat or electricity, compressibility, and so on, in different directions. Also known as aeolotropic. { ¦a,nī·sə¦träp·ik }

anisotropic magnetoresistance [SOLID STATE] A type of magnetoresistance displayed by all metallic magnetic materials, which arises because conduction electrons have more frequent collisions when they move parallel to the magnetization in the material than when they move perpendicular to it. { ˌan·ə·sə¦trō·pik ˌmag ˌned·ō·ri'sis·təns }

anisotropic membrane [CHEM ENG] An ultrafiltration membrane which has a thin skin at the separating surface and is supported by a spongy sublayer of membrane material. { ¦a,nī·sə¦träp·ik 'mem,brān }

anisotropy [PHYS] The characteristic of a substance for which a physical property, such as index of refraction, varies in value with the direction in or along which the measurement is made. Also known as aeolotropy; eolotropy. { ¦a,nī'sä·trə·pē }

anisotropy constant [ELECTROMAG] In a ferromagnetic material, temperature-dependent parameters relating the magnetization in various directions to the anisotropy energy. { ¦a ˌnī'sä·trə·pē ˌkän·stənt }

anisotropy energy [ELECTROMAG] Energy stored in a ferromagnetic crystal by virtue of the work done in rotating the magnetization of a domain away from the direction of easy magnetization. { ¦a,nī'sä·trə·pē ˌen·ər·jē }

anneal [ENG] To treat a metal, alloy, or glass with heat and then cool to remove internal stresses and to make the material less brittle. Also known as temper. { ə'nēl }

annealing furnace [ENG] A furnace for annealing metals or glass. Also known as annealing oven. { ə'nēl·iŋ ˌfər·nəs }

annealing oven *See* annealing furnace. { ə'nēl·iŋ ˌəv·ən }

annealing point [THERMO] The temperature at which the viscosity of a glass is $10^{13.0}$ poises. Also known as annealing temperature; 13.0 temperature. { ə'nēl·iŋ ˌpóint }

annealing temperature *See* annealing point. { ə'nēl·iŋ ˌtem·prə·chər }

annealing twin [MET] A twinned crystal that is formed as molten metal is cooled and solidified. { ə'nēl·iŋ ˌtwin }

anode [ELEC] The terminal at which current enters a primary cell or storage battery; it is positive with respect to the device, and negative with respect to the external circuit. [ELECTR] **1.** The collector of electrons in an electron tube. Also known as plate; positive electrode. **2.** In a semiconductor diode, the terminal toward which forward current flows from the external circuit. [PHYS CHEM] The positive terminal of an electrolytic cell. { 'a,nōd }

anode copper [MET] Slabs of refined blister copper used as anodes in the electrolytic refining of copper. { 'a,nōd ˌkäp·ər }

anode corrosion [MET] The disintegration of a metal acting as an anode. { 'a,nōd kə'rō·zhən }

anode-corrosion efficiency [PHYS CHEM] The ratio of actual weight loss of an anode due to corrosion to the theoretical loss as calculated by Faraday's law. { 'a,nōd kə¦rō·zhən i,fish·ən·sē }

anode effect [PHYS CHEM] A condition produced by polarization of the anode in the electrolysis of fused salts and characterized by a sudden increase in voltage and a corresponding decrease in amperage. { 'a,nōd i,fekt }

anode film [CHEM] The portion of solution in immediate contact with the anode. { 'a,nōd ˌfilm }

anode furnace [MET] A furnace in which blister copper or impure nickel is refined. { 'a,nōd ˌfər·nəs }

anode metal [MET] The metal used as anode in an electroplating process. { 'a,nōd ˌmed·əl }

anode mud [MET] An insoluble substance or mixture that collects at the anode in an electrolytic refining or plating process. Also known as anode slime. { 'a,nōd ˌməd }

anode scrap [MET] Portions of anode copper retrieved from electrolytic refining of the metal. { 'a,nōd ˌskrap }

anode slime *See* anode mud. { 'a,nōd ˌslīm }

anodic cleaning |MET| The removal of a foreign substance from a metallic surface by electrolysis with the metal as the anode. Also known as anodic pickling; reverse-current cleaning. { ə'näd·ik 'klēn·iŋ }

anodic coating |MET| A film of oxide produced on a metal by electrolysis with the metal as the anode. { ə'näd·ik 'kōd·iŋ }

anodic pickling See anodic cleaning. { ə'näd·ik 'pik·liŋ }

anodic polarization |PHYS CHEM| The change in potential of an anode caused by current flow. { ə'näd·ik pō·lə·rə'zā·shən }

anodic protection |MET| Reduction of the corrosion rate in an anode by polarizing it into a potential region where the dissolution rates low. { ə'näd·ik prə'tek·shən }

anodic reaction |MET| The reaction in the mechanism of electrochemical corrosion in which the metal forming the anode dissolves in the electrolyte in the form of positively charged ions. { ə'näd·ik rē'ak·shən }

anodize |MET| To form a decorative or protective passive film on a metal part by making it the anode of a cell and applying electric current. { 'an·ə‚dīz }

anodized aluminum |MET| Aluminum coated with a layer of aluminum oxide by an anodic process in a suitable electrolyte such as chromic acid or sulfuric acid solution. { 'an·ə‚dīzd ə'lüm·ə·nəm }

anodized dielectric film |ELEC| An insulating film produced on a conducting surface by anodizing; used for producing thin-film capacitors, trimming resistor values, and passivation in the manufacture of integrated circuits. { 'an·ə‚dīzd dī·ə‚lek·trik 'film }

anodized magnesium |MET| An anodic coating on magnesium produced in one of various electrolytes, mainly of fluorides, phosphates, or chromates. { 'an·ə‚dīzd ‚mag'nēz·ē·əm }

anomalous Barkhausen effect |ELECTROMAG| The occurrence of large steps in the magnetization of an iron-aluminum alloy at temperature above about 400°C (750°F). { ə¦näm·ə·ləs 'bark‚haủz·ən i‚fekt }

anomalous expansion |THERMO| An increase in the volume of a substance that results from a decrease in its temperature, such as is displayed by water at temperatures between 0 and 4°C (32 and 39°F). { ə'näm·ə·ləs ik'span·shən }

anomalous Hall effect |ELECTROMAG| **1.** In a current-carrying conductor in a magnetic field, development of a transverse voltage resulting from the deflection of positive charge carriers (hole states) by the Lorentz force. **2.** The Hall effect in ferromagnetic metals, which arises from the unsymmetrical scattering of conduction electrons at magnetic moments. { ə¦näm·ə·ləs 'hȯl i‚fekt }

anomalous skin effect |ELEC| The skin effect at very low temperatures and high frequencies at which the thickness of the conducting skin layer is less than the electron mean free path, so that the

classical theory of electrical conductivity breaks down. { ə¦näm·ə·ləs 'skin i‚fekt }

anomalous viscosity See non-Newtonian viscosity. { ə'näm·ə·ləs vis'käs·əd·ē }

anorthic crystal See triclinic crystal. { ə'nȯr·thik 'kris·təl }

antemedium |MATER| A substance, such as wax or resin solvent, used in tissue processing prior to infiltration for histological examination. { ‚an·tē'mēd·ē·əm }

anthracite duff |MATER| In Wales, fine anthracite screenings used in briquets or mixed with bituminous coal to fuel cement kilns on chain grate stokers. { 'an·thrə‚sīt ‚dəf }

antiacid bronze |MET| A high-lead, acid-resistant bronze used for casting chemical machine parts. { ¦an·tē¦as·əd 'bränz }

anticorrosive paint |MATER| A paint formulated with a corrosive-resistant pigment (such as lead chromate, zinc chromate, or red lead) and a chemical- and moisture-resistant binder; used to protect iron and steel surfaces. { ‚an·tē·kə'rō·siv 'pānt }

anticurl coating |GRAPHICS| A thin layer generally applied to or on the back of photographic material to prevent front warping. { 'an·tē‚kərl 'kōd·iŋ }

antidesiccant |MATER| Material applied to plants prior to transplanting to reduce the amount of moisture lost by transpiration. { ‚an·tē'des·ə‚kənt }

antiferroelectric crystal |SOLID STATE| A crystalline substance characterized by a state of lower symmetry consisting of two interpenetrating sublattices with equal but opposite electric polarization, and a state of higher symmetry in which the sublattices are unpolarized and indistinguishable. { ¦an·tē‚fer·ō·i'lek·trik 'kris·təl }

antiferromagnetic domain |SOLID STATE| A region in a solid within which equal groups of elementary atomic or molecular magnetic moments are aligned antiparallel. { ¦an·tē‚fer·ō ‚mag'ned·ik dō'mān }

antiferromagnetic resonance |ELECTROMAG| Magnetic resonance in antiferromagnetic materials which may be observed by rotating magnetic fields in either of two opposite directions. { ¦an·tē‚fer·ō‚mag'ned·ik 'rez·ə·nəns }

antiferromagnetic substance |ELECTROMAG| A substance that is composed of antiferromagnetic domains. { ¦an·tē‚fer·ō‚mag'ned·ik 'səb·stəns }

antiferromagnetic susceptibility |ELECTROMAG| The magnetic response to an applied magnetic field of a substance whose atomic magnetic moments are aligned in antiparallel fashion. { ¦an·tē‚fer·ō‚mag'ned·ik sə‚sep·tə'bil·əd·ē }

antiferromagnetism |SOLID STATE| A property possessed by some metals, alloys, and salts of transition elements by which the atomic magnetic moments form an ordered array which alternates or spirals so as to give no net total moment in zero applied magnetic field. { ¦an·tē ‚fer·ō'mag·nə‚tiz·əm }

antifoaming agent |ORG CHEM| A substance,

antifogging compound

such as silicones, organic phosphates, and alcohols, that inhibits the formation of bubbles in a liquid during its agitation by reducing its surface tension. { ¦an·tē¦fōm·iŋ ,ā·jənt }

antifogging compound [MATER] A compound of one or more basic chemicals with filler or extenders for preventing condensation of moisture on glass and other transparent material, such as lenses or windshields. { ¦an·tē¦fäg·iŋ ,käm·pȧu̇nd }

antifouling coating [MATER] A special paint containing copper used on ships' bottoms to prevent marine organisms from attaching themselves. { ,an·tē'fau̇l·iŋ ,kōd·iŋ }

antifreeze [CHEM] A substance added to a liquid to lower its freezing point. { 'an·tē,frēz }

antifreeze proteins [BIOMATER] Proteins that decrease the nonequilibrium freezing point of water without significantly affecting the melting point by directly binding to the surface of an ice crystal, thereby disrupting its normal structure and growth pattern and inhibiting further ice growth; found in a number of fish, insects, and plants. { 'an·ti,frēz ¦prō,tēnz }

antifriction [MECH] Making friction smaller in magnitude. { ,an·tē'frik·shən }

antifriction alloy [MET] An alloy generally having more than 50% tin as a base and cast as facings of machinery bearings, domestic equipment, and small parts in rolling contact. { ,an·tē'frik·shən 'al,ȯi }

antifriction material [ENG] A machine element made of Babbitt metal, lignum vitae, rubber, or a combination of a soft, easily deformable metal overlaid on a hard, resistant one. { ,an·tē'frik·shən mə'tir·ē·əl }

antifungal agent [MATER] A chemical compound that either destroys or inhibits the growth of fungi. { ,an·tē'fəŋ·gəl ,ā·jənt }

antihalation backing [GRAPHICS] A coating on the back of film to minimize reflection of light from the base into the emulsion. { ,an·tē·hə'lā·shən 'bak·iŋ }

antilogous pole [SOLID STATE] That crystal pole which becomes electrically negative when the crystal is heated or is expanded by decompression. { an'til·ə·gəs ¦pōl }

antimagnetic [ENG] Constructed so as to avoid the influence of magnetic fields, usually by the use of nonmagnetic materials and by magnetic shielding. { ,an·tē,mag'ned·ik }

antimonial lead [MET] A lead alloy containing up to 25% antimony and possessing greater hardness and tensile strength than lead; used for storage-battery plates, pipes, cable coverings, and roofing. { ¦an·tə¦mōn·ē·əl 'led }

antimonide [INORG CHEM] A binary compound of antimony with a more positive compound, for example, H₅Sb. Also known as stibide. { 'an·tə·mə,nīd }

antimony(III) oxide [INORG CHEM] Sb_2O_3 Colorless, rhombic crystals, melting at 656°C; insoluble in water; powerful reducing agent. { 'an·tə ,mō·nē ,thrē 'äk,sīd }

antimony [CHEM] A chemical element, symbol

Sb, atomic number 51, atomic weight 121.75. { 'an·tə,mō·nē }

antimony pentachloride [INORG CHEM] $SbCl_5$ A reddish-yellow, oily liquid; hygroscopic, it solidifies after moisture is absorbed and decomposes in excess water; soluble in hydrochloric acid and chloroform; used in analytical testing for cesium and alkaloids, for dyeing, and as an intermediary in synthesis. Also known as antimony perchloride. { 'an·tə,mō·nē ,pent·ə'klȯr,īd }

antimony pentafluoride [INORG CHEM] SbF_5 A corrosive, hygroscopic, moderately viscous fluid; reacts violently with water; forms a clear solution with glacial acetic acid; used in the fluorination of organic compounds. { 'an·tə,mō·nē ,pent·ə'flu̇r ,īd }

antimony pentasulfide [INORG CHEM] Sb_2S_5 An orange-yellow powder; soluble in alkali, soluble in concentrated hydrochloric acid, with hydrogen sulfide as a by-product, and insoluble in water; used as a red pigment. Also known as antimony persulfide; antimony red; golden antimony sulfide. { 'an·tə,mō·nē ,pent·ə'səl,fīd }

antimony perchloride *See* antimony pentachloride. { 'an·tə,mō·nē ,per'klȯr,īd }

antimony persulfide *See* antimony pentasulfide. { 'an·tə,mō·nē ,per'səl,fīd }

antimony red *See* antimony pentasulfide. { 'an·tə,mō·nē 'red }

antimony sodiate *See* sodium antimonate. { 'an·tə,mō·nē 'sō·dē·āt }

antimony sulfate [INORG CHEM] $Sb_2(SO_4)_3$ Antimony(III) sulfate, a white, deliquescent powder; soluble in acids. { 'an·tə,mō·nē 'səl,fāt }

antimony trichloride [INORG CHEM] $SbCl_3$ Hygroscopic, colorless, crystalline mass; fumes slightly in air, is soluble in alcohol and acetone, and forms antimony oxychloride in water; used as a mordant, as a chlorinating agent, and in fireproofing textiles. { 'an·tə,mō·nē ,trī'klȯr ,īd }

antimony trisulfide [INORG CHEM] Sb_2S_3 Black and orange-red rhombic crystals; soluble in concentrated hydrochloric acid and sulfide solutions, insoluble in water; melting point 546°C; used as a pigment, and in matches and pyrotechnics. { 'an·tə,mō·nē ,trī'səl,fīd }

antimony yellow *See* lead antimonite. { 'an·tə ,mō·nē 'ye·lō }

Antioch process [MET] A method of plaster molding in which a plaster-water mixture is poured over a pattern, after which the mold is steam-treated, allowed to set in air, dried in an oven, and cooled for use in casting certain alloys. { 'an·tē,äk 'präs·əs }

antioxidant [PHYS CHEM] A substance that, when present at a lower concentration than that of the oxidizable substrate, significantly inhibits or delays oxidative processes, while being itself oxidized. In primary antioxidants, antioxidative activity is implemented by the donation of an electron or hydrogen atom to a radical derivative, and in secondary antioxidants by the removal of an oxidative catalyst and the consequent prevention of the initiation of oxidation.

Antioxidants are used in polymers to prevent degradation, and in foods, beverages, and cosmetic products to inhibit deterioration and spoilage. { ¡an·tē'äk·sə·dənt }

antique finish |MATER| A paper finish, somewhat rougher than eggshell, obtained by operating wet presses and calender stacks at reduced pressures. { an¦tēk ¦fin·ish }

antirad |CHEM ENG| An inhibitor incorporated into rubber during manufacturing to reduce the degrading effects of radiation. { 'an·te¦rad }

antireflection coating |ENG| The application of a thin film of dielectric material to a surface to reduce its reflection and to increase its transmission of light or other electromagnetic radiation. { ¡an·tē·ri'flek·shən ¸kōd·iŋ }

antislip metal |MET| Metal, usually iron, bronze, or aluminum, containing abrasive grains cast with the metal or rolled out to the surface; used for car steps, stair treads, or floor plates. { ¦an·tē ¦slip ¦med·əl }

antislip paint |MATER| A paint with a high coefficient of friction, caused by addition of sand, wood flour, or cork dust; used on steps, porches, and walkways to prevent slipping. { ¦an·tē¦slip ¦pānt }

antistat See antistatic agent. { 'an·tē¸stat }

antistatic agent |MATER| A material used with textiles, plastics, paper products, or wax polishes to reduce static-electrical charges by allowing the charge to leak off. Also known as antistat. { ¡an·tē¦stad·ik 'ā·jənt }

antistripping agent |MATER| An additive used in an asphaltic binder to overcome the affinity of an aggregate for water instead of asphalt; assists the asphalt to adhere to wet surfaces. { ¸an·tē'strip·iŋ 'ā·jənt }

anvil |MET| **1.** A heavy wrought-iron, cast-iron, or steel block upon which metal is hammered in smith forging. **2.** The base of the hammer, holding the die bed and lower die part in drop forging. { 'an·vəl }

anyris oil |MATER| Sandalwood oil from West India sandalwood species. { ə'nī·rəs ¸oil }

aperture mask See shadow mask. { 'ap·ə¸chər ¸mask }

A phase See liquid A. { 'ā ¸fāz }

A₁ phase See liquid A. { ¦ā·səb¦wən ¸fāz }

apitong |MATER| A wood from the Philippine tree Dipterocarpus grandiflorus; sold as mahogany although it is not a true mahogany. { ə'pē¸tòn }

apparent density |MET| The weight per unit volume of a metal powder, in contrast to the weight per unit volume of the individual particles. { ə'pa·rənt 'dens·əd·ē }

apparent expansion |THERMO| The expansion of a liquid with temperature, as measured in a graduated container without taking into account the container's expansion. { ə'pa·rənt ik'span·shən }

apparent viscosity |FL MECH| The value obtained by applying the instrumental equations used in obtaining the viscosity of a Newtonian fluid to viscometer measurements of a non-Newtonian fluid. { ə¦par·ənt vi'skäs·əd·ē }

aprotic solvent |CHEM| A solvent that does not yield or accept a proton. { ā'präd·ik 'säl·vənt }

APW method See augmented plane-wave method. { ¦ā¦pē'dəb·əl¸yü ¸meth·əd }

aqua |CHEM| Latin for water. { 'äk·wə }

aqua ammonia See ammonium hydroxide. { 'äk·wə ə'mōn·ē·ə }

aquadag |ELECTR| Graphite coating on the inside of certain cathode-ray tubes for collecting secondary electrons emitted by the face of the tube. |MATER| A colloidal suspension of graphite in water. { 'ak·wə¸dag }

aquafortis See nitric acid. { ¦äk·wə'fórd·əs }

aqua regia |INORG CHEM| A fuming, highly corrosive, volatile liquid with a suffocating odor made by mixing 1 part concentrated nitric acid and 3 parts concentrated hydrochloric acid; reacts with all metals, including silver and gold. { ¦äk·wə 'rē·jə }

Ar See argon.

Ar₁ |MET| The temperature at which conversion of austenite to ferrite or to ferrite plus cementite is completed upon cooling a steel.

Ar₃ |MET| The temperature at which austenite begins to convert to ferrite upon cooling a steel.

Ar₄ |MET| The temperature at which delta ferrite is converted to gamma iron (austenite) upon cooling a steel.

Ar_cm |MET| The temperature at which austenite is converted to cementite upon cooling a hypereutectoid steel.

aralkyl |ORG CHEM| A radical in which an aryl group is substituted for an alkyl H atom. Derived from arylated alkyl. { a'ral¸kil }

arbitration bar |MET| A cast-iron specimen, in the form of a standard-sized bar, to be tested for conformity to specifications of the American Society for Testing and Materials. { ¸ar·bə'trā·shən ¸bär }

arbor |MET| A device which supports sand cores in molds. { 'ar·bər }

arborescent powder See dendritic powder. { ¦är·bə¦res·ənt 'paúd·ər }

arc See electric arc. { ärk }

arc blow |MET| The shifting of the arc in various directions in electric-arc welding due to the magnetic fields at the arc. { 'ärk ¸blō }

arc brazing |MET| Brazing with the use of an electric arc. { 'ärk ¸brāz·iŋ }

arc cutting |MET| A type of thermal cutting of metal using the temperature generated by an electric arc. { 'ärk ¸kəd·iŋ }

arc force |MECH| The force of a plasma arc through a nozzle or opening. { 'ärk ¸fórs }

arc furnace |MET| A furnace used to heat materials by the energy from an electric arc. Also known as electric-arc furnace. { 'ärk 'fər·nəs }

arc gouging |MET| The formation of a bevel or groove by an arc cutting process. { 'ärk ¸gaúj·iŋ }

arch brick |MATER| **1.** A wedge-shaped brick used in arches. **2.** An overburned brick, resulting from contact with the fire in the arch of a kiln. { 'ärch ¸brik }

arc heating |MET| The heating of a material by

the heat energy from an electric arc, which has a very high temperature and very high concentration of heat energy. Also known as electric-arc heating. { 'ärk ,hēd·iŋ }

Archimedes number |FL MECH| One of a dimensionless group of numbers denoting the ratio of gravitational force to viscous force. { är·kə¦mēd¸ēz 'nəm·bər }

architectural bronze |MET| An alloy containing 57% copper, 40% zinc, 2.75% lead, 0.25% tin; used for extruded moldings and forgings. { ¦är·kə ¦tek·chər·əl 'bränz }

architectural concrete |MATER| Concrete used for ornamentation or finish on exterior or interior surfaces of a building or other structure. { ¦är·kə ¦tek·chər·əl 'kän,krēt }

architectural terra-cotta |MATER| A hard-burnt, glazed or unglazed clay unit used in building construction. { ¦är·kə¦tek·chər·əl ,ter·ə'käd·ə }

arc melting |MET| Melting and purification of metal in an electric-arc furnace. { ¦ärk ¦melt·iŋ }

arc seam weld |MET| A linear weld or overlapping spot welds made by an arc welding process. { ¦ärk ¦sēm ,weld }

arc-spot weld |MET| A weld that covers a very small area of the surface in contact and was made by an arc welding process. { ¦ärk 'spät ,weld }

arc spraying |MET| Depositing on a surface a metal melted by an electric arc and blown at high speed in an atomized state. { 'ärk ,sprā·iŋ }

arc time |MET| The time interval during which an arc is maintained in the making of an arc weld. { 'ärk ,tīm }

arc voltage |MET| The voltage across a welding arc. { 'ärk ,vōl·tij }

arc welding See electric-arc welding. { 'ärk ,weld·iŋ }

arene See aromatic hydrocarbon. { 'a,rēn }

argentic oxide See silver suboxide. { är'jen·tik 'äk,sīd }

argentocyanides |INORG CHEM| Complexes formed, for example, in the cyanidation of silver ores and in electroplating, when silver cyanide reacts with solutions of soluble metal cyanides. Also known as dicyanoargentates. { är,jen·tō'sī·ə,nīdz }

argentum |CHEM| Latin for silver. { är'jen·təm }

argil See potter's clay. { 'är·jəl }

argon |CHEM| A chemical element, symbol Ar, atomic number 18, atomic weight 39.998. { 'ar ,gän }

armchair nanotube |PHYS CHEM| A carbon nanotube formed from a graphite sheet that is rolled up so that the edge is in the shape of armchairs. { ,ärm,chär 'nan·ō,tüb }

armor castings |MET| A type of armor made of high-alloy steel frequently used when complicated shapes are involved. { 'är·mər ,kast·iŋz }

armored wood |MATER| Wood which is faced on one or both sides with metal sheeting. { 'är·mərd 'wud }

armor plate |MET| Heavy, flat steel, either surface-hardened or hardened throughout, used as a sheathing for warships, tanks, and so forth

to resist penetration and deformation from heavy gunfire. { 'är·mər ¦plāt }

aromatic |ORG CHEM| **1.** Pertaining to or characterized by the presence of at least one benzene ring. **2.** Describing those compounds having physical and chemical properties resembling those of benzene. { ar·ə¦mad·ik }

aromatic alcohol |ORG CHEM| Any of the compounds containing the hydroxyl group in a side chain to a benzene ring, such as benzyl alcohol. { ¦ar·ə¦mad·ik 'al·kə,hól }

aromatic aldehyde |ORG CHEM| An aromatic compound containing the CHO radical, such as benzaldehyde. { ¦ar·ə¦mad·ik 'al·də,hīd }

aromatic amine |ORG CHEM| An organic compound that contains one or more amino groups joined to an aromatic structure. { ¦ar·ə¦mad·ik 'am,ēn }

aromatic hydrocarbon |ORG CHEM| A member of the class of hydrocarbons, of which benzene is the first member, consisting of assemblages of cyclic conjugated carbon atoms and characterized by large resonance energies. Also known as arene. { ¦ar·ə¦mad·ik ¦hī·drə'kär·bən }

aromatic ketone |ORG CHEM| An aromatic compound containing the —CO radical, such as acetophenone. { ¦ar·ə¦mad·ik 'kē,tōn }

aroyl |ORG CHEM| The radical RCO, where R is an aromatic (benzoyl, napthoyl) group. { 'ar·ə·wəl }

Arrhenius equation |PHYS CHEM| The relationship that the specific reaction rate constant k equals the frequency factor constant s times $\exp(-\Delta H_{act}/RT)$, where ΔH_{act} is the heat of activation, R the gas constant, and T the absolute temperature. { ar'rā·nē·əs i'kwā·zhən }

Arrhenius-Guzman equation |PHYS| The relation between the viscosity η of a liquid and the Kelvin temperature T at constant pressure: $\eta = A\exp(B/RT)$, where A and B are constants and R is the gas constant. { ar'rā·nē·əs ,güth·mən i,kwä·zhən }

Arrhenius viscosity formulas |PHYS| A series of three equations which relate the viscosity of a liquid to the temperature, the viscosity of a solution to its concentration and to the viscosity of the solvent, and the viscosity of a sol to the viscosity of the medium. { ar'rā·nē·əs vis'käs·əd·ē ,för·myə·ləz }

arsenate |INORG CHEM| **1.** $AsO_4{}^{3-}$ A negative ion derived from orthoarsenic acid, $H_3AsO_4 \cdot \frac{1}{2}H_2O$. **2.** A salt or ester of arsenic acid. { 'ärs·ən,āt }

arsenic |CHEM| A chemical element, symbol As, atomic number 33, atomic weight 74.9216. { 'ärs·ən·ik }

arsenic acid |INORG CHEM| $H_3AsO_4 \cdot \frac{1}{2}H_2O$ White, poisonous crystals, soluble in water and alcohol; used in manufacturing insecticides, glass, and arsenates and as a defoliant. Also known as orthoarsenic acid. { är'sen·ik 'as·əd }

arsenic disulfide |INORG CHEM| As_2S_2 Red, orange, or black monoclinic crystals, insoluble in water; used in fireworks; occurs naturally as realgar. { 'ärs·ən·ik dī'səl,fīd }

arsenic oxide |INORG CHEM| **1.** An oxide of

arsenic. **2.** *See* arsenic pentoxide;arsenic trioxide. { 'ärs·ən·ik 'äk‚sīd }

arsenic pentasulfide |INORG CHEM| As_2S_5 Yellow crystals that are insoluble in water and readily decompose to the trisulfide and sulfur; used as a pigment. { 'ärs·ən·ik ‚pent·ə'səl‚fīd }

arsenic pentoxide |INORG CHEM| As_2O_5 A white, deliquescent compound that decomposes by heat and is soluble in water. Also known as arsenic oxide. { 'ärs·ən·ik ‚pent'äk‚sīd }

arsenic trichloride |INORG CHEM| $AsCl_3$ An oily, colorless liquid that dissolves in water; used in ceramics, organic chemical syntheses, and in the preparation of pharmaceuticals. { 'ärs·ən·ik ‚tri'klȯr‚īd }

arsenic trioxide |INORG CHEM| As_2O_3 A toxic compound, slightly soluble in water; octahedral crystals change to the monoclinic form by heating at 200°C; occurs naturally as arsenolite and claudetite; used in small quantities in some medicinal preparations. Also known as arsenic oxide; arsenious acid. { 'ärs·ən·ik ‚tri'äk‚sīd }

arsenic trisulfide |INORG CHEM| As_2S_3 An acidic compound in the form of yellow or red monoclinic crystals with a melting point at 300°C; occurs as the mineral orpiment; used as a pigment. { 'ärs·ən·ik ‚tri'səl‚fīd }

arsenide [CHEM] A binary compound of negative, trivalent arsenic; for example, H_3As or GaAs. { 'ärs·ən‚īd }

arsenious acid *See* arsenic trioxide. { är'sēn·ē·əs 'as·əd }

arsenite |INORG CHEM| **1.** AsO_3^{3-} A negative ion derived from aqueous solutions of As_4O_6. **2.** A salt or ester of arsenious acid. { 'är·sə‚nīt }

arsenous oxide *See* arsenic trioxide. { 'är·sə·nəs 'äk‚sīd }

arsine |INORG CHEM| H_3As A colorless, highly poisonous gas with an unpleasant odor. { är'sēn }

arsinic acid |INORG CHEM| An acid of general formula R_2AsO_2H; derived from trivalent arsenic; an example is cacodylic acid, or dimethylarsinic acid, $(CH_3)_2AsO_2H$. { är'sin·ik 'as·əd }

arsonic acid |INORG CHEM| An acid derived from orthoarsenic acid, $OAs(OH)_3$; the type formula is generally considered to be $RAsO(OH)_2$; an example is *para*-aminobenzenearsonic acid, $NH_2C_6H_4AsO(OH)_2$. { är'sän·ik 'as·əd }

arsonium |INORG CHEM| $-AsH_4$ A radical which may be considered analogous to the ammonium radical in that a compound such as AsH_4OH may form. { är'sōn·ē·əm }

artificial aging [MET] The heat treatment of an alloy at moderately elevated temperatures to accelerate precipitation of a component from the supersaturated solid solution. { ¦ärd·ə¦fish·əl 'āj·iŋ }

artificial atom [ELECTR] A structure, typically 50–100 nanometers in diameter, that is fabricated in a semiconductor crystal and holds a small number of electrons which are trapped in a bowllike potential well. { ‚ärd·ə‚fish·əl 'ad·əm }

artificial crystal *See* superlattice. { ¦ärd·ə ¦fish·əl 'krist·əl }

artificial fiber [TEXT] A filament made from materials such as glass, rayon, or nylon. Also known as synthetic fiber. { ¦ärd·ə¦fish·əl 'fī·bər }

artificial gold *See* stannic sulfide. { ¦ärd·ə¦fish·əl 'gōld }

artificially layered structure *See* superlattice. { ¦ärd·ə¦fish·əl·ē ¦lā·ərd 'strək·chər }

artificial malachite *See* copper carbonate. { ¦ärd·ə¦fish·əl 'mal·ə‚kīt }

artificial nerve graft |BIOMATER| Used to enhance peripheral nerve regeneration, a porous or resorbable tube containing matrix material that may lead axons to grow in the desired direction. { ‚ärd·ə¦fish·əl 'nərv ‚graft }

artificial scheelite *See* calcium tungstate. { ¦ärd·ə¦fish·əl 'shā‚līt }

aryl |ORG CHEM| An organic group derived from an aromatic hydrocarbon by removal of one hydrogen. { 'ar·əl }

aryl compound |ORG CHEM| Molecules with the six-carbon aromatic ring structure characteristic of benzene or compounds derived from aromatics. { 'ar·əl ‚käm‚paúnd }

aryl halide |ORG CHEM| An aromatic derivative in which a ring hydrogen has been replaced by a halide atom. { 'ar·əl 'hal‚īd }

arylide |ORG CHEM| A compound formed from a metal and an aryl group, for example, PbR_4, where R is the aryl group. { 'ar·ə‚līd }

aryne |ORG CHEM| An aromatic species in which two adjacent atoms of a ring lack substituents, with two orbitals each missing an electron. Also known as benzyne. { 'a‚rīn }

As *See* arsenic.

asbestine [MATER] A material with the properties of asbestos. { as'be‚stēn }

asbestos blanket [MATER] Asbestos fibers (alone or in combination with other fibers) stitched, bonded, or woven into flexible blanket form; used for high-temperature insulation or for fire barriers. { as'bes·təs ‚blaŋ·kət }

asbestos board [MATER] A sheet of fire-resistant material made from asbestos fiber and portland cement. { as'bes·təs ‚bȯrd }

asbestos cement [MATER] A building material composed of a mixture of asbestos fiber, portland cement, and water made into plain sheets, corrugated sheets, tiles, and piping. { as'bes·təs si‚ment }

asbestos-cement pipe [MATER] A concrete pipe made of a mixture of portland cement and asbestos fiber and highly resistant to corrosion; used in drainage systems, waterworks systems, and gas lines. { as'bes·təs si‚ment 'pīp }

asbestos felt [MATER] A product made by saturating felted asbestos with asphalt or other suitable binder, such as a synthetic elastomer. { as'bes·təs 'felt }

asbestos insulation |MATER| A material composed of asbestos fibers bonded with mixtures of clay or sodium silicate; used as thermal insulation for temperatures above 1500°F (816°C). { as'bes·təs ‚in·sə'lā·shən }

asbestos plaster [MATER] A fireproof insulating material generally composed of asbestos with bentonite as the binder. { as'bes·təs 'plas·tər }

asbestos roofing [MATER] Roofing or wall cladding sheets made of asbestos cement. { as'bes·təs 'rüf·iŋ }

asbestos shingle [MATER] A shingle composed of asbestos cement formed under pressure; used on houses for roofing and siding that resist the destructive effects of time, weather, and fire. { as'bes·təs 'shiŋ·gəl }

as-brazed [MET] A brazement prior to any additional treatment such as thermal, chemical, or mechanical treatments. { ‚az 'brāzd }

ashlar brick [MATER] A brick having rough-hackled faces resembling stone. { 'ash·lər ‚brik }

asperomagnetic state [SOLID STATE] The condition of a rare-earth glass in which the spins are oriented in fixed directions, with most nearest-neighbor spins parallel or nearly parallel, so that the spin directions are distributed in one hemisphere. { a‚sper·ō‚mag'ned·ik 'stāt }

asphalt [MATER] A brown to black, hard, brittle, or plastic bituminous material composed principally of hydrocarbons; formed in oil-bearing rocks near the Dead Sea, and in Trinidad; prepared by pyrolysis from coal tar, certain petroleums, and lignite tar; melts on heating; insoluble in water but soluble in gasoline; used for paving and roofing and in paints and varnishes. { 'a‚sfólt }

asphalt base crude [MATER] A crude oil which yields napthenic asphalt residues when processed. { 'a‚sfólt ‚bās 'krüd }

asphalt block [MATER] A paving block composed of a mixture of 88–92% crushed stone with the balance asphalt cement. { 'a‚sfólt ‚bläk }

asphalt cement [MATER] Fluxed or unfluxed asphalt especially prepared for direct use in making bituminous pavements. { 'a‚sfólt si'ment }

asphalt emulsion [MATER] Asphalt cement in water containing a small amount of emulsifying agent. { 'a‚sfólt i'məl·shən }

asphalt-emulsion slurry seal [MATER] A mixture of slow-setting emulsified asphalt, fine aggregate, and mineral filler, with water added to produce a slurry consistency. { 'a‚sfólt i'məl·shən 'slə·rē ‚sēl }

asphalt enamel [MATER] A surface coating composed of a mixture of asphalt, finely pulverized mica, clay, soapstone, or talc; applied to pipe that will be buried. { 'a‚sfólt i'nam·əl }

asphaltene [MATER] Any of the dark, solid constituents of crude oils and other bitumens which are soluble in carbon disulfide but insoluble in paraffin naphthas; they hold most of the organic constituents of bitumens. { a'sfól‚tēn }

asphalt flux [MATER] An oil used to reduce the consistency or viscosity of hard asphalt to the point required for use. { 'a‚sfólt ‚fləks }

asphalt fog seal [MATER] An asphalt surface treatment consisting of a light application of liquid asphalt without a mineral aggregate cover. { 'a‚sfólt 'fäg ‚sēl }

asphaltic base oil [MATER] A crude oil that has asphaltum as the predominating solid residual. { a'sfólt·ik ‚bas ‚óil }

asphaltic concrete [MATER] A special concrete consisting of a mixture of graded aggregate and heated asphalt; must be applied and spread while hot. { a'sfólt·ik 'kän‚krēt }

asphaltic material [MATER] A solid, liquid, or semisolid mixture of heavy hydrocarbons and nonmetallic derivatives; obtained from naturally occurring bituminous deposits or from residues of petroleum refining. { a'sfól·tik mə'tir·ē·əl }

asphaltic road oil [MATER] A thick, fluid solution of asphalt. { a'sfólt·ik 'rōd ‚óil }

asphalt lamination [MATER] A laminate of sheet material, such as paper or felt, which uses asphalt as the adhesive. { 'a‚sfólt ‚lam·ə'nā·shən }

asphalt macadam [MATER] Pavement made with an asphalt binder rather than tar. { 'a ‚sfólt mə'kad·əm }

asphalt mastic [MATER] A mixture of asphalt with sand, asbestos, crushed rock, or similar material; used like cement. Also known as mastic asphalt. { 'a‚sfólt 'mas·tik }

asphalt paint [MATER] Asphaltic material dissolved in volatile solvent with or without pigments, drying oils, resins, and so on. { 'a‚sfólt 'pānt }

asphalt paper [MATER] A paper that is coated or impregnated with asphalt. { 'a‚sfólt 'pā·pər }

asphalt pavement [CIV ENG] A pavement consisting of a surface layer of mineral aggregate, coated and cemented together with asphalt cement on supporting layers. { 'a‚sfólt 'pāv·mənt }

asphalt primer [MATER] Low-viscosity, liquid asphaltic material applied to and absorbed by nonbituminous surfaces, as in waterproofing. { 'a‚sfólt 'prīm·ər }

asphalt roofing [MATER] A roofing material made by impregnating a dry roofing felt with a hot asphalt saturant, applying asphalt coatings to the weather and reverse sides, and embedding a mineral surfacing in the coating on the weather side. { 'a‚sfólt 'rüf·iŋ }

asphalt shingle [MATER] A roof shingle made of felt impregnated with asphalt and covered with mineral granules. { 'a‚sfólt 'shiŋ·gəl }

asphalt tile [MATER] Floor tile composed of asbestos fibers, mineral coloring pigments, and inert fillers bound together; used on rigid subfloors or hardwood floors. { 'a‚sfólt 'tīl }

asphaltum [MATER] Bituminous material in oil of turpentine; used in photomechanical work because of its ability to be rendered insoluble in light. { a'sfól·təm }

assay bar [MET] A bar of pure or nearly pure gold and silver; used by a government as a standard. { 'a‚sā ‚bär }

assembled stone [MATER] A stone made of two or more gem materials, whether genuine or imitation. { ə'sem·bəld 'stōn }

association [CHEM] Combination or correlation of substances or functions. { ə‚sō·sē'ā·shən }

A stage [POLYM CHEM] An early stage in a thermosetting resin reaction characterized by linear

structure, solubility, and fusibility of the material.
{ 'ā ‚stāj }

astatic [PHYS] Without orientation or directional characteristics; having no tendency to change position. { ā'stad·ik }

astatine [CHEM] A radioactive chemical element, symbol At, atomic number 85, the heaviest of the halogen elements. { 'as·tə‚tēn }

Aston process [MET] A process for making controlled-quality wrought iron synthetically. { 'as·tən 'präs·əs }

as-welded [MET] The condition of a weldment prior to any additional thermal, chemical, or mechanical treatment. { ‚az'weld·əd }

asymmetrical conductivity [ELEC] A variation in the conductivity of a conductor over its cross section that is not symmetric about the conductor's central axis. { ¦ā·sə¦me·tri·kəl ‚kän ‚dək'tiv·əd·ē }

asymmetric carbon atom [ORG CHEM] A carbon atom with four different atoms or groups of atoms bonded to it. Also known as chiral carbon atom; stereogenic center. { ¦ā·sə¦me·trik ¦kär·bən 'ad·əm }

asymmetric synthesis [ORG CHEM] Chemical synthesis of a pure enantiomer, or of an enantiomorphic mixture in which one enantiomer predominates, without the use of resolution. { ¦ā·sə¦me·trik 'sin·thə·səs }

asymmetry [PHYS CHEM] The geometrical design of a molecule, atom, or ion that cannot be divided into like portions by one or more hypothetical planes. Also known as molecular asymmetry. { ¦ā'sim·ə·trə }

asynchronous [PHYS] Not synchronous. { ā 'siŋ·krə·nəs }

At See astatine.

atactic [POLYM CHEM] Of the configuration for a polymer, having the opposite steric configurations for the carbon atoms of the polymer chain occur in equal frequency and more or less at random. { ā'tak·tik }

athermancy [ELECTROMAG] Property of a substance which cannot transmit infrared radiation. { ¦ā¦thər·mən·sē }

atm See atmosphere.

atmosphere [MECH] A unit of pressure equal to 101.325 kilopascals, which is the air pressure measured at mean sea level. Abbreviated atm. Also known as standard atmosphere. { 'at·mə ‚sfir }

atmospheric corrosion [MET] The gradual destruction or alteration of a metal or alloy by contact with substances present in the atmosphere, such as oxygen, carbon dioxide, water vapor, and sulfur and chlorine compounds. { ¦at·mə¦sfir·ik kə'rō·zhən }

atmospheric pressure [PHYS] The pressure at any point in an atmosphere due solely to the weight of the atmospheric gases above the point concerned. Also known as barometric pressure. { ¦at·mə¦sfir·ik 'presh·ər }

atom [CHEM] The individual structure which constitutes the basic unit of any chemical element. { 'ad·əm }

atom cluster [PHYS CHEM] An assembly of between three and a few thousand atoms or molecules that are weakly bound together and have properties intermediate between those of the isolated atom or molecule and the bulk or solid-state material. { 'ad·əm ‚kləs·tər }

atomic beam [PHYS] A stream of atoms, which may or may not be ionized. { ə'täm·ik 'bēm }

atomic constants See fundamental constants. { ə'täm·ik 'kän·stəns }

atomic force microscope [ENG] A device for mapping surface atomic structure by measuring the force acting on the tip of a sharply pointed wire or other object that is moved over the surface. { ə¦täm·ik ¦fòrs 'mī¦krə‚skōp }

atomic frequency [SOLID STATE] One of the vibrational frequencies of an atom in a crystal lattice. { ə¦täm·ik 'frē·kwən·sē }

atomic hydrogen [CHEM] Gaseous hydrogen whose molecules are dissociated into atoms. { ə'täm·ik 'hī·drə·jən }

atomic hydrogen welding [MET] An arc welding process in which hydrogen gas dissociated by the arc recombines outside the arc to provide intense heat and protection against oxidation for the weld. { ə'täm·ik 'hī·drə·jən 'weld·iŋ }

atomic mass [PHYS] The mass of a neutral atom usually expressed in atomic mass units. { ə'täm·ik 'mas }

atomic mass unit [PHYS] An arbitrarily defined unit in terms of which the masses of individual atoms are expressed; the standard is the unit of mass equal to one-twelfth the mass of the carbon atom, having as nucleus the isotope with mass number 12. Abbreviated amu. Also known as dalton. { ə'täm·ik 'mas 'yü·nət }

atomic paramagnetism [ELECTROMAG] The result of a permanent magnetic moment in an atom. { ə'täm·ik ‚par·ə'mag·nə‚tiz·əm }

atomic radius [PHYS CHEM] Also known as covalent radius. **1.** Half the distance between the nuclei of two like atoms that are covalently bonded. **2.** The experimentally determined radius of an atom in a covalently bonded compound. { ə'täm·ik 'rād·ē·əs }

atomic susceptibility [ELECTROMAG] The magnetization of a material per atom per unit of applied field; measured in ergs per oersted per atom. { ə'täm·ik sə‚sep·tə'bil·əd·ē }

atomic theory [CHEM] The assumption that matter is composed of particles called atoms and that these are the limit to which matter can be subdivided. { ə'täm·ik 'thē·ə·rē }

atomic weight [CHEM] The relative mass of an atom based on a scale in which a specific carbon atom (carbon-12) is assigned a mass value of 12. Abbreviated at. wt. Also known as relative atomic mass. { ə'täm·ik 'wät }

atomization [CHEM] A process in which the chemical bonds in a molecule are broken to yield separated (free) atoms. { ‚ad·ə·mə'zā· shən }

atom optics [PHYS] The use of laser light and nanofabricated structures to manipulate the motion of atoms in the same manner that rudimentary optical elements control light. { ¦ad·əm 'äp·tiks }

atom probe [ENG] An instrument for identifying a single atom or molecule on a metal surface; it consists of a field ion microscope with a probe hole in its screen opening into a mass spectrometer; atoms that are removed from the specimen by pulsed field evaporation fly through the probe hole and are detected in the mass spectrometer. { 'ad·əm ‚prōb }

attenuation [ELEC] The exponential decrease with distance in the amplitude of an electrical signal traveling along a very long uniform transmission line, due to conductor and dielectric losses. [ENG] A process by which a material is fabricated into a thin, slender configuration, such as forming a fiber from molten glass. [PHYS] The reduction in level of a quantity, such as the intensity of a wave, over an interval of a variable, such as the distance from a source. { ə‚ten·yə'wā·shən }

attrition [MATER] Wear caused by rubbing or friction. For metal surfaces, also known as scoring; scouring. { ə'trish·ən }

attritor [MET] A high-energy stirred-ball mill used for mechanically alloying metal powder particles. { ə'trid·ər }

at. wt *See* atomic weight.

Au *See* gold.

augmented plane-wave method [SOLID STATE] A method of approximating the energy states of electrons in a crystal lattice; the potential is assumed to be spherically symmetrical within spheres centered at each atomic nucleus and constant in the interstitial region, wave functions (the augmented plane waves) are constructed by matching solutions of the Schrödinger equation within each sphere with plane-wave solutions in the interstitial region, and linear combinations of these wave functions are then determined by the variational method. Abbreviated APW method. { ȯg¦ment·əd 'plān ‚wāv ‚meth·əd }

auric oxide *See* gold oxide. { 'ȯr·ik 'äk‚sīd }

aurora *See* corona discharge. { ə'rȯr·ə }

ausforging [MET] The forming of austenitic steel into required shapes by hammering or pressing after cooling. { ȯs'fȯrj·iŋ }

ausforming [MET] The processing of steel by subjecting it to deformation followed by quenching and tempering in order to increase strength, ductility, and resistance to fatigue. { 'aüs ‚fȯrm·iŋ }

ausrolling [MET] The working of austenitic steel by passing it, after cooling, between oppositely revolving rollers. { ȯs'rōl·iŋ }

austempering [MET] A process for the heat treatment of austenitic steel. { ‚ȯs'tem·pə·riŋ }

austenite [MET] Gamma iron with carbon in solution. { 'ȯs·tə‚nīt }

austenitic [MET] Composed mainly of austenite. { ¦ȯs·tə¦nid·ik }

austenitic cast iron [MET] The product resulting from changing the basic crystalline structure of gray or ductile iron by alloying it with substantial amounts of nickel, manganese, silicon, or other elements. { ¦ȯs·tə¦nid·ik ¦kast 'ī·ərn }

austenitic manganese steel *See* Hadfield manganese steel. { ¦ȯs·tə¦nid·ik ¦maŋ·ga‚nēs 'stēl }

austenitic stainless steel [MET] Stainless steel composed principally of austenite made stable by alloying with nickel. { ¦ȯs·tə¦nid·ik ¦stān·ləs 'stēl }

austenitic steel [MET] An alloy whose structure is typically that of austenite at room temperature. { ¦ȯs·tə¦nid·ik 'stēl }

austenitize [MET] To heat steel to the temperature range in which the crystalline form of the iron is austenite. { ¦ȯs·tə·nə‚tīz }

autocatalysis [CHEM] A catalytic reaction started by the products of a reaction that was itself catalytic. { ¦ȯd·ō·kə'tal·ə·səs }

autoclave curing [ENG] Steam curing of concrete products, sand-lime brick, asbestos cement products, hydrous calcium silicate insulation products, or cement in an autoclave at maximum ambient temperatures generally between 340 and 420°F (170 and 215°C). { 'ȯd·ō‚klāv 'kyúr·iŋ }

autoclave molding [ENG] A method of curing reinforced plastics that uses an autoclave with 50–100 pounds per square inch (345–690 kilopascals) steam pressure to set the resin. { 'ȯd·ō ‚klāv ¦mōld·iŋ }

autogenous ignition temperature *See* ignition temperature. { ȯ'täj·ə·nəs ig'nish·ən ‚tem· prə·chər }

autogenous volume change [MATER] The change in volume produced by continued hydration of cement, exclusive of effects of external forces or change of water content or temperature. { ȯ'täj·ə·nəs 'väl·yəm ‚chanj }

autogenous welding [MET] A fusion welding process using heat without the addition of filler metal to join two pieces of the same metal. { ȯ'täj·ə·nəs 'weld·iŋ }

autoignition temperature [CHEM] The temperature at which a material (solid, liquid, or gas) will self-ignite and sustain combustion in air without an external spark or flame. { ¦ȯd·ō·ig¦nish·ən 'tem·prə·chər }

automatic brazing [MET] Brazing by the use of either portable or stationary equipment which does not require constant supervision by the operator. { ¦ȯd·ə¦mad·ik 'brāz·iŋ }

automatic welding [MET] Electric-arc welding with automatic control of the arc movement along the welding line, the electrode feed, and the arc-gap length. { ¦ȯd·ə¦mad·ik 'weld·iŋ }

autooxidation *See* autoxidation. { ¦ȯd·ō‚äk·sə 'dā·shən }

autoprotolysis [CHEM] Transfer of a proton from one molecule to another of the same substance. { ‚ȯd·ō·prə'täl·ə·səs }

autoxidation [CHEM] **1.** The slow, flameless combustion of materials by reaction with oxygen. **2.** An oxidation reaction that is self-catalyzed and spontaneous. **3.** An oxidation reaction begun only by an inductor. Also known as autooxidation. { ȯ¦täk·sə'dā·shən }

elastic [MECH] Capable of sustaining deformation without permanent loss of size or shape. { i'las·tik }

elastica [MECH] The elastic curve formed by a uniform rod that is originally straight, then is bent in a principal plane by applying forces, and couples only at its ends. { i'las·tə·kə }

elastic aftereffect [MECH] The delay of certain substances in regaining their original shape after being deformed within their elastic limits. Also known as elastic lag. { i'las·tik 'af·tər·i,fekt }

elastic axis [MECH] The lengthwise line of a beam along which transverse loads must be applied in order to produce bending only, with no torsion of the beam at any section. { i'las·tik 'ak·səs }

elastic body [MECH] A solid body for which the additional deformation produced by an increment of stress completely disappears when the increment is removed. Also known as elastic solid. { i'las·tik 'bäd·ē }

elastic buckling [MECH] An abrupt increase in the lateral deflection of a column at a critical load while the stresses acting on the column are wholly elastic. { i'las·tik 'bək·liŋ }

elastic center [MECH] That point of a beam in the plane of the section lying midway between the flexural center and the center of twist in that section. { i'las·tik 'sen·tər }

elastic collision [MECH] A collision in which the sum of the kinetic energies of translation of the participating systems is the same after the collision as before. { i'las·tik kə'lizh·ən }

elastic constant See compliance constant; stiffness constant. { i'las·tik 'kän·stənt }

elastic curve [MECH] The curved shape of the longitudinal centroidal surface of a beam when the transverse loads acting on it produced wholly elastic stresses. { i'las·tik 'kərv }

elastic deformation [MECH] Reversible alteration of the form or dimensions of a solid body under stress or strain. { i'las·tik ,dē·fər'mā·shən }

elastic design [CIV ENG] In the design of a structural member, a method of analysis based on a linear stress-strain relationship, with the assumption that the working stresses constitute only a fraction of the elastic limit of the material. { i'las·tik di'zīn }

elastic equilibrium [MECH] The condition of an elastic body in which each volume element of the body is in equilibrium under the combined effect of elastic stresses and externally applied body forces. { i'las·tik ,ē·kwə'lib·rē·əm }

elastic failure [MECH] Failure of a body to recover its original size and shape after a stress is removed. { i'las·tik 'fāl·yər }

elastic flow [MECH] Return of a material to its original shape following deformation. { i'las·tik 'flō }

elastic force [MECH] A force arising from the deformation of a solid body which depends only on the body's instantaneous deformation and not on its previous history, and which is conservative. { i'las·tik 'fȯrs }

elastic hysteresis [MECH] Phenomenon exhibited by some solids in which the deformation of the solid depends not only on the stress applied to the solid but also on the previous history of this stress; analogous to magnetic hysteresis, with magnetic field strength and magnetic induction replaced by stress and strain respectively. { i'las·tik ,his·tə'rē·səs }

elasticity [MECH] **1.** The property whereby a solid material changes its shape and size under action of opposing forces, but recovers its original configuration when the forces are removed. **2.** The existence of forces which tend to restore to its original position any part of a medium (solid or fluid) which has been displaced. { i,las'tis·əd·ē }

elasticity modulus See modulus of elasticity. { i,las'tis·əd·ē ,mäj·ə·ləs }

elastic lag See elastic aftereffect. { i'las·tik 'lag }

elastic limit [MECH] The maximum stress a solid can sustain without undergoing permanent deformation. { i,las'tis·tik 'lim·ət }

elastic-limit effect [SOLID STATE] A phenomenon in which a material acquires a sharp elastic limit because of the introduction of foreign atoms. { i'las·tik 'lim·ət i,fekt }

elastic modulus See modulus of elasticity. { i,las·tik 'mäj·ə·ləs }

elasticoviscosity [FL MECH] That property of a fluid whose rate of deformation under stress is the sum of a part corresponding to a viscous Newtonian fluid and a part obeying Hooke's law. { i'las·tə·kō·vis'käs·əd·ē }

elastic potential energy [MECH] Capacity that a body has to do work by virtue of its deformation. { i'las·tik pə'ten·chəl 'en·ər·jē }

elastic ratio [MECH] The ratio of the elastic limit to the ultimate strength of a solid. { i'las·tik 'rā·shō }

elastic recovery [MECH] That fraction of a given deformation of a solid which behaves elastically. { i'las·tik ri'kəv·ə·rē }

elastic scattering [MECH] Scattering due to an elastic collision. { i'las·tik 'skad·ə·riŋ }

elastic solid See elastic body. { i'las·tik 'säl·əd }

elastic strain energy [MECH] The work done in deforming a solid within its elastic limit. { i'las·tik 'strān ,en·ər·jē }

elastic theory [MECH] Theory of the relations between the forces acting on a body and the resulting changes in dimensions. { i'las·tik 'thē·ə·rē }

elastic vibration [MECH] Oscillatory motion of a solid body which is sustained by elastic forces and the inertia of the body. { i'las·tik vī'brā·shen }

elastodynamics [MECH] The study of the mechanical properties of elastic waves. { i'las·tō·dī'nam·iks }

elastomer [POLYM CHEM] A polymeric material, such as a synthetic rubber or plastic, which at room temperature can be stretched under stress and, upon immediate release of the stress, will return with force to its approximate original length. { i'las·tə·mər }

elastoplasticity [MECH] State of a substance

subjected to a stress greater than its elastic limit but not so great as to cause it to rupture, in which it exhibits both elastic and plastic properties. { i¦las·tō·plə'stis·əd·ē }

elastoresistance |ELEC| The change in a material's electrical resistance as it undergoes a stress within its elastic limit. { i¦las·tō·ri'zis·təns }

electret |ELEC| A solid dielectric possessing persistent electric polarization, by virtue of a long time constant for decay of a charge instability. { i'lek,tret }

electric |ELEC| Containing, producing, arising from, or actuated by electricity; often used interchangeably with electrical. { i'lek·trik }

electrical |ELEC| Related to or associated with electricity, but not containing it or having its properties or characteristics; often used interchangeably with electric. { ə'lek·trə·kəl }

electrical axis |SOLID STATE| The x axis in a quartz crystal; there are three such axes in a crystal, each parallel to one pair of opposite sides of the hexagon; all pass through and are perpendicular to the optical, or z, axis. { ə'lek·trə·kəl 'ak·səs }

electrical conductance See conductance. { ə'lek·trə·kəl kən'dək·təns }

electrical conduction See conduction. { ə'lek·trə·kəl kən'dək·shən }

electrical conductivity See conductivity. { ə'lek·trə·kəl ¸kän,dək'tiv·əd·ē }

electrical discharge machining See electron discharge machining. { i'lek·trə·kəl 'dis,chärj mə ¸shēn·iŋ }

electrical disintegration |MET| Removing excess metal by using an electric spark in air. { i'lek·trə·kəl dis,in·tə'grā·shən }

electrical impedance |ELEC| **1.** The total opposition that a circuit presents to an alternating current, equal to the complex ratio of the voltage to the current in complex notation. Also known as complex impedance. **2.** The ratio of the maximum voltage in an alternating-current circuit to the maximum current; equal to the magnitude of the quantity in the first definition. Also known as impedance. { i'lek·trə·kəl im'pēd·əns }

electrical insulator See insulator. { i'lek·trə·kəl 'in·sə,lād·ər }

electrically active fluid |PHYS CHEM| A fluid whose properties are altered by either an electric field (electrorheological fluid) or a magnetic field (ferrofluid). { i'lek·trə·klē ¦ak·tiv 'flü·əd }

electrical oil See insulating oil. { i'lek·trə·kəl 'óil }

electrical porcelain See insulation porcelain. { i'lek·trə·kəl 'pòrs·lən }

electrical properties |ELEC| Properties of a substance which determine its response to an electric field, such as its dielectric constant or conductivity. { i'lek·trə·kəl 'präp·ərd·ēz }

electrical resistance See resistance. { i'lek·trə·kəl ri'zis·təns }

electrical resistivity |ELEC| The electrical resistance offered by a material to the flow of current, times the cross-sectional area of current flow and

per unit length of current path; the reciprocal of the conductivity. Also known as resistivity; specific resistance. { i'lek·trə·kəl ¸rē·zis'tiv·əd·ē }

electrical steel |MET| Low carbon-iron alloy containing 0.5–5% silicon, produced in an electric-arc furnace and used for the cores of transformers, alternators, and other iron-core electric machines. { i'lek·trə·kəl 'stēl }

electrical tape See insulating tape. { i'lek·trə·kəl ¸tāp }

electric arc |ELEC| A discharge of electricity through a gas, normally characterized by a voltage drop approximately equal to the ionization potential of the gas. Also known as arc. { i¦lek·trik 'ärk }

electric-arc furnace See arc furnace. { i'lek·trik ¸ärk 'fər·nəs }

electric-arc heating See arc heating. { i'lek·trik ¸ärk 'hēd·iŋ }

electric-arc spraying |MET| A thermal spraying process with an electric arc as a heat source and with compressed gas to propel the material. { i'lek·trik ¸ärk 'sprā·iŋ }

electric-arc welding |MET| Welding in which the joint is heated to fusion by an electric arc or by a large electric current. Also known as arc welding. { i'lek·trik ¸ärk 'weld·iŋ }

electric cell |ELEC| **1.** A single unit of a primary or secondary battery that converts chemical energy into electric energy. **2.** A single unit of a device that converts radiant energy into electric energy, such as a nuclear, solar, or photovoltaic cell. { i¦lek·trik 'sel }

electric charge See charge. { i¦lek·trik 'chärj }

electric circuit |ELEC| Also known as circuit. **1.** A path or group of interconnected paths capable of carrying electric currents. **2.** An arrangement of one or more complete, closed paths for electron flow. { i¦lek·trik 'sər·kət }

electric condenser See capacitor. { i¦lek·trik kən'den·sər }

electric conductor See conductor. { i¦lek·trik kən'dək·tər }

electric connection |ELEC| A direct wire path for current between two points in a circuit. { i¦lek·trik kə'nek·shən }

electric corona See corona discharge. { i¦lek·trik kə'rō·nə }

electric current See current. { i¦lek·trik 'kə·rənt }

electric current density See current density. { i¦lek·trik ¦kə·rənt ¸den·səd·ē }

electric dipole |ELEC| A localized distribution of positive and negative electricity, without net charge, whose mean positions of positive and negative charges do not coincide. { i¦lek·trik 'dī,pōl }

electric dipole moment |ELEC| A quantity characteristic of a charge distribution, equal to the vector sum over the electric charges of the product of the charge and the position vector of the charge. { i¦lek·trik 'dī,pōl ¸mō·mənt }

electric displacement |ELEC| The electric field intensity multiplied by the permittivity. Symbolized D. Also known as dielectric displacement; dielectric flux density; displacement; electric

in water; used as a pigment in overglazes. { 'bar·ē·əm 'krō,māt }

barium dioxide See barium peroxide. { 'bar·ē·əm dī'äk,sīd }

barium fluoride [INORG CHEM] BaF₂ Colorless, cubic crystals, slightly soluble in water; used in enamels. { 'bar·ē·əm flúr,īd }

barium fluosilicate [INORG CHEM] BaSiF₆H A white, crystalline powder; insoluble in water; used in ceramics and insecticides. { 'bar·ē·əm ,flü·ə'sil·ə,kāt }

barium glass [MATER] Glass which differs from ordinary lime-soda glass in that barium oxide replaces part of the calcium oxide. { 'bar·ē·əm 'glas }

barium hydroxide [INORG CHEM] Ba(OH)₂·8H₂O Colorless, monoclinic crystals, melting at 78°C; soluble in water, insoluble in acetone; used for fat saponification and fusing of silicates. { 'bar·ē·əm hī'dräk,sīd }

barium hyposulfite See barium thiosulfate. { 'bar·ē·əm ,hī·pō'səl,fīt }

barium manganate [INORG CHEM] BaMnO₄ A toxic, emerald-green powder which is used as a paint pigment. Also known as Cassel green; manganese green. { 'bar·ē·əm 'maŋ·gə,nāt }

barium mercury iodide See mercuric barium iodide. { 'bar·ē·əm 'mər·kyə·rē 'ī·ə,dīd }

barium molybdate [INORG CHEM] BaMoO₄ A toxic, white powder with a melting point of approximately 1600°C; used in electronic and optical equipment and as a paint pigment. { 'bar·ē·əm mə'lib,dāt }

barium monosulfide [INORG CHEM] BaS A colorless, cubic crystal that is soluble in water; used in pigments. { 'bar·ē·əm ,män·ō'səl,fīd }

barium monoxide See barium oxide. { 'bar·ē·əm mə'näk ,sīd }

barium nitrate [INORG CHEM] Ba(NO₃)₂ A toxic salt occurring as colorless, cubic crystals, melting at 592°C, and soluble in water; used as a reagent, in explosives, and in pyrotechnics. Also known as nitrobarite. { 'bar·ē·əm 'nī,trāt }

barium oxide [INORG CHEM] BaO A white to yellow powder that melts at 1923°C; it forms the hydroxide with water; may be used as a dehydrating agent. Also known as barium monoxide; barium protoxide. { 'bar·ē·əm 'äk,sīd }

barium perchlorate [INORG CHEM] Ba(ClO₄)₂·4H₂O Tetrahydrate variety which forms colorless hexagons; used in pyrotechnics. { 'bar·ē·əm pər'klór,āt }

barium permanganate [INORG CHEM] Ba(MnO₄)₂ Brownish-violet, toxic crystals; soluble in water; used as a disinfectant. { 'bar·ē·əm pər'maŋ·gə ,nāt }

barium peroxide [INORG CHEM] BaO₂ A compound formed as white toxic powder, insoluble in water; used as a bleach and in the glass industry. Also known as barium binoxide; barium dioxide; barium superoxide. { 'bar·ē·əm pər'äk,sīd }

barium plaster [MATER] A special mill-mixed gypsum plaster containing barium salts; used to plaster walls of x-ray rooms. { 'bar·ē·əm ,plas·tər }

barium protoxide See barium oxide. { 'bar·ē·əm prō'täk,sīd }

barium silicide [INORG CHEM] BaSi₂ A compound that has the appearance of metal-gray lumps; melts at white heat; used in metallurgy to deoxidize steel. { 'bar·ē·əm 'sil·ə,sīd }

barium-sodium niobate [MATER] Synthetic electrooptical crystals used to produce coherent green light in lasers and to manufacture devices such as electrooptical modulators and optical polarimetric oscillators. { 'bar·ē·əm 'sōd·ē·əm 'nī·ə,bāt }

barium sulfate [INORG CHEM] BaSO₄ A salt occurring in the form of white, rhombic crystals, insoluble in water; used as a white pigment, as an opaque contrast medium for roentgenographic processes, and as an antidiarrheal. { 'bar·ē·əm 'səl,fāt }

barium sulfite [INORG CHEM] BaSO₃ A toxic, white powder; soluble in dilute hydrochloric acid; used in paper manufacturing. { 'bar·ē·əm 'səl ,fīt }

barium superoxide See barium peroxide. { 'bar·ē·əm ,sü·pər'äk,sīd }

barium tetrasulfide [INORG CHEM] BaS₄·H₂O Red or yellow, rhombic crystals, soluble in water. { 'bar·ē·əm ,te·trə'səl,fīd }

barium thiocyanate [INORG CHEM] Ba(SCN)·2H₂O White crystals that deliquesce; used in dyeing and in photography. { 'bar·ē·əm ,thī·ō'sī·ə,nāt }

barium thiosulfate [INORG CHEM] BaS₂O₃·H₂O A white powder that decomposes upon heating; used to make explosives and in matches. Also known as barium hyposulfite. { 'bar·ē·əm ,thī·ō'səl,fāt }

barium titanate [INORG CHEM] BaTiO₃ A grayish powder that is insoluble in water but soluble in concentrated sulfuric acid; used as a ferroelectric ceramic. { 'bar·ē·əm 'tī·tə,nāt }

barium tungstate [INORG CHEM] BaWO₄ A toxic, white powder used as a pigment and in x-ray photography. Also known as barium white; barium wolframate; tungstate white; wolfram white. { 'bar·ē·əm 'təŋ,stāt }

barium white See barium tungstate. { 'bar·ē·əm 'wīt }

barium wolframate See barium tungstate. { 'bar·ē·əm 'wúl·frə,māt }

bark [MET] The decarburized layer formed beneath the scale on the surface of steel heated in air. { bärk }

Barker method [CRYSTAL] A method utilizing a number of convenient rules which allow two observers to choose the same reference system to describe the same noncubic crystal. { 'bär·kər ,meth·əd }

Barkhausen effect [ELECTROMAG] The succession of abrupt changes in magnetization occurring when the magnetizing force acting on a piece of iron or other magnetic material is varied. { 'bärk,haúz·ən i'fekt }

bar magnet [ELECTROMAG] A bar of hard steel that has been strongly magnetized and holds its magnetism, thereby serving as a permanent magnet.   { 'bär ,mag·nət }

barometric pressure   *See* atmospheric pressure.   { bar·ə'met·rik 'presh·ər }

barotropic phenomenon   [THERMO] The sinking of a vapor beneath the surface of a liquid when the vapor phase has the greater density.   { ,bar·ə'träp·ik fə'näm·ə,nän }

barrel-etch reactor   [ENG] A type of plasma reactor in which the specimens to be etched are placed in a quartz support stand and a plasma is generated that diffuses and contacts them.   { ¦bar·əl ¦ech rē'ak·tər }

barrel plating   [MET] An electroplating process by which articles are brought into contact with an electrolyte while rotating in a perforated hardwood barrel.   { 'bar·əl ,plād·iŋ }

barrier layer   *See* depletion layer.   { 'bar·ē·ər ,lā·ər }

barrier material   [MATER] Packing material impervious to moisture, vapor, or other liquids and gases.   { 'bar·ē·ər mə'tir·ē·əl }

barrier plastic   [MATER] A polymer that is impermeable to gases and can be fabricated into an object such as a container that will not permit aromas or other gases to pass through it in either direction.   { 'bar·ē·ər ,plas·tik }

bar steel   [MET] Steel formed into bars.   { 'bär ,stēl }

barstock   *See* bar.   { 'bär,stäk }

basal cleavage   [CRYSTAL] Cleavage parallel to the base of the crystal structure or to the lattice plane which is normal to one of the lattice axes.   { 'bā·səl 'klēv·ij }

basal orientation   [CRYSTAL] A crystal orientation in which the surface is parallel to the base of the lattice or to the lattice plane which is normal to one of the lattice axes.   { 'bā·səl ,ȯr·ē·ən'tā·shən }

basal plane   [CRYSTAL] The plane perpendicular to the long, or *c*, axis in all crystals except those of the isometric system.   { 'bā·səl ¦plān }

base   [CHEM] Any chemical species, ionic or molecular, capable of accepting or receiving a proton (hydrogen ion) from another substance; the other substance acts as an acid in giving of the proton. Also known as Brønsted base.   [GRAPHICS] A transparent plastic film on which a photographic emulsion is applied.   { bās }

base box   [MET] A unit of area used for tin-plated steel sheet; one base box is equivalent to 112 sheets, 14 by 20 inches (35.6 by 50.8 centimeters), or 62,720 square inches of surface, coated on two sides; 1 pound (0.454 kilogram) of tin per base box is equal to a coating of tin 0.000059 inch (0.0014986 millimeter) thick.   { 'bās ,bäks }

base bullion   [MET] Crude lead that has enough silver in it to make the extraction of silver worthwhile; gold may be present.   { 'bās ,bùl·yən }

base-centered lattice   [CRYSTAL] A space lattice in which each unit cell has lattice points at the centers of each of two opposite faces as well as at the vertices; in a monoclinic crystal, they are the faces normal to one of the lattice axes.   { 'bās ,sen·tərd 'lad·əs }

base metal   [CHEM] Any of the metals on the lower end of the electrochemical series.   [MET] **1.** The metal that is to be worked.   **2.** The principal metal of an alloy.   **3.** Any metal that will oxidize when heated in air.   **4.** The metal of parts to be welded. Also known as parent metal.   **5.** Metal to which cladding or plating is applied. Also known as basis metal.   { 'bās ¦med·əl }

basic converter   *See* basic-lined converter.   { 'bā·sik kən'vərd·ər }

basic copper carbonate   *See* copper carbonate.   { 'bā·sik 'käp·ər 'kär·bə,nāt }

basic group   [CHEM] A chemical group (for example, OH⁻) which, when freed by ionization in solution, produces a pH greater than 7.   { 'bā·sik ¦grüp }

basic-lined   [MET] Pertaining to the walls and bottom of a melting furnace made of refractory materials, such as lime or dolomite, that have a basic reaction in the melting process.   { 'bā·sik ¦līnd }

basic-lined converter   [MET] A converter, such as the Peir-Smith copper converter, which has a basic refractory lining. Also known as basic converter.   { 'bā·sik ¦līnd kən'vərd·ər }

basic open-hearth process   [MET] An open-hearth process for steelmaking under basic slag; used for pig iron and scrap with a phosphorus content too low for the Bessemer process and too high for the acid open-hearth process.   { 'bā·sik ¦ō·pən ¦härth 'präs·əs }

basic oxide   [INORG CHEM] A metallic oxide that is a base, or that forms a hydroxide when combined with water, such as sodium oxide to sodium hydroxide.   { 'bā·sik 'äk,sīd }

basic refractory   [MATER] Any heat-resistant material used for basic linings; examples are dolomite and magnesite.   { 'bā·sik ri'frak·trē }

basic salt   [INORG CHEM] A compound that is a base and a salt because it contains elements of both, for example, copper carbonate hydroxide, $Cu_2(OH)_2CO_3$.   { 'bā·sik 'sȯlt }

basic slag   [MET] A slag resulting from the steelmaking process; rich in phosphorus, it is ground and used as a nutrient in grasslands.   { 'bā·sik 'slag }

basic steel   [MET] Steel made by the basic process, in a furnace with a basic lining.   { 'bā·sik 'stēl }

basin   [MET] The mouth of a sprue in a gating system of castings into which the molten metal is first poured.   { 'bās·ən }

basis metal   *See* base metal.   { 'bā·səs 'med·əl }

bassora gum   [MATER] A type of high-colored gum similar to tragacanth gum; includes Indian gum.   { 'bas·ə·rə ,gəm }

Batchinsky relation   [FL MECH] The relation stating that the fluidity of a liquid is proportional to the difference between the specific volume and a characteristic specific volume, approximately equal to the specific volume appearing in the van der Waals equation.   { ba'chin·skē ri,lā·shən }

bathochromatic shift   [PHYS CHEM] The shift of the fluorescence of a compound toward the red part of the spectrum due to the presence of a bathochrome radical in the molecule. { ¦bath·ō ¦krō¦mad·ik 'shift }

batt [MATER] A blanket of insulating material usually 16 inches (41 centimeters) wide and 3 to 6 inches (8 to 15 centimeters) thick, used to insulate building walls and roofs. { bat }

battery [ELEC] A direct-current voltage source made up of one or more units that convert chemical, thermal, nuclear, or solar energy into electrical energy. { 'bad·ə·rē }

battery amalgamation [MET] Amalgamation by means of mercury placed in the mortar box of a stamp battery. { 'bad·ə·rē ə,mal·gə'mā·shən }

battery electrolyte [PHYS CHEM] A liquid, paste, or other conducting medium in a battery, in which the flow of electric current takes place by migration of ions. { 'bad·ə·rē i'lek·trə,līt }

battery manganese See manganese dioxide. { 'bad·ə·rē ,maŋ·gə,nēs }

Bauschinger effect [MET] A phenomenon by which the plastic deformation of a metal increases the tensile yield strength and decreases the compressive yield strength. { 'bau̇,shiŋ·ər i'fekt }

Baveno twin law [CRYSTAL] An uncommon twin law applicable in feldspar, in which the twin plane and composition surface are (021); a Baveno twin usually consists of two individuals. { bə'vē·nō ,twin ,lȯ }

b axis [CRYSTAL] A crystallographic axis that is oriented horizontally, right to left. { 'bē ,ak·səs }

Bayer process [MET] A method of producing alumina from bauxite by heating it in a sodium hydroxide solution. { 'bī·ər ,präs·əs }

BCS theory See Bardeen-Cooper-Schrieffer theory. { ¦bē¦sē¦es ,thē·ə·rē }

Be See beryllium.

bead [MET] **1.** The drop of precious metal obtained by cupellation in fire assaying. **2.** See weld bead. { bēd }

beading [MET] The placing of a bead on a piece of sheet metal for either decorative or strengthening purposes. { 'bēd·iŋ }

beam [CIV ENG] A body, with one dimension large compared with the other dimensions, whose function is to carry lateral loads (perpendicular to the large dimension) and bending movements. [PHYS] A concentrated, nearly unidirectional flow of particles, or a like propagation of electromagnetic or acoustic waves. { bēm }

beam splitting [OPTICS] The division of a beam of light into two beams by placing a special type of mirror in the path of the beam that reflects part of the light falling on it and transmits part. { 'bēm ,splid·iŋ }

beam spread [ENG] The angle of divergence from the central axis of an electromagnetic or acoustic beam as it travels through a material. { 'bēm ,spred }

beam test [CIV ENG] A test of the flexural strength (modulus of rupture) of concrete from measurements on a standard reinforced concrete beam. { 'bēm ,test }

bearing bronze [MET] A form of bronze usually having a high lead content and antifriction characteristics and used for fabricating bearings. { 'ber·iŋ ,bränz }

bearing capacity [MECH] Load per unit area which can be safely supported by the ground. { 'ber·iŋ kə'pas·əd·ē }

bearing pressure [MECH] Load on a bearing surface divided by its area. Also known as bearing stress. { 'ber·iŋ ,presh·ər }

bearing strain [MECH] The deformation of bearing parts subjected to a load. { 'ber·iŋ ,strān }

bearing strength [MECH] The maximum load that a column, wall, footing, or joint will sustain at failure, divided by the effective bearing area. { 'ber·iŋ ,strenkth }

bearing stress See bearing pressure. { 'ber·iŋ ,stres }

Beattie and Bridgman equation [THERMO] An equation that relates the pressure, volume, and temperature of a real gas to the gas constant. { ¦bēd·ē ən ¦brij·mən i'kwā·zhən }

beaumontage [MATER] A material that is made of a mixture of resin, beeswax, and shellac and used to fill small holes or cracks in wood or metal. { ,bō·män'tij }

bed charge [MET] The primary charge of fuel in a cupola to initiate the melting process. { 'bed ,chärj }

bed coke [MET] The primary layer of coke placed in a cupola at a preselected height above the tuyeres for the initial combustion and melting operation. { 'bed ,kōk }

beef stearin See oleostearin. { bēf ,stir·ən }

Beer's law [PHYS CHEM] The law which states that the absorption of light by a solution changes exponentially with the concentration, all else remaining the same. { 'bā·ərz ,lȯ }

beeswax [MATER] Yellow to grayish-brown solid wax obtained from bee honeycombs by boiling and straining; used in floor waxes, waxed paper, and textile finishes and in pharmacy. { 'bēz ,waks }

Beilby layer [MET] An amorphous layer formed on a polished metal surface. { 'bīl·bē ,lā·ər }

Belgian block [MATER] **1.** A stone block used for paving, having the shape of a truncated pyramid, with a depth of 7–8 inches (18–20 centimeters), a base of 5–6 inches (13–15 centimeters) square, and a face opposite the base that is 1 inch (2.5 centimeters) or less smaller than the base. **2.** Any stone block used for paving. { 'bel·jən ¦bläk }

Belgian retort process See horizontal retort process. { 'bel·jən ri'tȯrt ,präs·əs }

bell [ENG] A hollow metallic cylinder closed at one end and flared at the other; it is used as a fixed-pitch musical instrument or signaling device and is set vibrating by a clapper or tongue which strikes the lip. [MET] A conical device that seals the top of a blast furnace. { bel }

bellite [MATER] An explosive consisting of five parts of ammonium nitrate to one part of meta-dinitrobenzene with some potassium nitrate. { 'be,līt }

bell metal [MET] An alloy of copper and tin, containing 15–25% tin but 20–24% for best tonal quality. { 'bel ,med·əl }

belt dressing [MATER] A material, usually beeswax, applied to leather belts to increase friction between the belt and pulley surface. { 'belt ,dres·iŋ }

belt grinding [MET] Grinding with an abrasive-coated continuous belt. Also known as linishing. { 'belt ,grīn·diŋ }

belting [MATER] **1.** A sturdy fabric, usually of cotton, used in belts. **2.** A heavy leather, made from hides of cattle, used in power transmission belts. Also known as belting leather. { 'bel·tiŋ }

belting leather See belting. { 'bel·tiŋ ,leth·ər }

bending moment [MECH] Algebraic sum of all moments located between a cross section and one end of a structural member; a bending moment that bends the beam convex downward is positive, and one that bends it convex upward is negative. { 'ben·diŋ ,mō·mənt }

bending-moment diagram [MECH] A diagram showing the bending moment at every point along the length of a beam plotted as an ordinate. { 'ben·diŋ ¦mō·mənt ,dī·ə,gram }

bending stress [MECH] An internal tensile or compressive longitudinal stress developed in a beam in response to curvature induced by an external load. { 'ben·diŋ ,stres }

bend plane See tilt boundary. { 'bend ,plān }

bend test [MET] A ductility test in which a specimen is bent through an arc of known radius and angle. { 'bend ,test }

beneficiation [MET] Improving the chemical or physical properties of an ore so that metal can be recovered at a profit. Also known as mineral dressing. { ,ben·ə,fish·ē'ā·shən }

bentonite slurry [MATER] Slurry composed of water and clay powder consisting chiefly of montmorillonite minerals. { 'bent·ən,īt ,slər·ē }

benzene [ORG CHEM] C_6H_6 A colorless, liquid, flammable, aromatic hydrocarbon that boils at 80.1°C and freezes at 5.4–5.5°C; used to manufacture styrene and phenol. Also known as benzol. { 'ben,zēn }

benzil [ORG CHEM] $C_6H_5COCOC_6H_5$ A yellow powder; melting point 95°C; insoluble in water, soluble in ethanol, ether, and benzene; used in organic synthesis. { 'ben,zil }

benzol See benzene. { 'ben,zól }

benzyl [ORG CHEM] The radical $C_6H_5CH_2^-$, as found, for example, in benzyl alcohol, $C_6H_5CH_2OH$. { 'ben·zəl }

benzyne [ORG CHEM] C_6H_4 A chemical species whose structure consists of an aromatic ring in which four carbon atoms are bonded to hydrogen atoms and two adjacent carbon atoms lack substitutents; a member of a class of compounds known as arynes. { 'ben,zīn }

berkelium [CHEM] A radioactive element, symbol Bk, atomic number 97, the eighth member of the actinide series; properties resemble those of the rare-earth cerium. { 'bər·klē·əm }

Bernal chart [CRYSTAL] A chart used to determine the coordinates in reciprocal space of x-ray reflections that produce the spots on an x-ray diffraction photograph of a single crystal. { bər'nal ,chärt }

Bernoulli-Euler law [MECH] A law stating that the curvature of a beam is proportional to the bending moment. { ber,nü·lē ¦öil·ər ¦lö }

Berthelot method [THERMO] A method of measuring the latent heat of vaporization of a liquid that involves determining the temperature rise of a water bath that encloses a tube in which a given amount of vapor is condensed. { 'ber·tə ,lö ,meth·əd }

berthollide [CHEM] A compound whose solid phase exhibits a range of composition. { 'bər·thə,līd }

beryllate [INORG CHEM] **1.** $BeO_2{}^{2-}$ An ion containing beryllium and oxygen. **2.** A salt produced by the reaction of a strong alkali such as sodium hydroxide with beryllium oxide. { 'ber·ə ,lāt }

beryllia See beryllium oxide. { bə'ril·ē·ə }

beryllide [INORG CHEM] A chemical combination of beryllium with a metal, such as zirconium or tantalum. { bə'ril·ə,dē }

beryllium [CHEM] A chemical element, symbol Be, atomic number 4, atomic weight 9.0122. [MET] A rare metal, occurring naturally in combinations, with density about one-third of aluminum; used most commonly in the manufacture of beryllium-copper alloys which find numerous industrial and scientific applications. { bə'ril·ē·əm }

beryllium alloy [MET] Any dilute alloy of base metals containing a few percent of beryllium in a precipitation-hardening system. { bə'ril·ē·əm 'al,oi }

beryllium bronze See beryllium copper. { bə'ril·ē·əm 'bränz }

beryllium copper [MET] **1.** An alloy of copper and beryllium containing not more than about 3% beryllium; used for springs, tools, and plastic molds. Also known as beryllium bronze. **2.** An alloy of copper and beryllium specifically for addition to metals in the foundry. { bə'ril·ē·əm 'käp·ər }

beryllium fluoride [INORG CHEM] BeF_2 A hygroscopic, amorphous solid with a melting point of 800°C; soluble in water; used in beryllium metallurgy. { bə'ril·ē·əm ¦flür,īd }

beryllium monel [MET] A nickel-copper alloy containing beryllium. { bə'ril·ē·əm mō,nel }

beryllium nitrate [INORG CHEM] $Be(NO_3)_2 \cdot 3H_2O$ A compound that forms colorless, deliquescent crystals that are soluble in water; used to introduce beryllium oxide into materials used in incandescent mantles. { bə'ril·ē·əm 'nī,trāt }

beryllium nitride [INORG CHEM] Be_3N_2 Refractory, white crystals with a melting point of 2200±40°C; used in the manufacture of radioactive carbon-14 and in experimental rocket fuels. { bə'ril·ē·əm ¦nī,trīd }

beryllium oxide [INORG CHEM] BeO An amorphous white powder, insoluble in water; used to make beryllium salts and as a refractory; also known as beryllia. { bə'ril·ē·əm 'äk,sīd }

Bessemer converter [MET] A pear-shaped, basic-lined, cylindrical vessel for producing steel by the Bessemer process. { 'bes·ə·mər kən'vərd·ər }

Bessemer iron [MET] Pig iron with about 1% silicon and a sulfur and a phosphorus content below 0.10; used to make steel by the Bessemer or the acid open-hearth process. Also known as Bessemer pig iron. { 'bes·ə·mər 'ī·ərn }

Bessemer matte [MET] Product of the oxidation of furnace matte; contains nickel, copper, cobalt, precious metals, and about 22% sulfur. { 'bes·ə·mər 'mat }

Bessemer ore [MET] An iron ore containing very little phosphorus, considered suitable for refining by the Bessemer process. { 'bes·ə·mər 'òr }

Bessemer pig iron See Bessemer iron. { 'bes·ə·mər 'pig ,ī·ərn }

Bessemer process [MET] A steelmaking process in which carbon, silicon, phosphorus, and manganese contained in molten pig iron are oxidized by a strong blast of air. { 'bes·ə·mər 'präs·əs }

Bessemer steel [MET] Steel manufactured by the Bessemer process. { 'bes·ə·mər 'stēl }

beta brass [MET] A type of brass containing nearly equal proportions of copper and zinc. { 'bād·ə ,bras }

Betti reciprocal theorem [MECH] A theorem in the mathematical theory of elasticity which states that if an elastic body is subjected to two systems of surface and body forces, then the work that would be done by the first system acting through the displacements resulting from the second system equals the work that would be done by the second system acting through the displacements resulting from the first system. { 'bāt·tē ri'sip·rə·kəl ,thir·əm }

Betti's method [MECH] A method of finding the solution of the equations of equilibrium of an elastic body whose surface displacements are specified; it uses the fact that the dilatation is a harmonic function to reduce the problem to the Dirichlet problem. { 'bāt·tēz ,meth·əd }

Betts' process [MET] A refining process for electrolytically purifying lead. { 'bets ,präs·əs }

Bi See bismuth.

bias [MATER] In a reinforced composite material, the angle made by the reinforcing fibers with the longitudinal direction of the material. { 'bī·əs }

bias voltage [ELECTR] A voltage applied or developed between two electrodes as a bias. { 'bī·əs ,vōl·tij }

biaxial crystal [CRYSTAL] A crystal of low symmetry in which the index ellipsoid has three unequal axes. { bī'ak·sē·əl 'krist·əl }

biaxial indicatrix [CRYSTAL] An ellipsoid whose three axes at right angles to each other are proportional to the refractive indices of a biaxial crystal. { bī'ak·sē·əl in'dik·ə,triks }

biaxial stress [MECH] The condition in which there are three mutually perpendicular principal stresses; two act in the same plane and one is zero. { bī'ak·sē·əl ,stress }

bicarbonate [INORG CHEM] A salt obtained by the neutralization of one hydrogen in carbonic acid. { bī'kär·bə,nət }

bicarbonate of soda See sodium bicarbonate. { bī¦kär·bə·nət əv 'sō·də }

bichloride of mercury See mercuric chloride. { bī'klòr,īd əv 'mər·kyə·rē }

bichromate See dichromate. { ,bī'krō,māt }

bicomponent fiber [TEXT] A fiber manufactured from two different polymers by spinning and joining them in a simultaneous process from one spinneret. Also known as conjugate fiber. { bī·kəm'pō·nənt 'fī·bər }

biconstituent fiber [TEXT] A manufactured fiber composed of two or more dissimilar fibers combined before the extrusion process. Also known as matrix fiber. { ¦bī·kən'stich·ə·wənt 'fī·bər }

bidentate ligand [INORG CHEM] A chelating agent having two groups capable of attachment to a metal ion. { bī'den,tāt 'lig·ənd }

bifluoride [INORG CHEM] An acid fluoride whose formula has the form MHF_2; an example is sodium bifluoride, $NaHF_2$. { bī'flùr,īd }

bifunctional catalyst [CHEM] A catalytic substance that possesses two catalytic sites and thus is capable of catalyzing two different types of reactions. Also known as dual-function catalyst. { ¦bī¦fəŋk·shən·əl 'kad·ə,list }

bilayer [CHEM] A layer two molecules thick, such as that formed on the surface of the aqueous phase by phospholipids in aqueous solution. { 'bī,lā·ər }

billet [MET] A semifinished, short, thick bar of iron or steel in the form of a cylinder or rectangular prism produced from an ingot; limited to 1.5 inches (3.8 centimeters) in width and thickness with a cross-sectional area up to 36 square inches (232 square centimeters). { 'bil·ət }

Billet furnace [MET] A furnace for heating steel in sizes between a bloom and a bar. { 'bil·ət 'fər·nəs }

billet mill [MET] A rolling mill for making billets from ingots. { 'bil·ət ,mil }

bimetal [MATER] A laminate of two dissimilar metals, with different coefficients of thermal expansion. { ¦bī¦med·əl }

bimetallic corrosion [MET] Corrosion of two different metals exposed to an electrolyte while in contact. { ¦bī·mə'tal·ik kə'rōzh·ən }

bimetallic strip [ENG] A strip formed of two dissimilar metals welded together; different temperature coefficients of expansion of the metals cause the strip to bend or curl when the temperature changes. { ¦bī·mə'tal·ik ,strip }

bimolecular [CHEM] Referring to two molecules. { ¦bī·mə'lek·yə·lər }

bimolecular reaction [CHEM] A chemical transformation or change involving two molecules. { ¦bī·mə'lek·yə·lər rē'ak·shən }

binary alloy [MET] An alloy composed of two principal metallic components. { 'bīn·ə·rē 'a ,lòi }

binary compound [CHEM] A compound that has two elements; it may contain two or more atoms; examples are KCl and $AlCl_3$. { 'bīn·ə·rē 'käm ,paùnd }

binary magnetic core [SOLID STATE] A ferromagnetic core that can be made to take either of two stable magnetic states. { 'bīn·ə·rē mag'ned·ik 'kȯr }

binary optics [OPTICS] A technology that uses etching technology to produce optical elements with computer-generated microscopic surface relief patterns having two or more levels. { 'bī·ner·ē 'äp·tiks }

binder [MATER] 1. A resin or cementlike material used to hold particles together and provide mechanical strength or to ensure uniform consistency, solidification, or adhesion to a surface coating. 2. See binding agent. { 'bīn·dər }

binding agent [MATER] A liquid component of paint that hardens as it dries and thereby serves to bind the pigment particles and develop adhesion to a surface. Also known as binder. { 'bīn·diŋ ,ā·jənt }

binding energy [PHYS] Abbreviated BE. Also known as total binding energy (TBE). 1. The net energy required to remove a particle from a system. 2. The net energy required to decompose a system into its constituent particles. { 'bīn·diŋ ¦en·ər·jē }

binding site See active site. { 'bīn·diŋ ,sīt }

Bingham number [FL MECH] A dimensionless number used to study the flow of Bingham plastics. { 'biŋ·əm ,nəm·bər }

Bingham plastic [FL MECH] A non-Newtonian fluid exhibiting a yield stress which must be exceeded before flow starts; thereafter the rate of shear versus shear stress curve is linear. { 'biŋ·əm ,plas·tik }

bioartificial organs [BIOMATER] Devices, used for both short-term and long-term organ replacement, that are designed and manufactured for membrane biocompatibility, diffusion limitations, device retrieval in the event of failure, and mechanical stability. { ,bī·ō,ard·ə¦fish·əl 'ȯr·gənz }

biocatalyst [BIOMATER] A biochemical catalyst, especially an enzyme. { ,bī·ō'kad·əl·ist }

bioceramic [MATER] Biocompatible or osteoinductive (stimulating bone growth) ceramic material, such as hydroxyapatite or some other type of calcium phosphate ceramic, used for reconstructive bone surgery and dental implants. { ,bī·ō·sə'ram·ik }

biocide See pesticide. { 'bī·ə,sīd }

biocompatibility [BIOMATER] The condition of being compatible with living tissue by virtue of a lack of toxicity or ability to cause immunological rejection. { ,bī·ō·kəm,pad·ə'bil·əd·ē }

biocorrosion See biological corrosion. { ,bī·ō·kə'rō·zhən }

biodegradability [MATER] The characteristic of a substance that can be broken down by microorganisms. { ,bī·ō·di,grād·ə'bil·əd·ē }

biodeterioration [MATER] Decay of wood or other material caused by fungi, bacteria, insects, or marine boring organisms. { ,bī·ō·di ,tir·ē·ər'ā·shən }

bioelectronics [BIOMATER] 1. The application of electronic theories and techniques to the problems of biology. 2. The use of biotechnology in electronic devices such as biosensors, molecular electronics, and neuronal interfaces; more speculatively, the use of proteins in constructing circuits. { ,bī·ō,i,lek'trän·iks }

biofilm [BIOMATER] A microbial (bacterial, fungal, algal) community, enveloped by the extracellular biopolymer which these microbial cells produce, that adheres to the interface of a liquid and a surface. { 'bī·ō,film }

biological corrosion [MET] Deterioration of metals as a result of the metabolic activity of microorganisms. Also known as biocorrosion. { ¦bī·ə¦läj·ə·kəl kə'rō·zhən }

biomimetic catalyst [ORG CHEM] A synthetic compound that can simulate the mode of action of a natural enzyme by catalyzing a reaction at ambient conditions. { ¦bī·ō·mə'med·ik 'kad·ə ,list }

biomimetics [BIOMATER] A branch of science in which synthetic systems are developed by using information obtained from biological systems. { ¦bī·ō·mə¦med·iks }

biopolymer [BIOMATER] A biological macromolecule such as a protein or nucleic acid. { ¦bī·ō'päl·ə·mər }

Biot-Fourier equation [THERMO] An equation for heat conduction which states that the rate of change of temperature at any point divided by the thermal diffusivity equals the Laplacian of the temperature. { ¦byō ¦für·yā i'kwā·zhən }

Biot-Savart law [ELECTROMAG] A law that gives the intensity of the magnetic field due to a wire carrying a constant electric current. { ¦byō sə ¦vär 'lȯ }

biphenyl [ORG CHEM] $C_{12}H_{10}$ A white or slightly yellow crystalline hydrocarbon, melting point 70.0°C, boiling point 255.9°C, and density 1.9896, which gives plates or monoclinic prismatic crystals; used as a heat-transfer medium and as a raw material for chlorinated diphenyls. Also known as diphenyl; phenylbenzene. { bī'fen·əl }

bipolar electrode [ELEC] Electrode, without metallic connection with the current supply, one face of which acts as anode surface and the opposite face as a cathode surface when an electric current is passed through a cell. { bī'pō·lər i'lek,trōd }

bipyramid [CRYSTAL] A crystal having the form of two pyramids that meet at a plane of symmetry. Also known as dipyramid. { bī¦pir·ə,mid }

biradical [CHEM] A chemical species having two independent odd-electron sites. { bī'rad· ə·kəl }

birdseye [MATER] A small localized area in wood in which the fibers are indented and otherwise contorted to form few to many circular or elliptical figures on the tangential surface. { 'bərdz,ī }

birefringence [OPTICS] 1. Splitting of a light beam into two components, which travel at different velocities, by a material. 2. For a light beam that has been split into two components by a material, the difference in the indices of refraction of the components within the material. Also known as double refraction. { ,bī·ri'frin·jəns }

bis- [CHEM] A prefix indicating doubled or twice. { bis }

biscuit [MATER] **1.** A clay object that has been fired once prior to glazing. **2.** Pottery that is unglazed in its final form. [MET] An upset blank for drop forging. { 'bis·kət }

bisectrix [CRYSTAL] A line that is the bisector of the angle between the optic axes of a biaxial crystal. { ,bī'sek,triks }

bisilicate [MET] A type of slag whose silicate degree is 2. { ,bī'sil·ə·kāt }

bismanol [MET] A magnetic alloy of bismuth and manganese. { 'biz·mə,nól }

bismuth [CHEM] A metallic element, symbol Bi, of atomic number 83 and atomic weight 208.980. { 'biz·məth }

bismuth alloy [MET] A group of low-melting alloys (many below 100°C) of bismuth combined with lead, tin, and cadmium; used in automatic sprinklers, special solders, safety plugs in compressed-gas cylinders, automatic shutoffs for water-heating systems, castings, and type metal. { 'biz·məth 'al,ói }

bismuthate [INORG CHEM] A compound of bismuth in which the bismuth has a valence of +5; an example is sodium bismuthate, $NaBiO_3$. { 'biz·mə,thāt }

bismuth carbonate See bismuth subcarbonate. { 'bis·məth 'kär·bə,nāt }

bismuth chloride [INORG CHEM] $BiCl_3$ A deliquescent material that melts at 230–232°C and decomposes in water to form the oxychloride; used to make bismuth salts. Also known as bismuth trichloride. { 'biz·məth 'klór,īd }

bismuth chromate [INORG CHEM] $Bi_2O_3 \cdot Cr_2O_3$ An orange-red powder, soluble in alkalies and acids; used as a pigment. { 'biz·məth 'krō,māt }

bismuth hydroxide [INORG CHEM] $Bi(OH)_3$ A water-insoluble, white powder; precipitated by hydroxyl ion from bismuth salt solutions. { 'biz·məth hī'dräk,sīd }

bismuth iodide [INORG CHEM] BiI_3 A bismuth halide that sublimes in grayish-black hexagonal crystals melting at 408°C, insoluble in water; used in analytical chemistry. { 'biz·məth 'ī·ə,dīd }

bismuth nitrate [INORG CHEM] $Bi(NO_3)_3 \cdot 5H_2O$ White, triclinic crystals that decompose in water; used as an astringent and antiseptic. { 'biz·məth 'nī,trāt }

bismuth oxide See bismuth trioxide. { 'biz·məth 'äk,sīd }

bismuth oxycarbonate See bismuth subcarbonate. { 'biz·məth ,äk·sē'kär·bə,nāt }

bismuth oxychloride [INORG CHEM] $BiOCl$ A white powder; insoluble in water, soluble in acid; a toxic material if ingested; used in pigments and cosmetics. { 'biz·məth ,äk·sē'klór,īd }

bismuth subcarbonate [INORG CHEM] $(BiO)_2CO_3$ or $Bi_2O_3 \cdot CO_2 \cdot \frac{1}{2}H_2O$ A white powder; dissolves in hydrochloric or nitric acid, insoluble in alcohol and water; used as opacifier in x-ray

diagnosis, in ceramic glass, and in enamel fluxes. { 'biz·məth səb'kär·bə,nāt }

bismuth subnitrate [INORG CHEM] $4BiNO_3(OH)_2 \cdot BiO(OH)$ A white, hygroscopic powder; used in bismuth salts, perfumes, cosmetics, ceramic enamels, pharmaceuticals, and analytical chemistry. { 'biz·məth ,səb'ni,trāt }

bismuth subsalicylate [INORG CHEM] $Bi(C_7H_5O)_3 \cdot Bi_2O_3$ A white powder that is insoluble in ethanol and water; used in medicine and as a fungicide for tobacco crops. { 'biz·məth ,səb·sə'lis·ə,lāt }

bismuth telluride [INORG CHEM] Bi_2Te_3 Gray, hexagonal platelets with a melting point of 573°C; used for semiconductors, thermoelectric cooling, and power generation applications. { 'biz·məth 'tel·yə,rīd }

bismuth trichloride See bismuth chloride. { 'biz·məth ,trī·klór,īd }

bismuth trioxide [INORG CHEM] Bi_2O_3 A yellow powder; melting point 820°C; insoluble in water, dissolves in acid; used to make enamels and to color ceramics. Also known as bismuth oxide; bismuth yellow. { 'biz·məth trī'äk,sīd }

bismuth yellow See bismuth trioxide. { 'biz·məth 'yel·ō }

bisphenoid [CRYSTAL] A form apparently consisting of two sphenoids placed together symmetrically. { bī'sfē,nóid }

bisulfate [INORG CHEM] A compound that has the $HSO_4{}^-$ radical; derived from sulfuric acid. { bī'səl,fāt }

bit [MET] In soldering, the portion of the iron that transfers either heat or solder to the joint involved. { bit }

Bitter pattern [SOLID STATE] A pattern produced when a drop of a colloidal suspension of ferromagnetic particles is placed on the surface of a ferromagnetic crystal; the particles collect along domain boundaries at the surface. { 'bid·ər ,pad·ərn }

bitudobe [MATER] A sun-baked brick composed of adobe and a binder of emulsified asphalt. { 'bich·yü,dō·bē }

bitumastic [MATER] A combination of asphalt and filler used mainly to coat metals to protect them against corrosion and weathering. { ,bi·tyü'mas·tik }

bitumen [MATER] Naturally occurring or pyrolytically obtained substances of dark to black color consisting almost entirely of carbon and hydrogen, with very little oxygen, nitrogen, or sulfur. { bī'tü·mən }

bituminous [MATER] **1.** Containing much organic, or at least carbonaceous, matter, mostly in the form of tarry hydrocarbons which are usually described as bitumen. **2.** Similar to bitumen. **3.** Giving off volatile bituminous substances on heating, as in bituminous coal. { bī'tü·mə·nəs }

bituminous cement [MATER] A bituminous material suitable for use as a binder, having cementing qualities which are dependent mainly on its bituminous character. { bī'tü·mə·nəs si'ment }

bituminous coating [MATER] A coating made principally of bituminous material and used as

37

a surfacing for roads and as a water-repellent barrier in buildings. { bī'tü·mə·nəs 'kōd·iŋ }

bituminous concrete [MATER] A concrete made with bituminous material as a binder for sand and gravel. { bī'tü·mə·nəs käŋ'krēt }

bituminous grout [MATER] A mixture of bitumen and an aggregate, such as sand, that liquefies when heated and cures in air and is used as a sealant for joints or cracks. { bī'tüm·ə·nəs 'graút }

bituminous paint [MATER] Paint with a high proportion of bitumen; usually dark in color. { bī'tü·mə·nəs 'pānt }

Bk See berkelium.

black [CHEM] Fine particles of impure carbon that are made by the incomplete burning of carbon compounds, such as natural gas, naphthas, acetylene, bones, ivory, and vegetables. [OPTICS] Quality of an object which uniformly absorbs large percentages of light of all visible wavelengths. { blak }

black-and-white groups See Shubnikov groups. { ¦blak ən ¦wīt 'grüps }

black annealing [MET] A type of box annealing used to impart a black color to the metal surface; first process in tin plating. { ¦blak ə'nēl·iŋ }

blackbody [THERMO] An ideal body which would absorb all incident radiation and reflect none. Also known as hohlraum; ideal radiator. { 'blak ¦bäd·ē }

blackbody radiation [THERMO] The emission of radiant energy which would take place from a blackbody at a fixed temperature; it takes place at a rate expressed by the Stefan-Boltzmann law, with a spectral energy distribution described by Planck's equation. { 'blak¦bäd·ē ,rā·dē'ā·shən }

blackbody temperature [THERMO] The temperature of a blackbody that emits the same amount of heat radiation per unit area as a given object; measured by a total radiation pyrometer. Also known as brightness temperature. { 'blak¦bäd·ē ,tem·prə·chər }

black copper [MET] The more or less impure metallic copper (70–99% copper) produced in blast furnaces when running on oxide ores or roasted sulfide material. { ¦blak 'käp·ər }

black cyanide See calcium cyanide. { ¦blak 'sī·ə ,nīd }

black grease [MATER] Grease which has black coloration due to use of asphalt, either naturally occurring or from residues used in the manufacture of the grease. { ¦blak 'grēs }

blackheart malleable iron See whiteheart malleable iron. { 'blak,härt 'mal·yə·bəl 'ī·ərn }

blacking [MET] Carbonaceous material, such as powdered graphite, used to coat the inner surfaces of a dry-sand mold to improve the separation and finish of a casting. { 'blak·iŋ }

black iron oxide See ferrous oxide. { 'blak 'ī·ərn 'äk,sīd }

black light [OPTICS] Invisible light, such as ultraviolet rays which fall on fluorescent materials and cause them to emit visible light. { 'blak ,līt }

black liquor [MATER] 1. The liquid material remaining from pulpwood cooking in the soda or sulfate papermaking process. 2. See iron acetate liquor. { ¦blak 'lik·ər }

black-surface enclosure [THERMO] An enclosure for which the interior surfaces of the walls possess the radiation characteristics of a blackbody. { 'blak ,sər·fəs in'klozh·ər }

blacktop [MATER] A black bituminous material that is used to pave roadways; it is spread over a layer of crushed rocks and packed down into a level surface; it may be spread over small areas of roadways in need of repair. { 'blak,täp }

blanc fixe [INORG CHEM] $BaSO_4$ A commercial name for barium sulfate, with some use in pure form in the paint, paper, and pigment industries as a pigment extender. { ,bläŋk 'fēks }

blank [MET] 1. A semifinished piece of metal to be stamped or forged into a tool or implement. 2. A semifinished, pressed, compacted mass of powdered metal. 3. Metal sheet prepared for a forming operation. { blaŋk }

blank carburizing [MET] A simulated carburizing procedure carried out without a carburizing medium. Also known as pseudocarburizing. { 'blaŋk 'kär·bə,rīz·iŋ }

blanket insulation [MATER] Insulation in the form of a rolled sheet sometimes having a vapor-barrier treated paper backing. { 'blaŋ·kət ,in·sə'lā·shən }

blank holder [MET] A tool to prevent the edge of a sheet metal blank from wrinkling during deep-drawing operations. { 'blaŋk ,hōl·dər }

blank nitriding [MET] Simulation of nitriding without the introduction of nitrogen; achieved by use of an inert material or by application of a coating to the piece. { ¦blaŋk 'nī,trīd·iŋ }

blast furnace [MET] A tall, cylindrical smelting furnace for reducing iron ore to pig iron; the blast of air blown through solid fuel increases the combustion rate. { 'blast ,fər·nəs }

blast-furnace coke [MATER] Coke which supplies carbon monoxide to reduce the ore in a blast furnace and supplies heat to melt the iron. { 'blast ,fər·nəs ,kōk }

blast-furnace gas [MATER] The gas product from iron ore smelting when hot air passes over coke in blast ovens; contains carbon dioxide, carbon monoxide, hydrogen, and nitrogen and is used as fuel gas. { 'blast ,fər·nəs ,gas }

blast gate [MET] A sliding plate in a cupola blast pipe to channel airflow in the proper proportion. { 'blast ,gāt }

bleaching [GRAPHICS] An afterprocess in the production of direct positive photographs, in which an oxidizing solution dissolves the negative silver. [OPTICS] A decrease in the optical absorption of a medium, produced by radiation or by external forces. [TEXT] A process in which natural coloring matter is removed from a fiber, yarn, or fabric to make it white. { 'blēch·iŋ }

bleaching agent [CHEM] An oxidizing or reducing chemical such as sodium hypochlorite, sulfur dioxide, sodium acid sulfite, or hydrogen peroxide. { 'blēch·iŋ ,ā·jənt }

bleaching assistant |MATER| A material added to textile bleach baths to cause better penetration by the bleach; includes pine oils, borax, and sulfonated oils. { 'blēch·iŋ a'sis·tənt }

bleaching clay |MATER| Absorbent clay contacted with petroleum and vegetable and other oils to make decolorized products. { 'blēch·iŋ ‚klā }

bleb |MATER| A small bubble in a material that has solidified, such as glass. { bleb }

bleed |CHEM| Diffusion of coloring matter from a substance. { blēd }

bleeder |MET| An incomplete casting that results from some molten metal draining out of the mold cavity after pouring has ceased. { 'blēd·ər }

bleeder cloth |MATER| A layer of material placed over a composite material during the curing process to allow escape of excess gas and resin. { 'blēd·ər ‚klòth }

bleeding |CHEM ENG| The undesirable movement of certain components of a plastic material to the surface of a finished article. Also known as migration. |ENG| Natural separation of a liquid from a liquid-solid or semisolid mixture; for example, separation of oil from a stored lubricating grease, or water from freshly poured concrete. |MATER| **1.** The outward penetration of a coloring agent from a substrate through the surface coat of paint. **2.** The movement of grout through a pavement from below a road surfacing material to the outer surface. |TEXT| Referring to a fabric in which the dye is not fast and therefore comes out when the fabric is wet. { 'blēd·iŋ }

blend |MATER| A mixture so combined as to render the parts indistinguishable from one another. { blend }

blending |MATER| The process of mixing two or more substances having different properties to obtain a final product having characteristics different from those of the starting materials. { 'blen·diŋ }

blind riser |MET| An internal riser that does not extend to the outer surface of a mold. { ¦blīnd 'rī·zər }

blister |ENG| A raised area on the surface of a metallic or plastic object caused by the pressure of gases developed while the surface was in a partly molten state, or by diffusion of high-pressure gases from an inner surface. |MATER| A roughly circular or elliptic unbonded area between plies of a laminated material; usually caused by trapped moisture. Also known as steam blow. { 'blis·tər }

blister copper |MET| Copper having 96–99% purity and a blistered appearance and formed by forcing air through molten copper matte. { 'blis·tər ‚käp·ər }

blister furnace |MET| A furnace for smelting ore to produce blister copper. { 'blis·tər ‚fər·nəs }

blister gas |MATER| Any of several war gases, such as lewisite, which cause burning, inflammation, or tissue destruction internally or externally. { 'blis·tər ‚gas }

blistering |ENG| The appearance of enclosed or broken macroscopic cavities in a body or in a glaze or other coating during firing. { 'blis·tə·riŋ }

blister steel |MET| Raw steel, made from wrought iron by cementation followed by slow cooling, in which blisters are formed by gas attempting to escape from the metal. { 'blis·tər ‚stēl }

Bloch equations |SOLID STATE| Approximate equations for the rate of change of magnetization of a solid in a magnetic field due to spin relaxation and gyroscopic precession. { 'bläk i'kwā·zhənz }

Bloch function |SOLID STATE| A wave function for an electron in a periodic lattice, of the form $u(\mathbf{r})\exp[i\mathbf{k}\cdot\mathbf{r}]$ where $u(\mathbf{r})$ has the periodicity of the lattice. { 'bläk ‚fəŋk·shən }

Bloch theorem |SOLID STATE| The theorem that, in a periodic structure, every electronic wave function can be represented by a Bloch function. { 'bläk ‚thir·əm }

Bloch wall |SOLID STATE| A transition layer, with a finite thickness of a few hundred lattice constants, between adjacent ferromagnetic domains. Also known as domain wall. { 'bläk ‚wòl }

block brazing |MET| The process of joining metals by applying heated blocks to the joint and using a nonferrous filler metal with a melting point above 800°F (427°C). { 'bläk ‚brāz·iŋ }

block copolymer |POLYM CHEM| A copolymer in which the like monomer units occur in relatively long alternate sequences on a chain. Also known as block polymer. { 'bläk kō'päl·ə·mər }

block grease |MATER| High-melting-point grease that can be handled as a block or stick at normal temperatures; used for journal bearings. Also known as brick grease. { 'bläk ‚grēs }

blocking |ENG| Undesired adhesion between layers of plastic materials in contact during storage or use. |MET| **1.** A preliminary hot-forging operation which imparts an approximate shape to the rough stock. **2.** Reducing the oxygen content of the bath in an open-hearth furnace. |SOLID STATE| The hindering of motion of dislocations in a solid substance by small particles of a second substance included in the solid; results in hardening of the substance. { 'bläk·iŋ }

blocking group |ORG CHEM| In peptide synthesis, a group that is reacted with a free amino or carboxyl group on an amino acid to prevent its taking part in subsequent formation of peptide bonds. { 'bläk·iŋ ‚grüp }

blocking layer See depletion layer. { 'bläk·iŋ ‚lā·ər }

block polymer See block copolymer. { 'bläk 'päl·ə·mər }

block sequence |MET| A procedure by which a continuous multiple-pass joint is completed by alternating the deposition of intermittent cross-sectional buildup and intervening lengths of weld metal. { 'bläk ‚sē·kwəns }

bloom |ENG| **1.** Fluorescence in lubricating oils or a cloudy surface on varnished or enameled surfaces. **2.** To apply an antireflection coating to glass. |MATER| Crystals formed on the surface of treated wood by exudation and evaporation of the solvent in preservative solutions. |MET| **1.** A semifinished bar of metal formed from an ingot and having a rectangular cross section exceeding 36 square inches (232 square centimeters). **2.** To hammer or roll metal in order to make its surface bright. **3.** *See* slag. |OPTICS| Color of oil in reflected light, differing from its color in transmitted light. Also known as fluorescence. { blüm }

bloomer |MET| A furnace used to shape wrought iron directly from ore. { 'blü·mər }

blooming |MATER| The migration of sulfur or other substances to the surface of a sample of rubber, causing discoloration. { 'blüm·iŋ }

blooming mill |MET| A rolling mill for making blooms from ingots. Also known as cogging mill. { 'blü·miŋ ˌmil }

blotting paper |MATER| An unsized paper used to absorb excess ink from penned letters or signatures; also used for other applications where a soft, spongy paper is required. { 'bläd·iŋ ˌpā·pər }

blowhole |MET| A pocket of air or gas formed in a metal during solidification. { 'blō,hōl }

blow in |MET| To put a blast furnace into operation. { 'blō ˌin }

blowing agent |MATER| A chemical added to plastics and rubbers that generates inert gases on heating, causing the resin to assume a cellular structure. Also known as foaming agent. { 'blō·iŋ ˌā·jənt }

blowing-up furnace |MET| A furnace used for sintering ore and for the volatilization of lead and zinc. { 'blō·iŋ ˌəp ˌfər·nəs }

blown asphalt |MATER| Asphalt which is treated or heated with air or steam within a blowing still at relatively high temperature. { ¦blōn ¦as ˌfȯlt }

blown film |MATER| Film that has been produced by extruding a tube of plastic material which is then expanded in diameter and reduced in thickness by internal air pressure, and then slit along one side. { ¦blōn 'film }

blown foam |MATER| A cellular plastic consisting of an expanded matrix resembling natural foam. { ¦blōn 'fōm }

blown oil |MATER| A vegetable or animal oil that has been agitated and partially oxidized by a current of warm air or oxygen, including castor, whale, fish, rape, and linseed oils; used in paints, lubricants, and plasticizers. { ¦blōn 'ȯil }

blue |OPTICS| The hue evoked in an average observer by monochromatic radiation having a wavelength in the approximate range from 455 to 492 nanometers; however, the same sensation can be produced in a variety of other ways. { blü }

blue annealing |MET| Softening metal sheets by heating in an open furnace and cooling in air; bluish oxide forms on the metal surface. { ¦blü ə'nēl·iŋ }

blue brittleness |MET| Loss of ductility noted for

some steels when heated to 400–600°F (204–316°C), the blue heat range. { ¦blü 'brid·əl·nəs }

blue gas |MATER| A gas consisting chiefly of carbon monoxide and hydrogen, formed by the action of steam upon hot coke; used mainly as a source of hydrogen and in synthesis of other chemical compounds. Also known as blue water gas. { 'blü ˌgas }

blue glow |MET| Luminescence emitted by certain metallic oxides when heated. { 'blü ˌglō }

blueing *See* bluing. { 'blü·iŋ }

blue laser |OPTICS| A laser that emits bluish-purple light efficiently at room temperature from a semiconductor diode based on multiple quantum wells of III-V nitrides such as indium gallium nitride. { ¦blü ¦lā·zər }

blue vitriol |MATER| A hydrous solution of copper sulfate that is applied to the surface of a metal for layout purposes. { 'blü 'vit·rē,ȯl }

blue water gas *See* blue gas. { 'blü 'wȯd·ər ˌgas }

bluing |MET| Also spelled blueing. **1.** Formation of a bluish oxide film on polished steel; improves appearance and provides some corrosion resistance. **2.** Heating of formed springs to reduce internal stress. **3.** A blue oxide film formed on the polished surface of a metal due to extremely high temperatures. { 'blü·iŋ }

board |MATER| A piece of lumber whose dimensions are less than 2 inches (5 centimeters) thick and between 4 and 12 inches (10 and 30 centimeters) wide. { 'bȯrd }

bob |MET| A feeding device for providing molten metal to a casting during solidification to prevent shrinkage. { bäb }

Bobeck effect |SOLID STATE| The contraction of magnetic strip domains in a thin magnetic film to cylindrical domains called magnetic bubbles. { 'bäb·ək i,fekt }

body |MATER| The consistency or viscosity of fluid materials, such as lubricating oils, paints, and cosmetics. { 'bäd·ē }

body-centered lattice |CRYSTAL| A space lattice in which the point at the intersection of the body diagonals is identical to the points at the corners of the unit cell. { 'bäd·ē ,sen·tərd 'lad·əs }

body force |MECH| An external force, such as gravity, which acts on all parts of a body. { 'bäd·ē ,fȯrs }

Bohemian glass *See* hard glass. { bō'hem·ē·ən ˌglas }

bohrium |CHEM| A synthetic chemical element, symbolized Bh, atomic number 107; the fifteenth transuranium element. { 'bȯr·ē·əm }

boiler plate |MET| Flat-rolled steel, usually ¼ to ½ inch (6 to 13 millimeters) thick; used mainly for covering ships and making boilers and tanks. Also known as boiler steel. { 'bȯil·ər ,plāt }

boiler steel *See* boiler plate. { 'bȯil·ər ,stēl }

boiling point |PHYS CHEM| The temperature at which the transition from the liquid to the gaseous phase occurs in a pure substance at fixed pressure. Abbreviated bp. { 'bȯil·iŋ ,pȯint }

boiling-point elevation |CHEM| The raising of

the normal boiling point of a pure liquid compound by the presence of a dissolved substance, the elevation being in direct relation to the dissolved substance's molecular weight. { 'bȯil·iŋ ˌpȯint el·ə'vā·shən }

boiling range [CHEM] The temperature range of a laboratory distillation of an oil from start until evaporation is complete. { 'bȯil·iŋ ˌrānj }

boil-off [TEXT] Removal of impurities from fabric by boiling the material in a solution. [THERMO] The vaporization of a liquid, such as liquid oxygen or liquid hydrogen, as its temperature reaches its boiling point under conditions of exposure, as in the tank of a rocket being readied for launch. { 'bȯil,ȯf }

boil-off assistant [MATER] A material used to facilitate the removal of natural glue from silk during boiling in soap solution (degumming); examples are sulfonated oils, pine oils, and solvents and other wetting agents. { 'bȯil,ȯf ə'sis·tənt }

bolster [MET] A block of steel to which dropforging dies are attached. { 'bōl·stər }

bolt [MATER] In veneer production, a short log of a length suitable for peeling on a lathe. [TEXT] The entire length of cloth from a loom. { bōlt }

bolting cloth [MATER] A sieve cloth made of wire, hair, or silk or other thread used to remove lumps from flour or to make screen prints or needlework. { 'bōl·tiŋ ˌklȯth }

Boltzmann engine [THERMO] An ideal thermodynamic engine that utilizes blackbody radiation; used to derive the Stefan-Boltzmann law. { 'bōlts·mən ˌen·jən }

bond [CHEM] The strong attractive force that holds together atoms in molecules and crystalline salts. Also known as chemical bond. [MET] **1.** Material added to molding sand to impart bond strength. **2.** Junction of the base metal and filler metal, or the base metal beads, in a welded joint. { bänd }

bond angle [PHYS CHEM] The angle between bonds sharing a common atom. Also known as valence angle. { 'bänd ˌaŋ·gəl }

bond blister [MET] An unbonded area at the interface between the coating and the metal core of a cladded surface. { 'bänd ˌblis·tər }

bond clay [MATER] A type of clay with high plasticity and high dry strength used to bond nonplastic materials; may be refractory. { 'bänd ˌklā }

bond coat [MATER] **1.** A coat of bonding agent or plaster applied to a surface to provide a bond for succeeding coats of plaster. **2.** A coat of primer applied to a surface to act as a sealer or to ensure adhesion of paint to the surface. { 'bänd ˌkōt }

bond dissociation energy [PHYS CHEM] The change in enthalpy that occurs with the homolytic cleavage of a chemical bond under conditions of standard state. { ˌbänd di,sō·sē ¦ā·shən 'en·ər·jē }

bond distance [PHYS CHEM] The distance separating the two nuclei of two atoms bonded to each other in a molecule. Also known as bond length. { 'bänd ˌdis·təns }

bonded coating [MATER] A finishing or protecting layer of any compound affixed to a surface. { ¦bän·dəd ¦kōd·iŋ }

bond energy [PHYS CHEM] **1.** The average value of specific bond dissociation energies that have been measured from different molecules of a given type. **2.** *See* average bond dissociation energy. { 'bänd ˌen·ər·jē }

bonding [CHEM] The joining together of atoms to form molecules or crystalline aggregates. [ELEC] The use of low-resistance material to connect electrically a chassis, metal shield cans, cable shielding braid, and other supposedly equipotential points to eliminate undesirable electrical interaction resulting from high-impedance paths between them. [ENG] **1.** The fastening together of two components of a device by means of adhesives, as in anchoring the copper foil of printed wiring to an insulating baseboard. **2.** *See* cladding. [TEXT] The joining of two fabrics, usually a face fabric and a lining fabric. { 'bän·diŋ }

bonding agent [MATER] Any substance that fixes one material to another. { 'bän·diŋ ˌā·jənt }

bonding electron [PHYS CHEM] An electron whose orbit spans the entire molecule and so assists in holding it together. { 'bän·diŋ i'lek ˌträn }

bonding orbital [PHYS CHEM] A molecular orbital formed by a bonding electron whose energy decreases as the nuclei are brought closer together, resulting in a net attraction and chemical bonding. { 'bän·diŋ 'ȯr·bəd·əl }

bonding strength [MECH] Structural effectiveness of adhesives, welds, solders, glues, or of the chemical bond formed between the metallic and ceramic components of a cermet, when subjected to stress loading, for example, shear, tension, or compression. { 'bän·diŋ ˌstreŋkth }

bonding wire [ELEC] Wire used to connect metal objects so they have the same potential (usually ground potential). { 'bän·diŋ ˌwīr }

bond length *See* bond distance. { 'bänd ˌleŋkth }

bond paper [MATER] A paper used for writing paper, business forms, and typewriter paper; the less expensive bond papers are made from wood sulfite pulps; rag-content bonds contain 25, 50, 75, or 100% of pulp made from rags, and offer greater permanence and strength. { 'bänd ˌpā·pər }

bond strength [CHEM] The strength with which a chemical bond holds two atoms together; conventionally measured in terms of the amount of energy, in kilocalories per mole, required to break the bond. [ENG] The amount of adhesion between bonded surfaces measured in terms of the stress required to separate a layer of material from the base to which it is bonded. { 'bänd ˌstreŋkth }

bone ash [CHEM] A white ash consisting primarily of tribasic calcium phosphate obtained by burning bones in air; used in cleaning jewelry and in some pottery. { 'bōn ˌash }

bone black [MATER] A black substance made

by carbonizing crushed, defatted bones in closed vessels; used as a paint and varnish pigment, as a decolorizing absorbent in clarifying shellac, in cementation, and in gas masks. Also known as animal black; bone char. { 'bōn ˌblak }

bone char *See* bone black. { 'bōn ˌchär }

book mold [MET] A mold made in two halves hinged together like a book. { 'búk ˌmōld }

Boolean algebra [MATH] An algebraic system with two binary operations and one unary operation important in representing a two-valued logic. { 'bü·lē·ən 'al·jə·brə }

boracic acid *See* boric acid. { bə'ras·ik 'as·əd }

borane [INORG CHEM] **1.** A class of binary compounds of boron and hydrogen; boranes are used as fuels. Also known as boron hydride. **2.** A substance which may be considered a derivative of a boron-hydrogen compound, such as BCl_3 and $B_{10}H_{12}I_2$. { 'bȯ,rän }

borate [CHEM] **1.** A generic term referring to salts or esters of boric acid. **2.** Related to boric oxide, B_2O_3, or commonly to only the salts of orthoboric acid, H_3BO_3. { 'bȯ,rāt }

borax glass [MATER] A glassy, transparent solid formed by fusing borax. { 'bȯ,raks ˌglas }

borazole [INORG CHEM] $B_3N_3H_6$ A colorless liquid boiling at $53°C$; with water it hydrolyzes to form boron hydrides; the borazole molecule is the inorganic analog of the benzene molecule. { 'bȯr·ə,zōl }

borazon [INORG CHEM] A form of boron nitride with a zinc blende structure produced by subjecting the ordinary form to high pressure and temperature. { 'bȯr·ə,zän }

Bordini effect [MET] A phenomenon that takes place when metals having a close-packed crystal structure are subjected to an oscillating stress; there is a peak in the internal friction at a particular frequency of the stress. { bȯr'dēn·ē i'fekt }

boric acid [INORG CHEM] H_3BO_3 An acid derived from boric oxide in the form of white, triclinic crystals, melting at $185°C$, soluble in water. Also known as boracic acid; orthoboric acid. { ¦bȯr·ik 'as·əd }

boric oxide [INORG CHEM] B_2O_3 A trioxide of boron obtained as rhombic crystals melting at $460°C$; used as an intermediate in the production of boron halides and metallic borides and as a thermal neutron absorber in nuclear engineering. Also known as boron oxide. { ¦bȯr·ik 'äk,sīd }

boride [INORG CHEM] A binary compound of boron and a metal formed by heating a mixture of the two elements. { 'bȯr,īd }

Born-Haber cycle [SOLID STATE] A sequence of chemical and physical processes by means of which the cohesive energy of an ionic crystal can be deduced from experimental quantities; it leads from an initial state in which a crystal is at zero pressure and 0 K to a final state which is an infinitely dilute gas of its constituent ions, also at zero pressure and 0 K. { ¦bȯrn 'hä·bər ˌsī·kəl }

Born-Madelung model [SOLID STATE] A classical theory of cohesive energy, lattice spacing,

and compressibility of ionic crystals. { ¦bȯrn 'mäd·əl·əŋ 'mäd·əl }

Born-Mayer equation [SOLID STATE] An equation for the cohesive energy of an ionic crystal which is deduced by assuming that this energy is the sum of terms arising from the Coulomb interaction and a repulsive interaction between nearest neighbors. { ¦bȯrn 'mī·ər i'kwā·zhən }

Born-von Kármán theory [SOLID STATE] A theory of specific heat which considers an acoustical spectrum for the vibrations of a system of point particles distributed like the atoms in a crystal lattice. { ¦bȯrn fən'kär,män ,thē·ə·rē }

boron [CHEM] A chemical element, symbol B, atomic number 5, atomic weight 10.811; it has three valence electrons and is nonmetallic. { 'bȯ,rän }

boron alloy [MET] Alloy of boron and iron used to increase the high-temperature strength characteristics of alloy steel. { 'bȯ,rän 'al,ȯi }

boron fiber [CHEM] Fiber produced by vapor-deposition methods; used in various composite materials to impart a balance of strength and stiffness. Also known as boron filament. { 'bȯ,rän ,fī·bər }

boron filament *See* boron fiber. { 'bȯ,rän ,fil·ə·mənt }

boron fluoride [INORG CHEM] BF_3 A colorless pungent gas in a dry atmosphere; used in industry as an acidic catalyst for polymerizations, esterifications, and alkylations. Also known as boron trifluoride. { 'bȯ,rän 'flúr,īd }

boron hydride *See* borane. { 'bȯ,rän 'hī,drīd }

boron nitride [INORG CHEM] BN A binary compound of boron and nitrogen, especially a white, fluffy powder with high chemical and thermal stability and high electrical resistance. { 'bȯ,rän 'nī,trīd }

boron nitride fiber [INORG CHEM] Inorganic, high-strength fiber, made of boron nitride, that is resistant to chemicals and electricity but susceptible to oxidation above $1600°F$ ($870°C$); used in composite structures for yarns, fibers, and woven products. { 'bȯ,rän 'nī,trīd 'fī·bər }

boron oxide *See* boric oxide. { 'bȯ,rän 'äk,sīd }

boron steel [MET] Alloy steel with a small amount (as little as 0.0005) of boron added to increase hardenability; can be used to replace other alloys in short supply. { 'bȯ,rän 'stēl }

boron trichloride [INORG CHEM] BCl_3 A colorless liquid used as a catalyst and in refining of aluminum, magnesium, zinc, and copper. { 'bȯ,rän trī'klȯr,īd }

boron trifluoride *See* boron fluoride. { 'bȯ,rän trī'flúr,īd }

borosilicate glass [MATER] A type of glass containing at least 5% boric oxide; used in glassware that resists heat. { ¦bȯ·rō'sil·ə·kət 'glas }

Bosanquet's law [ELECTROMAG] The statement that, in analogy to Ohm's law for the resistance of an electric circuit, in a magnetic circuit the ratio of the magnetomotive force to the magnetic flux is a constant known as the reluctance. { 'bō·zən ,kets ,lȯ }

Bose-Einstein condensate [CRYO] The state of

matter of a gas of bosonic particles below a critical temperature such that a large number of particles occupy the ground state of the system. { ¦bōz ¦īn₁stīn 'kan·dən₁sāt }

Bose-Einstein condensation [CRYO] A phase transition that occurs when a gas of bosonic particles is cooled below a critical temperature very close to absolute zero, in which a large number of the particles come to occupy the ground state of the system and form a coherent matter wave. { ¦bōz ¦īn₁stīn kän·den'sā·shən }

bosh [MET] **1.** Tapering lower portion of a blast furnace, from the blast holes of the hearth up to the maximum internal diameter at the bottom of the stack. **2.** Quartz deposited on the furnace lining during the smelting of copper ore. { bäsh }

bottom tap [MET] A hole in the bottom of a furnace for draining out the slag. { 'bäd·əm ₁tap }

Bouin's solution [MATER] A picric acid-acetic acid-formaldehyde fixative and preserving fluid for contractile forms. { bü'anz sə'lü·shən }

boule [CRYSTAL] A pure crystal, such as silicon, having the atomic structure of a single crystal, formed synthetically by rotating a small seed crystal while pulling it slowly out of molten material in a special furnace. { bül }

bouncing putty [MATER] A silicone polymer in a soft elastic mass; the material's elasticity will increase as the applied force increases. { ¦baúns·iŋ 'pəd·ē }

boundary friction [MECH] Friction between surfaces that are neither completely dry nor completely separated by a lubricant. { 'baún·drē ₁frik·shən }

boundary-layer theory See film theory. { 'baún·drē ₁lā·ər ₁thē·ə·rē }

bound charge [ELEC] Electric charge which is confined to atoms or molecules, in contrast to free charge, such as metallic conduction electrons, which is not. Also known as polarization charge. { ¦baúnd 'chärj }

bound water [CHEM] Water that is a portion of a system such as tissues or soil and does not form ice crystals until the material's temperature is lowered to about −20°C. { ¦baúnd 'wód·ər }

Boussinesq approximation [FL MECH] The assumption (frequently used in the theory of convection) that the fluid is incompressible except insofar as the thermal expansion produces a buoyancy, represented by a term $g\alpha T$, where g is the acceleration of gravity, α is the coefficient of thermal expansion, and T is the perturbation temperature. { 'bü·si'nesk ə₁präk·sə'mā·shən }

Boussinesq's problem [MECH] The problem of determining the stresses and strains in an infinite elastic body, initially occupying all the space on one side of an infinite plane, and indented by a rigid punch having the form of a surface of revolution with axis of revolution perpendicular to the plane. Also known as Cerruti's problem. { 'bü·si'nesks ₁präb·ləm }

bow [MATER] The distortion of lumber in which there is a deviation from a straight line

in a direction perpendicular to the flat face. { baú }

Bower-Barff process [MET] A method of coating iron or steel with magnetic oxide, such as Fe_3O_4, in order to minimize atmospheric corrosion. { ¦baú·ər ¦bärf ₁präs·əs }

Bow's notation [MECH] A graphical method of representing coplanar forces and stresses, using alphabetical letters, in the solution of stresses or in determining the resultant of a system of concurrent forces. { 'bōz nō'tā·shən }

box annealing [MET] Slow heating of metal sheets in a closed metal box to prevent oxidation, followed by cooling; usually limited to iron-base alloys. { 'bäks ə'nēl·iŋ }

boxboard [MATER] Paperboard used for making cardboard boxes. { 'bäks₁bórd }

boxing [MET] Continuing a fillet weld around a corner. Also known as end turning. { 'bäks·iŋ }

Boyle's temperature [THERMO] For a given gas, the temperature at which the virial coefficient B in the equation of state $Pv = RT[1 + (B/v) + (C/v^2) + \cdots]$ vanishes. { 'bóilz 'tem·prə·chər }

bp See boiling point.

B phase See liquid B. { 'bē ₁fāz }

Br See bromine.

brachistochrone [MECH] The curve along which a smooth-sliding particle, under the influence of gravity alone, will fall from one point to another in the minimum time. { brə'kis·tə₁krōn }

brachyaxis [CRYSTAL] The shorter lateral axis, usually the a axis, of an orthorhombic or triclinic crystal. Also known as brachydiagonal. { ¦bra·kē'ak·səs }

brachydiagonal See brachyaxis. { ₁brak·i·dī'ag·ə·nəl }

Bragg angle [SOLID STATE] One of the characteristic angles at which x-rays reflect specularly from planes of atoms in a crystal. { 'brag ₁aŋ·gəl }

Bragg diffraction See Bragg scattering. { 'brag di'frak·shən }

Bragg reflection See Bragg scattering. { 'brag ri'flek·shən }

Bragg scattering [SOLID STATE] Scattering of x-rays or neutrons by the regularly spaced atoms in a crystal, for which constructive interference occurs only at definite angles called Bragg angles. Also known as Bragg diffraction; Bragg reflection. { 'brag ₁skad·ər·iŋ }

Bragg's equation See Bragg's law. { 'bragz i'kwā·zhən }

Bragg's law [SOLID STATE] A statement of the conditions under which a crystal will reflect a beam of x-rays with maximum intensity. Also known as Bragg's equation; Bravais' law. { 'bragz ₁lō }

brale [MET] A conical diamond indenter with an angle of 120° used in the Rockwell hardness test. { brāl }

branch See side chain. { branch }

branched chain See side chain. { 'brancht 'chān }

branched polymer [POLYM CHEM] A polymer chain having branch points that connect three or more chain segments; examples include graft copolymers, star polymers, comb polymers, and dendritic polymers. { ¦brancht 'päl·ə·mər }

brashness [MATER] Pertaining to wood that possesses relatively low shock resistance and therefore fails abruptly when broken during bending. { 'brash·nəs }

brass [MET] A copper-zinc alloy of varying proportions but typically containing 67% copper and 33% zinc. { bras }

Bravais indices [CRYSTAL] A modification of the Miller indices; frequently used for hexagonal and trigonal crystalline systems; they refer to four axes: the c axis and three others at $120°$ angles in the basal plane. { brə'vä 'in·də,sēz }

Bravais lattice [CRYSTAL] One of the 14 possible arrangements of lattice points in space such that the arrangement of points about any chosen point is identical with that about any other point. { brə'vä 'lad·əs }

Bravais' law See Bragg's law. { brə'väz ,lȯ }

braze [MET] To solder metals by melting a non-ferrous filler metal, such as brass or brazing alloy (hard solder), with a melting point lower than that of the base metals, at the point of contact. Also known as hard-solder. { brāz }

brazed joint [MET] The joining of two or more metallic components by brazing or braze welding. { ¦brāzd 'jȯint }

braze welding [MET] A method of welding in which coalescence is produced by heating above $800°F$ ($427°C$) and by using a nonferrous filler metal having a melting point below that of the base metals; in distinction to brazing, capillary attraction does not distribute the filler metal in the joint. { 'brāz ,wel·diŋ }

Brazil wax See carnauba wax. { brə'zil ,waks }

brazing alloy [MET] **1.** An alloy used as a filler metal for brazing; copper alloys and nickel alloys are used for brazing of steels. **2.** Solder that does not melt below red heat. Formerly known as hard solder. { 'brāz·iŋ ¦al,oi }

brazing brass See spelter solder. { 'brāz·iŋ ,bras }

brazing metal [MET] A nonferrous metal to be added to a joint in braze welding; can be a pure metal such as copper, zinc, or nickel, or a brazing alloy. { 'brāz·iŋ ,med·əl }

brazing sheet [MET] **1.** Brazing filler metal in sheet form. **2.** Flat-rolled metal clad with brazing filler metal on one or both sides. { 'brāz·iŋ ,shēt }

breakdown [MET] The initial process of rolling and drawing, or a series of such processes, which reduce a casting or extruded shape before its final reduction to desired size. { 'brāk,daùn }

breaking-down rolls [MET] A rolling mill unit used for breakdown operations. { 'brāk·iŋ ,daùn ,rōlz }

breaking load [MECH] The stress which, when steadily applied to a structural member, is just sufficient to break or rupture it. Also known as ultimate load. { 'brāk·iŋ ,lōd }

breaking radius [MATER] The limiting radius of curvature to which wood or plywood can be bent without breaking. { 'brāk·iŋ ,rād·ē·əs }

breaking strength [MECH] The ability of a material to resist breaking or rupture from a tension force. { 'brāk·iŋ ,streŋkth }

breaking stress [MECH] The stress required to fracture a material whether by compression, tension, or shear. { 'brāk·iŋ ,stres }

breast hole [MET] A hole for raking cinders out of a smelting cupola. { 'brest ,hōl }

breathing [MATER] Permeability of plastic sheeting to air, bubbles, voids, or trapped gas globules. { 'brēth·iŋ }

brick [MATER] A building material usually made from clay, molded as a rectangular block, and baked or burned in a kiln. { brik }

brick grease See block grease. { 'brik ,grēs }

bridge [ELEC] **1.** An electrical instrument having four or more branches, by means of which one or more of the electrical constants of an unknown component may be measured. **2.** An electrical shunt path. [ORG CHEM] A connection between two different parts of a molecule consisting of a valence bond, an atom, or an unbranched chain of atoms. { brij }

bridging [ELEC] **1.** Connecting one electric circuit in parallel with another. **2.** The action of a selector switch whose movable contact is wide enough to touch two adjacent contacts so that the circuit is not broken during contact transfer. [MET] **1.** Formation of arched cavities in a powder compact. **2.** Jamming of the charge in a blast or a cupola furnace due to adherence of fine ore particles to the inner walls. **3.** Formation of solidified metal over the top of the charge in a mold or crucible. { 'brij·iŋ }

bridging ligand [ORG CHEM] A ligand in which an atom or molecular species which is able to exist independently is simultaneously bonded to two or more metal atoms. { 'brij·iŋ ,lig·ənd }

bridging material [MATER] A fibrous, flaky, or granular substance added to a cement slurry or drilling fluid to seal a formation in which lost circulation has occurred. Also known as lost circulation material. { 'brij·iŋ mə,tir·ē·əl }

Bridgman effect [SOLID STATE] The phenomenon that when an electric current passes through an anisotropic crystal, there is an absorption or liberation of heat due to the nonuniformity in current distribution. { 'brij·mən i'fekt }

Bridgman relation [SOLID STATE] $P = QT\sigma$ in a metal or semiconductor, where P is the Etting-shausen coefficient, Q the Nernst-Ettingshausen coefficient, T the temperature, and σ the thermal conductivity in a transverse magnetic field. { 'brij·mən ri'lā·shən }

Bridgman technique [SOLID STATE] A method of growing single crystals in which a vertical cylinder that tapers conically to a point at the bottom and contains the substance to be crystallized in molten form is slowly lowered into a cold zone, resulting in crystallization beginning at the tip. { 'brij·mən tek'nēk }

bright |OPTICS| Attribute of an area that appears to emit a large amount of light. { brīt }

bright annealing |MET| Heating and cooling a metal in an inert atmosphere to inhibit oxidation; surface remains relatively bright. { ¦brīt ə'nēl·iŋ }

bright dipping |MET| Immersing metal into an acid solution to give a bright, clean surface. { brīt ¸dip·iŋ }

brightener |MET| Any of the agents which are employed in small concentrations in the electrolytic bath for electroplating metal to yield smoother or brighter coatings. { 'brīt·nər }

brightness |OPTICS| 1. The characteristic of light that gives a visual sensation of more or less light. 2. See luminance. { 'brīt·nəs }

brightness temperature See blackbody temperature. { 'brīt·nəs ¸tem·prə·chər }

bright plating |MET| An electroplating process resulting in a smooth, lustrous surface without polishing. { brīt ¸plād·iŋ }

bright stock |MATER| High-viscosity refined and dewaxed lubricating oils used in the compounding of motor oils. { brīt ¸stäk }

Brillouin function |SOLID STATE| A function of x with index (or parameter) n that appears in the quantum-mechanical theories of paramagnetism and ferromagnetism and is expressed as $|(2n+1)/2n|$ coth $|(2n+1)x/2n| - (1/2n)$ coth $(x/2n)$. { brēy·wan ¦fəŋk·shən }

Brillouin scattering |SOLID STATE| Light scattering by acoustic phonons. { brēy·wan ¦skad·ər·iŋ }

Brillouin zone |SOLID STATE| A fundamental region of wave vectors in the theory of the propagation of waves through a crystal lattice; any wave vector outside this region is equivalent to some vector inside it. { brēy·wan ¦zōn }

brine |MATER| A liquid used in a refrigeration system, usually an aqueous solution of calcium chloride or sodium chloride, which is cooled by contact with the evaporator surface and then goes to the space to be refrigerated. { brīn }

Brinkmann number |FL MECH| A dimensionless number used to study viscous flow. { 'briŋk·män ¸nəm·bər }

briquet |MATER| A block of some compressed substance, such as coal dust, metal powder, or sawdust, used as a fuel. Also spelled briquette. { bri'ket }

briquette See briquet. { bri'ket }

Bristol board |MATER| Cardboard with a surface smooth enough for painting or writing, usually at least 0.006 inch (0.15 millimeter) thick. { 'brist·əl ¸bȯrd }

britannia metal |MET| A silver-white tin alloy, similar to pewter, containing about 7% antimony, 2% copper, and often some zinc and bismuth; used in domestic utensils. { bri'tan·yə ¸med·əl }

British thermal unit |THERMO| Abbreviated Btu. 1. A unit of heat energy equal to the heat needed to raise the temperature of 1 pound of air-free water from 60 to 61°F at a constant pressure of 1 standard atmosphere; it is found experimentally to be equal to 1054.5 joules. Also known as sixty degrees Fahrenheit British thermal unit ($Btu_{60/61}$). 2. A unit of heat energy that is equal to 1/180 of the heat needed to raise 1 pound of air-free water from 32°F (0°C) to 212°F (100°C) at a constant pressure of 1 standard atmosphere; it is found experimentally to be equal to 1055.79 joules. Also known as mean British thermal unit (Btu_{mean}). 3. A unit of heat energy whose magnitude is such that 1 British thermal unit per pound equals 2326 joules per kilogram; it is equal to exactly 1055.05585262 joules. Also known as international table British thermal unit (Btu_{IT}). { 'brid·ish 'thər·məl ¸yü·nət }

brittle fracture |MET| A break in a brittle piece of metal which failed because stress exceeded cohesion. { ¦brid·əl 'frak·chər }

brittleness |MECH| That property of a material manifested by fracture without appreciable prior plastic deformation. { 'brid·əl·nəs }

brittle temperature |THERMO| The temperature point below which a material, especially metal, is brittle; that is, the critical normal stress for fracture is reached before the critical shear stress for plastic deformation. { 'brid·əl ¸tem·prə·chər }

bromate |CHEM| 1. BrO_3^- A negative ion derived from bromic acid, $HBrO_3$ 2. A salt of bromic acid. 3. $C_9H_9ClO_3$ A light brown solid with a melting point of 118-119°C; used as a herbicide to control weeds in crops such as flax, cereals, and legumes. { 'brō¸māt }

bromic acid |INORG CHEM| $HBrO_3$ A liquid, colorless to slightly yellow; boils with decomposition at 100°C; used in dyes and as a chemical intermediate. { 'brō·mik 'as·əd }

bromide |CHEM| A compound derived from hydrobromic acid, HBr, with the bromine atom in the −1 oxidation state. { 'brō¸mīd }

bromine |CHEM| A chemical element, symbol Br, atomic number 35, atomic weight 79.904; usually occurs as the molecule Br_2 used to test for unsaturation (by bleaching of an aqueous solution), to make ethylene dibromide, and in organic synthesis and plastics. { 'brō¸mēn }

bromo- |CHEM| A prefix that indicates the presence of bromine in a molecule. { 'brō·mō }

Brønsted acid |CHEM| A chemical species which can act as a source of protons. Also known as proton acid; protonic acid. { 'brən·steth or 'bren¸sted 'as·əd }

Brønsted base See base. { 'brən·steth ¸bās }

Brønsted-Lowry theory |CHEM| A theory that all acid-base reactions consist simply of the transfer of a proton from one base to another. Also known as Brønsted theory. { ¦brən·steth ¦laü·rē ¸thē·ə·rē }

Brønsted theory See Brønsted-Lowry theory. { 'brən·steth ¸thē·ə·rē }

bronze |MET| An alloy of copper and tin in varying proportions; other elements such as zinc, nickel, and lead may be added. { bränz }

bronzing liquid |MATER| A solvent, gloss oil, or varnish containing a bronze powder; used to produce bronze-colored finishes. { 'brän·ziŋ ¸lik·wəd }

45

Brown and Sharpe gage   *See* American wire gage. { ¦braůn ən ¦shärp ¸gāj }

brown coat |MATER| Mortar with about 1 to 1$\frac{1}{2}$ bushels (35 to 53 liters) of hair per 200 pounds (91 kilograms) quicklime; used to make a brown coat of plaster, which is covered with a finish coat, and often covers a scratch coat.   { ¦braůn ¦kōt }

brown lead oxide   *See* lead dioxide. { ¦braůn ¦led 'äk¸sīd }

brown petroleum |MATER| A solid or semisolid product formed by air acting on asphalt. { ¦braůn pə'trō·lē·əm }

brush plating |MET| Electroplating in which the anode with the solution is in the form of a brush or pad; used for plating equipment too large to be immersed.   { 'brəsh ¸plād·iŋ }

B stage |POLYM CHEM| An intermediate stage in a thermosetting resin reaction in which the plastic softens but does not fuse when heated, and swells but does not dissolve in contact with certain liquids.   { 'bē ¸stāj }

Btu   *See* British thermal unit.

bubble |PHYS| **1.** A small, approximately spherical body of fluid within another fluid or solid. **2.** A thin, approximately spherical film of liquid inflated with air or other gas. |SOLID STATE| *See* magnetic bubble.   { 'bəb·əl }

bubble raft |SOLID STATE| A visual demonstration for the structure of dislocations in metal lattices, showing slip propagation; it consists of many identical bubbles floating on a liquid surface in something like a crystalline array. { 'bəb·əl ¸raft }

buckle |MET| An up-and-down wrinkle on the surface of a metal bar or sheet.   { 'bək·əl }

buckling |ENG| Wrinkling or warping of fibers in a composite material. |MECH| Bending of a sheet, plate, or column supporting a compressive load.   { 'bək·liŋ }

buckling stress |MECH| Force exerted by the crippling load.   { 'bək·liŋ ¸stres }

buckminsterfullerene |CHEM| C_{60} The most abundant and most stable of the fullerenes, containing 60 carbon atoms in a highly spherical arrangement; named in honor of R. Buckminster Fuller, a practitioner of geodesic dome architecture. Also known as buckyball. { 'bək¸min·stər'fül·ə¸rēn }

buckyball   *See* buckminsterfullerene.   { 'bək·ē ¸ból }

Buerger precession method |CRYSTAL| The recording on film of a single level of the reciprocal lattice of an individual crystal, by means of x-ray diffraction, for the purpose of determining unit cell dimensions and space groups.   { 'bər·gər prē'sesh·ən ¸meth·əd }

buffer |CHEM| A solution selected or prepared to minimize changes in hydrogen ion concentration which would otherwise occur as a result of a chemical reaction. Also known as buffer solution. |ENG| A device, apparatus, or piece of material designed to reduce mechanical shock due to impact.   { 'bəf·ər }

buffer solution   *See* buffer.   { 'bəf·ər sə'lü·shən }

builder |MATER| An additive used in a detergent to improve the cleaning efficiency of the surfactant, principally by inactivating water-hardness ions. |MET| A fire-clay brick cull used for bottom construction in kilns or for boxing brick during burning.   { 'bil·dər }

building block |MATER| A hollow block, either of burned clay or concrete, used in constructing the walls of buildings, and often faced with brick or stone.   { 'bil·diŋ ¸bläk }

building paper |MATER| Heavy, waterproof paper used to cover sheathing and subfloors to prevent passage of air and water.   { 'bil·diŋ ¸pā·pər }

buildup |MET| Deposition of excess metal, either by electroplating or spraying, on worn or undersized machine components to restore required dimensions.   { 'bil¸dəp }

built-up mica |MATER| Large, laminated plates of mica made by bonding thin splittings of natural mica with shellac, glyptol, or some other suitable adhesive.   { 'bilt¸əp 'mī·kə }

built-up roofing |MATER| A seamless piece of flexible, waterproofed roofing material consisting of plies of felt mopped with asphalt or pitch. { 'bilt¸əp 'rüf·iŋ }

bulk density |ENG| The mass of powdered or granulated solid material per unit of volume. { ¦bəlk 'den·səd·ē }

bulked-down   *See* solid-piled.   { ¦bəlkt 'daůn }

bulk factor |ENG| The ratio of the volume of loose powdered or granulated solids to the volume of an equal weight of the material after consolidation into a voidless solid.   { 'bəlk ¸fak·tər }

bulk flow   *See* convection.   { ¦bəlk 'flō }

bulking |MATER| The difference in volume of a given mass of sand or other fine material in moist and dry conditions.   { 'bəl·kiŋ }

bulk lifetime |SOLID STATE| The average time that elapses between the formation and recombination of minority charge carriers in the bulk material of a semiconductor.   { ¦bəlk 'līf ¸tīm }

bulk modulus   *See* bulk modulus of elasticity. { ¦bəlk 'mäj·ə·ləs }

bulk modulus of elasticity |MECH| The ratio of the compressive or tensile force applied to a substance per unit surface area to the change in volume of the substance per unit volume. Also known as bulk modulus; compression modulus; hydrostatic modulus; modulus of compression; modulus of volume elasticity.   { ¦bəlk 'mäj·¸·ləs əv i¸las'tis·əd·ē }

bulk molding compound |MATER| A mixture of resin, inert fillers, reinforcements, and other additives which forms a puttylike rope, sheet, or preformed shape; used as a premix in composite manufacture.   { 'bəlk 'mōld·iŋ ¸käm¸paůnd }

bulk rheology |MECH| The branch of rheology wherein study of the behavior of matter neglects effects due to the surface of a system.   { ¦bəlk rē'äl·ə·jē }

bulk solid |MATER| An assembly of solid particles that is large enough for the statistical mean of any property to be independent of the number of particles. Also known as particulate solid; powder. { ¦bəlk 'säl·əd }

bulk strain |MECH| The ratio of the change in the volume of a body that occurs when the body is placed under pressure, to the original volume of the body. { 'balk ¸strān }

bulk strength |MECH| The strength per unit volume of a solid. { ¦bəlk 'streŋkth }

bull block |MET| Power-driven machine for drawing wire through a die. { 'bəl ¸bläk }

bullet |MATER| A small, lustrous, nearly spherical industrial diamond. { 'búl·ət }

bullion |MET| Gold or silver in bulk in the shape of bars or ingots. { 'búl·yən }

bull ladle |MET| A ladle used in foundry operations for carrying molten metal. { 'búl ¸lād·əl }

bull nose |MET| A ladle used in foundry operations for carrying molten metal. { 'búl ¸nōz }

Bulygen number |THERMO| A dimensionless number used in the study of heat transfer during evaporation. { 'bül·ə·jən ¸nəm·bər }

bumper |MET| A vibrating machine for ramming and consolidating sand into a mold. { 'bəm·pər }

bumping |CHEM| Uneven boiling of a liquid caused by irregular rapid escape of large bubbles of highly volatile components as the liquid mixture is heated. { 'bəm·piŋ }

Bunn chart |CRYSTAL| A chart for classifying x-ray diffraction powder photographs of substances whose crystals have tetragonal or hexagonal symmetry. { 'bən ¸chärt }

buoyancy |FL MECH| The resultant vertical force exerted on a body by a static fluid in which it is submerged or floating. { 'bói·ən·sē }

buoyancy parameter |FL MECH| The Grashof number divided by the square of the Reynolds number. { 'bói·ən·sē pə'ram·əd·ər }

buoyant force |FL MECH| The force exerted vertically upward by a fluid on a body wholly or partly immersed in it; its magnitude is equal to the weight of the fluid displaced by the body. { 'bói·ənt 'fòrs }

burden |MET| **1.** The material which is melted in a direct arc furnace. **2.** In an iron blast furnace, the ratio of iron and flux to coke and other fuels in the charge. { 'bərd·ən }

Burgers vector |CRYSTAL| A translation vector of a crystal lattice representing the displacement of the material to create a dislocation. { 'bər·gərz ¸vek·tər }

burl |MATER| In lumber or veneer, a localized severe distortion of the grain that is generally rounded in outline. { bərl }

burned sand |MET| The dissipated claying portion of a casting sand resulting from the heat of the metal. { ¦bərnd 'sand }

burning |ENG| The firing of clay products placed in a kiln. |MET| **1.** Permanent damage to a metal caused by heating beyond temperature limits of the treatment. **2.** Deep pitting of a metal caused by excessive pickling. { 'bər·niŋ }

burning rate |MATER| The tendency and rate of materials to burn at given temperatures, in contrast to melting or disintegrating. { 'bər·niŋ ¸rāt }

burnish |ENG| To polish or make shiny. |MET| To develop a smooth, lustrous surface finish by tumbling with steel balls or rubbing with a hard metal pad. { 'bər·nish }

burnisher |ENG| A tool with a hard, smooth rounded edge or surface; used for finishing the edges of scraper blades, for smoothing or polishing plastic or metal surfaces, or for other applications requiring manipulation by rubbing. { 'bər·nə·shər }

burnt deposit |MET| A dark, powdery deposit obtained by excessive current density in electroplating. { 'bərnt di'päz·ət }

burnt lime See calcium oxide. { ¦bərnt 'līm }

burnt-on sand |MET| A mixture of sand and metal on the surface of a casting due to metal penetration of the sand mold. { 'bərnt ¸on ¸sand }

burnt plate oil |MATER| A material used to thin etching ink too heavy in body; made by boiling linseed oil and, at a certain temperature, igniting it. { ¦bərnt ¸plāt ¸óil }

burr |MET| A thin, ragged fin left on the edge of a piece of metal by a cutting or punching tool. { bər }

bursting strength |MECH| A measure of the ability of a material to withstand pressure without rupture; it is the hydraulic pressure required to burst a vessel of given thickness. { 'bər·stiŋ ¸streŋkth }

burst pressure |MECH| The maximum inside pressure that a process vessel can safely withstand. { 'bərst ¸presh·ər }

buster |MET| A pair of dies used in press forging for barreling, for flattening a hot metal billet, or for loosening scale on hot, ferrous forging billets. { 'bəs·tər }

butadiene rubber See polybutadiene. { ¸byüd·ə'dī·ēn 'rəb·ər }

butadiene-styrene rubber |MATER| A synthetic rubber that is formed by copolymerization of butadiene and styrene. { ¸byüd·ə'dī·ēn 'stī¸rēn ¦rəb·ər }

Butler finish See satin finish. { 'bət·lər ¸fin·ish }

butter finish |MET| The semilustrous surface produced with a mildly abrasive wheel. { 'bəd·ər ¸fin·ish }

butterfly |MATER| A color imperfection in a lime-putty finish caused by lumps in the lime that were not broken up during mixing. { 'bəd·ər¸flī }

buttering |MET| Coating the faces of a weld joint prior to welding to prevent cross contamination of the weld metal and base metal. { 'bəd·ə·riŋ }

butt-off |MET| To supplement ramming in the production of castings by either manual or pneumatic jolting. { 'bət ¸óf }

button |MET| **1.** Mass of metal remaining in a crucible after fusion has been completed. **2.** That part of a weld which tears out in the destructive testing of spot-, seam-, or projection-welded specimens. { 'bət·ən }

butt weld [MET] A weld that joins the ends of two pieces of metal of similar cross section without overlapping. { 'bət ,weld }

butyl [ORG CHEM] Any of the four variations of the hydrocarbon radical C_4H_9: $CH_3CH_2CH_2CH_2-$, $(CH_3)_2CHCH_2-$, $CH_3CH_2CHCH_3-$, and $(CH_3)_3C-$. { 'byüd·əl }

butyl rubber [MATER] A synthetic rubber made by the polymerization of isoprene and isobutylene. { 'byüd·əl ,rəb·ər }

Byers process [MET] The main process for manufacturing wrought iron; pig iron is melted in a cupola, desulfurized in a ladle, and refined in a Bessemer converter. { 'bī·ərz ,präs·əs }

C

c *See* calorie.

C *See* capacitance; capacitor; carbon; coulomb.

Ca *See* calcium.

cabble [MET] To break up into pieces preparatory to the processes of fagoting, fusing, and rolling into bars, as is done with charcoal iron. { 'ka·bəl }

cable lacquer [MATER] Black, colored, or clear lacquer made from synthetic resins, having a high dielectric strength and being resistant to oils and heat; used to give a tough, flexible coating. { 'kā·bəl ,la·kər }

cable paper [MATER] A paper used to insulate electrical cables. { 'kā·bəl ,pā·pər }

CABRA numbers [MET] Copper and Brass Research Association number designations for various wrought copper and copper alloy grades. { 'kab·rə ,nəm·bərz }

cadmium [CHEM] A chemical element, symbol Cd, atomic number 48, atomic weight 112.40. [MET] A tin-white, malleable, ductile metal capable of high polish; principal use is in the plating of iron and steel, and to a much less extent of copper, brass, and other alloys, to protect them from corrosion and improve solderability and surface conductivity, and as a control absorber and shield in nuclear reactors. { 'kad·mē·əm }

cadmium bromate [INORG CHEM] $Cd(BrO_3)_2$ Colorless powder, soluble in water; used as an analytical reagent. { 'kad·mē·əm 'brō,māt }

cadmium bromide [INORG CHEM] $CdBr_2$ A compound produced as a yellow crystalline powder, soluble in water and alcohol; used in photography, process engraving, and lithography. { 'kad·mē·əm 'brō,mīd }

cadmium carbonate [INORG CHEM] $CdCO_3$ A white crystalline powder, insoluble in water, soluble in acids and potassium cyanide; used as a starting compound for other cadmium salts. { 'kad·mē·əm 'kär·bə,nāt }

cadmium chlorate [INORG CHEM] $CdClO_3$ White crystals, soluble in water; a highly toxic material. { 'kad·mē·əm 'klȯr,āt }

cadmium chloride [INORG CHEM] $CdCl_2$ A cadmium halide in the form of colorless crystals, soluble in water, methanol, and ethanol; used in photography, in dyeing and calico printing, and as a solution to precipitate sulfides. { 'kad·mē·əm 'klȯr,īd }

cadmium fluoride [INORG CHEM] CdF_2 A crystalline compound with a melting point of 1110°C; soluble in water and acids; used for electronic and optical applications and as a starting material for laser crystals. { 'kad·mē·əm 'flur,īd }

cadmium hydroxide [INORG CHEM] $Cd(OH)_2$ A white powder, soluble in dilute acids; used to prepare negative electrodes for cadmium-nickel storage batteries. { 'kad·mē·əm hī'dräk,sīd }

cadmium iodide [INORG CHEM] CdI_2 A cadmium halide that forms lustrous, white, hexagonal scales, consisting of two water-soluble allotropes; used in photography, in process engraving, and formerly as an antiseptic. { 'kad·mē·əm 'ī·ə,dīd }

cadmium metallurgy [MET] The extraction of cadmium from zinc ores, or from complex ores as a by-product of zinc, lead, and copper smelting. { 'kad·mē·əm 'med·əl·ər·jē }

cadmium oxide [INORG CHEM] CdO In the cubic form, a brown, amorphous powder, insoluble in water, soluble in acids and ammonia salts; used for cadmium plating baths and in the manufacture of paint pigments. { 'kad·mē·əm 'äk,sīd }

cadmium potassium iodide *See* potassium tetraiodocadmate. { 'kad·mē·əm pə'tas·ē·əm 'ī·ə,dīd }

cadmium red [MATER] A pigment composed of a mixture of cadmium sulfide, cadmium selenide, and barite; used as a red pigment. { 'kad·mē·əm 'red }

cadmium sulfate [INORG CHEM] $CdSO_4$ A compound that forms colorless, efflorescent crystals, soluble in water; used as an antiseptic and astringent, in the treatment of syphilis, gonorrhea, and rheumatism, and as a detector of hydrogen sulfide and fumaric acid. { 'kad·mē·əm 'səl,fāt }

cadmium sulfide [INORG CHEM] CdS A compound with two forms: orange, insoluble in water, used as a pigment, and also known as orange cadmium; light yellow, hexagonal crystals, insoluble in water, and also known as cadmium yellow. { 'kad·mē·əm 'səl,fīd }

cadmium telluride [INORG CHEM] CdTe Brownish-black, cubic crystals with a melting point of 1090°C; soluble, with decomposition, in nitric acid; used for semiconductors. { 'kad·mē·əm 'tel·yə,rīd }

cadmium tungstate

cadmium tungstate [INORG CHEM] $CdWO_4$ White or yellow crystals or powder; soluble in ammonium hydroxide and alkali cyanides; used in fluorescent paint, x-ray screens, and scintillation counters. { 'kad·mē·əm 'təŋ ,stāt }

cage [CRYSTAL] A void occurring in a crystal structure capable of trapping one or more foreign atoms. [PHYS CHEM] An aggregate of molecules in the condensed phase that surrounds fragments formed by thermal or photochemical dissociation or pairs of molecules in a solution that have collided without reacting. { kāj }

cage compound See clathrate. { ¦kāj ¦käm ,pau̇nd }

cage hydrocarbon [ORG CHEM] A compound composed of only carbon and hydrogen atoms that contains three or more rings arranged topologically so as to enclose a volume of space; in general, the space within a cage hydrocarbon is too small to accommodate even a proton. { ¦kāj ,hī·drə'kär·bən }

cake of gold [MET] Gold formed into a compact mass (though not melted) by distillation of the mercury from a mercury-gold amalgam. Also known as sponge gold. { ¦kāk əv ,gōld }

cal See calorie.

Cal See kilocalorie.

calamine [MET] An alloy composed of zinc, lead, and tin. { 'kal·ə,mīn }

calcimine [MATER] A thin paint, white or colored, made of pigment, glue, and water. Also known as kalsomine. { 'kal·sə,mīn }

calcine [ENG] **1.** To heat to a high temperature without fusing, as to heat unformed ceramic materials in a kiln, or to heat ores, precipitates, concentrates, or residues so that hydrates, carbonates, or other compounds are decomposed and the volatile material is expelled. **2.** To heat under oxidizing conditions. [MATER] Product of calcining or roasting. { 'kal,sīn }

calcined clay [MATER] Clay that has been heated to drive out volatile materials; a natural abrasive. { 'kal,sīnd 'klā }

calcined coke [MATER] Coke that has been heated to expel volatile material. { 'kal,sīnd ,kōk }

calcined gypsum See plaster of paris. { 'kal ,sīnd 'jip·səm }

calcined limestone [MATER] Limestone that has been heated in a vertical-shaft kiln to drive off carbon dioxide. { 'kal,sīnd 'līm,stōn }

calcined soda See soda ash. { 'kal,sīnd 'sō·də }

calcium [CHEM] A chemical element, symbol Ca, atomic number 20, atomic weight 40.08; used in metallurgy as an alloying agent for aluminum-bearing metal, as an aid in removing bismuth from lead, and as a deoxidizer in steel manufacture, and also used as a cathode coating in some types of photo tubes. { 'kal·sē·əm }

calcium arsenate [INORG CHEM] $Ca_3(AsO_4)_2$ An arsenic compound used as an insecticide to control cotton pests. { 'kal·sē·əm 'ärs·ən,āt }

calcium arsenite [INORG CHEM] $Ca_3(AsO_3)_2$

White granules that are soluble in water; used as an insecticide. { 'kal·sē·əm 'ärs·ən,īt }

calcium bisulfite [INORG CHEM] $Ca(HSO_3)_2$ A white powder, used as an antiseptic and in the sulfite pulping process. { 'kal·sē·əm bī'səl ,fīt }

calcium-bomb process [MET] A former process to produce pellets of pure titanium metal by mixing titanium chloride with calcium in a steel bomb and heating. { 'kal·sē·əm ,bäm ,präs·əs }

calcium bromide [INORG CHEM] $CaBr_2$ A deliquescent salt in the form of colorless hexagonal crystals that are soluble in water and absolute alcohol. { 'kal·sē·əm 'brō,mīd }

calcium carbide [INORG CHEM] CaC_2 An alkaline earth carbide obtained in the pure form as transparent crystals that decompose in water; used to make acetylene gas. { 'kal·sē·əm 'kär ,bīd }

calcium carbonate [INORG CHEM] $CaCO_3$ White rhombohedrons or a white powder; occurs naturally as calcite; used in paint manufacture, as a dentifrice, as an anticaking medium for table salt, and in manufacture of rubber tires. { 'kal·sē·əm 'kär·bə,nāt }

calcium chlorate [INORG CHEM] $Ca(ClO_3)_2 \cdot 2H_2O$ White monoclinic crystals, decomposed by heating. { 'kal·sē·əm 'klȯr,āt }

calcium chloride [INORG CHEM] $CaCl_2$ A colorless, deliquescent powder that is soluble in water and ethanol; used as an antifreeze and as an antidust agent. { 'kal·sē·əm 'klȯr,īd }

calcium chromate [INORG CHEM] $CaCrO_4 \cdot 2H_2O$ Yellow, monoclinic crystals that are slightly soluble in water; used to make other pigments. { 'kal·sē·əm 'krō,māt }

calcium cyanamide [INORG CHEM] $CaCN_2$ In pure form, colorless rhombohedral crystals, the commercial form being a gray material containing 55–70% $CaCN_2$; used as a fertilizer, weed killer, and defoliant. { 'kal·sē·əm sī'an·ə,mīd }

calcium cyanide [INORG CHEM] $Ca(CN)_2$ In pure form, a white powder that gives off hydrogen cyanide in air at normal humidity; prepared commercially in impure black or gray flakes; used as an insecticide and rodenticide. Also known as black cyanide. { 'kal·sē·əm 'sī·ə,nīd }

calcium dihydrogen phosphate See calcium phosphate. { 'kal·sē·əm dī'hī·drə·jən 'fäs,fāt }

calcium fluoride [INORG CHEM] CaF_2 Colorless, cubic crystals that are slightly soluble in water and soluble in ammonium salt solutions; used in etching glass and preparing hydrofluoric acid. { 'kal·sē·əm 'flu̇r,īd }

calcium hydride [INORG CHEM] CaH_2 In pure form, white crystals that are insoluble in water; used in the production of chromium, titanium, and zirconium in the Hydromet process. { 'kal·sē·əm 'hī,drīd }

calcium hydrogen phosphate See calcium phosphate. { ¦kal·sē·əm ¦hī·drə·jən 'fäs,fāt }

calcium hydroxide [INORG CHEM] $Ca(OH)_2$ White crystals, slightly soluble in water; used in cement, mortar, and manufacture of calcium

50

salts. Also known as hydrated lime. { 'kal·sē·əm hī'dräk,sīd }

calcium hypochlorite [INORG CHEM] $Ca(OCl)_2 \cdot 4H_2O$ A white powder, used as a bleaching agent and disinfectant for swimming pools. { 'kal·sē·əm hī·pō'klȯr,īt }

calcium iodide [INORG CHEM] CaI_2 A yellow, hygroscopic powder that is very soluble in water; used in photography. { 'kal·sē·əm 'ī·ə,dīd }

calcium nitrate [INORG CHEM] $Ca(NO_3)_2 \cdot 4H_2O$ Colorless, monoclinic crystals that are soluble in water; the anhydrous salt is very deliquescent; used as a fertilizer and in explosives. Also known as nitrocalcite. { 'kal·se·əm 'nī,trāt }

calcium oxalate [INORG CHEM] $CaC_2O_4 \cdot H_2O$ A salt of oxalic acid in the form of white crystals that are insoluble in water. { 'kal·se·əm 'äk·sə ,lāt }

calcium oxide [INORG CHEM] CaO A caustic white solid sparingly soluble in water; the commercial form is prepared by roasting calcium carbonate limestone in kilns until all the carbon dioxide is driven off; used as a refractory, in pulp and paper manufacture, and as a flux in manufacture of steel. Also known as burnt lime; calx; caustic lime. { 'kal·se·əm 'äk,sīd }

calcium peroxide [INORG CHEM] CaO_2 A cream-colored powder that decomposes in water; used as an antiseptic and a detergent. { 'kal·se·əm pə'räk,sīd }

calcium phosphate [INORG CHEM] **1.** Any phosphate of calcium. **2.** Any of the following three calcium orthophosphates, all of which are white or colorless in pure form: $Ca(H_2PO_4)_2$ is used as a fertilizer, as a plastics stabilizer, and in baking powder, and is also known as acid calcium phosphate, calcium dihydrogen phosphate, monobasic calcium phosphate, monocalcium phosphate; $CaHPO_4$ is used in pharmaceuticals, animal feeds, and toothpastes, and is also known as calcium hydrogen phosphate, dibasic calcium phosphate, dicalcium orthophosphate, dicalcium phosphate; $Ca_3(PO_4)_2$ is used as a fertilizer, and is also known as tribasic calcium phosphate, tricalcium phosphate. { 'kal·se·əm 'fäs,fāt }

calcium plumbate [INORG CHEM] $Ca(PbO_3)_2$ Orange crystals that are insoluble in cold water but decompose in hot water; used as an oxidizer in the manufacture of glass and matches. { 'kal·se·əm 'pləm,bāt }

calcium plumbite [INORG CHEM] $CaPbO_2$ Colorless crystals that are slightly soluble in water. { 'kal·se·əm 'pləm,bīt }

calcium pyrophosphate [INORG CHEM] $Ca_2P_2O_7$ White, abrasive powder, used in dentifrice polishes, in metal polishes, and as a food supplement. { 'kal·se·əm ,pī·rō'fäs,fāt }

calcium silicate [INORG CHEM] Any of three silicates of calcium: tricalcium silicate, Ca_3SiO_5; dicalcium silicate, Ca_2SiO_4; calcium metasilicate, $CaSiO_3$. { 'kal·se·əm 'sil·ə,kāt }

calcium sulfate [INORG CHEM] **1.** $CaSO_4$ A white crystalline salt, insoluble in water; used in Keene's cement, in pigments, as a paper filler, and as a drying agent. **2.** Either of two hydrated forms of the salt: the dihydrate, $CaSO_4 \cdot 2H_2O$, and the hemihydrate, $CaSO_4 \cdot \frac{1}{2}H_2O$. { 'kal·se·əm 'səl,fāt }

calcium sulfide [INORG CHEM] CaS In pure form, white cubic crystals, slightly soluble in water; used as a base for luminescent materials. Also known as hepar calcies; sulfurated lime. { 'kal·se·əm 'səl,fīd }

calcium sulfite [INORG CHEM] $CaSO_3 \cdot 2H_2O$ A white powder that is soluble in dilute sulfurous acid; may be dehydrated at $150°C$ to the anhydrous salt; used in the sulfite process for the manufacture of wood pulp. { 'kal·se·əm 'səl,fīt }

calcium tungstate [INORG CHEM] $CaWO_4$ White, tetragonal crystals, slightly soluble in water; used in manufacture of luminous paints. Also known as artificial scheelite; calcium wolframate. { 'kal·se·əm 'təŋ,stāt }

calcium wolframate See calcium tungstate. { 'kal·se·əm 'wu̇l·frə,māt }

calendered paper [MATER] Paper that has passed through the calenders of a paper machine. { 'kal·ən·dərd 'pā·pər }

californium [CHEM] A chemical element, symbol Cf, atomic number 98; all isotopes are radioactive. { ,kal·ə'fȯr·nē·əm }

calite [MET] A heat-resistant alloy of iron, nickel, and aluminum which resists oxidation up to $2200°F$ ($1204°C$) and is practically noncorrodible under ordinary conditions of exposure. { 'kā ,līt }

Callendar's equation [THERMO] **1.** An equation of state for steam whose temperature is well above the boiling point at the existing pressure, but is less than the critical temperature: $(V - b) = (RT/p) - (a/T^n)$, where V is the volume, R is the gas constant, T is the temperature, p is the pressure, n equals $^{10}/_3$, and a and b are constants. **2.** A very accurate equation relating temperature and resistance of platinum, according to which the temperature is the sum of a linear function of the resistance of platinum and a small correction term, which is a quadratic function of temperature. { 'kal·ən·dərz i'kwā·zhən }

calorescence [PHYS] The production of visible light by infrared radiation; the transformation is indirect, the light being produced by heat and not by any direct change of wavelength. { ,kal·ə'res·əns }

calorie [THERMO] Abbreviated cal; often designated c. **1.** A unit of heat energy, equal to 4.1868 joules. Also known as International Table calorie (IT calorie). **2.** A unit of energy, equal to the heat required to raise the temperature of 1 gram of water from $14.5°$ to $15.5°C$ at a constant pressure of 1 standard atmosphere; equal to 4.1855 ± 0.0005 joules. Also known as fifteen-degrees calorie; gram-calorie (g-cal); small calorie. **3.** A unit of heat energy equal to 4.184 joules; used in thermochemistry. Also known as thermochemical calorie. { 'kal·ə·rē }

calorimeter [ENG] An apparatus for measuring heat quantities generated in or emitted by materials in processes such as chemical reactions, changes of state, or formation of solutions. { ‚kal·ə'rim·əd·ər }

calorimetric test [ENG] The use of a calorimeter to determine the thermochemical characteristics of propellants and explosives; properties normally determined are heat of combustion, heat of explosion, heat of formation, and heat of reaction. { kə¦lȯr·ə¦me·trik 'test }

calorimetry [ENG] The measurement of the quantity of heat involved in various processes, such as chemical reactions, changes of state, and formations of solutions, or in the determination of the heat capacities of substances; fundamental unit of measurement is the joule or the calorie (4.184 joules). { kal·ə'rim·ə·trē }

calorize [MET] To treat by a process by which a coating of aluminum and aluminum-iron alloys is produced on iron and steel (and, less commonly, brass, copper, or nickel), which coating protects the metal from burning in temperatures up to 1800°F (982°C). { 'kal·ə‚rīz }

calorized steel See aluminized steel. { 'kal·ə‚rīzd ‚stēl }

calx See calcium oxide. { kalks }

Calzecchi-Onesti effect [ELEC] A change in the conductivity of a loosely aggregated metallic powder caused by an applied electric field. { ‚kält'se·kē ‚ȯ'nes·tē i'fekt }

Campbell process [MET] An open-hearth steel manufacturing process in which ore and pig iron are used as raw materials in a tilting furnace. { 'kam·əl ‚präs·əs }

candelilla wax [MATER] A wax obtained from the wax-coated stems of candelilla shrubs, especially Euphorbia antisyphilitica; used for varnishes and furniture and shoe polishes. { ‚kan·də'lē·ə ‚waks }

candlepower [OPTICS] Luminous intensity expressed in candelas. Abbreviated cp. { 'kan·dəl ‚paú·ər }

caoutchouc [MATER] Formerly, crude rubber which had been cured over a fire into a solid, dark mass for shipment. { kaú'chük }

capacitance [ELEC] The ratio of the charge on one of the conductors of a capacitor (there being an equal and opposite charge on the other conductor) to the potential difference between the conductors. Symbolized C. Formerly known as capacity. { kə'pas·ə·təns }

capacitor [ELEC] A device which consists essentially of two conductors (such as parallel metal plates) insulated from each other by a dielectric and which introduces capacitance into a circuit, stores electrical energy, blocks the flow of direct current, and permits the flow of alternating current to a degree dependent on the capacitor's capacitance and the current frequency. Symbolized C. Also known as condenser; electric condenser. { kə'pas·əd·ər }

capacitor color code [ELEC] A method of marking the value on a capacitor by means of dots or bands of colors as specified in the Electronic Industry Association color code. { kə'pas·əd·ər 'kəl·ər ‚kōd }

capacity See capacitance. { kə'pas·əd·ē }

capillary number [FL MECH] A dimensionless number associated with a liquid that compares the intensity of liquid viscosity and surface tension, equal to $\mu V/\sigma$, where μ is the viscosity, σ is the surface tension, and V is a fluid velocity such as the deposition velocity on a solid that is drawn out of the liquid. { 'kap·ə·ler·ē ‚nəm·bər }

capillary viscometer [ENG] A long, narrow tube that is used to measure the laminar flow of fluids. { 'kap·ə‚ler·ē vis'käm·əd·ər }

capped steel [MET] Partially deoxidized steel cast in an open-top mold, which is capped to solidify the top metal and enforce internal pressure, resulting in a surface condition similar to that of rimming steel. { 'kapt 'stēl }

caprolactam [ORG CHEM] $(CH_2)_5NH\cdot CO$ White flakes, melting point 68–69°C, made from cyclohexanone; used to make synthetic fiber, particularly nylon-6. { ‚ka·prō¦lak·təm }

ε-caprolactone [ORG CHEM] $CH_2(CH_2)_4NHCO$ White crystals, used to make synthetic fibers, plastics, films, coatings, and plasticizers; its vapors or fine crystals are respiratory irritants. { ¦ā·də ¦ka·prō¦lak‚tōn }

caranda wax [MATER] A wax similar to carnauba wax; obtained from the tropical palm caranda (Copernicia australis). { ¦ka·rən¦dä ‚waks }

Carathéodory's principle [THERMO] An expression of the second law of thermodynamics which says that in the neighborhood of any equilibrium state of a system, there are states which are not accessible by a reversible or irreversible adiabatic process. Also known as principle of inaccessibility. { ‚kär·ə‚tā·ə'dȯr·ēz 'prin·sə·pəl }

carbamide See urea. { 'kär·bə‚mīd }

carbanion [CHEM] An anion in which the negative charge is localized on a carbon atom, for example, CH_3^- or $HC\equiv C^-$. { ¦kärb'an‚ī·ən }

carbide [INORG CHEM] A binary compound of carbon with an element more electropositive than carbon; carbon-hydrogen compounds are excluded. [MATER] A cemented or compacted mixture of powdered carbides of heavy metals forming a hard material used in metal-cutting tools. Also known as cemented carbide. { 'kär‚bīd }

carbolic acid See phenol. { kär'bäl·ik 'as·əd }

carboloy [MET] A hard alloy, containing carbon, tungsten, and cobalt, used for cutting tools and as an abrasive. { 'kär·bə·lȯi }

carbon [CHEM] A nonmetallic chemical element, symbol C, atomic number 6, atomic weight 12.01115; occurs freely as diamond, graphite, and coal. { 'kär·bən }

carbon-arc brazing [MET] Brazing base metals by heating with an electric arc between a carbon electrode and the workpiece. { ¦kär·bən ¦ärk 'brāz·iŋ }

carbon-arc cutting [MET] An arc cutting process which generates heat in order to melt a base metal with a carbon electrode to eventually produce a cut. { ¦kär·bən ¦ärk 'kəd·iŋ }

carbon-arc welding [MET] Welding by maintaining an electric arc between a nonconsumable carbon electrode and the work. { ¦kär·bən ¦ärk 'wel·diŋ }

carbonate [CHEM] **1.** An ester or salt of carbonic acid. **2.** A compound containing the carbonate (CO_3^{2-}) ion. **3.** Containing carbonates. { 'kär·bə·nət }

carbon black [CHEM] An amorphous form of carbon produced commercially by thermal or oxidative decomposition of hydrocarbons and used principally in rubber goods, pigments, and printer's ink. { 'kär·bən ¦blak }

carbon dioxide [INORG CHEM] CO_2 A colorless, odorless, tasteless gas about 1.5 times as dense as air. { ¦kär·bən dī'äk,sīd }

carbon dioxide process [MET] A casting process in which the molding material is a mixture of sand and 1.5–6% liquid silicate as a binder, and the mixture is packed around the pattern and hardened by blowing carbon dioxide gas through it. { ¦kär·bən dī'äk,sīd ,präs·əs }

carbon electrode [MET] A nonfiller-metal electrode consisting of carbon or a graphite rod; sometimes contains copper powder for increased electrical conductivity; used in carbon arc welding. { ¦kär·bən i'lek·trōd }

carbon equivalent [MET] An empirical relationship of the total carbon (TC), silicon (Si), and phosphorus (P) content of gray iron: CE = %TC + 0.3(%Si + %P). { ¦kär·bən i'kwiv·ə·lənt }

carbon fiber [MATER] **1.** Commercial material made by pyrolyzing any spun, felted, or woven raw material to a char at temperatures from 700 to 1800°C. **2.** A filamentary form of carbon, usually with a diameter in the 6–10-micrometer range. { 'kär·bən ,fī·bər }

carbonic acid [INORG CHEM] H_2CO_3 The acid formed by combination of carbon dioxide and water. { kär'bän·ik 'as·əd }

carbonitrided steel [MET] Steel which is produced by carbonitriding. { ,kär·bə'nī,trīd·əd 'stēl }

carbonitriding [MET] A case-hardening process that maintains carbon and alloy steels in a hot gaseous atmosphere from which they absorb carbon and nitrogen simultaneously, producing a hard, wear-resistant surface. { ,kär·bə'nī ,trīd·iŋ }

carbonless copy paper [MATER] A sheet of paper, used to make duplicate copies of written or printed material, whose back is coated with a layer of microcapsules that contain a dye in colorless form in a hydrocarbon solvent; writing or printing pressure breaks the capsules and releases the dye, which reacts with a clay or phenolic resin coating on top of a second paper sheet, located directly below the first, to produce visible color. { 'kär·bən·ləs 'käp·ē ,pā·pər }

carbon monoxide [INORG CHEM] CO A colorless, odorless gas resulting from the incomplete oxidation of carbon; found, for example, in mines and automobile exhaust; poisonous to animals. { ¦kär·bən mə'näk,sīd }

carbon nanotubes [CHEM] Cylindrical molecules (sealed at both ends with a convex arrangement of atoms) composed of carbon with a diameter of around 1 nanometer and lengths up to a few micrometers. Single-walled nanotubes may be conducting or semiconducting, depending on the diameter and chirality of the tube. Multiwall nanotubes containing coaxial shells of the elemental single-wall nanotubes are also possible. { ,kär·bən 'nan·ō,tübz }

carbon paper [MATER] A paper, coated with dark waxy pigment, used to make duplicate copies while typewriting or handwriting; a sheet of carbon paper is sandwiched between two paper sheets, so that the impression made on the top sheet causes the carbon paper to transfer a pigmented impression onto the bottom sheet. { 'kär·bən ,pā·pər }

carbon potential [MET] Determination of the extent to which an environment containing active carbon can affect the carbon content of a steel. { 'kär·bən pə'ten·chəl }

carbon refractory [MATER] Carbon, generally in the form of graphite, used as a refractory in such equipment as crucibles and stopper nozzles for steel casting. { 'kär·bən ri'frak·trē }

carbon restoration [MET] Carburizing of the surface layer of a material to achieve the original level of carbon content which had been depleted during processing. { 'kär·bən res·tə'rā·shən }

carbon steel [MET] Steel containing carbon, to about 2%, as the principal alloying element. { 'kär·bən 'stēl }

carbon suboxide [INORG CHEM] C_3O_2 A colorless lacrimatory gas having an unpleasant odor with a boiling point of −6.8°C. { 'kär·bən səb'äk ,sīd }

carbonyl [ORG CHEM] A functional group found in organic compounds in which a carbon atom is doubly bonded to an oxygen atom (−CO−). { 'kär·bə,nil }

carbonyl process [MET] **1.** A process in powder metallurgy for the production of iron, nickel, and iron-nickel alloy powders for magnetic applications. **2.** A process used in putting a metallic coating on molybdenum tungsten and other metals. { 'kär·bə,nil ,präs·əs }

carborundum [MATER] A manufactured crystalline material (silicon carbide), prepared by fusing coke and sand in an electric furnace; used as an abrasive in the grinding of low-tensile-strength materials, and as a semiconductor with a maximum operating temperature of 1300°C, to rectify and detect radio waves. { ,kär·bə'rən·dəm }

carbosand [MATER] Sand treated with an organic solution and then roasted; used as a spray to disperse oil slicks. { 'kär·bō,sand }

carboxy group [ORG CHEM] −COOH The functional group of carboxylic acid. { kär'bäk·sē ,grüp }

carboxylic [CHEM] Having chemical properties resembling those of carboxylic acid. { ¦kär,bäk ¦sil·ik }

carboxylic acid [ORG CHEM] Any of a family of organic acids characterized by the presence of one or more carboxyl groups. { ¦kär,bäk¦sil·ik 'as·əd }

carboxymethylcellulose [POLYM CHEM] An acid

ether derivative of cellulose used as a sodium salt; a white, odorless, bulky solid used as a stabilizer and emulsifier; negatively charged resin used in ion-exchange chromatography as a cation exchanger. Also known as cellulose gum. { kär ¦bäk-sē¦meth·əl 'sel·yə¦lōs }

carburize |MET| To surface-harden steel by converting the outer layer of low-carbon steel to high-carbon steel by heating the steel above the transformation range in contact with a carbonaceous material. { 'kär·bə₁rīz }

carbyne |CHEM| Elemental carbon in a triply bonded form. { 'kär₁bīn }

carcerand |ORG CHEM| A macrocyclic compound capable of including organic guest molecules. { 'kär·sə·rənd }

cardboard |MATER| A good quality of chemical pulp or rag pasteboard made by combining two or more webs of paper, either with or without paste, while still wet; used for signs, printed material, and high-quality boxes. { 'kärd₁bȯrd }

Carinthian furnace |MET| A zinc distillation furnace with small, vertical retorts. { kə'rin·thē·ən 'fər·nəs }

Carlsbad law |CRYSTAL| A feldspar twin law in which the twinning axis is the *c* axis, the operation is rotation of 180°, and the contact surface is parallel to the side pinacoid. { 'kärlz₁bad ₁lȯ }

Carlsbad turn |CRYSTAL| A twin crystal in the monoclinic system with the vertical axis as the turning axis. { 'kärlz₁bad ₁tərn }

carnauba wax |MATER| The hardest natural wax, having a melting point of 85°C, exuded from the leaves of the carnauba palm (*Carnauba cerifera*); used for insulating purposes and in making candles, shoe polish, high-luster wax, varnishes, phonograph records, and surface coating of automobiles. Also known as Brazil wax. { kär'nȯ·bə 'waks }

Carnot-Clausius equation |THERMO| For any system executing a closed cycle of reversible changes, the integral over the cycle of the infinitesimal amount of heat transferred to the system divided by its temperature equals 0. Also known as Clausius theorem. { kär¦nōt 'klȯz·ē· əs i₁kwä·zhən }

Carnot cycle |THERMO| A hypothetical cycle consisting of four reversible processes in succession: an isothermal expansion and heat addition, an isentropic expansion, an isothermal compression and heat rejection process, and an isentropic compression. { kär'nō ₁si·kəl }

Carnot efficiency |THERMO| The efficiency of a Carnot engine receiving heat at a temperature absolute T_1 and giving it up at a lower temperature absolute T_2; equal to $(T_1 - T_2)/T_1$. { kär'nō i'fish·ən·sē }

Carnot number |THERMO| A property of two heat sinks, equal to the Carnot efficiency of an engine operating between them. { kär'nō ₁nəm·bər }

Carnot's theorem |THERMO| **1.** The theorem that all Carnot engines operating between two given temperatures have the same efficiency, and no cyclic heat engine operating between two given temperatures is more efficient than a

Carnot engine. **2.** The theorem that any system has two properties, the thermodynamic temperature T and the entropy S, such that the amount of heat exchanged in an infinitesimal reversible process is given by $dQ = T dS$; the thermodynamic temperature is a strictly increasing function of the empirical temperature measured on an arbitrary scale. { kär'noz 'thir·əm }

caroba *See* carob wood. { kə'rō·bə }

carob wood |MATER| The wood from a large Brazilian tree, *Jacaranda copaia*. Also known as caroba; jacaranda. { 'kar·əb ₁wu̇d }

Caro's acid |INORG CHEM| H_2SO_5 A white solid melting at about 45°C, formed during the acid hydrolysis of peroxydisulfates. { 'kä·roz 'as·əd }

carpincho |MATER| Processed capybara skin, noted for its elastic properties. { kär'pin·chō }

carrier |CHEM| A substance that, when associated with a trace of another substance, will carry the trace with it through a chemical or physical process. { 'kar·ē·ər }

carrier density |SOLID STATE| The density of electrons and holes in a semiconductor. { 'kar·ē·ər ₁den·səd·ē }

carrier gas |MET| The gas used in thermal spraying which transmits powder from the feeder to the spray gun. { 'kar·ē·ər ₁gas }

carrier mobility |SOLID STATE| The average drift velocity of carriers per unit electric field in a homogeneous semiconductor; the mobility of electrons is usually different from that of holes. { 'kar·ē·ər mō'bil·əd·ē }

cartridge brass |MET| An alloy containing 70% copper and 30% zinc; uses include cartridge cases, automotive radiator cores and tanks, lighting fixtures, eyelets, rivets, springs, screws, and plumbing products. { 'kär·trij ₁bras }

cascade liquefaction |CRYO| A method of liquefying gases in which a gas with a high critical temperature is liquefied by increasing its pressure; evaporation of this liquid cools a second liquid so that it can also be liquefied by compression, and so on. { ka'skäd lik·wə'fak·shən }

cascade molecule *See* dendrimer. { ka¦skäd 'mäl·ə₁kyül }

cascade sequence |MET| Combined longitudinal and buildup sequence in which weld beads are deposited in overlapping layers. { ka'skäd ₁sē·kwəns }

case |MET| Outer layer of a ferrous alloy which has been made harder than the core by case hardening. { kās }

cased glass |MATER| Glass composed of two or more layers of different colors. { 'kāst ₁glas }

case hardening |MATER| A condition of stress and set in dry lumber characterized by compressive stress in the outer layers and tensile stress in the center or core. |MET| Process of carburizing low-carbon steel or other ferrous alloy for making the outer layer (case) harder than the core. { 'kās ₁härd·ən·iŋ }

casein |ORG CHEM| The protein of milk; a white solid soluble in acids. { 'ka₁sēn }

casein-formaldehyde |POLYM CHEM| A modified natural polymer. { 'ka₁sēn fȯr'mal·də₁hīd }

casein glue [MATER] A produce of dried curds of milk, lime, and other chemical ingredients, mixed cold; used for both plywood and assembly work. { 'ka,sēn ,glü }

casein paint [MATER] A paint with casein substituted for linseed oil. { 'ka,sēn ¦pānt }

casein plastic [MATER] A plastic made with casein and used for buttons, beads, knitting needles, and novelties; thin sheets and rods of casein plastic are cured (hardened) in formaldehyde baths. { 'ka,sēn ¦plas·tik }

cashew gum [MATER] A gum obtained from the bark of the cashew tree; hard, yellowish-brown substance used for inks, insecticides, pharmaceuticals, varnishes, and bookbinders' gum. Also known as anacardium gum. { 'kash·ü ,gəm }

Casimir-du Pré theory [SOLID STATE] A theory of spin-lattice relaxation which treats the lattice and spin systems as distinct thermodynamic systems in thermal contact with one another. { 'kaz·ə,mir dyü'prā ,thē·ə·rē }

Cassel green See barium manganate. { 'kas·əl ,grēn }

cast [ENG] **1.** To form a liquid or plastic substance into a fixed shape by letting it cool in the mold. **2.** Any object which is formed by placing a castable substance in a mold or form and allowing it to solidify. Also known as casting. [OPTICS] A change in a color because of the adding of a different hue. { kast }

castable [MATER] A refractory aggregate mixed with a bonding agent such as aluminous hydraulic cement which, with addition of water, will develop structural strength and set in a mold. { 'ka·stə·bəl }

cast coated paper [MATER] A paper with a high-gloss enamel finish that has been produced by drying coated paper under pressure from a polished cylinder. { 'kast ,kōd·əd 'pā·pər }

Castigliano's principle See Castigliano's theorem. { ,kas·til'yä·nōz ,prin·sə·pəl }

Castigliano's theorem [MECH] The theorem that the component in a given direction of the deflection of the point of application of an external force on an elastic body is equal to the partial derivative of the work of deformation with respect to the component of the force in that direction. Also known as Castigliano's principle. { ,kas·til'yä·nōz ,thir·əm }

casting See cast. { 'kast·iŋ }

casting alloy [MET] An alloy which cannot be forged or rolled and can be shaped only as a casting. { 'kast·iŋ ,a,lȯi }

casting copper [MET] Copper used for making foundry castings; obtained from copper ores, and inferior to electrolytic copper. { 'kast·iŋ ,käp·ər }

casting ladle [MET] A refractory-lined steel ladle used to transport molten metal from the furnace to a mold. { 'kast·iŋ ,lād·əl }

casting plaster [MATER] A white plaster used for castings and carvings. { 'kast·iŋ ,plas·tər }

casting shrinkage [MET] **1.** Total reduction in volume of a casting due to partial reductions at each stage of solidification. **2.** Reduction in volume at each stage of solidification of a casting. { 'kast,iŋ ,shriŋ·kij }

casting slip [MATER] A slurry of clay and additives mixed in water with deflocculating agents and used for casting in molds. { 'kast·iŋ ,slip }

casting strain [MECH] Any strain that results from the cooling of a casting, causing casting stress. { 'kast·iŋ ,strān }

casting stress [MECH] Any stress that develops in a casting due to geometry and casting shrinkage. { 'kast·iŋ ,stres }

casting wheel [MET] A large turntable with molds mounted on the outer edge; used primarily in the base-metal industries for cast ingots, anodes, and so on. { 'kast·iŋ ,wēl }

cast iron [MET] Any carbon-iron alloy cast to shape and containing 1.8–4.5% carbon, that is, in excess of the solubility in austenite at the eutectic temperature. Abbreviated C.I. { ¦kast 'ī·ərn }

castor machine oil [MATER] Petroleum-base lubricating oil thickened with an aluminum-base soap, such as aluminum oleate. { 'kas·tər mə'shēn ,ȯil }

cast steel [MET] Steel shaped by casting. { 'kast ,stēl }

cast stone [MATER] Building stone molded from concrete so that it resembles natural stone. { 'kast ,stōn }

cast structure [MET] The microstructure of a casting. { 'kast ,strək·chər }

cast-weld [MET] Joining parts by pouring molten metal over them in a mold. { 'kast ,weld }

Catalan forge [MET] A furnace or forge for making wrought iron from ore, in which the ore loaded at front and charcoal at rear are covered with fine ore and charcoal dust moistened with water. { 'kad·ə,lan ,fȯrj }

catalysis [CHEM] A phenomenon in which a relatively small amount of substance augments the rate of a chemical reaction without itself being consumed. { kə'tal·ə·səs }

catalyst [CHEM] Substance that alters the velocity of a chemical reaction and may be recovered essentially unaltered in form and amount at the end of the reaction. { 'kad·əl·əst }

catalytic hydrogenation [CHEM ENG] Hydrogenating by means of catalysts such as nickel or palladium. { ¦kad·əl¦id·ik ,hī·dra·ja'nā·shən }

catalytic polymerization [POLYM CHEM] Polymerization of monomers to form high-molecular-weight molecules in the presence of catalysts. { ¦kad·əl¦id·ik pə,lim·ə·rə'zā·shən }

catalytic site See active site. { ¦kad·əl¦id·ik ,sīt }

cataphoresis See electrophoresis. { ,kad·ə·fə 'rē·səs }

catastrophic failure [ENG] **1.** A sudden failure without warning, as opposed to degradation failure. **2.** A failure whose occurrence can prevent the satisfactory performance of an entire assembly or system. { ,kad·ə'sträf·ik 'fāl·yər }

catenane [ORG CHEM] A supramolecular species consisting of mechanically interlocked macrocyclic rings. { 'kat·ən,ān }

catenary [MATER] In fiberglass-reinforced plastics, a measure of the sag in an assemblage of a number of strands, which have a minimal amount of twist, at the midpoint of a specified length. { 'kat·ə,ner·ē }

catenation |CHEM| Formation of a chain structure by the bonding of atoms of the same element, for example, carbon in the hydrocarbons. { ,kat·ən'ā·shən }

catgut |MATER| A thin cord made from the submucosa of sheep and other animal intestine; used for sutures and ligatures, for strings of musical instruments, and for tennis racket strings. { 'kat,gət }

cathedral glass |MATER| Unpolished translucent sheet glass. { kə'thē·drəl ,glas }

cathode |ELEC| The terminal at which current leaves a primary cell or storage battery; it is negative with respect to the device, and positive with respect to the external circuit. |ELECTR| **1.** The primary source of electrons in an electron tube; in directly heated tubes the filament is the cathode, and in indirectly heated tubes a coated metal cathode surrounds a heater. Designated K. Also known as negative electrode. **2.** The terminal of a semiconductor diode that is negative with respect to the other terminal when the diode is biased in the forward direction. |PHYS CHEM| The electrode at which reduction takes place in an electrochemical cell, that is, a cell through which electrons are being forced. { 'kath,ōd }

cathode cleaning |MET| Electrolytically removing soil from work connected as the cathode. { 'kath,ōd ,klēn·iŋ }

cathode copper |MET| Copper deposited at the cathode during electrolytic refining; it is melted and marketed as electrolytic copper. { 'kath,ōd ,käp·ər }

cathode corrosion |MET| **1.** Corrosion of the cathode of an electrochemical circuit, usually caused by production of alkaline reaction products. **2.** Corrosion of the cathodic member of a galvanic couple. { 'kath,ōd kə'rō·zhən }

cathode efficiency |CHEM ENG| The proportion of current used for completion of a given process at the cathode. { 'kath,ōd i,fish·ən·sē }

cathode-ray tube |ELECTR| An electron tube in which a beam of electrons can be focused to a small area and varied in position and intensity on a surface. Abbreviated CRT. Originally known as Braun tube; also known as electron-ray tube. { 'kath,ōd ¦rā ,tüb }

cathode sputtering See sputtering. { 'kath,ōd 'spəd·ə·riŋ }

cathodic coating |MET| Material forming a continuous film on a base metal, deposited by mechanical coating or electroplating. { kə'thäd·ik ,kōd·iŋ }

cathodic disbonding |MET| The destruction of adhesion between a cathodic coating and its substrate by products of a cathodic reduction reaction. { kə¦thōd·ik ,dis'bänd·iŋ }

cathodic inhibitor |CHEM ENG| A compound, such as calcium bicarbonate or sodium phosphate, which is deposited on a metal surface in a thin film that operates at the cathodes

to provide physical protection over the entire surface against corrosive attack in a conducting medium. { kə'thäd·ik in'hib·əd·ər }

cathodic polarization |PHYS CHEM| Portion of electric cell polarization occurring at the cathode. { kə'thäd·ik ,pō·lə·rə'zā·shən }

cathodic protection |MET| Protecting a metal from electrochemical corrosion by using it as the cathode of a cell with a sacrificial anode. Also known as electrolytic protection. { kə'thäd·ik prə'tek·shən }

cation |CHEM| A positively charged atom or group of atoms, or a radical which moves to the negative pole (cathode) during electrolysis. { 'kat,ī·ən }

cation exchange resin |ORG CHEM| A highly polymerized synthetic organic compound consisting of a large, nondiffusible anion and a simple, diffusible cation, which later can be exchanged for a cation in the medium in which the resin is placed. { 'kat,ī·ən iks'chānj 'rez·ən }

cationic detergent |CHEM| A member of a group of detergents that have molecules containing a quaternary ammonium salt cation with a group of 12 to 24 carbon atoms attached to the nitrogen atom in the cation; an example is alkyltrimethyl ammonium bromide. { ,kad·ē'än·ik di'tər·jənt }

cationic polymerization |POLYM CHEM| A type of polymerization in which Lewis acids act as initiators. { ,kad·ē'än·ik pə,lim·ə·rə'zā·shən }

Cauchy number |FL MECH| A dimensionless number used in the study of compressible flow, equal to the density of a fluid times the square of its velocity divided by its bulk modulus. Also known as Hooke number. { kō·shē ,nəm·bər }

Cauchy relations |SOLID STATE| A set of six relations between the compliance constants of a solid which should be satisfied provided the forces between atoms in the solid depend only on the distances between them and act along the lines joining them, and provided that each atom is a center of symmetry in the lattice. { kō·shē ri'lā·shənz }

caulk |ENG| To make a seam or point airtight, watertight, or steamtight by driving in caulking compound, dry pack, lead wool, or other material. |MATER| Material used to caulk seams. Also spelled calk. { kók }

caulking compound |MATER| A heavy paste, such as a synthetic, containing a polysulfide rubber and lead peroxide curing agent, or a natural product such as oakum, used for caulking. { kók·iŋ ,käm,paúnd }

caustic |CHEM| **1.** Burning or corrosive. **2.** A hydroxide of a light metal. { 'kó·stik }

caustic cracking See caustic embrittlement. { 'kó·stik 'krak·iŋ }

caustic dip |MET| Immersion of metal into a caustic solution such as sodium hydroxide. { 'kó·stik 'dip }

caustic embrittlement |MET| Intercrystalline cracking of steel caused by exposure to caustic solutions above 70°C while under tensile stress;

once common in riveted boilers. Also known as caustic cracking. { 'kȯ·stik im'brid·əl·mənt }

caustic lime See calcium oxide. { 'kȯ·stik'līm }

caustic potash See potassium hydroxide. { 'kȯ·stik 'päd,ash }

caustic soda [INORG CHEM] See sodium hydroxide. [MATER] Sodium hydroxide that contains 76–78% sodium oxide; the most important of the commercial caustic materials, used in chemical manufacture, petroleum refining, and pulp and paper manufacture. { 'kȯ·stik 'sōd·ə }

cavitation corrosion See cavitation erosion. { ,kav·ə'tā·shən kə'rō·zhən }

cavitation damage See cavitation erosion. { ,kav·ə'tā·shən ,dam·ij }

cavitation erosion [MET] Attack of metal surfaces caused by the collapse of cavitation bubbles on the surface of the liquid and characterized by pitting. Also known as cavitation corrosion; cavitation damage. { ,kav·ə'tā·shən i'rō·zhən }

cavity radiator [THERMO] A heated enclosure with a small opening which allows some radiation to escape or enter; the escaping radiation approximates that of a blackbody. { 'kav·əd·ē 'rād·ē,ād·ər }

c axis [CRYSTAL] A vertically oriented crystal axis, usually the principal axis; the unique symmetry axis in tetragonal and hexagonal crystals. { 'sē ,ak·səs }

CCD See charge-coupled device.

Cd See cadmium.

CD See circular dichroism.

Ce See cerium.

ceiling temperature [POLYM CHEM] For addition (chain) polymerization, the temperature at which the propagation and depropagation rates are equal, that is, above which the net rate of polymer formation is zero. Above the ceiling temperture, depolymerization, an unzipping reaction to reform monomer, occurs. { 'sēl·iŋ ,tem·prə·chər }

cell [ELEC] A single unit of a battery. [PHYS CHEM] A cup, jar, or vessel containing electrolyte solutions and metal electrodes to produce an electric current (conductiometric or potentiometric) or for electrolysis (electrolytic). { sel }

cell constant [PHYS CHEM] The ratio of distance between conductance-titration electrodes to the area of the electrodes, measured from the determined resistance of a solution of known specific conductance. { 'sel ,kän·stənt }

celloidin [MATER] A concentrated solution of pyroxylin used principally in microscopy for embedding specimens or for section-cutting. { se'lȯid·ən }

cellophane [MATER] A thin, transparent sheeting of regenerated cellulose; it is moisture-proof, and sometimes dyed, and used chiefly as food wrapping or as bags for dialysis. { 'sel·ə,fān }

cellular glass [MATER] Sheets or blocks of thermal insulating material for walls and roofs made from pulverized glass that is heated with a gasforming chemical at the flow temperature of glass. Also known as cellulated glass; foam glass. { 'sel·yə·lər 'glas }

cellular plastic [MATER] A type of plastic with apparent density decreased substantially by the presence of numerous cells disposed throughout its mass. { 'sel·yə·lər 'plas·tik }

cellular rubber See rubber sponge. { 'sel·yə·lər 'rəb·ər }

cellulated glass See cellular glass. { 'sel·yə ,lād·əd ,glas }

cellulose [BIOMATER] $(C_6H_{10}O_5)_n$ The main polysaccharide in living plants, forming the skeletal structure of the plant cell wall; a polymer of β-D-glucose units linked together, with the elimination of water, to form chains comprising 2000–4000 units. { 'sel·yə,lōs }

cellulose acetate [POLYM CHEM] An acetic acid ester of cellulose; a tough, flexible, slow-burning, and long-lasting thermoplastic material used as the base for magnetic tape and movie film, in acetate rayon, as a plastic film in food packaging, in lacquers, and for molded receiver cabinets. { 'sel·yə,lōs 'as·ə,tāt }

cellulose acetate butyrate [POLYM CHEM] An ester of cellulose formed by the action of a mixture of acetic acid and butyric acid and their anhydrides on purified cellulose; has high impact resistance, clarity, and weatherability; used in making plastic film, lacquer, lenses, and outdoor signs. { 'sel·yə,lōs 'as·ə,tāt 'byüd·ə,rāt }

cellulose acetate rayon [TEXT] The spun product of the acetic ester of cellulose. { 'sel·yə,lōs 'as·ə,tāt 'rā,än }

cellulose ester [POLYM CHEM] Cellulose in which the free hydroxyl groups have been replaced wholly or in part by acidic groups. { 'sel·yə,lōs 'es·tər }

cellulose ether [ORG CHEM] The product of the partial or complete etherification of the hydroxyl groups in a cellulose molecule. { 'sel·yə,lōs 'ē·thər }

cellulose fiber [POLYM CHEM] Any fiber based on esters or ethers of cellulose. { 'sel·yə,lōs 'fī·bər }

cellulose gum See carboxymethylcellulose. { 'sel·yə,lōs 'gəm }

cellulose methyl ether See methylcellulose. { 'sel·yə,lōs 'meth·əl ,ē·thər }

cellulose nitrate [POLYM CHEM] Any of several esters of nitric acid, produced by treating cotton or some other form of cellulose with a mixture of nitric and sulfuric acids; used as explosive and propellant. Also known as nitrocellulose; nitrocotton. { 'sel·yə,lōs 'nī,trāt }

cellulosic [POLYM CHEM] Any of the derivatives of cellulose, such as cellulose acetate. { ,sel·yə'lō·sik }

cellulosic resin [POLYM CHEM] Any resin based on cellulose compounds such as esters and ethers. { 'sel·yə,lōs 'rez·ən }

Celsius degree [THERMO] Unit of temperature interval or difference equal to the kelvin. { 'sel·sē·əs di'grē }

Celsius temperature scale [THERMO] Temperature scale in which the temperature Θ_c in degrees Celsius (°C) is related to the temperature T_k in kelvins by the formula

$\Theta_c = T_k - 273.15$; the freezing point of water at standard atmospheric pressure is very nearly 0°C and the corresponding boiling point is very nearly 100°C. Formerly known as centigrade temperature scale. { 'sel·se·əs 'tem·prə·chər ‚skāl }

cement [MATER] **1.** A dry powder made from silica, alumina, lime, iron oxide, and magnesia which hardens when mixed with water; used as an ingredient in concrete. **2.** An adhesive for the assembling of surfaces which are not in close contact. { si'ment }

cementation [CHEM] The setting of a plastic material. [ENG] **1.** Plugging a cavity or drill hole with cement. **2.** Consolidation of loose sediments or sand by injection of a chemical agent or binder. [MET] **1.** High-temperature impregnation of a metal surface with another material. **2.** Conversion of wrought iron into steel by packing layers of bars in charcoal sealed with clay and heating to 1000°C for 7–10 days. { ‚sē‚men'tā·shən }

cement brick [MATER] A type of brick made from a mixture of cement and sand, molded under pressure and cured under steam at 200°F (93°C); used as backing brick and where there is no danger of attack from acid or alkaline conditions. { si'ment ‚brik }

cement copper [MET] A precipitate of copper from copper sulfate solution by the addition of iron. { si'ment ‚käp·ər }

cemented carbide See carbide. { si'men·təd 'kär‚bīd }

cementite [MET] Fe_3C A hard, brittle, crystalline compound occurring as lamellae or plates in steel. Also known as iron carbide. { si'men‚tīt }

cementitious material [MATER] Any of various building materials which may be mixed with a liquid, such as water, to form a plastic paste, and to which an aggregate may be added; includes cements, limes, and mortar. { ¦sē‚men¦tish·əs mə'tir·ē·əl }

cement mortar [MATER] A mixture of approximately four parts of sand to one part of portland cement with a small amount of lime and enough water to make it plastic. { si'ment 'mórd·ər }

cement paint [MATER] A mixture based on portland cement, with filler, accelerator, and water repellent added, that is combined with water and applied to masonry, concrete, or brickwork; provides a waterproof coating. { si'ment ‚pānt }

cement paste [MATER] A mixture of cement and water, hardened or unhardened. { si'ment ‚pāst }

cement plaster [MATER] A gypsum plaster used in mortar mixtures for plastering interior surfaces. { si'ment ‚plas·tər }

cement-sand process [MET] A sand casting process in which portland cement is the binder for sand; typical mixtures have 11% portland cement and 89% silica sand. { si'ment 'sand ‚präs·əs }

cement temper [MATER] Use of portland cement as an additive in a lime plaster preparation to increase strength and durability. { sə'ment ‚tem·pər }

centered lattice [CRYSTAL] A crystal lattice in which the axes have been chosen according to the rules for the crystal system, and in which there are lattice points at the centers of certain planes as well as at the corners. { 'sen·tərd 'lad·əs }

center of inversion [CRYSTAL] A point in a crystal lattice such that the lattice is left invariant by an inversion in the point. { 'sen·tər əv in'vər·zhən }

centigrade heat unit [THERMO] A unit of heat energy, equal to 0.01 of the quantity of heat needed to raise 1 pound of air-free water from 0 to 100°C at a constant pressure of 1 standard atmosphere; equal to 1900.44 joules. Symbolized CHU; (more correctly) CHU_{mean}. { 'sent·ə‚grād 'hēt ‚yü·nət }

centigrade temperature scale See Celsius temperature scale. { 'sent·ə‚grād 'tem·prə·chər ‚skāl }

centipoise [FL MECH] A unit of viscosity which is equal to 0.01 poise. Abbreviated cp. { 'sent·ə ‚póiz }

centistoke [FL MECH] A cgs unit of kinematic viscosity in customary use, equal to the kinematic viscosity of a fluid having a dynamic viscosity of 1 centipoise and a density of 1 gram per cubic centimeter. Abbreviated cs. { 'sent·ə‚stōk }

central force [MECH] A force whose line of action is always directed toward a fixed point; the force may attract or repel. { 'sen·trəl 'fórs }

central mix concrete [MATER] A concrete prepared at a concrete mixing plant and transported to the building site. { 'sen·trəl ‚miks 'kän‚krēt }

centrifugal [MECH] Acting or moving in a direction away from the axis of rotation or the center of a circle along which a body is moving. { ‚sen'trif·i·gəl }

centrifugal casting [ENG] A method for casting metals or forming thermoplastic resins in which the molten material solidifies in and conforms to the shape of the inner surface of a heated, rapidly rotating container. { ‚sen'trif·i·gəl 'kast·iŋ }

centrifugal force [MECH] **1.** An outward pseudo-force, in a reference frame that is rotating with respect to an inertial reference frame, which is equal and opposite to the centripetal force that must act on a particle stationary in the rotating frame. **2.** The reaction force to a centripetal force. { ‚sen'trif·i·gəl 'fórs }

centrifugal moment [MECH] The product of the magnitude of centrifugal force acting on a body and the distance to the center of rotation. { ‚sen'trif·i·gəl 'mō·mənt }

centripetal [MECH] Acting or moving in a direction toward the axis of rotation or the center of a circle along which a body is moving. { ‚sen'trip·əd·əl }

centripetal force [MECH] The radial force required to keep a particle or object moving in a circular path, which can be shown to be directed toward the center of the circle. { ‚sen'trip·əd·əl 'fórs }

centrosymmetry [PHYS] Property of a body or system which is unchanged under space inversion through a specified point. { ¦sen·trō'sim·ə·trē }

ceramagnet |ELECTROMAG| A ferrimagnet composed of the hard magnetic material BaO·6Fe₂O₃. { 'se·rə,mag·nət }

ceramal *See* cermet. { sə'ram·əl }

ceramet *See* cermet. { sə'ram·ət }

ceramic |MATER| **1.** Inorganic, nonmetallic materials processed or used at high temperature, generally including oxides, nitrides, borides, carbides, silicides, and sulfides. Intermetallic compounds such as aluminides and beryllides are also considered ceramics, as are phosphides, antimonides, and arsenides. **2.** Consisting of such a product. { sə'ram·ik }

ceramic aggregate |MATER| **1.** Portland cement concrete containing lumps of ceramic material. **2.** Concrete made with porous clay to reduce its weight. { sə'ram·ik 'ag·rə·gət }

ceramic capacitor |ELEC| A capacitor whose dielectric is a ceramic material such as steatite or barium titanate, the composition of which can be varied to give a wide range of temperature coefficients. { sə'ram·ik kə'pas·əd·ər }

ceramic coating |MET| A nonmetallic, inorganic coating made of sprayed aluminum oxide or of zirconium oxide, or a cemented coating of an intermetallic compound, such as aluminum disilicide, of essentially crystalline nature, applied as a protective film on metal to protect against temperatures above 1100°C. { sə'ram·ik 'kōd·iŋ }

ceramic fiber |MATER| A small-dimension filament or thread composed of a ceramic material, usually alumina and silica, used in lightweight units for electrical, thermal, and sound insulation, filtration at high temperatures, packing, and reinforcing other ceramic materials. { sə'ram·ik 'fī·bər }

ceramic glaze |ENG| A glossy finish on a clay body obtained by spraying with metallic oxides, chemicals, and clays and firing at high temperature. { sə'ram·ik 'glāz }

ceramic magnet |ELECTROMAG| A permanent magnet made from pressed and sintered mixtures of ceramic and magnetic powders. Also known as ferromagnetic ceramic. { sə'ram·ik 'mag·nət }

ceramic mold casting |MET| A precision casting process using a ceramic body fired to high temperature as the mold, and carbon, low-alloy, or stainless steel as the casting. { sə'ram·ik 'mōld ,kast·iŋ }

ceramic rod flame spraying |MET| A method of flame spraying in which the ceramic rod is fed into a gun that utilizes an oxyfuel gas flame to atomize and airblast the rod material to the substrate. { sə'ram·ik 'räd ,flām ,sprā·iŋ }

ceramics |ENG| The art and science of making ceramic products. { sə'ram·iks }

ceramic tile |MATER| A burned-clay product composed of a clay body with a decorative surface glaze; used principally for decorative and sanitary effects. { sə'ram·ik 'tīl }

ceramic veneer |MATER| A burned clay non-load-bearing unit used in masonry construction. { sə'ram·ik və'nir }

ceramide |ORG CHEM| Any of a group of amides formed by linking a fatty acid to sphingosine. { 'ser·ə,mīd }

ceramoplastic |MATER| A high-temperature insulating material made by bonding synthetic mica with glass. { sə¦ram·ō¦plas·tik }

cereal binder |MATER| A binding material derived from flour; used for core mixtures in a casting process. { 'sir·ē·əl ,bīn·dər }

ceresin |MATER| **1.** A hydrocarbon wax refined from veins of wax shale known as ozocerite; used in manufacture of candles, shoe polish, electrical insulation, and floor waxes because of its great compatibility with other substances. Also known as ceresine; ozocerite. **2.** A mixture of paraffin wax and beeswax, or a mixture of ozocerite and paraffin. { 'se·rə·sən }

ceresine *See* ceresin. { 'se·rə,sēn }

ceria *See* ceric oxide. { 'ser·ē·ə }

ceric oxide |INORG CHEM| CeO₂ A pale-yellow to white powder; soluble in sulfuric acid, insoluble in dilute acid and water; used in ceramics and as a polish for optical glass. Also known as ceria; cerium dioxide; cerium oxide. { 'sir·ik 'äk ,sīd }

ceric sulfate |INORG CHEM| Ce(SO₄)₂·4H₂O Yellow needles forming a basic salt with excess water; used in waterproofing, mildew-proofing, and in dyeing and printing textiles. { 'sir·ik 'səl ,fāt }

cerium |CHEM| A chemical element, symbol Ce, atomic number 58, atomic weight 140.12; a rare-earth metal, used as a getter in the metal industry, as an opacifier and polisher in the glass industry, in Welsbach gas mantles, in cored carbon arcs, and as a liquid-liquid extraction agent to remove fission products from spent uranium fuel. { 'sir·ē·əm }

cerium dioxide *See* ceric oxide. { 'sir·ē·əm dī'äk ,sīd }

cerium fluoride |INORG CHEM| CeF₃ White hexagonal crystals, melting point 1460°C; used in arc carbons to increase the brilliance of carbon-arc lamps. { 'sir·ē·əm 'flur,īd }

cerium oxide *See* ceric oxide. { 'sir·ē·əm 'äk ,sīd }

cermet |MATER| Any of a group of composite materials made by mixing, pressing, and sintering metal with ceramic; examples are silicon-silicon carbide and chromium-alumina carbide. Also known as ceramal; ceramet; metal ceramic. { 'sər,met }

cermet resistor |ELEC| A metal-glaze resistor, consisting of a mixture of finely powdered precious metals and insulating materials fired onto a ceramic substrate. { 'sər,met ri'zis·tər }

Cerruti's problem *See* Boussinesq's problem. { se'rü·dēz ,präb·ləm }

cesium |CHEM| A chemical element, symbol Cs, atomic number 55, atomic weight 132.905. { 'sē·zē·əm }

cesium bromide |INORG CHEM| CsBr A colorless, crystalline powder with a melting point of 636°C; soluble in water; used in medicine, for infrared spectroscopy, and in scintillation counters. { 'sē·zē·əm 'brō,mīd }

cesium carbonate [INORG CHEM] Cs_2CO_3 A white, hygroscopic, crystalline powder; soluble in water; used in specialty glasses. { 'sē·zē·əm 'kär·bə,nāt }

cesium chloride [INORG CHEM] CsCl Colorless cuboid crystals, melting point 646°C; used in filaments of radio tubes to increase sensitivity, in photoelectric cells, and for photosensitive deposit on cathodes. { 'sē·zē·əm 'klȯr,īd }

cesium fluoride [INORG CHEM] CsF Toxic, irritating, deliquescent crystals with a melting point of 682°C; soluble in water and methanol; used in medicine, mineral water, and brewing. { 'sē·zē·əm 'flür,īd }

cesium hydroxide [INORG CHEM] CsOH Colorless or yellow, fused crystalline mass with a melting point of 272.3°C; soluble in water; used as electrolyte in alkaline storage batteries at subzero temperatures. { 'sē·zē·əm ,hī'dräk,sīd }

cesium iodide [INORG CHEM] CsI A colorless, deliquescent, crystalline powder with a melting point of 621°C; soluble in water and alcohol; crystals used for infrared spectroscopy. { 'sē·zē·əm 'ī·ə,dīd }

cesium perchlorate [INORG CHEM] $CsClO_4$ A crystalline solid with a melting point of 250°C; soluble in water; used in optics and for specialty glasses. { 'sē·zē·əm pər'klȯr,āt }

cesium sulfate [INORG CHEM] Cs_2SO_4 Colorless crystals with a melting point of 1010°C; soluble in water; used for brewing and in mineral waters. { 'sē·zē·əm 'səl,fāt }

Cf See californium.

chadacryst See xenocryst. { 'kad·ə,krist }

chafing corrosion See fretting corrosion. { 'chāf·iŋ kə'rō·zhən }

chafing fatigue [MET] Fatigue induced by corrosion damage between metal surfaces in close contact under pressure. { 'chāf·iŋ fə,tēg }

chain [CHEM] A structure in which similar atoms are linked by bonds. { chān }

chain intermittent fillet welding [MET] **1.** The forming of two lines of intermittent fillet welds in a T joint or lap joint so that the increments of welding in one line are approximately opposite those in the other line. **2.** The forming of two lines of equal-length fillet welds concurrently on opposite sides of the perpendicular member of a T joint at intermittent intervals. { 'chān ,in·tər'mit·ənt 'fil·ət ,wel·diŋ }

chain reaction [CHEM] A chemical reaction in which many molecules undergo chemical reaction after one molecule becomes activated. { ¦chān rē'ak·shən }

chain scission [POLYM CHEM] The cleavage of polymer chains, as in natural rubber as a result of heating. { 'chān ,sizh·ən }

chain structure [SOLID STATE] A crystalline structure in which forces between atoms in one direction are greater than those in other directions, so that the atoms are concentrated in chains. { 'chān ,strək·chər }

chain transfer [POLYM CHEM] The abstraction of an atom from another molecule (initiator, monomer, polymer, or solvent) by the radical end of a growing (addition) polymer, which simultaneously terminates the polymer chain and creates a new radical capable of chain polymerization; also occurs in cationic polymerization. { 'chān ,tranz·fər }

chalcogen [INORG CHEM] Any of the elements that form group 16 of the periodic table; included are oxygen, sulfur, selenium, tellurium, and polonium. { 'kal·kə·jən }

chalcogenide [INORG CHEM] A binary compound containing a chalcogen and a more electropositive element or radical. { 'kal·kə·jə ,nīd }

chalcogenide glass [MATER] A type of glass containing large amounts of one of the chalcogens tellurium, selenium, or sulfur; used in glass switches. { 'kal·kə·jə,nīd 'glas }

chalk [MATER] Artificially prepared pure calcium carbonate; used as the basis for pastels. { chȯk }

chalking [MET] Defect of coated metals in which a layer of powder forms between the coating and the base metal. { 'chȯk·iŋ }

chamber acid [INORG CHEM] Sulfuric acid made by the obsolete chamber process. { 'chām·bər 'as·əd }

chaplet [MET] Metal support used to space and hold the core in position within a sand mold. { 'chap·lət }

characteristic temperature See Debye temperature. { ,kar·ik·tə'ris·tik 'tem·prə·chər }

charge [ELEC] **1.** A basic property of elementary particles of matter; the charge of an object may be a positive or negative number or zero; only integral multiples of the proton charge occur, and the charge of a body is the algebraic sum of the charges of its constituents; the value of the charge may be inferred from the Coulomb force between charged objects. Also known as electric charge, quantity of electricity. **2.** To convert electrical energy to chemical energy in a secondary battery. **3.** To feed electrical energy to a capacitor or other device that can store it. [MET] Material introduced into a furnace for melting. { chärj }

charge carrier [SOLID STATE] A mobile conduction electron or mobile hole in a semiconductor. Also known as carrier. { 'chärj ,kar·ē·ər }

charge conservation See conservation of charge. { 'chärj ,kän·sər'vā·shən }

charge-coupled device [ELECTR] A semiconductor device wherein minority charge is stored in a spatially defined depletion region (potential well) at the surface of a semiconductor and is moved about the surface by transferring this charge to similar adjacent wells. Abbreviated CCD. { 'chärj ¦kəp·əld di'vīs }

charge-coupled image sensor [ELECTR] A device in which charges are introduced when light from a scene is focused on the surface of the device; image points are accessed sequentially to produce a television-type output signal. Also known as solid-state image sensor. { 'chärj ¦kəp·əld 'im·ij ,sen·sər }

charge density [ELEC] The charge per unit area

on a surface or per unit volume in space. { 'chärj ,den·səd·ē }

charge-density wave [SOLID STATE] The ground state of a metal in which the conduction-electron charge density is sinusoidally modulated in space. { 'chärj ,den·səd·ē ,wāv }

charged species [CHEM] A chemical entity in which the overall total of electrons is unequal to the overall total of protons. { 'chärjd 'spē·shēz }

charge-mass ratio [ELEC] The ratio of the electric charge of a particle to its mass. { ,chärj ,mas 'rā·shō }

charge neutrality [SOLID STATE] The condition in which electrons and holes are present in equal numbers in a semiconductor. { ¦chärj nü'tral·əd·ē }

charge-state process [SOLID STATE] A process involving the motion of preexisting crystal defects in a solid, following a change in the charges of the defects. { 'chärj ,stāt ,präs·əs }

charge transfer [PHYS CHEM] The process in which an ion takes an electron from a neutral atom, with a resultant transfer of charge. { 'chärj ,tranz·fər }

charge-transfer complexes [CHEM] Compounds in which electrons move between molecules. { 'chärj ,tranz·fər 'käm·plek·səs }

Charles' law [PHYS] The law that at constant pressure the volume of a fixed mass or quantity of gas varies directly with the absolute temperature; a close approximation. Also known as Gay-Lussac's first law. { 'chärlz ,lò }

Charlton white See lithopone. { 'chärl·tən 'wīt }

Charpy test [MET] An impact test to determine the ductility of a metal; a freely swinging pendulum is allowed to strike and break a notched specimen that has been laid loosely on a support; the work done by the pendulum is obtained by comparing the position of the pendulum before release with the position to which the pendulum swings after breaking the specimen. { 'shär·pē ,test }

charring ablator [MATER] An ablation material characterized by the formation of a carbonaceous layer at the heated surface which impedes heat flow into the material by its insulating and reradiating characteristics. { 'chär·iŋ ə'blād·ər }

chatter mark [MATER] One of a series of marks made in a crosswise direction on a material; caused by vibration during rolling, extrusion, cutting, or drawing. { 'chad·ər ,märk }

check [MATER] A lengthwise crack in a board. [MET] A minute crack occurring in steel that has been cooled too quickly. { chek }

check cracks See checking. { 'chek ,kraks }

checking [MATER] **1.** Fine, shallow cracks appearing on the surface of a material or in a film of a surface coating. Also known as check cracks. **2.** A defect in lumber characterized by separation along the fiber direction. [MET] Temporarily reducing the volume or temperature of the air blast in a blast furnace. { 'chek·iŋ }

cheek [MET] Portion of a three-part flask between the cope and the drag. { chēk }

cheese cement [MATER] A glue made from cheese or milk curd. { 'chēz si'ment }

chelate [ORG CHEM] A molecular structure in which a heterocyclic ring can be formed by the unshared electrons of neighboring atoms. { 'kē ,lāt }

chelating agent [ORG CHEM] An organic compound in which atoms form more than one coordinate bond with metals in solution. { 'ke ,lād·iŋ ,ā·jənt }

chelating resin [ORG CHEM] Any of the ion-exchange resins with unusually high selectivity for specific cations; for example, phenol-formaldehyde resin with 8-quinolinol replacing part of the phenol, particularly selective for copper, nickel, cobalt, and iron(III). { 'ke,lād·iŋ 'rez·ən }

chelation [ORG CHEM] A chemical process involving formation of a heterocyclic ring compound which contains at least one metal cation or hydrogen ion in the ring. { kē'lā·shən }

chemical [CHEM] **1.** Related to the science of chemistry. **2.** A substance characterized by definite molecular composition. { 'kem·i·kəl }

chemical affinity See affinity. { 'kem·i·kəl ə'fin·əd·ē }

chemical bond See bond. { 'kem·i·kəl ¦bänd }

chemical compound See compound. { 'kem·i·kəl 'käm,paund }

chemical conversion coating [MET] A protective or decorative coating formed on the surface of a metal as the result of chemical reaction of the metal with a selected environment. { 'kem·i·kəl kən'vər·zhən ,kōd·iŋ }

chemical crystallography [CRYSTAL] The geometric description, and study, of the internal arrangement of atoms in crystals formed from chemical compounds. { 'kem·i·kəl kris·tə'läg·rə·fē }

chemical element See element. { 'kem·i·kəl 'el·ə·mənt }

chemical equilibrium [CHEM] A condition in which a chemical reaction is occurring at equal rates in its forward and reverse directions, so that the concentrations of the reacting substances do not change with time. Also known as equilibrium. { 'kem·i·kəl ,ē·kwə'lib·rē·əm }

chemical etching [MET] Formation of characteristic surface features when a polished metal surface is etched by suitable reagents. { 'kem·i·kəl 'ech·iŋ }

chemical film dielectric [ELEC] An extremely thin layer of material on one or both electrodes of an electrolytic capacitor, which conducts electricity in only one direction and thereby constitutes the insulating element of the capacitor. { 'kem·i·kəl ,film ,dī·ə'lek·trik }

chemical flux cutting [MET] An oxygen cutting process in which metals are cut by using flux. { 'kem·i·kəl 'fləks ,kəd·iŋ }

chemical force microscope [ENG] A modification of the atomic force microscope in which an organic monolayer on the probe tip that terminates with specific chemical functional groups is sensitive to specific molecular interactions between these groups and those on the sample surface. { ¦kem·ə·kəl ¸förs 'mī·krə¸sköp }

chemical formula |CHEM| A notation utilizing chemical symbols and numbers to indicate the chemical composition of a pure substance; examples are CH_4 for methane and HCl for hydrogen chloride. { 'kem·i·kəl 'för·myə·lə }

chemical inhibitor |CHEM| A substance capable of stopping or retarding a chemical reaction. { 'kem·i·kəl in'hib·əd·ər }

chemically foamed plastic [MATER] A foamed plastic having its cellular structure produced by gases generated from chemical reaction of the components. { 'kem·ik·lē ¸fōmd 'plas·tik }

chemical machining [MET] Making of metal parts to specified dimensions by removing surface metal with chemicals (acids or alkalies). Also known as chemical milling. { 'kem·i·kəl mə'shēn·iŋ }

chemical metallurgy [MET] The science and technology of extracting metals from ores and refining them. Also known as process metallurgy. { 'kem·i·kəl 'med·əl·ər·jē }

chemical milling See chemical machining. { 'kem·i·kəl 'mil·iŋ }

chemical polishing [MET] Smoothing and brightening the surface of a metal by treatment with a chemical agent. { 'kem·i·kəl 'päl·ish·iŋ }

chemical porcelain [MATER] High-purity, nonporous grade of porcelain used to make laboratory analysis utensils, such as crucibles, retorts, and spatulas. { 'kem·i·kəl 'pörs·lən }

chemical pulp [MATER] Wood pulp made by separating the fibers of wood chips by the action of alkalies or acids. { 'kem·i·kəl 'pəlp }

chemical pulping [CHEM ENG] Separation of wood fiber for paper pulp by chemical treatment of wood chips to dissolve the lignin that cements the fibers together. { 'kem·i·kəl 'pəlp·iŋ }

chemical reaction [CHEM] A change in which a substance (or substances) is changed into one or more new substances. { 'kem·i·kəl rē'ak·shən }

chemical reactivity [CHEM] The tendency of two or more chemicals to react to form one or more products differing from the reactants. { 'kem·i·kəl rē¸ak'tiv·əd·ē }

chemical relaxation [CHEM] The readjustment of a chemical system to a new equilibrium after the equilibrium of a chemical reaction is disturbed by a sudden change, particularly in an external parameter such as pressure or temperature. { 'kem·ə·kəl ¸rē¸lak'sā·shən }

chemical resistance [MATER] Ability of solid materials to resist damage by chemical reactivity or solvent action. { 'kem·i·kəl ri'zis·təns }

chemical species See species. { 'kem·i·kəl 'spē ¸shēz }

chemical stoneware [MATER] Clay pottery material that resists acids and alkalies; used for ball mills, pipes, laboratory sinks and utensils, and so on. { 'kem·i·kəl 'stōn¸wer }

chemical symbol |CHEM| A notation for one of the chemical elements, consisting of letters; for example Ne, O, C, and Na represent neon, oxygen, carbon, and sodium. { 'kem·i·kəl 'sim·bəl }

chemical vapor deposition |SOLID STATE| The growth of thin solid films on a crystalline substrate as the result of thermochemical vapor-phase reactions. Abbreviated CVD. { 'kem·i·kəl ¦vā·pər ¸dep·ə'zish·ən }

chemiluminescence |PHYS CHEM| Emission of light as a result of a chemical reaction without an apparent change in temperature. { ¸kem·i ¸lüm·ə'nes·əns }

chemimechanical pulp [MATER] Plant material treated by the sulfite, soda, or sulfate process for papermaking. { ¦kem·i·mə¦kan·i·kəl 'pəlp }

chemionics |CHEM| The chemistry of molecular components and devices that operate on photons, electrons, and ions. { ¸kem·ē'än·iks }

chemiosmosis |CHEM| A chemical reaction occurring through an intervening semipermeable membrane. Also known as chemosmosis. { ¦kem·ē¸äs¦mō·səs }

chemonite [MATER] A wood preservative consisting of a water solution of small percentages of copper hydroxide, arsenic trioxide, ammonia, and acetic acid. { 'kem·ə¸nīt }

chemosmosis See chemiosmosis. { ¸kem¸äs 'mō·səs }

Chevrel phase |SOLID STATE| One of a series of ternary molybdenum chalcogenide compounds with unusual superconducting properties and the general formula $M_xMo_6X_8$, where M represents any one of a large number of metallic elements, x has values between 1 and 4, and X is a chalcogen (sulfur, selenium, or tellurium). { she'vrel ¸fāz }

chill [MET] **1.** A metal plate inserted in the surface of a sand mold or placed in the mold cavity to rapidly cool and solidify the casting, producing a hard surface. **2.** White or mottled iron occurring on the surface of a rapidly cooled gray iron casting. { chil }

chill-block melt spinning [MET] A rapid quenching process in which a jet of molten metal is directed onto a cold moving surface, such as a spinning disk, where the jet is shaped and solidified; quench rates are 1000 to 1,000,000 K per second. { 'chil ¸bläk 'melt ¸spin·iŋ }

chilled iron [MET] Cast iron made in iron- or steel-faced molds so the surface of the casting cools rapidly, retaining most of the carbon and becoming white and hard. { 'child 'ī·ərn }

chilled roll [MET] A roll consisting of an outer hard layer of white (chilled) iron with a middle transitional layer of mottled iron and a core of full gray iron. { 'child 'rōl }

chilled shot [MET] Lead shots containing 3–6% antimony. { 'child 'shät }

chilling [MET] Rapidly removing the heat from a casting. { 'chil·iŋ }

chimney rock [MATER] A porous phosphate rock used principally in chimney construction. { 'chim¸nē ¸räk }

china clay [MATER] A high-grade white kaolin composed principally of the mineral kaolinite, and often occurring as a lenticular-shaped body; used in the manufacture of ceramics, paper, rubber, catalysts, and ink. { 'chī·nə ¸klā }

China wood oil See tung oil. { 'chī·nə ¸wüd ¸oil }

Chinese ink *See* india ink. { chīn'nēz 'iŋk }

Chinese vermilion *See* mercuric sulfide. { chīn 'nēz vər'mil·yən }

Chinese wax [MATER] A white or yellowish crystalline wax formed on certain trees by the secretions of a scale insect, especially *Ceroplastes ceriferus*. { chīn'nēz 'waks }

Chinese white [CHEM] A term used in the paint industry for zinc oxide and kaolin used as a white pigment. Also known as zinc white. { chīn'nēz 'wīt }

chip [ELECTR] **1.** The shaped and processed semiconductor die that is mounted on a substrate to form a transistor, diode, or other semiconductor device. **2.** An integrated microcircuit performing a significant number of functions and constituting a subsystem. Also known as microchip. [MET] A section of metal that is removed as a workpiece is being machined. { chip }

chipboard [MATER] A low-density paper board made from mixed waste paper and used where strength and quality are needed. { 'chip,bȯrd }

chipping [MET] Removing seams, surface defects, or other excess fragments from semifinished metal products by using a manual or pneumatic chisel or a continuous machine. { 'chip·iŋ }

chiral carbon atom *See* asymmetric carbon atom. { ¦kī·rəl ¦kär·bən 'ad·əm }

chiral center [ORG CHEM] An atom in a molecule that is attached to four different groups. { 'kī·rəl 'sen·tər }

chirality [CHEM] The handedness of an asymmetric molecule. [PHYS] The characteristic of an object that cannot be superimposed upon its mirror image. { kī'ral·əd·ē }

chiral molecules [CHEM] Molecules which are not superposable with their mirror images. { 'kī·rəl 'mäl·ə,kyülz }

chiral nanotube [PHYS CHEM] A carbon nanotube formed from a graphite sheet that is rolled up so that the succession of hexagons of carbon atoms on a particular cylinder makes an angle with the axis of the nanotube. { ¦kī·rəl 'nan·ō ¦tüb }

chiral twinning *See* optical twinning. { 'kī·rəl 'twin·iŋ }

chlorate [INORG CHEM] ClO₃⁻. **1.** A negative ion derived from chloric acid. **2.** A salt of chloric acid. { 'klȯr,āt }

chloric acid [INORG CHEM] HClO₃ A compound that exists only in solution and as chlorate salts; breaks down at 40°C. { 'klȯr·ik 'as·əd }

chloride [CHEM] **1.** A compound which is derived from hydrochloric acid and contains the chlorine atom in the −1 oxidation state. **2.** In general, any binary compound containing chloride. { 'klȯr,īd }

chloride paper [MATER] A paper made with an emulsion of silver chloride; usually used in photography as contact paper or very-slow-speed enlarging paper. { 'klȯr,īd ,pā·pər }

chloridization [MET] Treatment of mineral ores with hydrochloric acid or chlorine to form

the chloride of the main metal present. { ,klȯr·ə·də'zā·shən }

chlorimide *See* dichloramine. { 'klȯr·ə,mīd }

chlorinated paraffin [ORG CHEM] One of a group of chlorine derivatives of paraffin compounds. { 'klȯr·ə,nād·əd 'par·ə·fən }

chlorinated rubber [MATER] A nonrubbery, incombustible rubber derivative produced by the action of chlorine on rubber in solution; used in corrosion-resistant paints and varnishes, and in inks and adhesives. { 'klȯr·ə,nād·əd 'rəb·ər }

chlorine [CHEM] A chemical element, symbol Cl, atomic number 17, atomic weight 35.453; used in manufacture of solvents, insecticides, and many non-chlorine-containing compounds, and to bleach paper and pulp. { 'klȯr,ēn }

chlorine dioxide [INORG CHEM] ClO₂ A green gas used to bleach cellulose and to treat water. { 'klȯr,ēn dī'äk,sīd }

chlorite [INORG CHEM] A salt of chlorous acid. { 'klȯr,īt }

chloro- [ORG CHEM] A prefix describing an organic compound which contains chlorine atoms substituted for hydrogen; for example, a hydrocarbon in which chlorine is substituted for one hydrogen atom. { 'klȯr·ō }

chlorobutadiene *See* chloroprene. { ,klȯr·ō ,byüd·ə'dī,ēn }

chlorochromic anhydride *See* chromyl chloride. { ¦klȯr·ō¦krō·mik an'hī,drīd }

chloroethene *See* vinyl chloride. { ,klȯr·ō'eth ,ēn }

chloropicrin [INORG CHEM] CCl₃NO₂ A colorless liquid with a sweet odor whose vapor is very irritating to the lungs and causes vomiting, coughing, and crying; used as a soil fumigant. Also known as trichloronitromethane. { ,klȯr·ō'pik·rən }

chloroplatinate [INORG CHEM] **1.** A double salt of platinic chloride and another chloride. **2.** A salt of chloroplatinic acid. Also known as platinochloride. { ,klȯr·ō'plat·ən,āt }

chloroplatinic acid [INORG CHEM] H₂PtCl₆ An acid obtained as red-brown deliquescent crystals; used in chemical analysis. Also known as platinic chloride. { ¦klȯr·ō·sə¦plat·ən·ik 'as·əd }

chloroprene [ORG CHEM] C₄H₅Cl A colorless liquid which polymerizes to chloroprene resin. Also known as chlorobutadiene. { 'klȯr·ə,prēn }

chloroprene resin [POLYM CHEM] A polymer of chloroprene used to form materials resembling natural rubber. { 'klȯr·ə,prēn 'rez·ən }

chlorosulfonic acid [INORG CHEM] ClSO₂OH A fuming liquid that decomposes in water to sulfuric acid and hydrochloric acid; used in pharmaceuticals, pesticides, and dyes, and as a chemical intermediate. { ¦klȯr·ō·səl'fän·ik 'as·əd }

chlorotrifluoroethylene polymer [POLYM CHEM] A colorless, noninflammable, heat-resistant resin, soluble in most organic solvents, and with a high impact strength; can be made into transparent filling and thin sheets; used for chemical piping, fittings, and insulation for wire and cables, and in electronic components. Also known as

fluorothene; polytrifluorochloroethylene resin. { ¦klór·ō·tri¦'flür·ō'eth·əl,en 'päl·ə·mər }

cholesteric material [PHYS CHEM] A liquid crystal material in which the elongated molecules are parallel to each other within the plane of a layer, but the direction of orientation is twisted slightly from layer to layer to form a helix through the layers. { kə'les·tə·rik mə'tir·ē·əl }

cholesteric phase [PHYS CHEM] A form of the nematic phase of a liquid crystal in which the molecules are spiral. { kə'les·tə·rik ,fāz }

chroma [OPTICS] **1.** The dimension of the Munsell system of color that corresponds most closely to saturation, which is the degree of vividness of a hue. Also known as Munsell chroma. **2.** See color saturation. { 'krō·mə }

chromadizing [MET] Treating the surface of aluminum or aluminum alloys with chromic acid to improve paint adhesion. { 'krō·mə,dīz·iŋ }

chromate [INORG CHEM] CrO_4^{2-}. **1.** An ion derived from the unstable acid H_2CrO_4. **2.** A salt or ester of chromic acid. { 'krō,māt }

chromate treatment [MET] Treatment of metal with a solution of a hexavalent chromium compound to produce a protective coating of metal chromate. { 'krō,māt ,trēt·mənt }

chromatic [OPTICS] Relating to color. { krō 'mad·ik }

chromatic aberration [OPTICS] An optical lens defect causing color fringes, because the lens material brings different colors of light to focus at different points. Also known as color aberration. { krō'mad·ik ab·ə'rā·shən }

chromaticity [OPTICS] The color quality of light that can be defined by its chromaticity coordinates; depends only on hue and saturation of a color, and not on its luminance (brightness). { ,krō·mə'tis·əd·ē }

chromatics [OPTICS] **1.** The branch of optics concerned with the properties of colors. **2.** The part of colorimetry concerned with hue and saturation. { krō'mad·iks }

chromatic sensitivity [OPTICS] The smallest change in wavelength of light that produces a change in hue which is just large enough to be detected by human vision. { krō'mad·ik sen·sə'tiv·əd·ē }

chromating [MET] Performing a chromate treatment. { 'krō,mād·iŋ }

chrome alum [INORG CHEM] $KCr(SO_4)_2 \cdot 12H_2O$ An alum obtained as purple crystals and used as a mordant, in tanning, and in photography in the fixing bath. Also known as potassium chromium sulfate. { 'krōm 'al·əm }

chrome brick See chrome refractory. { 'krōm ,brik }

chrome dye [CHEM] One of a class of acid dyes used on wool with a chromium compound as mordant. { ¦krōm ¦dī }

chrome green See chromic oxide. { ¦krōm ,grēn }

chrome plating [MET] A thin plate of chromium deposited by electrolysis on a corrodible metal, giving a bright, metallic surface which is highly resistant to tarnish; used to coat automobile trimming, bathroom fixtures, and many household and other articles. Also known as chromium coating; chromium plating. { ¦krōm ¦plād·iŋ }

chrome red [CHEM] **1.** A pigment containing basic lead chromate. **2.** Any of several mordant acid dyes. { ¦krōm 'red }

chrome refractory [MATER] A ceramic material made from chrome ore and used to line steel furnaces. Also known as chrome brick. { ¦krōm ri'frak·trē }

chrome steel See chromium steel. { ¦krōm 'stēl }

chrome-vanadium steel See chromium-vanadium steel. { ¦krōm və¦nād·ē·əm 'stēl }

chrome yellow [CHEM] **1.** A yellow pigment composed of normal lead chromate, $PbCrO_4$, or other lead compounds. **2.** Any of several mordant acid dyes. { ¦krōm 'yel·ō }

chromic acid [INORG CHEM] H_2CrO_4 The hydrate of CrO_3; exists only as salts or in solution. { ¦krō·mik 'as·əd }

chromic chloride [INORG CHEM] $CrCl_3$ Crystals that are pinkish violet shimmering plates, almost insoluble in water, but easily soluble in presence of minute traces of chromous chloride; used in calico printing, as a mordant for cotton and silk. { ¦krō·mik 'klór,īd }

chromic fluoride [INORG CHEM] $CrF_3 \cdot 4H_2O$ Crystals that are green, soluble in water; used in dyeing cottons. { ¦krō·mik 'flür,īd }

chromic hydroxide [INORG CHEM] $Cr(OH)_3 \cdot 2H_2O$ Gray-green, gelatinous precipitate formed when a base is added to a chromic salt; the precipitate dries to a bluish, amorphous powder; prepared as an intermediate in the manufacture of other soluble chromium salts. { ¦krō·mik hī'dräk,sīd }

chromic nitrate [INORG CHEM] $Cr(NO_3)_3 \cdot 9H_2O$ Purple, rhombic crystals that are soluble in water; used as a mordant in textile dyeing. { ¦krō·mik 'nī,trāt }

chromic oxide [INORG CHEM] Cr_2O_3 A dark green, amorphous powder, forming hexagonal crystals on heating that are insoluble in water or acids; used as a pigment to color glass and ceramic ware and as a catalyst. Also known as chrome green. { ¦krō·mik 'äk,sīd }

chrominance [OPTICS] The difference between any color and a specified reference color of equal brightness; in color television, this reference color is white having coordinates $x = 0.310$ and $y = 0.316$ on the chromaticity diagram. { 'krō·mə·nəns }

chromium [CHEM] A metallic chemical element, symbol Cr, atomic number 24, atomic weight 51.996. [MET] A blue-white, hard, brittle metal used in chrome plating, in chromizing, and in many alloys. { 'krō·mē·əm }

chromium carbide [INORG CHEM] Cr_3C_2 Orthorhombic crystals with a melting point of 1890°C; resistant to oxidation, acids, and alkalies; used for hot-extrusion dies, in spray-coating materials, and as a component for pumps and valves. { 'krō·mē·əm 'kär,bīd }

chromium chloride [INORG CHEM] A group

of compounds of chromium and chloride; chromium may be in the +2, +3, or +6 oxidation state. { 'krō·mē·əm 'klȯr‚īd }

chromium coating See chrome plating. { 'krō·mē·əm 'kōd·iŋ }

chromium dioxide [INORG CHEM] Cr_2O_2 Black, acicular crystals; a semiconducting material with strong magnetic properties used in recording tapes. { 'krō·mē·əm dī'äk‚sīd }

chromium-iron alloy [MET] Any of several acid- and corrosion-resistant alloys containing chromium and iron. { ¦krō·mē·əm ¦ī·ərn 'al‚ȯi }

chromium molybdenum steel [MET] Cast steel containing up to 1% carbon, 0.7–1.1% chromium, and 0.2–0.4% molybdenum; characterized by high strength and ductility. { ¦krō·mē·əm mə ¦lib·də·nəm 'stēl }

chromium-nickel alloy [MET] Any of several alloys containing chromium and nickel in various proportions together with small amounts of other metals. { ¦krō·mē·əm ¦nik·əl 'al‚ȯi }

chromium oxide [INORG CHEM] A compound of chromium and oxygen; chromium may be in the +2, +3, or +6 oxidation state. { 'krō·mē·əm 'äk ‚sīd }

chromium oxychloride See chromyl chloride. { 'krō·mē·əm äk·sē'klȯr‚īd }

chromium plating See chrome plating. { 'krō·mē·əm 'plād·iŋ }

chromium steel [MET] Hard, wear-resistant steel containing chromium as the predominating alloying element. Also known as chrome steel. { 'krō·mē·əm 'stēl }

chromium-vanadium steel [MET] Any of several strong, hard alloy steels containing 0.15–0.25% vanadium, 0.50–1% chromium, and 0.45–0.55% carbon. Also known as chrome-vanadium steel. { ¦krō·mē·əm və¦nād·ē·əm 'stēl }

chromizing [MET] Surface-alloying of metals in which an alloy is formed by diffusion of chromium into the base metal. { 'krō‚miz·iŋ }

chromophore [CHEM] An arrangement of atoms that gives rise to color in many organic substances. { 'krō·mə‚fȯr }

chromyl chloride [INORG CHEM] CrO_2Cl_2 A dark-red, toxic, fuming liquid that boils at 116°C; reacts with water to form chromic acid; used to make dyes and chromium complexes. Also known as chlorochromic anhydride; chromium oxychloride. { 'krō·məl 'klȯr‚īd }

CHU See centigrade heat unit.

CHU_mean See centigrade heat unit.

chuck [MET] A small bar between flask bars to secure the sand in the upper box (cope) of a flask. { chək }

chucking lug [MET] A projection forged or cast onto a piece of metal that functions as a location marker when the work is being machined. { 'chək·iŋ ‚ləg }

Chworinov rule [MET] The postulation that total freezing time for a casting is a function of the ratio of volume to surface area. { 'shvȯr·ə·nȯv ‚rül }

C.I. See cast iron.

cinder [MATER] Slag from a metal furnace. [MET] Scale cast off in forging metal. { 'sin·dər }

cinder block [MATER] A hollow block made of cinder concrete. [MET] A block which closes the front of a blast furnace, containing the cinder notch. { 'sin·dər ‚bläk }

cinder concrete [MATER] A concrete containing cinders as the aggregate. { 'sin·dər 'kän‚krēt }

cinder notch [MET] An opening in a blast furnace that allows molten slag to flow out. { 'sin·dər ‚näch }

cinder pig [MET] Pig iron produced from a mixture of slag in the furnace and crude metal or ore. { 'sin·dər ‚pig }

cinders [MATER] Incombustible residue from a burning process; in particular, small pieces of clinker from the burning of soft coal. { 'sin·dərz }

circuit [ELEC] See electric circuit. [ELECTROMAG] A complete wire, radio, or carrier communications channel. { 'sər·kət }

circuit analyzer See volt-ohm-milliammeter. { 'sər·kət ‚an·ə‚līz·ər }

circular birefringence [OPTICS] The phenomenon in which an optically active substance transmits right circularly polarized light with a different velocity from left circularly polarized light. { 'sər·kyə·lər ‚bī·rə¦frin·jəns }

circular dichroism [OPTICS] A change from planar to elliptic polarization when an initially plane-polarized light wave traverses an optically active medium. Abbreviated CD. { 'sər·kyə·lər 'dī·krō‚iz·əm }

circular polarization [PHYS] Attribute of a transverse wave (either of electromagnetic radiation, or in an elastic medium) whose electric or displacement vector is of constant amplitude and, at a fixed point in space, rotates in a plane perpendicular to the propagation direction with constant angular velocity. { 'sər·kyə·lər ‚pō·lə·rə'zā·shən }

circulating scrap [MET] At steelworks and foundries, scrap arising during the manufacture of finished iron and steel or of castings. { 'sər·kyə‚lād·iŋ 'skrap }

Cl See chlorine.

cladding [ENG] Process of covering one material with another and bonding them together under high pressure and temperature. Also known as bonding. { 'klad·iŋ }

clad metal [MET] A metal overlaid on one or both sides with a different metal. { 'klad ‚med·əl }

clapboard [MATER] A board, thicker at one edge than the other, used to cover exterior walls. { 'kla‚bȯrd }

Clapeyron-Clausius equation See Clausius-Clapeyron equation. { kla·pā·rōn ¦klōz·ē·əs i ‚kwā·zhən }

Clapeyron equation See Clausius-Clapeyron equation. { kla·pā·rōn i'kwā·zhən }

clarifying agent See fining. { 'klar·ə‚fī·iŋ ‚ā·jənt }

classical mechanics [MECH] Mechanics based on Newton's laws of motion. { 'klas·ə·kəl mə'kan·iks }

clathrate [CHEM] An inclusion compound in which the guest species is enclosed on all sides by the species forming the crystal lattice.

Also known as cage compound; inclusion compound. { 'klath,rāt }

clathrochelate [INORG CHEM] A type of coordination compound containing a metal ion both coordinately saturated and encapsulated by a single ligand. { ¦klath·rō'kē,lāt }

clausius [THERMO] A unit of entropy equal to the increase in entropy associated with the absorption of 1000 international table calories of heat at a temperature of 1 K, or to 4186.8 joules per kelvin. { 'klȯz·ē·əs }

Clausius-Clapeyron equation [THERMO] An equation governing phase transitions of a substance, $dp/dT = \Delta H/(T\Delta V)$, in which p is the pressure, T is the temperature at which the phase transition occurs, ΔH is the change in heat content (enthalpy), and ΔV is the change in volume during the transition. Also known as Clapeyron-Clausius equation; Clapeyron equation. { klȯz·ē·əs kla·pā,rōn i,kwä·zhən }

Clausius equation [THERMO] An equation of state in reference to gases which applies a correction to the van der Waals equation: $\{P + (n^2a/[T(V + c)^2])\}(V - nb) = nRT$, where P is the pressure, T the temperature, V the volume of the gas, n the number of moles in the gas, R the gas constant, a depends only on temperature, b is a constant, and c is a function of a and b. { 'klȯz·ē·əs i'kwä·zhən }

Clausius inequality [THERMO] The principle that for any system executing a cyclical process, the integral over the cycle of the infinitesimal amount of heat transferred to the system divided by its temperature is equal to or less than zero. Also known as Clausius theorem; inequality of Clausius. { 'klȯz·ē·əs in·i'kwäl·əd·ē }

Clausius law [THERMO] The law that an ideal gas's specific heat at constant volume does not depend on the temperature. { 'klȯz·ē·əs ,lȯ }

Clausius-Mosotti equation [ELEC] An expression for the polarizability γ of an individual molecule in a medium which has the relative dielectric constant ϵ and has N molecules per unit volume: $\gamma = (3/4\pi N) [(\epsilon - 1)/(\epsilon + 2)]$ (Gaussian units). { ¦klȯz·ē·əs mə'zäd·ē i'kwä·zhən }

Clausius number [THERMO] A dimensionless number used in the study of heat conduction in forced fluid flow, equal to $V^3L\rho/k\Delta T$, where V is the fluid velocity, ρ is its density, L is a characteristic dimension, k is the thermal conductivity, and ΔT is the temperature difference. { 'klȯz·ē·əs ,nəm·bər }

Clausius' statement [THERMO] A formulation of the second law of thermodynamics, stating it is not possible that, at the end of a cycle of changes, heat has been transferred from a colder to a hotter body without producing some other effect. { 'klȯz·ē·əs 'stāt·mənt }

Clausius theorem See Clausius inequality. { 'klȯz·ē·əs 'thir·əm }

clay [MATER] A special grade of absorbent clay used as a filtering medium in refineries for removing solids or colorizing matter from lubricating oils. { klā }

clay brick [MATER] Brick made from diverse types of clays and used for normal constructional purposes. { ¦klā 'brik }

clay slip [MATER] A slurry of clay and water used in glazing pottery. { ¦klā 'slip }

clay wash [MATER] A light oil such as naphtha or kerosine used to clean fuller's earth after it has been used in a filter. { 'klā ,wäsh }

clean room [ENG] A room in which elaborate precautions are employed to reduce dust particles and other contaminants in the air, as required for assembly of delicate equipment. { 'klēn ,rüm }

cleanser [MATER] Any material used to remove dirt, soil, and impurities from surfaces of all kinds. { 'klen·zər }

cleavage [CRYSTAL] Splitting, or the tendency to split, along planes determined by crystal structure and always parallel to a possible face. { 'klēv·ij }

cleavage crystal [CRYSTAL] A crystal fragment bounded by cleavage faces giving it a regular form. { 'klēv·ij ,kris·təl }

cleavage fracture [CRYSTAL] 1. Manner of breaking a crystalline substance along the cleavage plane. 2. The appearance of such a broken surface. { 'klēv·ij ,frak·chər }

cleavage plane [CRYSTAL] Plane along which a crystalline substance may be split. { 'klēv·ij ,plān }

cleft weld [MET] A weld in which a V-shaped projection on one piece is joined to a V-shaped groove in the other. { ¦kleft ¦weld }

climb cutting [MET] A milling technique in which the teeth of a cutting tool advance into the work in the same direction as the feed. Also known as climb milling; down cutting; down milling. { 'klīm ,kəd·iŋ }

climb milling See climb cutting. { 'klīm ,mil·iŋ }

clinker [MATER] An overburned brick. { 'kliŋ·kər }

clinoaxis [CRYSTAL] The inclined lateral axis that makes an oblique angle with the vertical axis in the monoclinic system. Also known as clinodiagonal. { ¦klī·nō'ak·səs }

clinodiagonal See clinoaxis. { ¦klī·nō·dī'ag·ən·əl }

clinohedral class [CRYSTAL] A rare class of crystals in the monoclinic system having a plane of symmetry but no axis of symmetry. Also known as domatic class. { ¦klī·nō¦hē·drəl ,klas }

clinopinacoid [CRYSTAL] A form of monoclinic crystal whose faces are parallel to the inclined and vertical axes. { ¦klī·nə'pin·ə,kȯid }

clip and shave [MET] Dual forging operation in which one cutter removes the flash and then another cutter shaves and sizes the piece. { ¦klip ən 'shāv }

clipping edge [MET] Area of a forging where flash is removed. { 'klip·iŋ ,ej }

closed-cell foam [MATER] A cellular plastic in which there is a predominance of noninterconnecting cells. { ¦klōzd ¦sel 'fōm }

closed cycle [THERMO] A thermodynamic cycle in which the thermodynamic fluid does not enter or leave the system, but is used over and over again. { ‚klōzd 'sī·kəl }

closed die |MET| A forming or forging die in which the flow of metal is restricted to the cavity of the die set. { ‚klōzd 'dī }

closed pass |MET| A metal-rolling operation in which the top roll has a collar that fits a groove on the bottom roll, allowing a flash-free shape to be formed in the rolled metal. { ¦klōzd 'pas }

closed system |THERMO| A system which is isolated so that it cannot exchange matter or energy with its surroundings and can therefore attain a state of thermodynamic equilibrium. Also known as isolated system. { ¦klōzd 'sis·təm }

close-grained |MATER| Consisting of fine, closely spaced particles, crystals, or other elements. { 'klōs ‚grānd }

close-packed crystal |CRYSTAL| A crystal structure in which the lattice points are centers of spheres of equal radius arranged so that the volume of the interstices between the spheres is minimal. { ¦klōs ¦pakt 'kris·təl }

close-tolerance forging |MET| Forging in which draft angles are on the order of 1–3°, tolerances are less than half of those for commercial designs, and there is little or no allowance for finish. { ¦klōs 'täl·ə·rəns 'fór·iŋ }

closure domain |SOLID STATE| A small ferromagnetic domain whose position and orientation ensure that the flux lines between adjacent larger domains close on themselves. Also known as flux-closure domain. { 'klō·zhər dō‚mān }

cloth |TEXT| **1.** A sheet of fibers assembled by weaving, knitting, felting, or some other similar process. **2.** A nonfibrous material of similar properties. { klòth }

cloudburst hardness test |MET| A procedure in which a shower of steel balls, dropped from a predetermined height, dulls the surface of a hardened part in proportion to its softness and thus reveals defective areas. { 'klaùd‚bərst 'härd·nəs ‚test }

cloudburst treatment |MET| Cold-working the surface of a metal by impingement of an avalanche of metal shot; a form of shot-peening. { 'klaùd‚bərst ‚trēt·mənt }

cluster mill |MET| A rolling mill in which small-diameter rolls are supported by larger rolls. { 'kləs·tər ‚mil }

Cm *See* curium.

CMOS device |ELECTR| A device formed by the combination of a PMOS (*p*-type-channel metal oxide semiconductor device) with an NMOS (*n*-type-channel metal oxide semiconductor device). Derived from complementary metal oxide semiconductor device. { 'se‚mós di'vīs }

Co *See* cobalt.

coacervate |CHEM| An aggregate of colloidal droplets bound together by the force of electrostatic attraction. { kō'as·ər‚vāt }

coacervation |CHEM| The separation, by addition of a third component, of an aqueous solution of a macromolecule colloid (polymer) into two liquid phases, one of which is colloid-rich (the

coacervate) and the other an aqueous solution of the coacervating agent (the equilibrium liquid). { kō‚as·ər'vā·shən }

coagulant |CHEM| An agent that causes coagulation. { kō'ag·yə‚lənt }

coalesced copper |MET| A mass, oxygen-free copper made by compacting and sintering cathode copper at high pressure and temperature. { ‚kō·ə'lest 'käp·ər }

coalescence |MET| The bonding of welded materials into one body. |PHYS| The uniting by growth in one body, as particles, gas, or a liquid. { ‚kō·ə'les·əns }

coalescent |CHEM| Chemical additive used in immiscible liquid-liquid mixtures to cause small droplets of the suspended liquid to unite, preparatory to removal from the carrier liquid. { ‚kō·ə'les·ənt }

coal tar |MATER| A tar obtained from carbonization of coal, usually in coke ovens or retorts, containing several hundred organic chemicals. { 'kōl ‚tär }

coal-tar enamel |MATER| Coal tar used as a paintlike coating for petroleum-product pipelines; provides protection from both water and cathodic corrosion. { 'kōl ¦tär i‚nam·əl }

coal-tar epoxy |MATER| A thermosetting resin produced as a by-product of the carbonization of bituminous coal. { 'kōl ¦tär ə'päk·sē }

coal-tar pitch |MATER| Dark-brown to black amorphous residue from the redistillation of coal tar; melts at 150°F (66°C); used as a thermoplastic. { 'kōl ¦tär ‚pich }

coarse aggregate |MATER| Crushed stone or gravel used in concrete; will not, when dry, pass through a sieve with $\frac{1}{4}$-inch-diameter (6-millimeter) holes. { 'kòrs 'ag·rə·gət }

coarse-grained |MATER| Having a coarse texture. { 'kòrs ¦grānd }

coated abrasive |MATER| An abrasive product having the abrasive particles attached to a backing material with glue or a synthetic resin. { 'kōd·əd ə'brā·siv }

coated electrode |MET| A wire covered with metal oxides and silicates and used as a filler-metal electrode in arc welding. Also known as covered electrode. { 'kōd·əd i'lek‚trōd }

coated fabric |TEXT| A fabric that has been coated, covered, or impregnated with substances such as lacquer, varnish, rubber, or polymers. { 'kōd·əd 'fab·rik }

coated paper |MATER| Paper with a surface coating of clay and other materials to produce a smooth, shiny surface; especially useful for fine, detailed, blur-free reproductions in color or black and white. Also known as enamel paper. { 'kōd·əd 'pā·pər }

coating |MATER| **1.** Any material that will form a continuous film over a surface. **2.** The film formed by the material. { 'kōd·iŋ }

coating density ratio |MET| In thermal spraying, the ratio of actual density to theoretical density of the coating material used. { 'kōd·iŋ ‚den·səd·ē ‚rā·shō }

coaxing |MET| Improving fatigue strength of a

cobalt

metal by increasing the stress range, beginning just below the fatigue limit. { kōks·iŋ }

cobalt [CHEM] A metallic element, symbol Co, atomic number 27, atomic weight 58.93; used chiefly in alloys. { 'kō‚bȯlt }

cobalt blue [CHEM] A green-blue pigment formed of alumina and cobalt oxide. Also known as cobalt ultramarine; king's blue. { ¦kō‚bȯlt ¦blü }

cobalt bromide See cobaltous bromide. { 'kō ‚bȯlt 'brō‚mīd }

cobalt chloride See cobaltous chloride. { 'kō ‚bȯlt 'klȯr‚īd }

cobaltic fluoride See cobalt trifluoride. { kə 'bȯl·tik 'flür‚īd }

cobalt nitrate See cobaltous nitrate. { 'kō‚bȯlt 'nī‚trāt }

cobaltous bromide [INORG CHEM] CoBr$_2$·6H$_2$O Red-violet crystals with a melting point of 47–48°C; soluble in water, alcohol, and ether; used in hygrometers. Also known as cobalt bromide. { kō'bȯl·təs 'brō‚mīd }

cobaltous chloride [INORG CHEM] CoCl$_2$ or CoCl$_2$·6H$_2$O A compound whose anhydrous form consists of blue crystals and sublimes when heated, and whose hydrated form consists of red crystals and melts at 86.8°C; both forms are used as an absorbent for ammonia in dyes and as a catalyst. Also known as cobalt chloride. { kō'bȯl·təs 'klȯr‚īd }

cobaltous fluorosilicate [INORG CHEM] CoSiF$_6$· H$_2$O A water-soluble, orange-red powder, used in toothpastes. { kō'bȯl·təs ¦flür·ō'sil·ə‚kāt }

cobaltous nitrate [INORG CHEM] Co(NO$_3$)$_2$· 6H$_2$O A red crystalline compound with a melting point of 56°C; soluble in organic solvents; used in sympathetic inks, as an additive to soils and animal feeds, and for vitamin preparations and hair dyes. Also known as cobalt nitrate. { kō'bȯl·təs 'nī‚trāt }

cobalt oxide [INORG CHEM] CoO A grayish brown powder that decomposes at 1935°C, insoluble in water; used as a colorant in ceramics and in manufacture of glass. { 'kō‚bȯlt 'äk‚sīd }

cobalt potassium nitrite [INORG CHEM] K$_3$Co(NO$_2$)$_6$ A yellow powder which decomposes at the melting point of 200°C; used in medicine and as a yellow pigment. Also known as cobalt yellow; Fischer's salt; potassium cobaltinitrite. { 'kō‚bȯlt pə'tas·ē·əm 'nī‚trīt }

cobalt sulfate [INORG CHEM] Any compound of either divalent or trivalent cobalt and the sulfate group; anhydrous cobaltous sulfate, CoSO$_4$, contains divalent cobalt, has a melting point of 96.8°C, is soluble in methanol, and is utilized to prepare pigments and cobalt salts; cobaltic sulfate, Co$_2$(SO$_4$)$_3$·18H$_2$O, contains trivalent cobalt, is soluble in sulfuric acid, and functions as an oxidizing agent. { 'kō‚bȯlt 'səl‚fāt }

cobalt trifluoride [INORG CHEM] CoF$_3$ A brownish powder that reacts with water to form a precipitate of cobaltic hydroxide; used as a fluorinating agent. Also known as cobaltic fluoride. { 'kō ‚bȯlt trī'flür‚īd }

cobalt ultramarine See cobalt blue. { 'kō‚bȯlt ‚əl·trə·mə'rēn }

cobalt yellow See cobalt potassium nitrite. { 'kō ‚bȯlt 'yel·ō }

cochineal [CHEM] A red dye made of the dried bodies of the female cochineal insect (*Coccus cacti*), found in Central America and Mexico; used as a biological stain and indicator. { 'käch·ə ‚nēl }

cockle finish [MATER] An irregular surface usually produced on rag bond and ledger paper, obtained by coating the paper with sizing and drying it in loop or festoon fashion in heated air. { 'käk·əl ‚fin·ish }

coefficient of absorption See absorption coefficient. { ¦kō·ə'fish·ənt əv ab'sȯrp·shən }

coefficient of capacitance [ELEC] One of the coefficients which appears in the linear equations giving the charges on a set of conductors in terms of the potentials of the conductors; a coefficient is equal to the ratio of the charge on a given conductor to the potential of the same conductor when the potentials of all the other conductors are 0. { ¦kō·ə'fish·ənt əv kə'pas·ə·təns }

coefficient of compressibility [MECH] The decrease in volume per unit volume of a substance resulting from a unit increase in pressure; it is the reciprocal of the bulk modulus. { ¦kō·ə'fish·ənt əv kəm‚pres·ə'bil·əd·ē }

coefficient of conductivity See thermal conductivity. { ¦kō·ə'fish·ənt əv ‚kän·dək'tiv·əd·ē }

coefficient of contraction [FL MECH] The ratio of the minimum cross-sectional area of a jet of liquid discharging from an orifice to the area of the orifice. Also known as contraction coefficient. { ¦kō·ə'fish·ənt əv kən'trak·shən }

coefficient of cubical expansion [THERMO] The increment in volume of a unit volume of solid, liquid, or gas for a rise of temperature of 1° at constant pressure. Also known as coefficient of expansion; coefficient of thermal expansion; coefficient of volumetric expansion; expansion coefficient; expansivity. { ¦kō·ə'fish·ənt əv 'kyüb·ə·kəl ik'span·shən }

coefficient of elasticity See modulus of elasticity. { ¦kō·ə'fish·ənt əv i‚las'tis·əd·ē }

coefficient of expansion See coefficient of cubical expansion. { ¦kō·ə'fish·ənt əv ik'span·shən }

coefficient of friction [MECH] The ratio of the frictional force between two bodies in contact, parallel to the surface of contact, to the force, normal to the surface of contact, with which the bodies press against each other. Also known as friction coefficient. { ¦kō·ə'fish·ənt əv 'frik·shən }

coefficient of friction of rest See coefficient of static friction. { ¦kō·ə'fish·ənt əv 'frik·shən əv 'rest }

coefficient of induction [ELEC] One of the coefficients which appears in the linear equations giving the charges on a set of conductors in terms of the potentials of the conductors; a coefficient is equal to the ratio of the charge on a given conductor to the potential on another conductor, when the potentials of all the other conductors equal 0. { ¦kō·ə'fish·ənt əv in'dək·shən }

coefficient of kinematic viscosity See kinematic viscosity. { ¦kō·ə'fish·ənt əv ¦kin·ə¦mad·ik vis'käs·əd·ē }

coefficient of kinetic friction [MECH] The ratio of the frictional force, parallel to the surface of contact, that opposes the motion of a body which is sliding or rolling over another, to the force, normal to the surface of contact, with which the bodies press against each other. { ¦kō·ə'fish·ənt əv kə'ned·ik 'frik·shən }

coefficient of linear expansion [THERMO] The increment of length of a solid in a unit of length for a rise in temperature of 1° at constant pressure. Also known as linear expansity. { ¦kō·ə'fish·ənt əv 'lin·ē·ər ik'span·shən }

coefficient of performance [THERMO] In a refrigeration cycle, the ratio of the heat energy extracted by the heat engine at the low temperature to the work supplied to operate the cycle; when used as a heating device, it is the ratio of the heat delivered in the high-temperature coils to the work supplied. { ¦kō·ə'fish·ənt əv pər'fȯr·məns }

coefficient of potential [ELEC] One of the coefficients which appears in the linear equations giving the potentials of a set of conductors in terms of the charges on the conductors. { ¦kō·ə'fish·ənt əv pə'ten·chəl }

coefficient of reflection See reflection coefficient. { ¦kō·ə'fish·ənt əv ri'flek·shən }

coefficient of restitution [MECH] The constant *e*, which is the ratio of the relative velocity of two elastic spheres after direct impact to that before impact; *e* can vary from 0 to 1, with 1 equivalent to an elastic collision and 0 equivalent to a perfectly elastic collision. Also known as restitution coefficient. { ¦kō·ə'fish·ənt əv ‚res·tə'tü·shən }

coefficient of rigidity See modulus of elasticity in shear. { ¦kō·ə'fish·ənt əv rə'jid·əd·ē }

coefficient of rolling friction [MECH] The ratio of the frictional force, parallel to the surface of contact, opposing the motion of a body rolling over another, to the force, normal to the surface of contact, with which the bodies press against each other. { ¦kō·ə'fish·ənt əv 'rōl·iŋ 'frik·shən }

coefficient of sliding friction [MECH] The ratio of the frictional force, parallel to the surface of contact, opposing the motion of a body sliding over another, to the force, normal to the surface of contact, with which the bodies press against each other. { ¦kō·ə'fish·ənt əv 'slīd·iŋ 'frik·shən }

coefficient of static friction [MECH] The ratio of the maximum possible frictional force, parallel to the surface of contact, which acts to prevent two bodies in contact, and at rest with respect to each other, from sliding or rolling over each other, to the force, normal to the surface of contact, with which the bodies press against each other. Also known as coefficient of friction of rest. { ¦kō·ə'fish·ənt əv 'stad·ik 'frik·shən }

coefficient of strain [MECH] For a substance undergoing a one-dimensional strain, the ratio of the distance along the strain axis between two points in the body, to the distance between the same points when the body is undeformed. { ¦kō·ə'fish·ənt əv 'strān }

coefficient of superficial expansion [THERMO] The increment in area of a solid surface per unit of area for a rise in temperature of 1° at constant pressure. Also known as superficial expansivity. { ¦kō·ə'fish·ənt əv ‚sü·pər'fish·əl ik'span·chən }

coefficient of thermal expansion See coefficient of cubical expansion. { ¦kō·ə'fish·ənt əv 'thər·məl ik'span·shən }

coefficient of viscosity See absolute viscosity. { ¦kō·ə'fish·ənt əv vis'käs·əd·ē }

coefficient of volumetric expansion See coefficient of cubical expansion. { ¦kō·ə'fish·ənt əv ¦väl·yə¦me·trik ik'span·chən }

coercive force [ELECTROMAG] The magnetic field H which must be applied to a magnetic material in a symmetrical, cyclicly magnetized fashion, to make the magnetic induction B vanish. Also known as magnetic coercive force. { 'kō·ər·siv 'fȯrs }

coercivity [ELECTROMAG] The coercive force of a magnetic material in a hysteresis loop whose maximum induction approximates the saturation induction. { ‚kō·ər'siv·əd·ē }

cogging mill See blooming mill. { 'käg·iŋ ‚mil }

coherence [PHYS] 1. The existence of a correlation between the phases of two or more waves, so that interference effects may be produced between them, or of a correlation between the phases of part of a single wave. 2. Property of moving in unison, such as is characteristic of the particles in a synchrotron. { kō'hir·əns }

coherence distance See coherence length. { kō'hir·əns ‚dis·təns }

coherence length [PHYS] For a beam of particles, the typical length of a wave packet along the beam; the more monochromatic the beam, the greater its coherence length. [SOLID STATE] A measure of the distance through which the effect of any local disturbance is spread out in a superconducting material. Also known as coherence distance. { kō'hir·əns ‚leŋkth }

coherent light [OPTICS] Radiant electromagnetic energy of the same, or almost the same, wavelength, and with definite phase relationships between different points in the field. { kō'hir·ənt 'līt }

coherent radiation [PHYS] Radiation in which there are definite phase relationships between different points in a cross section of the beam. { kō'hir·ənt ‚rād·ē'ā·shən }

cohesion [PHYS] The tendency of parts of a body of like composition to hold together, as a result of intermolecular attractive forces. { kō'hē·zhən }

cohesive energy [SOLID STATE] The difference between the energy per atom of a system of free atoms at rest far apart from each other, and the energy of the solid. { kō'hē·siv 'en·ər·jē }

cohesive strength [MECH] 1. Strength corresponding to cohesive forces between atoms. 2. Hypothetically, the stress causing tensile fracture without plastic deformation. { kō'hē·siv 'streŋkth }

coil breaks [MET] Creases across a metal strip transverse to the direction of coiling and representing areas of reduced thickness. { 'kȯil ‚brāks }

coil weld [MET] A butt weld joining the ends of two metal sheets; forms a continuous strip for coiling. { 'kȯil ‚weld }

coincidence boundary [CRYSTAL] A grain boundary separating crystal lattices which are rotated with respect to each other by an angle with a special value, resulting in a periodic grain boundary structure and an extension of a sublattice of the original lattice across the boundary. { kō'in·sə·dəns ‚baȯn·drē }

coin gold [MET] Gold of the legal fineness for coins. { 'kȯin ‚gōld }

coining [MET] **1.** A process of forming metals by squeezing between two dies so as to impress well-defined imprints on both surfaces of the work; usually performed cold. **2.** Final pressing of a sintered compact in powder metallurgy. { 'kȯin·iŋ }

coin silver [MET] An alloy of 90% silver, 10% copper; has been used for coining American currency. { 'kȯin ‚sil·vər }

coir [MATER] A coarse, brown fiber obtained from the husk of the coconut. { kȯir }

coke [MATER] A coherent, cellular, solid residue remaining from the dry (destructive) distillation of a coking coal or of pitch, petroleum, petroleum residues, or other carbonaceous materials; contains carbon as its principal constituent, together with mineral matter and volatile matter. { kōk }

Colburn j factor equation [THERMO] Dimensionless heat-transfer equation to calculate the natural convection movement of heat from vertical surfaces or horizontal cylinders to fluids (gases or liquids) flowing past these surfaces. { 'kol·bərn 'jā ‚fak·tər i'kwā·zhən }

colcothar [INORG CHEM] Red ferric oxide made by heating ferrous sulfate in the air; used as a pigment and as an abrasive in polishing glass. { 'käl·kə‚thär }

cold bending [MET] The bending of metal rods, especially concrete-reinforcing rods, without heat. { 'kōld ‚bend·iŋ }

cold-conductor effect [SOLID STATE] A sudden increase in resistivity, up to seven orders of magnitude, as the temperature increases over a narrow range; observed in certain semiconducting materials, particularly ferroelectric titanate ceramics. { ¦kōld kən'dək·tər i‚fekt }

cold drawing [MET] Drawing a tube or wire through a series of successively smaller dies, without the application of heat, to reduce its diameter. [TEXT] Drawing a textile, such as nylon, when cold. { 'kōld ‚drȯ·iŋ }

cold extrusion [MET] Shaping cold metal by striking a slug in a closed cavity with a punch so that the metal is forced up around the punch. Also known as cold forging; cold pressing; extrusion pressing; impact extrusion. { 'kōld ik'strü·zhən }

cold-finished steel [MET] Steel bars which have been cold-drawn, cold-rolled, centerless-ground, or turned smooth. { 'kōld ‚fin·isht 'stēl }

cold flow [MECH] Creep in polymer plastics. { 'kōld ‚flō }

cold forging See cold extrusion. { 'kōld ‚fȯrj·iŋ }

cold forming [MET] Any forging operation performed cold, such as cold extrusion, cold drawing, or coining, which enables close dimensional accuracy to be achieved. { 'kōld ‚fȯrm·iŋ }

cold galvanizing [MET] Painting iron with a suspension of zinc particles in an organic solvent, so that a zinc coating remains following evaporation of the solvent. { 'kōld ¦gal·və‚nīz·iŋ }

cold heading [MET] The cold working of metal in order to increase part or all of the cross-sectional area of the stock. { 'kōld ‚hed·iŋ }

cold inspection [MET] The inspection of a forging at room temperature by visible or nondestructive means to detect surface conditions or defects at room temperature. { 'kōld ‚in'spek·shən }

cold lap See cold shut. { 'kōld ‚lap }

cold plasma [CHEM ENG] Low-energy ionized gas. { 'kōld 'plaz·mə }

cold pressing See cold extrusion. { 'kōld ‚pres·iŋ }

cold rolling [MET] Rolling metal at room temperature to reduce thickness or harden the surface; results in a smooth finish and improved resistance to fatigue. { 'kōld ‚rōl·iŋ }

cold rubber [MATER] Butadiene-styrene type of synthetic rubber produced by polymerization at about 40°F (4°C), instead of the conventional 120°F (49°C); has improved strength and abrasion resistance. { ¦kōld ¦rəb·ər }

cold-setting adhesive [MATER] A synthetic resin that can harden at normal room temperature without the addition of a hardener. { ¦kōld ‚sed·iŋ əd'hē·ziv }

cold-short [MET] Pertaining to lack of ductility in some metals at temperatures below the recrystallization temperature. { 'kōld ‚shȯrt }

cold shot [MET] Intensely hard, globular portions of the surface of an ingot or casting formed by premature solidification upon first contact with the cold sand during pouring. { 'kōld ‚shät }

cold shut [MET] **1.** A surface defect of a metal casting in the form of a discontinuity where two streams failed to unite. Also known as cold lap. **2.** Freezing of the top surface of an ingot before the mold is full. { 'kōld ‚shət }

cold soldering [MET] Soldering of parts without heat. { 'kōld 'säd·ə·riŋ }

cold stress [MECH] Forces tending to deform steel, cement, and other materials, resulting from low temperatures. { 'kōld ‚stres }

cold treatment [MET] Subzero cooling, to a temperature of −100°F (−73°C). { 'kōld ‚trēt·mənt }

cold trimming [MET] The removal of excess metal from a forging at room temperature by means of a trimming press. { 'kōld ‚trim·iŋ }

cold welding [MET] Welding in which a molecular bond is obtained by a cold flow of metal under extremely high pressures, without heat; widely used for sealing transistors and quartz crystal holders. { 'kōld ‚weld·iŋ }

cold working [MET] Plastic deformation of a

metal below the annealing temperature to cause permanent strain hardening. { 'kōld ˌwərk·iŋ }

collagen [BIOMATER] A fibrous protein found in all multicellular animals, especially in connective tissue. { 'kä·lə·jən }

collapse [ENG] Contraction of plastic container walls during cooling; produces permanent indentation. { kə'laps }

collapse properties [MECH] Strength and dimensional attributes of piping, tubing, or process vessels, related to the ability to resist collapse from exterior pressure or internal vacuum. { kə'laps ˌpräp·ərd·ēz }

collapsing pressure [MECH] The minimum external pressure which causes a thin-walled body or structure to collapse. { kə'lap·siŋ ˌpresh·ər }

collective electron theory [SOLID STATE] A theory of ferromagnetism in which electrons responsible for ferromagnetism are supposed to move more or less freely throughout a crystal, and to align with one another as the result of an exchange interaction. { kə¦lek·tiv i'lek¸trän ˌthē·ə·rē }

collective paramagnetism [ELECTROMAG] Magnetization of a collection of extremely small ferromagnetic particles, each containing only one magnetic domain, that resembles paramagnetism of a collection of atoms or molecules. Also known as superparamagnetism. { kə'lek·tiv ˌpar·ə'mag·nəˌtiz·əm }

colligative properties [PHYS CHEM] Properties dependent on the number of molecules but not their nature. { kə'lig·ə·div ˌpräp·ərd·ēz }

collimated beam [PHYS] A beam of radiation or matter whose rays or particles are nearly parallel so that the beam does not converge or diverge appreciably. { 'käl·əˌmād·əd 'bēm }

Collins helium liquefier [CRYO] A machine which uses the Joule-Thomson effect and work done by helium gas in expansion against a movable piston to liquefy helium. { 'käl·ənz 'hē·lē·əm 'lik·wəˌfī·ər }

colloid [CHEM] The phase of a colloidal system made up of particles having dimensions of 10–10,000 angstroms (1–1000 nanometers) and which is dispersed in a different phase. { 'käl ˌoid }

colloidal crystal [CHEM] A periodic array of suspended colloidal particles that can arise spontaneously in a monodisperse colloidal system under appropriate conditions. { kə'lóid·əl 'krist·əl }

colloidal dispersion See colloidal system. { kə'lóid·əl dis'pər·zhən }

colloidal electrolyte [PHYS CHEM] An electrolyte that yields at least one type of ion in the colloidal size range. { kə'lóid·əl i'lek·trəˌlīt }

colloidal graphite [MATER] Extremely fine flakes of graphite suspended in water, petroleum oil, castor oil, glycerin, or other liquids; used to provide conductive shields on the inside or outside surfaces of electron tubes. { kə'lóid·əl 'graˌfīt }

colloidal silver [MATER] Finely divided particles of silver, sometimes used on terminals of elec-

tronic components to give a larger surface area for connections. { kə'lóid·əl 'sil·vər }

colloidal suspension See colloidal system. { kə'lóid·əl səs'pen·shən }

colloidal system [CHEM] An intimate mixture of two substances, one of which, called the dispersed phase (or colloid), is uniformly distributed in a finely divided state through the second substance, called the dispersion medium (or dispersing medium); the dispersion medium or dispersed phase may be a gas, liquid, or solid. Also known as colloidal dispersion; colloidal suspension. { kə'lóid·əl 'sis·təm }

colloid chemistry [PHYS CHEM] The scientific study of matter whose size is approximately 10 to 10,000 angstroms (1 to 1000 nanometers), and which exists as a suspension in a continuous medium, especially a liquid, solid, or gaseous substance. { ¦käl¸óid 'kem·ə·strē }

color [OPTICS] A general term that refers to the wavelength composition of light, with particular reference to its visual appearance. { 'kəl·ər }

color aberration See chromatic aberration. { 'kəl·ər ab·ə'rā·shən }

color attribute [OPTICS] Any of the visual qualities of hue, saturation, or brightness. { 'kəl·ər ˌa·traˌbyüt }

color center [SOLID STATE] A point lattice defect which produces optical absorption bands in an otherwise transparent crystal. { 'kəl·ər ˌsen·tər }

color circle [OPTICS] An arrangement of hues about the circumference of a circle in the order in which they appear in the electromagnetic spectrum, with pairs of complementary colors at opposite ends of diameters. { 'kəl·ər ˌsar·kəl }

color code [ELEC] A system of colors used to indicate the electrical value of a component or to identify terminals and leads. { 'kəl·ər ˌkōd }

color correction [OPTICS] The construction of an optical system so that the image positions of an object are the same for two or more wavelengths, and chromatic aberration is thus minimized. { 'kəl·ər kə'rek·shən }

color disk [OPTICS] A rotating circular disk having three filter sections to produce the individual red, green, and blue pictures in a field-sequential color television system. { 'kəl·ər ˌdisk }

color emissivity See monochromatic emissivity. { ¦kəl·ər ˌe·mi'siv·əd·ē }

color equation [OPTICS] An algebraic equation that expresses a specified color as an additive mixture of primary colors. { 'kəl·ər i'kwā·zhən }

colorfast [TEXT] Referring to a fabric that does not fade during normal wear. { 'kəl·ərˌfast }

color film [GRAPHICS] A sensitized film used for three-color photography; consists of three emulsions coated one above another and sensitive respectively to blue, green, and red light. { 'kəl·ər ˌfilm }

color filter [OPTICS] An optical element that partially absorbs incident light, consisting of a pane of glass or other partially transparent material, or of films separated by narrow layers; the absorption may be either selective or nonselective with respect to wavelength. Also known as light filter. { 'kəl·ər ,fil·tər }

colorimeter [OPTICS] An instrument that measures color by determining the intensities of the three primary colors that will give that color. { ,kəl·ə'rim·əd·ər }

colorimetric photometer [OPTICS] A photometer that can measure light intensities in several spectral regions, using color filters placed in the path of the light. { ,kəl·ə·rə'me·trik fō'täm·əd·ər }

colorimetry [OPTICS] Any technique by which an unknown color is evaluated in terms of standard colors; the technique may be visual, photoelectric, or indirect by means of spectrophotometry; used in chemistry and physics. { ,kəl·ə'rim·ə·trē }

color lake See lake. { 'kəl·ər ,lāk }

color medium [OPTICS] Any colored, transparent material that is placed in front of a lighting unit to color the light transmitted. { 'kəl·ər ,mēd·ē·əm }

color rendering [OPTICS] For a light source, the extent of the agreement between the perceived color of a surface illuminated by the source and that of the same surface illuminated by a reference source under specified viewing conditions, measured and expressed in terms of the chromaticity coordinates of the source and the luminance of the source in agreed spectral bands. { 'kəl·ər 'ren·dər·iŋ }

color saturation [OPTICS] The degree to which a color is mixed with white; high saturation means little white, low saturation means much white. Also known as chroma; saturation. { 'kəl·ər sach·ə'rā·shən }

color solid [OPTICS] A three-dimensional diagram which represents the relationship of three attributes of surface color: hue, saturation, and brightness. { 'kəl·ər ,säl·əd }

color stability [CHEM] Resistance of materials to change in color that can be caused by light or aging. { 'kəl·ər stə'bil·əd·ē }

color system [OPTICS] Any three-component coordinate system used to represent the attributes of colors. { 'kəl·ər ,sis·təm }

color tempering [MET] Reheating of hardened steel and observing color changes to determine quenching temperature and to obtain the desired hardness. { 'kəl·ər ¦tem·pə·riŋ }

color triangle [OPTICS] A triangle on a chromaticity diagram that represents the range of chromaticities that can be obtained as additive mixtures of three prescribed primary colors represented by the corners of the triangle. { 'kəl·ər 'trī,aŋ·gəl }

colossal magnetoresistance [SOLID STATE] A very large magnetoresistance associated with magnetic phase transitions in certain homogeneous materials, particularly a class of rare-earth perovskite manganites. { kə¦läs·əl mag,ned·ō·ri'zis·təns }

columbium See niobium. { kə'ləm·bē·əm }

combination die [MET] A die having more than one cavity for different castings. { ,käm·bə 'nā·shən 'dī }

combination mill [MET] A rolling mill arranged with continuous rolls for roughing and a guide or looping mill for shaping. { ,käm·bə'nā·shən ¦mil }

combinatorial chemistry [ORG CHEM] A method for reacting a small number of chemicals to produce simultaneously a very large number of compounds, called libraries, which are screened to identify useful products such as drug candidates. { kəm,bīn·ə¦tör·ē·əl 'kem·ə·strē }

combined flexure [MECH] The flexure of a beam under a combination of transverse and longitudinal loads. { kəm'bīnd 'flek·shər }

combined stresses [MECH] Bending or twisting stresses in a structural member combined with direct tension or compression. { kəm'bīnd 'stres·əz }

combining glass [OPTICS] A glass screen designed to reflect display imagery to the viewer, usually at selected wavelengths of light, while being sufficiently transmissive for the viewer to see the scene beyond. { kəm'bīn·iŋ ,glas }

comb polymer [POLYM CHEM] A macromolecule in which the main chain has one long branch per repeat unit. { ¦kōm 'päl·ə·mər }

combustion [CHEM] The burning of gas, liquid, or solid, in which the fuel is oxidized, evolving heat and often light. { kəm'bəs·chən }

common brick [MATER] Brick made from natural clay. { ¦käm·ən ¦brik }

common-ion effect [CHEM] The lowering of the degree of ionization of a compound when another ionizable compound is added to a solution; the compound added has a common ion with the other compound. { ¦käm·ən ¦ī,än i'fekt }

common salt See sodium chloride. { ¦käm·ən 'sólt }

commutator-controlled welding [MET] Spot or projection welding in which several electrodes in contact with the work simultaneously are operated under the control of an electrical commutating device. Also known as ultraspeed welding. { ¦käm·yə,täd·ər kən¦trōld 'weld·iŋ }

comonomer [CHEM] One of the compounds used to produce a specific polymeric product. { ¦kō'män·ə·mər }

compact [MET] A briquette made by the compression of metal powder, with or without the addition of nonmetallic constituents. { 'käm ,pakt }

compar [MATER] Generic term for a family of compounded, modified, and plasticized polyvinyl alcohol resins; used for making full- and solvent-resistant tubing and hose, and printing rolls. { 'käm,pär }

comparator method [THERMO] A method of determining the coefficient of linear expansion of a substance in which one measures the distance that each of two traveling microscopes must be moved in order to remain centered on scratches on a rod-shaped specimen when the temperature of the specimen is raised by a measured amount. { kəm'par·əd·ər ,meth·əd }

compatibility conditions [MECH] A set of six

differential relations between the strain components of an elastic solid which must be satisfied in order for these components to correspond to a continuous and single-valued displacement of the solid.   { kəm₁pad·ə'bil·əd·ē kən₁dish·ənz }

compensating impurity   [SOLID STATE] A semiconductor impurity that is of the opposite electrical type to a given impurity, and that reduces the concentration of charge carriers (electrons or holes) that resulted from the given impurity.   { 'käm·pən₁sād·iŋ im'pyür·əd·ē }

compensation effect   See Jaccarino-Peter effect.   { ₁käm·pən'sā·shən i₁fekt }

complementary colors   [OPTICS] Two colors which lie on opposite sides of the white point in the chromaticity diagram so that an additive mixture of the two, in appropriate proportions, can be made to yield an achromatic mixture.   { ₁käm·plə'men·trē 'kəl·ərz }

complementary metal oxide semiconductor device   See CMOS device.   { ₁käm·plə¦men·trē ¦med·əl ¦äk₁sīd 'sem·i·kən₁dək·tər di'vīs }

complementary wavelength   [OPTICS] The wavelength of light that, when combined with a sample color in suitable proportions, matches a reference color standard light.   { ₁käm·plə'men·trē 'wāv₁leŋkth }

complete fusion   [MET] Fusion which has occurred over all surfaces of the base metal exposed for welding.   { kəm'plēt 'fyü·zhən }

complete joint penetration   [MET] The fusion of weld metal to base metal throughout the entire thickness of the base metal, that is, the deposited weld metal occupies the entire groove.   { kəm'plēt 'jöint pen·ə'trā·shən }

complexation   See complexing.   { ₁käm₁plek'sā·shən }

complex impedance   See electrical impedance. See impedance.   { 'käm₁pleks im'pēd·əns }

complexing   [CHEM] Formation of a complex compound. Also known as complexation.   { 'käm₁plek·siŋ }

complexing agent   [CHEM] A substance capable of forming a complex compound with another material in solution.   { 'käm₁plek·siŋ ₁ā·jənt }

complex ion   [CHEM] A complex, electrically charged group of atoms or radical, for example, $Cu(NH_3)_2{}^{2+}$.   { 'käm₁pleks 'ī₁än }

complex permittivity   [ELEC] A property of a dielectric, equal to $\epsilon_0(C/C_0)$, where C is the complex capacitance of a capacitor in which the dielectric is the insulating material when the capacitor is connected to a sinusoidal voltage source, and C_0 is the vacuum capacitance of the capacitor.   { 'käm₁pleks ₁pər·mə'tiv·əd·ē }

complex salt   [INORG CHEM] A class of salts in which there are no detectable quantities of each of the metal ions existing in solution; an example is $K_3Fe(CN)_6$, which in solution has K^+ but no Fe^{3+} because Fe is strongly bound in the complex ion, $Fe(CN)_6{}^{3-}$.   { 'käm₁pleks 'sölt }

compliance   [MECH] The displacement of a linear mechanical system under a unit force.   { kəm'plī·əns }

compliance constant   [MECH] Any one of the coefficients of the relations in the generalized Hooke's law used to express strain components as linear functions of the stress components. Also known as elastic constant.   { kəm'plī·əns ₁kän·stənt }

compliant substrate   [ELECTR] A semiconductor substrate into which an artificially formed interface is introduced near the surface which makes the substrate more readily deformable and allows it to support a defect-free semiconductor film of essentially any lattice constant, with dislocations forming in the substrate instead of in the film. Also known as sacrificial compliant substrate.   { kəm¦plī·ənt 'səb₁strāt }

composite   [MATER] A material that results when two or more materials, each having its own, usually different characteristics, are combined, giving useful properties for specific applications. Also known as composite material.   { kəm'päz·ət }

composite beam   [CIV ENG] A structural member composed of two or more dissimilar materials joined together to act as a unit in which the resulting system is stronger than the sum of its parts. An example in civil structures is the steel-concrete composite beam in which a steel wide-flange shape (I or W shape) is attached to a concrete floor slab.   { kəm'päz·ət 'bēm }

composite column   [CIV ENG] A concrete column having a structural-steel or cast-iron core with a maximum core area of 20.   { kəm'päz·ət 'käl·əm }

composite compact   [MET] A powder compact composed of more than one layer of different components with each layer retaining its identity.   { kəm'päz·ət 'käm₁pakt }

composite electrode   [MET] A filler-metal electrode composed of more than one metal.   { kəm'päz·ət i'lek₁trōd }

composite I-beam bridge   [CIV ENG] A beam bridge in which the concrete roadway is mechanically bonded to the I beams by means of shear connectors.   { kəm'päz·ət 'ī ₁bēm ₁brij }

composite joint   [MET] A joint connected by welding in conjunction with one or more mechanical means.   { kəm'päz·ət 'jöint }

composite macromechanics   [ENG] The study of composite material behavior wherein the material is presumed homogeneous and the effects of the constituent materials are detected only as averaged apparent properties of the composite.   { kəm'päz·ət ¦mak·rō·mə'kan·iks }

composite material   See composite.   { kəm ¦päz·ət mə¦tir·ē·əl }

composite micromechanics   [ENG] The study of composite material behavior wherein the constituent materials are studied on a microscopic scale with specific properties being assigned to each constituent; the interaction of the constituent materials is used to determine the properties of the composite.   { kəm'päz·ət ¦mik·rō·mə'kan·iks }

composite pile   [CIV ENG] A pile in which the

upper and lower portions consist of different types of piles. { kəm'päz·ət 'pīl }

composite plate [MET] A layer of electrodeposited material consisting of at least two different constituents. { kəm'päz·ət 'plāt }

composite steel [MET] Bar steel machined along the entire length which is cast around an insert of tool steel welded to the backing of mild steel; used for shear blades and die parts. { kəm'päz·ət 'stēl }

composite truss [CIV ENG] A truss having compressive members and tension members. { kəm'päz·ət 'trəs }

composition [CHEM] The elements or compounds making up a material or produced from it by analysis. [MECH] The determination of a force whose effect is the same as that of two or more given forces acting simultaneously; all forces are considered acting at the same point. { ,käm·pə'zish·ən }

composition face See composition surface. { ,käm·pə'zish·ən ,fās }

composition metal [MET] A cast copper alloy having a composition of more than 80% copper, with tin, zinc, and lead. { ,käm·pə'zish·ən ,med·əl }

composition plane [CRYSTAL] A planar composition surface in a crystal uniting two individuals of a contact twin. { ,käm·pə'zish·ən,plān }

composition surface [CRYSTAL] The surface uniting individuals of a crystal twin; may or may not be planar. Also known as composition face. { ,käm·pə'zish·ən ,sər·fəs }

compound [CHEM] A substance whose molecules consist of unlike atoms and whose constituents cannot be separated by physical means. Also known as chemical compound. { 'käm ,paùnd }

compound compact [MET] A powder compact made from a mixture of metals, with each particle retaining its original composition. { 'käm ,paùnd 'käm,pakt }

compound die [MET] A die designed to perform more than one operation on the work with each stroke of the press. { 'käm,paùnd 'dī }

compound magnet [ELEC] A permanent magnet that is constructed from a number of thin magnets having the same shape. { 'käm,paùnd 'mag·nət }

compound twins [CRYSTAL] Individuals of one mineral group united in accordance with two or more different twin laws. { 'käm,paùnd 'twinz }

compressed straw slab See strawboard. { kəm ¦prest ¦strò 'slab }

compressibility [MECH] The property of a substance capable of being reduced in volume by application of pressure; quantitively, the reciprocal of the bulk modulus. { kəm,pres·ə'bil·əd·ē }

compressibility factor [THERMO] The product of the pressure and the volume of a gas, divided by the product of the temperature of the gas and the gas constant; this factor may be inserted in the ideal gas law to take into account the departure of true gases from ideal gas behavior. Also known as deviation factor; gas-deviation factor; su-

percompressibility factor. { kəm,pres·ə'bil·əd·ē ,fak·tər }

compression [MECH] Reduction in the volume of a substance due to pressure; for example in building, the type of stress which causes shortening of the fibers of a wooden member. { kəm'presh·ən }

compression modulus See bulk modulus of elasticity. { kəm'presh·ən ,mäj·ə·ləs }

compression ratio [MET] Ratio of the volume of loose metal powder to the volume of the compact made from it. { kəm'presh·ən ,rā·shō }

compression strength [MECH] Property of a material to resist rupture under compression. { kəm'presh·ən ,streŋkth }

compression test [ENG] A test to determine compression strength, usually applied to materials of high compression but low tensile strength, in which the specimen is subjected to increasing compressive forces until failure occurs. { kəm'presh·ən ,test }

compressive member [CIV ENG] A structural member subject to tension. { kəm'pres·iv 'mem·bər }

compressive strength [MECH] The maximum compressive stress a material can withstand without failure. { kəm'pres·iv 'streŋkth }

compressive stress [MECH] A stress which causes an elastic body to shorten in the direction of the applied force. { kəm'pres·iv 'stres }

concave fillet weld [MET] A fillet weld having a concave surface. { 'kän,kāv 'fil·ət ,weld }

concentrate [CHEM] To increase the amount of a dissolved substance by evaporation. { 'kän·sən ,trāt }

concentrated load [MECH] A force that is negligible because of a small contact area; a beam supported on a girder represents a concentrated load on the girder. { 'kän·sən,träd·əd 'lōd }

concentration [CHEM] In solutions, the mass, volume, or number of moles of solute present in proportion to the amount of solvent or total solution. { ,kän·sən'trā·shən }

concentration cell [PHYS CHEM] **1.** Electrochemical cell for potentiometric measurement of ionic concentrations where the electrode potential electromotive force produced is determined as the difference in emf between a known cell (concentration) and the unknown cell. **2.** An electrolytic cell in which the electromotive force is due to a difference in electrolyte concentrations at the anode and the cathode. { ,kän·sən'trā·shən ,sel }

concentration polarization [PHYS CHEM] That part of the polarization of an electrolytic cell resulting from changes in the electrolyte concentration due to the passage of current through the solution. { ,kän·sən'trā·shən ,pō·lə·rə'zā·shən }

concrete [MATER] A mixture of aggregate, water, and a binder, usually portland cement; hardens to stonelike condition when dry. { 'kän ,krēt }

concrete beam [CIV ENG] A structural member of reinforced concrete, placed horizontally over openings to carry loads. { 'käŋ,krēt 'bēm }

concrete block [MATER] A solid or hollow block of precast concrete. { 'käŋ,krēt 'bläk }

concrete bridge [CIV ENG] A bridge constructed of prestressed or reinforced concrete. { 'käŋ ,krēt 'brij }

concrete finish [MATER] The texture or smoothness on the surface of hardened concrete. { ¦käŋ ,krēt ¦fin·ish }

concrete form oil [MATER] Nonviscous, neutral mineral oil used on wooden or metal forms to allow easy removal from set concrete. { 'käŋ ,krēt ,fórm ,óil }

concrete hardener [MATER] An admixture such as calcium chloride, sodium chloride, or sodium hydroxide that hastens or decreases the hydration rate of cementing material; the concrete takes less time to set and has earlier higher strength. { 'käŋ,krēt ,härd·ən·ər }

concrete masonry [MATER] Building units composed of block, brick, or tile laid by masons. { 'käŋ,krēt 'mās·ən·rē }

concrete retarder [MATER] A material added to concrete that decreases the hydration rate of cement, thereby increasing the setting time and decreasing the strengthening rate during the early age. { 'käŋ,krēt ri'tärd·ər }

concrete steel [MET] Steel used in reinforced concrete, which should comply with standard specifications for prestressed concrete. { 'käŋ ,krēt ,stēl }

concurrent heating [MET] Application of supplemental heat in metal cutting or welding. { 'kän'kər·ənt 'hēd·iŋ }

condensation [CHEM] Transformation from a gas to a liquid. [CRYO] See Bose-Einstein condensation. [ELEC] An increase of electric charge on a capacitor conductor. [MECH] An increase in density. [OPTICS] Focusing or collimation of light. { ,kän·dən'sā·shən }

condensation polymer [POLYM CHEM] A high-molecular-weight compound formed by condensation polymerization. { ,kän·dən'sā·shən 'päl·ə·mər }

condensation polymerization [POLYM CHEM] The stepwise reaction between functional groups of reactants in which a high-molecular-weight polymer is formed only after a large number of steps, for example, the reaction of dicarboxylic acids with diamines to form a polyamide. { ,kän·dən'sā·shən pə,lim·ə·rə 'zā·shən }

condensation resin [POLYM CHEM] A resin formed by polycondensation. { ,kän·dən 'sā·shən 'rez·ən }

condensed matter [PHYS] Matter in the liquid or solid state. { kən¦denst 'mad·ər }

condensed phase [PHYS CHEM] Either the solid or liquid phase of a material. { kən'denst ,fāz }

condenser [ELEC] See capacitor. [MECH ENG] A heat-transfer device that reduces a thermodynamic fluid from its vapor phase to its liquid phase, such as in a vapor-compression refrigeration plant or in a condensing steam power plant. [OPTICS] A system of lenses or mirrors in an optical projection system, which gathers as much of the light from the source as possible and directs it through the projection lens. { kən'den·sər }

conductance See thermal conductance. { kən' dək·təns }

conducting polymer [MATER] A plastic having high conductivity. { kən'dək·tiŋ 'päl·ə·mər }

conduction [ELEC] The passage of electric charge, which can occur by a variety of processes, such as the passage of electrons or ionized atoms. [PHYS] Transmission of energy by a medium which does not involve movement of the medium itself. Also known as electrical conduction. { kən'dək·shən }

conduction band [SOLID STATE] An energy band in which electrons can move freely in a solid, producing net transport of charge. { kən'dək·shən ,band }

conduction current [SOLID STATE] A current due to a flow of conduction electrons through a body. { kən'dək·shən ,kər·ənt }

conduction electron [SOLID STATE] An electron in the conduction band of a solid, where it is free to move under the influence of an electric field. Also known as outer-shell electron; valence electron. { kən'dək·shən i'lek,trän }

conductive coating [MATER] A coating used to reduce surface resistance and thus prevent the accumulation of static electric charges. { kən'dək·tiv 'kōd·iŋ }

conductive elastomer [MATER] A rubberlike silicone material in which suspended metal particles conduct electricity. { kən'dək·tiv i'las·tə·mər }

conductive rubber [MATER] Rubber that contains suspended carbon or silver spheres; its electrical resistance decreases when it is compressed, making it useful as a contact sensor. { kən'dək·tiv 'rəb·ər }

conductive silver paste [MATER] Silver powder in a suitable vehicle for applying to ceramic or other insulating materials by silk-screening or other methods, then fixing or firing at appropriate temperatures to provide a hard conductive surface or joint. { kən'dək·tiv 'sil·vər ,pāst }

conductivity [ELEC] The ratio of the electric current density to the electric field in a material. Also known as electrical conductivity; specific conductance. { ,kän,dək'tiv·əd·ē }

conductivity bridge [ELEC] A modified Kelvin bridge for measuring very low resistances. { ,kän,dək'tiv·əd·ē ,brij }

conductivity cell [ELEC] A glass vessel with two electrodes at a definite distance apart and filled with a solution whose conductivity is to be measured. { ,kän,dək'tiv·əd·ē ,sel }

conductor [ELEC] A wire, cable, or other body or medium that is suitable for carrying electric current. Also known as electric conductor. { kən'dək·tər }

cone [MET] The part of an oxygen gas flame adjacent to the orifice of the tip. { kōn }

configuration [CHEM] The three-dimensional spatial arrangement of atoms in a stable or isolable molecule. { kən,fig·yə'rā·shən }

conformability [MET] The ability of a metal that

has been cast as a bearing surface to deform plastically to compensate for irregularities in bearing assembly. { kən,fȯr·mə'bil·əd·ē }

conformation [ORG CHEM] In a molecule, a specific orientation of the atoms that varies from other possible orientations by rotation or rotations about single bonds; generally in mobile equilibrium with other conformations of the same structure. Also known as conformational isomer; conformer. { kän·fər'mā·shən }

conformational isomer See conformation. { kän·fər'mā·shən·əl 'ī·sə·mər }

conformer See conformation. { kən'fȯr·mər }

congener [CHEM] A chemical substance that is related to another substance, such as a derivative of a compound or an element belonging to the same family as another element in the periodic table. { 'kän·jə·nər }

conglomerate See racemate. { kən'gläm·ə·rət }

congruent evaporation [MET] In laser deposition, the tendency of the deposited film to have the same composition or stoichiometry as the material that serves as the target of the laser radiation. { kən'grü·ənt i,vap·ə'rā·shən }

congruent melting point [THERMO] A point on a temperature composition plot of a nonstoichiometric compound at which the one solid phase and one liquid phase are adjacent. { kən'grü·ənt 'melt·iŋ ,pȯint }

congruent transformation [MET] An isothermal or isobaric phase change in an alloy where the integrity of both phases is maintained throughout the process. { kən'grü·ənt ,tranz·fər'mā·shən }

conical helimagnet [SOLID STATE] A helimagnet in which the directions of atomic magnetic moments all make the same angle (greater than 0° and less than 90°) with a specified axis of the crystal, moments of atoms in successive basal planes are separated by equal azimuthal angles, and all moments have the same magnitude. { 'kän·ə·kəl 'hel·ə,mag·nət }

conjugate acid-base pair [CHEM] An acid and a base related by the ability of the acid to generate the base by loss of a proton. { 'kän·jə·gət ¦as·əd ¦bās 'per }

conjugated diene [ORG CHEM] A hydrocarbon with a molecular structure containing two carbon-carbon double bonds separated by a single bond. { 'kän·jə,gād·əd 'dī,ēn }

conjugated polyene [ORG CHEM] An acyclic hydrocarbon with a molecular structure containing alternating carbon-carbon double and single bonds. { 'kän·jə,gād·əd 'päl·ē,ēn }

conjugate fiber See bicomponent fiber. { 'kän·jə·gət 'fī·bər }

conservation law [PHYS] A law which states that some physical quantity associated with an isolated system is constant. { ,kän·sər'vā·shən ,lȯ }

conservation of charge [ELEC] A law which states that the total charge of an isolated system is constant; no violation of this law has been discovered. Also known as charge conservation. { ,kän·sər'vā·shən əv 'chärj }

conservation of energy [PHYS] The principle that energy cannot be created or destroyed. { ,kän·sər'vā·shən əv 'en·ər·jē }

conservation of mass [PHYS] The notion that mass can neither be created nor destroyed; it is violated by many microscopic phenomena. { ,kän·sər'vā·shən əv 'mas }

conservation of matter [PHYS] The notion that matter can be neither created nor destroyed; it is violated by microscopic phenomena. { ,kän·sər'vā·shən əv 'mad·ər }

conservative property [THERMO] A property of a system whose value remains constant during a series of events. { kən'sər·və·tiv 'präp·ərd·ē }

consistency [MATER] The degree of solidity or fluidity of a material such as grease, pulp, or slurry. { kən'sis·tən·sē }

consolute temperature [THERMO] The upper temperature of immiscibility for a two-component liquid system. Also known as upper consolute temperature; upper critical solution temperature. { 'kän·sə,lüt 'tem·prə·chər }

constantan [MET] An alloy containing 45% nickel and 55% copper, used to form iron-constantan and copper-constantan thermocouples. { kən 'stan·tən }

constitutional unit [CHEM] An atom or group of atoms that is part of a chain in a polymer or oligomer. { ,kän·stə'tü·shən·əl 'yü·nət }

constitution diagram [MET] Graphical representation of phase-stability relationships in an alloy system as a function of temperature. Also known as phase diagram. { ,kän·stə'tü·shən 'dī·ə,gram }

constitutive property [CHEM] Any physical or chemical property that depends on the constitution or structure of the molecule. { 'kän·stə ,tüd·iv 'präp·ərd·ē }

constricting nozzle [MET] A copper nozzle which has a constricting orifice and envelopes the electrode in plasma arc processes. { kən'strik·tiŋ ,näz·əl }

construction paper [MATER] A heavy paper made from mechanical pulp and available in a large range of colors, most of which are not lightfast. { kən'strək·shən ,pā·pər }

constructive interference [PHYS] Phenomenon in which the phases of waves arriving at a specified point over two or more paths of different lengths are such that the square of the resultant amplitude is greater than the sum of the squares of the component amplitudes. { kən'strək·div ,in·tər'fir·əns }

consumable electrode [MET] A metal electrode that supplies the filler for welding. { kən'süm·ə·bəl i'lek,trōd }

consumable guide electroslag welding [MET] A modification of the wire process of electroslag welding in which the filler metal is supplied by a welding wire held by a stationary metal tube. { kən'süm·ə·bəl 'gīd i'lek·trō,slag 'weld·iŋ }

consumable insert [MET] Filler metal which is located at the root of a welded joint and which becomes part of the final weldment. { kən'süm·ə·bəl 'in·sərt }

contact acid [INORG CHEM] Sulfuric acid

produced by the contact process. { 'kän,takt 'as·əd }

contact angle See angle of contact. { 'kän,takt ,aŋ·gəl }

contact corrosion See crevice corrosion; galvanic corrosion. { 'kän,takt kə'rō·zhən }

contact electromotive force See contact potential difference. { 'kän,takt i¦lek·trə'mōd·iv 'fōrs }

contact material [MET] A metal having high electrical and thermal conductivity, low contact resistance, minimum sticking or welding tendencies, and high corrosion resistance. { 'kän,takt mə'tir·ē·əl }

contact potential See contact potential difference. { 'kän,takt pə'ten·chəl }

contact potential difference [ELEC] The potential difference that exists across the space between two electrically connected materials. Also known as contact electromotive force; contact potential; Volta effect. { 'kän,takt pə'ten·chəl 'dif·rəns }

contact-pressure resin See contact resin. { 'kän,takt ,presh·ər ,rez·ən }

contact resin [MATER] A liquid resin which thickens or polymerizes on heating and, when used for bonding laminates, requires little or no pressure for adherence. Also known as contact-pressure resin. { 'kän,takt ,rez·ən }

contact resistance [ELEC] The resistance in ohms between the contacts of a relay, switch, or other device when the contacts are touching each other. { 'kän,takt ri'zis·təns }

contact tube [MET] A device which provides electric current to a continuous electrode in a welding process. { 'kän,takt ,tüb }

contact twin [CRYSTAL] Twinned crystals whose members are symmetrically arranged about a twin plane. { 'kän,takt ,twin }

continuity bond [MET] A metallic connection that provides continuous electrical contact between metal structures. { ,känt·ən'ü·əd·ē ,bänd }

continuity of state [THERMO] Property of a transition between two states of matter, as between gas and liquid, during which there are no abrupt changes in physical properties. { ,känt·ən'ü·əd·ē əv 'stāt }

continuous casting [MET] A technique in which an ingot, billet, tube, or other shape is continuously solidified and withdrawn while it is being poured, so that its length is not determined by mold dimensions. { kən¦tin·yə·wəs 'kast·iŋ }

continuous furnace [MET] A type of reheating furnace in which the charge introduced at one end moves continuously through the furnace and is discharged at the other end. { kən¦tin·yə·wəs 'fər·nəs }

continuous mill [MET] A rolling mill in which metal is successively rolled thinner as it passes through a series of synchronized rolls in tandem. { kən¦tin·yə·wəs 'mil }

continuous phase [CHEM] The liquid in a disperse system in which solids are suspended or droplets of another liquid are dispersed. Also known as dispersion medium, external phase. [MET] The matrix or background phase of a multiphasic alloy. { kən¦tin·yə·wəs 'fāz }

continuous precipitation [MET] Precipitation that is characteristic of certain alloys, from a supersaturated solid solution, involving a gradual change of the lattice parameter of the matrix with aging time. { kən¦tin·yə·wəs prə ,sip·ə'tā·shən }

continuous sequence [MET] A welding sequence in which each succeeding pass is made longitudinally along the entire length of the joint. { kən¦tin·yə·wəs 'sē·kwəns }

continuous sintering [MET] Sintering process in which materials are moved through the furnace at a fixed rate without interruption. { kən ¦tin·yə·wəs 'sin·tə·riŋ }

continuous weld [MET] A weld that is continuous along the entire length of the joint. { kən ¦tin·yə·wəs 'weld }

contour forming [MET] Shaping a sheet of metal onto a shaped die. { 'kän,túr ,fórm·iŋ }

contour milling [MET] Milling of an irregular surface. { 'kän,túr ,mil·iŋ }

contraction [MECH] The action or process of becoming smaller or pressed together, as a gas on cooling. { kən'trak·shən }

contraction coefficient See coefficient of contraction. { kən'trak·shən ,kō·i'fish·ənt }

contraction crack [ENG] A crack resulting from restriction of metal in a mold while contracting. { kən'trak·shən ,krak }

contraction rule [MET] A measuring rule having larger divisions than standard measures to allow for shrinkage of a metal casting. Also known as shrinkage rule; shrink rule. { kən'trak·shən ,rül }

contrast sensitivity See threshold contrast. { 'kän,trast sen·sə'tiv·əd·ē }

contrast threshold See threshold contrast. { 'kän,trast ,thresh,hōld }

controlled cooling [MET] Process by which an object is cooled from an elevated temperature in a predetermined manner to avoid cracking, internal damage, or hardening, or to produce a desired microstructure. { kən'trōld 'kül·iŋ }

convection [FL MECH] Diffusion in which the fluid as a whole is moving in the direction of diffusion. Also known as bulk flow. [PHYS] Transmission of energy or mass by a medium involving movement of the medium itself. { kən'vek·shən }

convection coefficient See film coefficient. { kən'vek·shən ,kō·i'fish·ənt }

convection modulus [FL MECH] An intrinsic property of a fluid which is important in determining the Nusselt number, equal to the acceleration of gravity times the volume coefficient of thermal expansion divided by the product of the kinematic viscosity and the thermal diffusivity. { kən'vek·shən ,mäj·ə·ləs }

convection oven [ENG] An oven containing a fan that continuously circulates hot air. { kən'vek·shən ,əv·ən }

conventional milling |MET| Milling in which the cutter and feed move in opposite directions from the point of contact. { kən'ven·chən·əl 'mil·iŋ }

conventional spinning See manual spinning. { kən'ven·chən·əl 'spin·iŋ }

conversion coating |MET| A metal-surface coating consisting of a compound of the base metal. { kən'vər·zhən ,kōd·iŋ }

converter |MET| A type of furnace in which impurities are oxidized out by blowing air through or across a path of molten metal or matte. { kən'vərd·ər }

Conwell-Weisskopf equation |SOLID STATE| An equation for the mobility of electrons in a semiconductor in the presence of donor or acceptor impurities, in terms of the dielectric constant of the medium, the temperature, the concentration of ionized donors (or acceptors), and the average distance between them. { ¦kän ,wel ¦vīs,kȯpf i'kwā·zhən }

coolant |MATER| **1.** A cutting fluid for machine operations, which keeps the tool cool to prevent reduction in hardness and resistance to abrasion, and prevents distortion of the work. **2.** A substance, ordinarily fluid, used for cooling any part of a reactor in which heat is generated. **3.** In general, any cooling agent, usually a fluid. { 'kül·ənt }

cooling correction |THERMO| A correction that must be employed in calorimetry to allow for heat transfer between a body and its surroundings. Also known as radiation correction. { 'kül·iŋ kə'rek·shən }

cooling curve |THERMO| A curve obtained by plotting time against temperature for a solid-liquid mixture cooling under constant conditions. { 'kül·iŋ ,kərv }

cooling method |THERMO| A method of determining the specific heat of a liquid in which the times taken by the liquid and an equal volume of water in an identical vessel to cool through the same range of temperature are compared. { 'kül·iŋ ,meth·əd }

cooling stress |MECH| Stress resulting from uneven contraction during cooling of metals and ceramics due to uneven temperature distribution. { 'kül·iŋ ,stres }

cool time |MET| The period of time between successive heat times in pulsation and seam welding. { 'kül ,tīm }

cooperative phenomenon |SOLID STATE| A process that involves a simultaneous collective interaction among many atoms or electrons in a crystal, such as ferromagnetism, superconductivity, and order-disorder transformations. { kō'äp·rəd·iv fə'näm·ə,nän }

Cooper pairs |SOLID STATE| Pairs of bound electrons which occur in a superconducting medium according to the Bardeen-Cooper-Schrieffer theory. { 'kü·pər ,perz }

coordinated complex See coordination compound. { kō'ȯrd·ən,ād·əd 'käm,pleks }

coordination chemistry |CHEM| The chemistry of metal ions in their interactions with other molecules or ions. { kō,ȯrd·ən'ā·shən 'kem·ə·strē }

coordination compound |CHEM| A compound with a central atom or ion and a group of ions or molecules surrounding it. Also known as coordinated complex; Werner complex. { kō,ȯrd·ən'ā·shən ,käm,paúnd }

coordination lattice |CRYSTAL| The crystal structure of a coordination compound. { kō ,ȯrd·ən'ā·shən ,lad·əs }

coordination polymer |POLYM CHEM| Organic addition polymer that is neither free-radical nor simply ionic; prepared by catalysts that combine an organometallic (for example, triethyl aluminum) and a transition metal compound (for example, $TiCl_4$). { kō,ȯrd·ən'ā·shən ,päl·ə·mər }

copal |MATER| Hard, resinous substance exuded from certain trees in the East Indies, South America, and Africa and used in varnish and printing ink. { 'kō·pəl }

cope |MET| The upper portion of a flask, mold, or pattern. { kōp }

coping brick |MATER| A brick with a special shape that is used to cap the exposed top of a wall. { 'kōp·iŋ ,brik }

coplanar forces |MECH| Forces that act in a single plane; thus the forces are parallel to the plane and their points of application are in the plane. { kō'plān·ər ,fȯrs·əz }

copolymer |POLYM CHEM| A mixed polymer, the product of polymerization of two or more substances at the same time. { kō'päl·i·mər }

copolymerization |POLYM CHEM| A polymerization reaction that forms a copolymer. { ,kō·pə ,lim·ə·rə'zā·shən }

copper |CHEM| A chemical element, symbol Cu, atomic number 29, atomic weight 63.546. |MET| One of the most important nonferrous metals; a ductile and malleable metal found in various ores and used in industry, engineering, and the arts in both pure and alloyed form. { 'käp·ər }

copper alloy |MET| A solid solution of one or more metals in copper. { 'käp·ər 'al,ȯi }

copper amalgam |MET| An alloy of copper and mercury. { 'käp·ər ə'mal·gəm }

copper arsenate |INORG CHEM| $Cu_3(AsO_4)_2 \cdot 4H_2O$ or $Cu_5H_2(AsO_4)_4 \cdot 2H_2O$ Bluish powder, soluble in ammonium hydroxide and dilute acids, insoluble in water and alcohol; used as a fungicide and insecticide. { 'käp·ər 'ärs·ən ,āt }

copper arsenite |INORG CHEM| $CuHAsO_3$ A toxic, light green powder which is soluble in acids and decomposes at the melting point; used as a pigment and insecticide. Also known as copper orthoarsenite; cupric arsenite; Scheele's green. { 'käp·ər 'ärs·ən,īt }

copperas See ferrous sulfate. { 'käp·ə·rəs }

copper blue See mountain blue. { 'käp·ər ¦blü }

copper brazing |MET| Brazing by using copper as the filler metal. { 'käp·ər ,brāz·iŋ }

copper bromide See cupric bromide; cuprous bromide. { 'käp·ər 'brō,mīd }

copper carbonate |INORG CHEM| $Cu_2(OH)_2 \cdot CO_3$ A toxic, green powder; decomposes at

200°C and is soluble in acids; used in pigments and pyrotechnics and as a fungicide and feed additive. Also known as artificial malachite; cupric carbonate; mineral green. { 'käp·ər 'kär·bə‚nāt }

copper chloride *See* cupric chloride;cuprous chloride. { 'käp·ər 'klȯr‚īd }

copper chromate *See* cupric chromate. { 'käp·ər 'krō‚māt }

copper converter |MET| A converter for purifying copper. { 'käp·ər kən'vərd·ər }

copper cyanide *See* cupric cyanide. { 'käp·ər sī·ə‚nīd }

copper fluoride *See* cupric fluoride;cuprous fluoride. { 'käp·ər 'flur‚īd }

copper hydroxide *See* cupric hydroxide. { 'käp·ər hī'dräk‚sīd }

copper nitrite *See* cupric nitrate. { 'käp·ər 'nī‚trīt }

copper orthoarsenite *See* copper arsenite. { 'käp·ər ‚ȯr·thō'ärs·ən‚īt }

copper oxide *See* cupric oxide;cuprous oxide. { 'käp·ər 'äk‚sīd }

copper plating |MET| Coating of a substance with copper by an electrolytic process, to minimize corrosion. { 'käp·ər ‚plād·iŋ }

copper powder |MET| A bronzing powder made by saturating nitrous acid with copper and precipitating the latter by the addition of iron. { 'käp·ər 'paud·ər }

copper steel |MET| Low-carbon steel containing up to 0.25% copper. { 'käp·ər 'stēl }

copper sulfate *See* cupric sulfate. { 'käp·ər 'səl‚fāt }

copper sulfide |INORG CHEM| CuS Black, monoclinic or hexagonal crystals that break down at 220°C; used in paints on ship bottoms to prevent fouling. { 'käp·ər 'səl‚fīd }

copperweld |MET| Copper-covered steel, used as a conductor for high-voltage transmission spans where tensile strength is more important than high conductance. { 'käp·ər‚weld }

copper wire |MET| Wire commonly made from copper by drawing from a hot-rolled rod without annealing; however, the smaller sizes may involve intermediate anneals. { 'käp·ər 'wīr }

cord |MATER| **1.** A unit of measure for wood stacked for fuel or pulp; equals 4 × 4 × 8, or 128 cubic feet (approximately 3.6246 cubic meters). **2.** A long, flexible, cylindrical construction of natural or synthetic fibers twisted or woven together. **3.** Strands of material forming the plies in a motor vehicle tire. { kȯrd }

cordite |MATER| A trinitrate cellulose derivative prepared by treating cotton fiber or purified wood pulp with a mixture of nitric and sulfuric acids; an explosive powder. { 'kȯr‚dīt }

core |MATER| The center layers of a sheet of plywood. |MET| A specially formed part of a mold used to form internal holes in a casting. { kȯr }

core binder |MATER| A substance for binding core sand together; an example is core oil. { 'kȯr ‚bīnd·ər }

core blower |MET| A machine using compressed air to blow and pack sand into a core box. { 'kȯr ‚blō·ər }

core box |MET| A container for shaping a sand core for a casting. { 'kȯr ‚bäks }

cored bar |MET| A bar-shaped powder compact which has been heated by electricity to melt the interior. { ¦kȯrd ¦bär }

cored electrode |MET| An electrode made of metal with a core of flux or other material. { ¦kȯrd i'lek‚trōd }

core drier |MET| A light, skeleton cast-iron or aluminum box, whose internal shape conforms closely to the cope portion of a core for molding, used to support, during baking, a core which cannot be placed on a flat plate. { 'kȯr ‚drī·ər }

cored solder |MET| Soldering wire which has a core consisting of flux. { ¦kȯrd 'säd·ər }

core iron |MET| A grade of soft iron suitable for cores of chokes, transformers, and relays. { 'kȯr ‚ī·ərn }

core molding |MET| A molding process which makes use of assembled cores to construct the mold. { 'kȯr ‚mōld·iŋ }

core oil |MATER| An oil compound used with sand to make foundry core. { 'kȯr ‚ȯil }

core oven |MET| An oven used for baking cores for molding; the walls are constructed of inner and outer layers of sheet metal separated by rock wool or fiber glass insulation, with interlocked joints. { 'kȯr ‚əv·ən }

core print |MET| A projection on a cylindrical casting pattern which supports a core. { 'kȯr ‚print }

core rod |MET| The part of a die used to make a hole in a compact. { 'kȯr ‚räd }

core sand |MATER| Sand used in a core for molding, made from standard molding-sand mixtures or from silica sand, usually with a binder. Also known as foundry core sand. { 'kȯr ‚sand }

core wash |MATER| A suspension of fine clay or graphite that is applied to a core in a metal casting to improve that portion's cast surface. { 'kȯr ‚wäsh }

core wire |MET| Copper wire having a steel core, often used for antennas. { 'kȯr ‚wīr }

coring |MET| A variable composition of individual crystals across a casting, due to nonequilibrium growth over a range of temperature; the purest material is near the center. { 'kȯr·iŋ }

corkboard |MATER| Board made of compressed cork. { 'kȯrk‚bȯrd }

cork paint |MATER| A paint containing fine cork particles; used on steel parts on ships to prevent sweating. { ¦kȯrk ¦pānt }

cork tile |MATER| Floor tile made of compressed cork bound with phenolic or other resin binders; used on moisture-free rigid subfloors or on plywood or hardboard. { ¦kȯrk ¦tīl }

corona |ELEC| *See* corona discharge. |MET| An area sometimes surrounding the nugget at the faying surfaces of a spot weld which provides a degree of bond strength. { kə'rō·nə }

corona discharge |ELEC| A discharge of electricity appearing as a bluish-purple glow on the surface of and adjacent to a conductor when the voltage gradient exceeds a certain critical value; due to ionization of the surrounding air by the high voltage. Also known as aurora; corona; electric corona.   { kə'rō·nə 'dis,chärj }

coronizing   [MET] Process to electroplate zinc on nickel, thermally treated at 375°C; coating is used on ferrous and copper-based alloys to give resistance to sulfur dioxide, SO_2, and sulfur trioxide, SO_3.   { 'kȯr·ə,nīz·iŋ }

corpuscular theory of light   [OPTICS] Theory that light consists of a stream of particles; now considered a limiting case of the quantum theory. Also known as Newton's theory of light.   { kȯr'pəs·kyə·lər ,thē·ə·rē əv 'līt }

correlation energy   [SOLID STATE] The modification of the Coulomb energy of a crystal that results from the tendency of electrons to stay apart from each other.   { ,kär·ə'lā·shən ,en·ər·jē }

corroding lead   [MET] Lead that can be corroded to make white lead.   { kə'rōd·iŋ ,led }

corrosion   [MET] Gradual destruction of a metal or alloy due to chemical processes such as oxidation or the action of a chemical agent.   { kə'rō·zhən }

corrosion cell   [MET] A condition on a metal surface in which a flow of electric current occurs between the metal surface and an electrolyte with which it is in contact sufficient to cause the metal to degrade.   { kə'rō·zhən ,sel }

corrosion control   See corrosion protection.   { kə'rō·zhən kən,trōl }

corrosion fatigue   [MET] Damage to or failure of a metal due to corrosion combined with fluctuating fatigue stresses.   { kə'rō·zhən fə'tēg }

corrosion fatigue limit   [MET] The maximum stress that a corroded material can withstand for a given number of stress reversals.   { kə'rō·zhən fə'tēg 'lim·ət }

corrosion inhibitor   [PHYS CHEM] A compound or material deposited as a film on a metal surface that either provides physical protection against corrosive attack or reduces the open-circuit potential difference between local anodes and cathodes and increases the polarization of the former.   { kə'rōzh·ən in,hib·əd·ər }

corrosion number   See acid number.   { kə'rō·zhən ,nəm·bər }

corrosion potential   [MET] The measure of corroding surface potential in an electrolyte in relation to a reference electrode while the circuit is open.   { kə'rō·zhən pə'ten·chəl }

corrosion protection   [MET] The minimization of corrosion by coating with a protective metal, with an oxide or phosphide or similar substance, or with a protective paint, or by rendering the metal passive. Also known as corrosion control.   { kə'rō·zhən prə'tek·shən }

corrosion test   [MET] Any of various tests to determine the resistance of a metal to chemical attack.   { kə'rō·zhən ,test }

corrosive   [MATER] A substance that causes corrosion.   { kə'rō·siv }

corrosive flux   [MET] A soldering flux, usually composed of inorganic salts and acids, which provides oxide removal of the base metal upon application of solder; flux remaining on the base metal is corrosive and should be removed.   { kə'rō·siv 'fləks }

corrosiveness   [MET] The tendency of a metal to wear away another by chemical attack.   { kə'rō·siv·nəs }

corrosive sublimate   See mercuric chloride.   { kə'rō·siv 'səb·lə,māt }

Cosslett process   [MET] A process in which iron or steel articles immersed for 3 or 4 hours in a boiling solution, made by mixing iron filings with concentrated phosphoric acid, H_3PO_4 (sufficient to form a paste), and then adding to weak phosphoric acid, become coated with a rust-resisting deposit of basic ferrous phosphate.   { 'käs·lət ,präs·əs }

cotectic   [PHYS CHEM] Referring to conditions of pressure, temperature, and composition under which two or more solid phases crystallize at the same time, with no resorption, from a single liquid over a finite range of decreasing temperature.   { kō'tek·tik }

cotectic crystallization   [PHYS CHEM] Simultaneous crystallization of two or more solid phases from a single liquid over a finite range of falling temperature without resorption.   { kō'tek·tik ,krist·əl·ə'zā·shən }

Cotton-Mouton birefringence   See Cotton-Mouton effect.   { ¦kät·ən ¦mü·ton ,bī·ri'frin·jəns }

Cotton-Mouton constant   [OPTICS] A constant giving the strength of the Cotton-Mouton effect in a liquid; when multiplied by the path length and the square of the magnetic field, it gives the phase difference between the components of light parallel and perpendicular to the field.   { ¦kät·ən ¦mü·ton ,kän·stənt }

Cotton-Mouton effect   [OPTICS] The double refraction (birefringence) of light in a liquid in a magnetic field at right angles to the direction of light propagation. Also known as Cotton-Mouton birefringence.   { ¦kät·ən ¦mü·ton i¦fekt }

cotton oil   [MATER] The yellow, viscous fixed oil, containing principally linoleic acid, pressed from the seeds of various *Gossypium* species; the refined oil is colorless and used in foods and some pharmaceutical preparations. Also known as cottonseed oil; oleum gossypii seminis.   { 'kät·ən ,ȯil }

cottonseed oil   See cotton oil.   { 'kät·ən,sēd ,ȯil }

Cottrell atmosphere   [SOLID STATE] A cluster of impurity atoms surrounding a dislocation in a crystal.   { kä¦trel 'at·mə,sfir }

Cottrell hardening   [SOLID STATE] Hardening of a material caused by locking of its dislocations when impurity atoms whose size differs from that of the solvent cluster around them.   { 'kä·trəl ,härd·ən·iŋ }

coul   See coulomb.

coulomb   [ELEC] A unit of electric charge, defined as the amount of electric charge that crosses a surface in 1 second when a steady current of 1 absolute ampere is flowing across the surface; this is the absolute coulomb and has been the legal standard of quantity of electricity since 1950; the previous standard was the international coulomb, equal to 0.999835 absolute coulomb. Abbreviated coul. Symbolized C. { 'kü,läm }

Coulomb attraction |ELEC| The electrostatic force of attraction exerted by one charged particle on another charged particle of opposite sign. Also known as electrostatic attraction. { 'kü,läm ə'trak·shən }

Coulomb crystal |CRYO| A structure formed by electrons trapped at a liquid helium surface at sufficiently high electron densities and low temperatures, in which the electrons occupy the points of a two-dimensional hexagonal lattice. { 'kü,läm ,krist·əl }

Coulomb energy |PHYS| The part of the binding energy of a system of particles, such as an atomic nucleus of a solid, which is associated with electrostatic forces between the particles. |PHYS CHEM| The energy associated with the electrostatic interaction between two or more electron distributions in terms of which the actual electron distribution of a covalent bond is described. { 'kü,läm ,en·ər·jē }

Coulomb field |ELEC| The electric field created by a stationary charged particle. { 'kü,läm ,fēld }

Coulomb force |ELEC| The electrostatic force of attraction or repulsion exerted by one charged particle on another, in accordance with Coulomb's law. { 'kü,läm ,fórs }

Coulomb friction |MECH| Friction occurring between dry surfaces. { 'kü,läm ,frik·shən }

Coulomb interactions |ELEC| Interactions of charged particles associated with the Coulomb forces they exert on one another. Also known as electrostatic interactions. { 'kü,läm in·tər'ak·shənz }

Coulomb potential |ELEC| A scalar point function equal to the work per unit charge done against the Coulomb force in transferring a particle bearing an infinitesimal positive charge from infinity to a point in the field of a specific charge distribution. { kü'läm pə'ten·chəl }

Coulomb repulsion |ELEC| The electrostatic force of repulsion exerted by one charged particle on another charged particle of the same sign. Also known as electrostatic repulsion. { kü'läm ri'pəl·shən }

Coulomb's law |ELEC| The law that the attraction or repulsion between two electric charges acts along the line between them, is proportional to the product of their magnitudes, and is inversely proportional to the square of the distance between them. Also known as law of electrostatic attraction. { 'kü'lämz ,lö }

Coulomb's theorem |ELEC| The proposition that the intensity of an electric field near the surface of a conductor is equal to the surface charge density on the nearby conductor surface divided by the absolute permittivity of the surrounding medium. { kü'läm ,thir·əm }

coulometer |PHYS CHEM| An electrolytic cell for the precise measurement of electrical quantities or current intensity by quantitative determination of chemical substances produced or consumed. Also known as voltameter. { kü'läm·əd·ər }

coumarone |ORG CHEM| C_8H_6O A colorless liquid, boiling point 169°C. { 'kü·mə,rōn }

coumarone-indene resin |POLYM CHEM| A synthetic resin prepared by polymerization of coumarone and indene. { 'kü·mə,rōn 'in,dēn ,rez·ən }

counterion |PHYS CHEM| In a solution, an ion with a charge opposite to that of another ion included in the ionic makeup of the solution. { 'kaúnt·ər,ī,än }

couple |CHEM| Joining of two molecules. |ELECTR| Two metals placed in contact, as in a thermocouple. { 'kəp·əl }

coupled oscillators |MECH| A set of particles subject to elastic restoring forces and also to elastic interactions with each other. { 'kəp·əld 'äs·ə,lād·ərz }

coupled systems |PHYS| Mechanical, electrical, or other systems which are connected in such a way that they interact and exchange energy with each other. { 'kəp·əld 'sis·təmz }

covalence |CHEM| The number of covalent bonds which an atom can form. { kō'vā·ləns }

covalent bond |CHEM| A bond in which each atom of a bound pair contributes one electron to form a pair of electrons. Also known as electron pair bond. { kō'vā·lənt 'bänd }

covalent crystal |CRYSTAL| A crystal held together by covalent bonds. Also known as valence crystal. { kō'vā·lənt 'krist·əl }

covalent hydride |INORG CHEM| A compound formed from a nonmetal and hydrogen, for example, H_2S and NH_3. { kō'vā·lənt 'hī,drīd }

covalent radius See atomic radius. { kō'vā·lənt 'rād·ē·əs }

covered electrode See coated electrode. { 'kəv·ərd i'lek,trōd }

cover half |MET| The stationary portion of a die. { 'kəv·ər ,haf }

covering power |ENG| The degree to which a coating obscures the underlying material. |MET| The ability of an electroplating bath to produce a coating at a low current density. { 'kəv·riŋ ,paú·ər }

cp See candlepower; centipoise.

Cr See chromium.

crack |CHEM| To break a compound into simpler molecules. |ENG| To open something slightly, for instance, a valve. { krak }

cracking |CHEM ENG| A process that is used to reduce the molecular weight of hydrocarbons by breaking the molecular bonds by various thermal, catalytic, or hydrocracking methods. |ENG| Presence of relatively large cracks extending into the interior of a structure, usually produced by overstressing the structural material. { 'krak·iŋ }

crater |MET| A depression at the end of the weld head or under the electrode during welding. { 'krād·ər }

crawling |MATER| **1.** Separation and contraction of the glaze on the surface of a ceramic object during drying or firing so that unglazed areas

crazing

result. **2.** In a film of wet paint, redistribution of the paint so that it is no longer evenly spread, usually due to imperfect bonding of the paint with the surface. { 'kròl·iŋ }

crazing |ENG| A network of fine cracks on or under the surface of a material such as enamel, glaze, metal, or plastic. |MET| Development of a network of cracks on a metal surface. { 'krāz·iŋ }

creep |ELECTR| A slow change in a characteristic with time or usage. |MECH| A time-dependent strain of solids caused by stress. { krēp }

creepage |ELEC| The conduction of electricity across the surface of a dielectric. { 'krē·pij }

creep buckling |MECH| Buckling that may occur when a compressive load is maintained on a member over a long period, leading to creep which eventually reduces the member's bending stiffness. { 'krēp ,bək·liŋ }

creep limit |MECH| The maximum stress a given material can withstand in a given time without exceeding a specified quantity of creep. { 'krēp ,lim·ət }

creep recovery |MECH| Strain developed in a period of time after release of load in a creep test. { 'krēp ri'kəv·ə·rē }

creep rupture strength |MECH| The stress which, at a given temperature, will cause a material to rupture in a given time. { 'krēp 'rəp·chər ,streŋkth }

creep strength |MECH| The stress which, at a given temperature, will result in a creep rate of 1% deformation within 100,000 hours. { 'krēp ,streŋkth }

creep test |ENG| Any one of a number of methods of measuring creep, for example, by subjecting a material to a constant stress or deforming it at a constant rate. { 'krēp ,test }

creosol |ORG CHEM| $CH_3O(CH_3)C_6H_3OH$ A combination of isomers, derived from coal tar or petroleum; a yellowish liquid with a phenolic odor; used as a disinfectant, in the manufacture of resins, and in flotation of ore. { 'krē·ə,sōl }

creosote |MATER| A colorless or yellowish oily liquid containing a mixture of phenolic compounds obtained by distillation of tar; commercial creosote is distilled from coal tar, and pharmaceutical creosote is distilled from wood tar. { 'krē·ə,sōt }

creosote oil |MATER| A coal tar fraction, boiling between 240 and 270°F (116–132°C); used for producing materials such as creosote and tar acids and used directly as a germicide, insecticide, or pesticide. { 'krē·ə,sōt ,oil }

crepe paper |MATER| A lightweight, crinkled paper available in many colors and used for displays, floats, and decorations; it has no strength when wet, and the colors run. { 'krāp ,pā·pər }

cresol |ORG CHEM| $CH_3C_6H_4OH$ One of three poisonous, colorless isomeric methyl phenols: o-cresol, m-cresol, p-cresol; used in the production of phenolic resins, tricresyl phosphate, disinfectants, and solvents. { 'krē,sōl }

cressing See swaging. { 'kres·iŋ }

cresylic acid |MATER| **1.** A mixture of phenols containing varying amounts of xylenols, cresols, and other high-boiling fractions. **2.** A crude mixture of the three cresol isomers. { krə'sil·ik 'as·əd }

crevice corrosion |MET| Corrosive degradation of metal parts at the crevices left at rolled joints or from other forming procedures; common in stainless steel heat exchangers in contact with chloride-containing fluids or other dissolved corrosives. Also known as contact corrosion. { 'krev·əs kə'rō·zhən }

crinkling See wrinkling. { 'kriŋk·liŋ }

critical area See picture element. { 'krid·ə·kəl 'er·ē·ə }

critical condensation temperature |PHYS CHEM| The temperature at which the sublimand of a sublimed solid recondenses; used to analyze solid mixtures, analogous to liquid distillation. { 'krid·ə·kəl ,kän·dən'sā·shən ,tem·prə·chər }

critical cooling rate |MET| The minimum cooling rate that will suppress undesired transformations in a metal. { 'krid·ə·kəl 'kül·iŋ ,rāt }

critical current |SOLID STATE| The current in a superconductive material above which the material is normal and below which the material is superconducting, at a specified temperature and in the absence of external magnetic fields. { 'krid·ə·kəl 'kər·ənt }

critical current density |PHYS CHEM| The amount of current per unit area of electrode at which an abrupt change occurs in a variable of an electrolytic process. { 'krid·ə·kəl 'kər·ənt ,den·səd·ē }

critical density |CHEM| The density of a substance exhibited at its critical temperature and critical pressure. |THERMO| The density of a substance at the liquid-vapor critical point. { 'krid·ə·kəl 'den·səd·ē }

critical exponent |THERMO| A parameter n that characterizes the temperature dependence of a thermodynamic property of a substance near its critical point; the temperature dependence has the form $|T - T_c|^n$, where T is the temperature and T_c is the critical temperature. { 'krid·ə·kəl ik'spō·nənt }

critical field |ELECTR| The smallest theoretical value of steady magnetic flux density that would prevent an electron emitted from the cathode of a magnetron at zero velocity from reaching the anode. Also known as cutoff field. |SOLID STATE| The magnetic field strength below which magnetic flux is excluded from a type I superconductor. Symbolized H_c. { 'krid·ə·kəl 'fēld }

critical humidity |CHEM ENG| The humidity of a system's atmosphere above which a crystal of a water-soluble salt will always become damp (absorb moisture from the atmosphere) and below which it will always stay dry (release moisture to the atmosphere). |MET| The atmospheric humidity above which the corrosion rate increases rapidly for a particular metal. { 'krid·ə·kəl yü'mid·əd·ē }

critical isotherm |THERMO| A curve showing the relationship between the pressure and volume of a gas at its critical temperature. { 'krid·ə·kəl 'ī·sə,thərm }

critical magnetic field [SOLID STATE] The field below which a superconductive material is superconducting and above which the material is normal, at a specified temperature and in the absence of current. { 'krid·ə·kəl mag'ned·ik 'fēld }

critical magnetic scattering [SOLID STATE] Intense scattering of low-energy neutrons by a ferromagnetic crystal at temperatures near the Curie point. { 'krid·ə·kəl mag'ned·ik 'skad·ər·iŋ }

critical micelle concentration [PHYS CHEM] The concentration of a micelle (oriented molecular arrangement of an electrically charged colloidal particle or ion) at which the rate of increase of electrical conductance with increase in concentration levels off or proceeds at a much slower rate. { 'krid·ə·kəl mi'sel ,kän·sən'trā·shən }

critical phenomena [PHYS CHEM] Physical properties of liquids and gases at the critical point (conditions at which two phases are just about to become one); for example, critical pressure is that needed to condense a gas at the critical temperature, and above the critical temperature the gas cannot be liquefied at any pressure. { 'krid·ə·kəl fə'näm·ə·nə }

critical point [PHYS CHEM] 1. The temperature and pressure at which two phases of a substance in equilibrium with each other become identical, forming one phase. 2. The temperature and pressure at which two ordinarily partially miscible liquids are consolute. { 'krid·ə·kəl 'pòint }

critical pressure [THERMO] The pressure of the liquid-vapor critical point. { 'krid·ə·kəl 'presh·ər }

critical properties [PHYS CHEM] Physical and thermodynamic properties of materials at conditions of critical temperature, pressure, and volume, that is, at the critical point. { 'krid·ə·kəl 'präp·ərd·ēz }

critical range [MET] The temperature range for the reversible change of austenite to ferrite, pearlite, and cementite. { 'krid·ə·kəl 'rānj }

critical shear stress [SOLID STATE] The shear stress needed to cause slip in a given direction along a given crystallographic plane of a single crystal. { 'krid·ə·kəl 'shir,stres }

critical speed See critical velocity. { 'krid·ə·kəl 'spēd }

critical state [PHYS CHEM] Unique condition of pressure, temperature, and composition wherein all properties of coexisting vapor and liquid become identical. { 'krid·ə·kəl 'stāt }

critical strain [MET] The strain at which heating causes rapid growth of large grains in many metals and alloys; phase transformations do not occur. { 'krid·ə·kəl 'strān }

critical temperature [PHYS CHEM] The temperature of the liquid-vapor critical point, that is, the temperature above which the substance has no liquid-vapor transition. Symbolized T_c. { 'krid·ə·kəl 'tem·prə·chər }

critical velocity [CRYO] The velocity of a super-

fluid in very narrow channels (on the order of 10^{-5} centimeter), which is nearly constant. Also known as critical speed. { 'krid·ə·kəl və'läs·əd·ē }

crocodiling See alligatoring. { 'kräk·ə,dīl·iŋ }

crocus [MATER] Finely powdered oxide of iron, of dark red color, used for buffing and polishing. { 'krō·kəs }

croning process [MET] A shell-molding process. { 'krōn·iŋ ,präs·əs }

Crookes glass [MATER] A type of glass that contains cerium and other rare earths and has a high absorption of ultraviolet radiation; used in sunglasses. { 'krúks ,glas }

crop [MET] Defective end portion of an ingot which is removed for scrap before rolling the ingot. { kräp }

crossband [MATER] In plywood comprising three or more plies, a layer of veneer whose grain direction is at right angles to that of the face plies. { 'kròs,band }

cross-country mill [MET] A rolling mill in which the tables of the mill stands are parallel with a crossover table that connects them; used to produce special forms of bar stock. { ¦kròs ¦kən·trē 'mil }

crosshead [MET] A device generally employed in wire coating which is attached to the discharge end of the extruder cylinder; designed to facilitate extruding material at an angle. { 'kròs ,hed }

cross-lamination [MATER] Construction of a laminated composite material so that some layers are oriented at right angles to the other layers with respect to the grain or the strongest direction in terms of tension. { ¦kròs lam·ə'nā·shən }

crosslink [POLYM CHEM] The covalent bonds between adjacent polymer chains that lock the chains in place. { 'kròs ,liŋk }

crosslinking [POLYM CHEM] The setting up of chemical links between the molecular chains of polymers. { 'kròs ,liŋk·iŋ }

cross-rolling [MET] 1. Straightening metal sheets by passing them through rolls at right angles to the principal direction of rolling. 2. Straightening round bars or tubes by passing the work through parallel to the axes of rolls. { 'kròs ,rōl·iŋ }

cross-wire weld [MET] A weld made across wires or bars in order to make wire mesh or other similar products. { 'kròs ,wīr ,weld }

Crowe process [MET] In cyanidation, after extraction of the gold, separation of the solution from the ore tailings by filtration or countercurrent decantation and by passage through a vacuum chamber where deaeration occurs. { 'krō ,präs·əs }

crown [MET] That part of the sheet or roll where the thickness or diameter increases from edge to center. { kraún }

crown ether [ORG CHEM] A macrocyclic polyether whose structure exhibits a conformation with a so-called hole capable of trapping cations by coordination with a lone pair of electrons on the oxygen atoms. { kraún ,ē·thər }

crown glass |MATER| A soda-lime glass, typically having 72% SiO_2, 13% CaO, and 15% Na_2O, which is hard and will take a simple polish; highly transparent for visible light. { 'kraún ‚glas }

CRT *See* cathode-ray tube.

crucible melt extraction |MET| Melt extraction in which the molten metal is contained in a crucible. { 'krü·sə·bəl 'melt ik'strak·shən }

crucible steel *See* drill steel. { 'krü·sə·bəl ‚stēl }

crude scale *See* scale wax. { 'krüd ‚skāl }

crude yellow scale |MATER| The trade name for a low-grade paraffin wax. { 'krüd 'yel·ō ‚skāl }

crush |MET| Casting defect caused by damage to the mold before pouring the metal. { krəsh }

crushed steel |MATER| An abrasive used in the stone, brick, glass, and metal trades, made by heating high-grade crucible steel to white heat, quenching in a bath of cold water, and crushing the fragments. { ¦krəsht 'stēl }

crushing strain |MECH| Compression which causes the failure of a material. { 'krəsh·iŋ ‚strān }

crushing strength |MECH| The compressive stress required to cause a solid to fail by fracture; in essence, it is the resistance of the solid to vertical pressure placed upon it. { 'krəsh·iŋ ‚streŋkth }

crushing test |ENG| A test of the suitability of stone that might be mined for roads or building use. |MET| A test to determine quality of tubing, especially welded tubing, by applying compression parallel to the axis. { 'krəsh·iŋ ‚test }

cryogen *See* cryogenic fluid. { 'krī·ə·jən }

cryogenic coil |CRYO| A high-purity coil refrigerated to very low temperatures to reduce effective coil resistivity. { ‚krī·ə'jen·ik 'kȯil }

cryogenic conductor *See* superconductor. { ‚krī·ə'jen·ik kən'dək·tər }

cryogenic device |CRYO| A device whose operation depends on superconductivity as produced by temperatures near absolute zero. Also known as superconducting device. { ‚krī·ə'jen·ik di'vīs }

cryogenic fluid |CRYO| A liquid which boils at temperatures of less than about 110 K at atmospheric pressure, such as hydrogen, helium, nitrogen, oxygen, air, or methane. Also known as cryogen; cryogenic liquid. { ‚krī·ə'jen·ik 'flü·əd }

cryogenic liquid *See* cryogenic fluid. { ‚krī·ə'jen·ik 'lik·wəd }

cryogenic pump |CRYO| A high-speed vacuum pump that can produce an extremely low vacuum and has a low power consumption; to reduce the pressure, gases are condensed on surfaces within an enclosure at extremely low temperatures, usually attained by using liquid helium or liquid or gaseous hydrogen. Also known as cryopump. { ‚krī·ə'jen·ik 'pəmp }

cryogenics |PHYS| The production and maintenance of very low temperatures, and the study of phenomena at these temperatures. { ‚krī·ə'jen·iks }

cryogenic temperature |CRYO| A temperature within a few degrees of absolute zero. { ‚krī·ə'jen·ik 'tem·prə·chər }

cryomagnetic |CRYO| Pertaining to production of very low temperatures by adiabatic demagnetization of paramagnetic salts. { ¦krī·ō·mag'ned·ik }

cryophysics |CRYO| Physics as restricted to phenomena occurring at very low temperatures, approaching absolute zero. { ¦krī·ō'fiz·iks }

cryopump *See* cryogenic pump. { 'krī·ō‚pəmp }

cryptand |ORG CHEM| A macropolycyclic polyazo-polyether, where the three-coordinate nitrogen atoms provide the vertices of a three-dimensional structure. { 'krip‚tand }

cryptate |INORG CHEM| The adduct formed between a cryptand and a guest (cation, anion, or neutral species) molecular entity. { 'krip‚tāt }

crystal |CRYSTAL| A homogeneous solid made up of an element, chemical compound or isomorphous mixture throughout which the atoms or molecules are arranged in a regularly repeating pattern. |ELECTR| A natural or synthetic piezoelectric or semiconductor material whose atoms are arranged with some degree of geometric regularity. { 'krist·əl }

crystal axis |CRYSTAL| A reference axis used for the vectoral properties of a crystal. { ¦krist·əl 'ak·səs }

crystal base |CRYSTAL| The contents of a primitive cell of a crystal. { ¦krist·əl 'bās }

crystal chemistry |CRYSTAL| The study of the crystalline structure and properties of a mineral or other solid. { 'krist·əl 'kem·ə·strē }

crystal class |CRYSTAL| One of 32 categories of crystals according to the inversions, rotations about an axis, reflections, and combinations of these which leaves the crystal invariant. Also known as symmetry class. { 'krist·əl 'klas }

crystal defect |CRYSTAL| Any departure from crystal symmetry caused by free surfaces, disorder, impurities, vacancies and interstitials, dislocations, lattice vibrations, and grain boundaries. Also known as lattice defect. { 'krist·əl 'dē‚fekt }

crystal diffraction |SOLID STATE| Diffraction by a crystal of beams of x-rays, neutrons, or electrons whose wavelengths (or de Broglie wavelengths) are comparable with the interatomic spacing of the crystal. { 'krist·əl di'frak·shən }

crystal diode *See* semiconductor diode. { ¦krist·əl 'dī‚ōd }

crystal dynamics *See* lattice dynamics. { ¦krist·əl də'nam·iks }

crystal face |CRYSTAL| One of the outward planar surfaces which define a crystal and reflect its internal structure. Also known as face. { ¦krist·əl ¦fās }

crystal field theory |PHYS CHEM| The theory which assumes that the ligands of a coordination compound are the sources of negative charge which perturb the energy levels of the central metal ion and thus subject the metal ion to an electric field analogous to that within an ionic crystalline lattice. { ¦krist·əl 'fēld ‚thē·ə·rē }

crystal form |CRYSTAL| A collection of crystal faces generated by operating on a single face with a subgroup of the symmetry elements of the crystal class.   { ¦krist·əl 'fȯrm }

crystal glass |MATER| A water-clear lead glass which polishes readily and has a high index of refraction.   { 'krist·əl ˌglas }

crystal gliding |CRYSTAL| Slip along a crystal plane due to plastic deformation; often produces crystal twins. Also known as translation gliding.   { 'krist·əl ˌglīd·iŋ }

crystal growth |CRYSTAL| The growth of a crystal, which involves diffusion of the molecules of the crystallizing substance to the surface of the crystal, diffusion of these molecules over the crystal surface to special sites on the surface, incorporation of molecules into the surface at these sites, and diffusion of heat away from the surface.   { 'krist·əl ˌgrōth }

crystal habit |CRYSTAL| The size and shape of the crystals in a crystalline solid. Also known as habit.   { 'krist·əl ˌhab·ət }

crystal indices   See Miller indices.   { 'krist·əl 'in·dəˌsēz }

crystal laser |OPTICS| A laser that uses a pure crystal of ruby or other material for generating a coherent beam of output light.   { ¦krist·əl 'lā·zər }

crystal lattice |CRYSTAL| A lattice from which the structure of a crystal may be obtained by associating with every lattice point an assembly of atoms identical in composition, arrangement, and orientation.   { ¦krist·əl 'lad·əs }

crystalline |CRYSTAL| Of, pertaining to, resembling, or composed of crystals.   { 'kris·tə·lən }

crystalline alumina |MATER| An abrasive which consists of essentially the same mineral as corundum, but whose physical properties such as crystal structure, size, and shape of grain are so controlled as to produce the most desirable abrasives for specific types of grinding.   { 'kris·tə·lən ə'lüm·ə·nə }

crystalline anisotropy |SOLID STATE| The tendency of crystals to have different properties in different directions; for example, a ferromagnet will spontaneously magnetize along certain crystallographic axes.   { 'kris·tə·lən an·ə'sä·trə·pē }

crystalline double refraction |OPTICS| The splitting which a wavefront experiences when a wave disturbance propagates through an anisotropic crystal.   { 'kris·tə·lən ˌdəb·əl ri'frak·shən }

crystalline field |SOLID STATE| The internal electric field in a solid due to localized charges, especially ions, inside.   { 'kris·tə·lən 'fēld }

crystalline fracture |MET| A break in a polycrystalline metal, with the fractured surface having a grainy appearance.   { 'kris·tə·lən 'frak·chər }

crystalline laser |OPTICS| A solid laser in which the lasing material is a pure crystal like ruby or a doped crystal like neodymium-doped ruby or neodymium-doped yttrium aluminum garnet.   { 'kris·tə·lən 'lā·zər }

crystalline polymer |POLYM CHEM| A polymer whose sections of adjacent chains are packed in a regular array.   { 'kris·tə·lən 'päl·i·mər }

crystallinity |CRYSTAL| The quality or state of being crystalline. |POLYM CHEM| The degree

to which polymer molecules are oriented into repeating patterns.   { ˌkris·tə'lin·əd·ē }

crystallization |CRYSTAL| The formation of crystalline substances from solutions or melts.   { ˌkris·tə·lə'zā·shən }

crystallogram |CRYSTAL| A photograph of the x-ray diffraction pattern of a crystal.   { 'kris·tə·lō ˌgram }

crystallographic axis |CRYSTAL| One of three lines (sometimes four, in the case of a hexagonal crystal), passing through a common point, that are chosen to have definite relation to the symmetry properties of a crystal, and are used as a reference in describing crystal symmetry and structure.   { ¦kris·tə·lō¦graf·ik 'ak·səs }

crystallography |PHYS| The branch of science that deals with the geometric description of crystals and their internal arrangement.   { ˌkris·tə'läg·rə·fē }

crystallomagnetic |SOLID STATE| Pertaining to magnetic properties of crystals.   { ˌkris·tə·lō·mag'ned·ik }

crystal momentum |SOLID STATE| The product of Planck's constant and the wave vector associated with an elementary excitation in a crystal (the magnitude of the wave vector being taken as the reciprocal of the wavelength).   { ¦krist·əl mə'men·təm }

crystal optics |OPTICS| The study of the propagation of light, and associated phenomena, in crystalline solids.   { ¦krist·əl 'äp·tiks }

crystal phase |MET| A crystal structure formed by an alloy over a certain range of values of the relative proportions of its constituents.   { 'krist·əl ˌfāz }

crystal photoeffect |SOLID STATE| An electromotive force induced by illumination of natural cuprite crystals or transparent zinc sulfide, and having a direction dependent on that of the incident light beam.   { 'krist·əl 'fōd·ō·iˌfekt }

crystal plane |CRYSTAL| One of a set of parallel, equally spaced planes in a crystal structure, each of which contains an infinite periodic array of lattice points.   { ¦krist·əl 'plān }

crystal projection |CRYSTAL| Any method of displaying the positions of the poles of a crystal by projecting them on a plane.   { ¦krist·əl prə'jek·shən }

crystal pulling |CRYSTAL| A method of crystal growing in which the developing crystal is gradually withdrawn from a melt.   { 'krist·əl ˌpu̇l·iŋ }

crystal rectifier   See semiconductor diode.   { ¦krist·əl 'rek·tə ˌfī·ər }

crystal structure |CRYSTAL| The arrangement of atoms or ions in a crystalline solid.   { ¦krist·əl 'strək·chər }

crystal symmetry |CRYSTAL| The existence of nontrivial operations, consisting of inversions, rotations around an axis, reflections, and combinations of these, which bring a crystal into a position indistinguishable from its original position.   { ¦krist·əl 'sim·ə·trē }

crystal system |CRYSTAL| One of seven categories (cubic, hexagonal, tetragonal, trigonal, orthorhombic, monoclinic, and triclinic) into which a crystal may be classified according to the shape of the unit cell of its Bravais lattice, or according to the dominant symmetry elements of its crystal class. { ¦krist·əl 'sis·təm }

crystal twin See twin crystal. { 'krist·əl ˌtwin }

crystal whisker |CRYSTAL| A single crystal that has grown in a filamentary form. Also known as whisker. { ¦krist·əl 'wis·kər }

cs See centistoke.

Cs See cesium.

C stage |POLYM CHEM| The final stage in a thermosetting resin reaction in which the material is relatively insoluble and infusible; the resin in a fully cured thermoset molding is in this stage. Also known as resite. { 'sē ˌstāj }

Cu See copper.

cubical dilation |MECH| The isotropic part of the strain tensor describing the deformation of an elastic solid, equal to the fractional increase in volume. { ¦kyü·bə·kəl di'lā·shən }

cubical expansion |PHYS| The increase in volume of a substance with a change in temperature or pressure. { ¦kyü·bə·kəl ik'span·shən }

cubic cleavage |CRYSTAL| Isometric crystal cleavage occuring parallel to the faces of a cube. { ¦kyü·bik 'klē·vij }

cubic crystal |CRYSTAL| A crystal whose lattice has a unit cell with perpendicular axes of equal length. { ¦kyü·bik 'krist·əl }

cubic packing |CRYSTAL| The spacing pattern of uniform solid spheres in a clastic sediment or crystal lattice in which the unit cell is a cube. { ¦kyü·bik 'pak·iŋ }

cubic plane |CRYSTAL| A plane that is at right angles to any one of the three crystallographic axes of the cubic system. { ¦kyü·bik 'plān }

cubic system See isometric system. { ¦kyü·bik ˌsis·təm }

cup |MET| Sheet metal part formed during the first deep-drawing operation. { kəp }

cup-and-cone fracture See cup fracture. { ¦kəp ən 'kōn ˌfrak·chər }

cupel |MET| A cup made of bone ash or magnesite, used in assaying precious metals. { 'kyü·pel }

cupellation |MET| **1.** Method using a cupel for assaying precious metals. **2.** Process for refining gold and silver by alloying them with lead and then oxidizing the molten lead to separate the base metal from the precious metal. { ˌkyü·pə'lā·shən }

cup fracture |MET| A break in a ductile material under tensile stress in which the surface of failure on one piece has a central flat area with an exterior extended rim. Also known as cup-and-cone fracture. { 'kəp ˌfrak·chər }

cupola |MET| A vertical cylindrical furnace for melting gray iron for foundry use; the metal, coke, and flux are put into the top of the furnace onto a bed of coke through which air is blown. Also known as furnace cupola. { 'kyü·pə·lə }

cupola drop |MET| The bed and unmelted charges dropped from a cupola at the end of a heat. { 'kyü·pə·lə ˌdräp }

cupping |MET| **1.** First operation of a deep-drawing process. **2.** Fracture of a wire or rod in which one fracture surface is conical and the other concave. { 'kəp·iŋ }

cupric |CHEM| The divalent ion of copper. { 'kyü·prik }

cupric arsenite See copper arsenite. { 'kyü·prik ärs·ən,īt }

cupric bromide |INORG CHEM| CuBr₂ Black prismatic crystals; used in photography as an intensifier and in organic synthesis as a brominating agent. Also known as copper bromide. { 'kyü·prik 'brō,mīd }

cupric carbonate See copper carbonate. { 'kyü·prik 'kär·bə,nāt }

cupric chloride |INORG CHEM| Also known as copper chloride. **1.** CuCl₂ Yellowish-brown, deliquescent powder soluble in water, alcohol, and ammonium chloride. **2.** CuCl₂·H₂O A dihydrate of cupric chloride forming green crystals soluble in water; used as a mordant in dyeing and printing textile fabrics and in the refining of copper, gold, and silver. { 'kyü·prik 'klór,īd }

cupric chromate |INORG CHEM| CuCrO₄ A yellow liquid, used as a mordant. Also known as copper chromate. { 'kyü·prik 'krō,māt }

cupric cyanide |INORG CHEM| Cu(CN)₂ A green powder, insoluble in water; used in electroplating copper on iron. Also known as copper cyanide. { 'kyü·prik 'sī·ə,nīd }

cupric fluoride |INORG CHEM| CuF₂ White crystalline powder used in ceramics and in the preparation of brazing and soldering fluxes. Also known as copper fluoride. { 'kyü·prik 'flür ˌīd }

cupric hydroxide |INORG CHEM| Cu(OH)₂ Blue macro- or microscopic crystals; used as a mordant and pigment, in manufacture of many copper salts, and for staining paper. Also known as copper hydroxide. { 'kyü·prik hī'dräk,sīd }

cupric nitrate |INORG CHEM| Cu(NO₃)₂·3H₂O Green powder or blue crystals soluble in water; used in electroplating copper on iron. Also known as copper nitrate. { 'kyü·prik 'nī,trāt }

cupric oxide |INORG CHEM| CuO Black, monoclinic crystals, insoluble in water; used in making fibers and ceramics, and in organic and gas analyses. Also known as copper oxide. { 'kyü·prik 'äk,sīd }

cupric sulfate |INORG CHEM| CuSO₄ A water-soluble salt used in copper-plating baths; crystallizes as hydrous copper sulfate, which is blue. Also known as copper sulfate. { 'kyü·prik 'səl ˌfāt }

cupronickel |MET| A copper-base alloy with 10–30% nickel and small amounts of manganese and iron; used in industrial and marine installations as condenser and heat-exchanger tubing. { ¦kyü·prō'nik·əl }

cuprous bromide |INORG CHEM| Cu₂Br₂ White or gray crystals slightly soluble in cold water. Also known as copper bromide. { 'kyü·prəs 'brō ˌmīd }

cuprous chloride |INORG CHEM| CuCl or Cu₂Cl₂ Green, tetrahedral crystals, insoluble in water.

Also known as copper chloride; resin of copper. { 'kyü·prəs 'klȯr,īd }

cuprous fluoride [INORG CHEM] Cu_2F_2 Red, crystalline powder, melting point 908°C. Also known as copper fluoride. { 'kyü·prəs 'flür,īd }

cuprous oxide [INORG CHEM] Cu_2O An oxide of copper found in nature as cuprite and formed on copper by heat; used chiefly as a pigment and as a fungicide. Also known as copper oxide. { 'kyü·prəs 'äk,sīd }

cure [CHEM] To change the properties of a resin material by chemical polycondensation or addition reactions. [POLYM CHEM] See vulcanization. { kyür }

cure time [POLYM CHEM] The amount of time required for a rubber compound to reach maximum viscosity or modulus at a given temperature. { 'kyür ,tīm }

Curie constant [ELECTROMAG] The electric or magnetic susceptibility at some temperature times the difference of the temperature and the Curie temperature, which is a constant at temperatures above the Curie temperature according to the Curie-Weiss law. { 'kyür·ē ¦kän·stənt }

Curie point See Curie temperature. { 'kyür·ē ,pȯint }

Curie principle [THERMO] The principle that a macroscopic cause never has more elements of symmetry than the effect it produces; for example, a scalar cause cannot produce a vectorial effect. { 'kyür·ē ,prin·sə·pəl }

Curie's law [ELECTROMAG] The law that the magnetic susceptibilities of most paramagnetic substances are inversely proportional to their absolute temperatures. { 'kyür,ēz ,lȯ }

Curie temperature [ELECTROMAG] The temperature marking the transition between ferromagnetism and paramagnetism, or between the ferroelectric phase and paraelectric phase. Also known as Curie point. { 'kyür·ē ,tem·prə·chər }

Curie-Weiss law [ELECTROMAG] A relation between magnetic or electric susceptibilities and the absolute temperatures which is followed by ferromagnets, antiferromagnets, nonpolar ferroelectrics, antiferroelectrics, and some paramagnets. { ¦kyür·ē ¦vīs ,lȯ }

curing [POLYM CHEM] A process in which polymers or oligomers are chemically cross-linked to form polymer networks. { 'kyür·iŋ }

curing agent See hardener. { 'kyür·iŋ ,ā·jənt }

curing temperature [POLYM CHEM] That temperature at which a resin or adhesive is subjected to curing. { 'kyür·iŋ ,tem·prə·chər }

curing time [POLYM CHEM] The period of time in which a part is subjected to heat or pressure to cure the resin. { 'kyür·iŋ ,tīm }

curium [CHEM] An element, symbol Cm, atomic number 96; the isotope of mass 244 is the principal source of this artificially produced element. { 'kyür·ē·əm }

curl [MATER] A defect of paper caused by unequal alteration in the dimensions of the top and underside of the sheet. { kərl }

Curie scale of temperature [THERMO] A temperature scale based on the susceptibility of a paramagnetic substance, assuming that it obeys Curie's law; used at temperatures below about 1 kelvin. { ¦kyür·ē ¦skāl əv 'tem·prə·chər }

current [ELEC] The net transfer of electric charge per unit time; a specialization of the physics definition. Also known as electric current. [PHYS] **1.** The rate of flow of any conserved, indestructible quantity across a surface per unit time. **2.** See current density. { 'kər·ənt }

current decay [MET] In certain types of welding operations, controlled reduction of the welding impulse over a predetermined time interval to prevent rapid cooling of the weld nugget. { 'kər·ənt di,kā }

current density [ELEC] The current per unit cross-sectional area of a conductor; a specialization of the physics definition. Also known as electric current density. [PHYS] A vector quantity whose component perpendicular to any surface equals the rate of flow of some conserved, indestructible quantity across that surface per unit area per unit time. Also known as current. { 'kər·ənt ,den·səd·ē }

current efficiency [PHYS CHEM] The ratio of the amount of electricity, in coulombs, theoretically required to yield a given quantity of material in an electrochemical process, to the amount actually consumed. { 'kər·ənt i,fish·ən·sē }

current intensity [ELEC] The magnitude of an electric current. Also known as current strength. { 'kər·ənt in'ten· səd·ē }

current strength See current intensity. { 'kər·ənt ,streŋkth }

curtain coating [CHEM ENG] A method in which the substrate to be coated with low-viscosity resins or solutions is passed through, and is perpendicular to, a freely falling liquid curtain. { 'kərt·ən ,kōd·iŋ }

cut [CRYSTAL] A section of a crystal having two parallel major surfaces; cuts are specified by their orientation with respect to the axes of the natural crystal, such as X cut, Y cut, BT cut, and AT cut. See fraction. { kət }

cut-and-carry method [MET] A die-fabricating method in which the part remains attached to the strip or is forced back into the strip to be fed through the succeeding stations of a progressive die. { ¦kət ən 'kar·ē ,meth·əd }

cutback asphalt [MATER] Asphalt which has been softened or liquefied by blending with petroleum distillates. { 'kət,bak ,as,fȯlt }

cut glass [MATER] Flint glass ornamented with patterns cut into its surface. { ¦kət ¦glas }

cutoff field See critical field. { 'kət,ȯf ,fēld }

cutting fluid [MATER] A fluid flowed over the tool and work in metal cutting to reduce heat generated by friction, lubricate, prevent rust, and flush away chips. { 'kəd·iŋ ,flü·əd }

cutting oil [MATER] A type of cutting fluid used in machining metals to lubricate the tool and workpiece, reducing tool wear, increasing cutting speeds, and decreasing power needs; there are two types: active and inactive. { 'kəd·iŋ ,ȯil }

cutting process [MET] A process where metal is

severed by the application of a gas or an electric arc. { 'kəd·iŋ ,präs·əs }

CVD See chemical vapor deposition.

cyanamide |INORG CHEM| NHCNH An acidic compound that forms colorless needles, melting at 46°C, soluble in water. Also known as urea anhydride. { sī'an·ə,mid }

cyanate |INORG CHEM| A salt or ester of cyanic acid containing the radical CNO. { 'sī·ə,nāt }

cyanidation See cyanide process. { ,si·ə·nə 'dā·shən }

cyanide |INORG CHEM| Any of a group of compounds containing the CN group and derived from hydrogen cyanide, HCN. { 'sī·ə,nīd }

cyanide copper |MET| 1. An electrolytic solution containing a complex of copper and the cyanide radical. 2. Copper electrodeposited from the solution. { 'sī·ə,nīd 'käp·ər }

cyanide process |MET| Process of dissolving powdered gold and silver ores in a weak solution of sodium cyanide or potassium cyanide; the precious metals are precipitated from solution by zinc. Also known as cyanidation. { 'sī·ə,nīd ,präs·əs }

cyanide pulp |MET| The mixture resulting from grinding of gold and silver ore, then dissolving out the precious-metal content in a solution of sodium cyanide. { 'sī·ə,nīd ,pəlp }

cyanide slime |MET| Minute particles of precious metals precipitated from cyanide solutions used in extracting the metals from ore. { 'sī·ə ,nīd ,slīm }

cyaniding |MET| Introduction of carbon and nitrogen simultaneously into a ferrous alloy by heating while in contact with molten cyanide; usually followed by quenching to produce a hardened case. { 'sī·ə,nīd·iŋ }

cyano- |CHEM| Combining form indicating the radical CN. { 'sī·ə·nō }

cyanoacrylate adhesive |MATER| An adhesive having a base of an alkyl 2-cyanoacrylate compound and characterized by excellent polymerizing and bonding properties; used for rubber printing plates, tools, and rubber swimming masks. { ,sī·ə·nō'ak·rə,lāt ad'hē·ziv }

cyanogen |INORG CHEM| C_2N_2 A colorless, highly toxic gas with a pungent odor; a starting material for the production of complex thiocyanates used as insecticides. { sī'an·ə·jən }

cyanogen bromide |INORG CHEM| CNBr White crystals melting at 52°C, vaporizing at 61.3°C, and having toxic fumes that affect nerve centers; used in the synthesis of organic compounds and as a fumigant. { sī'an·ə·jən 'brō,mīd }

cyanogen chloride |INORG CHEM| ClCN A poisonous, colorless gas or liquid, soluble in water; used in organic synthesis. { sī'an·ə·jən 'klȯr,īd }

cyanogen fluoride |INORG CHEM| CNF A toxic, colorless gas, used as a tear gas. { sī'an·ə·jən 'flu̇r,īd }

cyanogen iodide See iodine cyanide. { sī'an· ə·jən 'i·ə,dīd }

cyanoplatinate See platinocyanide. { ¦sī·ə· nō'plat·ən,āt }

cyclane See alicyclic. { 'sī,klān }

cycle annealing |MET| Annealing at a controlled time-temperature cycle to achieve a specific microstructure. { 'sī·kəl ə'nēl·iŋ }

cycle per second See hertz. { 'sī·kəl pər 'sek· ənd }

cyclic amide |ORG CHEM| An amide arranged in a ring of carbon atoms. { 'sīk·lik 'a,mīd }

cyclic anhydride |ORG CHEM| A ring compound formed by the removal of water from a compound; an example is phthalic anhydride. { 'sīk·lik an'hī,drīd }

cyclic compound |ORG CHEM| A compound that contains a ring of atoms. { 'sīk·lik 'käm ,pau̇nd }

cyclic twinning |CRYSTAL| Repeated twinning of three or more individuals in accordance with the same twinning law but without parallel twinning axes. { 'sīk·lik 'twin·iŋ }

cyclic voltammetry |PHYS CHEM| An electrochemical technique for studying variable potential at an electrode involving application of a triangular potential sweep, allowing one to sweep back through the potential region just covered. { 'sīk·lik vōl'täm·ə·trē }

cyclization |ORG CHEM| Changing an open-chain hydrocarbon to a closed ring. { ,sī· klə'zā·shən }

cyclized rubber |MATER| A thermoplastic, non-rubbery, tough or hard rubber derivative formed by the action of certain agents, such as sulfonic acid and chlorostannic acid, on rubber; used in paints and adhesives and for insulation. { 'sī ,klīzd 'rəb·ər }

cycloaddition |ORG CHEM| A reaction in which unsaturated molecules combine to form a cyclic compound. { ¦sī·klō·ə'dish·ən }

cycloaliphatic See alicyclic. { ¦sī·klō·al·ə'fad· ik }

cycloalkane See alicyclic. { ¦sī·klō'al,kān }

cyclopean |MATER| Mass concrete with aggregate larger than 6 inches (15 centimeters); used for thick structures such as dams. { ¦sī·klə ¦pē·ən }

cyclophane |ORG CHEM| A molecule composed of an aromatic ring (most frequently a benzene ring) and an aliphatic unit which forms a bridge between two (or more) positions of the aromatic ring. { 'sī·klə,fān }

Czochralski process |CRYSTAL| A method of producing large single crystals by inserting a small seed crystal of germanium, silicon, or other semiconductor material into a crucible filled with similar molten material, then slowly pulling the seed up from the melt while rotating it. { chə'kräl·skē ,präs·əs }

D

dahoma *See* African greenheart. { də̇hō·mə }

dalton *See* atomic mass unit. { 'dȯl·tən }

Dalton's atomic theory [CHEM] Theory forming the basis of accepted modern atomic theory, according to which matter is made of particles called atoms, reactions must take place between atoms or groups of atoms, and atoms of the same element are all alike but differ from atoms of another element. { 'dȯl·tənz ə̇täm·ik 'thē·ə·rē }

Dalton's law [PHYS] The law that the pressure of a gas mixture is equal to the sum of the partial pressures of the gases composing it. Also known as law of partial pressures. { 'dȯl·tənz ˌlȯ }

Dalton's temperature scale [THERMO] A scale for measuring temperature such that the absolute temperature T is given in terms of the temperature on the Dalton scale τ by $T = 273.15(373.15/273.15)^{\tau/100}$. { 'dȯl·tənz 'tem·prə·chər ˌskāl }

damaging stress [MECH] The minimum unit stress for a given material and use that will cause damage to the member and make it unfit for its expected length of service. { ¦dam·ə·jiŋ 'stres }

Damköhler number V *See* Reynolds number. { 'däm,kər·lər ¦nəm·bər ¦fīv }

dammar [MATER] **1.** A type of hard resin obtained from evergreen trees of the genus *Agathis*. **2.** A type of soft, clear to yellow East Indian resin derived from several trees of the family Dipterocarpaceae. Also known as gum dammar. { 'dam·ər }

dammar varnish [MATER] Varnish made from East Indian dammar. { 'dam·ər ˌvär·nəsh }

damp down [MET] To stop the blast in a blast furnace by closing the openings in the furnace. { ¦damp 'daún }

damping capacity [MECH] A material's capability in absorbing vibrations. { 'dam·piŋ kə 'pas·əd·ē }

damping coefficient *See* resistance. { 'dam·piŋ ˌkō·i,fish·ənt }

damping constant *See* resistance. { 'dam·piŋ ˌkän·stənt }

dangler [MET] The flexible electrode used in barrel plating. { 'daŋ·glər }

dangling bond [SOLID STATE] A chemical bond associated with an atom in the surface layer of a solid that does not join the atom with a second atom but extends in the direction of the solid's exterior. { ¦daŋ·gliŋ 'bänd }

Daniell cell [PHYS CHEM] A primary cell with a constant electromotive force of 1.1 volts, having a copper electrode in a copper sulfate solution and a zinc electrode in dilute sulfuric acid or zinc sulfate, the solutions separated by a porous partition or by gravity. { 'dan·yəl ˌsel }

darcy [PHYS] A unit of permeability, equivalent to the passage of 1 cubic centimeter of fluid of 1 centipoise viscosity flowing in 1 second under a pressure of 1 atmosphere through a porous medium having a cross-sectional area of 1 square centimeter and a length of 1 centimeter. { 'där·sē }

Darcy number 1 [FL MECH] A dimensionless group, equal to four times the Fanning friction factor. Symbolized Da_1. Also known as Darcy-Weisbach coefficient; resistance coefficient 2. { 'där·sē ˌnəm·bər 'wən }

Darcy number 2 [FL MECH] A dimensionless group used in the study of the flow of fluids in porous media, equal to the fluid velocity times the flow path divided by the permeability of the medium. Symbolized Da_2. { 'där·sē ˌnəm·bər ˌtü }

Darcy's law [FL MECH] The law that the rate at which a fluid flows through a permeable substance per unit area is equal to the permeability, which is a property only of the substance through which the fluid is flowing, times the pressure drop per unit length of flow, divided by the viscosity of the fluid. { 'där·sēz ˌlȯ }

Darcy-Weisbach coefficient *See* Darcy number 1. { 'där·sē 'vīs,bäk ˌkō·i,fish·ənt }

dark plaster [MATER] A plaster made from calcined, unground gypsum. { ¦därk 'plas·tər }

Darwin curve [CRYSTAL] A plot of the intensity of diffracted x-rays from a perfect crystal as a function of angle. { 'där·win ˌkərv }

Dauphine law [CRYSTAL] A twin law in which the twinned parts are related by a rotation of 180° around the *c* axis. { dȯˌfēn ˌlȯ }

daylight glass [MATER] A glass that absorbs red light; used in incandescent lamps to remove excess red emission so that the light spectrum resembles natural daylight. { 'dā,līt ˌglas }

d-block element [CHEM] A transition element occupying groups 3 through 12 in the periodic table. { 'dē ˌbläk ˌel·ə·mənt }

dc *See* direct current.

dc casting *See* direct-chill casting. { 'dē,sē ,kast·iŋ }

D center *See* R center. { 'dē ,sen·tər }

d constant |SOLID STATE| The ratio of the induced strain in a piezoelectric material to the applied electric field that produces this strain. { 'dē ,kän·stənt }

deactivation |MET| Chemical removal of the active constituents of a corrosive liquid. { dē ,ak·tə'vā·shən }

dead-bright |MET| Of a metal, polished to remove tool marks. { ¦ded 'brīt }

dead-burnt gypsum *See* anhydrous calcium sulfate. { 'ded ,bərnt 'jip·səm }

deadhead |MET| The portion of a casting that fills up the ingate. { 'ded,hed }

dead load *See* static load. { 'ded ,lōd }

dead soft steel |MET| **1.** Steel very low in carbon. **2.** Steel annealed until it is very soft. { ¦ded ¦sóft 'stēl }

Deborah number |MECH| A dimensionless number used in rheology, equal to the relaxation time for some process divided by the time it is observed. Symbolized D. { də'bór·ə ,nəm·bər }

deboss |MET| To press a design into a metal surface. { dē'bós }

deburr |MET| To remove burrs, sharp edges, or fins from metal parts by placing them in a revolving barrel containing abrasives suspended in a liquid. { dē'bər }

debye |ELEC| A unit of electric dipole moment, equal to 10^{-18} Franklin centimeter. { də'bī }

Debye effect |ELECTROMAG| Selective absorption of electromagnetic waves by a dielectric, due to molecular dipoles. { də'bī i'fekt }

Debye equation |SOLID STATE| The equation for the Debye specific heat, which satisfies the Dulong and Petit law at high temperatures and the Debye T^3 law at low temperatures. { də'bī i'kwā·zhən }

Debye-Falkenhagen effect |PHYS CHEM| The increase in the conductance of an electrolytic solution when the applied voltage has a very high frequency. { də¦bī 'fäl·kən,häg·ən i,fekt }

Debye force *See* induction force. { də¦bī ,fórs }

Debye frequency |SOLID STATE| The maximum allowable frequency in the computation of the Debye specific heat. { də'bī ,frē·kwən·sē }

Debye-Hückel theory |PHYS CHEM| A theory of the behavior of strong electrolytes, according to which each ion is surrounded by an ionic atmosphere of charges of the opposite sign whose behavior retards the movement of ions when a current is passed through the medium. { də¦bī 'hik·əl ,thē·ə·rē }

Debye-Jauncey scattering |SOLID STATE| Incoherent background scattering of x-rays from a crystal in directions between those of the Bragg reflections. { də¦bī 'jón·sē ,skad·ə·riŋ }

Debye potentials |ELECTROMAG| Two scalar potentials, designated Π_e and Π_m, in terms of which one can express the electric and magnetic fields resulting from radiation or scattering of electromagnetic waves by a distribution of localized sources in a homogeneous isotropic medium. { də¦bī pə'ten·chəlz }

Debye relaxation time |PHYS CHEM| According to the Debye-Hückel theory, the time required for the ionic atmosphere of a charge to reach equilibrium in a current-carrying electrolyte, during which time the motion of the charge is retarded. { də'bī ,rē,lak'sā·shən ,tīm }

Debye-Scherrer method |SOLID STATE| An x-ray diffraction method in which the sample, consisting of a powder stuck to a thin fiber or contained in a thin-walled silica tube, is rotated in a monochromatic beam of x-rays, and the diffraction pattern is recorded on a cylindrical film whose axis is parallel to the axis of rotation of the sample. { də¦bī 'sher·ər ,meth·əd }

Debye specific heat |SOLID STATE| The specific heat of a solid under the assumption that the energy of the lattice arises entirely from acoustic lattice vibration modes which all have the same sound velocity, and that frequencies are cut off at a maximum such that the total number of modes equals the number of degrees of freedom of the solid. { də'bī spə,sif·ik 'hēt }

Debye temperature |SOLID STATE| The temperature θ arising in the computation of the Debye specific heat, defined by $k\theta = h\nu$, where k is the Boltzmann constant, h is Planck's constant, and ν is the Debye frequency. Also known as characteristic temperature. { də'bī 'tem·prə·chər }

Debye theory |ELEC| The classical theory of the orientation polarization of polar molecules in which the molecules have a single relaxation time, and the plot of the imaginary part of the complex relative permittivity against the real part is a semicircle. { də'bī ,thē·ə·rē }

Debye T^3 law |SOLID STATE| The law that the specific heat of a solid at constant volume varies as the cube of the absolute temperature T at temperatures which are small with respect to the Debye temperature. { də'bī ,tē'kyübd ,ló }

Debye-Waller factor |SOLID STATE| A reduction factor for the intensity of coherent (Bragg) scattering of x-rays, neutrons, or electrons by a crystal, arising from thermal motion of the atoms in the lattice. { də¦bī 'väl·ər ,fak·tər }

decaborane (14) |INORG CHEM| $B_{10}H_{14}$ A binary compound of boron and hydrogen that is relatively stable at room temperature; melting point 99.5°C, boiling point 213°C. { ¦dek·ə¦bór,ān 'fór ¦tēn }

decade |ELEC| A group or assembly of 10 units; for example, a decade counter counts 10 in one column, and a decade box inserts resistance quantities in multiples of powers of 10. { de'kād }

decalescence |MET| Darkening of a metal surface due to isothermal absorption of the latent heat of phase transformation. { ,de·kə'les·əns }

decarburize |MET| To remove carbon from the surface of a ferrous alloy, particularly steel, by heating in a medium that reacts with carbon. { dē'kär·bə,rīz }

decavanadate |INORG CHEM| A deep-orange

polyvanadate ($V_{10}O_{28}^{6-}$), composed of 10 fused VO_6 octahedra. { ,dē·kə 'van·ə,dāt }

deckle |MATER| In paper manufacturing, the width of the wet sheet as it comes off the wire of a paper machine. { 'dek·əl }

deckle edge |MATER| The unfinished edge of paper having a characteristic appearance as a result of leakage under the frame (deckle) in which the paper is made; handmade paper has a deckle edge on all four sides, machine-made paper only on two sides. { ,dek·ə'l ej }

decomposition |CHEM| The more or less permanent structural breakdown of a molecule into simpler molecules or atoms. { dē,käm·pə'zish·ən }

decomposition potential |PHYS CHEM| The electrode potential at which the electrolysis current begins to increase appreciably. Also known as decomposition voltage. { dē,käm·pə'zish·ən pə,ten·chəl }

decomposition voltage See decomposition potential. { dē,käm·pə'zish·ən ,vōl·tij }

decorative stone |MATER| A stone that serves for architectural decoration, as in mantles or store fronts. { ¦dek·rəd·iv 'stōn }

deep-draw |MET| To form shapes with large depth-diameter ratios in sheet or strip metal by considerable plastic distortion in dies. { ¦dēp 'drò }

deep-etch |MET| Severe etching of a metal surface to reveal gross features, such as abnormal grain size, segregation, or cracks, at magnifications of 10 diameters or less. Also known as macroetch. { ¦dēp 'ech }

defect chemistry |SOLID STATE| The study of the dynamic properties of crystal defects under particular conditions, such as raising of the temperature or exposure to electromagnetic particle radiation. { 'dē,fekt ,kem·ə·strē }

defect cluster |CRYSTAL| A macroscopic cluster of crystal defects which can arise from attraction among defects. { 'dē,fekt ,kləs·tər }

defect conduction |SOLID STATE| Electric conduction in a semiconductor by holes in the valence band. { 'dē,fekt kən'dək·shən }

defect motion |CRYSTAL| Movement of a point defect from one lattice point to another. { 'dē ,fekt 'mō·shən }

defect scattering |SOLID STATE| Scattering of particles or electromagnetic radiation by crystal defects. { di'fekt skad·ər·iŋ }

defect structure |SOLID STATE| A crystal structure in which some atomic positions are occupied by atoms other than those that would be found in a perfect crystal, or are unoccupied. { di'fekt ,strək·chər }

deflashing |ENG| Finishing technique to remove excess material (flash) from a plastic or metal molding. { dē'flash·iŋ }

deflocculant |CHEM| An agent that causes deflocculation; examples are sodium carbonate and other basic materials used to deflocculate clay slips. { dē'fläk·yə·lənt }

deflocculate |CHEM ENG| To break up and dis-

perse agglomerates and form a stable colloid. { dē'fläk·yə,lāt }

deformation |MECH| Any alteration of shape or dimensions of a body caused by stresses, thermal expansion or contraction, chemical or metallurgical transformations, or shrinkage and expansions due to moisture change. { ,def·ər'mā·shən }

deformation bands See Lüders' lines. { ,def·ər'mā·shən ,banz }

deformation curve |MECH| A curve showing the relationship between the stress or load on a structure, structural member, or a specimen and the strain or deformation that results. Also known as stress-strain curve. { ,def·ər'mā·shən ,kərv }

deformation potential |SOLID STATE| The effective electric potential experienced by free electrons in a semiconductor or metal resulting from a local deformation in the crystal lattice. { ,def·ər'mā·shən pə,ten·chəl }

deformeter |ENG| An instrument used to measure minute deformations in materials in structural models. { ,dē'fòr,mēd·ər }

degasifier |MET| An alloy added to molten metal to facilitate the removal of dissolved gases. { dē'gas·ə,fī·ər }

degauss |ELECTR| To remove, erase, or clear information from a magnetic tape, disk, drum, or core. |ELECTROMAG| To neutralize (demagnetize) a magnetic field of, for example, a ship hull or television tube; a direct current of the correct value is sent through a cable around the ship hull; a current-carrying coil is brought up to and then removed from the television tube. { dē'gaús }

degenerate conduction band |SOLID STATE| A band in which two or more orthogonal quantum states exist that have the same energy, the same spin, and zero mean velocity. { di'jen·ə·rət kən'dək·shən ,band }

degenerate semiconductor |SOLID STATE| A semiconductor in which the number of electrons in the conduction band approaches that of a metal. { di'jen·ə·rət 'sem·i·kən,dək·tər }

degradation |ORG CHEM| Conversion of an organic compound to one containing a smaller number of carbon atoms. { ,deg·rə'dā·shən }

degradation failure |ENG| Failure of a device because of a shift in a parameter or characteristic which exceeds some previously specified limit. { ,deg·rə'dā·shən ,fāl·yər }

degreaser |ENG| A machine designed to clean grease and foreign matter from mechanical parts and like items, usually metallic, by exposing them to vaporized or liquid solvent solutions confined in a tank or vessel. |MATER| A solvent, such as a polyhalogenated hydrocarbon, that removes fat or oil in many industrial processes. { dē'grēs·ər }

degree |FL MECH| One of the units in any of various scales of specific gravity, such as the Baumé scale. |THERMO| One of the units of temperature or temperature difference in any of various temperature scales, such as the Celsius, Fahrenheit, and Kelvin temperature scales (the Kelvin degree is now known as the kelvin). { di'grē }

degree of crystallinity |POLYM CHEM| In a fairly large sample of a polymer, the fraction that consists of regions showing long-range three-dimensional order. { di'grē əv ˌkris·tə'lin·əd·ē }

degree of freedom |PHYS CHEM| Any one of the variables, including pressure, temperature, composition, and specific volume, which must be specified to define the state of a system. { di'grē əv 'frē·dəm }

degree of polymerization |POLYM CHEM| The number of structural units in the average polymer molecule in a particular sample. { di'grē əv pə ˌlim·ə·rə'zā·shən }

de Haas-van Alphen effect |SOLID STATE| An effect occurring in many complex metals at low temperatures, consisting of a periodic variation in the diamagnetic susceptibility of conduction electrons with changes in the component of the applied magnetic field at right angles to the principal axis of the crystal. { də¦häs ˌvan'äl·fən i¸fekt }

dehydration |CHEM| Removal of water from any substance. |ORG CHEM| An elimination reaction in which a molecule loses both a hydroxyl group (OH) and a hydrogen atom (H) that was bonded to an adjacent carbon. { ˌdē·hī'drā·shən }

dehydrogenation |CHEM| Removal of hydrogen from a compound. { dē¦hī·drə·jə'nā·shən }

dehydrohalogenation |CHEM| Removal of hydrogen and a halogen from a compound. { dē ¦hī·dro¸hal·ə·jə'nā·shən }

deionization |CHEM| An ion-exchange process in which all charged species or ionizable organic and inorganic salts are removed from solution. { dē¸ī·ən·ə'zā·shən }

delamination |ENG| Separation of a laminate into its constituent layers. { dē¸lam·ə'nā·shən }

de Lavaud process |MET| A centrifugal casting process employing water-cooled metal molds, used to produce cast iron pipe, gun barrels, and other cylindrical objects. { də·lä'vō 'präs·əs }

delayed yield |MET| Time delay between the sudden application of a yield stress and the appearance of yielding. { di'lād 'yēld }

delocalized bond |CHEM| A type of molecular bonding in which the electron density of delocalized electrons is regarded as being spread over several atoms or over the whole molecule. Also known as nonlocalized bond. { dē'lō·kə ˌlīzd 'bänd }

delta E effect |ELECTROMAG| Magnetization of a ferromagnetic substance that is caused by elastic tension. { ¦del·tə 'ē i¸fekt }

delta ferrite See delta iron. { 'del·tə 'fe¸rīt }

delta iron |MET| The nonmagnetic polymorphic form of iron stable between about 1403°C and the melting point, about 1535°C. Also known as delta ferrite. { 'del·tə ˌī·ərn }

deltohedron |CRYSTAL| A polyhedron which has 12 quadrilateral faces, and is the form of a crystal belonging to the cubic system and having hemihedral symmetry. Also known as deltoid dodecahedron; tetragonal tristetrahedron. { ˌdel·tə'hē·drən }

deltoid dodecahedron See deltohedron. { 'del ˌtóid dō¸dek·ə'hē·drən }

demagnetization |ELECTROMAG| **1.** The process of reducing or removing the magnetism of a ferromagnetic material. **2.** The reduction of magnetic induction by the internal field of a magnet. { dē¸mag·nəd·ə'zā·shən }

denature |CHEM| To change a protein by heating it or treating it with alkali or acid so that the original properties such as solubility are changed as a result of the protein's molecular structure being changed in some way. { dē'nā·chər }

denatured alcohol |CHEM| Ethyl alcohol containing a poisonous substance, such as methyl alcohol or benzene, which makes it unfit for human consumption. { dē'nā·chərd 'al·kə ˌhól }

dendrimer |POLYM CHEM| **1.** A polymer with a well-defined core, an interior, and peripheral surface components constructed in a concentric ordered fashion around the core. **2.** A polymer having a regular branched structure. Also known as dendritic polymer; dendron; starburst polymer. { 'den·drə·mər }

dendrite |CRYSTAL| A crystal having a treelike structure. { 'den¸drīt }

dendritic macromolecule |POLYM CHEM| A macromolecule whose structure is characterized by a high degree of branching that originates from a single focal point (core). { den'drid·ik ˌmak·rō'mäl·ə¸kyül }

dendritic polymer See dendrimer. { den'drid·ik 'päl·ə·mər }

dendritic powder |MET| Fine metal particles having a dendritic structure; usually of electrolytic origin. Also known as arborescent powder. { den'drid·ik 'paúd·ər }

dendron See dendrimer. { 'den¸drän }

dense |GRAPHICS| Very opaque because of a concentration of material, as pertaining to a negative or transparency that has been overdeveloped or overexposed. { dens }

dense-air refrigeration cycle See reverse Brayton cycle. { ¦dens ¦er ri¸frij·ə'rā·shən ¸sī·kəl }

densification |MATER| In ceramic powder processes, the step where the porous powder shape is converted into a part; the three main processes are sintering, hot pressing, and hot isostatic pressing. { ˌden·si·fə'kā·shən }

densimeter |ENG| An instrument which measures the density or specific gravity of a liquid, gas, or solid. Also known as densitometer; density gage; density indicator; gravitometer. { den'sim·əd·ər }

densitometer |ENG| **1.** An instrument which measures optical density by measuring the intensity of transmitted or reflected light; used to measure photographic density. **2.** See densimeter. { ˌden·sə'täm·əd·ər }

density |MECH| The mass of a given substance per unit volume. |OPTICS| **1.** The degree of opacity of a translucent material. **2.** The common logarithm of opacity. |PHYS| The total amount of a quantity, such as energy, per unit of space. { 'den· səd·ē }

density gage *See* densimeter. { 'den·səd·ē ˌgāj }

density indicator *See* densimeter. { 'den·səd·ē ˌin·də,kād·ər }

density of states [SOLID STATE] A function of energy E equal to the number of quantum states in the energy range between E and E + dE divided by the product of dE and the volume of the substance. { 'den·səd·ē əv 'stāts }

density scale [GRAPHICS] A value for the range density for a photographic material that corresponds to the difference between the maximum density and the minimum density. Also known as net density. { 'den·səd·ē ˌskāl }

dental gold [MET] An alloy composed of 5 to 12% silver, 4 to 10% copper, with the balance gold. { 'dent·əl 'gōld }

dental materials science [BIOMATER] An interdisciplinary area that applies biology, chemistry, and physics to the development, understanding, and evaluation of materials used in the practice of dentistry; principally involved in restorative dentistry, prosthodontics, pedodontics, and orthodontics. { 'dent·əl mə,tir·ē·əlz 'sī·əns }

deoxidize [CHEM] **1.** To remove oxygen by any of several processes. **2.** To reduce from the state of an oxide. [MET] To remove an oxide film from a metal surface. { dē'äk·sə,dīz }

deoxidized copper [MET] Pure copper deoxidized with phosphorus to reduce cuprous oxide and eliminate porosity. { dē'äk·sə,dīzd 'käp·ər }

dephosphorize [MET] Removal of phosphorus from a molten metal such as steel. { dē'fäs·fə ˌrīz }

depletion layer [ELECTR] An electric double layer formed at the surface of contact between a metal and a semiconductor having different work functions, because the mobile carrier charge density is insufficient to neutralize the fixed charge density of donors and acceptors. Also known as barrier layer (deprecated); blocking layer (deprecated); space-charge layer. { də'plē·shən ˌlā·ər }

depolarization [ELEC] The removal or prevention of polarization in a substance (for example, through the use of a depolarizer in an electric cell) or of polarization arising from the field due to the charges induced on the surface of a dielectric when an external field is applied. { dē ˌpō·lə·rə'zā·shən }

depolymerization [POLYM CHEM] Decomposition of macromolecular compounds into relatively simple compounds. { ˌdē·pə,lim·ə·rə 'zā·shən }

deposit [MATER] Any material applied to a base by means of vacuum, electrical, chemical, screening, or vapor methods. { də'päz·ət }

deposit attack [MET] Corrosion under or around the edge of a noncontinuous local deposit on a metal surface. { də'päz·ət ə'tak }

deposited metal [MET] Molten metal used during a welding operation for the fusion of base metals. { də'päz·əd·əd 'med·əl }

deposition efficiency [MET] The ratio of the weight of deposited metal to the net weight of the consumed electrodes, exclusive of stubs, in welding. { ˌdep·ə'zish·ən ə'fish·ən·sē }

deposition potential [PHYS CHEM] The smallest potential when it can produce electrolytic deposition when applied to an electrolytic cell. { ˌdep·ə'zish·ən pə'ten·chəl }

deposition rate [MET] The amount of welding material deposited per unit of time, expressed in pounds per hour. { ˌdep·ə'zish·ən ˌrāt }

deposition sequence [MET] The order of deposition of weld-metal increments. { ˌdep·ə'zish·ən ˌsē·kwəns }

deproteinize [ORG CHEM] To remove protein from a substance. { dē'prō,tē,nīz }

depth of cut [MET] The thickness that is removed as a workpiece is being machined. { ¦depth əv 'kət }

depth of fusion [MET] The distance that fusion extends from the original surface into the base metal in a welding operation. { 'depth əv 'fyü·zhən }

derby [MET] A large, usually cylindrical piece of primary metal, whose weight may exceed 100 pounds (45 kilograms), formed by bomb reduction. { 'dər·bē }

derivative [CHEM] A substance that is made from another substance. { də'riv·əd·iv }

derivatized gelatin [MATER] A form of gelatin in which the amino groups have been acylated; used for photographic and microencapsulation applications. { də'riv·ə'tīzd 'jel·ət·ən }

descaling [ENG] Removing scale, usually oxides, from the surface of a metal or the inner surface of a pipe, boiler, or other object. { dē'skāl·iŋ }

deseaming [MET] Removing defects from the surfaces of ingots, blooms, or semifinished products, usually by means of a chipping hammer or an oxy-gas flame. { dē'sēm·iŋ }

desiccant *See* drying agent. { 'des·i·kənt }

desilverization [MET] The act or process of removing silver; specifically, the process used to remove silver and gold from ore after softening. { dē,sil·vər·ə'zā·shən }

desorption [PHYS CHEM] The process of removing a sorbed substance by the reverse of adsorption or absorption. { dē'sórp·shən }

Destriau effect [SOLID STATE] Sustained emission of light by suitable phosphor powders that are embedded in an insulator and subjected only to the action of an alternating electric field. { 'des·trē·aù i,fekt }

destructive interference [OPTICS] The interaction of superimposed light from two different sources when the phase relationship is such as to reduce or cancel the resultant intensity to less than the sum of the individual lights. { di'strək·tiv ,in·tər'fir·əns }

destructive testing [ENG] **1.** Intentional operation of equipment until it fails, to reveal design weaknesses. **2.** A method of testing a material that degrades the sample under investigation. { di'strək·tiv 'test·iŋ }

detergent [MATER] A synthetic cleansing agent resembling soap in the ability to emulsify oil and hold dirt, and containing surfactants which do not precipitate in hard water; may also contain protease enzymes and whitening agents. { di'tər·jənt }

detergent additive [MATER] A substance incorporated in lubricating oils which gives them the property of keeping insoluble material in suspension. { di'tər·jənt ‚ad·ə·tiv }

determinate structure [MECH] A structure in which the equations of statics alone are sufficient to determine the stresses and reactions. { də'tər·mə·nət 'strək·chər }

detonation flame spraying [MET] A flame-spraying method in which the combined mixture of fuel gas, oxygen, and powdered coating liquefies and explodes the material to the workpiece. { ‚det·ən'ā·shən 'flām ‚sprā·iŋ }

deuterium [CHEM] The isotope of the element hydrogen with one neutron and one proton in the nucleus; atomic weight 2.0144. Designated D, d, H^2, or ^2H. { dü'tir·ē·əm }

deuterium oxide See heavy water. { dü'tir·ē·əm 'äk‚sīd }

developed blank [MET] A blank requiring little or no trimming after being formed. { də'vel·əpt 'blaŋk }

developed dye [CHEM] A direct azo dye that can be further diazotized by a developer after application to the fiber; it couples with the fiber to form colorfast shades. Also known as diazo dye. { də'vel·əpt 'dī }

developer [CHEM] An organic compound which interacts on a textile fiber to develop a dye. [GRAPHICS] A chemical solution used to develop exposed photographic materials by reducing silver salts to metallic silver. { də'vel·əp·ər }

deviation factor See compressibility factor. { ‚dēv·ē'ā·shən ‚fak·tər }

deviatonic stress [MECH] The portion of the total stress that differs from an isostatic hydrostatic pressure; it is equal to the difference between the total stress and the spherical stress. { ‚dev·ē·ə'tän·ik 'stres }

device [ELECTR] An electronic element that cannot be divided without destroying its stated function; commonly applied to active elements such as transistors and transducers. { di'vīs }

devitrification [CHEM] The process by which the glassy texture of a material is converted into a crystalline texture. { dē‚vi·trə·fə'kā·shən }

devitrified glass [MATER] A glassy material which has been changed from a vitreous to a brittle crystalline state during manufacture. { dē'vi·trə‚frīd 'glas }

dewetting [MET] Flow of solder away from the soldered surface during reheating following initial soldering. { dē‚wed·iŋ }

dew point [CHEM] The temperature and pressure at which a gas begins to condense to a liquid. { 'dü ‚póint }

dew-point temperature See dew point. { 'dü ‚póint 'tem·prə·chər }

dextro See dextrorotatory. { 'dek‚strō }

dextrorotatory [OPTICS] Rotating clockwise the plane of polarization of a wave traveling through a medium in a clockwise direction, as seen by an eye observing the light. Abbreviated dextro. { ‚dek·strə'rōd·ə‚tór·ē }

dezincification [CHEM] Removal of zinc. [MET] Corrosion of brass in which both components of the alloy are dissolved and the copper is redeposited as a porous surface residue. { dē‚ziŋk·ə·fə'kā·shən }

D glass [MATER] A type of glass with a high boron content and a precisely controlled dielectric constant. { 'dē ‚glas }

diabatic [THERMO] A thermodynamic change of state of a system in which there is a transfer of heat across the boundaries of the system. Also known as nonadiabatic. { ¦dī·ə¦bad·ik }

diacid [CHEM] An acid that has two acidic hydrogen atoms; an example is oxalic acid. { dī'as·əd }

diad axis [CRYSTAL] A rotation axis whose multiplicity is equal to 2. { 'dī‚ad ‚aks·əs }

diadochy [CRYSTAL] Replacement or ability to be replaced of one atom or ion by another in a crystal lattice. { dī'ad·ə‚kē }

diamagnet [ELECTROMAG] A substance which is diamagnetic, such as the alkali and alkaline earth metals, the halogens, and the noble gases. { ¦dī·ə¦mag·nət }

diamagnetic [ELECTROMAG] Having a magnetic permeability less than 1; materials with this property are repelled by a magnet and tend to position themselves at right angles to magnetic lines of force. { ¦dī·ə·mag¦ned·ik }

diamagnetic susceptibility [ELECTROMAG] The susceptibility of a diamagnetic material, which is always negative and usually on the order of -10^{-5} cm^3/mole. { ¦dī·ə·mag¦ned·ik sə ‚sep·tə'bil·əd·ē }

diamagnetism [ELECTROMAG] The property of a material which is repelled by magnets. { ¦dī·ə¦mag·nə‚tiz·əm }

diamond paste [MATER] An abrasive consisting of diamond dust in a viscous material. { 'dī·mənd ‚pāst }

diamond-pyramid hardness number [MET] The quotient of the load applied in the diamond-pyramid hardness test divided by the pyramidal area of the impression. { 'dī·mənd ¦pir·ə·mid 'härd·nəs ‚nəm·bər }

diamond-pyramid hardness test [MET] An indentation hardness test in which a diamond-pyramid indenter, with a 136° angle between opposite faces, is forced under variable loads into the surface of a test specimen. Also known as Vickers hardness test. { 'dī·mənd ¦pir·ə·mid 'härd·nəs ‚test }

diamond structure [CRYSTAL] A crystal structure in which each atom is the center of a tetrahedron formed by its nearest neighbors. { 'dī·mənd ‚strək·chər }

diathermous envelope [THERMO] A surface enclosing a thermodynamic system in equilibrium that is not an adiabatic envelope; intuitively, this means that heat can flow through the surface. { ¦dī·ə¦thər·məs 'en·və‚lōp }

diatomic [CHEM] Consisting of two atoms. { ¦dī·ə'täm·ik }

diazo compound |ORG CHEM| An organic compound containing the radical –N=N–. { dī'a·zō 'käm,paủnd }

diazo dye See developed dye. { dī'a·zō ,dī }

diazo group |ORG CHEM| A functional group with the formula $=N_2$. { dī'a·zō ,grüp }

diazonium |ORG CHEM| The grouping $=N\equiv N$. { ,dī·ə'zō·nē·əm }

diazonium salts |ORG CHEM| Compounds of the type R·X·N:N, where R represents an alkyl or aryl group and X represents an anion such as a halide. { ,dī·ə'zō·nē·əm 'sȯls }

diazo oxide |ORG CHEM| An organic molecule or a grouping of organic molecules that have a diazo group and an oxygen atom joined to ortho positions of an aromatic nucleus. { dī'a·zō 'äk,sīd }

dibasic |CHEM| **1.** Compounds containing two hydrogens that may be replaced by a monovalent metal or radical. **2.** An alcohol that has two hydroxyl groups, for example, ethylene glycol. { dī'bās·ik }

dibasic acid |CHEM| An acid having two hydrogen atoms capable of replacement by two basic atoms or radicals. { dī'bās·ik }

dibasic calcium phosphate See calcium phosphate. { dī'bās·ik 'kal·sē·əm 'fäs,fāt }

diborane |INORG CHEM| B_2H_6 A colorless, volatile compound that is soluble in ether; boiling point $-92.5°C$, melting point $-165.5°C$; can be used to produce pentaborane and decaborane, proposed for use as rocket fuels; also used to synthesize organic boron compounds. { dī'bȯr,ān }

dicalcium orthophosphate See calcium phosphate. { dī'kal·sē·əm ,ȯr·thō'fäs,fāt }

dicalcium phosphate See calcium phosphate. { dī'kal·sē·əm 'fäs,fāt }

dicarboxylic acid |ORG CHEM| A compound with two carboxyl groups. { dī¦kär·bäk¦sil·ik 'as·əd }

dichloramine |INORG CHEM| **1.** NH_2Cl_2 An unstable molecule considered to be formed from ammonia by action of chlorine. Also known as chlorimide. **2.** Any chloramine with two chlorine atoms joined to the nitrogen atom. { dī'klȯr·ə ,mēn }

dichroism |OPTICS| In certain anisotropic materials, the property of having different absorption coefficients for light polarized in different directions. { 'dī·krō,iz·əm }

dichromate |INORG CHEM| A salt of dichromic acid, usually orange or red. Also known as bichromate. { dī'krō,māt }

dichromate treatment |MET| Processing technique involving the formation of a corrosion-resistant film on the surface of a magnesium alloy by boiling the alloy in a sodium dichromate solution. { dī'krō,māt ,trēt·mənt }

dichromatic dye |CHEM| Dye or indicator in which different colors are seen, depending upon the thickness of the solution. { dī·krə'mäd·ik 'dī }

dichromic acid |INORG CHEM| $H_2Cr_2O_7$ An acid known only in solution, especially in the form of dichromates. { dī'krō·mik 'as·əd }

dicyanoargentates l See argentocyanides. { dī ¦sī·ə·nō,är·jən'tād·ēz 'wən }

die |ELECTR| The tiny, sawed or otherwise machined piece of semiconductor material used in the construction of a transistor, diode, or other semiconductor device; plural is dice. { dī }

die drawing |MET| Reducing the diameter of wire or tubing by pulling it through a die. { 'dī ,drȯ·iŋ }

die forging |MET| Shaping metal by plastic deformation in a die. { 'dī ,fȯrj·iŋ }

die forming |MET| Shaping metal by means of a die under pressure. { 'dī ,fȯr·miŋ }

dielectric See dielectric material. { ,dī·ə'lek· trik }

dielectric absorption |ELEC| The persistence of electric polarization in certain dielectrics after removal of the electric field. |ELECTROMAG| See dielectric loss. { ,dī·ə'lek·trik əb'sȯrp·shən }

dielectric breakdown |ELECTR| Breakdown which occurs in an alkali halide crystal at field strengths on the order of 10^6 volts per centimeter. { ,dī·ə'lek·trik 'brāk,daủn }

dielectric constant |ELEC| **1.** For an isotropic medium, the ratio of the capacitance of a capacitor filled with a given dielectric to that of the same capacitor having only a vacuum as dielectric. **2.** More generally, $1 + \gamma\chi$, where γ is 4π in Gaussian and cgs electrostatic units or 1 in rationalized mks units, and χ is the electric susceptibility tensor. Also known as relative dielectric constant; relative permittivity; specific inductive capacity (SIC). { ,dī·ə'lek·trik 'kän·stənt }

dielectric crystal |ELEC| A crystal which is electrically nonconducting. { ,dī·ə'lek·trik 'krist· əl }

dielectric current |ELEC| The current flowing at any instant through a surface of a dielectric that is located in a changing electric field. { ,dī·ə'lek·trik 'kər·ənt }

dielectric displacement See electric displacement. { ,dī·ə'lek·trik di'splās·mənt }

dielectric ellipsoid |ELEC| For an anisotropic medium in which the dielectric constant is a tensor quantity **K**, the locus of points **r** satisfying $r \cdot K \cdot r = 1$. { ,dī·ə'lek·trik ə'lip,sȯid }

dielectric fatigue |ELECTR| The property of some dielectrics in which resistance to breakdown decreases after a voltage has been applied for a considerable time. { ,dī·ə'lek·trik fə'tēg }

dielectric field |ELEC| The average total electric field acting upon a molecule or group of molecules inside a dielectric. Also known as internal dielectric field. { ,dī·ə'lek·trik 'fēld }

dielectric film |ELEC| A film possessing dielectric properties; used as the central layer of a capacitor. { ,dī·ə'lek·trik 'film }

dielectric flux density See electric displacement. { ,dī·ə'lek·trik 'fləks ,den·səd·ē }

dielectric gas |ELEC| A gas having a high dielectric constant, such as sulfur hexafluoride. { ,dī·ə'lek·trik 'gas }

dielectric heating |ELEC| Heating of a nominally electrical insulating material due to its own electrical (dielectric) losses, when the material is placed in a varying electrostatic field. { ,dī·ə'lek·trik 'hēd·iŋ }

dielectric hysteresis *See* ferroelectric hysteresis. { ,dī·ə'lek·trik hi·stə'rē·səs }

dielectric imperfection levels |SOLID STATE| Energy levels that occur in the forbidden zone between the valence and conduction bands of a dielectric crystal, because of imperfections in the crystal. { ,dī·ə'lek·trik ,im·pər'fek·shən ,lev·əlz }

dielectric loss |ELECTROMAG| The electric energy that is converted into heat in a dielectric subjected to a varying electric field. Also known as dielectric absorption. { ,dī·ə'lek·trik 'lòs }

dielectric loss angle |ELEC| Difference between 90° and the dielectric phase angle. { ,dī·ə ¦lek·trik ,aŋ·gəl }

dielectric loss factor |ELEC| Product of the dielectric constant of a material and the tangent of its dielectric loss angle. { ,dī·ə¦lek·trik ¦los ,fak·tər }

dielectric material |MATER| **1.** Also known as dielectric. **2.** A material which is an electrical insulator or in which an electric field can be sustained with a minimum dissipation of power. **3.** In a more general sense, any material other than a condensed state of a metal. { ,dī·ə'lək·trik mə,tir·ē·əl }

dielectric phase angle |ELEC| Angular difference in phase between the sinusoidal alternating potential difference applied to a dielectric and the component of the resulting alternating current having the same period as the potential difference. { ,dī·ə'lek·trik 'fāz ,aŋ·gəl }

dielectric polarization *See* polarization. { ,dī·ə'lek·trik ,pō·lə·rə'zā·shən }

dielectric power factor |ELEC| Cosine of the dielectric phase angle (or sine of the dielectric loss angle). { ,dī·ə'lek·trik 'paùr ,fak·tər }

dielectric soak *See* absorption. { ,dī·ə'lek·trik 'sōk }

dielectric strength |ELEC| The maximum electrical potential gradient that a material can withstand without rupture; usually specified in volts per millimeter of thickness. Also known as electric strength. { ,dī·ə'lek·trik 'streŋkth }

dielectric susceptibility *See* electric susceptibility. { ,dī·ə'lek·trik sə,sep·tə'bil·əd·ē }

dielectric test |ELEC| A test involving application of a voltage higher than the rated value for a specified time, to determine the margin of safety against later failure of insulating materials. { ,dī·ə'lek·trik 'test }

die lubricant |MATER| Any material applied to a die to facilitate movement of the work in the die in certain die-forming operations. { 'dī ,lü·brə·kənt }

die match |MET| The proper alignment of forging dies in relation to each other in forging equipment. { 'dī ,mach }

diene |ORG CHEM| One of a class of organic compounds containing two ethylenic linkages (carbon-to-carbon double bonds) in the molecules. Also known as diolefin. { 'dī,ēn }

diene resin |ORG CHEM| Material containing the diene group of double bonds that may polymerize. { 'dī,ēn 'rez·ən }

dienophile |ORG CHEM| The alkene component of a reaction between an alkene and a diene. { dī'en·ə,fīl }

die opening |MET| The distance between electrodes in flash or upset welding; it is measured with parts in contact but before the beginning or immediately after completion of the weld cycle. { 'dī ,ō·pən·iŋ }

die radius |MET| The radius on the exposed edge of a deep-drawing die. { 'dī ,rād·ē·əs }

die scalping |MET| Drawing wire, tubing, bars, or rods through a sharp-edged die to remove surface layers containing defects. { 'dī ,skalp·iŋ }

die steel |MET| Plain carbon steel or alloy steel used in making tools for cutting, machining, shearing, stamping, punching, and chipping. { 'dī ,stēl }

diester |ORG CHEM| A compound containing two ester groupings. { ¦dī'es·tər }

Dieterici equation of state |THERMO| An empirical equation of state for gases, $pe^{a/VRT}(v - b) = RT$, where p is the pressure, T is the absolute temperature, v is the molar volume, R is the gas constant, and a and b are constants characteristic of the substance under consideration. { dē·də'rē·chē i'kwā·zhen əv 'stāt }

Dietert tester |MET| An apparatus for reading Brinell hardness directly from the impression made in the part being tested by means of a depth pin pressed into the depression. { 'dēd·ərt 'tes·tər }

diether |ORG CHEM| A molecule that has two oxygen atoms with ether bonds. { dī'ē·thər }

die welding |MET| Forge welding in which the weld is completed under pressure between dies. { 'dī ,weld·iŋ }

differential calorimetry |THERMO| Technique for measurement of and comparison (differential) of process heats (reaction, absorption, hydrolysis, and so on) for a specimen and a reference material. { ,dif·ə'ren·chəl ,kal·ə'rim·ə·trē }

differential heating |MET| A thermal gradient caused as heating takes place; can result in a distribution of stress in a material. { ,dif·ə'ren·chəl 'hēd·iŋ }

differential heat of solution |THERMO| The partial derivative of the total heat of solution with respect to the molal concentration of one component of the solution, when the concentration of the other component or components, the pressure, and the temperature are held constant. { ,dif·ə'ren·chəl 'hēt əv sə'lü·shən }

differential permeability |ELECTROMAG| The slope of the magnetization curve for a magnetic material. { ,dif·ə'ren·chəl ,pər·mē·ə'bil·əd·ē }

differential thermal analysis |THERMO| A method of determining the temperature at which thermal reactions occur in a material undergoing continuous heating to elevated temperatures; also involves a determination of the nature and intensity of such reactions. { ,dif·ə'ren·chəl 'thər·məl ə'nal·ə·səs }

differential thermogravimetric analysis [THERMO] Thermal analysis in which the rate of material weight change upon heating versus temperature is plotted; used to simplify reading of weight-versus-temperature thermogram peaks that occur close together. { ‚dif·ə'ren·chəl ¦thər·mō‚grav·ə¦me·trik ə'nal·ə·səs }

diffraction [PHYS] Any redistribution in space of the intensity of waves that results from the presence of an object causing variations of either the amplitude or phase of the waves; found in all types of wave phenomena. { di'frak·shən }

diffraction symmetry [CRYSTAL] Any symmetry in a crystal lattice which causes the systematic annihilation of certain beams in x-ray diffraction. { di'frak·shən ‚sim·ə·trē }

diffractometry [CRYSTAL] The science of determining crystal structures by studying the diffraction of beams of x-rays or other waves. { ‚di‚frak'täm·ə·trē }

diffusion [ELECTR] A method of producing a junction by difusing an impurity metal into a semiconductor at a high temperature. [OPTICS] **1.** The distribution of incident light by reflection. **2.** Transmission of light through a translucent material. [PHYS] **1.** The spontaneous movement and scattering of particles (atoms and molecules), of liquids, gases, and solids. **2.** In particular, the macroscopic motion of the components of a system of fluids that is driven by differences in concentration. [SOLID STATE] **1.** The actual transport of mass, in the form of discrete atoms, through the lattice of a crystalline solid. **2.** The movement of carriers in a semiconductor. { də'fyü·zhən }

diffusion annealing [MET] Heat treatment of metal to promote homogeneity by diffusion of components. { də'fyü·zhən ə‚nēl·iŋ }

diffusion area [PHYS] One-sixth of the mean-square displacement between the appearance and disappearance of a subatomic particle of a given type. { də'fyü·zhən ‚er·ē·ə }

diffusion bonding [MET] A solid-state process for joining metals by using only heat and pressure to achieve atomic bonding. { də'fyü·zhən ‚bänd·iŋ }

diffusion brazing [MET] A process which produces bonding of the faying surfaces by heating them to suitable temperatures; the filler metal is diffused with the base metal and approaches the properties of the base metal. Also known as transient liquid phase bonding. { də'fyü·zhən ‚brāz·iŋ }

diffusion coating [MET] An alloy coating produced by allowing the coating material to diffuse into the base at high temperature. { də'fyü·zhən ‚kōd·iŋ }

diffusion coefficient [PHYS] The weight of a material, in grams, diffusing across an area of 1 square centimeter in 1 second in a unit concentration gradient. Also known as diffusivity. { də'fyü·zhən ‚kō·i'fish·ənt }

diffusion constant [SOLID STATE] The diffusion current density in a hologeneous semiconductor divided by the charge carrier concentration gradient. { də'fyü·zhən ‚kän·stənt }

diffusion equation [PHYS] **1.** An equation for diffusion which states that the rate of change of the density of the diffusing substance, at a fixed point in space, equals the sum of the diffusion coefficient times the Laplacian of the density, the amount of the quantity generated per unit volume per unit time, and the negative of the quantity absorbed per unit volume per unit time. **2.** More generally, any equation which states that the rate of change of some quantity, at a fixed point in space, equals a positive constant times the Laplacian of that quantity. { də'fyü·zhən i'kwā·zhən }

diffusion flame [CHEM] A long gas flame that radiates uniformly over its length and precipitates free carbon uniformly. { də'fyü·zhən ‚flām }

diffusion gradient [PHYS] The graphed distance of penetration (diffusion) versus concentration of the material (or effect) diffusing through a second material; applies to heat, liquids, solids, or gases. { də'fyü·zhən ‚grād·ē·ənt }

diffusion length [PHYS] The average distance traveled by a particle, such as a minority carrier in a semiconductor or a thermal neutron in a nuclear reactor, from the point at which it is formed to the point at which it is absorbed. { də'fyü·zhən ‚leŋkth }

diffusion-limited aggregation [PHYS] A mathematical model for particle aggregation processes, such as the growth of a metal deposit on an electrochemical cell, in which particles move according to a random walk process until they arrive at a certain fixed distance from the current aggregate, where they stick to it. { də'fyü·zhən ¦lim·əd·əd ag·rə'gā·shən }

diffusion-limited current density [MET] The density corresponding to the maximum transfer rate that a material can sustain due to diffusion limits. { də¦fyü·zhən ¦lim·əd·əd 'kər·ənt ‚den·səd·ē }

diffusion potential [PHYS CHEM] A potential difference across the boundary between electrolytic solutions with different compositions. Also known as liquid junction potential. { də'fyü·zhən pə‚ten·chəl }

diffusion theory [ELEC] The theory that in semiconductors, where there is a variation of carrier concentration, a motion of the carriers is produced by diffusion in addition to the drift determined by the mobility and the electric field. { də'fyü·zhən ‚thē·ə·rē }

diffusion welding [MET] A welding process which utilizes high temperatures and pressures to coalesce the faying surfaces by solid-state bonding; there is no physical movement, visible deformation of the parts involved, or melting. { də'fyü·zhən ‚weld·iŋ }

diffusivity [PHYS] See diffusion coefficient. [THERMO] The quantity of heat passing normally through a unit area per unit time divided by the product of specific heat, density, and temperature gradient. Also known as thermal diffusivity; thermometric conductivity. { dif·yü'ziv·əd·ē }

difunctional molecule |ORG CHEM| An organic structure possessing two sites that are highly reactive. { ,dī¦fəŋk·shən·əl 'mäl·ə,kyül }

dihexagonal |CRYSTAL| Of crystals, having a symmetrical form with 12 sides. { dī·hek'sag·ən·əl }

dihexagonal-dipyramidal |CRYSTAL| Characterized by the class of crystals in the hexagonal system in which any section perpendicular to the sixfold axis is dihexagonal. { dī·hek'sag·ən·əl dī·pir·ə'mid·əl }

dihexahedron |CRYSTAL| A type of crystal that has 12 faces, such as a double six-sided pyramid. { dī¦hek·sə¦hē·drən }

dihydroxy alcohol See glycol. { ¦dī,hī¦dräk·sē 'al·kə,hól }

dilatancy |CHEM| The property of a viscous suspension which sets solid under the influence of pressure. { dī'lāt·ən·sē }

dilatant |CHEM| A material with the ability to increase in volume when its shape is changed. { dī'lāt·ənt }

dilatation |PHYS| The increase in volume per unit volume of any continuous substance, caused by deformation. { ,dil·ə'tā·shən }

dilatometer |ENG| An instrument for measuring thermal expansion and dilation of liquids or solids. { ,dil·ə'täm·əd·ər }

dilatometry |PHYS| The measurement of changes in the volume of a liquid or dimensions of a solid which occur in phenomena such as allotropic transformations, thermal expansion, compression, creep, or magnetostriction. { ,dil·ə'täm·ə·trē }

diluent |CHEM| An inert substance added to some other substance or solution so that the volume of the latter substance is increased and its concentration per unit volume is decreased. { 'dil·yə·wənt }

dilution |CHEM| Increasing the proportion of solvent to solute in any solution and thereby decreasing the concentration of the solute per unit volume. |MET| The use of a welding filler metal deposit with a base metal or a previously deposited weld material having a lower alloy content. |OPTICS| Reducing the intensity of a color by adding white. { də'lü·shən }

dimensional stability |GRAPHICS| The percentage of change in size of paper under two different conditions of temperature and humidity. |MATER| The ability of a material, such as a textile or plastic, to hold its shape over a period of time and under specific conditions. { də·'men·chən·əl stə'bil·əd·ē }

dimensions |PHYS| The product of powers of fundamental quantities (or of convenient derived quantities) which are used to define a physical quantity; the fundamental quantities are often mass, length, and time. { də'men·chənz }

dimension stone |MATER| Large, sound, relatively flawless blocks of stone used as building stone, monumental stone, paving stone, curbing, and flagging. { də'men·chən ,stōn }

dimer |CHEM| A molecule that results from a chemical combination of two entities of the same species, for example, the chlorine molecule (Cl_2) or cyanogen (NCCN). { 'dī·mər }

dimeric water |INORG CHEM| Water in which pairs of molecules are joined by hydrogen bonds. { dī'mer·ik 'wód·ər }

dimerization |CHEM| A chemical reaction in which two identical molecular entities react to form a single dimer. { ,dī·mər·ə'zā·shən }

dimorphism |CHEM| Having crystallization in two forms with the same chemical composition. { dī'mór,fiz·əm }

dimple crystal |CRYO| A periodic pattern of hexagons that forms on a liquid helium surface when electrons, forming an electric charge sheet, are trapped at the surface, and an intense electric field is applied; the hexagons are typically about 2 millimeters across and the interiors of the hexagons are depressions about 10 micrometers deep. { 'dim·pəl ,krist·əl }

dingot |MET| A massive derby, usually a ton or more, produced in a bomb reaction. { 'diŋ·gət }

dinitrogen |CHEM| N_2 The diatomic molecule of nitrogen. { dī'nī·trə·jən }

dinitrogen tetroxide See nitrogen dioxide. { dī'nī·trə·jən te'träk,sīd }

dioctahedral |CRYSTAL| Pertaining to a crystal structure in which only two of the three available octahedrally coordinated positions are occupied. { ,dī,äk·tə'hē·drəl }

diode |ELECTR| **1.** A two-electrode electron tube containing an anode and a cathode. **2.** See semiconductor diode. { 'dī,ōd }

diode laser See semiconductor laser. { 'dī,ōd ,lāz·ər }

diode theory |ELEC| The theory that in a semiconductor, when the barrier thickness is comparable to or smaller than the mean free path of the carriers, then the carriers cross the barrier without being scattered, much as in a vacuum tube diode. { 'dī,ōd ,thē·ə·rē }

diolefin See diene. { dī'ō·lə,fən }

dioxygen |CHEM| O_2 Molecular oxygen. { dī 'äk·si·jən }

dip brazing |MET| Soldering by dipping the work into a hot, molten salt or metal bath and by using a nonferrous metal with a melting point above 800°F (427°C). { 'dip ,brāz·iŋ }

dip coating |ENG| A coating applied to ceramic ware or metal by immersion into a tank of melted nonmetallic material, such as resin or plastic, then chilling the adhering melt. { 'dip ,kōd·iŋ }

diphase cleaning |MET| Removing soilage from metal surfaces in a cleaning tank incorporating a solvent phase and an aqueous phase. { 'dī,fāz 'klēn·iŋ }

diphenyl See biphenyl. { dī'fen·əl }

diploid |CRYSTAL| A crystal form in the isometric system having 24 similar quadrilateral faces arranged in pairs. { 'di,plóid }

dipolar ion |CHEM| An ion carrying both a positive and a negative charge. Also known as zwitterion. { 'dī,pō·lər 'ī,än }

dipole |ELECTROMAG| Any object or system that is oppositely charged at two points, or poles, such as a magnet or a polar molecule; more precisely, the limit as either charge goes to infinity, the separation distance to zero, while the product remains constant. { 'dī,pōl }

dipole moment |ELEC| *See* electric dipole moment. |ELECTROMAG| *See* magnetic dipole moment. |PHYS CHEM| The vector sum of the bond moments in a molecule, a measure of the polarity of the molecule. { 'dī,pōl ,mō·mənt }

dipole polarization *See* orientation polarization. { 'dī,pōl ,pō·lə·rə'zā·shən }

dipole relaxation |ELEC| The process, occupying a certain period of time after a change in the applied electric field, in which the orientation polarization of a substance reaches equilibrium. { 'dī,pōl ,rē,lak'sā·shən }

dipping acid *See* sulfuric acid. { 'dip·iŋ ,as·əd }

dip plating *See* immersion plating. { 'dip ,plāt·iŋ }

diprotic |CHEM| Pertaining to a chemical structure that has two ionizable hydrogen atoms. { dī'präd·ik }

diprotic acid |CHEM| An acid that has two ionizable hydrogen atoms in each molecule. { di'präd·ik 'as·əd }

dip soldering |MET| A method similar to dip brazing but using a filler metal having a melting point below 800°F (427°C). { 'dip ,säd·ər·iŋ }

dipyramid *See* bipyramid. { dī'pir·ə,mid }

direct-band-gap semiconductor |SOLID STATE| A semiconductor material in which the state of minimum energy in the conduction band and the state of maximum energy in the valence band have the same momentum, so that optical transitions between free electrons and holes are allowed. { də'rekt 'band ,gap 'sem·i·kən ,dək·tər }

direct-chill casting |MET| A continuous ingot-or billet-casting process in which metal is poured into short molds on a platform and then cooled when the platform is lowered into a water bath. Abbreviated dc casting. { də¦rekt ¦chil 'kast·iŋ }

direct current |ELEC| Electric current which flows in one direction only, as opposed to alternating current. Abbreviated dc. { də¦rekt 'kə·rənt }

direct-current electrode negative |MET| In direct-current arc welding, the arrangement of leads where the surface to be welded is the positive and the electrode is the negative relative to the welding arc. { də¦rekt ¦kə·rənt i¦lek,trōd 'neg·əd·iv }

direct-current electrode positive |MET| In direct-current arc welding, the arrangement of leads where the surface to be welded is the negative and the electrode is the positive relative to the welding arc. { də¦rekt ¦kə·rənt i¦lek,trōd 'päz·əd·iv }

direct-current Josephson effect |CRYO| The current flow resulting from the tunneling of electron pairs through a thin insulating barrier between two superconductors in the absence of a voltage drop across the barrier. { di¦rekt ¦kər·ənt 'jō·sef·sən i,fekt }

direct dye |MATER| A group of coal tar dyes that act without mordants, for example, benzidine

dyes. Also known as substantive dye. { də¦rekt 'dī }

direct-gap semiconductor |SOLID STATE| A semiconductor in which the minimum of the conduction band occurs at the same wave vector as the maximum of the valence band, and recombination radiation consequently occurs with relatively large intensity. { də¦rekt ¦gap 'sem·i·kən,dək·tər }

directional property |MET| Any property of a metal whose magnitude varies with the orientation of the test axis to a specific direction within the metal. { də'rek·shən·əl 'präp·ərd·ē }

directional solidification |MET| Controlled solidification of molten metal in a casting so as to provide feed metal to the solidifying front of the casting. { də'rek·shən·əl sə,lid·ə·fə'kā·shən }

directly heated cathode *See* filament. { də ¦rek·lē ¦hēd·əd 'kā,thōd }

direct piezoelectricity |SOLID STATE| Name sometimes given to the piezoelectric effect in which an electric charge is developed on a crystal by the application of mechanical stress. { də'rekt pē¦ā·zō,i,lek'tris·əd·ē }

direct process |MET| A process in which the metal is produced from the ore in a single step (for example, steel without intermediate pig iron). { də'rekt 'präs·əs }

direct quenching |MET| Rapid cooling of carburized parts directly from the carburizing process. { də'rekt 'kwench·iŋ }

direct-reduction process |MET| Any of several methods for extracting iron ore below the melting point of iron, to produce solid reduced iron that may be converted to steel with little further refining. { də¦rekt ri'dək·shən ,präs·əs }

discharge |ELEC| To remove a charge from a battery, capacitor, or other electric-energy storage device. { 'dis,chärj }

discontinuity |MET| The place where the structural nature of a weldment is interfered with because of the materials involved or where the mechanical, physical, or metallurgical aspects are not homogeneous. |PHYS| A break in the continuity of a medium or material at which a reflection of wave energy can occur. { dis ,känt·ən'ü·əd·ē }

discontinuous phase *See* disperse phase. { ,dis·kən'tin·yə·wəs 'fāz }

discontinuous precipitation |MET| Precipitation principally at and away from the grain boundaries in a supersaturated solid solution; diffraction patterns show two lattice parameters, the solute in solution and the precipitate. { ,dis·kən'tin·yə·wəs prə,sip·ə'tā·shən }

discontinuous yielding |MET| The nonuniform plastic deformation of a metal along the length strained in tension. { ,dis·kən'tin·yə·wəs 'yēld·iŋ }

disilane |INORG CHEM| Si_2H_6 A spontaneously flammable compound of silicon and hydrogen; it exists as a liquid at room temperature. { dī'si ,lān }

disinclination |CRYSTAL| A type of crystal imperfection in which one part of the crystal is rotated and therefore displaced relative to the rest of the crystal; observed in liquid crystals and protein coats of viruses. { dis₁in·klə'nā·shən }

dislocation |CRYSTAL| A defect occurring along certain lines in the crystal structure and present as a closed ring or a line anchored at its ends to other dislocations, grain boundaries, the surface, or other structural feature. Also known as line defect. { ₁dis·lō'kā·shən }

dislocation line |CRYSTAL| A curve running along the center of a dislocation. { ₁dis·lō'kā·shən ₁līn }

disodium hydrogen phosphate See disodium phosphate. { dī'sōd·ē·əm 'hī·drə·jən 'fäs₁fāt }

disodium phosphate |INORG CHEM| Na₂HPO₄ Transparent crystals, soluble in water; used in the textile processing and other industries to control pH in the range 4-9, as an additive in processed cheese to maintain spreadability, and as a laxative and antacid. Also known as disodium hydrogen phosphate. { dī'sōd·ē·əm 'fäs₁fāt }

disorder |CRYSTAL| Departures from regularity in the occupation of lattice sites in a crystal containing more than one element. { dis'ȯrd·ər }

disordered crystalline alloy |SOLID STATE| A mixture of two elements in which the atoms of the mixture are found at more or less random positions on a crystal lattice. { dis'ȯrd·ərd ¦krist·əl·ən 'al₁ȯi }

dispersant |MATER| Also known as dispersing agent. **1.** An additive that can hold finely ground materials in suspension; used as a thinning agent for a slurry. **2.** A material added to a paste, mortar, or concrete to improve the flow properties. { di'spərs·ənt }

disperse dye |MATER| A very slightly water-soluble, colored material for use on cellulose acetate and other synthetic fibers; color is transferred to the fiber as extremely finely divided particles, resulting in a solution of the dye in the solid fiber. { də'spərs ₁dī }

disperse phase |CHEM| The phase of a disperse system consisting of particles or droplets of one substance distributed through another system. Also known as discontinuous phase; internal phase. { də'spərs ₁fāz }

disperser |MATER| Material added to solid-in-liquid or liquid-in-liquid suspensions to separate the individual suspended particles; used in pigment grinding and dye dispersion. Also known as dispersing agent; emulsifier; emulsifying agent. { də'spər·sər }

disperse system |CHEM| A two-phase system consisting of a dispersion medium and a disperse phase. { də'spərs ₁sis·təm }

dispersing agent See dispersant; disperser. { də'spərs·iŋ ₁ā·jənt }

dispersion |CHEM| A distribution of finely divided particles in a medium. { də'spər·zhən }

dispersion force |PHYS CHEM| The force of attraction that exists between molecules that have no permanent dipole. { də'spər·zhən ₁fȯrs }

dispersion medium See continuous phase. { də'spər·zhən ₁mēd·ē·əm }

dispersion strengthening |SOLID STATE| The reduction of plastic deformation of a solid by the presence of a uniform dispersion of another substance which inhibits the motion of plastic dislocations. { də'spər·zhən ₁streŋk·thən·iŋ }

disphenoid |CRYSTAL| **1.** A crystal form with four similar triangular faces combined in a wedge shape; can be tetragonal or orthorhombic. **2.** A crystal form with eight scalene triangles combined in pairs. { dī'sfē₁nȯid }

displacement |ELEC| See electric displacement. |MECH| **1.** The linear distance from the initial to the final position of an object moved from one place to another, regardless of the length of path followed. **2.** The distance of an oscillating particle from its equilibrium position. { dis'plās·mənt }

disruptive strength |MET| Failure stress caused by hydrostatic tension. { dis¦rəp·tiv 'streŋkth }

dissipation |PHYS| Any loss of energy, generally by conversion into heat; quantitatively, the rate at which this loss occurs. Also known as energy dissipation. { ₁dis·ə'pā·shən }

dissipation factor |ELEC| The inverse of Q, the storage factor. { ₁dis·ə'pā·shən ₁fak·tər }

dissipative tunneling |SOLID STATE| Quantum-mechanical tunneling of individual electrons, rather than pairs, across a thin insulating layer separating two superconducting metals when there is a voltage across this layer, resulting in partial disruption of cooperative motion. { ₁dis·ə'pād·iv 'tən·əl·iŋ }

dissociation |PHYS CHEM| Separation of a molecule into two or more fragments (atoms, ions, radicals) by collision with a second body or by the absorption of electromagnetic radiation. { də₁sō·sē'ā·shən }

dissolution |CHEM| Dissolving of a material. { ₁dis·ə'lü·shən }

dissolve |CHEM| **1.** To cause to disperse. **2.** To cause to pass into solution. { də'zälv }

dithionic acid |INORG CHEM| H₂S₂O₆ A strong acid formed by the oxidation of sulfurous acid, and known only by its salts and in solution. { ¦dī₁thī'än·ik 'as·əd }

ditungsten carbide |INORG CHEM| W₂C A gray powder having hardness approaching that of diamond; forms hexagonal crystals with specific gravity 17.2; melting point 2850°C. { ¦dī₁təŋ·stən 'kär₁bīd }

divalent metal |CHEM| A metal whose atoms are each capable of chemically combining with two atoms of hydrogen. { dī'vā·lənt 'med·əl }

divariant system |THERMO| A system composed of only one phase, so that two variables, such as pressure and temperature, are sufficient to define its thermodynamic state. { di¦ver·ē·ənt 'sis·təm }

divinyl ether See vinyl ether. { dī'vīn·əl 'ē·thər }

divinyl oxide See vinyl ether. { dī'vīn·əl 'äk₁sīd }

djalmaite See microlite. { 'jal·mə₁īt }

D nickel |MET| A nickel-manganese (4.75%) alloy of medium strength that resists spark erosion and hence is used for spark-plug electrodes and for some electronic applications. { dē ₁nik·əl }

doctor bar See doctor blade. { 'däk·tər ₁bär }

doctor blade |ENG| A device for regulating the amount of liquid material on the rollers of a spreader. Also known as doctor bar; doctor knife; doctor roll. { 'däk·tər ,blād }

doctor knife *See* doctor blade. { 'däk·tər ,nīf }

doctor roll *See* doctor blade. { 'däk·tər ,rōl }

domain |SOLID STATE| A region in a solid within which elementary atomic or molecular magnetic or electric moments are uniformly arrayed. { dō'mān }

domain growth |SOLID STATE| A stage in the process of magnetization in which there is a growth of those magnetic domains in a ferromagnet oriented most nearly in the direction of an applied magnetic field. { dō'mān ,grōth }

domain rotation |SOLID STATE| The stage in the magnetization process in which there is rotation of the direction of magnetization of magnetic domains in a ferromagnet toward the direction of a magnetic applied field and against anisotropy forces. { dō'mān rō'tā·shən }

domain theory |SOLID STATE| A theory of the behavior of ferromagnetic and ferroelectric crystals according to which changes in the bulk magnetization and polarization arise from changes in size and orientation of domains that are each polarized to saturation but which point in different directions. { dō'mān ,thē·ə·rē }

domain wall *See* Bloch wall. { dō'mān ,wôl }

domatic class *See* clinohedral class. { dō'mad·ik 'klas }

dome |CRYSTAL| An open crystal form consisting of two faces astride a symmetry plane. { dōm }

domestic coke |MATER| Coke for residential heating, which must have as low an ash content and as high a softening temperature (preferably above 2300°F, or 1260°C) for the ash as possible. { də'mes·tik 'kōk }

Donohue equation |THERMO| Equation used to determine the heat-transfer film coefficient for a fluid on the outside of a baffled shell-and-tube heat exchanger. { 'dän·ə·hü i,kwā·zhən }

donor |SOLID STATE| An impurity that is added to a pure semiconductor material to increase the number of free electrons. Also known as donor impurity; electron donor. { 'dō·nər }

donor impurity *See* donor. { 'dō·nər im,pyúr·əd·ē }

donor level |SOLID STATE| An intermediate energy level close to the conduction band in the energy diagram of an extrinsic semiconductor. { 'dō·nər ,lev·əl }

dopant *See* doping agent. { 'dō·pənt }

dope |ELECTR| *See* doping agent. |MATER| A cellulose ester lacquer used as an adhesive or a coating. { dōp }

doped junction |ELECTR| A junction produced by adding an impurity to the melt during growing of a semiconductor crystal. { ¦dōpt 'jəŋk·shən }

doped solder |MET| Solder having an element added to it to ensure retention of a quality of the base metal on which it is used. { ¦dōpt 'säd·ər }

doping |ELECTR| The addition of impurities to a semiconductor to achieve a desired characteristic, as in producing an *n*-type or *p*-type material.

Also known as semiconductor doping. |ENG| Coating the mold or mandrel with a substance which will prevent the molded plywood part from sticking to it and will facilitate removal. { 'dōp·iŋ }

doping agent |ELECTR| An impurity element added to semiconductor materials used in crystal diodes and transistors. Also known as dopant; dope. { 'dōp·iŋ ,ā·jənt }

dore |MET| Gold and silver bullion remaining in a cupeling furnace after removal of the oxidized lead. { də'rā }

double-acting hammer |MET| A forging hammer in which the ram is raised and forced down by a charge of air or steam. { ¦dəb·əl ¦ak·tiŋ 'ham·ər }

double-action die |MET| A die designed to perform more than one operation with each stroke of the press. { ¦dəb·əl ¦ak·shən 'dī }

double-action forming |MET| Forming in which more than one shape is imparted by each stroke of the press. { ¦dəb·əl ¦ak·shən 'fôr·miŋ }

double aging |MET| Introduction of a primary (stabilizing) and a secondary aging treatment to control the precipitate formed from a supersaturated alloy and to achieve specific properties in the material. { ¦dəb·əl 'āj·iŋ }

double arcing |MET| An occurrence in plasma arc welding and cutting where a secondary electric arc displaces the main arc at the outlet of a welding nozzle. { ¦dəb·əl 'ärk·iŋ }

double-bevel groove weld |MET| A type of groove weld in which one member has a joint edge beveled on both sides. Also known as double-V groove weld. { ¦dəb·əl ¦bev·əl 'grüv ,weld }

double bond |PHYS CHEM| A type of linkage between atoms in which two pair of electrons are shared equally. { ¦dəb·əl 'bänd }

double-J groove weld |MET| A groove weld in which one member has a joint edge in the form of a double J or two half U's, one from each side. Also known as double-U groove weld. { ¦dəb·əl 'jā ,grüv ,weld }

double layer *See* electric double layer. { ¦dəb·əl 'lā·ər }

double nickel salt *See* nickel ammonium sulfate. { ¦dəb·əl ¦nik·əl 'sôlt }

doubler |MATER| A localized area of extra layers of reinforcement on a section of composite material to provide extra strength to a specified site. { 'dəb·lər }

double refraction *See* birefringence. { ¦dəb·əl ri'frak·shən }

double salt |INORG CHEM| **1.** A salt that upon hydrolysis forms two different anions and cations. **2.** A salt that is a molecular combination of two other salts. { ¦dəb·əl 'sôlt }

double tempering |MET| A technique for ensuring the stability of the microstructure of a quench-hardened steel by subjecting it to two tempering cycles at approximately the same temperature. { ¦dəb·əl 'tem·pər·iŋ }

double-U groove weld *See* double-J groove weld. { ¦dəb·əl 'yü 'grüv ,weld }

double-V groove weld *See* double-bevel groove weld. { ¦dəb·əl 've 'grüv ,weld }

double-welded joint [MET] Any joint that has been welded from both sides. { ¦dəb·əl ,weld·əd 'jȯint }

douse [MET] To thrust a hot piece of metal into a liquid during the hardening process. { daús }

dovetailing [MET] In thermal spraying, roughening the surface by angular cutting prior to the deposit of sprayed material. { 'dəv,tāl·iŋ }

down cutting *See* climb cutting. { 'daún ,kəd·iŋ }

downhand welding *See* flat-position welding. { 'daún,hand 'weld·iŋ }

down milling *See* climb cutting. { 'daún ,mil·iŋ }

downslope time [MET] Time necessary for current decrease when using slope control in resistance welding. { 'daún,slōp ,tīm }

draft [MET] **1.** The act or process of drawing, with dies. **2.** The work or quantity of work drawn. { draft }

drafting paper [MATER] A fine white or cream-colored paper that is hard-surfaced and has good erasing characteristics. { 'draf·tiŋ ,pā·pər }

drag [MET] **1.** The bottom part of a flask used in casting. **2.** In thermal cutting, the distance deviating from the theoretical vertical line of cutting measured along the bottom surface of the material. { drag }

drag-in [MET] Solution carried by the work and handling equipment to another solution. { 'drag ,in }

drag-out [MET] Solution taken from a bath by the work or handling equipment. { 'drag ,aút }

drag technique [MET] An arc-welding method in which the electrode is in contact with the joint being welded without being in short circuit. { 'drag tek,nēk }

drain [ELECTR] The region into which majority carriers flow in a field-effect transistor; it is comparable to the collector of a bipolar transistor and the anode of an electron tube. { drān }

draw [MET] **1.** A fissure or pocket in a casting formed when the supply of molten metal is inadequate during solidification. **2.** To remove a pattern from a foundry flask. { drȯ }

drawability [MET] The ability of a metal to be deep-drawn. { ¦drȯ·ə'bil·əd·ē }

draw bead [MET] A projection on the surface of a metal sheet to control its flow during drawing. { 'drȯ ,bēd }

drawbench [MET] A stand on which metal is drawn through dies; used in wire-making, or for drawing of rods and tubing. { 'drȯ,bench }

drawhead [MET] A group of rollers through which strip tubing or solid stock is drawn to form angled sections. { 'drȯ,hed }

drawing [CHEM ENG] Removing ceramic ware from a kiln after it has been fired. [MET] **1.** Pulling a wire or tube through a die to reduce the cross section. **2.** Forcing plastic deformation of metal in a die to form recessed parts. { 'drȯ·iŋ }

drawing back [MET] Reheating hardened steel to a temperature below the critical temperature in order to change its hardness. { 'drȯ·iŋ 'bak }

drawing bristol [MATER] A cardboard made of 100% cotton in the higher grades; has good characteristics of permanence, strength, and erasability. { 'drȯ·iŋ 'brist·əl }

drawing cloth [MATER] A linen cloth that is specially treated to be smooth and translucent so that it may be used for ink tracings. { 'drȯ·iŋ ,klȯth }

drawing compound [MET] A material applied to the work during drawing or pressing operations to eliminate draw marks by preventing direct contact between the work and die. { 'drȯ·iŋ ,käm,paúnd }

drawing die [MET] A die that forms sheet metal into cuplike, wrinkle-free shapes. { 'drȯ·iŋ ,dī }

drawing of temper [MET] The process of heating steel to red heat and then letting it cool slowly; opposite of hardening or tempering. { 'drȯ·iŋ əv 'tem·pər }

drawing out [MET] Lengthening of a piece of metal through a heating and hammering process, resulting in a proportional reduction in section area. [TEXT] The action of pulling staple textile fibers lengthwise over each other, producing longer and thinner slivers. { ¦drȯ·iŋ 'aút }

drawing paper [MATER] One of a wide variety of papers used for pen- and-pencil drawing by artists and architects. { 'drȯ·iŋ ,pā·pər }

draw mark [MET] An impairment of the die or metal surface caused during drawing due to friction or a defect in the die; examples are scoring and die lines. { 'drȯ ,märk }

drawn finish [MET] A smooth, bright finish on metal fabrications such as tubing or wire that is obtained by drawing the metal through a die. { 'drȯn 'fin·ish }

drawn glass [MATER] Glass made automatically by drawing the molten material through rollers. { ¦drȯn 'glas }

draw piece [MET] Any part made by drawing. { 'drȯ ,pēs }

drawplate [MET] A circular plate having a central hole through which wire is drawn by a punch. Also known as draw ring. { 'drȯ,plāt }

draw radius [MET] A measure of cutting edge of a die or punch over which the metal is drawn. { 'drȯ,rād·ē·əs }

draw ring *See* drawplate. { 'drȯ ,riŋ }

drier [MATER] **1.** A substance that absorbs water. **2.** A substance that is used to hasten solidification. **3.** Material, such as salts of lead, manganese, and cobalt, which facilitates the oxidation of oils; used in paints and varnishes to speed drying. { 'drī·ər }

drift [SOLID STATE] The movement of current carriers in a semiconductor under the influence of an applied voltage. { drift }

drift mobility [SOLID STATE] The average drift velocity of carriers per unit electric field in a homogeneous semiconductor. Also known as mobility. { 'drift mō'bil·əd·ē }

drift velocity [SOLID STATE] The average velocity of a carrier that is moving under the influence of an electric field in a semiconductor, conductor, or electron tube. { 'drift və'läs·əd·ē }

drilled extrusion ingot [MET] A hollow extrusion ingot made from a solid cast extrusion ingot by drilling. { ¦drild ik'strü·zhən ‚iŋ·gət }

drill steel [MET] Steel with at least 0.85% carbon content made by the electric furnace process. Formerly known as crucible steel, when made by the crucible process. { 'dril ‚stēl }

drip [MATER] **1.** Oil which comes through the cloth of a paraffin wax press. **2.** Filter drainings too dark to be included in filter stock. { drip }

driving force [CHEM] In a chemical reaction, the formation of products such as an insoluble compound, a gas, a nonelectrolyte, or a weak electrolyte that enable the reaction to go to completion. { 'drīv·iŋ ‚fórs }

drop [FL MECH] The quantity of liquid that coalesces into a single globule; sizes vary according to physical conditions and the properties of the fluid itself. [MET] A casting defect due to the falling of a portion of sand from an overhanging section of the mold. { dräp }

drop black [MATER] Black pigment shaped into droplets. { 'dräp ‚blak }

drop forging [MET] Plastic deformation of hot metal under a falling weight, such as a drop hammer. { 'dräp ‚fórj·iŋ }

drop hammer [MET] A hammer used in forging that is raised and then dropped on the metal resting on an anvil or on a die. { 'dräp ‚ham·ər }

droplet condensation [THERMO] The formation of numerous discrete droplets of liquid on a wall in contact with a vapor, when the wall is cooled below the local vapor saturation temperature and the liquid does not wet the wall. { 'dräp·lət ‚kän·dən'sā·shən }

dropping point [CHEM] The temperature at which grease changes from a semisolid to a liquid state under standardized conditions. { 'dräp·iŋ ‚póint }

dropping test [MET] A chemical method for determining thickness of zinc and cadmium plated coatings on metal in which a reagent is dropped on the surface until the basis metal is exposed. { 'dräp·iŋ ‚test }

dropwise condensation [THERMO] Condensation of a vapor on a surface in which the condensate forms into drops. { 'dräp‚wīz ‚kän·dən'sā·shən }

dross [MET] An impurity, usually an oxide, formed on the surface of a molten metal. { drós }

drossing [MET] A process used in nonferrous pyrometallurgy for removing solid oxide deposits on the surface of a molten metal. { 'drós·iŋ }

Drude's theory of conduction [SOLID STATE] A theory which treats the electrons in a metal as a gas of classical particles. { 'drüdz ‚thē·ə·rē əv kən'dək·shən }

dry acid [CHEM] Nonaqueous acetic acid used for oil-well reservoir acidizing treatment. { ¦drī 'as·əd }

dry assay [MET] Determination of the amount of a desired constituent in ores, metallurgical

residues, and alloys by methods other than those involving liquid means of separation. { ¦drī 'a ‚sā }

dry box [CHEM] A container or chamber filled with a controlled, low-humidity, or an inert atmosphere in which manipulation of very reactive chemicals is carried out in the laboratory. { 'drī ‚bäks }

dry coloring [CHEM ENG] A plastics coloring method in which uncolored particles of the plastic material are tumble-blended with selected dyes and pigments. [ENG] A method to color plastics by tumbleblending colorless plastic particles with dyes and pigments. [MATER] A powdered form of pigment. { 'drī ‚kəl·ə·riŋ }

dry corrosion [MET] Destruction of a metal or alloy by chemical processes resulting from attack by gases in the atmosphere above the dew point. { ¦drī kə'rō·zhən }

dry friction [MECH] Resistance between two dry solid surfaces, that is, surfaces free from contaminating films or fluids. { ¦drī 'frik·shən }

dry ice [INORG CHEM] Carbon dioxide in the solid form, usually made in blocks to be used as a coolant; changes directly to a gas at $-78.5°C$ as heat is absorbed. { ¦drī 'īs }

drying [CHEM] **1.** An operation in which a liquid, usually water, is removed from a wet solid in equipment termed a dryer. **2.** A process of oxidation whereby a liquid such as linseed oil changes into a solid film. { 'drī·iŋ }

drying agent [CHEM] Soluble or insoluble chemical substance that has such a great affinity for water that it will abstract water from a great many fluid materials; soluble chemicals are calcium chloride and glycerol, and insoluble chemicals are bauxite and silica gel. Also known as desiccant. { 'drī·iŋ ‚ā·jənt }

drying oil [MATER] Relatively highly unsaturated oil, such as cottonseed, soybean, and linseed oil, that is easily oxidized and polymerized to form a hard, dry film on exposure to air; used in paints and varnish. { 'drī·iŋ ‚óil }

dry ink [MATER] A finely powdered mixture of resin and pigment that is deposited to form an image in electrophotography. { 'drī 'iŋk }

dry mortar [MATER] A mortar that is significantly stiffer and has lower water content than standard mortar but contains sufficient water for hydration. { ¦drī 'mórd·ər }

dry plasma etching See plasma etching. { ¦drī 'plaz·mə 'ech·iŋ }

dry sand mold [MET] A mold made of greensand and then dried in an oven to increase its strength. { ¦drī ¦sand ‚mōld }

dry strength [ENG] The strength of an adhesive joint determined immediately after drying under specified conditions or after a period of conditioning in the standard laboratory atmosphere. { 'drī ‚streŋkth }

dry wire drawing [MET] The drawing of dry steel process wire, pretreated by acid cleaning, lime coating, and baking, through a lubricant and wire drawing frame. { 'drī ‚wīr ‚dró·iŋ }

dual-function catalyst   See bifunctional catalyst. { ¦dül ¦fəŋk·shən 'kad·ə·list }

dubnium   [CHEM] A chemical element, symbolized Db, atomic number 105, a synthetic element; the thirteenth transuranium element. { 'düb·nē·əm }

ductile fracture   See fibrous fracture. { ¦dək·təl 'frak·chər }

ductile iron   See nodular cast iron. { ¦dək·təl 'ī·ərn }

ductility   [MATER] The ability of a material to be plastically deformed by elongation, without fracture. { dək'til·əd·ē }

Dufour effect   [THERMO] Energy flux due to a mass gradient occurring as a coupled effect of irreversible processes. { ¦dü·fȯr i'fekt }

Dufour number   [THERMO] A dimensionless number used in studying thermodiffusion, equal to the increase in enthalpy of a unit mass during isothermal mass transfer divided by the enthalpy of a unit mass of mixture. Symbol Du_2. { ¦dü·fȯr ₁nəm·bər }

Duhem-Margules equation   [THERMO] An equation showing the relationship between the two constituents of a liquid-vapor system and their partial vapor pressures:

$$\frac{d \ln p_A}{d \ln x_A} = \frac{d \ln p_B}{d \ln x_B}$$

where x_A and x_B are the mole fractions of the two constituents, and p_A and p_B are the partial vapor pressures. { dü'em 'mär·gyə·lēz i₁kwā·zhən }

Dulong-Petit law   [THERMO] The law that the product of the specific heat per gram and the atomic weight of many solid elements at room temperature has almost the same value, about 6.3 calories (264 joules) per degree Celsius. { də'lȯŋ pə'tē ₁lȯ }

dummy   [MET] A cathode that undergoes electroplating at low current densities. { 'dəm·ē }

dummy block   [MET] A thick disk positioned between the ram and billet in extrusion working to prevent the ram from overheating. { 'dəm·ē ₁bläk }

Duovac method   [MET] Technique for testing for defects in magnetic parts; a moving magnetic field magnetizes the part in many directions, and the part is then sprayed with fluorescent magnetic particles and examined under ultraviolet light so that defects become apparent. { 'dü·ə ₁vak ₁meth·əd }

duplexing   See duplex process. { 'dü₁pleks·iŋ }

duplex iron   [MET] Cast iron heated in an electric furnace after it has been melted in a cupola. { ¦dü₁pleks ₁ī·ərn }

duplex practice   See duplex process. { 'dü₁pleks ₁prak·təs }

duplex process   [MET] A two-step procedure in which steel is refined by one process (usually the

Bessemer process) and finished by another process (usually open-hearth or electric-furnace). Also known as duplexing; duplex practice. { 'dü ₁pleks ₁präs·əs }

Dupré equation   [THERMO] The work W_{LS} done by adhesion at a gas-solid-liquid interface, expressed in terms of the surface tensions γ of the three phases, is $W_{LS} = \gamma_{GS} + \gamma_{GL} - \gamma_{LS}$. { dü'prā i₁kwä·zhən }

durometer   [ENG] An instrument consisting of a small drill or blunt indenter point under pressure; used to measure hardness of metals and other materials. { də'räm·əd·ər }

durometer hardness   [ENG] The hardness of a material as measured by a durometer. { də'räm·əd·ər ₁härd·nəs }

Durville process   [MET] A casting process involving the attachment of an inverted mold to the top of a crucible; the metal is melted in the bottom of the crucible, and then the molten metal is decanted into the mold by inverting the entire apparatus. { 'dər·vil ₁präs·əs }

dusting   [MET] Spontaneous disintegration of a material on cooling due to expansion or inversion. { 'dəst·iŋ }

Dutch metal   [MET] An alloy of 80% copper and 20% zinc that is ductile, is easily drawn, and takes a high polish; used for low-priced jewelry. { 'dəch ₁med·əl }

duty cycle   [MET] The percentage of time that current flows in equipment over a specific period during electric resistance welding. { 'düd·ē ₁sī·kəl }

Dy   See dysprosium.

dye   [CHEM] A colored substance which imparts more or less permanent color to other materials. Also known as dyestuff. { dī }

dye penetrant   [MET] A dye-containing liquid used for detecting cracks or other surface defects in nonmagnetic materials. { ¦dī 'pen·ə·trənt }

dye-retarding agent   [MATER] Materials that decrease the rate of dye absorption, preventing rapid exhaustion of dye baths. { 'dī ri₁tärd·iŋ ₁ā·jənt }

dyestuff   See dye. { 'dī₁stəf }

dynamic creep   [MECH] Creep resulting from fluctuations in a load or temperature. { dī ¦nam·ik 'krēp }

dynamic load   [CIV ENG] A force exerted by a moving body on a resisting member, usually in a relatively short time interval. { dī¦nam·ik 'lōd }

dyne-centimeter   See erg. { ¦dīn 'sen·tə₁mēd·ər }

dyne-cm   See erg. { ¦dək·təl 'frak·chər }

dysprosium   [CHEM] A metallic rare-earth element, symbol Dy, atomic number 66, atomic weight 162.50. { dis'prō·zē·əm }

dystetic mixture   [PHYS CHEM] A mixture of two or more substances that has the highest possible melting point of all mixtures of these substances. { di'sted·ik 'miks·chər }

E

earing |MET| Formation of scallops around the top edge of a deep-drawn product due to differences in the directional properties of the metal sheet. { 'ir·iŋ }

Earnshaw's theorem |ELEC| The theorem that a charge cannot be held in stable equilibrium by an electrostatic field. { 'ərn,shȯz ,thir·əm }

earth See ground. { ərth }

earth detector See leakage indicator. { 'ərth di'tek·tər }

earthquake-resistant |CIV ENG| Of a structure or building, able to withstand lateral seismic stresses at the base. { 'ərth,kwāk ri,zis·tənt }

easy glide |SOLID STATE| A large increase in plastic deformation of a single crystal accompanying a small increase in stress as the result of the passage of many thousands of dislocations through the crystal along a single glide system. { 'ē·zē ,glīd }

EBM See electron-beam machining.

ebonite See hard rubber. { ēb·ə,nīt }

ebulliometer |PHYS CHEM| The instrument used for ebulliometry. { ə,bú·lē'äm·əd·ər }

ebulliometry |PHYS CHEM| The precise measurement of the absolute or differential boiling points of solutions. { ə,bú·lē'äm·ə·trē }

E center See R center. { 'ē ,sen· tər }

ECM See electrochemical machining.

eddy conduction See eddy heat conduction. { 'ed·ē kən,dək·shən }

eddy conductivity |THERMO| The exchange coefficient for eddy heat conduction. { 'ed·ē ,kän ,dək·tiv·əd·ē }

eddy current |ELECTROMAG| An electric current induced within the body of a conductor when that conductor either moves through a nonuniform magnetic field or is in a region where there is a change in magnetic flux. Also known as Foucault current. { 'ed·ē ,kə·rənt }

eddy-current heating See induction heating. { 'ed·ē ,kə·rənt ,hēd·iŋ }

eddy-current loss |ELECTROMAG| Energy loss due to undesired eddy currents circulating in a magnetic core. { 'ed·ē ,kə·rənt ,lȯs }

eddy-current test |ELECTROMAG| A nondestructive test in which the change of impedance of a test coil brought close to a conducting specimen indicates the eddy currents induced by the coil, and thereby indicates certain properties or defects of the specimen. { 'ed·ē ,kə·rənt ,test }

eddy heat conduction |THERMO| The transfer of heat by means of eddies in turbulent flow, treated analogously to molecular conduction. Also known as eddy heat flux; eddy conduction. { 'ed·ē 'hēt kən'dək·shən }

eddy heat flux See eddy heat conduction. { 'ed· ē 'hēt ,fləks }

eddy stress See Reynolds stress. { 'ed·ē ,stres }

edge dislocation |CRYSTAL| A dislocation which may be regarded as the result of inserting an extra plane of atoms, terminating along the line of the dislocation. Also known as Taylor-Orowan dislocation. { 'ej ,dis·lō,kā·shən }

edge effect |ELEC| An outward-curving distortion of lines of force near the edges of two parallel metal plates that form a capacitor. { 'ej i,fekt }

edge excitation |CRYO| An excitation of a droplet of incompressible quantum Hall liquid in which a surface wave propagates along the edge of the droplet in the same direction that electrons drift along the edge, as determined by the direction of the magnetic field. { 'ej ,ek· sī'tā·shən }

edge grain |MATER| The grain pattern produced when soft wood is cut so that the tree's annular rings form an angle of more than 45° with the board's surface. { 'ej ,grān }

edge joint |MET| A joint between the edges of welded members which are essentially parallel to each other. { 'ej ,jȯint }

edger |MET| The part of a forging die which portions out the quantity of metal needed for shaping. { 'ej·ər }

edging |MET| Controlling the plate width or the edge shape during rolling operations. { 'ej·iŋ }

EDM See electron discharge machining.

EDTC See electrochemical machining.

eff See efficiency.

effective ampere |ELEC| The amount of alternating current flowing through a resistance that produces heat at the same average rate as 1 ampere of direct current flowing in the same resistance. { ə¦fek·tiv 'am,pir }

effective capacitance |ELEC| Total capacitance existing between any two given points of an electric circuit. { ə¦fek·tiv kə'pas·əd·əns }

effective mass |SOLID STATE| A parameter with the dimensions of mass that is assigned to electrons in a solid; in the presence of an external electromagnetic field the electrons behave in many respects as if they were free, but with a mass equal to this parameter rather than the true mass. { ə|fek·tiv 'mas }

effective resistance *See* high-frequency resistance. { ə|fek·tiv ri'zis·təns }

effervescence [CHEM] The bubbling of a solution of an element or chemical compound as the result of the emission of gas without the application of heat; for example, the escape of carbon dioxide from carbonated water. { ,ef·ər'ves·əns }

efficiency Abbreviated eff. [CHEM] In an ion-exchange system, a measurement of the effectiveness of a system expressed as the amount of regenerant required to remove a given unit of adsorbed material. [PHYS] The ratio, usually expressed as a percentage, of the useful power output to the power input of a device. [THERMO] The ratio of the work done by a heat engine to the heat energy absorbed by it. Also known as thermal efficiency. { ə'fish·ən·sē }

efflorescence [CHEM] The property of hydrated crystals to lose water of hydration and crumble when exposed to air. [MATER] A crust of salts, usually white, that forms on the surface of stone, brick, plaster, or mortar because of leaching of free alkalies from adjacent concrete or mortar. { ,ef·lə'res·əns }

Egerton's effusion method [THERMO] A method of determining vapor pressures of solids at high temperatures, in which one measures the mass lost by effusion from a sample placed in a tightly sealed silica pot with a small hole; the pot rests at the bottom of a tube that is evacuated for several hours, and is maintained at a high temperature by a heated block of metal surrounding it. { |ej·ər·tənz ə'fyü·zhən ,meth·əd }

Eggertz's method [MET] A colorimetric estimation of the carbon content of steel by dissolving the metal in nitric acid and comparing the color with that produced by a similar metal of known carbon content. { 'e·gərts·əz ,meth·əd }

E glass [MATER] A type of borosilicate glass used to produce glass fibers for reinforced plastics designed for applications requiring high electrical resistivity. Also known as electric glass. { 'ē ,glas }

Ehrenfest's equations [THERMO] Equations which state that for the phase curve P(T) of a second-order phase transition the derivative of pressure P with respect to temperature T is equal to $(C fp − C i p)/TV(\gamma^f − \gamma^i) = (\gamma^f − \gamma^i)/(K^f − K^i)$, where i and f refer to the two phases, γ is the coefficient of volume expansion, K is the compressibility, C_p is the specific heat at constant pressure, and V is the volume. { 'er·ən,fests i,kwā·zhənz }

eicosane [MATER] A mixture of saturated hydrocarbons mostly straight-chained and averaging 20 carbons in the chain; for this reason, the formula $C_{20}H_{42}$ is given to the technical mixture; used in lubricants and plasticizers. { 'ī·kə ,sān }

einstein [PHYS] A unit of light energy used in photochemistry, equal to Avogadro's number times the energy of one photon of light of the frequency in question. { 'īn,stīn }

Einstein characteristic temperature [SOLID STATE] A temperature, characteristic of a substance, that appears in Einstein's equation for specific heat; it is equal to the product of Planck's constant and the Einstein frequency divided by Boltzmann's constant. { |īn,stīn ,kar·ik·tə|ris·tik 'tem·prə·chər }

Einstein condensation *See* Bose-Einstein condensation. { 'īn,stīn ,kän·dən'sā·shən }

Einstein-de Haas effect [ELECTROMAG] A freely suspended body consisting of a ferromagnetic material acquires a rotation when its magnetization changes. { 'īn,stīn də'häs i,fekt }

Einstein-de Haas method [ELECTROMAG] Method of measuring the gyromagnetic ratio of a ferromagnetic substance; one measures the angular displacement induced in a ferromagnetic cylinder suspended from a torsion fiber when magnetization of the object is reversed, and the magnetization change is measured with a magnetometer. { 'īn,stīn də'häs ,meth·əd }

Einstein frequency [SOLID STATE] Single frequency with which each atom vibrates independently of other atoms, in a model of lattice vibrations; equal to the frequency observed in infrared absorption studies. { 'īn,stīn ,frē·kwən·sē }

Einstein frequency condition [SOLID STATE] The assumption that all vibrations of a crystal lattice are harmonic with the same characteristic frequency. { 'īn,stīn 'frē·kwən·se kən,dish·ən }

einsteinium [CHEM] Synthetic radioactive element, symbol Es, atomic number 99; discovered in debris of 1952 hydrogen bomb explosion; now made in cyclotrons. { īn'stīn·ē·əm }

Einstein photochemical equivalence law [PHYS CHEM] The law that each molecule taking part in a chemical reaction caused by electromagnetic radiation absorbs one photon of the radiation. Also known as Stark-Einstein law. { 'īn,stīn |fōd·ō|kem·ə·kəl i'kwiv·ə·ləns ,ló }

Einstein relation [PHYS] The relation in which the mobility of charges in an ionic solution or semiconductor is equal to the magnitude of the charge times the diffusion coefficient divided by the product of the Boltzmann constant and the absolute temperature. { 'īn,stīn ri,lā·shən }

Einstein's equation for specific heat [SOLID STATE] The earliest equation based on quantum mechanics for the specific heat of a solid; uses the assumption that each atom oscillates with the same frequency. { 'īn,stīnz i'kwä·zhən fər spə,si·fik 'hēt }

Einstein viscosity equation [PHYS CHEM] An equation which gives the viscosity of a sol in terms of the volume of dissolved particles divided by the total volume. { 'īn,stīn vis'käs·əd·ē i,kwä·zhən }

Einthoven galvanometer *See* string galvanometer. { 'īnt,hō·vən ,gal·və'näm·əd·ər }

ejector half [MET] The movable half of a casting die. { ē'jek·tər ,haf }

elastance [ELEC] The reciprocal of capacitance. { i'las·təns }

auxiliary anode |MET| A supplementary anode that alters the current distribution in electroplating to give a more uniform plating thickness. { òg'zil·yə·rē 'an,ōd }

auxiliary electrode |PHYS CHEM| An electrode in an electrochemical cell used for transfer of electric current to the test electrode. { òg'zil·yə·rē i'lek,trōd }

auxochrome |CHEM| Any substituent group such as $-NH_2$ and $-OH$ which, by affecting the spectral regions of strong absorption in chromophores, enhance the ability of the chromogen to act as a dye. { 'òk·sə,krōm }

average bond dissociation energy |PHYS CHEM| The average value of the bond dissociation energies associated with the homolytic cleavage of several bonds of a set of equivalent bonds of a molecule. Also known as bond energy. { 'av·rij ¦bänd di·sō·sē'ā·shən ,en·ər·jē }

average molecular weight |POLYM CHEM| The calculated number to average the molecular weights of the varying-length polymer chains present in a polymer mixture. { 'av·rij mə'lek·yə·lər 'wāt }

Avogadro's hypothesis See Avogadro's law. { ¦a·və¦gäd·drōz hī'päth·ə·səs }

Avogadro's law |PHYS| The law which states that under the same conditions of pressure and temperature, equal volumes of all gases contain equal numbers of molecules; for example, 359 cubic feet at 32°F and 1 atmosphere for a perfect gas. Also known as Avogadro's hypothesis. { ¦a·və¦gäd·drōz ,lò }

Avogadro's number |PHYS| The number (6.02×10^{23}) of molecules in a gram-molecular weight of a substance. { ¦a·və¦gäd·drōz ,nəm·bər }

AWG See American wire gage.

axial angle |CRYSTAL| **1.** The acute angle between the two optic axes of a biaxial crystal. Also known as optic angle; optic-axial angle. **2.** In air, the larger angle between the optic axes after refraction on leaving the crystal. { 'ak·sē·əl 'aŋ·gəl }

axial element |CRYSTAL| The lengths, length ratios, and angles which define a crystal's unit cell. { 'ak·sē·əl 'el·ə·mənt }

axial load |MECH| A force with its resultant passing through the centroid of a particular section and being perpendicular to the plane of the section. { 'ak·sē·əl 'lōd }

axial modulus |MECH| The ratio of a simple tension stress applied to a material to the resulting strain parallel to the tension when the sides of the sample are restricted so that there is no lateral deformation. Also known as modulus of simple longitudinal extension. { 'ak·sē·əl 'mäj·ə·ləs }

axial plane |CRYSTAL| **1.** A plane that includes two of the crystallographic axes. **2.** The plane of the optic axis of an optically biaxial crystal. { 'ak·sē·əl 'plān }

axial ratio |CRYSTAL| The ratio obtained by comparing the length of a crystallographic axis with one of the lateral axes taken as unity. { 'ak·sē·əl 'rā·shō }

axial surface See axial plane. { 'ak·sē·əl 'sər·fəs }

axial winding |MATER| A winding used in filament-wound fiberglass-reinforced plastic construction in which the filaments run along the axis at a zero helix angle. { 'ak·sē·əl 'wind·iŋ }

axis |MECH| A line about which a body rotates. { 'ak·səs }

axis of weld |MET| A line along a weld used to describe the positions of the localized welds. { 'ak·səs əv 'weld }

azeotrope See azeotropic mixture. { ā'zē·ə ,trōp }

azeotropic mixture |CHEM| A solution of two or more liquids, the composition of which does not change upon distillation. Also known as azeotrope. { ¦a,zē·ə¦träp·ik 'miks·chər }

azide |ORG CHEM| One of several types of compounds containing the $-N_3$ group and derived from hydrazoic acid, HN_3. { 'ā,zīd }

azimuthal orthomorphic projection See stereographic projection. { ,az·ə'məth·əl ¦òrth·ə ¦mòr·fik prə'jek·shən }

azine |ORG CHEM| A compound of six atoms in a ring; at least one of the atoms is nitrogen, and the ring structure resembles benzene; an example is pyridine. { 'ā,zēn }

azo- |ORG CHEM| A prefix indicating the group $-N=N-$. { 'a·zō }

azo compound |ORG CHEM| A compound having two organic groups separated by an azo group ($-N=N-$). { 'ā·zō ,käm,pau̇nd }

azo dyes |ORG CHEM| Widely used commercial dyestuffs derived from amino compounds, with the $-N-$ chromophore group; can be made as acid, basic, direct, or mordant dyes. { 'a·zō ,dīz }

B

B *See* boron.

Ba *See* barium.

babbitt metal [MET] Any of the white alloys composed primarily of tin or lead and of lesser amounts of antimony, copper, and perhaps other metals, and used for bearings. { 'bab·ət ,med·əl }

back bond [SOLID STATE] A chemical bond between an atom in the surface layer of a solid and an atom in the second layer. { 'bak ,bänd }

back draft [MET] A reversed taper given to a casting model or pattern to prevent its withdrawal from the mold. { 'bak ,draft }

back gouging [MET] The elimination of excess material from both weld metal and base metal on the opposite side of a partly welded joint; a groove or bevel is formed in order to facilitate complete joint penetration. { 'bak ,gaüj·iŋ }

backhand welding [MET] A welding technique in which the flame is directed back against the completed weld. Also known as backward welding. { 'bak,hand 'weld·iŋ }

backing [ELECTR] Flexible material, usually cellulose acetate or polyester, used on magnetic tape as the carrier for the oxide coating. [MET] *See* backing strip. { 'bak·iŋ }

backing strip [MET] A piece of metal, asbestos, or other nonflammable material placed behind a joint to facilitate welding. Also known as backing. { 'bak·iŋ ,strip }

backout [MET] Process of nullifying the effect of positive electrical potentials occurring in an anodic area in a cathodic protection system. { 'bak,aüt }

back-reflection photography [CRYSTAL] A method of studying crystalline structure by x-ray diffraction in which the photographic film is placed between the source of x-rays and the crystal specimen. { 'bak ri'flek·shən fə'täg·rə·fē }

backstep sequence [MET] Sequential deposition of weld beads in the direction opposite to the direction of welding. { 'bak,step ,sē·kwəns }

backup [MET] A support used to balance the upsetting force in the workpieces during flash welding. { 'bak,əp }

back veneer [MATER] In veneer plywood, the layer of veneer on the side of a plywood sheet which is opposite the face veneer; usually of lower quality. { 'bak və,nir }

backward welding *See* backhand welding. { 'bak·wərd 'weld·iŋ }

backweld [MET] A weld placed behind a single groove weld. { 'bak,weld }

bainite [MET] Steel formed by austempering, having an acicular structure of ferrite and carbides, exhibiting considerable toughness, and combining high strength with high ductility. { 'bā,nīt }

baked core [MET] In sand castings, a core which has been heated in core stoves or by dielectric heating to attain uniform physical properties. { 'bākt ¦kór }

baked permeability [MET] The property of a molded mass of sand which permits passage of gases through it when molten metal is poured in a mold, baked above 230°F (110°C), and then cooled. { 'bākt ,per·mē·ə'bil·əd·ē }

baked strength [MET] The strength of a molded sand mixture when baked above 230°F (110°C) and then cooled to ambient temperature. { 'bākt ¦streŋkth }

baking [ENG] The use of heat on fresh paint films to speed the evaporation of thinners and to promote the reaction of binder components so as to form a hard polymeric film. [MET] Heating metal at low temperatures to remove gases. { 'bāk·iŋ }

baking finish [MATER] Varnish or paint that must be baked at temperatures greater than 150°F (66°C) to develop desired final properties of strength and hardness. { 'bāk·iŋ ,fin·ish }

baking soda *See* sodium bicarbonate. { 'bāk·iŋ ,sōd·ə }

baking varnish [MATER] A chemical-resistant varnish made of synthetic resins that requires baking to be dried. { 'bāk·iŋ ,vär·nəsh }

balata [MATER] A hard substance, similar to gutta-percha, used mainly in golf balls and belting, which is made by drying the milky juice of the bully tree (*Manilkara bidentata*). Also known as gutta-balata. { bə'läd·ə }

ballast [ELEC] A circuit element that serves to limit an electric current or to provide a starting voltage, as in certain types of lamps, such as in fluorescent ceiling fixtures. [MATER] Coarse gravel used as an ingredient in concrete. { 'bal·əst }

ball burnishing [MET] A method of giving small stainless steel parts a lustrous finish by rotating them in a wood-lined barrel with water, burnishing soap, and hardened steel balls. { 'bȯl ˌbər·nish·iŋ }

ball clay [MATER] A clay used in ceramics that is characterized by strong binding properties, a tendency to ball, and excellent plasticity. { 'bȯl ˌklā }

balling-up [MET] Formation of balls of molten brazing filler metal when the base material has not been sufficiently wetted. { ¦bȯl·iŋ ¦əp }

ball sizing [MET] Finishing a hole to a precise diameter and burnishing the surface by forcing a steel ball through it. { 'bȯl ˌsīz·iŋ }

band [SOLID STATE] A restricted range in which the energies of electrons in solids lie, or from which they are excluded, as understood in quantum-mechanical terms. Also known as energy band. { band }

banded structure [MET] The appearance of a metal showing light and dark parallel bands in the direction of rolling or working. { 'ban·dəd 'strək·chər }

band gap [SOLID STATE] An energy difference between two allowed bands of electron energy in a metal. { 'band ˌgap }

band scheme [SOLID STATE] The identification of energy bands of a solid with the levels of independent atoms from which they arise as the atoms are brought together to form the solid, together with the width and spacing of the bands. { 'band ˌskēm }

B and S gage See American wire gage. { ¦bē ən ¦es ˌgāj }

band theory of ferromagnetism [SOLID STATE] A theory according to which ferromagnetism is caused by electrons in the unfilled energy bands of a crystal. { ¦band ˌthē·ə·rē əv ˌfer·ō'mag·nə ˌtiz·əm }

band theory of solids [SOLID STATE] A quantum-mechanical theory of the motion of electrons in solids that predicts certain restricted ranges or bands for the energies of these electrons. Also known as energy-band theory of solids. { 'band ˌthē·ə·rē əv ¦säl·ədz }

bank gravel See bank-run gravel. { 'baŋk ˌgrav·əl }

bank-run gravel [MATER] Aggregate taken directly from natural deposits; contains both large and small stones. Also known as bank gravel; run-of-bank gravel. { 'baŋk ˌrən 'grav·əl }

bar [MECH] A unit of pressure equal to 10^5 pascals, or 10^5 newtons per square meter, or 10^6 dynes per square centimeter. [MET] An elongated piece of metal of simple uniform cross-section dimensions, usually rectangular, circular, or hexagonal, produced by forging or hot rolling. Also known as barstock. { bär }

barberite [MET] A nonferrous alloy with good resistance to sulfuric acid, sea water, and mine waters; 88.5% copper, 5% nickel, 5% tin, 1.5% silicon. { 'bär·bə ˌrīt }

Bardeen-Cooper-Schrieffer theory [SOLID STATE] A theory of super-conductivity that describes quantum-mechanically those states of the system in which conduction electrons cooperate

in their motion so as to reduce the total energy appreciably below that of other states by exploiting their effective mutual attraction; these states predominate in a superconducting material. Abbreviated BCS theory. { ¦bär ˌdēn ¦kü·pər ¦shrē·fər ˌthē·ə·rē }

bar drawing [MET] An operation in which a metallic bar is pulled through a die so that the cross-sectional area of the bar is reduced. { 'bär ˌdrȯ·iŋ }

bare electrode [MET] An uncoated electrode used in submerged arc automatic welding with a gas-shielded arc or a granular flux deposited in an elongated mound over a joint. { 'ber i'lek ˌtrōd }

bar folder [MET] A machine used to bend a metal sheet into a sharp, narrow, and accurate fold, or a rounded fold, along the edge. { 'bär ˌfōld· ər }

bar iron [MET] Wrought iron formed into bars. { 'bär ˌi·ərn }

barium [CHEM] A chemical element, symbol Ba, with atomic number 56 and atomic weight of 137.34. { 'bar·ē·əm }

barium acetate [INORG CHEM] $Ba(C_2H_3O_2)_2 \cdot H_2O$ A barium salt made by treating barium sulfide or barium carbonate with acetic acids; it forms colorless, triclinic crystals that decompose upon heating; used as a reagent for sulfates and chromates. { 'bar·ē·əm 'as·ə ˌtāt }

barium azide [INORG CHEM] $Ba(N_3)_2$ A crystalline compound soluble in water; used in high explosives. { 'bar·ē·əm 'ā ˌzīd }

barium-base grease [MATER] A lubricating material made from lubricating oil and barium soap. { 'bar·ē·əm ˌbās 'grēs }

barium binoxide See barium peroxide. { 'bar· ē·əm bī'näk ˌsīd }

barium bromate [INORG CHEM] $Ba(BrO_3)_2 \cdot H_2O$ A poisonous compound that forms colorless, monoclinic crystals, decomposing at 260°C; used for preparing other bromates. { 'bar·ē·əm 'brō ˌmāt }

barium bromide [INORG CHEM] $BaBr_2 \cdot 2H_2O$ Colorless crystals soluble in water and alcohol; used in photographic compounds. { 'bar·ē·əm 'brō ˌmīd }

barium carbonate [INORG CHEM] $BaCO_3$ A white powder with a melting point of 174°C; soluble in acids (except sulfuric acid); used in rodenticides, ceramic flux, optical glass, and television picture tubes. { 'bar·ē·əm 'kär·bə nət }

barium chlorate [INORG CHEM] $Ba(ClO_3)_2 \cdot H_2O$ A salt prepared by the reaction of barium chloride and sodium chlorate; it forms colorless, monoclinic crystals, soluble in water; used in pyrotechnics. { 'bar·ē·əm 'klȯr ˌāt }

barium chloride [INORG CHEM] $BaCl_2$ A toxic salt obtained as colorless, water-soluble cubic crystals, melting at 963°C; used as a rat poison, in metal surface treatment, and as a laboratory reagent. { 'bar·ē·əm 'klȯr ˌīd }

barium chromate [INORG CHEM] $BaCrO_4$ A toxic salt that forms yellow, rhombic crystals, insoluble

displacement density; electric flux density; electric induction. { i'lek·trik dis'plās·mənt }

electric displacement density See electric displacement. { i'lek·trik dis'plās·mənt ‚den·səd·ē }

electric double layer |PHYS CHEM| A phenomenon found at a solid-liquid interface; it is made up of ions of one charge type which are fixed to the surface of the solid and an equal number of mobile ions of the opposite charge which are distributed through the neighboring region of the liquid; in such a system the movement of liquid causes a displacement of the mobile counterions with respect to the fixed charges on the solid surface. Also known as double layer. { i'lek·trik ¦dəb·əl 'lā·ər }

electric field |ELEC| **1.** One of the fundamental fields in nature, causing a charged body to be attracted to or repelled by other charged bodies; associated with an electromagnetic wave or a changing magnetic field. **2.** Specifically, the electric force per unit test charge. { i¦lek·trik 'fēld }

electric flux density See electric displacement. { i¦lek·trik 'fləks ‚den·səd·ē }

electric-furnace steel |MET| Steel produced in an electric furnace. { i¦lek·trik ‚fər·nəs 'stēl }

electric glass See E glass. { i'lek·trik ‚glas }

electric hysteresis See ferroelectric hysteresis. { i¦lek·trik ‚his·tə'rē·səs }

electric induction See electric displacement. { i ¦lek·trik in'dək·shən }

electricity |PHYS| Physical phenomenon involving electric charges and their effects when at rest and when in motion. { i‚lek'tris·əd·ē }

electric moment |ELEC| One of a series of quantities characterizing an electric charge distribution; an l-th moment is given by integrating the product of the charge density, the l-th power of the distance from the origin, and a spherical harmonic Y^*_{lm} over the charge distribution. { i¦lek·trik 'mō·mənt }

electric monopole |ELEC| A distribution of electric charge which is concentrated at a point or is spherically symmetric. { i¦lek·trik 'män·ə‚pōl }

electric polarizability |ELEC| Induced dipole moment of an atom or molecule in a unit electric field. { i¦lek·trik ‚pō·lə‚rī·zə'bil·əd·ē }

electric polarization See polarization. { i¦lek·trik ‚pō·lə·rə'zā·shən }

electric potential |ELEC| The work which must be done against electric forces to bring a unit charge from a reference point to the point in question; the reference point is located at an infinite distance, or, for practical purposes, at the surface of the earth or some other large conductor. Also known as electrostatic potential; potential. Abbreviated V. { i¦lek·trik pə'ten·chəl }

electric quadrupole |ELEC| A charge distribution that produces an electric field equivalent to that produced by two electric dipoles whose dipole moments have the same magnitude but point in opposite directions and which are separated from each other by a small distance. { i¦lek·trik 'kwä·drə‚pōl }

electric quadrupole moment |ELEC| A quantity characterizing an electric charge distribution, obtained by integrating the product of the charge density, the second power of the distance from the origin, and a spherical harmonic Y^*_{2m} over the charge distribution. { i¦lek·trik 'kwä·drə‚pōl ‚mō·mənt }

electric resistance See resistance. { i¦lek·trik ri'zis·təns }

electric spark machining See electron discharge machining. { i¦lek·trik ¦spärk mə'shēn·iŋ }

electric steel |MET| Steel melted in an electric furnace which permits close control and the addition of alloying elements directly into the furnace. { i¦lek·trik 'stēl }

electric strength See dielectric strength. { i¦lek·trik 'streŋkth }

electric susceptibility |ELEC| A dimensionless parameter measuring the ease of polarization of a dielectric, equal (in meter-kilogram-second units) to the ratio of the polarization to the product of the electric field strength and the vacuum permittivity. Also known as dielectric susceptibility. { i¦lek·trik sə‚sep·tə'bil·əd·ē }

electric twinning |SOLID STATE| A defect occurring in natural quartz crystals, in which adjacent regions of quartz have their electric axes oppositely poled. { i¦lek·trik 'twin·iŋ }

electric wire See wire. { i¦lek·trik 'wīr }

electride |INORG CHEM| A member of a class of ionic compounds in which the anion is believed to be an electron. { i'lek‚trīd }

electrocaloric effect |SOLID STATE| A temperature change in certain crystals caused by alteration of the permanent polarization by application of an external electric field. { i¦lek·trō·kə ¦lȯr·ik i'fekt }

electrocatalysis |CHEM| Any one of the mechanisms which produce a speeding up of half-cell reactions at electrode surfaces. { i‚lek·trō· kə'tal·ə·səs }

electroceramics |MATER| Ceramic materials having electrical and other properties which make them suitable for use as insulators for power lines and in electrical components. { i‚lek· trō·sə'ram·iks }

electrochemical cell |PHYS CHEM| A combination of two electrodes arranged so that an overall oxidation-reduction reaction produces an electromotive force; includes dry cells, wet cells, standard cells, fuel cells, solid-electrolyte cells, and reserve cells. { i‚lek·trō'kem·ə·kəl 'sel }

electrochemical cleaning |MET| Removing soil by the chemical action caused or sustained by a current of electricity in an electrolyte. Also known as electrolytic cleaning. { i‚lek·trō'kem·ə·kəl 'klēn·iŋ }

electrochemical coating |MET| A coating formed by chemical action on the metal surface and effected by a current of electricity through an electrolyte. { i‚lek·trō'kem·ə·kəl 'kōd·iŋ }

electrochemical corrosion |MET| Corrosion of a metal associated with the flow of electric current in an electrolyte. Also known as electrolytic corrosion. { i‚lek·trō'kem·ə·kəl kə'rō·zhən }

electrochemical effect |PHYS CHEM| Conversion of chemical to electric energy, as in electrochemical cells; or the reverse process, used to produce elemental aluminum, magnesium, and bromine from compounds of these elements. { i,lek·trō'kem·ə·kəl i'fekt }

electrochemical emf |PHYS CHEM| Electrical force generated by means of chemical action, in manufactured cells (such as dry batteries) or by natural means (galvanic reaction). { i,lek·trō'kem·ə·kəl ¦e̱¦em'ef }

electrochemical equivalent |PHYS CHEM| The weight in grams of a substance produced or consumed by electrolysis with 100% current efficiency during the flow of a quantity of electricity equal to 1 faraday (96,485.34 coulombs). { i,lek·trō'kem·ə·kəl i'kwiv·ə·lənt }

electrochemical machining |MET| Removing excess metal by electrolytic dissolution, effected by the tool acting as the cathode against the workpiece acting as the anode. Abbreviated ECM. { i,lek·trō'kem·ə·kəl mə'shēn·iŋ }

electrochemical potential |PHYS CHEM| The difference in potential that exists when two dissimilar electrodes are connected through an external conducting circuit and the two electrodes are placed in a conducting solution so that electrochemical reactions occur. { i,lek·trō'kem·ə·kəl pə'ten·chəl }

electrochemical process |PHYS CHEM| 1. A chemical change accompanying the passage of an electric current, especially as used in the preparation of commercially important quantities of certain chemical substances. 2. The reverse change, in which a chemical reaction is used as the source of energy to produce an electric current, as in a battery. { i,lek·trō'kem·ə·kəl 'präs·əs }

electrochemical recording |ELECTR| Recording by means of a chemical reaction brought about by the passage of signal-controlled current through the sensitized portion of the record sheet. { i,lek·trō'kem·ə·kəl ri'kórd·iŋ }

electrochemical reduction cell |PHYS CHEM| The cathode component of an electrochemical cell, at which chemical reduction occurs (while at the anode, chemical oxidation occurs). { i,lek·trō'kem·ə·kəl ri'dək·shən ,sel }

electrochemical series |PHYS CHEM| A series in which the metals and other substances are listed in the order of their chemical reactivity or electrode potentials, the most reactive at the top and the less reactive at the bottom. Also known as electromotive series. { i,lek·trō'kem·ə·kəl 'sir·ēz }

electrochemical thermodynamics |THERMO| The application of the laws of thermodynamics to electrochemical systems. { i,lek·trō'kem·ə·kəl ,thərm·ō·dī'nam·iks }

electrochemical valve |ELEC| Electric valve consisting of a metal in contact with a solution or compound, across the boundary of which current flows more readily in one direction than in the other direction, and in which the valve action is accompanied by chemical changes. { i,lek·trō'kem·ə·kəl 'valv }

electrochemiluminescence |PHYS CHEM| Emission of light produced by an electrochemical reaction. Also known as electrogenerated chemiluminescence. { i,lek·trō,kem·ē·ə,lüm·ə'nes·əns }

electrochromic display |ELECTR| A solid-state passive display that uses organic or inorganic insulating solids which change color when injected with positive or negative charges. { i¦lek·trō¦krō·mik di'splā }

electrochromic material |MATER| An organic or inorganic substance that can interconvert between two or more colored states upon oxidation or reduction, that is, upon electrolytic loss or gain of electrons. { i,lek·trə¦krōm·ik mə'tir·ē·əl }

electrode |ELEC| 1. An electric conductor through which an electric current enters or leaves a medium, whether it be an electrolytic solution, solid, molten mass, gas, or vacuum. 2. One of the terminals used in dielectric heating or diathermy for applying the high-frequency electric field to the material being heated. { i'lek,trōd }

electrode capacitance |ELECTR| Capacitance between one electrode and all the other electrodes connected together. { i'lek,trōd kə'pas·əd·əns }

electrode current |ELECTR| Current passing to or from an electrode, through the interelectrode space within a vacuum tube. { i'lek,trōd ,kə·rənt }

electrode efficiency |PHYS CHEM| The ratio of the amount of metal actually deposited in an electrolytic cell to the amount that could theoretically be deposited as a result of electricity passing through the cell. { i'lek,trōd ə,fish·ən·sē }

electrode force |MET| The force that occurs between electrodes during seam, spot, and projection welding. Also known as welding force. { i'lek,trōd ,fórs }

electrode impedance |ELECTR| Reciprocal of the electrode admittance. { i'lek,trōd im'pēd·əns }

electrodeposition |MET| Electrolytic process in which a metal is deposited at the cathode from a solution of its ions; includes electroplating and electroforming. Also known as electrolytic deposition. { i¦lek·trō,dep·ə'zish·ən }

electrode potential |ELECTR| The instantaneous voltage of an electrode with respect to the cathode of an electron tube. |PHYS CHEM| The voltage existing between an electrode and the solution or electrolyte in which it is immersed; usually, electrode potentials are referred to a standard electrode, such as the hydrogen electrode. Also known as electrode voltage. { i'lek ,trōd pə'ten·chəl }

electrode skid |MET| Sliding of an electrode over the work surface in spot, seam, or projection welding. { i'lek,trōd ,skid }

electrode voltage See electrode potential. { i'lek,trōd ,vōl·tij }

electrodynamics |ELECTROMAG| The study of the relations between electrical, magnetic, and mechanical phenomena. { i,lek·trō·dī'nam·iks }

electroerosive machining See electron discharge machining. { i̧lek·trō·ə'rō·siv mə'shēn·iŋ }

electrofluid [FL MECH] Newtonian (or shear-thinning) fluid whose rheological or flow properties are changed into those of a viscoplastic type by the addition of electric-field modulation. { i̧lek·trō'flü·əd }

electroformed mold [MET] A mold made by electroplating metal on the reverse pattern on the cavity; molten steel may be then sprayed on the back of the mold to increase its strength. { i'lek·trə̧fórmd 'mōld }

electroforming [MET] Shaping components by electrodeposition of the metal on a pattern. { i'lek·trə̧fór·miŋ }

electrogalvanizing [MET] Coating of a metal, especially iron or steel, with zinc by electroplating. { i̧lek·trō'gal·və̧nīz·iŋ }

electrogas flux-cored welding [MET] A modification of the flux-cored welding process in which there is an externally supplied source of gas or gas mixture. { i̧lek·trō̧gas 'fləks ̧kórd 'weld·iŋ }

electrogenerated chemiluminescence See electrochemiluminescence. { i̧lek·trō̧jen·ə̧rād·əd ̧kem·ȩ̄lüm·ə'nes·əns }

electrohydraulic effect [PHYS CHEM] Generation of shock waves and highly reactive species in a liquid as the result of application of very brief but powerful electrical pulses. { i̧lek·trō·hī̧dról·ik i'fekt }

electrokinetic phenomena [PHYS CHEM] The phenomena associated with movement of charged particles through a continuous medium or with the movement of a continuous medium over a charged surface. { i̧lek·trō·kə'ned·ik fə'näm·ə·nə }

electroless plating [MET] Deposition of a metal coating by immersion of a metal or nonmetal in a suitable bath containing a chemical reducing agent. { i'lek·trə·ləs'plād·iŋ }

electroluminescence [ELECTR] The emission of light, not due to heating effects alone, resulting from application of an electric field to a material, usually solid. { i̧lek·trō̧lü·mə'nes·əns }

electroluminescent display [ELECTR] A display in which various combinations of electroluminescent segments may be activated by applying voltages to produce any desired numeral or other character. { i̧lek·trō̧lü·mə'nes·ənt di'splā }

electroluminescent phosphor [MATER] Zinc sulfide powder, with small additions of copper or manganese, which emits light when suspended in an insulator in an intense alternating electric field. Also known as electroluminor. { i̧lek·trō ̧lü·mə'nes·ənt 'fäs·fər }

electroluminor See electroluminescent phosphor. { i̧lek·trō'lü·mə̧nór }

electrolysis [PHYS CHEM] A method by which chemical reactions are carried out by passage of electric current through a solution of an electrolyte or through a molten salt. { i̧lek'trä·lə·səs }

electrolyte [PHYS CHEM] A chemical compound which when molten or dissolved in certain solvents, usually water, will conduct an electric current. { i'lek·trə̧līt }

electrolytic acid See sulfuric acid. { i'lek·trə̧līt 'as·əd }

electrolytic brightening See electropolishing. { i'lek·trə̧lid·ik 'brīt·ən·iŋ }

electrolytic cell [PHYS CHEM] A cell consisting of electrodes immersed in an electrolyte solution, for carrying out electrolysis. { i'lek·trə ̧lid·ik 'sel }

electrolytic cleaning See electrochemical cleaning. { i'lek·trə̧lid·ik 'klēn·iŋ }

electrolytic conductance [PHYS CHEM] The transport of electric charges, under electric potential differences, by charged particles (called ions) of atomic or larger size. { i'lek·trə̧lid·ik kən'dək·təns }

electrolytic conductivity [PHYS CHEM] The conductivity of a medium in which the transport of electric charges, under electric potential differences, is by particles of atomic or larger size. { i'lek·trə̧lid·ik ̧kän·dək'tiv·əd·ē }

electrolytic copper [MET] Metallic copper produced by electrochemical deposition from a copper ion-containing electrolyte. { i'lek·trə ̧lid·ik 'käp·ər }

electrolytic corrosion See electrochemical corrosion. { i'lek·trə̧lid·ik kə'rō·zhən }

electrolytic deposition See electrodeposition. { i'lek·trə̧lid·ik dep·ə'zish·ən }

electrolytic etching [MET] Engraving the surface of a metal by electrolysis. { i'lek·trə̧lid·ik 'ech·iŋ }

electrolytic migration [PHYS CHEM] The motions of ions in a liquid under the action of an electric field. { i̧lek·trə̧lid·ik mī'grā·shən }

electrolytic pickling [MET] Removal of metal by electrolysis using the metal as an electrode in a suitable electrolyte. { i'lek·trə̧lid·ik 'pik·liŋ }

electrolytic polarization [PHYS CHEM] The existence of a minimum potential difference necessary to cause a steady current to flow through an electrolytic cell, resulting from the tendency of the products of electrolysis to recombine. { i̧lek·trə̧lid·ik pō·lər·ə'zā·shən }

electrolytic polishing See electropolishing. { i'lek·trə̧lid·ik 'päl·ish·iŋ }

electrolytic potential [PHYS CHEM] Difference in potential between an electrode and the immediately adjacent electrolyte, expressed in terms of some standard electrode difference. { i'lek·trə̧lid·ik pə'ten·chəl }

electrolytic powder [MET] Metal powder produced directly or indirectly by electrodeposition. { i'lek·trə̧lid·ik 'paúd·ər }

electrolytic process [PHYS CHEM] An electrochemical process involving the principles of electrolysis, especially as relating to the separation and deposition of metals. { i'lek·trə̧lid·ik 'präs·əs }

electrolytic protection See cathodic protection. { i'lek·trə̧lid·ik prə'tek·shən }

electrolytic refining See electrorefining. { i'lek·trə̧lid·ik rə'fīn·iŋ }

electrolytic separation |PHYS CHEM| Separation of isotopes by electrolysis, based on differing rates of discharge at the electrode of ions of different isotopes. { i'lek·trə‚lid·ik ‚sep·ə'rā·shən }

electrolytic solution |PHYS CHEM| A solution made up of a solvent and an ionically dissociated solute; it will conduct electricity, and ions can be separated from the solution by deposition on an electrically charged electrode. { i'lek·trə‚lid·ik sə'lü·shən }

electrolytic tough pitch |MET| Copper which has been refined electrolytically, containing mostly oxygen as an impurity. { i'lek·trə‚lid·ik 'təf 'pich }

electromagnet |ELECTROMAG| A magnet consisting of a coil wound around a soft iron or steel core; the core is strongly magnetized when current flows through the coil, and is almost completely demagnetized when the current is interrupted. { i¦lek·trō'mag·nət }

electromagnetic |PHYS| Pertaining to phenomena in which electricity and magnetism are related. { i¦lek·trō·mag'ned·ik }

electromagnetic crack detector |MET| An instrument that detects cracks in iron or steel objects by applying a strong magnetizing force and measuring the resulting magnetic flux through the object. { i¦lek·trō·mag'ned·ik 'krak di‚tek·tər }

electromagnetic damping |ELEC| Retardation of motion that results from the reaction between eddy currents in a moving conductor and the magnetic field in which it is moving. { i¦lek·trō·mag'ned·ik 'damp·iŋ }

electromagnetic field |ELECTROMAG| An electric or magnetic field, or a combination of the two, as in an electromagnetic wave. { i¦lek·trō·mag'ned·ik 'fēld }

electromagnetic induction |ELECTROMAG| The production of an electromotive force either by motion of a conductor through a magnetic field so as to cut across the magnetic flux or by a change in the magnetic flux that threads a conductor. Also known as induction. { i¦lek·trō·mag'ned·ik in'dək·shən }

electromagnetic interference |ELEC| Interference, generally at radio frequencies, that is generated inside systems, as contrasted to radio-frequency interference coming from sources outside a system. Abbreviated emi. { i¦lek·trō·mag'ned·ik ‚in·tər'fir·əns }

electromagnetic mixing |MET| Mixing of molten alloys by exposing the melt to a strong magnetic field while passing direct current between electrodes at opposite ends of the crucible; stirring action results from interaction of the magnetic field of the current-carrying molten alloy with the external transverse magnetic field. { i¦lek·trō·mag'ned·ik 'mik·siŋ }

electromagnetic noise |ELEC| Noise in a communications system resulting from undesired electromagnetic radiation. Also known as radiation noise. { i¦lek·trō·mag'ned·ik 'nȯiz }

electromagnetic radiation |ELECTROMAG| Electromagnetic waves and, especially, the associated electromagnetic energy. { i¦lek·trō·mag'ned·ik ‚rād·ē'ā·shən }

electromagnetic spectrum |ELECTROMAG| The total range of wavelengths or frequencies of electromagnetic radiation, extending from the longest radio waves to the shortest known cosmic rays. { i¦lek·trō·mag'ned·ik 'spek·trəm }

electromagnetic system of units |ELECTROMAG| A centimeter-gram-second system of electric and magnetic units in which the unit of current is defined as the current which, if maintained in two straight parallel wires having infinite length and being 1 centimeter apart in vacuum, would produce between these conductors a force of 2 dynes per centimeter of length; other units are derived from this definition by assigning unit coefficients in equations relating electric and magnetic quantities. Also known as electromagnetic units (emu). { i¦lek·trō·mag'ned·ik ¦sis·təm əv 'yü·nəts }

electromagnetic theory of light |ELECTROMAG| Theory according to which light is an electromagnetic wave whose electric and magnetic fields obey Maxwell's equations. { i¦lek·trō·mag'ned·ik ¦thē·ə·rē əv 'līt }

electromagnetic units See electromagnetic system of units. { i¦lek·trō·mag'ned·ik 'yü·nəts }

electromechanical coupling coefficient |SOLID STATE| The ratio of the mutual elasto-dielectric energy density in a piezoelectric material to the square root of the product of the stored elastic and dielectric energy densities. { i‚lek·trō·mi¦kan·ə·kəl 'kəp·liŋ ‚kō·ə‚fish·ənt }

electrometallurgy |MET| Industrial recovery and processing of metals by electrical and electrolytic procedures. { i'lek·trō'med·əl·ər‚jē }

electromigration |PHYS CHEM| The movement of ions under the influence of an electrical potential difference. { i¦lek·trō·mī'grā·shən }

electromotance See electromotive force. { i¦lek·trō'mōt·əns }

electromotive force |PHYS CHEM| **1.** The difference in electric potential that exists between two dissimilar electrodes immersed in the same electrolyte or otherwise connected by ionic conductors. **2.** The resultant of the relative electrode potential of the two dissimilar electrodes at which electrochemical reactions occur. Abbreviated emf. Also known as electromotance. { i¦lek·trə'mōd·iv 'fȯrs }

electromotive series See electrochemical series. { i¦lek·trə'mōd·iv 'sir·ēz }

electron |PHYS| **1.** A stable elementary particle which is the negatively charged constituent of ordinary matter, having a mass of about 9.11×10^{-28} gram (equivalent to 0.511 MeV), a charge of about -1.602×10^{-19} coulomb, and a spin of $\frac{1}{2}$. **2.** Collective name for the electron, as in the first definition, and the positron. { i'lek‚trän }

electron acceptor |PHYS CHEM| **1.** An atom or part of a molecule joined by a covalent bond to an electron donor. **2.** See electrophile. |SOLID STATE| See acceptor. { i'lek‚trän ak'sep·tər }

electron-acoustic microscopy |PHYS| A technique for producing images that show variations in an object's thermal and elastic properties; an electron beam generates ultrasonic waves in the specimen which are detected by a piezoelectric transducer whose output controls the brightness of a spot sweeping a cathode-ray tube in synchronism with the electron beam. { i'lek,trän ə'küs·tik mī'kräs·kə·pē }

electron beam |ELECTR| A narrow stream of electrons moving in the same direction, all having about the same velocity. { i'lek,trän ,bēm }

electron-beam cutting |MET| A process which uses high-velocity electrons to heat the workpieces to be cut. { i'lek,trän ,bēm 'kəd·iŋ }

electron-beam drilling |ELECTR| Drilling of tiny holes in a ferrite, semiconductor, or other material by using a sharply focused electron beam to melt and evaporate or sublimate the material in a vacuum. { i'lek,trän ,bēm 'dril·iŋ }

electron-beam laser |OPTICS| A semiconductor laser in which the electron beam that provides pumping action in a thin plate of cadmium sulfide or other material is swept electrically in two dimensions by a deflection yoke, much as in a cathode-ray tube. { i'lek,trän ,bēm 'lā·zər }

electron-beam lithography |ELECTR| Lithography in which the radiation-sensitive film or resist is placed in the vacuum chamber of a scanning-beam electron microscope and exposed by an electron beam under digital computer control; after exposure, the film is removed from the vacuum chamber for conventional development and other production processes. { i'lek,trän ,bēm li'thäg·rə·fē }

electron-beam machining |MET| A machining process in which heat is produced by a focused electron beam at a sufficiently high temperature to volatilize and thereby remove metal in a desired manner; takes place in a vacuum. Abbreviated EBM. { i'lek,trän ,bēm mə'shēn·iŋ }

electron-beam melting |MET| A melting process in which an electron beam provides the necessary heat. { i'lek,trän ,bēm 'melt·iŋ }

electron-beam welding |MET| A technique for joining materials in which highly collimated electron beams are used at a pressure below 10^{-3} mmHg (0.1333 pascal) to produce a highly concentrated heat source; used in outer space. { i'lek,trän ,bēm 'weld·iŋ }

electron charge |PHYS| The charge carried by an electron, equal to about -1.602×10^{-19} coulomb, or -4.803×10^{-10} statcoulomb. { i'lek,trän ,chärj }

electron compound |MET| Alloy of two metals in which a progressive change in composition is accompanied by a progression of phases, differing in crystal structure. Also known as Hume-Rothery compound; intermetallic compound. { i'lek,trän ,käm,paůnd }

electron conduction |ELEC| Conduction of electricity resulting from motion of electrons, rather than from ions in a gas or solution, or holes in a solid. |THERMO| The transport of energy in highly ionized matter primarily by electrons

of relatively high temperature moving in one direction and electrons of lower temperature moving in the other. { i'lek,trän kən,dək·shən }

electron density |PHYS| **1.** The number of electrons in a unit volume. **2.** When quantum-mechanical effects are significant, the total probability of finding an electron in a unit volume. { i'lek,trän 'den·səd·ē }

electron diffraction |PHYS| The phenomenon associated with the interference processes which occur when electrons are scattered by atoms in crystals to form diffraction patterns. { i'lek,trän di'frak·shən }

electron discharge machining |MET| A process by which materials that conduct electricity can be removed from a metal by an electric spark; used to form holes of varied shapes in materials of poor machinability. Abbreviated EDM. Also known as electrical discharge machining; electric spark machining; electroerosive machining; electrospark machining. { i'lek,trän 'dis,chärj mə,shēn·iŋ }

electron donor |PHYS CHEM| **1.** An atom or part of a molecule which supplies both electrons of a duplet forming a covalent bond. **2.** See nucleophile. |SOLID STATE| See donor. { i'lek,trän ,dō·nər }

electronegative |ELEC| **1.** Carrying a negative electric charge. **2.** Capable of acting as the negative electrode in an electric cell. |PHYS CHEM| Pertaining to an atom or group of atoms that has a relatively great tendency to attract electrons to itself. { i¦lek·trō'neg·əd·iv }

electronegative potential |PHYS CHEM| Potential of an electrode expressed as negative with respect to the hydrogen electrode. { i¦lek·trō'neg·əd·iv pə'ten·chəl }

electron exchanger See redox polymer. { i'lek ,trän iks,chān·jər }

electron flow |ELEC| A current produced by the movement of free electrons toward a positive terminal; the direction of electron flow is opposite to that of current. { i'lek,trän ,flō }

electron hole See hole. { i'lek,trän ¦hōl }

electron-hole droplets |SOLID STATE| A form of electronic excitation observed in germanium and silicon at sufficiently low cryogenic temperatures; it is associated with a liquid-gas phase transition of the charge carriers, and consists of regions of conducting electron-hole Fermi liquid coexisting with regions of insulating exciton gas. { i'lek ,trän ¦hōl 'dräp·ləts }

electron-hole recombination |SOLID STATE| The process in which an electron, which has been excited from the valence band to the conduction band of a semiconductor, falls back into an empty state in the valence band, which is known as a hole. { i'lek,trän 'hōl rē ,käm·bə'nā·shən }

electronic polarization |ELEC| Polarization arising from the displacement of electrons with respect to the nuclei with which they are associated, upon application of an external electric field. { i,lek'trän·ik ,pō·lə·rə'zā·shən }

electronics |PHYS| Study, control, and application of the conduction of electricity through gases or vacuum or through semiconducting or conducting materials. { i,lek'trän·iks }

electronic specific heat [SOLID STATE] Contribution to the specific heat of a metal from the motion of conduction electrons. { i,lek'trän·ik spə¦sif·ik 'hēt }

electronic structure [PHYS] The arrangement of electrons in an atom, molecule, or solid, specified by their wave functions, energy levels, or quantum numbers. { i,lek'trän·ik 'strək·chər }

electronic work function [SOLID STATE] The energy required to raise an electron with the Fermi energy in a solid to the energy level of an electron at rest in vacuum outside the solid. { i,lek'trän·ik 'wərk ,faŋk·shən }

electron mass [PHYS] The mass of an electron, equal to about 9.11×10^{-28} gram, equivalent to 0.511 MeV. Also known as electron rest mass. { i,lek'trän 'mas }

electron metallography [MET] The study of the microscopic structure of metals employing an electron microscope. { i'lek,trän med·əl'äg·rə·fē }

electron microprobe [PHYS] An x-ray machine in which electrons emitted from a hot-filament source are accelerated electrostatically, then focused to an extremely small point on the surface of a specimen by an electromagnetic lens; nondestructive analysis of the specimen can then be made by measuring the backscattered electrons, the specimen current, the resulting x-radiation, or any other resulting phenomenon. Also known as electron probe. { i'lek,trän 'mī·krō,prōb }

electron microscope [ELECTR] A device for forming greatly magnified images of objects by means of electrons, usually focused by electron lenses. { i'lek,trän 'mī·krə,skōp }

electron mobility [SOLID STATE] The drift mobility of electrons in a semiconductor, being the electron velocity divided by the applied electric field. { i'lek,trän mō'bil·əd·ē }

electron orbit [PHYS] The path described by an electron. { i'lek,trän 'ȯr·bət }

electron paramagnetic resonance [PHYS] Magnetic resonance arising from the magnetic moment of unpaired electrons in a paramagnetic substance or in a paramagnetic center in a diamagnetic substance. Abbreviated EPR. Also known as electron spin resonance (ESR); paramagnetic resonance. { i'lek,trän ¦par·ə·mag¦ned·ik 'rez·ən·əns }

electron paramagnetism [PHYS] Paramagnetism in a substance whose atoms or molecules possess a net electronic magnetic moment; arises because of the tendency of a magnetic field to orient the electronic magnetic moments parallel to itself. { i'lek,trän ¦par·ə'mag·nə,tiz·əm }

electron probe See electron microprobe. { i'lek,trän ,prōb }

electron radius [PHYS] The classical value r of $2.8179403 \times 10^{-13}$ centimeter for the radius of an electron; obtained by equating mc^2 for the electron to e^2/r, where e and m are the charge and mass of the electron respectively; any classical model for an electron will have approximately this radius. { i'lek,trän 'rād·ē·əs }

electron rest mass See electron mass. { i'lek ,trän 'rest ,mas }

electron spin density [PHYS] The vector sum of the spin angular momenta of electrons at each point in a substance per unit volume. { i'lek ,trän 'spin ,den·səd·ē }

electron spin echo [SOLID STATE] A net magnetization of a material that is sometimes observed at a particular time following the application of two or more short, intense pulses of microwave radiation; in the simplest case, two pulses separated by a time interval t are followed by a net magnetization at time t' after the second pulse. { i,lek,trän 'spin ,ek·ō }

electron spin resonance See electron paramagnetic resonance. { i'lek,trän 'spin ,rez·ən·əns }

electron stain [MATER] A substance such as phosphotungstic acid or osmic acid which scatters large numbers of electrons and can therefore be used to stain objects to be examined by an electron microscope. { i'lek,trän ,stān }

electron transfer [PHYS] The passage of an electron from one constituent of a system to another. { i'lek,trän 'trans·fər }

electron trap [SOLID STATE] A defect or chemical impurity in a semiconductor or insulator which captures mobile electrons in a special way. { i'lek,trän ,trap }

electronvolt [PHYS] A unit of energy which is equal to the energy acquired by an electron when it passes through a potential difference of 1 volt in a vacuum; it is equal to $(1.60217646 \pm 0.00000006) \times 10^{-19}$ volt. Abbreviated eV. { i'lek,trän ,vōlt }

electrooptic material [OPTICS] A material in which the indices of refraction are changed by an applied electric field. { i,lek·trō'äp·tik mə'tir·ē·əl }

electrooptics [OPTICS] The study of the influence of an electric field on optical phenomena, as in the electrooptical Kerr effect and the Stark effect. Also known as optoelectronics. { i,lek·trō'äp·tiks }

electrophile [PHYS CHEM] An electron-deficient ion or molecule that takes part in an electrophilic process. { i'lek·trō,fīl }

electrophilic [PHYS CHEM] **1.** Pertaining to any chemical process in which electrons are acquired from or shared with other molecules or ions. **2.** Referring to an electron-deficient species. { i¦lek·trō'fil·ik }

electrophoresis [PHYS CHEM] An electrochemical process in which colloidal particles or macromolecules with a net electric charge migrate in a solution under the influence of an electric current. Also known as cataphoresis. { i,lek·trō·fə'rē·səs }

electrophoretic coating [MET] A surface coating on a metal deposited by electric discharge of particles from a colloidal solution. { i,lek·trō·fə'red·ik 'kōd·iŋ }

electrophotography [GRAPHICS] An electrostatic image-forming process in which light, x-rays, or gamma rays form an electrostatic image on a photoconductive, insulating medium; the charged image areas attract and hold a fine powder called a toner, and the powder image is then transferred to paper or fused there by heat.   { i¦lek·trō·fə'täg·rə·fē }

electrophotophoresis |PHYS| Helical motion of small particles suspended in a gas along the direction of an electric field when exposed to a beam of light.   { i,lek·trō,fōd·ə·fə'rē·səs }

electroplating |MET| Electrodeposition of a metal or alloy from a suitable electrolyte solution; the article to be plated is connected as the cathode in the electrolyte solution; direct current is introduced through the anode which consists of the metal to be deposited.   { i'lek·trō,plād·iŋ }

electropolishing |MET| Smoothing and enhancing the appearance of a metal surface by making it an anode in a suitable electrolyte. Also known as electrolytic brightening; electrolytic polishing.   { i¦lek·trō'pä·lə·shiŋ }

electropositive |ELEC| **1.** Carrying a positive electric charge. **2.** Capable of acting as the positive electrode in an electric cell. |PHYS CHEM| Pertaining to elements, ions, or radicals that tend to give up or lose electrons.   { i,lek·trə'päz·əd·iv }

electropositive potential |PHYS CHEM| Potential of an electrode expressed as positive with respect to the hydrogen electrode.   { i¦lek·trə ¦päz·əd·iv pə'ten·chəl }

electrorefining |MET| Purifying metals by electrolysis using an impure metal as anode from which the pure metal is dissolved and subsequently deposited at the cathode. Also known as electrolytic refining.   { i¦lek·trō·ri'fīn·iŋ }

electrorheological fluid |PHYS CHEM| A colloidal suspension of finely divided particles in a carrier liquid, usually an insulating oil, whose rheological properties are changed through an increase in resistance when an electric field is applied.   { i¦lek·trō,rē·ə¦läj·ə·kəl 'flü·əd }

electrorheological material |MATER| A material possessing rheological properties that are controlled by an imposed electric field.   { i,lek·trō ,rē·ə¦läj·ə·kəl mə'tir·ē·əl }

electrosensitive paper |MATER| A conductive paper that darkens when electric current is sent through it.   { i¦lek·trō'sen·səd·iv 'pā·pər }

electroslag welding |MET| A welding process in which consumable electrodes are fed into a joint containing flux; the current melts the flux, and the flux in turn melts the faces of the joint and the electrodes, allowing the weld metal to form a continuously cast ingot between the joint faces.   { i'lekm̂trō,slag 'weldm̂iŋ }

electrospark machining   See electron discharge machining.   { i'lek·trō,spärk mə'shēn·iŋ }

electrostatic |ELEC| Pertaining to electricity at rest, such as an electric charge on an object.   { i,lek·trə'stad·ik }

electrostatic attraction   See Coulomb attraction.   { i,lek·trə'stad·ik ə'trak·shən }

electrostatic bond |PHYS CHEM| A valence

bond in which two atoms are kept together by electrostatic forces caused by transferring one or more electrons from one atom to the other.   { i,lek·trə'stad·ik 'bänd }

electrostatic energy |ELEC| The potential energy which a collection of electric charges possesses by virtue of their positions relative to each other.   { i,lek·trə'stad·ik 'en·ər·jē }

electrostatic field |ELEC| A time-independent electric field, such as that produced by stationary charges.   { i,lek·trə'stad·ik 'fēld }

electrostatic force |ELEC| Force on a charged particle due to an electrostatic field, equal to the electric field vector times the charge of the particle.   { i,lek·trə'stad·ik ' fórs }

electrostatic induction |ELEC| The process of charging an object electrically by bringing it near another charged object, then touching it to ground. Also known as induction.   { i,lek· trə'stad·ik in'dak·shən }

electrostatic interactions   See Coulomb interactions.   { i,lek·trə'stad·ik int·ə'rak·shənz }

electrostatic memory   See electrostatic storage.   { i'lek·trə,stad·ik 'mem·rē }

electrostatic potential   See electric potential.   { i'lek·trə,stad·ik pə'ten·chəl }

electrostatic repulsion   See Coulomb repulsion.   { i'lek·trə,stad·ik ri'pəl·shən }

electrostatics |ELEC| The study of electric charges at rest, their electric fields, and potentials.   { i,lek·trə'stad·iks }

electrostatic storage |ELECTR| A storage in which information is retained as the presence or absence of electrostatic charges at specific spot locations, generally on the screen of a special type of cathode-ray tube known as a storage tube. Also known as electrostatic memory.   { i'lek·trə ,stad·ik 'stór·ij }

electrostatic stress |ELEC| An electrostatic field acting on an insulator, which produces polarization in the insulator and causes electrical breakdown if raised beyond a certain intensity.   { i'lek·trə,stad·ik 'stres }

electrostriction |MECH| A form of elastic deformation of a dielectric induced by an electric field, associated with those components of strain which are independent of reversal of field direction, in contrast to the piezoelectric effect. Also known as electrostrictive strain.   { i¦lek· trō'strik·shən }

electrostrictive strain   See electrostriction.   { i¦lek·trō'strik·tiv 'strān }

electrothermal |PHYS| **1.** Pertaining to both heat and electricity. **2.** In particular, pertaining to conversion of electrical energy into heat energy.   { i¦lek·trō'thər·məl }

electrothermal recording |ELECTR| Type of electrochemical recording, used in facsimile equipment, wherein the chemical change is produced principally by signal-controlled thermal action.   { i¦lek·trō'thər·məl ri'kórd·iŋ }

electrotinning |MET| Electroplating an object with tin.   { i¦lek·trō'tin·iŋ }

electrovalent bond   See ionic bond.   { i¦lek·trō ¦vā·lənt 'bänd }

electroviscous effect |FL MECH| Change in a liquid's viscosity induced by a strong electrostatic field. { i¦lek·trō'vis·kəs i'fekt }

electrowinning |MET| Extracting metal from solutions by electrochemical processes. { i¦lek·trō'win·iŋ }

electrum |MET| A naturally occurring alloy of gold and silver. { i'lek·trəm }

element 110 |CHEM| A synthetic chemical element, atomic number 110; the eighteenth transuranium element. { ¦el·ə·mənt ,wən'ten }

element 111 |CHEM| A synthetic chemical element, atomic number 111; the nineteenth transuranium element. { ¦el·ə·mənt wən·i'lev·ən }

element 112 |CHEM| A synthetic chemical element, atomic number 112; the twentieth transuranium element. { ¦el·ə·mənt wən'twelv }

element |CHEM| A substance made up of atoms with the same atomic number; common examples are hydrogen, gold, and iron. Also known as chemical element. |ELEC| A part of an electronic or semiconductor device that contributes directly to the electrical performance. { 'el·ə·mənt }

elemental area See picture element. { ,el·ə 'ment·əl 'er·ē·ə }

elementary charge |PHYS| An electric charge such that the electric charge of any body is an integral multiple of it, equal to the electron charge. { ,el·ə'men·trē ,chärj }

elementary particle |PARTIC PHYS| A particle which, in the present state of knowledge, cannot be described as compound, and is thus one of the fundamental constituents of all matter. Also known as fundamental particle; particle; subnuclear particle. { ,el·ə'men·trē 'pärd·i·kəl }

elemi |MATER| A soft resin obtained from tropical trees of the family Burseraceae in the Philippines; used as a plasticizer, in cements and printing inks, and for perfumery and waterproofing. { 'el·ə·mē }

elinvar |MET| A nickel-chromium steel alloy containing manganese and tungsten in varying amounts and having a low thermal expansion and almost invariable modulus of elasticity; used for chronometer balances and springs for gages and other instruments. { 'el·in,vär }

elongation |MECH| The fractional increase in a material's length due to stress in tension or to thermal expansion. { ē,loŋ'gā·shən }

eluant |CHEM| A liquid used to extract one material from another, as in chromatography. { 'el·yə·wənt }

eluate |CHEM| The solution that results from the elution process. { 'el·yə,wāt }

elution |CHEM| The removal of adsorbed species from a porous bed or chromatographic column by means of a stream of liquid or gas. { ē'lü·shən }

embedability |MET| The ability of a metal to enclose foreign particles within itself. { em ,bed·ə'bil·əd·ē }

embrittlement |MECH| Reduction or loss of ductility or toughness in a metal or plastic with little change in other mechanical properties. { ,em'brid·əl·mənt }

emery |MATER| An abrasive which is composed of pulverized, impure corundum; used in polishing and grinding. { 'em·ə·rē }

emery cake |MATER| Caked, powdered emery in a binding material. { 'em·ə·rē ,kāk }

emery paper |MATER| An abrasive paper or cloth with an adherent surface layer of emery powder; used for polishing and cleaning metal. { 'em·ə·rē ,pā·pər }

emery stone |MATER| **1.** A sharpening stone. **2.** A mixture of powdered emery and a binder which can be molded into grinding devices. { 'em·ə·rē ,stōn }

emf See electromotive force.

emi See electromagnetic interference.

emissive power See emittance. { i¦mis·iv 'paü·ər }

emissivity |THERMO| The ratio of the radiation emitted by a surface to the radiation emitted by a perfect blackbody radiator at the same temperature. Also known as thermal emissivity. { ,ē·mə'siv·əd·ē }

emittance |THERMO| The power radiated per unit area of a radiating surface. Also known as emissive power; radiating power. { i'mit·əns }

empire cloth |MATER| Cotton cloth coated with oxidized oil; used as an electrical insulator. { 'em,pīr ,klòth }

empirical formula |CHEM| A chemical formula that indicates the composition of a compound in terms of the relative numbers and kinds of atoms in the simplest ratio; for example, the empirical formula for fluorobenzene is C_6H_5F. { em'pirmərkəl 'fórm̄myəm̄lə }

emulsan |BIOMATER| A lipopolysaccharide produced by a strain of *Acinetobacter calcoaceticus*, used to stabilize oil-in-water emulsions. { i'məl·sən }

emulsification |CHEM| The process of dispersing one liquid in a second immiscible liquid; the largest group of emulsifying agents are soaps, detergents, and other compounds. { ə,məl·sə·fə'kā·shən }

emulsified asphalt |MATER| An emulsion of asphalt cement and water with a small quantity of an emulsifying agent. { ə'məl·sə,fīd 'as,fòlt }

emulsifier See disperser. { ə'məl·sə,fī·ər }

emulsifying agent See disperser. { ə'məl·sə ,fī·iŋ ,ā·jənt }

emulsion |CHEM| A stable dispersion of one liquid in a second immiscible liquid, such as milk (oil dispersed in water). |GRAPHICS| In photography, the photosensitized material on film, plates, and various photographic papers. { ə'məl·shən }

emulsion breaking |CHEM| In an emulsion, the combined sedimentation and coalescence of emulsified drops of the dispersed phase so that they will settle out of the carrier liquid; can be accomplished mechanically (in settlers, cyclones, or centrifuges) with or without the aid of chemical additives to increase the surface tension of the droplets. { ə'məl·shən ,brāk·iŋ }

emulsion paint |MATER| Paint whose vehicle is

an emulsion of a binder (oil, resin, latex, and so on) in water.   { ə'məl·shən ,pānt }

emulsion polymerization   [POLYM CHEM] A polymerization reaction that occurs in one phase of an emulsion.   { ə'məl·shən pə,lim·ə·rə'zā·shən }

emulsion speed   [GRAPHICS] Sensitivity of a photographic emulsion to light, under standard conditions of exposure and development.   { ə'məl·shən ,spēd }

enamel   [MATER] A finely ground, resin-containing oil paint that dries relatively harder, smoother, and glossier than ordinary paint.   { i'nam·əl }

enamel clay   [MATER] A ball clay able to float nonplastic enamel slips to make them spray or dip more evenly.   { i'nam·əl ,klā }

enameled brick   [MATER] Brick with a smooth hard surface acquired from the application of a special wash before burning.   { i'nam·əld 'brik }

enamel oxide   [MATER] Any of the mixtures of calcined oxides used to color vitreous enamels used on sheet steel or cast iron.   { i'nam·əl 'äk ,sīd }

enamel paper   See coated paper.   { i'nam·əl ,pā·pər }

enantiomer   See enantiomorph.   { ə¦nan·tē¦ō·mər }

enantiomerically pure   [ORG CHEM] Referring to a sample of molecules having the same chirality. IUPAC discourages the use of homochiral as a synonym.   { ə¦nan·tē·ō¦mer·ə·klē 'pyür }

enantiomeric excess   [ORG CHEM] In an asymmetric synthesis, a chemical yield that contains more of the desired enantiomer than other products.   { ə¦nan·tē·ō¦mer·ik ek'ses }

enantiomorph   [CHEM] One of an isomeric pair of either crystalline forms or chemical compounds whose molecules are nonsuperimposable mirror images. Also known as enantiomer; optical antipode; optical isomer.   { ə'nan·tē·ə,mórf }

enantioselective reaction   See stereoselective reaction.   { ə¦nan·tē·ə·si¦lek·tiv rē'ak·shən }

enantiotropy   [CHEM] The relation of crystal forms of the same substance in which one form is stable above the transition-point temperature and the other stable below it, so that the forms can change reversibly one into the other.   { ə,nan·te'ä·trə·pē }

end-construction tile   [MATER] A type of structural clay tile designed to receive its principal stress parallel to the axis of the cells.   { 'end kən,strək·shən ,tīl }

end loss   [MET] That portion remaining after designated lengths of bar have been cut into multiples.   { 'end ,lós }

endo-   [ORG CHEM] Prefix that denotes inward-directed valence bonds of a six-membered ring in its boat form.   { 'en·dō }

endotherm   [PHYS CHEM] In differential thermal analysis, a graph of the temperature difference between a sample compound and a thermally inert reference compound (commonly aluminum oxide) as the substances are simultaneously heated to elevated temperatures at a pre-termined rate, and the sample compound undergoes endothermal or exothermal processes.   { 'en·də,thərm }

end turning   See boxing.   { 'end ,tərn·iŋ }

endurance limit   See fatigue limit.   { in'dúr·əns ,lim·ət }

endurance ratio   See fatigue ratio.   { in'dúr·əns ,rā·shō }

endurance strength   See fatigue strength.   { in'dúr·əns ,streŋkth }

energy   [PHYS] The capacity for doing work.   { 'en·ər·jē }

energy absorption   [PHYS] Conversion of mechanical or radiant energy into the internal potential energy or heat energy of a system.   { 'en·ər·jē ab,sórp·shən }

energy band   See band.   { 'en·ər·jē ,band }

energy-band theory of solids   See band theory of solids.   { 'en·ər·jē ,band ¦thē·ə·rē əv 'säl·ədz }

energy conservation   See conservation of energy.   { 'en·ər·jē ,kän·sər'vā·shən }

energy dissipation   See dissipation.   { 'en·ər·jē dis·ə'pā·shən }

energy gap   [SOLID STATE] A range of forbidden energies in the band theory of solids.   { 'en·ər·jē ,gap }

energy of a charge   [ELEC] Charge energy measured in ergs according to the equation $E = QV$, where Q is the charge and V is the potential in electrostatic units.   { 'en·ər·jē əv ə 'chärj }

engine cycle   [THERMO] Any series of thermodynamic phases constituting a cycle for the conversion of heat into work; examples are the Otto cycle, Stirling cycle, and Diesel cycle.   { 'en·jən ,sī·kəl }

engineering plastics   [POLYM CHEM] A class of polymers, based on aromatic backbones, having high strength, stiffness, and toughness together with high thermal and oxidative stability, low creep, and the ability to be processed by standard techniques for thermoplastics; examples include polyacetal, polyamide, polycarbonate, and polysulfone resins.   { ,en·jə'nir·iŋ 'plas·tiks }

English red   [MATER] Pigment consisting mostly of red iron oxide.   { 'iŋ·glish 'red }

English vermilion   [INORG CHEM] Bright vermilion pigment of precipitated mercury sulfide; in paints, it tends to darken when exposed to light.   { 'iŋ·glish vər'mil·yən }

engobe   [MATER] A thin layer of fluid clay applied to a piece of earthenware to support a glaze or enamel or to cover blemishes.   { än'gōb }

enol   [ORG CHEM] An organic compound with a hydroxide group adjacent to a double bond; varies with a ketone form in the effect known as enol-keto tautomerism; an example is the compound $CH_3COH{=}CHCO_2C_2H_5$.   { 'ē,nól }

enthalpy   [THERMO] The sum of the internal energy of a system plus the product of the system's volume multiplied by the pressure exerted on the system by its surroundings. Also known as heat content; sensible heat; total heat.   { en'thal·pē }

enthalpy-entropy chart   [THERMO] A graph of the enthalpy of a substance versus its entropy at various values of temperature, pressure, or specific volume; useful in making calculations about a machine or process in which this substance is the working medium. { en¦thal·pē 'en·trə·pē ¸chärt }

enthalpy of vaporization See heat of vaporization. { en'thal·pē əv ¸vā·pə·rə'zā·shən }

entropy |THERMO| Function of the state of a thermodynamic system whose change in any differential reversible process is equal to the heat absorbed by the system from its surroundings divided by the absolute temperature of the system. Also known as thermal charge. { 'en·trə·pē }

envenomation |MATER| The process by which the surface of a plastic close to or in contact with another surface is deteriorated. { in ¸ven·ə'mā·shən }

environmental stress cracking |MECH| The susceptibility of a material to crack or craze in the presence of surface-active agents or other factors. { in¦vī-ərn¦mənt·əl 'stres ¸krak·iŋ }

environmental test |ENG| A laboratory test conducted to determine the functional performance of a component or system under conditions that simulate the real environment in which the component or system is expected to operate. { in¦vī-ərn¦mənt·əl 'test }

eolotropy See anisotropy. { ¸ē·ə'lä·trə·pē }

Eötvös rule |THERMO| The rule that the rate of change of molar surface energy with temperature is a constant for all liquids; deviations are encountered in practice. { 'ət·vəsh ¸rül }

EPDM See ethylene-propylene terpolymer.

epi- |ORG CHEM| A prefix used in naming compounds to indicate the presence of a bridge or intramolecular connection. { 'ep·ē }

epichlorohydrin |ORG CHEM| C_3H_5OCl A colorless, unstable liquid, insoluble in water; used as a solvent for resins, and to make epoxy resins. { ¸ep·ə¸klòr·ō'hī·drən }

epitaxial diffused-junction transistor |ELECTR| A junction transistor produced by growing a thin, high-purity layer of semiconductor material on a heavily doped region of the same type. { ¸ep·ə'tak·sē·əl də¦fyüzd ¦jəŋk·shən tran'zis·tər }

epitaxial layer |SOLID STATE| A semiconductor layer having the same crystalline orientation as the substrate on which it is grown. { ¸ep·ə'tak·sē·əl ¸lā·ər }

epitaxial transistor |ELECTR| Transistor with one or more epitaxial layers. { ¸ep·ə'tak·sē·əl tran'zis·tər }

epitaxy |CRYSTAL| Growth of one crystal on the surface of another crystal in which the growth of the deposited crystal is oriented by the lattice structure of the substrate. { 'ep·ə¸tak·sē }

EP lubricant |MATER| A lubricating oil or grease that contains additives to improve ability to adhere to the surfaces of metals under high bearing pressures. Derived from extreme-pressure lubricant. { ¦ē¦pē'lü·brə·kənt }

epoxidation |ORG CHEM| Reaction yielding an epoxy compound, such as the conversion of ethylene to ethylene oxide. { e¸päk·sə'dā·shən }

epoxide |ORG CHEM| **1.** A reactive group in which an oxygen atom is joined to each of two carbon atoms which are already bonded to each other. **2.** A three-membered cyclic ether. Also known as oxirane. { e'päk¸sīd }

epoxy- |ORG CHEM| A prefix indicating presence of an epoxide group in a molecule. { ə'päk·sē }

epoxy adhesive |MATER| An adhesive material made of epoxy resin. { ə'päk·sē ad'hē·siv }

1,2-epoxyethane See ethylene oxide. { ¦wən ¦tü ə¦päk·sē'e¸thān }

epoxy resin |POLYM CHEM| A polyether resin formed originally by the polymerization of a diol, such as bisphenol A, and epichlorohydrin, having high strength, and low shrinkage during curing; used as a coating, adhesive, casting, or foam. { ə'päk·sē 'rez·ən }

epsilon structure |SOLID STATE| The hexagonal close-packed structure of the ϵ-phase of an electron compound. { 'ep·sə¸län ¸strək·chər }

EPT See ethylene-propylene terpolymer.

equation of piezotropy |THERMO| An equation obeyed by certain fluids which states that the time rate of change of the fluid's density equals the product of a function of the thermodynamic variables and the time rate of change of the pressure. { i'kwā·zhən əv pē·ə'zä·trə·pē }

equilibrium |CHEM| See chemical equilibrium. |MECH| Condition in which a particle, or all the constituent particles of a body, are at rest or in unaccelerated motion in an inertial reference frame. Also known as static equilibrium. |PHYS| Condition in which no change occurs in the state of a system as long as its surroundings are unaltered. { ¸ē·kwə'lib·rē·əm }

equilibrium diagram |PHYS CHEM| A phase diagram of the equilibrium relationship between temperature, pressure, and composition in any system. { ¸ē·kwə'lib·rē·əm 'dī·ə¸gram }

equilibrium potential |PHYS CHEM| A point in which forward and reverse reaction rates are equal in an electrolytic solution, thereby establishing the potential of an electrode. { ¸ē·kwə'lib·rē·əm pə'ten·chəl }

equilibrium prism |PHYS CHEM| Three-dimensional (solid) diagram for multicomponent mixtures to show the effects of composition changes on some key property, such as freezing point. { ¸ē·kwə'lib·rē·əm ¸priz·əm }

equipotential surface |ELEC| A surface on which the electric potential is the same at every point. |MECH| A surface which is always normal to the lines of force of a field and on which the potential is everywhere the same. { ¦e·kwə·pə'ten·chəl 'sər·fəs }

equivalent bending moment |MECH| A bending moment which, acting alone, would produce in a circular shaft a normal stress of the same magnitude as the maximum normal stress produced by a given bending moment and a given twisting moment acting simultaneously. { i'kwiv·ə·lənt 'bend·iŋ ¸mō·mənt }

equivalent blackbody temperature |THERMO| For a surface, the temperature of a blackbody which emits the same amount of radiation per unit area as does the surface. { i'kwiv·ə·lənt 'blak,bäd·ē ,tem·prə·chər }

equivalent conductance |PHYS CHEM| Property of an electrolyte, equal to the specific conductance divided by the number of gram equivalents of solute per cubic centimeter of solvent. { i'kwiv·ə·lənt kən'dək·təns }

equivalent resistance |ELEC| Concentrated or lumped resistance that would cause the same power loss as the actual small resistance values distributed throughout a circuit. { i'kwiv·ə·lənt ri'zis·təns }

equivalent temperature |THERMO| A term used in British engineering for that temperature of a uniform enclosure in which, in still air, a sizable blackbody at 75°F (23.9°C) would lose heat at the same rate as in the environment. { i'kwiv·ə·lənt 'tem·prə·chər }

equivalent twisting moment |MECH| A twisting moment which, if acting alone, would produce in a circular shaft a shear stress of the same magnitude as the shear stress produced by a given twisting moment and a given bending moment acting simultaneously. { i'kwiv·ə·lənt 'twist·iŋ ,mō·mənt }

equivalent viscous damping |MECH| An assumed value of viscous damping used in analyzing a vibratory motion, such that the dissipation of energy per cycle at resonance is the same for the assumed or the actual damping force. { i'kwiv·ə·lənt ¦vis·kəs 'damp·iŋ }

equivalent weight |CHEM| The number of parts by weight of an element or compound which will combine with or replace, directly or indirectly, 1.008 parts by weight of hydrogen, 8.00 parts of oxygen, or the equivalent weight of any other element or compound. { i'kwiv·ə·lənt 'wāt }

Er *See* erbium.

erbia *See* erbium oxide. { 'ər·bē·ə }

erbium |CHEM| A trivalent metallic rare-earth element, symbol Er, of the yttrium subgroup, found in euxenite, gadolinite, fergusonite, and xenotine; atomic number 68, atomic weight 167.26, specific gravity 9.051; insoluble in water, soluble in acids; melts at 1400–1500°C. { 'ər·bē·əm }

erbium halide |INORG CHEM| A compound of erbium and one of the halide elements. { 'ər·bē·əm 'hal,īd }

erbium nitrate |INORG CHEM| $Er(NO_3)_3 \cdot 5H_2O$ Pink crystals that are soluble in water, alcohol, and acetone; may explode if it is heated or shocked. { 'ər·bē·əm 'nī,trāt }

erbium oxide |INORG CHEM| Er_2O_3 Pink powder that is insoluble in water; used as an actuator for phosphors and in manufacture of glass that absorbs in the infrared. Also known as erbia. { 'ər·bē·əm 'äk,sīd }

erbium sulfate |INORG CHEM| $Er_2(SO_4)_3 \cdot 8H_2O$ Red crystals that are soluble in water. { 'ər·bē·əm 'səl,fāt }

erg |PHYS| A unit of energy or work in the centimeter-gram-second system of units, equal to the work done by a force of magnitude of 1 dyne when the point at which the force is applied is displaced 1 centimeter in the direction of the force. Also known as dyne-centimeter (dyne-cm). { ərg }

Erichsen test |MET| A cupping test to measure the ductility of a piece of sheet metal and to determine its suitability for deep drawing. { 'er·ik·sən ,test }

Erichsen value |MET| The depth of impression in millimeters required to fracture a cupped sheet metal supported on a ring and deformed at the center by a spherically shaped tool. { 'er·ik·sən ,val·yü }

Ericsson cycle |THERMO| An ideal thermodynamic cycle consisting of two isobaric processes interspersed with processes which are, in effect, isothermal, but each of which consists of an infinite number of alternating isentropic and isobaric processes. { 'er·ik·sən ,sī·kəl }

erosion-corrosion |MET| Attack on a metal surface resulting from the combined effects of erosion and corrosion. { ə¦rō·zhən kə¦rō·zhən }

erosion scab |MET| A defect in the form of a solid mass of sand and metal that occurs when molten metal has been agitated or boiled or has partially eroded in the sand of a metal-casting mold. { ə'rō·zhən ,skab }

Es *See* einsteinium.

Eshelby twist |SOLID STATE| A torsional deformation of a crystal whisker resulting from a screw dislocation along the whisker axis. { 'esh·əl·bē 'twist }

ester |ORG CHEM| The compound formed by the elimination of water and the bonding of an alcohol and an organic acid. { 'es·tər }

esterification |ORG CHEM| A chemical reaction whereby esters are formed. { e,ster·ə·fə'kā·shən }

etch |MET| To corrode the surface of a metal in order to reveal its composition and structure. { ech }

etch cleaning |MET| Removing soil by electrolytic or chemical action; removes some of the surface metal along with the dirt. { 'ech ,klēn·iŋ }

etch cracks |MET| Shallow cracks in the surface of hardened steel that result from reaction with an acid, causing hydrogen cracking. { 'ech ,kraks }

etch figures |MET| Minute, faceted pits or surfaces produced on a metal surface by chemical reaction with an etchant. { 'ech ,fig·yərz }

ethene *See* ethylene. { 'e,thēn }

ethenol *See* vinyl alcohol. { 'eth·ə,nȯl }

ether |ORG CHEM| **1.** One of a class of organic compounds characterized by the structural feature of an oxygen linking two hydrocarbon groups (such as R–O–R). **2.** $(C_2H_5)_2O$ A colorless liquid, slightly soluble in water; used as a reagent, intermediate, anesthetic, and solvent. Also known as ethyl ether. { 'e·thər }

ethyl |ORG CHEM| **1.** The hydrocarbon radical C_2H_5. **2.** Trade name for the tetraethyllead antiknock compound in gasoline. { 'eth·əl }

ethyl acrylate |ORG CHEM| $C_5H_8O_2$ A colorless liquid, boiling at 99°C; used to manufacture chemicals and resins. { 'eth·əl 'ak·rə,lāt }

ethyl carbamate

ethyl carbamate *See* urethane. { ¦eth·əl 'kär·bə ¸māt }

ethyl cellulose [POLYM CHEM] The ethyl ether of cellulose; it has film-forming properties and is inert to alkalies and dilute acids; used in adhesives, lacquers, and coatings. { ¦eth·əl'sel·yə ¸lōs }

ethylene [ORG CHEM] C_2H_4 A colorless, flammable gas, boiling at $-102.7°C$; used as an agricultural chemical, in medicine, and for the manufacture of organic chemicals and polyethylene. Also known as ethene; olefiant gas. { 'eth·ə¸lēn }

ethylene glycol *See* glycol. { 'eth·ə·lēn 'glī¸kól }

ethylene oxide [ORG CHEM] $(CH_2)_2O$ A colorless gas, soluble in organic solvents and miscible in water, boiling point $11°C$; used in organic synthesis, for sterilizing, and for fumigating. Also known as 1,2-epoxyethane. { 'eth·ə¸lēn 'äk ¸sīd }

ethylene-propylene terpolymer [POLYM CHEM] An elastomer which is based on ethylene and propylene terpolymers with small amounts of a nonconjugated diene and which can be vulcanized; used for automotive parts, cable coating, hose, footwear, and other products. Abbreviated EPDM; EPT. { ¦eth·ə¸lēn ¦prō·pə ¸lēn tər'päl·ə·mər }

ethylene resin [POLYM CHEM] A thermoplastic material composed of polymers of ethylene; the resin is synthesized by polymerization of ethylene at elevated temperatures and pressures in the presence of catalysts. Also known as polyethylene; polyethylene resin. { 'eth·ə¸lēn 'rez·ən }

ethyl ether *See* ether. { ¦eth·əl 'ē·thər }

ethyl urethane *See* urethane. { ¦eth·əl 'yùr·ə ¸thān }

Eu *See* europium.

European porcelain *See* porcelain. { ¦yùr·ə ¦pē·ən 'pórs·lən }

europium [CHEM] A member of the rare-earth elements in the cerium subgroup, symbol Eu, atomic number 63, atomic weight 151.96, steel gray and malleable, melting at $1100–1200°C$. { yù'rō·pē·əm }

europium halide [INORG CHEM] Any of the compounds of the element europium and the halogen elements; for example, europium chloride, $EuCl_3·xH_2O$. { yù'rō·pē·əm 'ha¸līd }

europium oxide [INORG CHEM] Eu_2O_3 A white powder, insoluble in water; used in red- and infrared-sensitive phosphors. { yù'rō·pē·əm 'äk ¸sīd }

eutectic [MET] The microstructure that results when a metal of eutectic composition solidifies. [PHYS CHEM] An alloy or solution that has the lowest possible constant melting point. { yù'tek·tik }

eutectic crystallization [MET] Simultaneous crystallization of the constituents of a eutectic alloy during cooling of the melt. { yù'tek·tik krist·əl·ə'zā·shən }

eutectic melting [MET] Melting of isolated, microscopic areas of an alloy that correspond to the location of the eutectic of the system. { yù'tek·tik 'melt·iŋ }

eutectic mixture *See* eutectic system. { yù¦tek· tik 'miks·chər }

eutectic point [PHYS CHEM] The point in the constitutional diagram indicating the composition and temperature of the lowest melting point of a eutectic. { yù'tek·tik 'póint }

eutectic system [PHYS CHEM] The particular composition and temperature of materials at the eutectic point. Also known as eutectic mixture. { yù'tek·tik 'sis·təm }

eutectic temperature [PHYS CHEM] The temperature at the lowest melting point of a eutectic. { yù'tek·tik 'tem·prə·chər }

eutectogenic system [PHYS CHEM] A multicomponent liquid-solid mixture in which pure solid phases of each component are in equilibrium with the remaining liquid mixture at a specific (usually minimum) temperature for a given composition, that is, the eutectic point. { yù¦tek·tə¦jen·ik 'sis·təm }

eutectoid [MET] A mixture of phases whose composition is determined by the eutectoid point in the solid region of an equilibrium diagram and whose constituents are formed by the eutectoid reaction. [PHYS CHEM] The point in an equilibrium diagram for a solid solution at which the solution on cooling is converted to a mixture of solids. { yù'tek¸tóid }

evaporable water [MATER] Water present in capillaries or held by surface forces in cement that has set; measured as the amount of water that can be removed by drying under specified conditions. { i¦vap·rə·bəl 'wód·ər }

evaporation [PHYS] Conversion of a liquid to the vapor state by the addition of latent heat. { i¸vap·ə'rā·shən }

Evjen method [SOLID STATE] Method of calculating lattice sums in which groups of charges whose total charge is zero are taken together, so that the contribution of each group is small and the series rapidly converges. { 'ev·yən ¸meth·əd }

Ewald-Kornfeld method [SOLID STATE] An extension of the Ewald method to calculate Coulomb energies of dipole arrays. { ¦ē·valt ¦kórn¸feld ¸meth·əd }

Ewald method [SOLID STATE] Method of calculating lattice sums in which certain mathematical techniques are employed to make series converge rapidly. { 'ē·valt ¸meth·əd }

Ewald sphere [SOLID STATE] A sphere superimposed on the reciprocal lattice of a crystal, used to determine the directions in which an x-ray or other beam will be reflected by a crystal lattice. { 'ē·valt ¸sfir }

Ewing theory of ferromagnetism [SOLID STATE] Theory of ferromagnetic phenomena which assumes each atom is a permanent magnet which can turn freely about its center under the influence of applied fields and other magnets. { 'yü· iŋ ¸thē·ə·rē əv ¸fe·rō'mag·nə¸tiz·əm }

excelsior [MATER] Fine, curled wood shavings, used as packing material. { ek'sel·sē·ər }

excess conduction |SOLID STATE| Electrical conduction by excess electrons in a semiconductor. { 'ek,ses kən'dək·shən }

excess electron |SOLID STATE| Electron introduced into a semiconductor by a donor impurity and available for conduction. { 'ek,ses i'lek ,trän }

exchange anisotropy |ELECTROMAG| Phenomenon observed in certain mixtures of magnetic materials under certain conditions, in which magnetization is favored in some direction (rather than merely along some axis); thought to be caused by exchange coupling across the interface between compounds when one is ferromagnetic and one is antiferromagnetic. { iks'chānj ,an·ə'sä·trə·pē }

excimer |CHEM| An excited diatomic molecule where both atoms are of the same species and are dissociated in the ground state. { 'ek· sə·mər }

excimer laser |OPTICS| A laser containing a noble gas, such as helium or neon, which is based on a transition between an excited state in which a metastable bond exists between two gas atoms and a rapidly dissociating ground state. { 'ek·sə·mər ,lā·zər }

exciplex |CHEM| An excited electron donoracceptor complex which is dissociated in the ground state. { 'ek·sə,pleks }

excited-state effect |SOLID STATE| The motion of a crystal defect through a process in which the defect is first raised into an excited state and then decays, together with the surroundings, into a state in which motion of the defect readily occurs. { ek¦sīd·əd 'stāt i,fekt }

exciton |SOLID STATE| An excited state of an insulator or semiconductor which allows energy to be transported without transport of electric charge; may be thought of as an electron and a hole in a bound state. { 'ek·sə,tän }

exergy |THERMO| The portion of the total energy of a system that is available for conversion to useful work; in particular, the quantity of work that can be performed by a fluid relative to a reference condition, usually the surrounding ambient condition. { 'eks·ər,jē }

exfoliation |MET| Peeling off or separation of metal at its surface in the form of thin, parallel scales or lamellae. { eks,fō·lē'ā·shən }

exfoliation corrosion |MET| A type of corrosion that progresses parallel to the metal surface in such a manner that underlying layers are gradually separated. { eks,fō·lē'ā·shən kə,rō·zhən }

exo- |ORG CHEM| A conformation of carbon bonds in a six-membered ring such that the molecule is boat-shaped with one or more substituents directed outward from the ring. { 'ek·sō }

exoergic See exothermic. { ¦ek·sō¦wər·jik }

exothermic |PHYS| Indicating liberation of heat. Also known as exoergic. { ¦ek·sō'thər·mik }

expanded clay |MATER| A material made from common brick clays by grinding, screening, and then feeding through a gas burner at about 2700°F (1482°C), thus changing the ferric oxide

to ferrous oxide and causing the formation of bubbles. { ik'spand·əd 'klā }

expanded metal |MET| An alloy which has expanded following cooling and solidification. { ik'spand·əd 'med·əl }

expanded perlite |MATER| Perlite that has been finely ground and subjected to extreme heat, causing the particles to become considerably expanded and porous because of release of water. { ik'spand·əd 'pər,līt }

expanded plastic |MATER| A light, spongy plastic made by introducing pockets of air or gas. Also known as foamed plastic; plastic foam. { ik'spand·əd 'plas·tik }

expanded slag |MATER| Slag formed by running slag from phosphate rock onto a forehearth at about 2000°F (1093°C) and then treating it with water, high-pressure steam, and air; used to make lightweight concrete blocks. { ik'spand·əd 'slag }

expanding |MET| A process used to increase the inside diameter of a hollow piece, such as a tube, cup, or shell. { ik'spand·in }

expansion |ELECTR| A process in which the effective gain of an amplifier is varied as a function of signal magnitude, the effective gain being greater for large signals than for small signals; the result is greater volume range in an audio amplifier and greater contrast range in facsimile. |PHYS| Process in which the volume of a constant mass of a substance increases. { ik'span·shən }

expansion coefficient See coefficient of cubical expansion. { ik'span·shən kō·ə'fish·ənt }

expansion ellipsoid |SOLID STATE| An ellipsoid whose axes have lengths which are proportional to the coefficient of linear expansion in the corresponding direction in a crystal. { ik'span·shən ə'lip,sóid }

expansion scab |MET| A defect in the form of a rough thin layer of metal that is partially separated from the body of a metal casting by a thin layer of sand and held in place by a thin vein of metal. { ik'span·shən ,skab }

expansive cement |MATER| A type of hydraulic cement, usually of high sulfate and alumina content, that expands after hardening to compensate for drying shrinkage. { ek'span·siv si'ment }

expansivity See coefficient of cubical expansion. { ,ek,span'siv·əd·ē }

expendable pattern |MET| A pattern that is destroyed in the metal-casting process. { ik ¦spen·də·bəl 'pad·ərn }

explosion method |THERMO| Method of measuring the specific heat of a gas at constant volume by enclosing the gas with an explosive mixture, whose heat of reaction is known, in a chamber closed with a corrugated steel membrane which acts as a manometer, and by deducing the maximum temperature reached on ignition of the mixture from the pressure change. { ik'splō·zhən ,meth·əd }

explosion welding |MET| A solid-state process wherein bonding is produced by a controlled detonation, resulting in rapid movement together of the members to be joined. Also known as explosive welding. { ik'splō·zhən ,weld·iŋ }

explosive cladding [MET] Bonding of a metal coating or metal cladding to a base metal by using the force of an explosive charge. { ik'splō·siv 'klad·iŋ }

explosive forming [MET] Shaping metal parts in dies by using an explosive charge to generate forming pressure. { ik'splō·siv 'fōr·miŋ }

explosive limits [CHEM ENG] The upper and lower limits of percentage composition of a combustible gas mixed with other gases or air within which the mixture explodes when ignited. { ik'splō·siv 'lim·əts }

explosive welding See explosion welding. { ik ¦splō·siv 'weld·iŋ }

extended dislocation [CRYSTAL] A dislocation in a close-packed structure consisting of a strip of stacking fault edged by two lines across which slip through a fraction of a lattice constant, into one of the alternative stacking positions, has occurred. { ik'stend·əd ,dis ,lō'kā·shən }

extender [CHEM] A material used to dilute or extend or change the properties of resins, ceramics, paints, rubber, and so on. { ik'sten·dər }

extender plasticizer See secondary plasticizer. { ik'sten·dər 'plas·tə,sīz·ər }

extensibility [MATER] The extent to which a material can be stretched without causing it to tear or break. [MECH] The amount to which a material can be stretched or distorted without breaking. { ik,sten·sə'bil·əd·ē }

external force [MECH] A force exerted on a system or on some of its components by an agency outside the system. { ek¦stərn·əl 'fōrs }

external phase See continuous phase. { ek 'stərn·əl 'fāz }

external photoelectric effect See photoemission. { ek¦stərn·əl ,fō·dō·i'lek·trik i,fekt }

external work [THERMO] The work done by a system in expanding against forces exerted from outside. { ek¦stərn·əl 'wərk }

extract [CHEM] Material separated from liquid or solid mixture by a solvent. [MET] To separate a metal or a mineral from an ore by various chemical or mechanical methods. { 'ek,strakt (noun) or ik'strakt (verb) }

extraction [CHEM] A method of separation in which a solid or solution is contacted with a liquid solvent (the two being essential mutually insoluble) to transfer one or more components into the solvent. { ik'strak·shən }

extractive metallurgy [MET] Extraction of metals from ore by various chemical and mechanical methods. { ik'strak·tiv 'med·əl·ər·jē }

extreme-pressure lubricant See EP lubricant. { ek¦strēm ¦presh·ər 'lü·brə·kənt }

extrinsic sol [PHYS CHEM] A colloid whose stability is attributed to electric charge on the surface of the colloidal particles. { ek¦strinz·ik 'säl }

extrusion billet [MET] A slug of heated metal that is forced through a die by a hydraulic ram in direct extrusion operations. { ek'strü·zhən ,bil·ət }

extrusion defect [MET] Impaired flow of an extrusion product due to surface oxidation of the ingot or billet. { ek'strü·zhən di,fekt }

extrusion ingot [MET] A cylindrical casting used to form extruded products. { ek'strü·zhən ,iŋ·gət }

extrusion metal [MET] Any of numerous nonferrous metals, alloys, and other materials used in extrusion operations. { ek'strü·zhən ,med·əl }

extrusion pressing See cold extrusion. { ek 'strü·zhən ,pres·iŋ }

Eyring formula [FL MECH] A formula, based on the Eyring theory of rate processes, which relates shear stress acting on a liquid and the resulting rate of shear. { 'ī·riŋ ,fōr·myə·lə }

Eyring molecular system [FL MECH] Theory to account for liquid properties; assumes that each liquid molecule can move freely within a certain free volume. Also known as Eyring theory. { 'ī·riŋ mə'lek·yə·lər ,sis·təm }

Eyring theory See Eyring molecular system. { 'ī·riŋ ,thē·ə·rē }

F

F *See* farad; fluorine.

Faber flaw |SOLID STATE| A deformation in a superconducting material that acts as a nucleation center for the growth of a superconducting region. { 'fā·bər ,flȯ }

fabric |TEXT| A thin, flexible material made of any combination of cloth, fiber, polymeric film, sheet, or foam. { 'fab·rik }

fabricable |MATER| Capable of being shaped, such as an alloy. { 'fab·rik·ə·bəl }

fabric laminate |MATER| Layers of fabric alternating with plastic, used as insulation in electrical equipment. { 'fab·rik 'lam·ə,nāt }

fabric weight |TEXT| The number of ounces per square yard of a fabric. { 'fab·rik ,wāt }

face |CRYSTAL| *See* crystal face. |MATER| The veneer on the exposed surface of a sheet of plywood. |TEXT| The side of a fabric which is more attractive than the other side because of features such as weave, luster, or finish. { fās }

face brick |MATER| A brick of some esthetic quality to be used on the exposed surface of a building wall or other structure. { 'fās ,brik }

face-bridging ligand |ORG CHEM| A ligand that forms a bridge over one triangular face of the polyhedron of a metal cluster structure. { 'fās ,brij·iŋ 'līg·ənd }

face-centered cubic lattice |CRYSTAL| A lattice whose unit cells are cubes, with lattice points at the center of each face of the cube, as well as at the vertices. Abbreviated fcc lattice. { 'fās ,sen·tərd ¦kyüb·ik 'lad·əs }

face-centered orthorhombic lattice |CRYSTAL| An orthorhombic lattice which has lattice points at the center of each face of a unit cell, as well as at the vertices. { 'fās ,sen·tərd ¦ȯr·thō¦räm·bik 'lad·əs }

face feed |MET| In brazing or soldering, the deposition of filler metal to the joint, usually by hand. { 'fās ,fēd }

facet |MATER| The plane surface of a crystal, a cut precious stone, or other fractured surface. { 'fas·ət }

face tile |MATER| Tile with one finished surface, intended for use on a face. { 'fās ,tīl }

face veneer |MATER| Wood veneer selected for its decorative qualities rather than its strength. { 'fās və,nir }

facing |MET| A fine molding sand applied to the face of a mold. { 'fās·iŋ }

factor of safety |MECH| **1.** The ratio between the breaking load on a member, appliance, or hoisting rope and the safe permissible load on it. Also known as safety factor. **2.** *See* factor of stress intensity. { 'fak·tər əv 'sāf·tē }

factor of stress concentration |MECH| Any irregularity producing localized stress in a structural member subject to load. Also known as fatigue-strength reduction factor. { 'fak·tər əv 'stres ,käns·ən,trā·shən }

factor of stress intensity |MECH| The ratio of the maximum stress to which a structural member can be subjected, to the maximum stress to which it is likely to be subjected. Also known as factor of safety. { 'fak·tər əv 'stres in,ten·səd·ē }

factory lumber |MATER| Softwood lumber graded and used in the factory for the manufacture of such items as doors, sashes, moldings, and so on. { 'fak·trē ,ləm·bər }

Fahrenheit scale |THERMO| A temperature scale; the temperature in degrees Fahrenheit (°F) is the sum of 32 plus $\frac{9}{5}$ the temperature in degrees Celsius; water at 1 atmosphere (101,325 pascals) pressure freezes very near 32°F and boils very near 212°F. { 'far·ən,hīt ,skāl }

failure |MECH| Condition caused by collapse, break, or bending, so that a structure or structural element can no longer fulfill its purpose. { 'fāl·yər }

fake set *See* false set. { ¦fāk ¦set }

Falconbridge process |MET| Recovery of nickel from a nickel-copper matte; the matte is first crushed and roasted to remove sulfur, and the copper is acid-leached, filtered off, and electrolyzed; the residual solids are melted, cast as anodes, and refined electrolytically to produce nickel. { 'fȯk·ən,brij ,präs· əs }

falling-ball viscometer *See* falling-sphere viscometer. { 'fȯl· iŋ ¦bȯl vi'skäm·əd·ər }

falling film |FL MECH| A theoretical liquid film that moves downward in even flow on a vertical surface in laminar flow; the concept is used for heat-and mass-transfer calculations. { 'fȯl·iŋ ,film }

falling-sphere viscometer |ENG| A viscometer which measures the speed of a spherical body falling with constant velocity in the fluid whose viscosity is to be determined. Also known as falling-ball viscometer. { 'fȯl·iŋ ,sfir vi'skäm·əd·ər }

false body [PHYS CHEM] The property of certain colloidal substances, such as paints and printing inks, to become very viscous and not flow when left standing; they should flow if a shear stress is applied. { ¦fóls 'bäd·ē }

false bottom [MET] An insert put in either member of a die set to increase the strength and improve the life of the die. { ¦fóls 'bäd·əm }

false pyroelectricity *See* tertiary pyroelectricity. { ¦fóls ¦pī· rō·i¸lek'tri·səd·ē }

false set [MATER] Rapid hardening of freshly mixed cement paste, mortar, or concrete with minimum evolution of heat; plasticity can be restored by mixing without addition of water. { ¦fóls 'set }

farad [ELEC] The unit of capacitance in the meter-kilogram-second system, equal to the capacitance of a capacitor which has a potential difference of 1 volt between its plates when the charge on one of its plates is 1 coulomb, there being an equal and opposite charge on the other plate. Symbolized F. { 'fa¸rad }

faradaic current *See* faradic current. { ¸far·ə¦dā·ik ¦kər·ənt }

faraday [PHYS] The electric charge required to liberate 1 gram-equivalent of a substance by electrolysis; experimentally equal to 96,485.3415 ± 0.0039 coulombs. Also known as Faraday constant. { 'far·ə¸dā }

Faraday birefringence [OPTICS] Difference in the indices of refraction of left and right circularly polarized light passing through matter parallel to an applied magnetic field; it is responsible for the Faraday effect. { 'far·ə¸dā ¸bī·ri'frin·jəns }

Faraday constant *See* faraday. { 'far·ə¸dā¸kän·stənt }

Faraday effect [OPTICS] Rotation of polarization of a beam of linearly polarized light when it passes through matter in the direction of an applied magnetic field; it is the result of Faraday birefringence. Also known as Faraday rotation; Kundt effect; magnetic rotation. { 'far·ə¸dā i'fekt }

Faraday rotation *See* Faraday effect. { 'far·ə¸dā rō'tā·shən }

Faraday's law of electromagnetic induction [ELECTROMAG] The law that the electromotive force induced in a circuit by a changing magnetic field is equal to the negative of the rate of change of the magnetic flux linking the circuit. Also known as law of electromagnetic induction. { 'far·ə¸dāz 'ló əv i¦lek·trō¸mag¦ned·ik in'dək·shən }

Faraday's laws of electrolysis [PHYS CHEM] **1.** The amount of any substance dissolved or deposited in electrolysis is proportional to the total electric charge passed. **2.** The amounts of different substances dissolved or desposited by the passage of the same electric charge are proportional to their equivalent weights. { 'far·ə¸dāz 'lóz əv i¸lek'träl·ə·səs }

faradic current [CHEM] An electric current that corresponds to the reduction or oxidation of a chemical species. Also spelled faradaic current. { fə'rad·ik ¸kə·rənt }

far-infrared radiation [ELECTROMAG] Infrared radiation the wavelengths of which are the longest of those in the infrared region, about 50–1000 micrometers; requires diffraction gratings for spectroscopic analysis. { ¦fär in·frə'red ¸rād·ē'ā·shən }

far-ultraviolet radiation [ELECTROMAG] Ultraviolet radiation in the wavelength range of 200–300 nanometers; germicidal effects are greatest in this range. Abbreviated FUV radiation. { 'fär ¸əl·trə'vī·lət ¸rād·ē'ā·shən }

fast [CHEM] Of dye or pigment, being stable to heat, light, acid, or alkali. { fast }

fast break [MET] Interruption of the current in the magnetizing coil during nondestructive testing of magnetic particles; induces eddy currents and strong magnetization. { ¦fast 'brāk }

fatigue [ELECTR] The decrease of efficiency of a luminescent or light-sensitive material as a result of excitation. [MECH] Failure of a material by cracking resulting from repeated or cyclic stress. { fə'tēg }

fatigue life [MECH] The number of applied repeated stress cycles a material can endure before failure. { fə'tēg ¸līf }

fatigue limit [MECH] The maximum stress that a material can endure for an infinite number of stress cycles without breaking. Also known as endurance limit. { fə'tēg ¸lim·ət }

fatigue notch factor [MET] A notch, scratch, or other impairment on the surface of a metal resulting in premature failure of the metal. { fə'tēg ¦näch ¸fak·tər }

fatigue notch sensitivity [MET] A measure of the reduction of fatigue strength of a metal resulting from a notch. { fə'tēg 'näch sen·sə¸tiv·əd·ē }

fatigue ratio [MECH] The ratio of the fatigue limit or fatigue strength to the static tensile strength. Also known as endurance ratio. { fə'tēg ¸rā·shō }

fatigue strength [MECH] The maximum stress a material can endure for a given number of stress cycles without breaking. Also known as endurance strength. { fə'tēg ¸streŋkth }

fatigue-strength reduction factor *See* factor of stress concentration. { fə'tēg ¸streŋkth ri'dək·shən ¸fak·tər }

fatigue test [ENG] Test to determine the range of alternating stress which a material can withstand without breaking. { fə'tēg ¸test }

fat mortar [MATER] Mortar that adheres to the trowel. { ¦fat 'mór·dər }

fatty acid [ORG CHEM] An organic monobasic acid of the general formula $C_nH_{2n+1}COOH$ derived from the saturated series of aliphatic hydrocarbons; examples are palmitic acid, stearic acid, and oleic acid; used as a lubricant in cosmetics and nutrition, and for soaps and detergents. { ¦fad·ē 'as·əd }

fatty-acid pitch *See* packing house pitch. { ¦fad·ē ¸as·əd 'pich }

fatty alcohol [ORG CHEM] A high-molecular-weight, straight-chain primary alcohol derived from natural fats and oils; includes lauryl, stearyl, oleyl, and linoleyl alcohols; used in pharmaceuticals, cosmetics, detergents, plastics, and lube oils and in textile manufacture. { 'fad·ē 'al·kə,hȯl }

fatty amine [ORG CHEM] RCH_2NH_2 A normal aliphatic amine from oils and fats; used as a plasticizer, in medicine, as a chemical intermediate, and in rubber manufacture. { 'fad·ē 'am,ēn }

fatty ester [ORG CHEM] RCOOR' A fatty acid in which the alkyl group (R') of a monohydric alcohol replaces the active hydrogen; for example, $RCOOCH_3$ from reaction of RCOOH with methanol. { 'fad·ē 'es·tər }

F band [SOLID STATE] The optical absorption band arising from F centers. { 'ef ,band }

fcc lattice See face-centered cubic lattice. { ¦ef ¦sē¦sē 'lad·əs }

F center [SOLID STATE] A color center consisting of an electron trapped by a negative ion vacancy in an ionic crystal, such as an alkali halide or an alkaline-earth fluoride or oxide. { ¦ef ¦sen·tər }

F' center [SOLID STATE] A color center that gives rise to a broad absorption band at longer wavelengths than the band of the F center; probably an F center that has trapped an additional electron. { ¦ef,prīm ,sen·tər }

Fe See iron.

feeder [MET] A runner or riser so placed that it can feed molten metal to the contracting mass of the casting as it cools in its flask, therefore preventing formation of cavities or porous structure. { 'fēd·ər }

feedhead [MET] A reservoir of molten metal that is left above a casting in order to supply additional metal as the casting solidifies and shrinks. Also known as riser; sinkhead. { 'fēd,hed }

feeding rod [MET] A rod used by working up and down to keep the passage clear between riser and casting. { 'fēd·iŋ ,räd }

feed lines [MET] The pattern produced on the surface of a piece of metal by machine grinding. { 'fēd ,līnz }

felt [MATER] A fibrous, watertight heavy paper of organic or asbestos fibers impregnated with asphalt and used as an overlining or an underlining for roofs. [TEXT] A compressed, densely matted unwoven fabric of wool, sometimes with rayon or hair. { felt }

felt side [MATER] The upper side of a sheet of paper which was not in contact with the wire in a papermaking machine. { 'felt ,sīd }

Fermi distribution [SOLID STATE] Distribution of energies of electrons in a semiconductor or metal as given by the Fermi-Dirac distribution function; nearly all energy levels below the Fermi level are filled, and nearly all above this level are empty. { 'fer·mē ,dis·trə,byü·shən }

Fermi hole [SOLID STATE] A region surrounding an electron in a solid in which the energy band theory predicts that the probability of finding other electrons is less than the average over the volume of the solid. { 'fer·mē ,hōl }

Fermi liquid [CRYO] A liquid of particles which have Fermi-Dirac statistics; an example is the liquid phase of helium-3, in which the atoms belong to the isotope with mass number 3. { 'fer·mē ,lik·wəd }

Fermi surface [SOLID STATE] A constant-energy surface in the space containing the wave vectors of states of members of an assembly of independent fermions, such as electrons in a semiconductor or metal, whose energy is that of the Fermi level. { 'fer·mē ,sər·fəs }

fermium [CHEM] A synthetic radioactive element, symbol Fm, with atomic number 100; discovered in debris of the 1952 hydrogen bomb explosion, and now made in nuclear reactors. { 'fer·mē·əm }

fernico [MET] An iron-nickel-cobalt alloy used for metal-to-glass seals. { fər'nī·kō }

ferrate [INORG CHEM] A multiple iron oxide with another oxide, for example, Na_2FeO_4. { 'fe,rāt }

ferric [INORG CHEM] The term for a compound of trivalent iron, for example, ferric bromide, $FeBr_3$. { 'fer·ik }

ferric ammonium alum See ferric ammonium sulfate. { 'fer·ik ə'mōn·ē·əm 'al·əm }

ferric ammonium sulfate [INORG CHEM] $FeNH_4$-$(SO_4)_2 \cdot 12H_2O$ Efflorescent, water-soluble crystals; used in medicine, in analytical chemistry, and as a mordant in textile dyeing. Also known as ferric ammonium alum; iron ammonium sulfate. { 'fer·ik ə'mōn·ē·əm 'səl,fāt }

ferric arsenate [INORG CHEM] $FeAsO_4 \cdot 2H_2O$ A green or brown powder, insoluble in water, soluble in dilute mineral acids; used as an insecticide. { 'fer·ik 'ärs·ən,āt }

ferric bromide [INORG CHEM] $FeBr_3$ Red, deliquescent crystals that decompose upon heating; soluble in water, ether, and alcohol; used in medicine and analytical chemistry. Also known as ferric sesquibromide; ferric tribromide; iron bromide. { 'fer·ik 'brō,mīd }

ferric chloride [INORG CHEM] $FeCl_3$ Brown crystals, melting at 300°C, that are soluble in water, alcohol, and glycerol; used as a coagulant for sewage and industrial wastes, as an oxidizing and chlorinating agent, as a disinfectant, in copper etching, and as a mordant. Also known as anhydrous ferric chloride; ferric trichloride; flores martis; iron chloride. { 'fer·ik 'klȯr,īd }

ferric dichromate [INORG CHEM] $Fe_2(CrO_4)_3$ A red-brown, granular powder, miscible in water; used as a mordant. { 'fer·ik dī'krō,māt }

ferric ferrocyanide [INORG CHEM] $Fe_4[Fe(CN)_6]_3$ Dark-blue crystals, used as a pigment, and with oxalic acid in blue ink. Also known as iron ferrocyanide. { 'fer·ik ,fer·ə'sī·ə,nīd }

ferric fluoride [INORG CHEM] FeF_3 Green, rhombohedral crystals, soluble in water and acids; used in porcelain and pottery manufacture. Also known as iron fluoride. { 'fer·ik 'flür,īd }

ferric hydrate See ferric hydroxide. { 'fer·ik 'hī ,drāt }

ferric hydroxide [INORG CHEM] $Fe(OH)_3$ A brown powder, insoluble in water; used as arsenic poisoning antidote, in pigments, and in pharmaceutical preparations. Also known as ferric hydrate; iron hydroxide. { 'fer·ik hī'dräk,sīd }

ferric nitrate [INORG CHEM] $Fe(NO_3)_3 \cdot 9H_2O$ Colorless crystals, soluble in water and decomposed by heat; used as a dyeing mordant, in tanning, and in analytical chemistry. Also known as iron nitrate.   { 'fer·ik 'nī,trāt }

ferric oxide |INORG CHEM| Fe_2O_3 Red, hexagonal crystals or powder, insoluble in water and soluble in acids, melting at 1565°C; used as a catalyst and pigment for metal polishing, in metallurgy, and in medicine. Also known as ferric oxide red; jeweler's rouge; red ocher.   { 'fer·ik 'äk,sīd }

ferric oxide red   See ferric oxide.   { 'fer·ik ¦ak,sīd 'red }

ferric phosphate |INORG CHEM| $FePO_4 \cdot 2H_2O$ Yellow, rhombohedral crystals, insoluble in water, soluble in acids; used in medicines and fertilizers. Also known as iron phosphate.   { 'fer·ik 'fäs,fāt }

ferric sesquibromide   See ferric bromide.   { 'fer·ik ,ses·kwə'brō,mīd }

ferric sulfate |INORG CHEM| $Fe_2(SO_4)_3 \cdot 9H_2O$ Yellow, water-soluble, rhombohedral crystals, decomposing when heated; used as a chemical intermediate, disinfectant, soil conditioner, pigment, and analytical reagent, and in medicine. Also known as iron sulfate.   { 'fer·ik 'səl,fāt }

ferric tribromide   See ferric bromide.   { 'fer·ik trī'brō,mīd }

ferric trichloride   See ferric chloride.   { 'fer·ik trī'klor,īd }

ferric vanadate |INORG CHEM| $Fe(VO_3)_3$ Grayish-brown powder, insoluble in water and alcohol; used in metallurgy. Also known as iron metavanadate.   { 'fer·ik 'van·ə,dāt }

ferricyanic acid |INORG CHEM| $H_3Fe(CN)_6$ A red-brown unstable solid.   { ,fer·i·sī'an·ik 'as·əd }

ferricyanide |INORG CHEM| A salt containing the radical $Fe(CN)_6{}^{3-}$.   { fer·i'sī·ə,nīd }

ferrimagnet   See ferrimagnetic material.   { 'fe·ri ,mag·nət }

ferrimagnetic garnet |MATER| A synthetic ferrimagnetic oxide material that has the garnet structure and the chemical formula $X_3Fe_5O_{12}$, where the trivalent X ion is yttrium, or any of the rare-earth ions with an atomic number greater than 61.   { ,fe·ri·mag'ned·ik 'gär·nət }

ferrimagnetic material |SOLID STATE| A material displaying ferrimagnetism; the ferrites are the principal example. Also known as ferrimagnet.   { ,fe·ri·mag'ned·ik mə'tir·ē·əl }

ferrimagnetic resonance |PHYS| Magnetic resonance of a ferrimagnetic material.   { ,fe·ri·mag'ned·ik 'rez·ən·əns }

ferrimagnetism |SOLID STATE| A type of magnetism in which the magnetic moments of neighboring ions tend to align nonparallel, usually antiparallel, to each other, but the moments are of different magnitudes, so there is an appreciable resultant magnetization. Also known as Néel ferromagnetism.   { ,fe·ri'mag·nə,tiz·əm }

ferrisulphas   See ferrous sulfate.   { ¦fe·ri'səl·fəs }

ferrite |INORG CHEM| An unstable compound of a strong base and ferric oxide which exists in alkaline solution, such as $NaFeO_2$. |MET| Iron that has not combined with carbon in pig iron or steel. |SOLID STATE| Any ferrimagnetic material having high electrical resistivity which has a spinel crystal structure and the chemical formula XFe_2O_4, where X represents any divalent metal ion whose size is such that it will fit into the crystal structure.   { 'fe,rīt }

ferrite banding |MET| The formation of faint bands (flow lines) of free ferrite in rolled steel, running in the direction of working. Also known as ferrite ghosts; ferrite streaks; ghost lines; ghost structure.   { 'fe,rīt 'ban·diŋ }

ferrite device |ELEC| An electrical device whose principle of operation is based upon the use of ferrites in powdered, compressed, sintered form, making use of their ferrimagnetism and their high electrical resistivity, which makes eddy-current losses extremely low at high frequencies.   { 'fe ,rīt di,vīs }

ferrite ghosts   See ferrite banding.   { 'fe,rīt 'gōsts }

ferrite number |MET| The standard value assigned to austenitic stainless steel to denote a specific ferrite content; used instead of percent ferrite.   { 'fe,rīt 'nəm·bər }

ferrite streaks   See ferrite banding.   { 'fe,rīt 'strēks }

ferritic stainless steel |MET| Any magnetic iron alloy containing more than 12% chromium having a body-centered cubic structure. Also known as stainless iron.   { fe,rid·ik ¦stān·ləs 'stēl }

ferroalloy |MET| Any alloy containing iron, usually in major amount. Also known as ferrous alloy.   { ¦fe·rō'al,ói }

ferroaluminum |MET| An alloy of iron and aluminum; added to molten steel as a deoxidizer or as an alloying component.   { ¦fe·rō· ə'lü·mə·nəm }

ferroboron |MET| An alloy of iron and boron that is added to steel to form hardened special steels; two grades are used, 10% boron and 17% boron.   { ¦fe·rō'bó,rän }

ferrocarbon titanium |MET| An alloy of iron with 15–20% titanium and 3–8% carbon; may be added to molten steel as a component of low-alloy steel.   { ¦fe·rō'kär·bən tī'tān·ē·əm }

ferrocerium |MET| An alloy of iron with a high percentage of cerium; used to make cigarette lighter flints.   { ¦fe·rō'sir·ē·əm }

ferrochromium |MET| A crude ferroalloy containing chromium.   { ¦fe·rō'krō·mē·əm }

ferrocolumbium |MET| An alloy of iron and columbium (niobium); used to add columbium to certain alloy steels.   { ¦fe·rō·kə'ləm·bē·əm }

ferrocyanic acid |INORG CHEM| $H_4Fe(CN)_6$ A white solid obtained by treating ferrocyanides with acid.   { ¦fe·rō·sī'an·ik 'as·əd }

ferrocyanide |INORG CHEM| A salt containing the radical $Fe(CN)_6{}^{4-}$.   { ¦fe·rō'sī·ə,nīd }

ferroelectric |SOLID STATE| A crystalline substance displaying ferroelectricity, such as barium titanate, potassium dihydrogen phosphate, and Rochelle salt; used in ceramic capacitors, acoustic transducers, and dielectric amplifiers. Also known as Rochelle-electric; Seignette-electric.   { ¦fe·rō·i'lek·trik }

ferroelectric Barkhausen effect |SOLID STATE| A series of abrupt changes in the dielectric polarization of a ferroelectric material that occurs when the external electric field acting on the material is varied. { ǀfeˈrō·iǀlek·trik ˈbärk ˌhaúz·ən iˌfekt }

ferroelectric crystal [SOLID STATE] A crystal of a ferroelectric material. { ǀfeˈrō·iˈlek·trik ˈkrist·əl }

ferroelectric domain [SOLID STATE] A region of a ferroelectric material within which the spontaneous polarization is constant. { ǀfeˈrō·iˈlek·trik dəˈmān }

ferroelectric hysteresis [ELEC] The dependence of the polarization of ferroelectric materials not only on the applied electric field but also on their previous history; analogous to magnetic hysteresis in ferromagnetic materials. Also known as dielectric hysteresis; electric hysteresis. { feˈrō·iˈlek·trik ˌhis·təˈrē·səs }

ferroelectricity [SOLID STATE] Spontaneous electric polarization in a crystal; analogous to ferromagnetism. { ǀfeˈrò·iˈlek·trisˈəd·ē }

ferrofluid [MATER] A colloidal suspension of ultramicroscopic magnetic particles in a carrier liquid, used as a lubricant or damping liquid. [PHYS CHEM] A colloidal suspension that becomes magnetized in a magnetic field because of a disperse phase consisting of ferromagnetic or ferrimagnetic particles. { ˈfeˈrō ˌflü·əd }

ferrograph analyzer [ENG] An instrument used for ferrography; a pump delivers a small sample of the fluid to a microscope slide mounted above a magnet that generates a high-gradient magnetic field, causing particles to be deposited in a gradient of sizes along the slide. { ˈferˈəˌgraf ˈanˈəˌlīzˈər }

ferrography [ENG] Wear analysis of machine bearing surfaces by collection of ferrous (or nonferrous) wear particles from lubricating oil in a ferrograph analyzer; the method can be applied to human joints by collecting fragments of cartilage, bone, or prosthetic materials from synovial fluid. { feˈräg· rəˈfē }

ferromagnetic ceramic See ceramic magnet. { ǀfeˈrō·magǀned·ik səˈram·ik }

ferromagnetic crystal [SOLID STATE] A crystal of a ferromagnetic material. Also known as polar crystal. { ǀfeˈrō·magǀned·ik ˈkrist·əl }

ferromagnetic domain [SOLID STATE] A region of a ferromagnetic material within which atomic or molecular magnetic moments are aligned parallel. Also known as magnetic domain. { ǀfeˈrō·magǀned·ik dəˈmān }

ferromagnetic film See magnetic thin film. { ǀfeˈrō·magǀned·ik ˈfilm }

ferromagnetic material [SOLID STATE] A material displaying ferromagnetism, such as the various forms of iron, steel, cobalt, nickel, and their alloys. { ǀfeˈrō·magǀned·ik məˈtirˈē·əl }

ferromagnetic resonance [SOLID STATE] Magnetic resonance of a ferromagnetic material. { ǀfeˈrō·magǀned·ik ˈrezˈənˈəns }

ferromagnetics [ELECTR] The science that deals with the storage of binary information and the logical control of pulse sequences through the utilization of the magnetic polarization properties of materials. { ǀfeˈrō· magǀned·iks }

ferromagnetism [SOLID STATE] A property, exhibited by certain metals, alloys, and compounds of the transition (iron group) rare-earth and actinide elements, in which the internal magnetic moments spontaneously organize in a common direction; gives rise to a permeability considerably greater than that of vacuum, and to magnetic hysteresis. { ǀfeˈrōˈmag·nəˌtiz·əm }

ferromanganese [MET] A ferroalloy containing about 80% manganese and used in steelmaking. { ǀfeˈrōˈmaŋ·gəˌnēs }

ferrometer [ENG] An instrument used to make permeability and hysteresis tests of iron and steel. { fəˈräm·ad·ər }

ferromolybdenum [MET] A molybdenum-iron alloy produced in the electric furnace or by a thermite process; used to introduce molybdenum into iron or steel alloys, and as a coating material on welding rods. { ǀfeˈrō·məˈlib·dəˌnəm }

ferronickel [MET] A crude ferroalloy containing nickel. { ǀfeˈrōǀnikˈəl }

ferrophosphorus [MET] A by-product formed in the heating of iron, phosphate rock, silica, and coke; this alloy is used to increase fluidity in steel casting. { ǀfeˈrōˈfäsˈfəˈrəs }

ferroprussiate paper [MATER] A paper used in a blueprint process to reproduce plans and drawings. { ǀfeˈrōˈprəshˈēˌāt ˌpāˈpər }

ferrosilicon [MET] A crude ferroalloy containing 15–95% silicon and used in steelmaking. { ǀfeˈrōˈsilˈəˌkən }

ferrosilicon process See Pidgeon process. { ǀfeˈrōˈsilˈəˈəkən ˌpräsˈəs }

ferrotitanium [MET] A ferroalloy containing 15–45% titanium and used in steelmaking. { ǀfeˈrōˈtīˈtänˈēˌəm }

ferrotungsten [MET] A crude ferroalloy containing tungsten and used in steelmaking. { ǀfeˈrōˈtəŋˈstən }

ferrouranium [MET] An alloy of iron and uranium. { ǀfeˈrōˈyüˈränˈēˈəm }

ferrous [CHEM] The term or prefix used to denote compounds of iron in which iron is in the divalent (2+) state. { ˈferˈəs }

ferrous alloy See ferroalloy. { ˈferˈəs ˈalˌòi }

ferrous ammonium sulfate [INORG CHEM] $Fe(SO_4)·(NH_4)_2SO_4·6H_2O$ Light-green, water-soluble crystals; used in medicine, analytical chemistry, and metallurgy. Also known as iron ammonium sulfate; Mohr's salt. { ˈferˈəs əˈmōnˈēˈəm ˈsəlˌfāt }

ferrous arsenate [INORG CHEM] $Fe_3(AsO_4)_2·6H_2O$ Water-insoluble, toxic green amorphous powder, soluble in acids; used in medicine and as an insecticide. Also known as iron arsenate. { ˈferˈəs ˈärsˈənˌāt }

ferrous carbonate [INORG CHEM] $FeCO_3$ Green rhombohedral crystals that are soluble in carbonated water and decompose when heated; used in medicine. { ˈferˈəs ˈkärˈbəˌnāt }

ferrous chloride [INORG CHEM] $FeCl_2·4H_2O$ Green, monoclinic crystals, soluble in water; used as a mordant in dyeing, for sewage treatment, in metallurgy, and in pharmaceutical preparations. Also known as iron chloride; iron dichloride. { 'fer·əs 'klȯr‚īd }

ferrous hydroxide [INORG CHEM] Fe(OH)$_2$ A white, water-insoluble, gelatinous solid that turns reddish-brown as it oxidizes to ferric hydroxide. { 'fer·əs hī'dräk‚sīd }

ferrous oxide [INORG CHEM] FeO A black powder, soluble in water, melting at 1419°C. Also known as black iron oxide; iron monoxide. { 'fer·əs 'äk‚sīd }

ferrous sulfate [INORG CHEM] FeSO$_4$·7H$_2$O Blue-green, water-soluble, monoclinic crystals; used as a mordant in dyeing wool, in the manufacture of ink, and as a disinfectant. Also known as copperas; ferrisulphas; green copperas; green vitriol; iron sulfate. { 'fer·əs 'səl‚fāt }

ferrous sulfide [INORG CHEM] FeS Black crystals, insoluble in water, soluble in acids, melting point 1195°C; used to generate hydrogen sulfide in ceramics manufacture. Also known as iron sulfide. { 'fer·əs 'səl‚fīd }

ferrovanadium [MET] An iron alloy high in vanadium (35–55); used to add 0.1–2.5% vanadium during the manufacture of engineering steels and high-strength steels. { ‚fe·rō·və'nād·ē·əm }

ferrum [CHEM] Latin term for iron; derivation of the symbol Fe. { 'fer·əm }

FET See field-effect transistor.

Feynman's superfluidity theory [CRYO] Microscopic theory of superfluid helium which accounts for the spectrum of elementary excitations assumed by Landau's superfluidity theory. { 'fīn·mənz ‚sü·pər‚flü'id·əd·ē ‚thē·ə·rē }

FG achromatism See actinic achromatism. { ‚ef¦jē ‚ā'krō·mə‚tiz·əm }

fiber [MET] **1.** The characteristic of wrought metal that indicates directional properties as revealed by etching or by fracture appearance. **2.** The pattern of preferred orientation of metal crystals after a deformation process, usually wiredrawing. [OPTICS] A transparent threadlike object made of glass or clear plastic, used to conduct light along selected paths. [TEXT] An extremely long, pliable, cohesive natural or manufactured threadlike object from which yarns are spun to be woven into textiles. { 'fī·bər }

fiberboard [MATER] A hard isotropic board made by compressing wood chips or other vegetable fibers. { 'fī·bər ‚bȯrd }

fiber bundle [OPTICS] A flexible bundle of glass or other transparent fibers, parallel to each other, used in fiber optics to transmit a complete image from one end of the bundle to the other. { 'fī·bər ‚bən·dəl }

fiber-composite material [MATER] A composite material in which a fibrous phase that retains its physical identity is dispersed in a continuous matrix phase. { 'fī·bər kəm‚päz·ət mə‚tir·ē·əl }

fiber diagram [SOLID STATE] The x-ray diffraction pattern of a collection of crystallites that have one crystallographic axis approximately parallel to a common direction but are otherwise randomly oriented. { 'fī·bər ‚dī·ə‚gram }

fiberglass See glass fiber. { 'fī·bər‚glas }

fiber grease [MATER] Solid-base lubricating grease; contains soap fibers 1–1000 micrometers long and 0.1–1.0 micrometer wide, which stabilize lubricating action by immobilizing fluid lubricating components. { 'fī·bər ‚grēs }

fiber metal [MET] Any material composed of metal fibers that are pressed or sintered together or infiltrated with resin or other material. { 'fī·bər ‚med·əl }

fiber metallurgy [MET] A branch of metallurgy concerned with the study of metal fibers. { 'fī·bər 'med·əl·ər·jē }

fiber-optic imaging [OPTICS] The formation of optical images by transmission through precisely aligned bundles of optical fibers; each fiber transmits one element of the image. { 'fī·bər ¦äp·tik 'im·əj·iŋ }

fiber optics [OPTICS] The technique of transmitting light through long, thin, flexible fibers of glass, plastic, or other transparent materials; bundles of parallel fibers can be used to transmit complete images. { 'fī·bər ‚äp·tiks }

fiber plaster [MATER] Gypsum plaster containing hair or wood fiber as a binder. { 'fī·bər ‚plas·tər }

fiber-reinforced concrete [MATER] A portland-cement concrete or mortar that is reinforced with dispersed, randomly oriented, unconnected fibers of metallic, mineral, or organic materials. { ¦fī·bər ‚rē·in¦fȯrst 'kan‚krēt }

fiber-reinforced polymer [MATER] Lightweight, high-strength composite material composed of fibers (glass, carbon, or silicon carbide) embedded in a polymeric (epoxy, phenolic, or polyester) matrix. Abbreviated FRP. { ¦fī·bər ‚rē·ən¦fȯrst 'päl·ə·mər }

fiber stress [MECH] **1.** The tensile or compressive stress on the fibers of a fiber metal or other fibrous material, especially when fiber orientation is parallel with the neutral axis. **2.** Local stress through a small area (a point or line) on a section where the stress is not uniform, as in a beam under bending load. { 'fī·bər ‚stres }

fibril [MATER] One of the minute threadlike elements of a natural or synthetic fiber. { 'fī·brəl }

fibrous composite [MATER] A composite material consisting of fibers embedded in a matrix. { 'fī·brəs kəm'päz·ət }

fibrous fracture [MECH] Failure of a material resulting from a ductile crack; broken surfaces are dull and silky. Also known as ductile fracture. { 'fī·brəs 'frak·chər }

fibrous plaster [MATER] Gypsum plaster reinforced or backed with sisal or canvas. { 'fī·brəs 'plas·tər }

fibrous structure [MATER] A ropy surface on a fractured material. [MET] **1.** A lamination on an etched section of a forging. **2.** In wrought iron, a structure consisting of slag fibers embedded in ferrite. { 'fī·brəs 'strək·chər }

Fick's law [PHYS] The law that the rate of diffusion of matter across a plane is proportional to the negative of the rate of change of the concentration of the diffusing substance in the direction perpendicular to the plane. { 'fiks ¦lò }

fiducial temperature [THERMO] Any of the temperatures assigned to a number of reproducible equilibrium states on the International Practical Temperature Scale; standard instruments are calibrated at these temperatures. { fə'dü·shəl 'tem·prə·chər }

field [PHYS] **1.** An entity which acts as an intermediary in interactions between particles, which is distributed over part or all of space, and whose properties are functions of space coordinates and, except for static fields, of time; examples include gravitational field, sound field, and the strain tensor of an elastic medium. **2.** The quantum-mechanical analog of this entity, in which the function of space and time is replaced by an operator at each point in space-time. { fēld }

field desorption [SOLID STATE] A technique which tears atoms from a surface by an electric field applied at a sharp dip to produce very well-ordered, clean, plane surfaces of many crystallographic orientations. { ¦fēld dē'sòrp·shən }

field effect [ELECTR] The local change from the normal value that an electric field produces in the charge-carrier concentration of a semiconductor. { fēld i¦fekt }

field-effect transistor [ELECTR] A transistor in which the resistance of the current path from source to drain is modulated by applying a transverse electric field between grid or gate electrodes; the electric field varies the thickness of the depletion layer between the gates, thereby reducing the conductance. Abbreviated FET. { 'fēld i¦fekt tran'zis·tər }

field quenching [MET] The quench cooling and tempering of a heated metal object at the site of construction or operation by using portable equipment rather than fixed manufacturing facilities. [SOLID STATE] Decrease in the emission of light of a phosphor excited by ultraviolet radiation, x-rays, alpha particles, or cathode rays when an electric field is simultaneously applied. { fēld ¦kwench·iŋ }

field weld [MET] A weld made at the construction site. { 'fēld ¦weld }

fifteen-degrees calorie See calorie. { ¦fif·tēn di ¦grēz ¦kal·ə·rē }

fifth sound [CRYO] A temperature oscillation which propagates in helium II contained in a porous material such as a tightly packed powder, where the normal component is immobilized by its viscosity. { ¦fifth ¦saùnd }

figure [MATER] The natural grain of wood, especially when it is cut as a veneer. { 'fig·yər }

figure of merit [ELECTR] A performance rating that governs the choice of a device for a particular application; for example, the figure of merit of a magnetic amplifier is the ratio of usable power gain to the control time constant. { 'fig·yər əv 'mer·ət }

filament [ELEC] Metallic wire or ribbon which is heated in an incandescent lamp to produce light,

by passing an electric current through the filament. [ELECTR] A cathode made of resistance wire or ribbon, through which an electric current is sent to produce the high temperature required for emission of electrons in a thermionic tube. Also known as directly heated cathode; filamentary cathode; filament-type cathode. [MET] A long, flexible metal wire drawn very fine. [TEXT] A single continuous manufactured fiber which is extruded from a spinneret and joined with others to make a thread. { 'fil·ə·mənt }

filamentary cathode See filament. { ¦fil·ə'ment·ə·rē 'kath¦ōd }

filament drawing [MET] Reducing the cross section of wire by pulling it through a die to form a filament. { 'fil·ə·mənt ¦drò·iŋ }

filament lamp See incandescent lamp. { 'fil·ə·mənt ¦lamp }

filament-type cathode See filament. { 'fil·ə·mənt ¦tīp 'kath¦ōd }

file hardness [ENG] Hardness of a material as determined by testing with a file of standardized hardness; a material which cannot be cut with the file is considered as hard as or harder than the file. { 'fīl ¦härd·nəs }

filiform corrosion [MET] A random threadlike deterioration of a painted or lacquered metal caused by superficial corrosion of the base metal. { 'fil·ə¦fórm kə'rō·zhən }

filled band [SOLID STATE] An energy band, each of whose energy levels is occupied by an electron. { ¦fild 'band }

filled composite [MATER] Mixture (composite) of thermoplastic or thermosetting resin and granular or short-strand fiber fill. { ¦fild kəm 'päz·ət }

filled insulation [MATER] A loose insulating material that is poured or blown into walls. { ¦fild in·sə'lā·shən }

filled thermoplastic [MATER] A thermoplastic resin material that has been extended (filled) with an inert filler powder or fibers before curing. { ¦fild ¦thər·mə¦plas·tik }

filled thermoset [MATER] Thermosetting resin material that has been extended (filled) with an inert filler powder or fibers before curing. { ¦fild 'thər·mə¦set }

filler [MATER] **1.** An inert material added to paper, resin, bituminous material, and other substances to modify their properties and improve quality. **2.** A material used to fill holes in wood, plaster, or other surfaces before applying a coating such as paint or varnish. [MET] The rod used to deposit metal in a joint in brazing, soldering, or welding. Also known as filler metal. { 'fil·ər }

filler metal See filler. { 'fil·ər ¦med·əl }

filler specks [MATER] In a cast plastic object, visible specks of a filler such as wood flour or asbestos that stand out in color contrast against the surface of the object. { 'fil·ər ¦speks }

fillet weld [MET] A weld joining two edges at right angles; cross-sectional configuration is approximately triangular. { 'fil·ət ¦weld }

filling factor [CRYO] The ratio of the electron

density of a quantum Hall liquid to the density of magnetic flux quanta. { 'fil·iŋ ˌfak·tər }

film |GRAPHICS| Plastic material, such as cellulose acetate or cellulose nitrate, coated with a light-sensitive emulsion, used to make negatives or transparencies in radiography or photography. |MATER| A flat section of material that is extremely thin in comparison to its other dimensions and has a nominal maximum thickness of 250 micrometers and a lower limit of thickness of about 25 micrometers. |MET| Oxide coating on a metal. { film }

film base |GRAPHICS| The celluloid component which supports the emulsion of photographic film. { 'film ˌbās }

film boiling |PHYS CHEM| A stage in the boiling process in which the heater surface is totally covered by a film of vapor and the liquid does not contact the solid. |THERMO| Boiling in which a continuous film of vapor forms at the hot surface of the container holding the boiling liquid, reducing heat transfer across the surface. { 'film ˌbȯil·iŋ }

film coefficient |THERMO| For a fluid confined in a vessel, the rate of flow of heat out of the fluid, per unit area of vessel wall divided by the difference between the temperature in the interior of the fluid and the temperature at the surface of the wall. Also known as convection coefficient. { 'film ˌkō·iˌfish·ənt }

filmogen |MATER| The film-forming material or binder in paint which imparts continuity. { 'fil·məˌjen }

film pressure |PHYS| The difference between the surface tension of a pure liquid and the surface tension of the liquid with a unimolecular layer of a given substance adsorbed on it. Also known as surface pressure. { 'film ˌpresh·ər }

film resistor |ELEC| A fixed resistor in which the resistance element is a thin layer of conductive material on an insulated form; the conductive material does not contain binders or insulating material. { 'film riˌzis·tər }

film strength |MATER| **1.** The measurement of a lubricant's ability to keep an unbroken film over surfaces. **2.** The resistance to disruption by films of all types, such as plastic films or surface-coating films. { 'film ˌstreŋkth }

film tension |PHYS CHEM| The contractile force per unit length that is exerted by an equilibrium film in contact with a supporting substrate. { 'film ˌten·chən }

film theory |PHYS| A theory of the transfer of material or heat across a phase boundary, where one or both of the phases are flowing fluids, the main controlling factor being resistance to heat conduction or mass diffusion through a relatively stagnant film of the fluid next to the surface. Also known as boundary-layer theory. { 'film ˌthē·ə·rē }

filter capacitor |ELEC| A capacitor used in a power-supply filter system to provide a low-reactance path for alternating currents and thereby suppress ripple currents, without affecting direct currents. { 'fil·tər kəˌpas·əd·ər }

final set |MATER| Hardening of a mixture of water and cement, concrete, or mortar to a greater degree than the hardening attained at the initial set; generally measured as the time required for the mixture to stiffen sufficiently to resist the penetration of a test needle. { ¦fin·əl 'set }

fine gold |MET| Almost pure gold; the value of bullion gold depends on its percentage of fineness. { ¦fīn 'gōld }

fineness |MET| Degree of purity of gold or silver in parts per thousand. { 'fīn·nəs }

fineness of grind |MATER| In inks and paint, the pigment particle size, size range, and population, which indirectly influence color strength, gloss, and rheology. { ¦fīn·nəs əv 'grīnd }

fines |MATER| **1.** Particles smaller than average in a mixture of particles varying in size. **2.** Fine material which passes through a standard screen on which coarser fragments are retained. |MET| That portion of a metal powder consisting of particles smaller than a specified size. { fīnz }

fine silver |MET| Silver having a minimum fineness of 999; considered to be pure silver. { ¦fīn 'sil·vər }

fining |CHEM ENG| A process in which molten glass is cleared of bubbles, usually by the addition of chemical agents. |MATER| A material such as gelatin, egg white, or bentonite that is used to clarify a liquid. Also known as clarifying agent. { 'fīn·iŋ }

finish |MATER| **1.** A chemical or other material applied to surfaces to protect them, to alter their appearances, or to modify their physical properties; finishes can be physically, chemically, or electrolytically applied and have value for fabrics and fibers, metals, paper products, plastics, woods, and so on. **2.** The ultimate quality, condition, or appearance of the surface of a material. { 'fin·ish }

finished steel |MET| Steel that has undergone final processing and is ready for market. { ¦fin·isht 'stēl }

finishing compound |MATER| A substance used to impart surface properties to textiles or leather, such as softness, flexibility, or fire resistance. { 'fin·ish·iŋ ˌkäm·paùnd }

finishing hydrated lime |MATER| Any hydrated lime suitable for use in the finishing coat of plaster; characterized by a high degree of whiteness and plasticity. { 'fin·ish·iŋ ¦hī.drād·əd ˌlīm }

finishing mill |MET| A rolling mill in which sheet, plate, and other mill products are subjected to final rolling operations. { 'fin·ish·iŋ ˌmil }

finishing temperature |MET| The temperature at which hot-working is completed. { 'fin·ish·iŋ ˌtem·prə·chər }

finite elasticity theory See finite strain theory. { ¦fī.nīt iˌlas'tis·əd·ē ˌthē·ə·rē }

finite strain theory |MECH| A theory of elasticity, appropriate for high compressions, in which it is not assumed that strains are infinitesimally small. Also known as finite elasticity theory. { 'fī.nīt 'strān ˌthē·ə·rē }

Fior process |MET| The prereduction of

130

high-grade iron particles or concentrates in a hot gaseous reactor to produce low-oxygen fines for partially metallized briquettes suitable for electric-arc steelmaking furnaces. { 'fyȯr ˌpräs·əs }

fire assay [MATER] Analysis of a metal-bearing material, especially gold and silver, by heating a sample with a suitable flux and weghing the resulting metal beads. { ¦fīr 'as˒ā }

firebrick [MATER] A refractory brick, often made of fireclay, that is able to withstand high temperature (up to 1500–1600°C) without fusion; used to line furnaces, fireplaces, and chimneys. { 'fīr ˌbrik }

fire point [CHEM] The lowest temperature at which a volatile combustible substance vaporizes rapidly enough to form above its surface an air-vapor mixture which burns continuously when ignited by a small flame. { 'fīr ˌpȯint }

fireproof [MATER] The property of being relatively resistant to combustion. { 'fīr˒prüf }

fireproofing compound See fire retardant. { 'fīr ˌprüf·iŋ ˌkäm˒pau̇nd }

fire refining [MET] The refining of blister copper by treatment in a furnace under oxidizing conditions to remove the impurities and under reducing conditions to remove the excess oxygen. { ¦fīr ri'fīn·iŋ }

fire retardant [MATER] A chemical used as a coating for or a component of a combustible material to reduce or eliminate a tendency to burn; used with textiles, plastics, rubbers, paints, and other materials. { ¦fīr ri'tärd·ənt }

fire-retardant paint [MATER] A paint applied as a thin coating to reduce the rate of flame spread of a combustible material; based on silicone, casein, polyvinylchloride, or other substance. { ¦fīr ri ¦tärd·ənt 'pānt }

fire scale [MET] Copper oxide remaining below the surface of silver-copper alloys after annealing and pickling. { 'fīr ˌskāl }

fire welding See forge welding. { 'fīr ˌweld·iŋ }

firmoviscosity [MECH] Property of a substance in which the stress is equal to the sum of a term proportional to the substance's deformation, and a term proportional to its rate of deformation. { ˌfər·mō·vis¦käs·əd·ē }

first fire composition [MATER] A pyrotechnic composition (readily ignitable and easily pressed into a strong, solid mass), compounded to produce a high temperature, preferably with creation of slag to give heat capacity. { ¦fərst ¦fīr ˌkäm·pə'zish·ən }

first law of thermodynamics [THERMO] The law that heat is a form of energy, and the total amount of energy of all kinds in an isolated system is constant; it is an application of the principle of conservation of energy. { 'fərst ˌlȯ əv ˌthər·mō·dī'nam·iks }

first-order transition [THERMO] A change in state of aggregation of a system accompanied by a discontinuous change in enthalpy, entropy, and volume at a single temperature and pressure. { ¦fərst ˌȯrd·ər trans'zish·ən }

first sound [CRYO] Ordinary sound in helium II,

in which pressure and density variations are propagated; in contrast to second sound. { 'fərst ˌsau̇nd }

Fischer's salt See cobalt potassium nitrite. { ¦fish·ərz 'sȯlt }

fisheye [MATER] A small globular mass which has not blended completely into the surrounding material and is particularly evident in a transparent or translucent material, such as a plastic coating or surface coating. [MET] See flake. { 'fish˒ī }

fish gelatin See isinglass. { fish ˌjel·ət·ən }

fish glue [MATER] An adhesive obtained from the skin of certain fish, principally cod; used in gummed tape, letterpress printing plates, and blueprint paper. { fish ˌglü }

fishmouthing See alligatoring. { 'fish˒mau̇th·iŋ }

fish paper [MATER] A type of fiber used in sheet form for insulating purposes where high mechanical strength is required, as in insulating transformer windings from the transformer core. { 'fish ˌpā·pər }

fishtail [MET] Excess metal trailing on the end of a roll forging. { 'fish˒tāl }

fissure [MET] A small cracklike discontinuity with a slight opening or displacement of the fracture surfaces. { 'fish·ər }

fitch [MATER] **1.** A thin sheet of wood, such as a veneer. **2.** A bundle of veneers arranged in the same order as they were cut from a log. { fich }

five-fourths power law [THERMO] The proposition that the rate of heat loss from a body by free convection is proportional to the five-fourths power of the difference between the temperature of the body and that of its surroundings. { ¦fīv ¦fȯrths 'pau̇·ər ˌlȯ }

fixative [MATER] **1.** A chemical or a mixture of chemicals used to treat biological specimens before preservation so as to retain a reasonable facsimile of their appearance when alive. **2.** A substance used to increase the durability of another substance; used to fix dye mordants, hold textile dyes and pigments, and slow the rate of perfume evaporation. Also known as fixing agent. { 'fik·səd·iv }

fixed-position welding [MET] A welding operation in which the work is stationary. { ¦fikst ˌpȯint 'weld·iŋ }

fixing agent See fixative. { 'fik·siŋ ˌā·jənt }

Flade potential [MET] The potential of a passive metal immediately preceding a final steep fall from the passive to the active region. { 'flād pə'ten·chəl }

flake [MATER] **1.** Dry, unplasticized, cellulosic plastics base. **2.** Plastic chip used as feed in molding operations. **3.** A small, flat wood particle of predetermined dimensions and uniform thickness, with fiber direction essentially in the plane of the flake. [MET] **1.** Discontinuous, internal cracks formed in steel during cooling due usually to the release of hydrogen. Also known as fisheye; shattercrack; snowflake. **2.** Fish-scale, flat particles in powder metallurgy. Also known as flake powder. { flāk }

flakeboard [MATER] Particleboard composed of wood flakes. { 'flāk,bòrd }

flake powder *See* flake. { 'flak ,paùd·ər }

flame annealing [MET] The careful heating of a metal part by flames, before or after working. { 'flām ə,nēl·iŋ }

flame cleaning [MET] Removing scale, rust, and dirt from metal surfaces by using a broad flame. { 'flām ,klēn·iŋ }

flame coating *See* flame plating. { 'flām ,kōd·iŋ }

flame cutting [MET] Use of an oxyacetylene, oxyhydrogen, or oxycoal gas flame to cut thick metal sections. { 'flām ,kəd·iŋ }

flame gouging [MET] A form of oxygen cutting by means of a cutting torch with a slightly curved tip, enabling the flame to strike the metal surface at a low angle and making shallow cuts possible. Also known as oxygen gouging. { 'flām ,gaùj·iŋ }

flame hardening [MET] A method for local surface hardening of steel by passing an oxyacetylene or similar flame over the work at a predetermined rate. { 'flām ,härd·ən·iŋ }

flame plating [MET] Coating a thin layer of refractory material on a surface by exploding a mixture of plating powder, oxygen, and acetylene. Also known as flame coating. { 'flām ,plād·iŋ }

flameproofing [CHEM ENG] The process of treating materials chemically so that they will not support combustion. { 'flām,prüf·iŋ }

flame retardant [CHEM ENG] A substance that can suppress, reduce, or delay the propagation of a flame through a polymer material; may be inserted chemically into the polymer molecule or blended in after polymerization. { 'flām ri ,tärd·ənt }

flame-retarded resin [MATER] A resin which is compounded with certain chemicals to reduce or eliminate its tendency to burn. { 'flām ri,tärd·əd 'rez·ən }

flame spraying [ENG] **1.** A method of applying a plastic coating onto a surface in which finely powdered fragments of the plastic, together with suitable fluxes, are projected through a cone of flame. **2.** Deposition of a conductor on a board in molten form, generally through a metal mask or stencil, by means of a spray gun that feeds wire into a gas flame and drives the molten particles against the work. { 'flām ,sprā·iŋ }

flame straightening [MET] Correcting distorted structural metal to a straight form by local application of a gas-flame heat. { 'flām ,strāt·ən·iŋ }

flammability [CHEM] A measure of the extent to which a material will support combustion. Also known as inflammability. { ,flam·ə'bil·əd·ē }

flammable [MATER] Of a material, capable of supporting combustion. { 'flam·ə·bəl }

flapping [MET] Striking through the surface of molten copper to hasten oxidation by increasing the exposure to air. { 'flap·iŋ }

flaring [MET] Increasing the diameter at the end of a pipe or tube. { 'fler·iŋ }

flash [ENG] In plastics or rubber molding or in metal casting, that portion of the charge which overflows from the mold cavity at the joint line. [MET] A fin of excess metal along the mold joint line of a casting, occurring between mating die faces of a forging or expelled from a joint in resistance welding. { flash }

flash butt welding [MET] Resistance welding to produce a butt joint by passing an electric current through two pieces of metal in light contact to create an arc which causes flashing and consequent heating; the weld is completed by applying pressure at the joint. { 'flash ,bət ,weld·iŋ }

flash coat [MET] A thin coating that is forced from a flash-welded joint after the abutting surfaces are forced together. { 'flash ,kōt }

flashing [ENG] Burning brick in an intermittent air supply in order to impart irregular color to the bricks. [MET] The violent expulsion of small metal particles due to arcing during flash butt welding. { 'flash·iŋ }

flash magnetization [ELECTROMAG] Magnetization of a ferromagnetic object by a current impulse of short duration. { ¦flash ,mag·nə·tə'zā·shən }

flashover [ELEC] An electric discharge around or over the surface of an insulator. { 'flash,ō·vər }

flash plating [MET] Electrodeposition of a thin film of metal. { 'flash ,plād·iŋ }

flash point [CHEM] The lowest temperature at which vapors from a volatile liquid will ignite momentarily upon the application of a small flame under specified conditions; test conditions can be either open- or closed-cup. { 'flash ,pòint }

flash set [MATER] Rapid hardening of freshly mixed cement paste, mortar, or concrete with considerable evolution of heat; plasticity cannot be restored. { ¦flash 'set }

flash smelting [MET] Production of molten metal or matte in a vertical furnace in which concentrates are reacted with hot gases; the molten product is collected in a horizontal refractory-lined accumulator at the base of the furnace. { 'flash ,smel·tiŋ }

flash test [ELEC] A method of testing insulation by applying momentarily a voltage much higher than the rated working voltage. { 'flash ,test }

flash welding [MET] A form of resistance butt welding used to weld wide, thin members or members with irregular faces, and tubing to tubing. { 'flash ,weld·iŋ }

flask [MET] A frame used to hold molding sand in foundry work. { flask }

flat die forging [MET] Die forging in which the metal is worked between simple contour dies. { 'flat 'dī ,fór·jiŋ }

flat fillet weld [MET] A fillet weld having a face that is relatively flat. { 'flat ,fil·ət 'weld }

flat grain [MATER] The grain pattern formed when soft wood is cut so that the tree's annular rings form an angle of 45° or less with the board's surface. { ¦flat 'grān }

flat-position welding [MET] Welding above the joint with the face of the weld in the horizontal reference plane. Also known as downhand welding. { 'flat pə,zish·ən 'weld·iŋ }

flattening [MET] Straightening of metal sheet by

passing it through special rollers which flatten it without changing its thickness. Also known as roll flattening. { 'flat·ən·iŋ }

flattening test [MET] Quality test performed by flattening metal tubing between parallel plates that are a specified distance from each other. { 'flat·ən·iŋ ,test }

flatting agent [MATER] Additive substance for paints or varnishes to disperse incident light rays to give the dried surface a nonglossy matte finish. { 'flat·ən·iŋ ,ā·jənt }

flaw [MATER] A discontinuity in a material beyond acceptable established limits. { flȯ }

flexibility [MECH] The quality or state of being able to be flexed or bent repeatedly. { ,flek·sə'bil·əd·ē }

flexibilizer [POLYM CHEM] An additive that gives an otherwise rigid plastic flexibility. Also known as plasticizer. { 'flek·sə·bə,līz·ər }

flexible collodion [MATER] A collodion which has two additives (2% camphor and 3% castor oil) to make a pliable film. { ,flek·sə·bəl kə'lōd·ē·ən }

flexible glue [MATER] A type of glue used for pliable molds and printers' rollers, for example, a mixture of glue, glycerol, and water. { ,flek·sə·bəl 'glü }

flexible pavement [CIV ENG] A road or runway made of bituminous material which has little tensile strength and is therefore flexible. { ,flek·sə·bəl 'pāv·mənt }

flexometer [ENG] An instrument for measuring the flexibility of materials. { flek'säm·əd·ər }

flexural modulus [MECH] A measure of the resistance of a beam of specified material and cross section to bending, equal to the product of Young's modulus for the material and the square of the radius of gyration of the beam about its neutral axis. { 'flek·shə·rəl 'mäj·ə·ləs }

flexural rigidity [MECH] The ratio of the sideward force applied to one end of a beam to the resulting displacement of this end, when the other end is clamped. { 'flek·shə·rəl ri'jid·əd·ē }

flexural strength [MECH] Strength of a material in blending, that is, resistance to fracture. { 'flek·shə·rəl 'streŋkth }

flexure [MECH] **1.** The deformation of any beam subjected to a load. **2.** Any deformation of an elastic body in which the points originally lying on any straight line are displaced to form a plane curve. { 'flek·shər }

flexure theory [MECH] Theory of the deformation of a prismatic beam having a length at least 10 times its depth and consisting of a material obeying Hooke's law, in response to stresses within the elastic limit. { 'flek·shər ,thē·ə·rē }

flint glass [MATER] **1.** Heavy, colorless, brilliant glass that contains lead oxide. **2.** Any high-quality glass. { 'flint ,glas }

flip-over process See Umklapp process. { 'flip ,ō·vər ,präs·əs }

flitch [MATER] **1.** A section that is cut from a log and subsequently manufactured into veneer or lumber. **2.** Sheets of veneer that are stacked in sequence as they have been cut from a log. **3.** A longitudinal section of a log, sometimes having bark on one or more edges. { flich }

float glass [MATER] Flat glass with a nearly true optical surface produced by floating a continuous sheet of molten glass on a bed of molten tin until the glass cools and hardens. { 'flōt ,glas }

floating plug [MET] A plug or mandrel attached to a rod and used in plug drawing. Also known as a plug die. { ¦flōd·iŋ 'pləg }

floating zone refining [MET] A variation of the zone-refining technique in which the molten zone is held in place by its own surface tension between two collinear rods; since no container is needed, contamination of the pure metal is avoided. { ¦flōd·iŋ 'zōn ri,fīn·iŋ }

floc [CHEM] Small masses formed in a fluid through coagulation, agglomeration, or biochemical reaction of fine suspended particles. { fläk }

flocculant See flocculating agent. { 'fläk·yə·lənt }

flocculate [CHEM] To cause to aggregate or coalesce into a flocculent mass. { 'fläk·yə,lət (adjective) or 'fläk·yə,lāt (verb) }

flocculating agent [CHEM] A reagent added to a dispersion of solids in a liquid to bring together the fine particles to form flocs. Also known as flocculant. { 'fläk·yə,lād·iŋ ,ā·jənt }

flores martis See ferric chloride. { 'flȯr·ēz 'märd·əs }

flospinning [MET] Power-spinning or flowing metal over a rotating bar for shaping into cylindrical, conical, and curvilinear parts. { 'flō,spin·iŋ }

floss [MET] Molten or solid slag floating on the surface of a metal melt. { fläs }

floss hole [MET] A small door or opening of the bottom of a smokestack or flue for removal of ash. { 'fläs ,hōl }

flour gold [MET] The finest-size gold dust, much of which will float on water. { 'flau̇·ər ,gōld }

flow [FL MECH] The forward continuous movement of a fluid, such as gases, vapors, or liquids, through closed or open channels or conduits. [PHYS] The movement of electric charges, gases, liquids, or other materials or quantities. { flō }

flow brazing [MET] A brazing process in which coalescence is produced by the heat of molten filler metal that is poured over a joint. { 'flō ,brāz·iŋ }

flow brightening [MET] In a soldering process, the melting of a chemical or mechanical metallic coating on the base metal to be soldered. { 'flō ,brīt·ən·iŋ }

flow curve [FL MECH] A graph of the total shear of a fluid as a function of time. [MECH] The stress-strain curve of a plastic material. { 'flō ,kərv }

flowers of tin See stannic oxide. { 'flau̇·ərz əv 'tin }

flowing furnace [MET] A furnace from which molten metal can be tapped or drawn. { ¦flō·iŋ ¦fər·nəs }

flow marks [MATER] Wavy surface marks on a thermoplastic resin molding due to improper flow of material into the mold. { 'flō ,märks }

flow resistance [FL MECH] Any factor within a

conduit or channel that impedes the flow of fluid, such as surface roughness or sudden bends, contractions, or expansions. See viscosity. { 'flō ri₁zis·təns }

flow stress [MECH] The stress along one axis at a given value of strain that is required to produce plastic deformation. { 'flō ,stres }

flow welding [MET] A welding process in which coalescence is produced by heating with molten filler metal, which is poured over the joint until the welding temperature is attained and the required amount of filler metal is added. { 'flō ,weld·iŋ }

flue dust [MET] Fine particles of metal or alloy emitted with the gases of a smelter or metallurgical furnace. { 'flü ,dəst }

fluence [ELECTROMAG] The total energy per unit area carried by a pulse of electromagnetic radiation. [PHYS] A measure of time-integrated particle flux, expressed in particles per square centimeter. { 'flü·əns }

fluid [PHYS] An aggregate of matter in which the molecules are able to flow past each other without limit and without fracture planes forming. { 'flü·əd }

fluid-compressed [MET] Pertaining to steel that has been compressed while still fluid to remove gases and make the material more homogeneous. { 'flü·əd kəm,prest }

fluid density [FL MECH] The mass of a fluid per unit volume. { 'flü·əd ¦den·səd·ē }

fluid dynamics [FL MECH] The science of fluids in motion. { ¦flü·əd dī'nam·iks }

fluid friction [FL MECH] Conversion of mechanical energy in fluid flow into heat energy. { ¦flü·əd 'frik·shən }

fluidity [FL MECH] The reciprocal of viscosity; expresses the ability of a substance to flow. { flü'id·əd·ē }

fluid resistance [FL MECH] The force exerted by a gas or liquid opposing the motion of a body through it. Also known as resistance. { ¦flü·əd ri'zis·təns }

fluoborate See fluoroborate. { ,flü·ə'bór,āt }

fluophor See luminophor. { 'flü·ə,fór }

fluorescence See bloom. { flu'res·əns }

fluorescent dye [CHEM] A highly reflective dye that serves to intensify color and add to the brilliance of a fabric. { flü¦res·ənt 'dī }

fluorescent lamp [ELECTR] A tubular discharge lamp in which ionization of mercury vapor produces radiation that activates the fluorescent coating on the inner surface of the glass. { flü ¦res·ənt 'lamp }

fluorescent pigment [CHEM] A pigment capable of absorbing both visible and nonvisible electromagnetic radiations and releasing them quickly as energy of desired wavelength; examples are zinc sulfide or cadmium sulfide. { flü¦res·ənt 'pig·mənt }

fluoride [INORG CHEM] A salt of hydrofluoric acid, HF, in which the fluorine atom is in the −1 oxidation state. { 'flúr,īd }

fluorine [CHEM] A gaseous or liquid chemical element, symbol F, atomic number 9, atomic weight 18.998403; a member of the halide family, it is the most electronegative element and the most chemically energetic of the nonmetallic elements; highly toxic, corrosive, and flammable; used in rocket fuels and as a chemical intermediate. { 'flúr,ēn }

fluoroborate [INORG CHEM] **1.** Any of a group of compounds related to the borates in which one or more oxygens have been replaced by fluorine atoms. **2.** The BF_4^- ion, which is derived from fluoroboric acid, HBF_4. Also known as fluoborate. { ,flúr·ə'bór,āt }

fluoroboric acid [INORG CHEM] HBF_4 Colorless, clear, water-miscible acid; used for electrolytic brightening of aluminum and for forming stabilized diazo salts. { ,flúr·ə,bór·ik 'as·əd }

fluorocarbon fiber [POLYM CHEM] Fiber made from a fluorocarbon resin, such as polytetrafluoroethylene resin. { flúr·ō'kär·bən 'fī·bər }

fluorocarbon resin [POLYM CHEM] Polymeric material made up of carbon and fluorine with or without other halogens (such as chlorine) or hydrogen; the resin is extremely inert and more dense than corresponding fluorocarbons such as Teflon. { ¦flúr·ō'kär·bən 'rez·ən }

fluoroelastomer [POLYM CHEM] A partially fluorinated polymer or a copolymer; it is the most chemically resistant of the elastomers and has good mechanical properties at high and low temperatures. { ,flúr·ō·i'las·tə·mər }

fluorophosphoric acid [INORG CHEM] H_2PO_3F A colorless, viscous liquid that is miscible with water; used in metal cleaners and as a catalyst. { flúr·ō,fäs'fór·ik 'as·əd }

fluoroplastics [POLYM CHEM] A family of plastics based on fluorine replacement of hydrogen atoms in hydrocarbon molecules; includes polytetrafluoroethylene (PTFE), polychlorotrifluoroethylene (PCTFE), polyvinylidene fluoride (PVDF or PVF_2), and fluorinated ethylene propylene (FEP). Also known as fluoropolymers. { ¦flúr·ō¦plas·tiks }

fluoropolymers See fluoroplastics. { ¦flúr·ō 'päl·ə·mərz }

fluorothene See chlorotrifluoroethylene polymer. { 'flúr·ə,thēn }

fluosilicate [INORG CHEM] A salt derived from fluosilicic acid, H_2SiF_6, and containing the SiF_6^{-2} ion. { ¦flü·ə'sil·ə,kāt }

fluosilicic acid [INORG CHEM] H_2SiF_6 A colorless acid, soluble in water, which attacks glass and stoneware; highly corrosive and toxic; used in water fluoridation and electroplating. Also known as hydrofluorosilicic acid; hydrofluosilicic acid. { ¦flü·ə·sə'lis·ik 'as·əd }

fluosolids system [MET] In pyrometallurgy, a roasting method for finely divided solids, in which air under pressure is blown through a heated bed of mineral to keep it fluid. { ¦flü·ō ¦säl·ədz 'sis·təm }

fluosulfonic acid [INORG CHEM] HSO_3F Colorless, corrosive, fuming liquid; soluble in water with partial decomposition; used as organic synthesis catalyst and in electroplating. { ¦flü·ə·səl'fän·ik 'as·əd }

flushing oil |MATER| A solvent oil designed to remove used lubricating oil, decomposition products, and accumulated dirt from lubrication passages, crankcase surfaces, and lubricated moving parts of automotive engines. { 'fləsh·iŋ ,ȯil }

flux |ELECTROMAG| The electric or magnetic lines of force in a region. |MATER| **1.** In soldering, welding, and brazing, a material applied to the pieces to be united to reduce the melting point of solders and filler metals and to prevent the formation of oxides. **2.** A substance used to promote the fusing of minerals or metals. **3.** Additive for plastics composition to improve flow during physical processing. **4.** In enamel work, a substance composed of silicates and other materials that forms a colorless, transparent glass when fired. |PHYS| **1.** The integral over a given surface of the component of a vector field (for example, the magnetic flux density, electric displacement, or gravitational field) perpendicular to the surface; by definition, it is proportional to the number of lines of force crossing the surface. **2.** The amount of some quantity flowing across a given area (often a unit area perpendicular to the flow) per unit time; the quantity may be, for example, mass or volume of fluid, electromagnetic energy, or number of particles. { fləks }

flux-closure domain *See* closure domain. { 'fləks ,klō·zhər dō,mān }

flux-cored welding |MET| Welding with a metal electrode that has a flux core. { ¦fləks ¦kȯrd 'weld·iŋ }

flux density |PHYS| Any vector field whose flux is a significant physical quantity; examples are magnetic flux density, electric displacement, gravitational field, and the Poynting vector. { 'fləks ,den·səd·ē }

flux density threshold *See* threshold illuminance. { 'fləks ,den·səd·ē 'thresh,hōld }

flux factor |MET| A factor for assessing the quality of steelworks-grade silica refractories. { 'fləks ,fak·tər }

flux guide |MET| A shaped piece of magnetic material used to guide magnetic flux in induction heating; may be used either to direct the flux to preferred locations or to prevent the flux from spreading beyond definite regions. { 'fləks ,gīd }

fluxing |MET| The development of the liquid phase in a ceramic body under heat treatment by the melting of low-fusion components; used in steel manufacture, metal smelting, and assaying. { 'fləks·iŋ }

fluxing ore |MET| An ore containing usually an appreciable amount of valuable metal, but smelted mainly because it contains fluxing agents which are required in the reduction of other ores. { 'fləks·iŋ ,ȯr }

flux jumping *See* Meissner effect. { 'fləks ,jəmp·iŋ }

flux lattice |SOLID STATE| The regular array of fluxoids in a type II superconductor in the mixed state, when the superconductor is sufficiently pure and free of defects. { 'fləks ,lad·əs }

flux-lattice melting |SOLID STATE| A pheno-menon in high-temperature superconductors in the mixed state, in which the regular ordering of fluxoids breaks down above a certain temperature. { 'fləks,lad·əs ,melt·iŋ }

flux line *See* fluxoid. { 'fləks ,līn }

flux-line pinning |SOLID STATE| The introduction, in a type II superconductor, of microscopic crystalline defects that suppress the motion of flux lines which would otherwise occur in the presence of a magnetic field. { 'fləks ,līn ,pin·iŋ }

flux of energy |PHYS| The energy which passes through a surface per unit area per unit time. { 'fləks əv 'en·ər·jē }

fluxoid |SOLID STATE| One of the microscopic filaments of magnetic flux that penetrates a type II superconductor in the mixed state, consisting of a normal core in which the magnetic field is large, surrounded by a superconducting region in which flows a vortex of persistent supercurrent which maintains the field in the core. Also known as flux line; fluxon; vortex. { 'fluk,sȯid }

flux oil |MATER| An oil suitable for blending with bitumen or asphalt to form a product of greater fluidity or softer consistency. { 'fləks ,ȯil }

fluxon *See* fluxoid. { 'flək,sän }

flux oxygen cutting |MET| Oxygen cutting of metal with the aid of a flux to reduce the temperature. { 'fləks 'äk·sə·jən ,kəd·iŋ }

flux pump |CRYO| A cryogenic direct-current generator that converts a small alternating-current input to a large direct-current output when cooled to about 4 K; the output current builds up in a series of steps, much like the action of a pump. { 'fləks ,pəmp }

flux refraction |ELECTROMAG| The abrupt change in direction of magnetic flux lines at the boundary between two media having different permeabilities, or of the electric flux lines at the boundary between two media having different dielectric constants, when these lines are oblique to the boundary. { 'fləks ri,frak·shən }

flying shear |MET| A machine which cuts lengths of rolled products and allows for continuous production by reciprocating with the product while cutting. { ¦flī·iŋ 'shir }

Fm *See* fermium.

foam |CHEM| An emulsionlike two-phase system where the dispersed phase is gas or air. |FL MECH| A collection of bubbles on the surface of a liquid, often stabilized by organic contaminants, as found at sea or along shore. { fōm }

foamed plastic *See* expanded plastic. { ¦fōmd 'plas·tik }

foam glass |MATER| A light, black, opaque, cellular glass made by adding powdered carbon to crushed glass and firing the mixture. { 'fōm ,glas }

foaming agent *See* blowing agent. { 'fōm·iŋ ,ā·jənt }

foam metal |MET| Cast metal with finely divided gas bubbles evenly distributed throughout the body of the metal; an example is foam aluminum. { ¦fōm 'med·əl }

foam rubber *See* rubber sponge. { ¦fōm 'rəb·ər }

focal power |OPTICS| A measure of the ability of a lens, mirror, prism, or optical system to converge a parallel beam of light; equals the reciprocal of the focal length. Also known as power. { 'fō·kəl ,paů·ər }

focal spot |MET| In electron-beam or laser welding, the spot where the beam has the highest concentrated energy level. { 'fō·kəl ,spät }

fogged metal |MET| A metal whose luster has been highly reduced by corrosion products. { ¦fägd ¦med·əl }

fog quenching |MET| Rapid cooling of a metal piece in a fine vapor or mist. { 'fäg ,kwench·iŋ }

foil |MET| A thin sheet of metal, usually less than 0.006 inch (0.15 millimeter) thick. { fȯil }

fold *See* lap. { fōld }

foliation |MET| Beating metal into thin sheets. { ,fō·lē'ā·shən }

footcandle |OPTICS| A unit of illumination, equal to the illumination of a surface, 1 square foot in area, on which there is a luminous flux of 1 lumen uniformly distributed, or equal to the illumination of a surface all points of which are at a distance of 1 foot from a uniform point source of 1 candela; equal to approximately 10.7639 lux. Abbreviated ftc. { 'fůt ,kand·əl }

Forbes bar |THERMO| A metal bar which has one end immersed in a crucible of molten metal and thermometers placed in holes at intervals along the bar; measurement of temperatures along the bar together with measurement of cooling of a short piece of the bar enables calculation of the thermal conductivity of the metal. { 'fȯrbz ,bär }

forbidden band |SOLID STATE| A range of unallowed energy levels for an electron in a solid. { fər¦bid·ən 'band }

force |MECH| That influence on a body which causes it to accelerate; quantitatively it is a vector, equal to the body's time rate of change of momentum. { fȯrs }

force constant |MECH| The ratio of the force to the deformation of a system whose deformation is proportional to the applied force. { 'fȯrs ,kän·stənt }

forced convection |THERMO| Heat convection in which fluid motion is maintained by some external agency. { ¦fȯrst kən'vek·shən }

forehand welding |MET| Welding in which the flame is directed against the base metal ahead of the weld and is moved in the direction of welding. Also known as forward welding. { ¦fȯr ,hand 'weld·iŋ }

forehearth |MET| 1. A bay in front of the hearth of a furnace. 2. A receptacle in front of a hearth to receive the molten products. { 'fȯr,härth }

forge |MET| 1. To form a metal, usually hot, into desirable shapes by employing compressive forces. 2. A machine or place in which metal is formed hot, or where iron is produced from its ore. { fȯrj }

forgeability |MET| Suitability of a material for forging. { 'fȯr·jə'bil·əd·ē }

forge delay time |MET| The time between the start of weld time and the time when forging pressure is reached by the electrode force. { 'fȯrj di'lā ,tīm }

forge welding |MET| A group of welding processes in which the parts to be joined, usually iron, are heated to about 1000°C and then hammered or pressed together. Also known as fire welding. { 'fȯrj ,weld·iŋ }

forging |MET| 1. Using compressive force to shape metal by plastic deformation; dies may be used. 2. A piece of work made by forging. { 'fȯrj·iŋ }

forging brass |MET| Brass composed of 60% copper, 38% zinc, and 2% lead, used for hot forgings, hardware, and plumbing supplies; it is extremely plastic when hot, is corrosion-resistant, and has excellent mechanical properties. { 'fȯrj·iŋ ,bras }

forging hammer |MET| A hammer used to pound metal into forgings. { 'fȯrj·iŋ ,ham·ər }

forging plane |MET| The plane of the principal die face when oriented normal to the direction of ram travel. { 'fȯrj·iŋ ,plān }

forging press |MET| A press designed to operate dies in die forging. { 'fȯrj·iŋ ,pres }

forging range |MET| Optimum temperature range in which a metal can be forged. { 'fȯrj·iŋ ,rānj }

forging rolls |MET| A machine used in making forgings by rolling the metal. { 'fȯrj·iŋ ,rōlz }

forging stock |MET| A section or a piece of metal used to make a forging. { 'fȯrj·iŋ ,stäk }

formability |MATER| Capability of a material to be shaped by plastic deformation. { ,fȯr·mə'bil·əd·ē }

formaldehyde |ORG CHEM| HCHO The simplest aldehyde; a gas at room temperature, and a poisonous, clear, colorless liquid solution with pungent odor; used to make synthetic resins by reaction with phenols, urea, and melamine, as a chemical intermediate, as an embalming fluid, and as a disinfectant. Also known as formol; methanal; methylene oxide. { fȯr'mal·də,hīd }

forming |ELEC| Application of voltage to an electrolytic capacitor, electrolytic rectifier, or semiconductor device to produce a desired permanent change in electrical characteristics as a part of the manufacturing process. { 'fȯrm·iŋ }

form oil |MATER| An oil utilized on the contact surface of wooden or metal concrete forms to prevent concrete from sticking. { 'fȯrm ,ȯil }

formol *See* formaldehyde. { 'fȯr,mȯl }

formonitrile *See* hydrocyanic acid. { ¦fȯr·mō ¦nī·trəl }

formulation |CHEM| The particular mixture of base chemicals and additives required for a product. { ,fȯr·myə'lā·shən }

formula weight |CHEM| 1. The gram-molecular weight of a substance. 2. In the case of a substance of uncertain molecular weight such as certain proteins, the molecular weight calculated from the composition, assuming that the element present in the smallest proportion is represented by only one atom. { 'fȯr·myə·lə ,wāt }

forward extrusion |MET| A cold extrusion

process in which a formed blank is placed in a die cavity and struck by a punch; the metal is extruded through an annular space between the die and the end of the punch, moving in the same direction as the punch. { ¦fȯr·wərd ik'strü·zhən }

forward welding See forehand welding. { ¦fȯr·wərd ¦weld·iŋ }

Foucault current See eddy current. { fü'kō ¦kə·rənt }

founding [MET] The art and science of melting and casting metals. { 'fau̇nd·iŋ }

foundry [ENG] A building where metal or glass castings are produced. { 'faun·drē }

foundry alloy See master alloy. { 'faun·drē ‚ȯi }

foundry core sand See core sand. { ¦faun·drē 'kȯr ‚sand }

foundry engineering [ENG] The science and practice of melting and casting glass or metal. { 'faun·drē ‚en·jə‚nir·iŋ }

foundry facing [MET] A material applied to a sand mold to improve the surface quality of a casting. { 'faun·drē ‚fās·iŋ }

foundry return [MET] A scrapped casting that is returned to the furnace for remelting. { 'faun·drē ri‚tərn }

foundry sand [MET] Sand used in foundries to make molds for the casting of metal shapes. Also known as molding sand. { 'faun·drē ‚sand }

four-ball tester [ENG] A machine designed to measure the efficiency of lubricants by driving one ball against three stationary balls clamped together in a cup filled with the lubricant; performance is evaluated by measuring wear-scar diameters on the stationary balls. { ¦fȯr ¦bȯl 'tes·tər }

fourier See thermal ohm. { fur·ē‚ā }

Fourier heat equation See Fourier law of heat conduction; heat equation. { ‚fur·ē‚ā 'hēt i‚kwā·zhən }

Fourier law of heat conduction [THERMO] The law that the rate of heat flow through a substance is proportional to the area normal to the direction of flow and to the negative of the rate of change of temperature with distance along the direction of flow. Also known as Fourier heat equation. { 'fur·ē‚ā ‚lȯ əv 'hēt kən‚dək·shən }

Fourier number [FL MECH] A dimensionless number used in unsteady-state flow problems, equal to the product of the dynamic viscosity and a characteristic time divided by the product of the fluid density and the square of a characteristic length. Symbolized Fo_f. [PHYS] A dimensionless number used in the study of unsteady-state mass transfer, equal to the product of the diffusion coefficient and a characteristic time divided by the square of a characteristic length. Symbolized N_{Fom}. [THERMO] A dimensionless number used in the study of unsteady-state heat transfer, equal to the product of the thermal conductivity and a characteristic time, divided by the product of the density, the specific heat at constant pressure, and the distance from the midpoint of the body

through which heat is passing to the surface. Symbolized N_{Foh}. { ‚fur·ē‚ā ‚nəm·bər }

fourth sound [CRYO] A pressure wave which propagates in helium II contained in a porous material such as a tightly packed powder, and which results entirely from motion of the super-fluid component, the normal component being immobilized by its viscosity. { ¦fȯrth 'saund }

Fowler-DuBridge theory [SOLID STATE] Theory of photoelectric emission from a metal based on the Sommerfeld model, which takes into account the thermal agitation of electrons in the metal and predicts the photoelectric yield and the energy spectrum of photoelectrons as functions of temperature and the frequency of incident radiation. { ¦faul·ər dü'brij ‚thē·ə·rē }

Fowler function [SOLID STATE] A mathematical function used in the Fowler-DuBridge theory to calculate the photoelectric yield. { 'faulər ‚faŋk·shən }

fp See freezing point.

Fr See francium.

fraction [CHEM] One of the portions of a volatile liquid within certain boiling point ranges, such as petroleum naphtha fractions or gas-oil fractions. [MET] In powder metallurgy, that portion of sample that lies between two stated particle sizes. Also known as cut. { 'frak·shən }

fractography [MET] The microscopic examination of fractured metal surfaces. { frak'täg·rə·fē }

fracture strength See fracture stress. { 'frak·shər ‚streŋkth }

fracture stress [MECH] The minimum tensile stress that will cause fracture. Also known as fracture strength. { 'frak·shər ‚stres }

fracture test [ENG] 1. Macro- or microscopic examination of a fractured surface to determine characteristics such as grain pattern, composition, or the presence of defects. 2. A test designed to evaluate fracture stress. { 'frak·shər ‚test }

fracture wear [MECH] The wear on individual abrasive grains on the surface of a grinding wheel caused by fracture. { 'frak·shər ‚wer }

framework structure [SOLID STATE] A crystalline structure in which there are strong interatomic bonds which are not confined to a single plane, in contrast to a layer structure. { 'frām ‚wərk ‚strək·chər }

francium [CHEM] A radioactive alkali-metal element, symbol Fr, atomic number 87, atomic weight distinguished by nuclear instability; exists in short-lived radioactive forms, the chief isotope being francium-223. { 'fran·sē·əm }

frangible [MECH] Breakable, fragile, or brittle. { 'fran·jə·bəl }

Frank partial dislocation [CRYSTAL] A partial dislocation whose Burger's vector is not parallel to the fault plane, so that it can only diffuse and not glide, in contrast to a Schockley partial dislocation. { ¦fräŋk ¦pär·shəl ‚dis·lō'kā·shən }

Frank-Read source [MET] 1. The creation of dislocations by application of shear stress to an edge dislocation anchored terminally, causing formation of an unstable loop form followed by formation of a closed dislocation line and the establishment of the original condition. **2.** One of the sources of dislocations in a plastically deforming metal. { ¦fräŋk ¦rēd ˌsórs }

Frary metal [MET] Metal containing 97–98% lead alloyed with 1–2% barium and calcium; used for bearings. { 'frer·ē ˌmed·əl }

Fraude's reagent See perchloric acid. { 'frȯdz rē ˌā·jənt }

free carbon [MET] Elemental carbon present in a metal in an uncombined state. { ¦frē 'kär·bən }

free charge [ELEC] Electric charge which is not bound to a definite site in a solid, in contrast to the polarization charge. { ¦frē 'chärj }

free convection See natural convection. { ¦frē kən'vek·shən }

free-cutting steel [MET] Steel that contains a higher percentage of sulfur than carbon steel, making it very easy to machine. { ¦frē ˌkəd·iŋ 'stēl }

free-electron laser [OPTICS] A device in which a beam of relativistic electrons passes through a static periodic magnetic field to amplify a superimposed coherent optical wave and thereby produce a powerful beam of coherent light. { 'frē i¦lek,trän 'lā·zər }

free-electron theory of metals [SOLID STATE] A model of a metal in which the free electrons, that is, those giving rise to the conductivity, are regarded as moving in a potential (due to the metal ions in the lattice and to all the remaining free electrons) which is approximated as constant everywhere inside the metal. Also known as Sommerfeld model; Sommerfeld theory. { ¦frē i'lek,trän ¦thē·ə·rē əv 'med·əlz }

free energy [THERMO] **1.** The internal energy of a system minus the product of its temperature and its entropy. Also known as Helmholtz free energy; Helmholtz function; Helmholtz potential; thermodynamic potential at constant volume; work function. **2.** See Gibbs free energy. { ¦frē 'en·ər·jē }

free enthalpy See Gibbs free energy. { ¦frē 'en ˌthal·pē }

free ferrite [MET] Relatively pure metallic iron phase present in steel or cast iron. { ¦frē 'fe,rīt }

free-flight melt spinning [MET] A rapid-quenching process in which the molten metal is forced through an orifice under pressure and the jet is solidified while in free flight; quench rates reach 1000 K per second. { 'frē 'flīt 'melt ˌspin·iŋ }

free gold [MET] Gold that is in the free state, that is, not combined with other substances. { ¦frē ¦gōld }

free hole [SOLID STATE] Any hole which is not bound to an impurity or to an exciton. { ¦frē ¦hōl }

free-machining steel [MET] Steel to which impurities have been added to improve machinability. { 'frē mə¦shēn·iŋ 'stēl }

free-milling gold [MET] Gold that has a clean surface so that it readily amalgamates with mercury (by gravity or on blankets) after liberation by comminution. { 'frē ˌmil·iŋ 'gōld }

free radical [CHEM] An atom or a diatomic or polyatomic molecule which possesses one unpaired electron. Also known as a radical. { ¦frē 'rad·ə·kəl }

free-radical reaction See homolytic cleavage. { ¦frē ¦rad·ə·kəl rē'ak·shən }

freeze etching [CRYO] A method using cryogenics to prepare specimens for study with a microscope. { 'frēz ˌech·iŋ }

freezing point [PHYS CHEM] The temperature at which a liquid and a solid may be in equilibrium. Abbreviated fp. { 'frēz·iŋ ˌpȯint }

freezing-point depression [PHYS CHEM] The lowering of the freezing point of a solution compared to the pure solvent; the depression is proportional to the active mass of the solute in a given amount of solvent. { 'frēz·iŋ ˌpȯint di ˌpresh·ən }

French chalk [MATER] Finely ground talc. { ¦french 'chȯk }

French polish [MATER] Shellac dissolved in methylated spirits. { ¦french 'päl·ish }

Frenkel defect [SOLID STATE] A crystal defect consisting of a vacancy and an interstitial which arise when an atom is plucked out of a normal lattice site and forced into an interstitial position. Also known as Frenkel pair. { 'freŋk·əl ¦dē¦fekt }

Frenkel exciton [SOLID STATE] A tightly bound exciton in which the electron and the hole are usually on the same atom, although the pair can travel anywhere in the crystal. { 'freŋk·əl 'ek·sə ˌtän }

Frenkel pair See Frenkel defect. { 'freŋk·əl ˌper }

frequency [PHYS] The number of cycles completed by a periodic quantity in a unit time. { 'frē·kwən·sē }

frequency mixing [OPTICS] The combination of two or more electromagnetic waves in a nonlinear medium to form another wave whose frequency is a sum or difference of the frequencies of the incident waves. { 'frē·kwən·sē ˌmik·siŋ }

frequency spectrum [PHYS] A plot of the distribution of the intensity of some type of electromagnetic or acoustic radiation as a function of frequency. { 'frē·kwən·sē ˌspek·trəm }

Fresnel theory of double refraction [OPTICS] The theory which explains double refraction of a crystal in terms of nonspherical wave surfaces. { frā'nel ˌthē·ə·rē əv ˌdəb·əl ri'frak·shən }

fretting corrosion [MET] Surface damage usually in an air environment between two surfaces, one or both of which are metals, in close contact under pressure and subject to a slight relative motion. Also known as chafing corrosion. { 'fred·iŋ kə'rō·zhən }

friability [MATER] The ease with which a material is crumbled, pulverized, or reduced to powder. { ˌfrī·ə'bil·əd·ē }

friable [MATER] Referring to the property of a substance capable of being easily rubbed, crumbled, or pulverized into powder. { 'frī·ə·bəl }

friction [MECH] A force which opposes the relative motion of two bodies whenever such motion exists or whenever there exist other forces which tend to produce such motion. { 'frik·shən }

frictional electricity [ELEC] The electric charges produced on two different objects, such as silk and glass, by rubbing them together. Also known as triboelectricity. { 'frik·shən·əl i,lek'tri·səd·ē }

friction bonding [ENG] Soldering of a semiconductor chip to a substrate by vibrating the chip back and forth under pressure to create friction that breaks up oxide layers and helps alloy the mating terminals. { 'frik·shən ,bänd·iŋ }

friction coefficient See coefficient of friction. { 'frik·shən ,kō·i'fish·ənt }

friction damping [MECH] The conversion of the mechanical vibrational energy of solids into heat energy by causing one dry member to slide on another. { 'frik·shən ,damp·iŋ }

friction force microscopy [ENG] The use of an atomic force microscope to measure the frictional forces on a surface. { ¦frik·shən ¦fors mī'krä·skə·pē }

friction loss [MECH] Mechanical energy lost because of mechanical friction between moving parts of a machine. { 'frik·shən ,lōs }

friction tape [MATER] Cotton tape impregnated with a sticky moisture-repelling compound; used chiefly to hold rubber-tape insulation in position over a joint or splice. { 'frik·shən ,tāp }

friction-tube viscometer [ENG] Device to determine liquid viscosity by measurement of pressure drop through a friction tube with the liquid in viscous flow; gives direct solution to Poiseuille's equation. { 'frik·shən ,tüb vi'skäm·əd·ər }

friction welding [ENG] A welding process for metals and thermoplastic materials in which two members are joined by rubbing the mating faces together under high pressure. { 'frik·shən ,weld·iŋ }

Friedel's law [CRYSTAL] The law that x-ray or electron diffraction measurements cannot determine whether or not a crystal has a center of symmetry. { frē'delz ,lò }

frigorie [THERMO] A unit of rate of extraction of heat used in refrigeration, equal to 1000 fifteen-degree calories per hour, or 1.16264 ± 0.00014 watts. { 'frig·ə·rē }

frit [MATER] Fusible ceramic mixture used to make glazes and enamels for dinnerware and metallic surfaces, as on stoves and metal-base basins and tubs. { frit }

frit seal [ENG] A seal made by fusing together metallic powders with a glass binder, for such applications as hermetically sealing ceramic packages for integrated circuits. { 'frit ,sēl }

fritting [ENG] Fusing materials for glass by application of heat. [MET] The pasty condition, usually occurring a little below the melting point, of the powdered ore, flux, and other reagents in fire assaying. { 'frid·iŋ }

front pinacoid [CRYSTAL] The {100}pinacoid in an orthorhombic, monoclinic, or triclinic crystal. Also known as macropinacoid; orthopinacoid. { ¦frənt 'pin·ə,kòid }

front-to-back ratio [SOLID STATE] Ratio of resistance of a crystal to current flowing in the normal

direction to current flowing in the opposite direction. { ¦frənt tə ¦bak 'rā·shō }

frosted glass [MATER] Glass that has been etched with sand, or appears to have been so treated. { ¦fròs·təd 'glas }

frost line [MATER] In polyethylene film extrusion, a ring-shaped area with a frosty appearance at the point where the film reaches its final diameter. { 'fròst ,līn }

FRP See fiber-reinforced polymer.

frustration [SOLID STATE] In spin glasses, a phenomenon in which individual magnetic moments receive competing ordering instructions via different routes, because of the variation of the interaction between pairs of atomic moments with separation. { frəs'trā·shən }

ftc See footcandle.

fuel cell [PHYS CHEM] An electrochemical device in which the reaction between a fuel, such as hydrogen, and an oxidant, such as oxygen or air, converts the chemical energy of the fuel directly into electrical energy without combustion. { 'fyül ,sel }

fuel-cell catalyst [CHEM] A substance, such as platinum, silver, or nickel, from which the electrodes of a fuel cell are made, and which speeds the reaction of the cell; it is especially important in a fuel cell which does not operate at high temperatures. { 'fyül ,sel 'kad·ə,list }

fuel-cell electrolyte [CHEM] The substance which conducts electricity between the electrodes of a fuel cell. { 'fyül ,sel i'lek·trə,līt }

fuel-cell fuel [CHEM] A substance, such as hydrogen, carbon monoxide, sodium, alcohol, or a hydrocarbon, which reacts with oxygen to generate energy in a fuel cell. { 'fyül ,sel 'fyül }

fugacity [THERMO] A function used as an analog of the partial pressure in applying thermodynamics to real systems; at a constant temperature it is proportional to the exponential of the ratio of the chemical potential of a constituent of a system divided by the product of the gas constant and the temperature, and it approaches the partial pressure as the total pressure of the gas approaches zero. { fyü'gas·əd·ē }

fugacity coefficient [THERMO] The ratio of the fugacity of a gas to its pressure. { fyü'gas·əd·ē ,kō·ə,fish·ənt }

fugitive dye [CHEM] A dye whose color is unstable, that is, not fast to light or washing. { ¦fyü·jəd·iv 'dī }

full annealing [MET] Heating steel to a high temperature and then cooling to ambient or near-ambient temperatures. { ¦ful ə'nēl·iŋ }

full automatic plating [MET] Electroplating a piece of work that is carried through the full cycle automatically. { ¦ful ,òd·ə,mad·ik 'plād·iŋ }

fuller [MET] A die or portion of a die used in preliminary forging operations to reduce the cross section somewhere between the ends of a piece of stock. { 'fül·ər }

fullerene [CHEM] A large molecule composed entirely of carbon, with the chemical formula C_n, where n is any even number from 32 to over 100; believed to have the structure of a hollow spheroidal cage with a surface network of carbon atoms connected in hexagonal and pentagonal rings. { 'fúl·ə,rēn }

full wires |MATER| In wire-mesh cloth, wires running the short way of the cloth as woven. { 'fúl ,wīrz }

fuming nitric acid |INORG CHEM| Concentrated nitric acid containing dissolved nitrogen dioxide; may be prepared by adding formaldehyde to concentrated nitric acid. { ¦fyüm·iŋ 'nī,trik ,as·əd }

fuming sulfuric acid |INORG CHEM| Concentrated sulfuric acid containing dissolved sulfur trioxide. Also known as oleum. { ¦fyüm·iŋ səl'fyúr·ik ,as·əd }

functional group |ORG CHEM| An atom or group of atoms, acting as a unit, that has replaced a hydrogen atom in a hydrocarbon molecule and whose presence imparts characteristic properties to this molecule. Also known as functionality. { ¦fəŋk·shən·əl 'grüp }

functionality See functional group. { ,fəŋk·shə'nal·əd·ē }

fundamental constants |PHYS| The physical constants which play a fundamental role in the basic theories of physics, including the speed of light, electronic charge, electronic mass, Planck's constant, and the fine-structure constant. Also known as atomic constants; universal constants. { ,fən·də'ment·əl 'kän·stəns }

fundamental interval |THERMO| **1.** The value arbitrarily assigned to the difference in temperature between two fixed points (such as the ice point and steam point) on a temperature scale, in order to define the scale. **2.** The difference between the values recorded by a thermometer at two fixed points; for example, the difference between the resistances recorded by a resistance thermometer at the ice point and steam point. { ¦fən·də¦men·təl 'int·ər·vəl }

fundamental particle See elementary particle. { ¦fən·də¦men·təl 'pärd·ə·kəl }

fungicide |MATER| An agent that kills or destroys fungi. { 'fən·jə,sīd }

fungistat |MATER| A compound that inhibits or prevents growth of fungi. { 'fən·jə,stat }

furan |ORG CHEM| **1.** One of a group of organic heterocyclic compounds containing a diunsaturated ring of four carbon atoms and one oxygen atom. **2.** $C_4H_4O_4$ The simplest furan type of molecule; a colorless, mildly toxic liquid, boiling at 32°C, insoluble in water, soluble in alcohol and ether; used as a chemical intermediate. Also known as furfuran. { 'fyür,an }

furan cement |MATER| A strong adhesive that is made from furfural-alcohol resins and is highly resistant to chemicals. { 'fyür,an si ,ment }

furan resin |POLYM CHEM| A liquid, thermosetting resin in which the furan ring is an integral part of the polymer chain, made by the condensation of furfuryl alcohol; used as a cement and

adhesive, casting resin, coating, and impregnant. { 'fyür,an ,rez·ən }

furfuran See furan. { 'fər·fə,ran }

furnace black |CHEM| A carbon black formed by partial combustion of liquid and gaseous hydrocarbons in a closed furnace with a deficiency of oxygen; used as a reinforcing filler for synthetic rubber. { 'fər·nəs ,blak }

furnace brazing |MET| Joining two metals by mechanical union of the filler metal and joint, then heating the composite in a furnace. { 'fər·nəs ,brāz·iŋ }

furnace refining |MET| Purification of molten metal by treatment in a reverberatory furnace. { ¦fər·nəs ri'fīn·iŋ }

furnace soldering |MET| Soldering by heating clamped members to the appropriate temperature in a furnace. { ¦fər·nəs 'säd·ə·riŋ }

furring brick |MATER| Hollow brick grooved for plastering. { 'fər·iŋ ,brik }

furring tile |MATER| Non-load-bearing clay tile used for lining interior walls. { 'fər·iŋ ,tīl }

fused aromatic ring |ORG CHEM| A molecular structure in which two aromatic rings have two carbon atoms in common. { ¦fyüzd ar·ə¦mad·ik 'riŋ }

fused junction See alloy junction. { ¦fyüzd 'jəŋk·shən }

fused-junction diode See alloy-junction diode. { ¦fyüzd ¦jəŋk·shən 'dī,ōd }

fused potassium sulfide See potassium sulfide. { 'fyüzd pə'tas·ē·əm 'səl,fīd }

fused quartz |MATER| A glasslike insulating material made by melting crushed crystals of natural quartz or a certain type of quartz sand. { 'fyüzd 'kwórts }

fused silica See silica glass. { ¦fyüzd 'sil·ə·kə }

fused silver nitrate See lunar caustic. { 'fyüzd 'sil·vər 'nī,trāt }

fused spray deposit |MET| In thermal spraying, deposit which is sprayed on a preheated substrate and has the capability to coalesce within itself as well as to the substrate. { ¦fyüzd ¦sprā di'päz·ət }

fusibility |THERMO| The quality or degree of being capable of being liquefied by heat. { ,fyü·zə'bil·əd·ē }

fusible alloy |MET| A low melting alloy, usually of bismuth, tin, cadmium, and lead, which melts at temperatures as low as 70°C (160°F). { ¦fyü·zə·bəl 'al,ói }

fusion face |MET| That portion of a metal surface that will be fused in a welding operation. { 'fyü·zhən ,fās }

fusion welding |MET| Any welding operation involving melting of the base or parent metal. { 'fyü·zhən ,weld·iŋ }

fusion zone |MET| The volume of base or parent metal melted during a welding operation. { 'fyü·zhən ,zōn }

FUV radiation See far-ultraviolet radiation. { ¦ef ¦yü¦vē ,räd·ē'ā·shən }

G

G *See* conductance.

Ga *See* gallium.

gadolinium |CHEM| A rare-earth element, symbol Gd, atomic number 64, atomic weight 157.25; highly magnetic, especially at low temperatures. { ,gad·əl'in·ē·əm }

gaged brick |MATER| Brick which has been ground or otherwise produced to accurate dimensions. { ¦gājd ¦brik }

gagger |MET| An irregular-shaped piece of metal used in a sand mold to reinforce and support a metal casting. { 'gag·ər }

gall |MET| Damage to metal surfaces resulting from friction and improper lubrication. { gȯl }

galling |MET| Surface damage on mating, moving metal parts due to friction caused by local welding of high spots. { gȯl·iŋ }

gallium |CHEM| A chemical element, symbol Ga, atomic number 31, atomic weight 69.72. |MET| A silvery-white metal, melting at 29.7°C, boiling at 1983°C. { 'gal·ē·əm }

gallium arsenide |INORG CHEM| GaAs A crystalline material, melting point 1238°C; frequently alloys of this material are formed with gallium phosphide or indium arsenide. { 'gal·ē·əm 'ärs·ən,īd }

gallium arsenide laser |OPTICS| A laser that emits light at right angles to a junction region in gallium arsenide, at a wavelength of 9000 angstroms (900 nanometers); can be modulated directly at microwave frequencies; cryogenic cooling is required. { ¦gal·ē·əm ¦ärs·ən,īd 'lā·zər }

gallium arsenide semiconductor |SOLID STATE| A semiconductor having a forbidden-band gap of 1.4 electronvolts and a maximum operating temperature of 400°C when used in a transistor. { ¦gal·ē·əm ¦ärs·ən,īd 'sem·i·kən,dək·tər }

gallium halide |INORG CHEM| A compound formed by bonding of gallium to either chlorine, bromine, iodine, fluorine, or astatine. { 'gal·ē·əm 'ha,līd }

gallium phosphide |INORG CHEM| GaP Transparent crystals made by reacting phosphorus and gallium suboxide at low temperature. { 'gal·ē·əm 'fäs,fīd }

galvanic |ELEC| Pertaining to electricity flowing as a result of chemical action. { gal'van·ik }

galvanic battery |ELEC| A galvanic cell, or two or more such cells electrically connected to produce energy. { gal'van·ik 'bad·ə·rē }

galvanic cell |ELEC| An electrolytic cell that is capable of producing electric energy by electrochemical action. { gal'van·ik 'sel }

galvanic corrosion |MET| Electrochemical corrosion associated with the current in a galvanic cell, caused by dissimilar metals in an electrolyte because of the difference in potential (emf) of the two metals. Also known as contact corrosion. { gal'van·ik kə'rō·zhən }

galvanic current |ELEC| A steady direct current. { gal'van·ik 'kə·rənt }

galvanic series |CHEM| The relative hierarchy of metals arranged in order from magnesium (least noble) at the anodic, corroded end through platinum (most noble) at the cathodic, protected end. { gal'van·ik 'sir·ēz }

galvanize |MET| To deposit zinc on the surface of iron or steel by the processes of hot dipping, sherardizing, or sometimes electroplating. { 'gal·və ,nīz }

gamene *See* madder. { 'ga,mēn }

gamma |CHEM| The gamma position (the third carbon atom in an aliphatic carbon chain) on a chemical compound. |ELECTROMAG| A unit of magnetic field strength, equal to 10 microoersteds, or 0.00001 oersted. |MECH| A unit of mass equal to 10^{-6} gram or 10^{-9} kilogram. { 'gam·ə }

gamma iron |MET| Iron having a face-centered cubic lattice structure, stable between 910 and 1400°C. { 'gam·ə ,ī·ərn }

gamma structure |SOLID STATE| A Hume-Rothery designation for structurally analogous phases or intermetallic phases having 21 valence electrons to 13 atoms, analogous to the γ-brass structure. { 'gam·ə ,strək·chər }

gamma transition *See* glass transition. { 'gam·ə tran'zish·ən }

ganister |MATER| A fine mixture of quartz and fireclay which is used to line certain furnaces for metallurgical processes. { 'gan·ə·stər }

gap |ELEC| The spacing between two electric contacts. |ELECTROMAG| A break in a closed magnetic circuit, containing only air or filled with a nonmagnetic material. |MET| An opening at the point of closest approach between faces of members in a weld joint. { gap }

gap-graded aggregate |MATER| Aggregate in

which certain size particles are entirely or substantially absent. { 'gap ,grād·əd 'ag·rə·gət }

garbage pitch |MATER| Dark-brown to black pitch material obtained as a by-product residue from the burning of garbage; properties are analogous to complex hydrocarbons; used to make paints, varnishes, tarred paper, and waterproofing compound. { 'gär·bij ,pich }

garnet paper |MATER| Paper with a layer of crushed garnet on one side; used as an abrasive or polisher. { 'gär·nət ,pā·pər }

gas |PHYS| A phase of matter in which the substance expands readily to fill any containing vessel; characterized by relatively low density. { gas }

gas brazing See gas-flame brazing. { ¦gas 'brāz·iŋ }

gas carburizing |MET| Surface hardening by heating a metal in gas of high carbon content in order to introduce carbon into the surface layers. { ¦gas 'kär·bə,riz·iŋ }

gas constant |THERMO| The constant of proportionality appearing in the equation of state of an ideal gas, equal to the pressure of the gas times its molar volume divided by its temperature. Also known as gas-law constant; universal gas constant. { gas ,kän·stənt }

gas cutting |MET| Cutting metal with the heat of an oxyacetylene flame. { 'gas ,kəd·iŋ }

gas cyaniding See carbonitriding. { 'gas 'sī·ə ,nīd·iŋ }

gas cycle |THERMO| A sequence in which a gaseous fluid undergoes a series of thermodynamic phases, ultimately returning to its original state. { 'gas ,sī·kəl }

gas-deviation factor See compressibility factor. { ¦gas ,dē·vē'ā·shən ,fak·tər }

gas-flame brazing |MET| A brazing process for which the heat is supplied by a gas flame. Also known as gas brazing. { 'gas ,flām 'brāz·iŋ }

gas law |THERMO| Any law relating the pressure, volume, and temperature of a gas. { 'gas ,lò }

gas-law constant See gas constant. { 'gas ,lò ,kän·stənt }

gas-metal arc welding |MET| A welding procedure that employs an electric arc to heat the joint to fusion, and an inert gas to prevent oxidation of the weld. { 'gas 'med·əl 'ärk ,weld·iŋ }

gaspar |MATER| A mixture of finely ground glass and quartz; a feldspar substitute in some applications and a hard-rubber filler. { 'ga,spär }

gas pocket |MET| A cavity in a metal which contains trapped gases. { 'gas ,päk·ət }

gas-shielded arc welding |MET| Use of a gas atmosphere to shield the molten metal from air in arc welding. { 'gas ,shēl·dəd 'ärk ,weld·iŋ }

gas-tungsten arc welding |MET| Gas-metal arc welding in which tungsten is the agent of fusion. { 'gas 'təŋ·stən 'ärk ,weld·iŋ }

gas welding |MET| A welding process in which metals are joined by the heat of an oxyacetylene flame. { 'gas ,weld·iŋ }

gate |ELECTR| 1. A circuit having an output and a multiplicity of inputs and so designed that the output is energized only when a certain combination of pulses is present at the inputs. 2. A circuit in which one signal, generally a square wave, serves to switch another signal on and off. 3. One of the electrodes in a field-effect transistor. 4. An output element of a cryotron. 5. To control the passage of a pulse or signal. |MET| The opening in a casting mold through which molten metal enters the cavity. Also known as in-gate. { gāt }

gauss |ELECTROMAG| Unit of magnetic induction in the electromagnetic and Gaussian systems of units, equal to 1 maxwell per square centimeter, or 10^{-4} weber per square meter. Also known as abtesla (abt). { gaús }

gaussmeter |ENG| A magnetometer whose scale is graduated in gauss or kilogauss, and usually measures only the intensity, and not the direction, of the magnetic field. { 'gaús,mēd·ər }

gauze |MATER| 1. A sheer, loosely woven textile fabric similar to cheesecloth; used for surgical dressings. 2. A plastic or wire mesh. { góz }

Gay-Lussac's first law See Charles' law. { ,gā·lù,säks 'fərst ,lò }

Gay-Lussac's second law |THERMO| The law that the internal energy of an ideal gas is independent of its volume. { ,gā·lù,säks 'sek·ənd ,lò }

g-cal See calorie. { 'jē,kal }

G center See N center. { 'jē ,sen·tər }

g constant |SOLID STATE| The ratio of the induced electric field in a piezoelectric material to the applied force that produces this field. { 'jē ,kän·stənt }

Gd See gadolinium.

Ge See germanium.

gear oil |MATER| A lubricating oil for use in transmissions, most types of differential gears, and gears in gear boxes. { 'gir ,óil }

gel |CHEM| A two-phase system consisting of a continuous solid network enveloped in a continuous liquid phase. In terms of their network structure, gels can consist of agglomerated particles (formed, for example, by destabilization of a colloidal suspension; plates (as in a clay) or fibers; polymers joined by small crystalline regions; or cross-linked polymers. { jel }

gelatin |ORG CHEM| A protein derived from the skin, white connective tissue, and bones of animals; used as a food and in photography, the plastics industry, metallurgy, and pharmaceuticals. { 'jel·ət·ən }

gelatinobromide |MATER| A preparation of gelatin silver bromide that is light-sensitive and used in photography. { jə¦lat·ən·ō¦brō,mīd }

gelation |CHEM| Formation of a gel from a sol. { jə'lā·shən }

gelation time |POLYM CHEM| In the manufacture of a thermosetting resin, the time interval between the addition of the catalyst into a liquid adhesive system and the formation of a gel. { jə'lā·shən ,tīm }

gel cement |MATER| Cement containing a small percentage of bentonite, which makes the mixture more homogeneous, increases the water-cement ratio, and reduces loss of water to the formation. { 'jel si'ment }

gel coat [MATER] A resin applied to the surface of a mold and gelled prior to placing plastic material in the mold in position for production; the gel coat becomes an integral part of the finished laminate and improves surface appearance. { 'jel ¦kōt }

gel electrophoresis [CHEM] Electrophoresis performed in silica gel, which is a porous, inert medium. { ¦jel i‚lek·trō·fə'rē·səs }

gelfoam [BIOMATER] Absorbable gelatin sponge partially insolubilized by crosslinking, used for arresting hemorrhage during surgery. { 'jel ‚fōm }

gel paint [MATER] Paint formulation made thixotropic by the reaction of a small amount of polyamide resin with an alkyd resin vehicle. { 'jel ‚pānt }

gel point [PHYS CHEM] Stage at which a liquid begins to exhibit elastic properties and increased viscosity. { 'jel ‚pȯint }

gel strength [FL MECH] Of a colloid, the ability or a measure of its ability to form gels. { 'jel ‚streŋkth }

general formula [CHEM] A formula that can apply not only to one specific compound but to a series of related compounds; for example, the general formula for an aldehyde RCHO, where R is hydrogen in formaldehyde (the simplest aldehyde) and is a hydrocarbon radical for other aldehydes in the series such as CH_3 for acetaldehyde and C_2H_5 for proprionaldehyde. { ¦jen·rəl 'fȯr·myə·lə }

genetic transformation See transformation. { jə¦ned·ik ‚tranz·fər'mā·shən }

geomembrane [CIV ENG] Any impermeable membrane (usually made of synthetic polymers in sheets) used with soils, rock, earth, or other geotechnical material in order to block the migration of fluids. { ‚jē·ō'mem‚brān }

geosynthetic [CIV ENG] Any synthetic material used in geotechnical engineering, such as geotextiles and geomembranes. { ‚jē·ō·sin'thed·ik }

geotextiles [CIV ENG] Woven or nonwoven fabrics used with foundations, soils, rock, earth, or other geotechnical material as an integral part of a manufactured project, structure, or system. { ¦jē·ō¦tek‚stīlz }

German cupellation [MET] A refining method using a large reverberatory furnace with a fixed bed and a movable roof; the bullion to be cupelled is all charged at once, and the silver is not refined in the same furnace where the cupellation is carried on. { ¦jər·mən ‚kyü·pə'lā·shən }

germane [INORG CHEM] **1.** A hydride of germanium whose general formula is Ge_nH_{2n+2}. **2.** The compound GeH_4, a hydride of germanium, a colorless gas that is combustible in air and burns with a blue flame. { ¦jər¦mān }

germanide [INORG CHEM] A compound of an alkaline earth or alkali metal with germanium; an example is magnesium germanide, Mg_2Ge; the germanides are reactive with water. { 'jər·mə ‚nīd }

germanium [CHEM] A brittle, water-insoluble, silvery-gray metallic element in the carbon family, symbol Ge, atomic number 32, atomic weight 72.59, melting at 959°C. [MET] A rare metal used in semiconductors, alloys, and glass. { jər'mān·ē·əm }

germanium halide [INORG CHEM] A dihalide or tetrahalide of fluorine, chlorine, bromine, or iodine with germanium. { jər'mān·ē·əm 'ha‚līd }

germanium oxide [INORG CHEM] The monoxide GeO or dioxide GeO_2; a study of GeO indicates it exists in polymeric form; GeO_2 is a white powder, soluble in alkalies; used in special glass and in medicine. { jər'mān·ē·əm 'äk‚sīd }

German silver See nickel silver. { ¦jər·mən 'sil·vər }

germicide [MATER] An agent that destroys germs. { 'jər·mə‚sīd }

gesso [MATER] A material made from chalk and gelatin or casein glue; painted on panels to furnish a surface for tempera work or for polymer-based paints. { 'je·sō }

getter [ELEC] See scavenger. [PHYS CHEM] **1.** A substance that binds gases on its surface and is used to maintain a high vacuum in a vacuum tube. **2.** A special metal alloy, such as barium-aluminum, that is used in vacuum tubes, lamps, and special devices to absorb residual gases. { 'ged·ər }

getter sputtering [ELECTR] The deposition of high-purity thin films at ordinary vacuum levels by using a getter to remove contaminants remaining in the vacuum. { 'gəd·ər ‚spəd·ə·riŋ }

ghatti gum [MATER] A water-soluble gum from the tree *Anogeissus latifolia*, forming a viscous glue in water; used as an emulsifier. { 'gäd·ē ‚gəm }

ghedda wax [MATER] A beeswax that is obtained from African and Indian bees. { 'ged·ə ‚waks }

ghost crystal See phantom crystal. { 'gōst ‚krist·əl }

ghost lines See ferrite banding. { 'gōst ‚līnz }

ghost structure See ferrite banding. { 'gōst ‚strək·chər }

giant magnetoresistance [SOLID STATE] A very large decrease in electrical resistance upon application of a magnetic field in certain structures composed of alternating layers of magnetic and nonmagnetic metals. { ¦jī·ənt ‚mag·nē‚d·ō·ri'zis·təns }

Giaque-Debye method See adiabatic demagnetization. { ¦zhyäk də'bī ‚meth·əd }

Giaque's temperature scale [THERMO] The internationally accepted scale of absolute temperature, in which the triple point of water is defined to have a temperature of 273.16 K. { ¦zhyäks 'tem·prə·chər ‚skāl }

gibbs [PHYS] A unit of amount of adsorption, equal to a surface concentration of 10^{-6} mole per square meter. { gibz }

Gibbs elasticity [PHYS] The elasticity of a film of liquid, equal to twice the product of the surface area and the derivative of the surface tension with respect to surface area. { 'gibz i ‚las'tis·əd·ē }

Gibbs free energy [THERMO] The thermodynamic function $G = H - TS$, where H is enthalpy,

T absolute temperature, and S entropy. Also known as free energy; free enthalpy; Gibbs function. { 'gibz ¦frē 'en·ər·jē }

Gibbs function *See* Gibbs free energy. { 'gibz ,fəŋk·shən }

Gibbs-Helmholtz equation [PHYS CHEM] An expression for the influence of temperature upon the equilibrium constant of a chemical reaction, $(d \ln K^0/dT)_P = \Delta H^0/RT^2$, where K^0 is the equilibrium constant, ΔH^0 the standard heat of the reaction at the absolute temperature T, and R the gas constant. [THERMO] **1.** Either of two thermodynamic relations that are useful in calculating the internal energy U or enthalpy H of a system; they may be written $U = F - T(\partial F/\partial T)_V$ and $H = G - T(\partial G/\partial T)_P$, where F is the free energy, G is the Gibbs free energy, T is the absolute temperature, V is the volume, and P is the pressure. **2.** Any of the similar equations for changes in thermodynamic potentials during an isothermal process. { 'gibz 'helm,hōlts i,kwā·zhən }

gilding metal [MET] A copper alloy (about 90% copper, 10% zinc) used to jacket small-arms bullets, to form detonator or primer cups, and to form rotating bands for artillery projectiles; it can be readily engraved by the lands as the projectile moves down the bore. { 'gild·iŋ ,med·əl }

Ginzburg-Landau theory [CRYO] A phenomenological theory of superconductivity which accounts for the coherence length; the ordered state of a superconductor is described by a complex order parameter which is similar to a Schrödinger wave function, but describes all the condensed superelectrons, rather than a single charged particle. Also known as Landau-Ginzburg theory. { 'ginz·bərg 'lan·daù ,thē·ə·rē }

Ginzburg-London superconductivity theory [SOLID STATE] A modification of the London superconductivity theory to take into account the boundary energy. { 'ginz·bərg 'lən·dən 'sü·pər,kän,dək'tiv·əd·ē ,thē·ə·rē }

glair [MATER] A sizing liquid made of egg white beaten with vinegar; used to prepare a surface of a book binding for gilding. { gler }

glass [MATER] A hard, amorphous, inorganic, usually transparent, brittle substance made by fusing silicates, sometimes borates and phosphates, with certain basic oxides and then rapidly cooling to prevent crystallization. { glas }

glass-bonded mica [MATER] An insulating material made by compressing a mixture of powdered glass and powdered natural or synthetic mica at high temperatures. { 'glas ,bän·dəd 'mī·kə }

glass brick [MATER] A hollow block of translucent glass with patterns molded on the faces; used in partitions. { ¦glas 'brik }

glass-ceramic [MATER] Hard, strong, nucleated glass with a nonporous crystalline structure; has high flexural strength and shock resistance; used for coatings, molded mechanical and electrical parts, heat-exchanger tubes, missile cones, and cookware and dinnerware. { ¦glas sə 'ram·ik }

glassed steel [CHEM ENG] Process piping or vessels lined with glass; a glass-steel composite has structural strength of steel and corrosion resistance of glass. { ¦glast ¦stēl }

glass fiber [MATER] A glass thread less than a thousandth of an inch (25 micrometers) thick, used loosely or in woven form as an acoustic, electrical, or thermal insulating material and as a reinforcing material in laminated plastics. Also known as fiberglass. { ¦glas ¦fī·bər }

glass former [MATER] **1.** An oxide that can readily form a glass. **2.** An oxide that can contribute to the network of a silica glass. { 'glas ,förm·ər }

glassine [MATER] A thin, dense, transparent, supercalendered paper from highly refined sulfite pulp, used for envelope windows, for sanitary wrapping, and as an insulating paper between layers of iron-core transformer windings. { ¦gla ¦sēn }

glass insulator [MATER] An insulator for a power transmission line made of annealed or toughened (tempered) glass. { ¦glas 'in·sə,lād·ər }

glass ionomer [BIOMATER] The only dental restorative material that forms a durable chemical bond to dentin; it is formed by the reaction of aluminosilicate glass with polyacrylic acid. { ¦glas ī'än·ə·mər }

glassmakers' soap [MATER] A substance such as manganese dioxide added to glass to remove the green color created by iron salts. { 'glas ,māk·ərz ,sōp }

glass paper [MATER] **1.** Paper with a layer of pulverized glass; used as an abrasive. **2.** Paper made of glass fibers. { 'glas ,pā·pər }

glass sand [MATER] High-quartz sand used in glassmaking; contains small amounts of aluminum oxide, iron oxide, calcium oxide, and magnesium oxide. { 'glas ,sand }

glass seal [ENG] An airtight seal made by molten glass. { 'glas ,sēl }

glass textile [TEXT] Fabric woven from glass fibers; used for electrical insulation, as filter cloth, and in plastic laminates. { ¦glas 'tek·stīl }

glass-to-metal seal [ELECTR] An airtight seal between glass and metal parts of an electron tube, made by fusing together a special glass and special metal alloy having nearly the same temperature coefficients of expansion. { ¦glas tə ¦med·əl 'sēl }

glass transition [PHYS] The transition that occurs when a liquid is cooled to an amorphous or glassy solid. [POLYM CHEM] The change in an amorphous region of a partially crystalline polymer from a viscous or rubbery condition to a hard and relatively brittle one; usually brought about by changing the temperature. Also known as gamma transition; glassy transition. { 'glas tran,zish·ən }

glass transition temperature [PHYS CHEM] The temperature at which a liquid changes to an amorphous or glassy solid. { ¦glas ,tran'zish·ən ,tem·prə·chər }

glassware [MATER] Articles, especially tableware, made of glass. { 'glas,wer }

glass wool [MATER] A mass of glass fibers resembling wool and used as insulation, packing, and air filters. { ¦glas 'wul }

glassy alloy [MET] An alloy having an amorphous or glassy structure. Also known as metallic glass. { ¦glas·ē 'al,ói }

glassy state See vitreous state. { ¦glas·ē 'stāt }

glassy transition See glass transition. { ¦glas·ē tran'zish·ən }

Glauber's salt [INORG CHEM] $Na_2SO_4 \cdot 10H_2O$ Crystalline hydrated sodium sulfate; loses water when exposed to air; water soluble, alcohol insoluble; used in textile dyeing and medicine. { 'glau·bərz ,sòlt }

glaze stain [INORG CHEM] Colorant for ceramic glazes; made of a finely ground calcined oxide, such as of cobalt, copper, manganese, or iron. { 'glāz ,stān }

glazing compound [MATER] A caulking compound used to seal and support a glass pane in place. { 'glāz·iŋ ,käm,paund }

glide See slip. { glīd }

glide plane [CRYSTAL] A lattice plane in a crystal on which translation or twin gliding occurs. Also known as slip plane. { 'glīd ,plān }

glissile dislocation See Shockley partial dislocation. { 'glis·əl ,dis·lō'kā·shən }

glitter [MATER] A group of decorative materials consisting of flakes large enough so that each flake produces a plainly visible sparkle or reflection; incorporated into plastic during compounding. { 'glid·ər }

globular transfer [MET] In electric arc welding, the transfer of weld metal across the arc in large drops. { 'gläb·yə·lər 'tranz·fər }

gloss [OPTICS] The ratio of the light specularly reflected from a surface to the total light reflected. { gläs }

gloss oil [MATER] Low-grade varnish composed of rosin dissolved in solvent naphtha. { 'gläs ,òil }

glossy [OPTICS] Pertaining to a surface from which much more light is specularly reflected than is diffusely reflected. { 'gläs·ē }

glue [MATER] A crude, impure, amber-colored form of commercial gelatin of unknown detailed composition produced by the hydrolysis of animal collagen; gelatinizes in aqueous solutions and dries to form a strong, adhesive layer. { glü }

glulam [MATER] A material fabricated by joining two or more layers of wood with an adhesive so that the grains of all the layers are approximately parallel. { 'glü,läm }

glycol [ORG CHEM] **1.** $C_nH_{2n}(OH)_2$ An organic chemical with two hydroxyl groups on an essentially aliphatic carbon chain. Also known as dihydroxy alcohol. **2.** $HOCH_2CH_2OH$ A colorless dihydroxy alcohol used as an antifreeze, in hydraulic fluids, and in the manufacture of dynamites and resins. { 'glī,kòl }

glycopeptide See glycoprotein. { ,glī·kō'pep ,tīd }

glycoprotein [BIOMATER] Any of a class of conjugated proteins containing both carbohydrate and protein units. Also known as glycopeptide. { ¦glī·kō'prō,tēn }

glycyl [ORG CHEM] NH_2CH_2COO- or $NHCH_2-COO=$ The radical from glycine, NH_2CH_2COOH; found in peptides. { 'glī·səl }

glyptal resin [ORG CHEM] A phthalic anhytxride glycerol made from an emulsion of an alkyd resin; used in lacquers and insulation. { 'glipt·əl 'rez·ən }

gnomonic projection [CRYSTAL] A projection for displaying the poles of a crystal in which the poles are projected radially from the center of a reference sphere onto a plane tangent to the sphere. { nō'män·ik prə'jek·shən }

gold [CHEM] A chemical element, symbol Au, atomic number 79, atomic weight 196.96765; soluble in aqua regia; melts at 1065°C. [MET] The native metallic element; a deep-yellow, very dense, soft, isometric metal, usually found alloyed with silver or copper; used in jewelry, dentistry, gilding, anodes, alloys, and solders. { gōld }

gold alloy [MET] Any alloy containing gold. { ¦gōld 'al,ói }

goldbeater's skin [MATER] The treated outside membrane of the large intestine of cattle; used between leaves of metal in goldbeating, and sometimes in hygrometers. { 'gōl,bēd·ərz ,skin }

goldbeating [MET] The process of producing gold leaf. { 'gōl,bēd·iŋ }

gold bronze [MET] A powdered alloy of copper used to simulate gold, or an alloy of copper containing 3–5% aluminum. { ¦gōld ¦bränz }

gold chloride [INORG CHEM] $AuCl_3$ A red, soluble compound made by reaction of gold and chlorine or by reaction of $HAuCl_4$ with chlorine; decomposes by heat; soluble in water, alcohol, and ether; used in photography, plating, inks, medicine, and ceramics. { ¦gōld 'klòr,īd }

golden antimony sulfide See antimony pentasulfide. { 'gōl·dən 'ant·ə,mō·nē 'səl,fíd }

gold-filled [MET] Covered by a layer of gold alloy. { ¦gōld 'fild }

gold foil [MET] A thin sheet of gold, thicker than gold leaf, formed by rolling or hammering. { ¦gōld 'fòil }

gold hydroxide [INORG CHEM] $Au(OH)_3$ A yellow-brown, light-sensitive, water-insoluble powder; dissolves in most acids; easily reduced to metallic gold; used in medicine, porcelain, gold plating, and daguerreotypes. { ¦gōld hī'dräk,sīd }

gold leaf [MET] Gold beaten or rolled into extremely thin sheets or leaves (10^{-6} inch or 25 nanometers thick); leaves are stored in books (a book consists of 25 leaves), the paper of which is rubbed with chalk to keep the leaves from sticking. { ¦gōld 'lēf }

gold metallurgy [MET] The science and technology of gold recovery and refining. { ¦gōld 'med·əl,ər·jē }

gold oxide [INORG CHEM] Au_2O_3 Water-insoluble, heat-decomposable, brownish-black powder; soluble in hydrochloric acid; used

gold plate

to gild, in medicine and porcelain, and for daguerreotypes. Also known as auric oxide; gold trioxide. { 'gōld 'äk,sīd }

gold plate |MET| Gold which has been electroplated on a material in a thin layer of controlled thickness; used on electric contacts for corrosion resistance and solderability and on jewelry and oraments. { ¦gōld 'plāt }

gold point |THERMO| The temperature of the freezing point of gold at a pressure of 1 standard atmosphere (101,325 pascals); used to define the International Temperature Scale of 1940, on which it is assigned a value of 1337.33 K or 1064.18°C. { 'gold ,pòint }

gold potassium chloride See potassium gold chloride. { ¦gōld pə'tas·ē·əm 'klòr,īd }

gold potassium cyanide See potassium gold cyanide. { ¦gōld pə'tas·ē·əm 'sī·ə,nīd }

gold salt See sodium gold chloride. { 'gōld ,sòlt }

Goldschmidt process |MET| 1. The thermite process of welding. 2. A process by which dry chlorine is employed to remove tin from scrap tinplate. { 'gōl,shmit ,präs·əs }

Goldschmidt's law |SOLID STATE| The law that crystal structure is determined by the ratios of the numbers of the constituents, the ratios of their sizes, and their polarization properties. { 'gōl ,shmits ,lò }

gold size |MATER| An adhesive used to fix gold leaf to a surface. { ¦gōld ¦sīz }

gold sodium chloride See sodium gold chloride. { ¦gōld 'sōd·ē·əm 'klòr,īd }

gold solder |MET| Solder composed of 60% gold, 20% silver, and 20% copper. { ¦gōld 'säd·ər }

gold tin precipitate See gold tin purple. { ¦gōld ¦tin prə'sip·ə,tát }

gold tin purple |ORG CHEM| A brown powder which is a mixture of gold chloride and brown tin oxide, soluble in ammonia; used in coloring enamels, manufacturing ruby glass, and painting porcelain. Also known as gold tin precipitate; purple of Cassius. { ¦gōld ¦tin 'pər·pəl }

gold trioxide See gold oxide. { ¦gōld trī'äk,sīd }

goniometer |ENG| 1. An instrument used to measure the angles between crystal faces. 2. An instrument which uses x-ray diffraction to measure the angular positions of the axes of a crystal. 3. Any instrument for measuring angles. { ,gō·nē'äm·əd·ər }

Gorsky effect |MET| For hydrogen dissolved in a metal, its migration from the compressed side to the stretched side when the metal is bent. { 'gòr·skē i,fekt }

grade |MATER| Any of the various purity standards for chemicals and chemical products that have been established for specific applications. { grād }

graduated coating |MET| In thermal spraying, a deposit consisting of a number of layers which vary in material composition. { 'graj·ə,wād·əd 'kōd·iŋ }

Graetz number |THERMO| A dimensionless number used in the study of streamline flow,

equal to the mass flow rate of a fluid times its specific heat at constant pressure divided by the product of its thermal conductivity and a characteristic length. Also spelled Grätz number. Symbolized N_{Gz}. { 'grets ,nəm·bər }

graft copolymer |POLYM CHEM| Any high polymer composed of two or more different polymeric entities chemically united. { ¦graft kō'päl·ə·mər }

grain |GRAPHICS| A small particle of metallic silver remaining in a photographic emulsion after developing and fixing; these grains together form the dark areas of a photographic image. |MATER| 1. The appearance and texture of wood due to the arrangement of constituent fibers. 2. The woodlike appearance or texture of a rock, metal, or other material. 3. The direction in which most fibers lie in a sample of paper, which corresponds with the way the paper was made on the manufacturing machine. |TEXT| The direction in a piece of fabric which is parallel with the selvage. { grān }

grain boundary |SOLID STATE| An interface between individual crystals in a polycrystalline solid. { 'grān ,baún·drē }

grain fineness number |MATER| Average grain size of a granular material. { ¦grān 'fīn·nəs ,nəm·bər }

grain flow |MET| The fiber patterns appearing on the surface of a forging and resulting from the alignment of the crystalline structure of the base metal in the direction of working. { 'grān ,flō }

grain growth |MET| Enlargement of grains in a metal, usually through heat treatment. { 'grān ,grōth }

grain size |GRAPHICS| Average size of silver halide grains in a photosensitive material. |MET| Average size of grains in a metal expressed as average diameter, or grains per unit area or volume. { 'grān ,sīz }

gram-atomic weight |CHEM| The atomic weight of an element expressed in grams, that is, the atomic weight on a scale on which the atomic weight of carbon-12 isotope is taken as 12 exactly. { ¦gram ə¦tam·ik 'wāt }

gram-calorie See calorie. { 'gram ¦kal·ə·rē }

gram-equivalent weight |CHEM| The equivalent weight of an element or compound expressed in grams on a scale in which carbon-12 has an equivalent weight of 3 grams in those compounds in which its formal valence is 4. { ¦gram i¦kwiv·ə·lənt 'wāt }

gram-molecular volume |CHEM| The volume occupied by a gram-molecular weight of a chemical in the gaseous state at 0°C and 760 millimeters of pressure (101,325 pascals). { ¦gram mə¦lek·yə·lər 'väl·yəm }

gram-molecular weight |CHEM| The molecular weight of compound expressed in grams, that is, the molecular weight on a scale on which the atomic weight of carbon-12 isotope is taken as 12 exactly. { ¦gram mə¦lek·yə·lər 'wāt }

granular fracture |MET| Grain-like or crystalline surface appearance of a broken metal. { 'gran·yə·lər 'frak·chər }

146

granular powder |MET| Equidimensional metal particles that are not spherical. { 'gran·yə·lər 'paúd·ər }

granular structure |MATER| Nonuniform appearance of molded or compressed material due to presence of particles of composition, either within the material or on the surface. { 'gran·yə·lər 'strək·chər }

granulate |CHEM| To form or crystallize into grains, granules, or small masses. { 'gran·yə ‚lāt }

granulated metal |MET| Small pellets produced by pouring molten metal through a screen or similar device and chilling the droppings in water. { 'gran·yə‚lād·əd 'med·əl }

graphite flake |MET| A curved graphite particle in gray cast iron. { 'gra‚fīt ¦flāk }

graphite grease |MATER| A lubricating grease that contains 2–10% amorphous graphite; used for bearings, especially in damp places. { 'gra ‚fīt ¦grēs }

graphite oil |MATER| A deflocculated suspension of graphite in oil; used as a lubricant. { 'gra ‚fīt ‚oil }

graphite rosette |MET| Graphite flakes which extend radially outward from the centers of crystallization. { 'gra‚fīt rō'zet }

graphitic carbon |MET| Carbon in iron or steel present in the form of graphite. { grə'fid·ik 'kär·bən }

graphitic corrosion |MET| Corrosion of gray cast iron in which the metallic iron constituent is converted into corrosion products which cement together the residual graphite. { grə'fid·ik kə'rō·zhən }

graphitic steel |MET| Alloy steel containing graphitic carbon. { grə'fid·ik 'stēl }

graphitizing |MET| Annealing cast iron to convert all or some of the combined carbon to graphitic carbon. { 'graf·ə‚tiz·iŋ }

grappier cement |MATER| A cement composed of finely ground lumps of leftover underburned and overburned slaked lime. { 'grap·ē‚ā si ‚ment }

Grätz number See Graetz number. { 'grets ‚nəm·bər }

gravitometer See densimeter. { grav·ə'täm·əd· ər }

graybody |THERMO| An energy radiator which has a blackbody energy distribution, reduced by a constant factor, throughout the radiation spectrum or within a certain wavelength interval. Also known as nonselective radiator. { 'grā ‚bäd·ē }

gray casting |MET| A casting of gray iron. { 'grā ‚kast·iŋ }

gray iron |MET| Pig or cast iron in which the carbon not contained in pearlite is present in the form of graphitic carbon. { 'grā ‚ī·ərn }

gray scale |OPTICS| A series of achromatic tones having varying proportions of white and black, to give a full range of grays between white and black; a gray scale is usually divided into 10 steps; however, electronic scanners can typically differentiate 16 to 256 levels. { 'grā ‚skāl }

grease |MATER| **1.** Rendered, inedible animal fat that is soft at room temperature and is obtained from lard, tallow, bone, raw animal fat, and other waste products. **2.** A lubricant in the form of a solid to semisolid dispersion of a thickening agent in a fluid lubricant, such as petroleum oil thickened with metallic soap. { grēs }

greasepaint |MATER| Makeup made of melted tallow or grease used by theatrical performers. { 'grēs‚pānt }

green |MET| Pertaining to an unsintered powder. |OPTICS| The hue evoked in an average observer by monochromatic radiation having a wavelength in the approximate range from 492 to 577 nanometers; however, the same sensation can be produced in a variety of other ways. { grēn }

green chemistry |CHEM| The use of chemical products and processes that reduce or eliminate substances hazardous to human health or the environment. { grēn 'kem·ə·strē }

green concrete |MATER| Concrete that has set but not hardened. { ¦grēn 'kaŋ‚krēt }

green copperas See ferrous sulfate. { ¦grēn 'käp·ə·rəs }

green forming |MATER| The ceramic fabrication step in which powders are formed into useful shapes; used in casting, extrusion, die pressing, tape casting, and injection molding. { 'grēn fórm·iŋ }

green glass |MATER| Glass given a blue-green hue by substituting cupric oxide for the chromium compound used in ordinary glass. { 'grēn ¦glas }

green gold |MET| A greenish alloy of gold obtained by using silver, silver and cadmium, or silver and copper as the alloying metal. { 'grēn ¦gōld }

greenheart |MATER| Wood from *Octolea rodioei*; resistant to fungi and termites and used for shipbuilding, docks, and marine planking. { 'grēn ‚härt }

green lumber |MATER| Freshly sawed lumber, before drying. { 'grēn ¦ləm·bər }

green mortar |MATER| Mortar that has set but not dried. { ¦grēn ¦mórd·ər }

green nickel oxide See nickel oxide. { 'grēn ¦nik·əl 'äk‚sīd }

green salt See uranium tetrafluoride. { 'grēn ¦sólt }

greensand |MATER| A highly siliceous sand that contains small amounts of magnesia and alumina, with about 8% of its bulk in powdered coal or charcoal; dampened with water to make foundry molds. { 'grēn‚sand }

green strength |MET| The mechanical strength which a compacted powder must have in order to withstand mechanical operations to which it is subjected after pressing and before sintering, without damaging its fine details and sharp edges. { 'grēn ¦streŋkth }

green vitriol See ferrous sulfate. { 'grēn ¦vi·trē ‚ól }

greenware |MATER| Ceramic ware that has not yet been fired. { 'grēn‚wer }

Greninger chart |CRYSTAL| A chart that enables angular relations between planes and zones in a crystal to be read directly from an x-ray diffraction photograph. { 'gren·iŋ·ər ,chärt }

grid metal |MET| An alloy of lead with 5–12% antimony and sometimes with a smaller amount of tin; used for grids in storage batteries. { 'grid ,med·əl }

Griebe-Schiebe method |SOLID STATE| A method of observing the piezoelectric behavior of small crystals, in which the crystals are placed between two electrodes connected to the resonant circuit of an oscillator, and tuning of the resonant circuit results in jumps in the oscillator frequency which produce clicks in headphones or a loudspeaker attached to the plate circuit of the oscillator. { 'grē·bə 'shē·bə ,meth·əd }

Griffith crack |MET| Any small flaw in a metal theorized as creating a low order of fracture strength. { 'grif·əth ,krak }

Griffiths' method |THERMO| A method of measuring the mechanical equivalent of heat in which the temperature rise of a known mass of water is compared with the electrical energy needed to produce this rise. { 'grif·əths ,meth·əd }

Griffith's white See lithopone. { 'grif·əths ¦wīt }

grindability |MATER| Relative ease with which a material can be ground. { ˌgrīn·də'bil·əd·ē }

grindability index |MATER| A numerical indication of the capacity of a material to be ground. { ˌgrīn·də'bil·əd·ē ,in,deks }

grind gage See grindometer. { 'grīnd gāj }

grinding cracks |MATER| Cracks in a workpiece resulting from grinding. { 'grīn·diŋ ,kraks }

grinding fluid |MATER| Cutting fluid used in metal-grinding operations. { 'grīn·diŋ ,flü·əd }

grinding relief |MET| A groove at the edge of a metal surface which permits overhang of the corner of the grinding wheel. { 'grīn·diŋ ri,lēf }

grinding sensitivity |MATER| Susceptibility of a material to the formation of grinding cracks. { 'grīn·diŋ ,sen·sə,tiv·əd·ē }

grinding stress |MECH| Residual tensile or compressive stress, or a combination of both, on the surface of a material due to grinding. { 'grīn·diŋ ,stres }

grinding-type resin |POLYM CHEM| Vinyl or other resin that requires grinding before dispersal into plastisols or organosols. { 'grīn· diŋ ,tīp ,rez·ən }

grindometer |MATER| A device used to measure the size of pigment particles in the inks and paint. Also know as grind gage. { grīn'däm·əd·ər }

grit |MATER| An abrasive material composed of angular grains. { grit }

grog |MATER| Fired refractory material that is used in the manufacture of products which must withstand extreme heat. { gräg }

groove weld |MET| A weld in the groove between a pair of members. { 'grüv ,weld }

gross porosity |MET| Gas pockets or pores of undesirable size and quantity in a casting or a weld metal. { ¦grōs pə'räs·əd·ē }

ground |ELEC| **1.** A conducting path, intentional or accidental, between an electric circuit or equipment and the earth, or some conducting body serving in place of the earth. Abbreviated gnd. Also known as earth. **2.** To connect electrical equipment to the earth or to some conducting body which serves in place of the earth. { graủnd }

ground bed |MET| In cathodic protection, a device that consists of an interconnected group of impressed-current anodes which absorbs the damage caused by generated electric current. { 'graủnd ,bed }

ground lead See work lead. { 'graủnd ,lēd }

groundwood pulp See mechanical pulp. { 'graủnd,wủd ¦pəlp }

group |CHEM| **1.** A family of elements with similar chemical properties. **2.** A combination of bonded atoms that behave as a unit under certain conditions, for example, the sulfate group, $SO_4{}^{2-}$. { grüp }

grout |MATER| **1.** A fluid mixture of cement and water, or a mixture of cement, sand, and water. **2.** Waste material of all sizes obtained in quarrying stone. { graủt }

growth spiral |CRYSTAL| A structure on a crystal surface, observed after growth, consisting of a growth step winding downward and outward in an Archimedean spiral which may be distorted by the crystal structure. { 'grōth ,spī·rəl }

growth step |CRYSTAL| A ledge on a crystal surface, one or more lattice spacings high, where crystal growth can take place. { 'grōth ,step }

Grüneisen constant |SOLID STATE| Three times the bulk modulus of a solid times its linear expansion coefficient, divided by its specific heat per unit volume; it is reasonably constant for most cubic crystals. Also known as Grüneisen gamma. { 'grü·nīz·ən ,kän·stənt }

Grüneisen gamma See Grüneisen constant. { 'grü·nīz·ən ,gam·ə }

Grüneisen relation |SOLID STATE| The relation stating that the electrical resistivity of a very pure metal is proportional to a mathematical function which depends on the ratio of the temperature to a characteristic temperature. { 'grü·nīz·ən ri,lā·shən }

guaiac |MATER| A resin obtained from the trees *Guaiacum santum* and *G. officinale*; soluble in alcohol, ether, and chloroform; used in medicine and varnish. { 'gwī,ak }

guar gum |MATER| A mucilage formed from seeds of the guar plant; light-gray powder dispersible in water; used as a thickening agent in paper, foods, pharmaceuticals, and cosmetics. { 'gwär ,gəm }

guest |CHEM| Cationic, anionic, or neutral organic, inorganic, or biological substance, bound by means of various interactions (electrostatic, hydrogen bonding, van der Waals, donor-acceptor) within a crystalline or molecular structure. Also known as guest molecule; guest substance. { gest }

guest molecule See guest. { 'gest ,mäl·ə,kyül }

guest substance See guest. { 'gest ,səb·stəns }

guide coat |MATER| A thin coat of paint applied

to a surface over a sealer or filler to indicate the locations of bumps or imperfections and thereby to serve as a guide for removing them.   { 'gīd ‚kōt }

guided bend test   [MET] A bend test in which the specimen is bent to a predetermined shape.   { 'gīd·əd ¦bend ‚test }

guide mill   [MET] A hand rolling mill with a series of stands and guides at the entrance to the rolls.   { 'gīd ‚mil }

Guinier-Preston zones   [MET] The initially formed zones of a precipitate as it comes out of solid solution.   { gēn'yā 'pres·tən ‚zōnz }

Gukhman number   [THERMO] A dimensionless number used in studying convective heat transfer in evaporation, equal to $(t_0 - t_m)/T_0$, where t_0 is the temperature of a hot gas stream, t_m is the temperature of a moist surface over which it is flowing, and T_0 is the absolute temperature of the gas stream. Symbolized Gu; N_{Gu}.   { 'gúk̲·mən ‚nəm·bər }

gum   [MATER] A hydrophilic plant polysaccharide or derivative that swells to produce a viscous dispersion or solution when added to water. Also known as hydrocolloid.   { gəm }

gum accroides   See acaroid resin.   { 'gəm ə'krói·dēz }

gum arabic   [MATER] A water-soluble gum obtained from acacia trees in Africa and Australia; produced commercially as a white powder; used in the manufacture of inks and adhesives, in textile finishing, and as the principal binder in watercolor and gouache. Also known as acacia gum; gum Kordofan; gum Senegal.   { 'gəm 'ar·ə·bik }

gum dammar   See dammar.   { 'gəm 'da‚mär }

gum Kordofan   See gum arabic.   { 'gəm ‚kòr·də'fan }

gum resin   [MATER] A group of oleoresinous substances obtained from plants; mixtures of true gums and resins less soluble in alcohol than natural resins; examples are rubber, gutta-percha, gamboge, myrrh, and olibanum; used to make certain pharmaceuticals.   { 'gəm ¦rez·ən }

gum Senegal   See gum arabic.   { 'gəm ¦sen·ə·gəl }

gunmetal   [MET] **1.** Bronze composed of copper and tin in proportions of 9:1, formerly used to make cannons.   **2.** Any metal or alloy from which guns are made.   **3.** Any metal or alloy treated to give the appearance of black, tarnished copper-alloy gunmetal.   { 'gən‚med·əl }

Gurevich effect   [SOLID STATE] An effect observed in electric conductors in which phonon-electron collisions are important, in the presence of a temperature gradient, in which phonons carrying a thermal current tend to drag the electrons with them from hot to cold. Also known as phonon-drag effect.   { 'gúr·ə·vich i‚fekt }

gutta-balata   See balata.   { ¦gəd·ə·bə'läd·ə }

gutta-percha   [MATER] A leathery, thermoplastic substance consisting of gutta hydrocarbon with some resin obtained from the latex of certain Malaysian sapotaceous trees; used as insulation for submarine cables, and in golf balls and other products.   { ¦gəd·ə'pər·chə }

gutter   [MET] A groove along the periphery of a die impression to allow for excess flash during forging.   { 'gəd·ər }

Gutzkow's process   [MET] A modification of the sulfuric acid parting process for bullion containing large amounts of copper; a large excess of acid is used, and the silver sulfate is then reduced with charcoal, or, in the original process, ferrous sulfate.   { 'gúts·kōz ‚präs·əs }

gypsum board   [MATER] A plaster board covered with paper.   { 'jip·səm ‚bórd }

gypsum cement   See gypsum plaster.   { 'jip·səm si¦ment }

gypsum lath   [MATER] Lath consisting of a core of set gypsum surfaced with paper that is treated to receive plaster.   { 'jip·səm ¦lath }

gypsum plank   [MATER] A structural precast unit consisting of gypsum core reinforced with welded galvanized steel mesh and bounded on all four edges with a tongue-and-groove steel form; used as the roof deck of steel-frame buildings, and sometimes for the floor system.   { 'jip·səm ¦plaŋk }

gypsum plaster   [MATER] Plaster made principally from gypsum. Also known as gypsum cement.   { 'jip·səm ¦plas·tər }

gypsum wallboard   [MATER] Wallboard consisting of a core of set gypsum surfaced with paper or other fibrous material suitable to receive paint or paper.   { 'jip·səm 'wòl‚bòrd }

gyration tensor   [SOLID STATE] A tensor characteristic of an optically active crystal, whose product with a unit vector in the direction of propagation of a light ray gives the gyration vector.   { ji'rā·shən ¦ten·sər }

H

H See hydrogen.

Ha See hahnium.

habit See crystal habit. { 'hab·ət }

habit plane [CRYSTAL] The crystallographic plane or system of planes along which certain phenomena such as twinning occur. { 'hab·ət ˌplān }

Hadfield manganese steel [MET] Austenitic steel (face-centered cubic structure) containing 11–14% manganese; resistant to shock and wear. Also known as austenitic manganese steel; manganese steel. { 'had,fēld ¦maŋ·gə ˌnēs 'stēl }

HAF black See high-abrasion furnace black. { ¦āch¦ā¦ef 'blak }

hafnium [CHEM] A metallic element, symbol Hf, atomic number 72, atomic weight 178.49; melting point 2000°C, boiling point above 5400°C. { 'haf·nē·əm }

hafnium carbide [INORG CHEM] HfC Gray powder, melting at 3887°C; used in the control rods of nuclear reactors. { 'haf·nē·əm 'kär,bīd }

hahnium [CHEM] The name suggested by workers in the United States for element 105. Symbolized Ha. { 'hän·ē·əm }

Hahn technique [SOLID STATE] A method of studying changes in solids under various treatments that involves incorporating small amounts of radium into the solid and measuring the emanating power. { 'hän tek,nēk }

hair cracks [MATER] Fine, random cracks in the surface of the top coat of paint or other coating material. { 'her ˌkraks }

hair felt [MATER] Felt made of cattle hair; used as insulation in buildings. { 'her ˌfelt }

halation [OPTICS] A halo on a photographic image of a bright object caused by light reflected from the back of the film or plate. { hā'lā·shən }

half-and-half solder [MET] Solder composed of tin and lead in equal parts. { ¦haf ən ¦haf ¦säd·ər }

half bat [MATER] One half of a brick, cut across the length. { 'haf ¦bat }

half-cell [PHYS CHEM] A single electrode immersed in an electrolyte. { 'haf ¦sel }

half-cell potential [PHYS CHEM] In electrochemical cells, the electrical potential developed by the overall cell reaction; can be considered, for calculation purposes, as the sum of the potential developed at the anode and the potential

developed at the cathode, each being a half-cell. { 'haf ¦sel pə'ten·chəl }

half-hard [MET] A rolled-metal product of intermediate hardness or temper. { 'haf ,härd }

half-life [CHEM] The time required for one-half of a given material to undergo a chemical or nuclear reaction. { 'haf ,līf }

half thickness [PHYS] The thickness of a sheet of material which reduces the intensity of a beam of radiation passing through it to one-half its initial value. Also known as half-value layer; half-value thickness. { 'haf ¦thik·nəs }

half-value layer See half thickness. { 'haf ¦val·yü ,lā·ər }

half-value period See half-life. { 'haf ¦val·yü ,pir·ē·əd }

half-value thickness See half thickness. { 'haf ¦val·yü ,thik·nəs }

halide [CHEM] A compound of the type MX, where X is a halogen and M is another element or group. { 'ha,līd }

Hall effect [ELECTROMAG] The development of a transverse electric field in a current-carrying conductor placed in a magnetic field; ordinarily the conductor is positioned so that the magnetic field is perpendicular to the direction of current flow and the electric field is perpendicular to both. { 'hól i,fekt }

Hall mobility [SOLID STATE] The product of conductivity and the Hall constant for a conductor or semiconductor; a measure of the mobility of the electrons or holes in a semiconductor. { 'hól mō'bil·əd·ē }

Hall process [MET] An electrolytic recovery process for aluminum employing a fused-bauxite (aluminum oxide), cryolite electrolyte. { 'hól ,prä·səs }

halo [OPTICS] A ring around the photographic image of a bright source caused by light scattering in any one of a number of possible ways. { 'hā·lō }

halocarbon [ORG CHEM] A compound of carbon and a halogen, sometimes with hydrogen. { ¦ha·lō¦kär·bən }

halocarbon plastic [POLYM CHEM] Plastic made from halocarbon resins. { ¦ha·lō¦kär·bən 'plas·tik }

halocarbon resin [POLYM CHEM] Resin produced by the polymerization of monomers made of halogenated hydrocarbons, such

halogen

as tetrafluoroethylene, C_2F_4, and trifluorochloroethylene, C_2F_3Cl. { ¦ha·lō¦kär·bən 'rez·ən }

halogen |CHEM| Any of the elements of the halogen family, consisting of fluorine, chlorine, bromine, iodine, and astatine. { 'hal·ə·jən }

halogen acid |INORG CHEM| A compound composed of hydrogen bonded to a halogen element, for example, hydrochloric acid. { 'hal·ə·jən ¸as·əd }

hammer forging |MET| Forging by means of repeated blows of a hammer. { 'ham·ər ¦fȯrj·iŋ }

hammer test |MET| An impact test conducted by dropping weights from increasing heights until a specified deflection of the weight is produced. { 'ham·ər ¸test }

hammer welding |MET| Forge welding by means of a hammer. { 'ham·ər ¸weld·iŋ }

Hampson process |CRYO| A process for liquefying gases which resembles the Linde process except that the Joule-Thomson expansion reduces the gas pressure to approximately atmospheric pressure. { 'ham·sən ¸prä·səs }

handedness |PHYS| A division of objects, such as coordinate systems, screws, and circularly polarized light beams, into two classes (right and left), which distinguishes an object from a mirror image but not from a rotated object. { 'han·dəd·nəs }

hand forging |MET| Plastic deformation of a metal by manual force. { 'hand ¦fȯrj·iŋ }

hanging |MET| Sticking or wedging of part of the charge in a blast furnace. { 'haŋ·iŋ }

hard |MATER| Quality of a material that is compact, solid, and difficult to deform. { härd }

hard acid |CHEM| A Lewis acid of low polarizability, small size, and high positive oxidation state; it does not have easily excitable outer electrons; some examples are H^+, Li^+, and Al^+. { 'härd 'as·əd }

hard base |CHEM| A Lewis base (electron donor) that has high polarizability and low electronegativity, is easily oxidized, or possesses lowlying empty orbitals; some examples are H_2O, HO^-, OCH_3^-, and F^-. { 'härd ¦bās }

hardboard |MATER| A fiberboard formed to a density of 30–50 pounds per cubic foot (480–800 kilograms per cubic meter) and having one textured and one smooth face. { 'härd ¸bȯrd }

hard bronze |MET| A high-tensile-strength alloy containing 88% copper, 7% tin, 3% zinc, and 2% lead. { 'härd ¦bränz }

hard-burned brick |MATER| A brick that has been fired and sintered at high temperature. { 'härd ¸bərnd 'brik }

hard chromium |MET| A thick coating of electrodeposited chromium on a base metal; increases wear resistance. { 'härd ¦krō·mē·əm }

hard detergent |CHEM| A nonbiodegradable detergent. { 'härd di'tər·jənt }

hard-drawn wire |MET| Cold-drawn metal wire, usually of high tensile strength. { 'härd ¸drȯn 'wīr }

hardenability |MET| In a ferrous alloy, the property that determines the depth and distribution of hardness induced by quenching from elevated temperatures. { ¸härd·ən·ə'bil·əd·ē }

hardened steel |MET| Steel hardened by quenching from high temperatures. { 'härd·ənd ¦stēl }

hardener |MET| A master alloy added to a melt to control hardness. |POLYM CHEM| Compound reacted with a resin polymer to harden it, such as the amines or anhydrides that react with epoxides to cure or harden them into plastic materials. Also known as curing agent. { 'härd·ən·ər }

hardening |MET| **1.** Imparting hardness to carbon steel by abrupt cooling (quenching) through a critical temperature range. **2.** Heat-treating an age-hardening or precipitation-hardening alloy at intermediate temperatures. { 'härd·ən·iŋ }

hard-face |MET| To apply a layer of hard, abrasion-resistant metal to a less resistant metal part by plating, welding, spraying, or other techniques. Also known as hard-surface. { 'härd ¸fās }

hard fiber |MATER| Indicating vulcanization with zinc chloride; used of paper or boards. { 'härd ¸fī·bər }

hard glass |MATER| A potash-lime glass with a high silica content, used for making brilliant glassware. Also known as Bohemian glass. { 'härd ¦glas }

hard grease |MATER| A lubricating grease that flows at a temperature of about 90°C. { 'härd ¦grēs }

hardhead |MET| A hard white deposit formed during tin refining by liquation. { 'härd¸hed }

hard iron |MET| Iron or steel which is not readily magnetized by induction, but which retains a high percentage of the magnetism acquired. { 'härd ¦ī·ərn }

hard lac |MATER| Solvent-extracted shellac. Also known as hard-lac resin. { 'härd ¦lak }

hard-lac resin See hard lac. { 'härd ¸lak ¸rez·ən }

hard lead |MET| Lead alloy with reduced malleability due to the presence of impurities, usually antimony. { 'härd ¦led }

hard magnetic material |MET| A metal having a high coercive force which gives a large magnetic hysteresis. { 'härd mag¦ned·ik mə'tir·ē·əl }

hardness |ELECTROMAG| That quality which determines the penetrating ability of x-rays; the shorter the wavelength, the harder and more penetrating the rays. |MATER| Resistance of a metal or other material to indentation, scratching, abrasion, or cutting. { 'härd·nəs }

hardness number |ENG| A number representing the relative hardness of a mineral, metal, or other material as determined by any of more than 30 different hardness tests. { 'härd·nəs ¸nəm·bər }

hardness test |ENG| A test to determine the relative hardness of a metal, mineral, or other material according to one of several scales, such as Brinell, Mohs, or Shore. { 'härd·nəs ¸test }

hard paste porcelain See porcelain. { 'härd ¸pāst 'pȯrs·lən }

hard porcelain |MATER| A ceramic material

having good resistance to thermal shock. { 'härd ¦pȯrs·lən }

hard rubber |MATER| Rubber that has been vulcanized at high temperatures and pressures to give hardness; used as an electrical insulating material and in tool handles. Also known as ebonite. { 'härd ¦rəb·ər }

hard solder See brazing alloy. { 'härd 'säd·ər }

hard-solder See braze. { ¦härd ¦säd·ər }

hard superconductor |CRYO| A superconductor that requires a strong magnetic field, over 1000 oersteds (79,577 amperes per meter), to destroy superconductivity; niobium and vanadium are examples. { 'härd 'sü·pər·kən¦däk·tər }

hard-surface See hard-face. { 'härd ¦sər·fəs }

hardwood |MATER| Dense, close-grained wood of an angiospermous tree, such as oak, walnut, cherry, and maple. { 'härd¸wu̇d }

Harker-Kasper inequalities |SOLID STATE| Inequalities used in the analysis of crystal structure by x-ray diffraction which relate the structure factors and help to determine their phase factors. { 'härk·ər 'kas·pər ¸in·i'kwäl·əd·ēz }

Harris process |MET| A method for refining lead in which the liquid bullion is sprayed through molten caustic soda and molten sodium nitrate; arsenic, antimony, and tin are oxidized, converted into sodium salts, and skimmed from the bath. { 'har·əs ¸prä·səs }

Hartmann lines See Lüders' lines. { 'härt·män ¸līnz }

hassium |CHEM| A chemical element, symbolized Hs, atomic number 108, a synthetic element; the sixteenth transuranium element. { 'hä·sē·əm }

Haüy law |CRYSTAL| The law that for a given crystal there is a set of ratios such that the ratios of the intercepts of any crystal plane on the crystal axes are rational fractions of these ratios. { ä'wē ¸lȯ }

Hauzeur furnace |MET| A double furnace for the distillation of zinc wherein waste heat from one set of retorts is utilized for heating the second set. { ō'zür ¸fər·nəs }

Hayden process |MET| A method of electrolytic copper refining; anodes of unrefined copper are suspended in an acid electrolyte, and one side of each then acts as an anode and the other as a cathode. { 'hād·ən ¸prä·səs }

haydite |MATER| Expanded shale, slate, or clay characterized by low unit weight and satisfactory structural properties; used as an aggregate to produce lightweight structural concrete. { 'hā ¸dīt }

hazardous material |MATER| A poison, corrosive agent, flammable substance, explosive, radioactive chemical, or any other material which can endanger human health or well-being if handled improperly. { 'haz·ərd·əs mə'tir·ē·əl }

haze |OPTICS| The degree of cloudiness in a solution, cured plastic material, or coating material. { hāz }

hcp structure See hexagonal close-packed structure. { ¦āch¦sē¦pē 'strək·chər }

HDPE See high-density polyethylene.

He See helium.

heap leaching |MET| A process used for the recovery of copper from weathered ore and material from mine dumps; material is laid to a thickness of 20 feet (6 meters) in alternately fine and coarse beds and treated with water at intervals during which oxidation occurs; liquor that runs off is treated with scrap iron to precipitate copper. { 'hēp ¸lēch·iŋ }

hearth |MET| The floor of a reverberatory, open-hearth, cupola, or blast furnace; it is made of refractory material able to support the charge and to collect the molten products. { härth }

hearth furnace |MET| A furnace designed to heat the charge, resting on the hearth, by passing hot gases over it. { 'härth ¸fər·nəs }

heat |THERMO| Energy in transit due to a temperature difference between the source from which the energy is coming and a sink toward which the energy is going; other types of energy in transit are called work. { hēt }

heat-affected zone |MET| The zone within a base metal that undergoes structural changes but does not melt during welding, cutting, or brazing. { 'hēt ə¦fek·təd 'zōn }

heat balance |MET| The calculation used in fluidization roasting so that the addition or removal of heat can be controlled to maintain the optimum temperature in the reacting vessel. |THERMO| The equilibrium which is known to exist when all sources of heat gain and loss for a given region or body are accounted for. { 'hēt ¸bal·əns }

heat budget |THERMO| The statement of the total inflow and outflow of heat for a planet, spacecraft, biological organism, or other entity. { 'hēt ¸bəj·ət }

heat capacity |THERMO| The quantity of heat required to raise a system one degree in temperature in a specified way, usually at constant pressure or constant volume. Also known as thermal capacity. { 'hēt kə¸pas·əd·ē }

heat check |MET| Parallel surface cracks forming a pattern on the surface of a metal as a result of thermal fatigue. { 'hēt ¸chek }

heat conduction |THERMO| The flow of thermal energy through a substance from a higher-to a lower-temperature region. { 'hēt kən¸dək·shən }

heat conductivity See thermal conductivity. { 'hēt ¸kän·dək'tiv·əd·ē }

heat content See enthalpy. { 'hēt ¦kän·tent }

heat convection |THERMO| The transfer of thermal energy by actual physical movement from one location to another of a substance in which thermal energy is stored. Also known as thermal convection. { 'hēt kən¦vek·shən }

heat cycle See thermodynamic cycle. { 'hēt ¸sī·kəl }

heat death |THERMO| The condition of any isolated system when its entropy reaches a maximum, in which matter is totally disordered and at a uniform temperature, and no energy is available for doing work. { 'hēt ¸deth }

heat dissipation See heat loss. { 'hēt ,dis·ə¦pā·shən }

heat energy See internal energy. { 'hēt ,en·ər·jē }

heat engine |THERMO| A thermodynamic system which undergoes a cyclic process during which a positive amount of work is done by the system; some heat flows into the system and a smaller amount flows out in each cycle. { 'hēt ,en·jən }

heat equation |THERMO| A parabolic second-order differential equation for the temperature of a substance in a region where no heat source exists: $\partial t/\partial \tau = (k/\rho c)(\partial^2 t/\partial x^2 + \partial^2 t/\partial y^2 + \partial t^2/\partial z^2)$, where x, y, and z are space coordinates, τ is the time, $t(x,y,z,\tau)$ is the temperature, k is the thermal conductivity of the body, ρ is its density, and c is its specific heat; this equation is fundamental to the study of heat flow in bodies. Also known as Fourier heat equation; heat flow equation. { 'hēt i,kwā·zhən }

heat flow |THERMO| Heat thought of as energy flowing from one substance to another; quantitatively, the amount of heat transferred in a unit time. Also known as heat transmission. { 'hēt ,flō }

heat flow equation See heat equation. { 'hēt ¦flō i,kwā·zhən }

heat flux |THERMO| The amount of heat transferred across a surface of unit area in a unit time. Also known as thermal flux. { 'hēt ,fləks }

heat insulator |MATER| A substance having relatively low heat conductivity. { 'hēt ,ins·ə,lād·ər }

heat loss |PHYS| Energy or power transmitted out of a system in the form of heat. Also known as heat dissipation. { 'hēt ,lòs }

heat of ablation |THERMO| A measure of the effective heat capacity of an ablating material, numerically the heating rate input divided by the mass loss rate which results from ablation. { 'hēt əv ə'blā·shən }

heat of adsorption |THERMO| The increase in enthalpy when 1 mole of a substance is adsorbed upon another at constant pressure. { 'hēt əv ad'sòrp·shən }

heat of aggregation |THERMO| The increase in enthalpy when an aggregate of matter, such as a crystal, is formed at constant pressure. { 'hēt əv ,ag·rə'gā·shən }

heat of compression |THERMO| Heat generated when air is compressed. { 'hēt əv kəm 'presh·ən }

heat of condensation |THERMO| The increase in enthalpy accompanying the conversion of 1 mole of vapor into liquid at constant pressure and temperature. { 'hēt əv ,künd·ən'sā·shən }

heat of cooling |THERMO| Increase in enthalpy during cooling of a system at constant pressure, resulting from an internal change such as an allotropic transformation. { 'hēt əv 'kül·iŋ }

heat of crystallization |THERMO| The increase in enthalpy when 1 mole of a substance is transformed into its crystalline state at constant pressure. { 'hēt əv ,krist·əl·ə'zā·shən }

heat of evaporation See heat of vaporization. { 'hēt əv i,vap·ə'rā·shən }

heat of fusion |THERMO| The increase in enthalpy accompanying the conversion of 1 mole, or a unit mass, of a solid to a liquid at its melting point at constant pressure and temperature. Also known as latent heat of fusion. { 'hēt əv 'fyü·zhən }

heat of mixing |THERMO| The difference between the enthalpy of a mixture and the sum of the enthalpies of its components at the same pressure and temperature. { 'hēt əv 'mik·siŋ }

heat of solidification |THERMO| The increase in enthalpy when 1 mole of a solid is formed from a liquid or, less commonly, a gas at constant pressure and temperature. { 'hēt əv sə,lid·ə·fə'kā·shən }

heat of sublimation |THERMO| The increase in enthalpy accompanying the conversion of 1 mole, or unit mass, of a solid to a vapor at constant pressure and temperature. Also known as latent heat of sublimation. { 'hēt əv ,səb·lə'mā·shən }

heat of transformation |THERMO| The increase in enthalpy of a substance when it undergoes some phase change at constant pressure and temperature. { 'hēt əv ,tranz·fər'mā·shən }

heat of vaporization |THERMO| The quantity of energy required to evaporate 1 mole, or a unit mass, of a liquid, at constant pressure and temperature. Also known as enthalpy of vaporization; heat of evaporation; latent heat of vaporization. { 'hēt əv ,vā·pə·rə'zā·shən }

heat of wetting |THERMO| 1. The heat of adsorption of water on a substance. 2. The additional heat required, above the heat of vaporization of free water, to evaporate water from a substance in which it has been absorbed. { 'hēt əv 'wed·iŋ }

heat quantity |THERMO| A measured amount of heat; units are the small calorie, normal calorie, mean calorie, and large calorie. { 'hēt ¦kwän·əd·ē }

heat radiation |THERMO| The energy radiated by solids, liquids, and gases in the form of electromagnetic waves as a result of their temperature. Also known as thermal radiation. { 'hēt ,rād·ē'ā·shən }

heat release |THERMO| The quantity of heat released by a furnace or other heating mechanism per second, divided by its volume. { 'hēt ri,lēs }

heat resistance See thermal resistance. { 'hēt ri,zis·təns }

heat-resistant alloy |MET| An oxidation-resistant alloy. { 'hēt ri,zis·tənt 'al,ói }

heat-resistant glass |MATER| Glass, such as borosilicate glass, that is heat-treated or leached to remove alkali so that it withstands high heat and sudden cooling without shattering. { 'hēt ri,zis·tənt 'glas }

heat shield |MATER| Any protective layer that gives protection from heat; used on the front of a reentry capsule. { 'hēt ,shēld }

heat-shrinkable tubing |MATER| A type of plastic tubing that can be heated and shrink-fitted over terminals and other objects of varying sizes and shapes, for insulating and other purposes. { 'hēt ¦shriŋk·ə·bəl 'tüb·iŋ }

heat shunt |MET| A heatsink placed in contact with the lead of a delicate component to prevent overheating during soldering. { 'hēt ‚shənt }

heatsink |ELEC| A mass of metal that is added to a device for the purpose of absorbing and dissipating heat; used with power transistors and many types of metallic rectifiers. |THERMO| Any (gas, solid, or liquid) region where heat is absorbed. { 'hēt‚siŋk }

heat source |THERMO| Any device or natural body that supplies heat. { 'hēt ‚sórs }

heat time |MET| Duration of a single current impulse in pulsation welding. { 'hēt ‚tīm }

heat tinting |MET| Oxidation of a polished metal surface by heating to reveal the microstructure. { 'hēt ‚tint·iŋ }

heat transfer |THERMO| The movement of heat from one body to another (gas, liquid, solid, or combinations thereof) by means of radiation, convection, or conduction. { 'hēt ¦tranz·fər }

heat-transfer coefficient |THERMO| The amount of heat which passes through a unit area of a medium or system in a unit time when the temperature difference between the boundaries of the system is 1 degree. { 'hēt ¦tranz·fər ‚kō·i'fish·ənt }

heat-transfer oil |MATER| An oil used to transport heat or cold between two areas of process-equipment surface, and especially compounded to avoid heat degradation in the temperature range of application. { 'hēt ¦tranz·fər ‚óil }

heat transmission See heat flow. { 'hēt tranz ‚mish·ən }

heat transport |THERMO| Process by which heat is carried past a fixed point or across a fixed plane, as in a warm current. { 'hēt ¦tranz‚pòrt }

heat-treatable alloy |MET| An alloy that can be hardened by thermal treatment. { 'hēt ¦trēd·ə·bəl 'al‚ói }

heat-treating film |MET| An oxide coating formed on a metal surface by heat treating. { 'hēt ‚trēd·iŋ ‚film }

heat treatment |MET| Heating and cooling a metal or alloy to obtain desired properties or conditions. { 'hēt ‚trēt·mənt }

heavy acid See phosphotungstic acid. { ¦hev·ē 'as·əd }

heavy alloy |MET| A tungsten-nickel alloy produced by pressing and sintering the metallic powders; used for screens for x-ray tubes and radioactivity units and for contact surfaces of circuit breakers. { ¦hev·ē 'al‚ói }

heavy concrete |MATER| Concrete in which some or all rock aggregate is replaced by metal aggregate. { 'hev·ē käŋ'krēt }

heavy-fermion superconductor |SOLID STATE| A superconductor in which the superconducting electrons have unusually large effective masses, more than 100 times the mass of a free electron. { ¦hev·ē 'fər·mē‚än 'sü·pər·kən‚dək·tər }

heavy-fermion system |SOLID STATE| A lanthanide-based or actinide-based interme-tallic compound in which the low-energy excitations (quasiparticles) of the conduction electron system have effective masses at low temperatures that are several hundred times the free-electron mass. { ‚hev·ē 'fər·mē‚än ‚sis·təm }

heavy metal |MET| A metal whose specific gravity is approximately 5.0 or higher. { 'hev·ē 'med·əl }

heavy water |INORG CHEM| A compound of hydrogen and oxygen containing a higher proportion of the hydrogen isotope deuterium than does naturally occurring water. Also known as deuterium oxide. { 'hev·ē 'wód·ər }

Hedvall effect I |SOLID STATE| A discontinuous change in the temperature dependence of the chemical reaction rate of certain substances at the Curie temperture. { 'hed·vòl i‚fekt 'wən }

Hedvall effect II |SOLID STATE| A discontin-uous change in the activation energy of certain substances at the Curie temperature. { 'hed·vòl i‚fekt 'tü }

heel |MET| **1.** A quantity of molten metal remain-ing in the ladle after pouring a metal cast-ing. **2.** A quantity of metal retained in an induction furnace during a stand-by period. { hēl }

Hegeler furnace |MET| A muffle furnace having seven tiers of hearths; lower hearths are heated by gas burned in flues beneath them. { 'heg·lər ‚fər·nəs }

Heisenberg exchange coupling |SOLID STATE| The exchange forces between electrons in neigh-boring atoms which give rise to ferromag-netism in the Heisenberg theory. { 'hīz·ən·bərg iks'chānj ‚kəp·liŋ }

Heisenberg theory of ferromagnetism |SOLID STATE| A theory in which exchange forces be-tween electrons in neighboring atoms are shown to depend on relative orientations of electron spins, and ferromagnetism is explained by the assumption that parallel spins are favored so that all the spins in a lattice have a tendency to point in the same direction. { 'hīz·ən·bərg 'thē·ə·rē əv ‚fer·ō'mag·nə‚tiz·əm }

Heliarc welding See inert gas-shielded arc weld-ing. { 'hēl·ē‚ärk ‚weld·iŋ }

helicate |ORG CHEM| Any member of a group of synthetic, helical arrays of molecules formed by the chemical recognition and organization of metals and organic bases. { 'hel·i‚kāt }

helimagnet |SOLID STATE| A metal, alloy, or salt that possesses helimagnetism. { 'hel·ə ‚mag·nət }

helimagnetism |SOLID STATE| A property pos-sessed by some metals, alloys, and salts of transition elements or rare earths, in which the atomic magnetic moments, at sufficiently low temperatures, are arranged in ferromag-netic planes, the direction of the magnetism varying in a uniform way from plane to plane. { ‚hel·ə'mag·nə‚tiz·əm }

helium |CHEM| A gaseous chemical element, symbol He, atomic number 2, and atomic weight 4.0026; one of the noble gases in group 0 of the periodic table. { 'hē·lē·əm }

helium film |CRYO| A superfluid film that covers

any surface in contact with helium II. Also known as Rollin film. { 'hē·lē·əm ,film }

helium I [CRYO] The phase of liquid helium-4 which is stable at temperatures above the lambda point (about 2.2 K) and has the properties of a normal liquid, except low density. { 'hē·lē·əm 'wən }

helium II [CRYO] The phase of liquid helium-4 which is stable at temperatures between absolute zero and the lambda point (about 2.2 K), and has many remarkable properties such as vanishing viscosity, extremely high heat conductivity, and the fountain effect. { 'hē·lē·əm 'tü }

helium liquefier [CRYO] Any one of several machines which liquefy helium by causing it to undergo adiabatic expansion and to do external work. { 'hē·lē·əm 'lik·wə,fī·ər }

helmholtz [ELEC] A unit of dipole moment per unit area, equal to 1 Debye unit per square angstrom, or approximately 3.335×10^{-10} coulomb per meter. { 'helm,hōlts }

Helmholtz equation [OPTICS] An equation which relates the linear and angular magnifications of a spherical refracting interface. Also known as Lagrange-Helmholtz equation. [PHYS CHEM] The relationship stating that the emf (electromotive force) of a reversible electrolytic cell equals the work equivalent of the chemical reaction when charge passes through the cell plus the product of the temperature and the derivative of the emf with respect to temperature. { 'helm ,hōlts i,kwā·zhən }

Helmholtz free energy See free energy. { 'helm ,hōlts ¦frē 'en·ər·jē }

Helmholtz function See free energy. { 'helm ,hōlts ,fəŋk·shən }

Helmholtz potential See free energy. { 'helm ,hōlts pə¦ten·chəl }

Helmholtz's theorem [FL MECH] The theorem that in the isentropic flow of a nonviscous fluid which is not subject to body forces, individual vortices always consist of the same fluid particles. { 'helm,hōlt·səz ,thir·əm }

helve hammer [MET] A belt-driven trip hammer with the hammer face or swage carried on the end of a beam; used for welding, forging, plating, drawing, and other metal-working operations. { 'helv ,ham·ər }

hemihedral symmetry [CRYSTAL] The possession by a crystal of only half of the elements of symmetry which are possible in the crystal system to which it belongs. { ¦he·mē¦hē·drəl 'sim·ə·trē }

hemiholohedral [CRYSTAL] Of hemihedral form but with half of the octants having the full number of planes. { ¦he·mē,hō·lə'hē·drəl }

hemimorphic crystal [CRYSTAL] A crystal with no transverse plane of symmetry and no center of symmetry; composed of forms belonging to only one end of the axis of symmetry. { ¦he·mē ¦mór·fik 'krist·əl }

hemiprism [CRYSTAL] A pinacoid that cuts two crystallographic axes. { 'he·mē,priz·əm }

hemitropic [CRYSTAL] Pertaining to a twinned structure in which, if one part were rotated 180°, the two parts would be parallel. { ¦he·mē ¦träp·ik }

Henderson process [MET] The treatment of copper sulfide ores by roasting with salt to form chlorides, which are then leached out and precipitated. { 'hen·dər·sən ,prä·səs }

henequen [MATER] A hard plant fiber, obtained from the leaves of the American agave (*Agave fourcroydes*) and other agave species; used to make rope, twine, and cord. { 'hen·ə·kən }

hepar calcies See calcium sulfide. { 'hē,pär 'kal ,sēz }

hepar sulfuris See potassium sulfide. { 'hē,pär səl'fyúr·əs }

Herbert cloudburst test [MET] A hardness test in which a shower of steel balls, dropped from a predetermined height, dulls the surface of a hardened part in proportion to its softness and thus reveals defective areas. { 'hər·bərt 'klaúd ,bərst ,test }

Herbert pendulum method [MET] Hardness testing in which a 1-millimeter steel or jewel ball resting on the surface to be tested acts as the fulcrum for a 4-kilogram compound pendulum of 10-second period; the swinging of the pendulum causes a rolling indentation in the material, and several hardness factors, such as work hardenability, are determined. { 'hər·bərt 'pen·jə·ləm ,meth·əd }

hereditary mechanics [MECH] A field of mechanics in which quantities, such as stress, depend not only on other quantities, such as strain, at the same instant but also on integrals involving the values of such quantities at previous times. { hə'red·ə,ter·ē mi'kan·iks }

Hermann-Mauguin symbols [CRYSTAL] Symbols representing the 32 symmetry classes, consisting of series of numbers giving the multiplicity of symmetry axes in descending order, with other symbols indicating inversion axes and mirror planes. { 'her·män 'mō,gan ,sim·bəlz }

Herreshoff furnace [MET] **1.** A rectangular-shaft blast furnace for smelting copper ore. **2.** A mechanical, cylindrical, multiple-deck muffle furnace of the McDougall type. { 'her·əs·hóf ,fər·nəs }

hertz [PHYS] Unit of frequency; a periodic oscillation has a frequency of n hertz if in 1 second it goes through n cycles. Also known as cycle per second (cps). Symbolized Hz. { hərts }

Hertz's law [MECH] A law which gives the radius of contact between a sphere of elastic material and a surface in terms of the sphere's radius, the normal force exerted on the sphere, and Young's modulus for the material of the sphere. { 'hərt·səs ,ló }

hetero- [CHEM] Prefix meaning different; for example, a heterocyclic compound is one in which the ring is made of more than one kind of atom. { 'hed·ə·rō }

heteroatom [ORG CHEM] In an organic compound, any atom other than carbon or hydrogen. { 'hed·ə·rō,ad·əm }

heterocyclic compound [ORG CHEM] Compound in which the ring structure is a combination of more than one kind of atom; for example, pyridine, C_5H_5N. { ‚hed·ə·rō'sī·klik ¦käm‚paúnd }

heterodesmic [CRYSTAL] Pertaining to those atoms bonded in more than one way in crystals. { ‚hed·ə·rō'dez·mik }

heterodyne [ELECTR] To mix two alternating-current signals of different frequencies in a nonlinear device for the purpose of producing two new frequencies, the sum of and difference between the two original frequencies. { 'hed·ə·rə‚dīn }

heterogeneous [CHEM] Pertaining to a mixture of phases such as liquid-vapor, or liquid-vapor-solid. { ‚hed·ə'räj·ə·nəs }

heterogeneous catalysis [CHEM] A chemical process in which the catalyst is in a separate phase, usually the reactants and products are in gaseous or liquid phases and the catalyst is a solid, and the catalytic reaction occurs on the surface of the solid. { ‚hed·ə·rə'jē·nē·əs kə'tal·ə·səs }

heterogeneous strain [MECH] A strain in which the components of the displacement of a point in the body cannot be expressed as linear functions of the original coordinates. { ‚hed·ə·rə¦jē·nē·əs 'strān }

heterojunction [ELECTR] The boundary between two different semiconductor materials, usually with a negligible discontinuity in the crystal structure. { ¦hed·ə·rō'jəŋk·shən }

heterolysis See heterolytic cleavage. { ‚hed·ə'räl·ə·səs }

heterolytic cleavage [ORG CHEM] The breaking of a single (two-electron) chemical bond in which both electrons remain on one of the atoms. Also known as heterolysis. { ‚hed·ə·rō¦lid·ik 'klēv·ij }

heteromorphic transformation [THERMO] A change in the values of the thermodynamic variables of a system in which one or more of the component substances also undergo a change of state. { ‚hed·ə·rə¦mòr·fik ‚tranz·fər'mā·shən }

heteropoly acid [INORG CHEM] Complex acids of metals, whose specific gravity is greater than 4, with phosphoric acid; an example is phospho-molybdic acid. { ‚hed·ə'räp·ə·lē 'as·əd }

heteropoly compound [INORG CHEM] Polymeric compounds of molybdates with anhydrides of other elements such as phosphorus; the yellow precipitate $(NH_4)_3P(Mo_3O_{10})_4$ is such a compound. { ‚hed·ə'räp·ə·lē 'käm‚paúnd }

heteropolymer [CHEM] A compound comprising two or more molecules that are different from one another. { ‚hed·ə·rə'päl·ə·mər }

Heusler alloy [MET] Any of a group of ferromagnetic nonferrous alloys typically composed of 18–25% manganese, 10–25% aluminum, and the balance copper. { 'hòis·lər ‚al‚òi }

hevea rubber [MATER] Rubber made from latex obtained from the rubber tree (*Hevea brasiliensis*); used for electrical insulation. { 'hē·vē·ə ‚rəb·ər }

hexad axis [CRYSTAL] A rotation axis whose multiplicity is equal to 6. { 'hek·sad ‚ax·səs }

hexadentate ligand [INORG CHEM] A chelating agent having six groups capable of attachment to a metal ion. { ‚hek·sə'den‚tāt 'līg·ənd }

hexagonal boron nitride [MATER] A synthetic material composed of boron and nitrogen with the atoms combined in a hexagonal lattice such that it has a structure similar to graphite. Also known as white graphite. { hek¦sag·ən·əl ‚bòr·än 'nī‚trīd }

hexagonal close-packed structure [CRYSTAL] Close-packed crystal structure characterized by the regular alternation of two layers; the atoms in each layer lie at the vertices of a series of equilateral triangles, and the atoms in one layer lie directly above the centers of the triangles in neighboring layers. Abbreviated hcp structure. { hek'sag·ə·nəl ¦klōs ¦pakt 'strək·chər }

hexagonal lattice [CRYSTAL] A Bravais lattice whose unit cells are right prisms with hexagonal bases and whose lattice points are located at the vertices of the unit cell and at the centers of the bases. { hek'sag·ə·nəl 'lad·əs }

hexagonal system [CRYSTAL] A crystal system that has three equal axes intersecting at 120° and lying in one plane; a fourth, unequal axis is perpendicular to the other three. { hek'sag·ə·nəl 'sis·təm }

hexatic phase [PHYS] A phase of matter that is intermediate between the normal solid and the isotropic liquid phases, and corresponds to a two-dimensional fluid with sixfold orientational order but no positional order. { hek¦sad·ik 'fāz }

hexoctahedron [CRYSTAL] A cubic crystal form that has 48 equal triangular faces, each of which cuts the three crystallographic axes at different distances. { ‚hek'säk·tə‚hē·drən }

hextetrahedron [CRYSTAL] A 24-faced form of crystal in the tetrahedral group of the isometric system. { ‚heks‚te·trə'hē·drən }

Heyn's reagent [MET] Double chloride of copper and ammonia; used in microanalysis of carbon steels. { 'hīnz rē‚ā·jənt }

Hf See hafnium.

HFS See type II superconductor.

Hg See mercury.

high-abrasion furnace black [MATER] A variety of carbon black made by burning oil in a deficiency of air and then quenching to stop the reaction short of equilibrium; particles are 26–30 nanometers in diameter; used in tires and mechanical rubber goods. Abbreviated HAF black. { 'hī ə‚brā·zhən 'fər·nəs ‚blak }

high-alloy steel [MET] Steel containing large percentages of elements other than carbon. { ¦hī 'al‚òi ¦stēl }

high-alumina brick [MATER] Refractory brick made from raw materials rich in alumina, such as diaspore and bauxite; when well fired, they contain a large amount of mullite; used for unusually severe temperature or load conditions. { ¦hī ə'lüm·ə·nə ¦brik }

high-alumina cement See aluminate cement. { ¦hī ə'lüm·ə·nə si'ment }

high boiler [MATER] A solvent added to lacquer thinner to slow the rate of evaporation. { 'hī ‚bòil·ər }

high brass [MET] The most common commercial wrought brass containing about 35% zinc and 65% copper. { 'hī ¦bras }

high-carbon chromium [MET] Chromium containing at least 86% chromium, 8–11% carbon, and a maximum of 0.5% each of iron and silicon. { 'hī ¦kär·bən 'krō·mē·əm }

high-carbon steel [MET] A cast or forged steel containing more than 0.5% carbon. { 'hī ¦kär·bən 'stēl }

high-density polyethylene [POLYM CHEM] A thermoplastic polyolefin with a density of 0.941–0.960 gram per cubic centimeter (0.543–0.555 ounce per cubic inch). Abbreviated HDPE. { ¦hī ¦den·səd·ē ¦päl·ē'eth·ə¦lēn }

high-energy fuel [MATER] Fuel that upon combustion provides greater energy than that from conventional carbonaceous fuels; specifically, a hydroboron. { 'hī ¦en·ər·jē 'fyül }

high-energy-rate forging [MET] The production of a forging by the use of a machine which utilizes a sudden surge of kinetic energy from compressed gas against a piston, causing a high ram velocity against the work. { 'hī ¦en·ər·jē ¦rāt 'förj·iŋ }

high-expansion alloy [MET] An alloy possessing a high coefficient of expansion. { 'hī ik 'span·chən 'al,öi }

high-expansion foam [MATER] Noncombustible foam made from ammonium lauryl sulfate; used in underground mine fire fighting. { 'hī ik'span·chən 'fōm }

high-field superconductor See type II superconductor. { hī ,fēld 'sü·pər·kən,dək·tər }

high-frequency resistance [ELEC] The total resistance offered by a device in an alternating-current circuit, including the direct-current resistance and the resistance due to eddy current, hysteresis, dielectric, and corona losses. Also known as alternating-current resistance; effective resistance; radio-frequency resistance. { 'hī ¦frē·kwən·sē ri'zis·təns }

high-frequency welding [MET] Resistance welding in which the heat is produced by the current flow induced by a high-frequency electromagnetic field. Also known as radio-frequency welding. { 'hī ¦frē·kwən·sē 'weld·iŋ }

high heat [THERMO] Heat absorbed by the cooling medium in a calorimeter when products of combustion are cooled to the initial atmospheric (ambient) temperature. { 'hī ¦hēt }

high-heat cement [MATER] A type of cement which releases a large amount of heat during curing. { 'hī ,hēt si'ment }

high-K capacitor [ELEC] A capacitor whose dielectric material is a ferroelectric having a high dielectric constant, up to about 6000. { 'hī ,kā kə'pas·əd·ər }

high-lead bronze [MET] Bronze containing high percentages of lead to give a soft matrix alloy. { 'hī ,led 'bränz }

high-modulus furnace black [MATER] A variety of carbon black made by burning oil in a deficiency of air; particle diameter is 49–60 nanometers; production is now negligible, but it was used in tire carcasses. Abbreviated HMF black. { 'hī ,mäj·ə·ləs 'fər·nəs ,blak }

high polymer [POLYM CHEM] A large molecule (of molecular weight greater than 10,000) usually composed of repeat units of low-molecular-weight species; for example, ethylene or propylene. { 'hī ¦päl·ə·mər }

high-pressure laminate [MATER] A plastic-substrate laminate molded and cured at pressures of, commonly, 1200–2000 pounds per square inch (8–14 × 10⁶ pascals). { 'hī ¦presh·ər 'lam·ə,nāt }

high-residual-phosphorus copper [MET] Deoxidized copper having reduced conductivity due to the presence of residual phosphorus, usually less than 0.1. { ¦hī ri¦zij·yə·wəl 'fäs·fə·rəs 'käp·ər }

high-solvency naphtha [MATER] Any of the petroleum-based solvents that boil in the naphtha range (95–650°F; 35–343°C) and have a high aromatic-chemical content to give high-solvency power for nitrocelluloses, dry paints, and certain resins. { 'hī ,säl·vən·sē 'naf·thə }

high-speed cement See aluminate cement. { 'hī ,spēd si'ment }

high-speed steel [MET] An alloy steel that remains hard and tough at red heat. { 'hī ,spēd 'stēl }

high-strength alloy See high-tensile alloy. { 'hī ,streŋkth 'al,öi }

high-strength low-alloy steel [MET] Steel containing small amounts of niobium or vanadium and having higher strength, better low-temperature impact toughness, and, in some grades, better atmospheric corrosion resistance than carbon steel. Abbreviated HSLA steel. { 'hī ,streŋkth ¦lō ¦al,öi 'stēl }

high-strength steel See high-tensile steel. { 'hī ,streŋkth 'stēl }

high-temperature alloy [MET] An alloy suitable for use at temperatures of 500°C and above. { 'hī ,tem·prə·chər 'al,öi }

high-temperature cement [MATER] A cement that resists fusing, softening, or spalling at elevated temperatures; used to bond refractory materials. { 'hī ,tem·prə·chər si'ment }

high-temperature coke [MATER] Coke produced at temperatures of 900–1150°C; used mainly for metallurgical purposes. { 'hī ,tem·prə·chər 'kōk }

high-temperature material [MATER] A material with high-temperature capability, including the superalloys, refractory alloys, and ceramics; used in structures such as spacecraft subjected to extreme thermal environments. { 'hī ,tem·prə·chər mə'tir·ē·əl }

high-temperature superconductor [SOLID STATE] A ceramic material that displays superconductivity at temperatures of 90 K (−298°F) or more. { 'hī 'tem·prə·chər 'sü·pər·kən,dək·tər }

high-tensile alloy [MET] An alloy having a high tensile strength. Also known as high-strength alloy. { 'hī ,ten·səl 'al,öi }

high-tensile steel [MET] Low-alloy steel having a yield strength range of 50,000 to 100,000 pounds per square inch (3.4–6.9×10^8 pascals). Also known as high-strength steel.   { 'hī ‚ten·səl 'stēl }

Hildebrand function |THERMO| The heat of vaporization of a compound as a function of the molal concentration of the vapor; it is nearly the same for many compounds.   { 'hil·də‚brand ‚faŋk·shən }

hindered contraction |MET| Thermal contraction of a casting that is hindered locally due to the particular geometry.   { 'hin·dərd kən'trak·shən }

H Monel |MET| A Monel containing 3.2% silicon available in cast form; it is harder and stronger than Monel.   { 'āch mō‚nel }

Ho   See holmium.

hobbing steel |MET| A high-speed steel used to make gear teeth cutters.   { 'häb·iŋ ‚stēl }

Hoepfner process   See Hopfner process.   { 'hepf·nər ‚präs·əs }

hog gum |MATER| A sticky gum, an exudate from various species of Sterculia trees, whose chief constituent is galactan; the gum is dried and marketed either as a powder or as white flakes; used in the food, cosmetic, and textile industry.   { 'häg ‚gəm }

hohlraum   See blackbody.   { 'hōl‚raùm }

holddown |MET| A device that holds the outer portion of a metal sheet in place during deep-drawing operations in order to keep it from becoming wrinkled.   { 'hōl‚daùn }

holding furnace |MET| A heated reservoir to hold molten metal preparatory to casting.   { 'hōl·diŋ ‚fər·nəs }

hold time |MET| In resistance welding, the time that pressure is applied to the electrodes after the welding current is cut off.   { 'hōl ‚tīm }

hole |SOLID STATE| A vacant electron energy state near the top of an energy band in a solid; behaves as though it were a positively charged particle. Also known as electron hole.   { hōl }

hollow block   See hollow tile.   { 'häl·ō 'bläk }

hollow tile |MATER| A hollow building block of concrete or burnt clay used for making partitions, exterior walls, or suspended floors or roofs. Also known as hollow block.   { 'häl·ō 'tīl }

holmium |CHEM| A rare-earth element belonging to the yttrium subgroup, symbol Ho, atomic number 67, atomic weight 164.9304, melting point 1400–1525°C.   { 'hōl·mē·əm }

holoaxial |CRYSTAL| Having all possible axes of symmetry.   { ¦häl·ō'ak·sē·əl }

hologram |OPTICS| The special photographic plate used in holography; when this negative is developed and illuminated from behind by a coherent gas-laser beam, it produces a three-dimensional image in space.   { 'häl·ə‚gram }

holography |PHYS| A technique for recording, and later reconstructing, the amplitude and phase distributions of a wave disturbance; widely used as a method of three-dimensional optical image formation, and also with acoustical and radio waves; in optical image formation, the technique is accomplished by recording on a photographic plate the pattern of interference between coherent light reflected from the object

of interest, and light that comes directly from the same source or is reflected from a mirror.   { hə'läg·rə·fē }

holohedral |CRYSTAL| Pertaining to a crystal structure having the highest symmetry in each crystal class. Also known as holosymmetric; holosystemic.   { ¦häl·ō¦hē·drəl }

holohedron |CRYSTAL| A crystal form of the holohedral class, having all the faces needed for complete symmetry.   { ¦häl·ō'hē·drən }

holosymmetric   See holohedral.   { ¦häl·ō·si'me·trik }

holosystemic   See holohedral.   { ¦häl·ō·si'stem·ik }

homenergic flow |THERMO| Fluid flow in which the sum of kinetic energy, potential energy, and enthalpy per unit mass is the same at all locations in the fluid and at all times.   { 'häm·ə‚nər·jik 'flō }

homeomorph |CRYSTAL| A crystal that displays a form similar to that of a crystal with a different chemical composition.   { 'hō·mē·ə‚mórf }

homo- |ORG CHEM| **1.** Indicating the homolog of a compound differing in formula from the latter by an increase of one CH_2 group. **2.** Indicating a homopolymer made up of a single type of monomer, such as polyethylene from ethylene. **3.** Indicating that a skeletal atom has been added to a well-known structure.   { 'hō·mō }

homodesmic |CRYSTAL| Of a crystal, having atoms bonded in a single way.   { ‚hä·mə 'dez·mik }

homogeneous |CHEM| Pertaining to a substance having uniform composition or structure.   { ‚hä·mə'jē·nē·əs }

homogeneous catalysis |CHEM| Catalysis occurring within a single phase, usually a gas or liquid.   { ‚hä·mə'jē·nē·əs kə'tal·ə·səs }

homogeneous strain |MECH| A strain in which the components of the displacement of any point in the body are linear functions of the original coordinates.   { ¦hō·mə‚jē·nē·əs 'strān }

homogenize |MET| To hold metal at a high temperature long enough to eliminate by diffusion any chemical segregation of the components.   { hə'mäj·ə‚nīz }

homology |CHEM| The relation among elements of the same group, or family, in the periodic table. |ORG CHEM| That state, in a series of organic compounds that differ from each other by a CH_2 such as the methane series C_nH_{2n+2}, in which there is a similarity between the compounds in the series and a graded change of their properties.   { hə'mäl·ə·jē }

homolysis   See homolytic cleavage.   { hə'mäl·ə·sis }

homolytic cleavage |ORG CHEM| The breaking of a single (two-electron) bond in which one electron remains on each of the atoms. Also known as free-radical reaction; homolysis.   { ¦häm·ə‚lid·ik 'klēv·ij }

homometric pair |CRYSTAL| A pair of crystal structures whose x-ray diffraction patterns are identical.   { ¦hä·mə¦me·trik 'per }

homomorphous transformation |THERMO| A change in the values of the thermodynamic variables of a system in which none of the component substances undergoes a change of state. { ˌhō·məˌmȯr·fəs ˌtranz·fər'mā·shən }

homonuclear molecule |CHEM| A diatomic molecule, both of whose atoms are of the same element. { ˌhō·mōˌnü·klē·ər 'mäl·ə‚kyül }

homopolar crystal |SOLID STATE| A crystal in which the bonds are all covalent. { ˌhä·mə 'pō·lər 'krist·əl }

homopolymer |POLYM CHEM| A polymer formed from a single monomer; an example is polyethylene, formed by polymerization of ethylene. { ˌhä·mō'päl·ə·mər }

hone |MATER| A fine-grit stone that is used for sharpening a cutting tool. { hōn }

honeycombing |MATER| **1.** Internal fiber separation in drying timber. **2.** Local roughness and weakening on the face of a concrete wall due to segregation of the concrete, with the result that there is little sand to fill in between the stone aggregate. { 'hən·ē‚kōm·iŋ }

Hookean deformation |MECH| Deformation of a substance which is proportional to the force applied to it. { 'hük·ē·ən ‚def·ər'mā·shən }

Hookean solid |MECH| An ideal solid which obeys Hooke's law exactly for all values of stress, however large. { 'hük·ē·ən 'säl·əd }

Hooke number See Cauchy number. { 'hük ‚nəm·bər }

Hooker process |MET| A forming process in which pierced slugs or cups are punched through a die to produce tubing and other shapes. { 'hük·ər ‚prä·səs }

Hooke's law |MECH| The law that the stress of a solid is directly proportional to the strain applied to it. { 'hüks ‚lȯ }

Hope's apparatus |THERMO| An apparatus consisting of a vessel containing water, a freezing mixture in a tray surrounding the vessel, and thermometers inserted in the water at points above and below the freezing mixture; used to show that the maximum density of water lies at about 4°C. { 'hōps ‚ap·ə‚rad·əs }

Hopfner process |MET| A process for the recovery of copper from its sulfide ores by leaching with a solution of cuprous chloride in sodium or calcium chloride and electrolyzing the resulting solution in tanks that are protected by diaphragms. Also spelled Hoepfner process. { 'häpf·nər ‚prä·səs }

horizontal-position welding |MET| **1.** Making a fillet weld on the upper side of the intersection of a vertical surface and a horizontal surface. **2.** Making a horizontal groove weld on a vertical surface. { ‚här·ə'zänt·əl pə‚zish·ən ‚weld·iŋ }

horizontal retort |MET| An intermittent unit made from a siliceous fireclay and formerly used for zinc smelting. { ‚här·ə'zänt·əl 'rē‚tȯrt }

horizontal retort process |MET| A zinc-smelting process that employs vast, honeycomblike batteries of fireclay or silicon-carbide retorts set horizontally in a gas- or coal-fired furnace. Also known as Belgian retort process. { ‚här·ə'zänt·əl 'rē‚tȯrt ‚prä·səs }

horizontal-rolled-position welding |MET| Top-

side welding of a butt joint connecting two horizontal pieces of rotating pipe. { ‚här·ə'zänt·əl ¦rōld ·pə¦zish·ən 'weld·iŋ }

horn |MET| Holding arm for the electrode of a resistance spot-welding machine. { hȯrn }

horn spacing |MET| Unobstructed work space in a resistance-welding machine between horns at right angles to the throat depth. { 'hȯrn ‚spās·iŋ }

host |CHEM| A crystalline lattice or receptor molecule for the strong and selective binding of a cationic, anionic, or neutral organic, inorganic, or biological substance (guest) by means of electrostatic, hydrogen-bonding, van der Waals, or donor-acceptor interactions. Examples include clathrates, crown ethers, cryptands, cyclodextrins, calixarenes, cavitands, cyclophanes, and cryptophanes. { hōst }

host-guest complexation chemistry |ORG CHEM| The design, synthesis, and study of highly structured organic molecular complexes that mimic biological complexes. { 'hōst 'gest ‚käm·plek'sā·shən ‚kem·ə·strē }

hot-air soldering |MET| Soldering with a narrow blast of air whose temperature is closely controlled at the value required for soldering individual joints on printed circuit boards. { 'hä ¦der 'säd·ə·riŋ }

hotbed |MET| An area where hot-rolled metal is placed to cool. Also known as cooling table. { 'hät‚bed }

hot-blast stove |MET| A retracting device for preheating the incoming air in an iron blast furnace by using heat from the burning gases of the furnace. { 'hät ¦blast ‚stōv }

hot-die steel |MET| A high-temperature, shock-resistant alloy steel used in forging machines. { 'hät ¦dī 'stēl }

hot dipping |MET| Coating metal components by immersion in a molten metal bath, such as tin or zinc. { 'hät 'dip·iŋ }

hot extrusion |MET| The process of extruding metal at very high temperatures. { 'hät ik'strü·zhən }

hot forming |MET| Shaping operations performed at temperatures above the recrystallization temperature of the metal. { 'hät ¦fȯr·miŋ }

hot plate |MET| A heated surface on which joints are brought to soldering temperature. { 'hät ‚plāt }

hot press forge |MET| A press in which metal parts are formed by forcing hot metal into dies under high pressure. { 'hät ‚pres 'fȯrj }

hot-pressure welding |MET| A pressure welding process in which macrodeformation of the base material to produce coalescence results from the application of heat and pressure. { 'hät ‚presh·ər 'weld·iŋ }

hot-quenching |MET| Quenching from high temperatures into a medium of lower but still high temperature. { 'hät ¦kwench·iŋ }

hot rolling |MET| Rolling of metal bars or sheets when hot. { 'hät 'rōl·iŋ }

hot shortness |MET| Brittleness, usually of steel

or wrought iron, when the metal is hot, due to a high sulfur content. { 'hät 'shȯrt·nəs }

hot strength *See* tensile strength. { 'hät ˌstreŋkth }

hot-swage [MET] To reduce the cross section of a hot metal tube or rod. { 'hät ˌswāj }

hot tear [MET] A separation either internally or externally in a casting due to loadings or internal stresses or both; it results from improper solidification, and shrinkage near the temperature at which the casting is completely solid. { 'hät ˌter }

hot trim [MET] Removal of flash in a heated forging. { 'hät ˌtrim }

hot working [MET] Plastic deformation of a metal at a rate and temperature such that strain hardening cannot occur. { 'hät ˌwərk·iŋ }

HSLA steel *See* high-strength low-alloy steel. { ¦āch¦es¦el¦ā 'stēl }

hub [MET] A steel punch used in making a working die for a coin or medal. { həb }

hubbing [MET] Forcing a male die into a blank to form a female die. { 'həb·iŋ }

hue [OPTICS] The name of a color, such as red, yellow, green, blue, or purple, as perceived subjectively. { hyü }

humectant [CHEM] A substance which absorbs or retains moisture; examples are glycerol, propylene glycol, and sorbitol; used in preparing confectioneries and dried fruit. { hyü'mek·tənt }

Hume-Rothery compound *See* electron compound. { 'hyüm 'rȯth·ə·rē ˌkäm,paùnd }

Hume-Rothery rule [SOLID STATE] The rule that the ratio of the number of valence electrons to the number of atoms in a given phase of an electron compound depends only on the phase, and not on the elements making up the compounds. { 'hyüm 'rȯth·ə·rē ˌrül }

humidity indicator [INORG CHEM] Cobalt salt (for example, cobaltous chloride) that changes color as the surrounding humidity changes; changes from pink when hydrated, to greenishblue when anhydrous. { hyü'mid·əd·ē ˌin·də ˌkād·ər }

humidity test [MET] Corrosion test in which a specimen is exposed to an environment of controlled humidity and temperature. { hyü'mid·əd·ē ˌtest }

Humphries equation [THERMO] An equation which gives the ratio of specific heats at constant pressure and constant volume in moist air as a function of water vapor pressure. { 'həm·frēz i,kwā·zhən }

Hunt and Douglas process [MET] Smelting process involving the roasting of matte carrying copper, lead, gold, and silver to form copper sulfate and oxide (but not silver sulfate); this product is leached with sulfuric acid for copper; the resulting solution is treated with calcium chloride by passing sulfur dioxide through it; the cuprous chloride is then reduced to cuprous oxide by milk of lime, (regenerating calcium chloride), and the cuprous oxide is smelted. { ¦hənt ən 'dəg·ləs ˌprä·səs }

Huttig equation [THERMO] An equation which

states that the ratio of the volume of gas adsorbed on the surface of a nonporous solid at a given pressure and temperature to the volume of gas required to cover the surface completely with a unimolecular layer equals $(1 + r) c'/(1 + c')$, where r is the ratio of the equilibrium gas pressure to the saturated vapor pressure of the adsorbate at the temperature of adsorption, and c is the product of a constant and the exponential of $(q - q_l)/RT$, where q is the heat of adsorption into a first layer molecule, q_l is the heat of liquefaction of the adsorbate, T is the temperature, and R is the gas constant. { 'həd·ik i,kwä·zhən }

Hybinette process [MET] A process used for refining of crude nickel anodes; anodes are placed in asphalt-lined, reinforced concrete tanks and dissolved electrochemically so that impurities such as copper and iron pass into solution while pure nickel electrolyte is continuously added. { ¦hī·bə¦net ˌprä·səs }

hybrid composite [MATER] A composite material in which two or more high-performance reinforcements are combined. { 'hī·brəd kəm 'päz·ət }

hybrid magnet [CRYO] A type of superconducting magnet consisting of a large-bore NbTi (niobium-titanium) external coil, which provides an external field of about 5 teslas, and an inner Nb_3Sn (niobium-tin) coil which provides additional field strength. [ELECTROMAG] An aircooled magnet consisting of a large-volume superconducting magnet surrounding a watercooled normal-conductor magnet that operates at the highest field. { 'hī·brəd 'mag·nət }

hydrargyrum *See* mercury. { hī'drär·jə·rəm }

hydrate [CHEM] **1.** A form of a solid compound which has water in the form of H_2O molecules associated with it; for example, anhydrous copper sulfate is a white solid with the formula $CuSO_4$, but when crystallized from water a blue crystalline solid with formula $CuSO_4·5H_2O$ results, and the water molecules are an integral part of the crystal. **2.** A crystalline compound resulting from the combination of water and a gas; frequently a constituent of natural gas that is under pressure. { 'hī,drāt }

hydrate aluminum oxide *See* alumina trihydrate. { 'hī,drāt ə'lü·mə·nəm 'äk,sīd }

hydrated alumina *See* alumina trihydrate. { 'hī ˌdrād·əd ə'lü·mə·nə }

hydrated cellulose *See* hydrocellulose. { 'hī ˌdrād·əd 'sel·yə,lōs }

hydrated grease [MATER] Grease made with a soap containing a hydrated alkali. { 'hī,drād·əd 'grēs }

hydrated lime *See* calcium hydroxide. { 'hī ˌdrād·əd 'līm }

hydrated manganic hydroxide *See* manganic hydroxide. { 'hī,drād·əd maŋ'gan·ik hī'dräk,sīd }

hydrated mercurous nitrate [INORG CHEM] $Hg_2(NO_3)_2·2H_2O$ Poisonous, light-sensitive crystals, soluble in warm water, decomposes at 70°C; used as an analytical reagent and in cosmetics and medicine. { 'hī,drād·əd mər 'kyùr·əs 'nī,trāt }

hydrated silica *See* silicic acid. { 'hī,drād·əd 'sil·ə·kə }

hydration [CHEM] The incorporation of molecular water into a complex molecule with the molecules or units of another species; the complex may be held together by relatively weak forces or may exist as a definite compound. { hī'drā·shən }

hydraulic cement [MATER] Cement that hardens in the presence of water. { hī'drȯ·lik si'ment }

hydraulic fluid [MATER] A low-viscosity fluid used in operating a hydraulic mechanism. { hī'drȯ·lik 'flü·əd }

hydraulic lime [MATER] A type of limestone which has been heated and pulverized, and absorbs water without swelling or heating, yielding a cement that hardens under water. { hī'drȯ·lik 'līm }

hydraulic limestone [MATER] Limestone, containing silica and alumina, which produces lime that hardens in water. { hī'drȯ·lik 'līm,stōn }

hydraulic spraying *See* airless spraying. { hi 'drȯ·lik 'sprā·iŋ }

hydrazide [INORG CHEM] An acyl hydrazine; a compound of the formula

$$R-\overset{\overset{\text{O}}{\|}}{C}-NH=NH_2$$

where R may be an alkyl group. { 'hī·drə,zīd }

hydrazine [INORG CHEM] H_2NNH_2 A colorless, hygroscopic liquid, boiling point $114°C$, with an ammonialike odor; it is reducing, decomposable, basic, and bifunctional; used as a rocket fuel, in corrosion inhibition in boilers, in the synthesis of biologically active materials, explosives, antioxidants, and photographic chemicals. { 'hī·drə,zēn }

hydrazoic acid [INORG CHEM] NHN:N Explosive liquid, a strong protoplasmic poison boiling at $37°C$. { hī·drə¦zō·ik 'as·əd }

hydride [INORG CHEM] A compound containing hydrogen and another element; examples are H_2S, which is a hydride although it may be properly called hydrogen sulfide, and lithium hydride, LiH. { 'hī,drīd }

hydride descaling [MET] Removing surface deposits of oxides from a metal by immersion in molten alkali that contains hydrides. { 'hī,drīd dē'skāl·iŋ }

hydriodic acid [INORG CHEM] A yellow liquid that is a water solution of the gas hydrogen iodide; a solution of 59% hydrogen iodide produces a liquid that is constant-boiling; it is a strong acid used in organic synthesis and as a reagent in analytical chemistry. { 'hī·drē,äd·ik 'as·əd }

hydriodic acid gas *See* hydrogen iodide. { 'hī·drē,äd·ik ¦as·əd 'gas }

hydrobromic acid [INORG CHEM] HBr A solution of hydrogen bromide in water, usually 40%; a clear, colorless liquid; used in medicine, analytical chemistry, and synthesis of organic compounds. { ¦hī·drə'brō·mik 'as·əd }

hydrocarbon [ORG CHEM] One of a very large group of chemical compounds composed only of carbon and hydrogen; the largest source of hydrocarbons is from petroleum crude oil. { ¦hī·drə'kär·bən }

hydrocarbon resins [POLYM CHEM] Brittle or gummy materials prepared by the polymerization of several unsaturated constituents of coal tar, rosin, or petroleum; they are inexpensive and find uses in rubber and asphalt formulations and in coating and caulking compositions. { ¦hī·drə'kär·bən 'rez·ənz }

hydrocellulose [MATER] A gelatinous mass formed from the reaction of cellulose with water either by grinding cellulose and mixing with water, or by using strong salt solutions, acids, or alkalies; used in the manufacture of artifical fibers such as rayon, mercerized cotton, paper, and vulcanized fiber. Also known as hydrated cellulose. { ¦hī·drō'sel·yə,lōs }

hydrochloric acid [INORG CHEM] HCl A solution of hydrogen chloride gas in water; a poisonous, pungent liquid forming a constant-boiling mixture at 20% concentration in water; widely used as a reagent, in organic synthesis, in acidizing oil wells, ore reduction, food processing, and metal cleaning and pickling. Also known as muriatic acid. { ¦hī·drə'klȯr·ik 'as·əd }

hydrocolloid *See* gum. { ,hī·drə'käl,ȯid }

hydrocyanic acid [INORG CHEM] HCN A highly toxic liquid that has the odor of bitter almonds and boils at $25.6°C$; used to manufacture cyanide salts, acrylonitrile, and dyes, and as a fumigant in agriculture. Also known as formonitrile; hydrogen cyanide; prussic acid. { ¦hī·drō·sī'an·ik 'as·əd }

hydroelasticity [FL MECH] **1.** Theory of elasticity of a fluid. **2.** The interaction between the flow of water or other liquid and the elastic behavior of a body immersed in it. { ,hī·drō,ē·las'tis·əd·ē }

hydrofluoric acid [INORG CHEM] An aqueous solution of hydrogen fluoride, HF; colorless, fuming, poisonous liquid; extremely corrosive, it is a weak acid as compared to hydrochloric acid, but will attack glass and other silica materials; used to polish, frost, and etch glass, to pickle copper, brass, and alloy steels, to clean stone and brick, to acidize oil wells, and to dissolve ores. { ¦hī·drə'flur·ik 'as·əd }

hydrofluorosilicic acid *See* fluosilicic acid. { ¦hī·drō¦flur·ō·sə'lis·ik 'as·əd }

hydrogel [CHEM] The formation of a colloid in which the disperse phase (colloid) has combined with the continuous phase (water) to produce a viscous jellylike product; for example, coagulated silicic acid. { 'hī·drə,jel }

hydrogen [CHEM] The first chemical element, symbol H, in the periodic table, atomic number 1, atomic weight 1.00797; under ordinary conditions it is a colorless, odorless, tasteless gas composed of diatomic molecules, H_2; used in manufacture of ammonia and methanol, for hydrofining, for desulfurization of petroleum products, and to reduce metallic oxide ores. { 'hī·drə·jən }

hydrogenation [ORG CHEM] Catalytic reaction of hydrogen with other compounds, usually unsaturated; for example, unsaturated oils are hydrogenated to form solid fats. For example, corn, soybean, or cottonseed oils are made into margarine. { 'hī‚dräj·ə'nā·shən }

hydrogen blistering [MET] Cracks or blisters caused when atomic hydrogen penetrates steel via submicroscopic discontinuities or voids and becomes molecular hydrogen and develops internal pressures. { 'hī·drə·jən 'blis·tə·riŋ }

hydrogen bond [PHYS CHEM] A type of bond formed when a hydrogen atom bonded to atom A in one molecule makes an additional bond to atom B either in the same or another molecule; the strongest hydrogen bonds are formed when A and B are highly electronegative atoms, such as fluorine, oxygen, or nitrogen. { 'hī·drə·jən 'bänd }

hydrogen brazing [MET] Brazing in an atmosphere rich in hydrogen. { 'hī·drə·jən 'brāz·iŋ }

hydrogen bromide [INORG CHEM] HBr A hazardous, toxic gas used as a chemical intermediate and as an alkylation catalyst; forms hydrobromic acid in aqueous solution. { 'hī·drə·jən 'brō‚mīd }

hydrogen chloride [INORG CHEM] HCl A fuming, highly toxic, colorless gas soluble in water, alcohol, and ether; used in the production of vinyl chloride and alkyl chlorides, and in polymerization, isomerization, and other reactions. { 'hī·drə·jən 'klōr‚īd }

hydrogen cyanide See hydrocyanic acid. { 'hī·drə·jən 'sī·ə‚nīd }

hydrogen damage [MET] Corrosion, common in boilers, caused by diffusion of hydrogen through steel reacting with carbon to form methane, which builds up local stresses at the interfaces between grains, forming voids that ultimately produce failure. { 'hī·drə·jən ‚dam·ij }

hydrogen disulfide See hydrogen sulfide. { 'hī·drə·jən dī'səl‚fīd }

hydrogen electrode [PHYS CHEM] A noble metal (such as platinum) of large surface area covered with hydrogen gas in a solution of hydrogen ion saturated with hydrogen gas; metal is used in a foil form and is welded to a wire sealed in the bottom of a hollow glass tube, which is partially filled with mercury; used as a standard electrode with a potential of zero to measure hydrogen ion activity. { 'hī·drə·jən i'lek‚trōd }

hydrogen embrittlement See acid brittleness. { 'hī·drə·jən em'brid·əl·mənt }

hydrogen fluoride [INORG CHEM] HF The hydride of fluoride; anhydrous HF is a mobile, colorless, liquid that fumes in air, melts at −83°C, boils at 19.8°C; used to make fluorine-containing refrigerants (such as Freon) and organic fluorocarbon compounds, as a catalyst in alkylate gasoline manufacture, as a fluorinating agent, and in preparation of hydrofluoric acid. { 'hī·drə·jən 'flur‚īd }

hydrogen iodide [INORG CHEM] HI A water-soluble, colorless gas that may be used in organic synthesis and as a reagent. Also known as hydriodic acid gas. { 'hī·drə·jən 'ī·ə‚dīd }

hydrogen ion See hydronium ion. { 'hī·drə·jən 'ī‚än }

hydrogen loss [MET] Loss of weight by a compact or a metal powder when heated in a hydrogen atmosphere; used as a measure of oxygen content of the sample. { 'hī·drə·jən ‚lòs }

hydrogenous [CHEM] Of, pertaining to, or containing hydrogen. { hī'dräj·ə·nəs }

hydrogen overvoltage [MET] An overvoltage occurring at an electrode as a result of the liberation of hydrogen gas. { 'hī·drə·jən ¦ō·vər¦vōl·tij }

hydrogen oxide See water. { 'hī·drə·jən 'äk‚sīd }

hydrogen peroxide [INORG CHEM] H_2O_2 Unstable, colorless liquid, boiling at 150°C; soluble in water and alcohol; used (indifferent concentrations) as a bleach, chemical intermediate, rocket fuel, and antiseptic. Also known as peroxide. { 'hī·drə·jən pə'räk‚sīd }

hydrogen phosphide See phosphine. { 'hī·drə·jən 'fäs‚fīd }

hydrogen-reduced powder [MET] Metal powder produced by hydrogen-reduction of a metal, metallic compound, or surface-contaminated metal particles. { 'hī·drə·jən ri¦düst 'pau̇d·ər }

hydrogen selenide [INORG CHEM] H_2Se Toxic, colorless gas, soluble in water, carbon disulfide, and phosgene; used to make metallic selenides and organoselenium compounds and in the preparation of semiconductor materials. { 'hī·drə·jən 'sel·ə‚nīd }

hydrogen sulfide [INORG CHEM] H_2S Flammable, toxic, colorless gas with offensive odor, boiling at −60°C; soluble in water and alcohol; used as an analytical reagent, as a sulfur source, and for purification of hydrochloric and sulfuric acids. Also known as hydrogen disulfide. { 'hī·drə·jən 'səl‚fīd }

hydrogen tellurate See telluric acid. { 'hī·drə·jən 'tel·yə‚rāt }

hydrokinematics [FL MECH] The study of the motion of a liquid apart from the cause of motion. { 'hī·drə‚kin·ə'mad·iks }

hydrokinetics [FL MECH] The study of the forces produced by a liquid as a consequence of its motion. { 'hī·drə·kə'ned·iks }

hydrometallurgy [MET] Treatment of metals and metal-containing materials by wet processes. { ¦hī·drō'med·əl‚ər·jē }

hydrometry [FL MECH] The science and technology of measuring specific gravities, particularly of liquids. { hī'dräm·ə·trē }

hydronium ion [INORG CHEM] H_3O^+ An oxonium ion consisting of a proton combined with a molecule of water; found in pure water and in all aqueous solutions. Also known as hydrogen ion. { hī'drō·nē·əm ¦ī‚än }

hydrophile-lipophile balance [ORG CHEM] The relative simultaneous attraction of an emulsifier for two phases of an emulsion system; for example, water and oil. { 'hī·drə‚fīl 'lip·ə‚fīl ‚bal·əns }

hydrophilic [CHEM] Having an affinity for, attracting, adsorbing, or absorbing water. { ‚hī·drə'fil·ik }

hydrophobic |CHEM| Lacking an affinity for, repelling, or failing to adsorb or absorb water. { ¦hī·drə¦fō·bik }

hydrosol |CHEM| A colloidal system in which the dispersion medium is water, and the dispersed phase may be a solid, a gas, or another liquid. { 'hī·drə,sól }

hydrostatic balance |MECH| An equal-arm balance in which an object is weighed first in air and then in a beaker of water to determine its specific gravity. { ,hī·drə¦stad·ik i'kwä·zhən }

hydrostatic forging |MET| Forging a metal part by using pressure supplied by a liquid. { ,hī·drə¦stad·ik 'fórj·iŋ }

hydrostatic modulus See bulk modulus of elasticity. { ,hī·drə'stad·ik 'mäj·ə·ləs }

hydrostatic strength |MECH| The ability of a body to withstand hydrostatic stress. { ,hī·drə¦stad·ik 'streŋkth }

hydrostatic stress |MECH| The condition in which there are equal compressive stresses or equal tensile stresses in all directions, and no shear stresses on any plane. { ,hī·drə¦stad·ik 'stres }

hydrotrope |CHEM| Polar organic substances having both hydrogen-bonding capabilities and hydrophobic character; used to increase the solubility of partially soluble compounds in water. { 'hī·drə,trōp }

hydrous |CHEM| Indicating the presence of an indefinite amount of water. { 'hī·drəs }

hydroxide |CHEM| Compound containing the OH⁻ group; the hydroxides of metals are usually bases and those of nonmetals are usually acids; a hydroxide can be organic or inorganic. { hī'dräk ,sīd }

hydroximino See nitroso. { ¦hī,dräk'sim·ə·nō }

hydroxy- |ORG CHEM| Chemical prefix indicating the OH⁻ group in an organic compound, such as hydroxybenzene for phenol, C_6H_5OH; the use of just oxy- for the prefix is incorrect. Also spelled hydroxyl-. { hī'dräk·sē }

hydroxyethylcellulose |MATER| A white powder made from cellulose, used for textile finishes and as a thickener for water-base paints. { hī ¦dräk·sē¦eth·əl'sel·yə,lōs }

hydroxylamine |INORG CHEM| NH_2OH A colorless, crystalline compound produced commercially by acid hydrolysis of nitroparaffins, decomposes on heating, melts at 33°C; used in organic synthesis and as a reducing agent. { ,hī ,dräk'sil·ə,mēn }

hygroscopic |CHEM| **1.** Possessing a marked ability to accelerate the condensation of water vapor; applied to condensation nuclei composed of salts which yield aqueous solutions of a very low equilibrium vapor pressure compared with that of pure water at the same temperature. **2.** Pertaining to a substance whose physical characteristics are appreciably altered by effects of water vapor. **3.** Pertaining to water absorbed by dry soil minerals from the atmosphere; the amounts depend on the physicochemical character of the surfaces, and increase with rising relative humidity. { ¦hī·grə¦skäp·ik }

hypereutectic alloy |MET| Any binary alloy whose composition lies to the right of the eutectic on an equilibrium diagram and which contains some eutectic structure. { ¦hī·pər· yü'tek·tik 'al,ói }

hypereutectoid steel |MET| Steel containing more than 0.8% carbon. { ¦hī·pər·yü'tek,tóid 'stēl }

hyperfine enhanced nuclear cooling |CRYO| A version of adiabatic demagnetization in which a sample containing magnetic ions, embedded in a suitable crystal which quenches the hyperfine fields, is cooled in a moderate external field which reinduces the hyperfine fields, and is then thermally isolated and removed from the external field. { 'hī·pər,fīn in'hanst ¦nü·klē·ər 'kül·iŋ }

hypergolic |CHEM| Capable of igniting spontaneously upon contact. { ¦hī·pər¦gäl·ik }

hypervalent atom |CHEM| A central atom in a single-bonded structure that imparts more than eight valence electrons in forming covalent bonds. { ,hī·pər'vā·lənt 'ad·əm }

hypervalent compounds |CHEM| Stable compounds of the main group elements in the third row of the periodic table, such as silicon, phosphorus, and sulfur, that have more than eight valence shell electrons, for example, PF_5 and SF_4. { ,hī·pər¦vā·lənt 'käm,paúnz }

hypo |GRAPHICS| In photography, the common fixing agent sodium thiosulfate, which formerly was incorrectly called hyposulfate of soda. See sodium thiosulfate. { 'hī·pō }

hypochlorite |INORG CHEM| OC^- The ion derived from hypochlorous acid, $HClO$; it is an oxidizing agent and a constituent of bleaching agents. { ,hī·pə'klór,īt }

hypochlorous acid |INORG CHEM| $HOCl$ Weak, unstable acid existing in solution only; its salts (such as calcium hypochlorite) are used as bleaching agents. { ¦hī·pə'klór·əs 'as·əd }

hypoeutectic alloy |MET| Any binary alloy whose composition lies to the left of the eutectic. { ¦hī·pō·yü'tek·tik 'al,ói }

hypoeutectoid steel |MET| Steel containing less than 0.8% carbon. { ¦hī·pō·yü'tek,tóid 'stēl }

hypohalous acid |INORG CHEM| An oxyacid of a halogen (fluorine, chlorine, bromine, iodine, or astatine) possessing the general chemical formula HOX, where X is the halogen atom. { hī ¦pöt·ə·ləs 'as·əd }

hypoiodous acid |INORG CHEM| HIO A very weak unstable acid that occurs as the result of the weak hydrolysis of iodine in water. { ¦hī·pō ,ī'ōd·əs 'as·əd }

hysteresis |ELECTR| An oscillator effect wherein a given value of an operating parameter may result in multiple values of output power or frequency. |ELECTROMAG| See magnetic hysteresis. |PHYS| The dependence of the state of a system on its previous history, generally in the form of a lagging of a physical effect behind its cause. { ,his·tə'rē·səs }

hysteresis coefficient |PHYS| A constant, characteristic of a particular material, in a formula for hysteresis loss. { ,his·tə'rē·səs ,kō·i'fish·ənt }

hysteresis damping [MECH] Damping of a vibration due to energy lost through mechanical hysteresis. { ‚his·tə'rē·səs 'dam·piŋ }

hysteresis error [PHYS] The maximum separation due to hysteresis between upscale-going and downscale-going indications of a measured variable. { ‚his·tə'rē·səs ‚er·ər }

hysteresis heating [PHYS] **1.** Supply of heat to a material through hysteresis loss. **2.** In particular, supply of a controlled amount of heat to a thermally isolated paramagnetic sample at temperatures below 1 kelvin by taking it through a magnetic hysteresis loop. { ‚his·tə'rē·səs ‚hēd·iŋ }

hysteresis loop [PHYS] The closed curve followed by a material displaying hysteresis (such as a ferromagnet or ferroelectric) on a graph of a driven variable (such as magnetic flux density or electric polarization) versus the driving variable (such as magnetic field or electric field). { ‚his·tə'rē·səs ‚lüp }

hysteresis loss [PHYS] The energy converted to heat in a material because of magnetic or other hysteresis, accompanying cyclic variation of the magnetic field or other driving variable. { ‚his·tə'rē·səs ‚lós }

hysteretic damping [MECH] Damping of a vibrating system in which the retarding force is proportional to the velocity and inversely proportional to the frequency of the vibration. { ‚his·tə'red·ik ¦damp·iŋ }

Hz *See* hertz.

I

I *See* iodine.

IC *See* integrated circuit.

ice |PHYS CHEM| **1.** The dense substance formed by the freezing of water to the solid state; has a melting point of $32°F$ ($0°C$) and commonly occurs in the form of hexagonal crystals. **2.** A layer or mass of frozen water. { īs }

ice line |THERMO| A graph of the freezing point of water as a function of pressure. { 'īs ‚līn }

ice point |PHYS CHEM| The true freezing point of water; the temperature at which a mixture of air-saturated pure water and pure ice may exist in equilibrium at a pressure of 1 standard atmosphere (101,325 pascals). { 'īs ‚point }

ichthyocolla *See* isinglass. { ‚ik·thē·ə'käl·ə }

icositetrahedron *See* trapezohedron. { ī‚kä·sə ‚te·tra'hē·drən }

ideal crystal *See* perfect crystal. { ‚ī‚dēl 'krist·əl }

ideal dielectric |ELEC| Dielectric in which all the energy required to establish an electric field in the dielectric is returned to the source when the field is removed. Also known as perfect dielectric. { ī'dēl ‚dī·i'lek·trik }

ideal fluid |FL MECH| **1.** A fluid which has ideal flow. **2.** *See* inviscid fluid. { ī'dēl 'flü·əd }

ideal gas |THERMO| **1.** Also known as perfect gas. **2.** A gas whose molecules are infinitely small and exert no force on each other. **3.** A gas that obeys Boyle's law (the product of the pressure and volume is constant at constant temperature) and Joule's law (the internal energy is a function of the temperature alone). { ī'dēl 'gas }

ideal gas law |THERMO| The equation of state of an ideal gas which is a good approximation to real gases at sufficiently high temperatures and low pressures; that is, $PV = RT$, where P is the pressure, V is the volume per mole of gas, T is the temperature, and R is the gas constant. { ī'dēl 'gas ‚lô }

ideal radiator *See* blackbody. { ī'dēl 'rād·ē‚ād·ər }

ideal solution |CHEM| A solution that conforms to Raoult's law over all ranges of temperature and concentration and shows no internal energy change on mixing and no attractive force between components. { ī'dēl sə'lü·shən }

ideal transducer |ELEC| Hypothetical passive transducer which transfers the maximum possi-

ble power from the source to the load. { ī'dēl tranz'dü·sər }

ID grinding |MET| The grinding of the inner diameter of a piece of tubing or piping. { ¦ī'de ‚grīn·diŋ }

Igewsky's solution |MET| Etchant used to prepare carbon steels for microscopic analysis; consists of 5% picric acid in absolute alcohol. { ē'gev·skēz sə‚lü·shən }

ignite |CHEM| To start a fuel burning. { ig'nīt }

ignition point *See* ignition temperature. { ig'nish·ən ‚point }

ignition temperature |CHEM| The lowest temperature at which combustion begins and continues in a substance when it is heated in air. Also known as autogenous ignition temperature; ignition point. { ig'nish·ən ‚tem·pra·chər }

illuminance |OPTICS| The density of the luminous flux on a surface. Also known as illumination; luminous flux density. { ə'lü·mə·nəns }

illuminating gas |MATER| Flammable mixture of gases suitable for illuminating purposes; contains hydrogen, methane, ethane, carbon monoxide, and some nitrogen and oxygen. { ə'lü·mə ‚nād·iŋ ‚gas }

illumination |OPTICS| **1.** The science of the application of lighting. **2.** *See* illuminance. { ə‚lü· mə‚nā·shən }

illumination distribution |OPTICS| The manner in which light is dispersed on a surface. { ə‚lü· mə‚nā·shən ‚di·strə'byü·shən }

illuminometer |OPTICS| A portable photometer which is used in the field or outside the laboratory and yields results of lower accuracy than a laboratory photometer. { ə‚lü·mə'näm·əd·ər }

image force |ELEC| The electrostatic force on a charge in the neighborhood of a conductor, which may be thought of as the attraction to the charge's electric image. { 'im·ij ‚fors }

image potential |ELEC| The potential set up by an electric image. { 'im·ij pə‚ten·chəl }

imide |ORG CHEM| **1.** A compound derived from acid anhydrides by replacing the oxygen (O) with the =NH group. **2.** A compound that has either the =NH group or a secondary amine in which R is an acyl functional group, as R_2NH. { 'i‚mīd }

imine |ORG CHEM| A class of compounds that are the product of condensation reactions of aldehydes or ketones with ammonia or amines; they have the NH radical attached to the carbon with the double bond, as R—HC=NH; an example is benzaldimine. { 'i,mēn }

imino compound [ORG CHEM] A compound that has the =NH radical attached to one or two carbon atoms. { 'im·ə,nō ,käm,paùnd }

immersion cleaning [MET] Removing surface dirt from metal by dipping into a cleaning liquid. { ə'mər·zhən ¦klēn·iŋ }

immersion coating [ENG] Applying material to the surface of a metal or ceramic by dipping into a liquid. { ə'mər·zhən ¦kōd·iŋ }

immersion plating [MET] Applying an adherent layer of more-noble metal to the surface of a metal object by dipping in a solution of more-noble metal ions; a replacement reaction. Also known as dip plating; metal replacement. { ə'mər·zhən ,plād·iŋ }

immiscible [CHEM] Pertaining to liquids that will not mix with each other. { i'mis·ə·bəl }

immobilized catalyst [CHEM] A molecular catalyst that is bound without substantial change in its structure to an insoluble solid to prevent solution of the catalyst in the contacting liquid. Also known as anchored catalyst. { i¦mō·bə,līzd 'kad·ə,list }

immunity [MET] The ability of metal to resist corrosion as a result of thermodynamic stability. { i'myü·nəd·ē }

impact [MECH] A forceful collision between two bodies which is sufficient to cause an appreciable change in the momentum of the system on which it acts. Also known as impulsive force. { 'im ,pakt }

impact energy [MECH] The energy necessary to fracture a material. Also known as impact strength. { 'im,pakt ,en·ər·jē }

impact extrusion [MET] A cold extrusion process for producing tubular components by striking a slug of the metal, which has been placed in the cavity of the die, with a punch moving at high velocity. { 'im,pakt ik,strü·zhən }

impact forging [MET] Plastic deformation of a metal using an impactive force. { 'im,pakt ,fórj·iŋ }

impact modifier [MATER] A material added to a substance during manufacture to improve resistance to deformation or breaking. { 'im ,pakt ,mäd·ə,fī·ər }

impact strength [MECH] 1. Ability of a material to resist shock loading. 2. See impact energy. { 'im,pakt ,streŋkth }

impact stress [MECH] Force per unit area imposed on a material by a suddenly applied force. { 'im,pakt ,stres }

impact test [ENG] Determination of the degree of resistance of a material to breaking by impact, under bending, tension, and torsion loads; the energy absorbed is measured in breaking the material by a single blow. { 'im ,pakt ,test }

impedance [ELEC] See electrical impedance. [PHYS] 1. The ratio of a sinusoidally varying quantity to a second quantity which measures the response of a physical system to the first, both being considered in complex notation; examples

are electrical impedance, acoustic impedance, and mechanical impedance. Also known as complex impedance. 2. The ratio of the greatest magnitude of a sinusoidally varying quantity to the greatest magnitude of a second quantity which measures the response of a physical system to the first; equal to the magnitude of the quantity in the first definition. { im'pēd·əns }

impedance drop [ELEC] The total voltage drop across a component or conductor of an alternating-current circuit, equal to the phasor sum of the resistance drop and the reactance drop. { im'pēd·əns ,dräp }

imperfect crystal [CRYSTAL] A crystal in which the regular, periodic structure is interrupted by various defects. { im¦pər,fekt 'krist·əl }

imperfect gas See real gas. { im'pər·fikt 'gas }

imperial red [INORG CHEM] Any of the red varieties of ferric oxide used as pigment. { im'pir·ē·əl 'red }

imperial smelting process [MET] A pyrometallurgical process which treats a complex concentrated feed to a single furnace to recover zinc, copper, lead, cadmium, silver, gold, and other metals in one pass. Abbreviated ISP. { im'pir·ē·əl 'smel·tiŋ ,präs·əs }

impervious carbon [MATER] Carbon compressed with bituminous binder, then carbonized by sintering to produce a dense, impervious material; used as brick to line chemical process and storage vessels. { im'pər·vē·əs 'kär·bən }

impingement attack [MET] Accelerated corrosive attack on a metal by moving liquids, resulting usually from erosion of a protective surface layer. { im'pinj·mənt ə¦tak }

implant [BIOMATER] 1. A quantity of radioactive material in a suitable container, intended to be embedded in a tissue or tumor for therapeutic purposes. 2. A tissue graft placed in depth in the body. { 'im,plant }

implosion [CHEM] The sudden reduction of pressure by chemical reaction or change of state which causes an inrushing of the surrounding medium. { im'plō·zhən }

impregnated timber [MATER] Timber which has been made flame-resistant, fungi-resistant, or insect-proof by forcing into it under vacuum or pressure a flame retardant or a fungal or insect poison. { im'preg,nād·əd 'tim·bər }

impression [MET] A machined cavity in a forging die for production of a specific geometric shape in the workpiece. { im'presh·ən }

impulse [MECH] The integral of a force over an interval of time. [MET] A single pulse or several pulses in welding current used in resistance welding. { 'im,pəls }

impulsive force See impact. { im'pəl·siv 'fórs }

impulsive stimulated thermal scattering [ENG] An optical, noncontacting method for characterizing the high-frequency acoustic behavior of surfaces, thin membrane, coatings, and multilayer assemblies, in which picosecond pulses of light from an excitation laser stimulate motions which are then detected with a continuous-wave probing laser. Abbreviated ISTS. Also known as transient grating photoacoustics. { im¦pəl·siv ¦stim·yə¦lād·əd ¦thərm·əl 'skad·ər·iŋ }

impurity |SOLID STATE| A substance that, when diffused into semiconductor metal in small amounts, either provides free electrons to the metal or accepts electrons from it. { im'pyúr·əd·ē }

impurity band |SOLID STATE| The impurity levels in a semiconductor, occupying a certain range of energies. { im'pyúr·əd·ē ,band }

impurity level |SOLID STATE| An energy level in the band gap of a semiconductor that results from the presence of an impurity atom. { im'pyúr·əd·ē ,lev·əl }

impurity scattering |SOLID STATE| Scattering of electrons by holes or phonons in the crystal. { im'pyúr·əd·ē ¦skad·ə·riŋ }

impurity semiconductor |SOLID STATE| A semiconductor whose properties are due to impurity levels produced by foreign atoms. { im'pyúr·əd·ē ¦sem·i·kən¦dək·tər }

In See indium.

incandescence |OPTICS| The emission of visible radiation by a hot body. { ,in·kən'des·əns }

incandescent lamp |ELEC| An electric lamp that produces light when a metallic filament is heated white-hot in a vacuum by passing an electric current through the filament. Also known as filament lamp; light bulb. { ,in·kən'des·ənt 'lamp }

incident light |OPTICS| The direct light that falls on a surface. { 'in·sə·dənt ¦līt }

inclusion |CRYSTAL| **1.** A crystal or fragment of a crystal found in another crystal. **2.** A small cavity filled with gas or liquid in a crystal. |MET| An impure particle, such as sand, trapped in molten metal during solidification. { in'klü·zhən }

inclusion complex |CHEM| An unbonded association in which the molecules of one component are contained wholly or partially within the crystal lattice of the other component. { in'klü·zhən 'käm,pleks }

inclusion compound See clathrate. { in'klü·zhən 'käm,paúnd }

incoherent light |OPTICS| Electromagnetic radiant energy not all of the same phase, and possibly also consisting of various wavelengths. { ,in·kō'hir·ənt 'līt }

incompressibility |MECH| Quality of a substance which maintains its original volume under increased pressure. { ¦in·kəm,pres·ə'bil·əd·ē }

incremental hysteresis loss |ELECTROMAG| Hysteresis loss when a magnetic material is subjected to a pulsating magnetizing force. { ,iŋ·krə'ment·əl ,his·tə'rē·səs ,lòs }

incremental induction |ELECTROMAG| The quantity lying between the highest and lowest value of a magnetic induction at a point in a polarized material, when subjected to a small cycle of magnetization. { ,iŋ·krə'ment·əl in'dək·shən }

incremental permeability |ELECTROMAG| The ratio of a small cyclic change in magnetic induction to the corresponding cyclic change in magnetizing force when the average magnetic

induction is greater than zero. { ,iŋ·krə'ment·əl ,pər·mē·ə'bil·əd·ē }

indan |ORG CHEM| $C_6H_4(CH_2)_3$ Colorless liquid boiling at 177°C; soluble in alcohol and ether, insoluble in water; derived from coal tar. { 'in ,dan }

indelible ink |MATER| An ink that cannot be removed. { in'del·ə·bəl 'iŋk }

indene |ORG CHEM| C_9H_8 A colorless, liquid, polynuclear hydrocarbon; boils at 181°C and freezes at −2°C; derived from coal tar distillates; copolymers with benzofuran have been manufactured on a small scale for use in coatings and floor coverings. { 'in,dēn }

indentation hardness |MET| The resistance of a metal surface to indention when subjected to pressure by a hard pointed or rounded tool. Also known as penetration hardness. { ,in ,den'tā·shən ¦hard·nəs }

index of refraction |OPTICS| The ratio of the phase velocity of light in a vacuum to that in a specified medium. Also known as absolute index of refraction; absolute refractive constant; refractive constant; refractive index. { 'in,deks əv ri'frak·shən }

india ink |MATER| A permanent black ink made of lampblack and blue binder; some varieties are waterproof. Also known as Chinese ink; sumi ink. { 'in·dē·ə 'iŋk }

Indian gum |MATER| Any of the gums, such as ghatti gum and sterculia gum, with mucilage consistency from trees in the forests in India and Ceylon. { 'in·dē·ən 'gəm }

Indian red |MATER| Iron-oxide-base, maroon pigment; used to polish gold, silver, and other metals. Also known as iron saffron. { 'in·dē·ən 'red }

Indian tragacanth See karaya gum. { 'in·dē·ən 'trag·ə,kanth }

Indian yellow |MATER| A yellow pigment which may be aureolin, made of cobalt and potassium nitrates; or puree, the impure basic magnesium salt of euxanthic acid; or the synthetic dye primuline. { 'in·dē·ən 'yel·ō }

indirect-band-gap semiconductor |SOLID STATE| A semiconductor material in which the state of minimun energy in the conduction band and the state of maximum energy in the valence band have different momenta, and consequently optical transitions between free electrons and holes are forbidden. { ,in·də'rekt 'band ,gap 'sem·i·kən¦dək·tər }

indirect extrusion |MET| An extrusion process in which the billet remains stationary while a hollow die stand forces the die back into the cylinder. { ,in·də'rekt ik'strü·zhən }

indium |CHEM| A metallic element, symbol In, atomic number 49, atomic weight 114.82; soluble in acids; melts at 156°C, boils at 1450°C. |MET| A ductile, silver-white, shiny metal that resists tarnishing and is used in precious-metal alloys for jewelry and dentistry, in glass-sealing alloys, lubricants, and bearing metals, and as an atomic-pile neutron indicator. { 'in·dē·əm }

indium antimonide [INORG CHEM] InSb Crystals that melt at 535°C; an intermetallic compound having semiconductor properties and the highest room-temperature electron mobility of any known material; used in Hall-effect and magnetoresistive devices and as an infrared detector. { 'in·dē·əm ,an'tim·ə,nīd }

indium arsenide [INORG CHEM] InAs Metallic crystals that melt at 943°C; an intermetallic compound having semiconductor properties; used in Hall-effect devices. { 'in·dē·əm 'ärs·ən ,īd }

indium chloride [INORG CHEM] InCl₃ Hygroscopic white powder, soluble in water and alcohol. { 'in·dē·əm 'klȯr,īd }

indium phosphide [INORG CHEM] InP A metallic mass that is brittle and melts at 1070°C; an intermetallic compound having semiconductor properties. { 'in·dē·əm 'fäs,fīd }

indium sulfate [INORG CHEM] In₂(SO₄)₃ Deliquescent, water-soluble, grayish powder; decomposes when heated. { 'in·dē·əm 'səl,fāt }

induced anisotropy [SOLID STATE] A type of uniaxial anisotropy in a magnetic material produced by annealing the magnetic material in a magnetic field. { in'düst ,an·ə'sä·trə·pē }

induced current [ELECTROMAG] A current produced in a conductor by a time-varying magnetic field, as in induction heating. { in'düst 'kə·rənt }

induced dipole [ELEC] An electric dipole produced by application of an electric field. { in'düst 'dī,pōl }

induced electromotive force [ELECTROMAG] An electromotive force resulting from the motion of a conductor through a magnetic field, or from a change in the magnetic flux that threads a conductor. { in'düst i,lek·trə¦mōd·iv 'fȯrs }

induced magnetism [ELECTROMAG] The magnetism acquired by magnetic material while it is in a magnetic field. { in'düst 'mag·nə,tiz·əm }

induced moment [ELEC] The average electric dipole moment per molecule which is produced by the action of an electric field on a dielectric substance. { in'düst 'mō·mənt }

induced potential See induced voltage. { in 'düst pə'ten·chəl }

induced voltage [ELECTROMAG] A voltage produced by electromagnetic or electrostatic induction. Also known as induced potential. { in'düst 'vōl·tij }

inductance [ELECTROMAG] **1.** That property of an electric circuit or of two neighboring circuits whereby an electromotive force is generated (by the process of electromagnetic induction) in one circuit by a change of current in itself or in the other. **2.** Quantitatively, the ratio of the emf (electromotive force) to the rate of change of the current. { in'dək·təns }

induction See electromagnetic induction; electrostatic induction. { in'dək·shən }

induction brazing [MET] A brazing process in which coalescence is produced by heat generated within the work by an induced electric current. { in'dək·shən ,brāz·iŋ }

induction force [PHYS CHEM] A type of van der Waals force resulting from the interaction of the dipole moment of a polar molecule and the induced dipole moment of a nonpolar molecule. Also known as Debye force. { in'dək·shən ,fȯrs }

induction furnace [ENG] An electric furnace in which heat is produced in a metal charge by electromagnetic induction. { in'dək·shən ,fər·nəs }

induction hardening [MET] A quench-hardening technique in which the required elevated temperature is obtained by electromagnetic induction. { in'dək·shən ,härd·ən·iŋ }

induction heating [ENG] Increasing the temperature in a material by induced electric current. Also known as eddy-current heating. { in'dək·shən ¦hēd·iŋ }

induction melting [MET] Converting a solid metal to the molten state in an induction furnace. { in'dək·shən ¦melt·iŋ }

induction soldering [MET] A soldering process in which the metals are heated by an induced electric current. { in'dək·shən ¦säd·ə·riŋ }

induction welding [MET] A process of welding by means of heat generated within the work by induced electric currents. { in'dək·shən ¦weld·iŋ }

inductive charge [ELEC] The charge that exists on an object as a result of its being near another charged object. { in'dək·tiv 'chärj }

inductive effect [PHYS CHEM] In a molecule, a shift of electron density due to the polarization of a bond by a nearby electronegative or electropositive atom. { in'dək·tiv ə'fekt }

industrial glass [MATER] Any glass molded into shapes for product parts, for example, lime glass and lead glass. { in'dəs·trē·əl ¦glas }

inelastic [MECH] Not capable of sustaining a deformation without permanent change in size or shape. { ,in·ə'las·tik }

inelastic buckling [MECH] Sudden increase of deflection or twist in a column when compressive stress reaches the elastic limit but before elastic buckling develops. { ,in·ə'las·tik 'bək·liŋ }

inelastic collision [MECH] A collision in which the total kinetic energy of the colliding particles is not the same after the collision as before it. { ,in·ə'las·tik kə'lizh·ən }

inelastic stress [MECH] A force acting on a solid which produces a deformation such that the original shape and size of the solid are not restored after removal of the force. { ,in·ə'las·tik 'stres }

inequality of Clausius See Clausius inequality. { ,in·i'kwäl·əd·ē əv 'klau·zē·əs }

inert gas See noble gas. { i'nərt 'gas }

inert-gas cutting [MET] Cutting of metal while inert gas flows around the cutting area to prevent oxidation. { i'nərt ¦gas 'kəd·iŋ }

inert gas-shielded arc welding [MET] An arc-welding process in which the weld area is shielded by an inert gas to prevent oxidation. Also known as Heliarc welding. { i'nərt ¦gas ¦shēld·əd 'ärk ,weld·iŋ }

inertia welding [MET] A form of friction welding which utilizes kinetic energy stored in a flywheel system to supply the power required for all of the heating and much of the forging.   { i'nər·shə 'weld·iŋ }

inextensional deformation   [MECH] A bending of a surface that leaves unchanged the length of any line drawn on the surface and the curvature of the surface at each point.   { in,ek'sten·chən·əl ,def·ər'mā·shən }

infiltration   [MET] **1.** Filling the pores of a metal powder compact with metal having a lower melting point.   **2.** Movement of molten metal into the pores of a fiber or foam metal.   { ,in·fil'trā·shən }

inflammability   *See* flammability.   { in,flam·ə'bil·əd·ē }

influence line   [MECH] A graph of the shear, stress, bending moment, or other effect of a movable load on a structural member versus the position of the load.   { 'in,flü·əns ,līn }

infrared brazing   [MET] A brazing process in which coalescence is produced by heat generated by infrared radiation.   { ¦in·frə¦red 'brāz·iŋ }

infrared heating   [ENG] Heating by means of infrared radiation.   { ¦in·frə¦red 'hēd·iŋ }

infrared laser   [PHYS] A laser which emits infrared radiation, especially in the near- and intermediate-infrared regions.   { ¦in·frə¦red 'lā·zər }

infrared optical material   [ELECTROMAG] A material which is transparent to infrared radiation.   { ¦in·frə¦red 'äp·tə·kəl mə,tir·ē·əl }

infrared phosphor   [SOLID STATE] A phosphor which, when exposed to infrared radiation during or even after decay of luminescence resulting from its usual or dominant activator, emits light having the same spectrum as that of the dominant activator; sulfide and selenide phosphors are the most important examples.   { ¦in·frə¦red 'fäs·fər }

infrared radiation   [ELECTROMAG] Electromagnetic radiation whose wavelengths lie in the range from 0.75 or 0.8 micrometer (the long-wavelength limit of visible red light) to 1000 micrometers (the shortest microwaves).   { ¦in·frə ¦red ,rād·ē'ā·shən }

infrared soldering   [MET] Soldering in which infrared radiation furnishes the required heat.   { ¦in·frə¦red 'säd·ə·riŋ }

infrared-transparent material   [MATER] An optical material that transmits infrared radiation; examples include sodium chloride (0.25 to 16 micrometers), cesium iodide (1 to 50 micrometers), and high-density polyethylene (16 to 300 micrometers).   { ¦in·frə¦red tranz¦par·ənt mə ,tir·ē·əl }

in-gate   *See* gate.   { 'in,gāt }

Ingen-Hausz apparatus   [THERMO] An apparatus for comparing the thermal conductivities of different conductors; specimens consisting of long wax-coated rods of equal length are placed with one end in a tank of boiling water covered with a radiation shield, and the lengths along the rods from which the wax melts are compared.   { ¦iŋ·gən 'haüs ,ap·ə,rad·əs }

ingot   [MET] **1.** A solid metal casting suitable for remelting or working.   **2.** A bar of gold or silver.   { 'iŋ·gət }

ingot iron   [MET] Relatively pure iron produced in an open-hearth furnace.   { 'iŋ·gət ,ī·ərn }

inhibiting pigment   [MATER] A paint additive that inhibits or prevents rust and corrosion of metals or the formation of mildew, for example, lead chromate.   { in'hib·əd·iŋ 'pig·mənt }

inhibitor   [CHEM] A substance which is capable of stopping or retarding a chemical reaction; to be technically useful, it must be effective in low concentration.   { in'hib·əd·ər }

in-house scrap   [MET] Metal that has been shaved, cropped, or slit off in various stages of casting and rolling of ingots into sheets. Also known as runaround scrap.   { ¦in,haüs 'skrap }

initial permeability   [ELECTROMAG] The limit of the normal permeability as the magnetic induction and magnetic field strength approach 0.   { i¦nish·əl ,pər·mē·ə'bil·əd·ē }

initial set   [MATER] The onset of hardening after water has been added to concrete, cement, or plaster.   { i'nish·əl 'set }

initiating agent   [MATER] An explosive material which has the necessary sensitivity to heat, friction, or percussion to make it suitable for use as the initial element in an explosive train.   { i'nish·ē,ād·iŋ ,ā·jənt }

initiation step   [CHEM] The reaction that causes a chain reaction to begin but is not itself the principal source of products.   { i,nish·ē'ā·shən ,step }

initiator   [CHEM] The substance or molecule (other than reactant) that initiates a chain reaction, as in polymerization; an example is acetyl peroxide.   { i'nish·ē,ād·ər }

ink   [MATER] A dispersion of a pigment or a solution of a dye in a carrier vehicle, yielding a fluid, paste, or powder to be applied to and dried on a substrate; writing, marking, drawing, and printing inks are applied by several methods to paper, metal, plastic, wood, glass, fabric, or other substrate.   { 'iŋk }

inkometer   [ENG] An instrument for measuring adhesion of liquids by rotating drums in contact with the liquid.   { iŋ'käm·əd·ər }

inlay cladding   [MET] A mechanical process in which a groove, ¹⁄₁₀₀–¹⁄₈ inch (1.778–3.175 millimeters) wide, is cut into a base metal and filled with cladding metal; mechanical bonding of the metals is accomplished by passing them through the pressure rolls of a bonding mill.   { 'in,lā 'klad·iŋ }

inner potential   [SOLID STATE] The average value of the electrostatic potential, taken over the volume of a crystal.   { ¦in·ər pə'ten·chəl }

inoculant   [MET] A substance which augments a melt, usually in the latter part of the melting operation, thus altering the solidification structure of the cast metal, as in grain refinement of aluminum alloys.   { i'näk·yə·lənt }

inoculation   [MET] Treating a molten material with another material before casting in order to nucleate crystals.   { i,näk·yə'lā·shən }

inorganic   [INORG CHEM] Pertaining to or composed of chemical compounds that do not contain carbon as the principal element (excepting carbonates, cyanides, and cyanates), that is, matter other than plant or animal. { ¦in·ȯr ¦gan·ik }

inorganic acid [INORG CHEM] A compound composed of hydrogen and a nonmetal element or radical; examples are hydrochloric acid, HCl, sulfuric acid, H_2SO_4, and carbonic acid, H_2CO_3. { ¦in·ȯr¦gan·ik 'as·əd }

inorganic peroxide [INORG CHEM] An inorganic compound containing an element at its highest state of oxidation (such as perchloric acid, $HClO_4$), or having the peroxy group, $-O-O-$ (such as perchromic acid, $H_3CrO_8 \cdot 2H_2O$). { ¦in·ȯr¦gan·ik pə'räk,sīd }

inorganic pigment [INORG CHEM] A natural or synthetic metal oxide, sulfide, or other salt used as a coloring agent for paints, plastics, and inks. { ¦in·ȯr¦gan·ik 'pig·mənt }

inorganic polymer [POLYM CHEM] A macromolecule tht contains metals, metalloids, or other elements; the inorganic element may be ligand-coordinated, or the inorganic element may be pendant to the polymer chain. { ¦in·ȯr ¦gan·ik 'päl·ə·mər }

inquartation [MET] A step in bullion assay that uses nitric acid to dissolve silver from associated gold. Also known as quartation. { ,in ,kwȯr'tā·shən }

insert [MET] **1.** The part of a die or mold that can be removed. **2.** A part, usually metal, which is placed in a mold and appears as an integral part of the final casting. { 'in,sərt }

insol See insoluble. { 'in,säl }

insoluble [CHEM] Incapable of being dissolved in another material; usually refers to solid-liquid or liquid-liquid systems. Abbreviated insol. { in'säl·yə·bəl }

insoluble anode [CHEM] An anode that resists dissolution during electrolysis. { in'säl·yə·bəl 'an,ōd }

instantaneous recovery [MECH] The immediate reduction in the strain of a solid when a stress is removed or reduced, in contrast to creep recovery. { ¦in·stən¦tā·nē·əs ri'kəv·ə·rē }

instantaneous strain [MECH] The immediate deformation of a solid upon initial application of a stress, in contrast to creep strain. { ¦in·stən ¦tā·nē·əs 'strān }

instrumental analysis [ENG] The use of an instrument to measure a component, to detect the completion of a quantitative reaction, or to detect a change in the properties of a system. { ,in·strə'ment·əl ə'nal·ə·səs }

instrument oil [MATER] Special grade of lubricating oil that has been refined to have oxidation resistance and gum resistance, that has compatibility with electrical insulation, and that prevents tarnish or oxidation of contacted metal surfaces; used to lubricate instruments and other intricate mechanisms. { 'in·strə·mənt ,ȯil }

insulated [ELEC] Separated from other conducting surfaces by a nonconducting material. { 'in·sə,lād·əd }

insulating board [MATER] Any board used in a

wall or ceiling to provide insulation. { 'in·sə ,lād·iŋ ,bȯrd }

insulating compound [MATER] A liquid, at low temperatures, which is poured into joint boxes and allowed to solidify; as a poor conductor of heat and electricity, it provides good insulation. { 'in·sə,lād·iŋ ,käm,paünd }

insulating concrete [MATER] Concrete with insulating properties, often made with asbestos fibers and in the form of blocks, corrugated slabs, or sheathing. { 'in·sə,lād·iŋ ¦käŋ,krēt }

insulating oil [MATER] A chlorinated hydrocarbon, such as trichlorobenzene, mixed with fluorinated hydrocarbons, whose high dielectric strength and high flash point allow it to be used in switches, circuit breakers, and transformers as an insulator and cooling medium. Also known as electrical oil. { 'in·sə,lād·iŋ ,ȯil }

insulating tape [MATER] Tape impregnated with insulating material, and usually adhesive; used to cover joints in insulated wires or cables. Also known as electrical tape. { 'in·sə,lād·iŋ ,tāp }

insulation porcelain [MATER] Any of the various insulating materials consisting of molded silica, molded steatite, or specially compounded ceramics, often containing zirconia or beryllia. Also known as electrical porcelain. { ,in·sə'lā·shən ¦pȯr·slən }

insulation resistance [ELEC] The electrical resistance between two conductors separated by an insulating material. { ,in·sə'lā·shən ri¦zis·təns }

insulator [ELEC] A device having high electrical resistance and used for supporting or separating conductors to prevent undesired flow of current from them to other objects. Also known as electrical insulator. [MATER] A material that is a poor conductor of heat, sound, or electricity. [SOLID STATE] A substance in which the normal energy band is full and is separated from the first excitation band by a forbidden band that can be penetrated only by an electron having an energy of several electronvolts, sufficient to disrupt the substance. { 'in·sə,lād·ər }

integrated circuit [ELECTR] An interconnected array of active and passive elements integrated with a single semiconductor substrate or deposited on the substrate by a continuous series of compatible processes, and capable of performing at least one complete electronic circuit function. Abbreviated IC. { 'int·ə,grād·əd 'sər·kət }

intensity [PHYS] **1.** The strength or amount of a quantity, as of electric field, current, magnetization, radiation, or radioactivity. **2.** The power transmitted by a light or sound wave across a unit area perpendicular to the wave. { in'ten·səd·ē }

intensive properties [CHEM] Properties independent of the quantity or shape of the substance under consideration; for example, temperature, pressure, or composition. { in 'ten·siv 'präp·ərd·ēz }

intercalated graphite [MATER] An electrically conductive material made by impregnating graphite fiber or powder with metal-rich compounds that lodge between the stacked layers of the graphite. { ,in·tər¦ka,lād·əd 'gra ,fīt }

intercept |CRYSTAL| One of the distances cut off a crystal's reference axis by planes. { ¦in·tər ¦sept }

intercept method |MET| A method for determining grain size or the quantity of a phase in a microstructure by measuring the number of grains or phase particles per unit length intersected by straight lines. { 'in·tər,sept ,meth·əd }

intercommunicating porosity |MET| The type of porosity in a sintered metal powder compact that allows fluid to pass from pore to pore. { ¦in·tər·kə'myü·nə,kād·iŋ pə'räs·əd·āe }

interconnection |ELEC| A link between power systems enabling them to draw on one another's reserves in time of need and to take advantage of energy cost differentials resulting from such factors as load diversity, seasonal conditions, time-zone differences, and shared investment in larger generating units. { ¦in·tər·kə'nek·shən }

intercrystalline corrosion |MET| Localized attack occurring along the crystal boundaries of a metal or alloy. Also known as intergranular corrosion. { ¦in·tər'krist·əl·ən kə'rō·zhən }

interdendritic attack See interdendritic corrosion. { ¦in·tər,den'drid·ik ə'tak }

interdendritic corrosion |MET| Preferential corrosion of the metal immediately surrounding dendrites in unworked or slightly worked alloys caused by composition gradients. Also known as interdendritic attack. { ¦in·tər,den'drid·ik kə'rō·zhən }

interface |PHYS CHEM| The boundary between any two phases: among the three phases (gas, liquid, and solid), there are five types of interfaces: gas-liquid, gas-solid, liquid-liquid, liquid-solid, and solid-solid. { 'in·tər,fās }

interface resistance |THERMO| **1.** Impairment of heat flow caused by the imperfect contact between two materials at an interface. **2.** Quantitatively, the temperature difference across the interface divided by the heat flux through it. { 'in·tər,fās ri'zis·təns }

interfacial angle |CRYSTAL| The angle between two crystal faces. { 'in·tər,fā·shəl ¦aŋ·gəl }

interfacial energy |PHYS| The free energy of the surfaces at an interface, resulting from differences in the tendencies of each phase to attract its own molecules; equal to the surface tension. Also known as surface energy. { 'in·tər,fā·shəl ¦en·ər·jē }

interfacial force See surface tension. { 'in·tər ,fā·shəl ¦fórs }

interfacial layer |PHYS CHEM| A one- or two-molecules-thick boundary between any two bulk phases (gas, liquid, or solid) in contact where the properties differ from the properties of the bulk phases. { ,in·tər¦fā·shəl 'lā·ər }

interfacial polarization See space-charge polarization. { 'in·tər,fā·shəl ,pō·lə·rə'zā·shən }

interfacial tension See surface tension. { 'in·tər ,fā·shəl 'ten·chən }

interference |PHYS| The variation with distance or time of the amplitude of a wave which results from the superposition (algebraic or vector addition) of two or more waves having the same, or nearly the same, frequency. { ,in·tər'fir·əns }

interference colors |OPTICS| Colors formed by interference of a beam of light passed through a thin section of a mineral placed in a polarizing microscope. { ,in·tər'fir·əns ,kəl·ərz }

interference pattern |PHYS| Resulting space distribution of pressure, particle density, particle velocity, energy density, or energy flux when progressive waves of the same frequency and kind are superimposed. { ,in·tər'fir·əns ¦pad·ərn }

intergranular corrosion See intercrystalline corrosion. { ¦in·tər'gran·yə·lər kə'rō·zhən }

intergranular fracture |MET| Propagation of a crack along the grain boundaries of a metal or alloy. { ¦in·tər'gran·yə·lər 'frak·chər }

interhalogen |INORG CHEM| Any of the compounds formed from the elements of the halogen family that react with each other to form a series of binary compounds; for example, iodine monofluoride. { ¦in·tər'hal·ə·jən }

intermediate |CHEM| A precursor to a desired product; ethylene is an intermediate for polyethylene, and ethane is an intermediate for ethylene. { ,in·tər'mēd·ē·ət }

intermediate annealing |MET| Softening of a metal by heat treatment at one or more stages during cold working and before final treatment. { ,in·tər'mēd·ē·ət ə'nēl·iŋ }

intermediate flux |MET| A flux consisting of organic halide compounds whose residues are decomposed by the heat of soldering; fluxing action approaches that of corrosive flux. { ,in·tər'mēd·ē·ət 'fləks }

intermediate phase |MET| In an alloy system, a distinct phase whose composition ranges do not extend to any of the pure constituents of the system. { ,in·tər'mēd·ē·ət ¦fāz }

intermediate state |CRYO| A state of partial superconductivity that occurs when a magnetic field slightly less than the critical field is applied to a superconducting material below its critical temperature. { ,in·tər'mēd·ē·ət ¦stāt }

intermetallic alloys |MET| Ordered alloys having a superlattice crystal structure. Unlike conventional alloys, they have a strong chemical arrangement that reduces the mobility of atoms and results in good structural stability, higher melting temperatures, and lower densities. However, most are brittle due to their complex crystal structure, resulting in poor fracture resistance. { ,in·tər·me'tal·ik 'a,lóis }

intermetallic compound See electron compound. { ¦in·tər·me'tal·ik 'kam,paùnd }

intermittent weld |MET| A weld in which the continuity is broken by recurring unwelded spaces. { ¦in·tər¦mit·ənt 'weld }

intermolecular force |PHYS CHEM| The force between two molecules; it is that negative gradient of the potential energy between the interacting molecules, if energy is a function of the distance between the centers of the molecules. { ,in·tər·mə'lek·yə·lər 'fōrs }

internal dielectric field See dielectric field. { in'tərn·əl ˌdī·ə'lek·trik 'fēld }

internal energy [THERMO] A characteristic property of the state of a thermodynamic system, introduced in the first law of thermodynamics; it includes intrinsic energies of individual molecules, kinetic energies of internal motions, and contributions from interactions between molecules, but excludes the potential or kinetic energy of the system as a whole; it is sometimes erroneously referred to as heat energy. { in'tərn·əl 'en·ər·jē }

internal force [MECH] A force exerted by one part of a system on another. { in'tərn·əl 'fōrs }

internal friction [FL MECH] See viscosity. [MECH] **1.** Conversion of mechanical strain energy to heat within a material subjected to fluctuating stress. **2.** In a powder, the friction that is developed by the particles sliding over each other; it is greater than the friction of the mass of solid that comprises the individual particles. { in'tərn·əl 'frik·shən }

internal oxidation [MET] The subsurface oxidation of components of an alloy due to oxygen diffusion into the metal. { in'tərn·əl ˌäk·sə'dā·shən }

internal phase See disperse phase. { in'tərn·əl 'fāz }

internal photoelectric effect [SOLID STATE] A process in which the absorption of a photon in a semiconductor results in the excitation of an electron from the valence band to the conduction band. { in'tərn·əl ˌfōd·ō·ə'lek·trik i,fekt }

internal reflection [OPTICS] The reflection of electromagnetic radiation in a given medium from the boundary with a less dense medium. { inˌtərn·əl ri'flek·shən }

internal stress [MECH] A stress system within a solid that is not dependent on external forces. Also known as residual stress. { in'tərn·əl 'stres }

internal work [THERMO] The work done in separating the particles composing a system against their forces of mutual attraction. { in'tərn·əl 'wərk }

international ampere [ELEC] The current that, when flowing through a solution of silver nitrate in water, deposits silver at a rate of 0.001118 gram per second; it has been superseded by the ampere as a unit of current, and is equal to approximately 0.999850 ampere. { ¦in·tər¦nash·ən·əl 'am,pir }

international candle [OPTICS] A unit of luminous intensity, now replaced by the candela.o { ¦in·tər¦nash·ən·əl 'kand·əl }

international ohm [ELEC] A unit of resistance, equal to that of a column of mercury of uniform cross section that has a length of 160.3 centimeters and a mass of 14.4521 grams at the temperature of melting ice; it has been superseded by the ohm, and is equal to 1.00049 ohms. { ¦in·tər ¦nash·ən·əl 'ōm }

international practical temperature scale [THERMO] Temperature scale based on six points: the water triple point, the boiling points of oxygen, water, sulfur, and the solidification points of silver and gold; designated as °C, degrees Celsius, or t_{int}; replaced in 1990 by the international temperature scale. { ¦in·tər ¦nash·ən·əl ¦prak·tə·kəl 'tem·prə·chər ,skāl }

international standard annealed copper [MET] An annealed pure copper having a resistivity of 1.7241 microhm-centimeter at 20°C, which is taken as 100% conductivity. { ¦in·tər¦nash·ən·əl ¦stan·dərd ə¦nēld 'käp·ər }

international system of electrical units [ELEC] System of electrical units based on agreed fundamental units for the ohm, ampere, centimeter, and second, in use between 1893 and 1947, inclusive; in 1948, the Giorgi, or meter-kilogram-second-absolute system, was adopted for international use. { ¦in·tər¦nash·ən·əl ¦sistəm əv i¦lek·tra·kəl 'yü·nəts }

international table British thermal unit See British thermal unit. { ¦in·tər¦nash·ən·əl ¦tā·bəl ¦brid·ish 'thər·məl ,yü·nət }

international table calorie See calorie. { ¦in·tər ¦nash·ən·əl ¦tā·bəl 'kal·ə·rē }

international temperature scale [THERMO] A standard temperature scale, adopted in 1990, that approximates the thermodynamic scale, based on assigned temperature values of 17 thermodynamic equilibrium fixed points and prescribed thermometers for interpolation between them. Abbreviated ITS-90. { ¦in·tər¦nash·ən·əl 'tem·prə·chər ,skāl }

international volt [ELEC] A unit of potential difference or electromotive force, equal to 1/1.01858 of the electromotive force of a Weston cell at 20°C; it has been superseded by the volt, and is equal to 1.00034 volts. { ¦in·tər¦nash·ən·əl 'vōlt }

interpass temperature [MET] In a multiple-pass weld, the lowest temperature of the deposited weld metal before the next run is started. { 'in·tər,pas ,tem·prə·chər }

interpenetrating polymer network [POLYM CHEM] Two or more polymer components, each of which is a crosslinked three-dimensional network, one of which is formed (crosslinked) in the presence of the other. The polymer networks are physically entangled with, but not covalently bonded to, each other. Characteristically, these networks do not dissolve in solvent or flow when heated. Abbreviated IPN. { ,in·tər,pen·ə,trād·iŋ ,päl·ə·mər 'net,wərk }

interpenetration twin [CRYSTAL] Two or more individual crystals so twinned that they appear to have grown through one another. Also known as penetration twin. { ¦in·tər,pen·ə'trā·shən ¦twin }

interphase [CHEM] A region between the two phases of a newly created interface that contains particles of both phases. { 'in·tər,fāz }

interplanar spacing [CRYSTAL] The perpendicular distance between successive parallel planes of atoms in a crystal. { ¦in·tər,plā·nər 'spās·iŋ }

interpolymer [POLYM CHEM] A mixed polymer made from two or more starting materials. { ¦in·tər'päl·ə·mər }

interpulse time [MET] In resistance welding, the

time between successive pulses of an impulse. { 'in·tər,pəls ,tīm }

interrupted aging |MET| A technique for aging material in several steps; the material is brought to room temperature after each step. { 'int·ə ,rəp·təd 'āj·iŋ }

interrupted quenching |MET| Quenching in which a material is intermittently removed from the quenching medium while it is still at a higher temperature than the medium. { 'int·ə,rəp·təd 'kwench·iŋ }

interstice |SOLID STATE| A space or volume between atoms of a lattice, or between groups of atoms or grains of a solid structure. { in'tərs·təs }

interstitial |CRYSTAL| A crystal defect in which an atom occupies a position between the regular lattice positions of a crystal. { ¦in·tər¦stish·əl }

interstitial atom |CRYSTAL| A displaced atom which is forced into a nonequilibrium site within a crystal lattice. { ¦in·tər¦stish·əl 'ad·əm }

interstitial compound |CHEM| A compound of a transition metal and hydrogen, boron, carbon, or nitrogen whose crystals have a close-packed structure of the metal ions, with the nonmetal atoms being located in the interstices. |SOLID STATE| A binary compound in which atoms of one element (usually a light, nonmetallic element) occupy spaces between atoms of the crystal lattice formed by the other element (usually a heavy, metallic element). { ¦in·tər¦stish·əl 'käm,paúnd }

interstitial-free steel |MET| An aluminum-killed steel with an extra-low carbon content, nominally 0.005%, in which the residual carbon is combined with niobium (columbium), titanium, or some similar element with a strong affinity for carbon. { ,int·ər'stish·əl 'frē 'stēl }

interstitial impurity |SOLID STATE| An atom which is not normally found in a solid, and which is located at a position in the lattice structure where atoms or ions normally do not exist. { ¦in·tər¦stish·əl im'pyúr·əd·ē }

intracrystalline See transcrystalline. { ¦in·trə 'krist·əl·ən }

intrinsic conductivity |SOLID STATE| The conductivity of a semiconductor or metal in which impurities and structural defects are absent or have a very low concentration. { in'trin·sik ,kän ,dək'tiv·əd·ē }

intrinsic contact potential difference |ELEC| True potential difference between two perfectly clean metals in contact. { in'trin·sik ¦kän,takt pə¦ten·chəl 'dif·ərns }

intrinsic electric strength |ELEC| The extremely high dielectric strength displayed by a substance at low temperatures. { in¦trin·sik i¦lek·trik ,streŋkth }

intrinsic layer |ELECTR| A layer of semiconductor material whose properties are essentially those of the pure undoped material. { in'trin·sik 'lā·ər }

intrinsic mobility |SOLID STATE| The mobility of the electrons in an intrinsic semiconductor. { in'trin·sik mō'bil·əd·ē }

intrinsic photoconductivity |SOLID STATE| Photoconductivity associated with excitation of charge carriers across the band gap of a material. { in'trin·sik ¦fōd·ō,kän,dək'tiv·əd·ē }

intrinsic photoemission |SOLID STATE| Photoemission which can occur in an ideally pure and perfect crystal, in contrast to other types of photoemission which are associated with crystal defects. { in'trin·sik ¦fōd·ō·i'mish·ən }

intrinsic property |SOLID STATE| A property of a substance that is not seriously affected by impurities or imperfections in the crystal structure. { in'trin·sik 'präp·ərd·ē }

intrinsic semiconductor |SOLID STATE| A semiconductor in which the concentration of charge carriers is characteristic of the material itself rather than of the content of impurities and structural defects of the crystal. Also known as *i*-type semiconductor. { in'trin·sik ¦sem·i·kən ¦dək·tər }

intrinsic temperature range |SOLID STATE| In a semiconductor, the temperature range in which its electrical properties are essentially not modified by impurities or imperfections within the crystal. { in'trin·sik 'tem·prə·chər ,rānj }

intrinsic viscosity |PHYS CHEM| The ratio of a solution's specific viscosity to the concentration of the solute, extrapolated to zero concentration. Also known as limiting viscosity number. { in'trin·sik vi'skäs·əd·ē }

intumescence |MATER| The property of a material to swell when heated; intumescent materials in bulk and sheet form are used as fireproofing agents. { ,in·tü'mes·əns }

in vacuo |PHYS| In a vacuum. { in 'vak·yə·wō }

invar |MATER| An alloy (64% iron–36% nickel) that exhibits almost no thermal expansion over the temperature range of -50 to $150°C$ (-58 to $302°F$). { 'in,vär }

inverse magnetostriction See magnetic tension effect. { ¦in,vərs mag,ned·ō'strik·shən }

inverse micelle See inverted micelle. { 'in,vərs mī'sel }

inverse piezoelectric effect |SOLID STATE| The contraction or expansion of a piezoelectric crystal under the influence of an electric field, as in crystal headphones; also occurs at *pn* junctions in some semiconductor materials. { 'in,vərs pē ¦ā·zō·i¦lek·trik i,fekt }

inverse segregation |MET| Segregation in a cast metal in which an excess of lower-melting metal occurs in the earlier-freezing portions because liquid metal enters cavities developed in the earlier-solidified metal. { 'in,vərs ,seg·rə'gā·shən }

inversion |CRYSTAL| A change from one crystal polymorph to another. Also known as transformation. |OPTICS| The formation of an inverted image by an optical system. |SOLID STATE| The production of a layer at the surface of a semiconductor which is of opposite type from that of the bulk of the semiconductor, usually as the result of an applied electric field. |THERMO| A reversal of the usual direction of a variation or process, such as the change in sign of the expansion coefficient of water at 4°C, or a change in sign in the Joule-Thomson coefficient at a certain temperature. { in'vər·zhən }

inversion axis See rotation-inversion axis. { in'vər·zhən ,ak·səs }

inversion temperature [ENG] The temperature to which one junction of a thermocouple must be raised in order to make the thermoelectric electromotive force in the circuit equal to zero, when the other junction of the thermocouple is held at a constant low temperature. [THERMO] The temperature at which the Joule-Thomson effect of a gas changes sign. { in'vər·zhən ,tem·prə·chər }

inverted micelle [PHYS CHEM] An aggregate of colloidal dimension in which the polar groups are concentrated in the interior and the lipophilic groups extend outward into the solvent. Also known as inverse micelle. { in¦vərd·əd mī'sel }

investment casting [MET] A casting method designed to achieve high dimensional accuracy for small castings by making a mold of refractory slurry, which sets at room temperature, surrounding a wax pattern which is then melted out to leave a mold without joints. { in'ves·mənt ,kast·iŋ }

investment compound [MATER] A mixture containing a refractory filler, a binder, and a liquid vehicle which is used to make molds for investment casting. { in'ves·mənt ,käm,paùnd }

investment process See lost-wax process. { in'ves·mənt ,präs·əs }

inviscid fluid [FL MECH] A fluid which has no viscosity; it therefore can support no shearing stress, and flows without energy dissipation. Also known as ideal fluid; nonviscous fluid; perfect fluid. { in'vis·əd 'flü·əd }

iodate [INORG CHEM] A salt of iodic acid containing the IO_3^- radical; sodium and potassium iodates are the most important salts and are used in medicine. { 'ī·ə,dāt }

iodic acid [INORG CHEM] HIO_3 Water-soluble, moderately strong acid; colorless or white powder or crystals; decomposes at 110°C; used in analytical chemistry and medicine. { ī'äd·ik 'as·əd }

iodic acid anhydride See iodine pentoxide. { ī'äd·ik 'as·əd an'hī,drīd }

iodide [CHEM] **1.** A compound which contains the iodine atom in the −1 oxidation state and which may be considered to be derived from hydriodic acid (HI); examples are KI and NaI. **2.** A compound of iodine, such as CH_3CH_2I, in which the iodine has combined with a more electropositive group. { 'ī·ə,dīd }

iodide process [MET] A refining process in which a metal, such as titanium or zirconium, is combined with iodine vapor and then the iodide volatilized and decomposed at high temperatures to yield a pure solid metal. { ī·ə,dīd ,prä·səs }

iodine [CHEM] A nonmetallic halogen element, symbol I, atomic number 53, atomic weight 126.9045; melts at 114°C, boils at 184°C; the

poisonous, corrosive, dark plates or granules are readily sublimed; insoluble in water, soluble in common solvents; used as germicide and antiseptic, in dyes, tinctures, and pharmaceuticals, in engraving lithography, and as a catalyst and analytical reagent. { 'ī·ə,dīn }

iodine bisulfide See sulfur iodine. { 'ī·ə,dīn bī'səl,fīd }

iodine cyanide [INORG CHEM] ICN Poisonous, colorless needles with pungent aroma and acrid taste; melts at 147°C; soluble in water, alcohol, and ether; used in taxidermy as a preservative. Also known as cyanogen iodide. { 'ī·ə,dīn 'sī·ə ,nīd }

iodine disulfide See sulfur iodine. { 'ī·ə,dīn dī'səl,fīd }

iodine pentoxide [INORG CHEM] I_2O_5 White crystals, decomposing at 275°C, very soluble in water, insoluble in absolute alcohol, ether, and chloroform; used as an oxidizing agent to oxidize carbon monoxide to dioxide at ordinary temperatures, and in organic synthesis. Also known as iodic acid anhydride. { 'ī·ə,dīn pen'täk,sīd }

iodonium [INORG CHEM] A halonium ion such as H_2I^+ or R_2I^+, it may be open-chain or cyclic. { ī·ə'dōn·ē·əm }

Ioffe effect [SOLID STATE] An effect in which the simultaneous exposure of an ionic crystal such as rock salt to a mechanical stress and a solvent results in an increase in its plasticity. { 'yäf·ē i ,fekt }

ion [CHEM] An isolated electron or positron or an atom or molecule which by loss or gain of one or more electrons has acquired a net electric charge. { 'ī,än }

ion backscattering [SOLID STATE] Large-angle elastic scattering of monoenergetic ions in a beam directed at a metallized film on silicon or some other thin multilayer system. { 'ī,än 'bak ,skad·ə·riŋ }

ion-beam thinning See ion machining. { 'ī,än ,bēm ¦thin·iŋ }

ion concentration See ion density. { 'ī,än ,kän· sən'trā·shən }

ion current [PHYS] The electric current resulting from motion of ions. { 'ī,än ,kə·rənt }

ion density [PHYS] The number of ions per unit volume. Also known as ion concentration. { 'ī ,än ,den·səd·ē }

ion exchange [PHYS CHEM] A chemical reaction in which mobile hydrated ions of a solid are exchanged, equivalent for equivalent, for ions of like charge in solution; the solid has an open, fishnetlike structure, and the mobile ions neutralize the charged, or potentially charged, groups attached to the solid matrix; the solid matrix is termed the ion exchanger. { 'ī,än iks ,chānj }

ion-exchange resin [MATER] A synthetic resin that can combine or exchange ions with a solution; such a resin produces the exchange of sodium for calcium ions in the softening of hard water. { 'ī,än iks,chānj ,rez·ən }

ionic bond [PHYS CHEM] A type of chemical

bonding in which one or more electrons are transferred completely from one atom to another, thus converting the neutral atoms into electrically charged ions; these ions are approximately spherical and attract one another because of their opposite charge. Also known as electrovalent bond. { ī'än·ik 'bänd }

ionic charge [PHYS] **1.** The total charge of an ion. **2.** The charge of an electron; the charge of any ion is equal to this electron charge in magnitude, or is an integral multiple of it. { ī'än·ik 'chärj }

ionic conductance [PHYS CHEM] The contribution of a given type of ion to the total equivalent conductance in the limit of infinite dilution. { ī'än·ik kən'dək·təns }

ionic conduction [SOLID STATE] Electrical conduction of a solid due to the displacement of ions within the crystal lattice. { ī'än·ik kən'dək·shən }

ionic conductivity [SOLID STATE] The portion of the electrical conductivity of a solid that results from ionic conduction. { i'än·ik ‚kän ‚dək'tiv·əd·ē }

ionic crystal [CRYSTAL] A crystal in which the lattice-site occupants are charged ions held together primarily by their electrostatic interaction. { ī'än·ik 'krist·əl }

ionic gel [CHEM] A gel with ionic groups attached to the structure of the gel; the groups cannot diffuse out into the surrounding solution. { ī'än·ik 'jel }

ionicity [CHEM] The ionic character of a solid. { ‚ī·ə'nis·əd·ē }

ionic lattice [CRYSTAL] The lattice of an ionic crystal. { ī'än·ik 'lad·əs }

ionic mobility [PHYS] The ratio of the average drift velocity of an ion in a liquid or gas to the electric field. { ī'än·ik mō'bil·əd·ē }

ionic polymerization [POLYM CHEM] Polymerization that proceeds via ionic intermediates (carbonium ions or carbanions) rather than through neutral species or free radicals. { ī'än·ik pə‚lim·ə·rə'zā·shən }

ionic semiconductor [SOLID STATE] A solid whose electrical conductivity is due primarily to the movement of ions rather than that of electrons and holes. { ī'än·ik ‚sem·i·kən ¦dək·tər }

ionic solid [SOLID STATE] A solid made up of ions held together primarily by their electrostatic interaction. { i'än·ik 'säl·əd }

ion implantation [ENG] A process of introducing impurities into the near-surface regions of solids by directing a beam of ions at the solid. { 'ī‚än ‚im‚plan'tā·shən }

ionization [CHEM] A process by which an electron is removed from an atom, molecule, or ion. { ‚ī·ə·nə'zā·shən }

ionized atom [CHEM] An atom with an excess or deficiency of electrons, so that it has a net charge. { 'ī·ə‚nīzd 'ad·əm }

ion machining [ENG] Use of a high-velocity ion beam to remove material from a surface. Also known as ion-beam thinning, ion milling. { 'ī‚än mə'shēn·iŋ }

ion migration [ELEC] Movement of ions produced in an electrolyte, semiconductor, and so on, by the application of an electric potential between electrodes. { 'ī‚än mī'grā·shən }

ion milling See ion machining. { 'ī‚än ‚mil·iŋ }

ionomer [POLYM CHEM] Polymer with covalent bonds between the elements of the chain, and ionic bonds between the chains. { ī'än·ə·mər }

ionomer resin [POLYM CHEM] A polymer which has ethylene as the major component, but which contains both covalent and ionic bonds. { ī'än·ə·mər 'rez·ən }

ion-permeable membrane [MATER] A film or sheet of a substance which is preferentially permeable to some species or types of ions. { ‚ī‚än ¦pər·mē·ə·bəl 'mem‚brän }

ion-solid interaction [SOLID STATE] An atomic process that occurs as a result of the collision of energetic ions, atoms, or molecules with condensed matter. { 'ī‚än ¦säl·əd ‚in·tər'ak·shən }

IPN See interpenetrating polymer network.

Ir See iridium.

IR drop See resistance drop. { ¦ī¦är 'dräp }

iridescence [OPTICS] A rainbow color effect exhibited in various bodies as a result of interference in a thin film (as of soap bubbles or mother of pearl) or of diffraction of light reflected from a ribbed surface (as of the plumage of some birds). { ‚ir·ə'des·əns }

iridic chloride [INORG CHEM] IrCl₄ A hygroscopic brownish-black mass, soluble in water and alcohol; used to analyze for nitric acid, HNO_3, and in analytical microscopic work. Also known as iridium chloride; iridium tetrachloride. { i'rid·ik 'klȯr‚īd }

iridium [CHEM] A metallic element, symbol Ir, atomic number 77, atomic weight 192.2, in the platinum group; insoluble in acids, melting at 2454°C. [MET] A silver-white, brittle, hard metal used in jewelry, electric contacts, electrodes, resistance wires, and pen tips. { i'rid·ē·əm }

iridium chloride See iridic chloride. { i'rid·ē·əm 'klȯr‚īd }

iridium tetrachloride See iridic chloride. { i'rid·ē·əm ‚te·trə'klȯr‚īd }

iridosmine [MET] A natural iridium-osmium alloy composed of 10–77% iridium, 17–80% osmium, 0-10% platinum, 0–17% rhodium, 0–9% ruthenium, 0–2% iron, and 0–1% copper; used for surgical needles and compass bearings and for hardening platinum. { ‚i·rə'däz‚mēn }

iron [CHEM] A silvery-white metallic element, symbol Fe, atomic number 26, atomic weight 55.847, melting at 1530°C. [MET] A heavy, magnetic, malleable and ductile metal occurring in meteorites and combined in a wide range of ores and most igneous rocks; it is one of the most widely used metals, and plays a role in biological processes. { 'ī·ərn }

iron acetate liquor [MATER] Black liquor containing 5–5.5% iron and sometimes copperas or tannin; results from pyroligneous acid attack on iron filings; used for dyeing and printing with logwood, and as a mordant for alizarine and nitroso dyes. Also known as black liquor; iron liquor. { 'ī·ərn 'as·ə,tāt 'lik·ər }

iron alloy |MET| An alloy having iron as the principal component. { 'ī·ərn 'al,ȯi }

iron ammonium sulfate *See* ferric ammonium sulfate;ferrous ammonium sulfate. { 'ī·ərn ə'mō·nē·əm 'səl,fāt }

Ironarc process |MET| An ultra-high-temperature smelting process employing plasma chemistry to process refractory materials, such as zirconia. { 'ī·ərn,ärk ,prä·səs }

iron arsenate *See* ferrous arsenate. { 'ī·ərn 'ärs· ən,āt }

iron black |CHEM| Fine black antimony powder used to give a polished-steel look to papier-maché and plaster of paris; made by reaction of zinc with acid solution of an antimony salt and precipitation of black antimony powder. { 'ī·ərn 'blak }

iron blue |INORG CHEM| Ferric ferrocyanide used as blue pigment by the paint industry for permanent body and trim paints; also used in blue ink, in paper dyeing, and as a fertilizer ingredient. { 'ī·ərn 'blü }

iron bromide *See* ferric bromide. { 'ī·ərn 'brō ,mīd }

iron carbide *See* cementite. { 'ī·ərn 'kär,bīd }

iron carbonyl *See* iron pentacarbonyl. { 'ī·ərn 'kär·bə,nil }

iron castings |MET| Shapes cast in molds from iron. { 'ī·ərn 'kast·iŋz }

iron cement |MATER| A mixture of small iron pieces with ammonium chloride, used to join iron or steel surfaces. { 'ī·ərn si'ment }

iron chloride *See* ferric chloride;ferrous chloride. { 'ī·ərn 'klȯr,īd }

iron-Constantan |MET| Dual-metal combination for thermocouple junctions, used for temperature measurement in oxidizing or reducing atmospheres. { 'ī·ərn kän'stant·ən }

iron dichloride *See* ferrous chloride. { 'ī·ərn dī'klȯr,īd }

iron ferrocyanide *See* ferric ferrocyanide. { 'ī·ərn ,fer·ə'sī·ə,nīd }

iron fluoride *See* ferric fluoride. { 'ī·ərn 'flu̇r,īd }

iron foundry |MET| A building in which iron castings are made. { 'ī·ərn ,fau̇n·drē }

iron hydroxide *See* ferric hydroxide. { 'ī·ərn hī'dräk,sīd }

ironing |MET| Reducing the wall thickness of a deep-drawn object by reducing the clearance between punch and die. { 'ī·ərn·iŋ }

iron liquor *See* iron acetate liquor. { 'ī·ərn ⁝lik·ər }

iron metavanadate *See* ferric vanadate. { 'ī·ərn ,med·ə'van·ə,dāt }

iron monoxide *See* ferrous oxide. { 'ī·ərn mə'näk,sīd }

iron-nickel alloy |MET| An iron alloy containing 20–80% nickel; it has high permeability and low hysteresis losses at low flux densities and is more readily rolled into thin laminations than silicon steels. { 'ī·ərn 'nik·əl 'al,ȯi }

iron nitrate *See* ferric nitrate. { 'ī·ərn 'nī,trāt }

iron nonacarbonyl |INORG CHEM| $Fe_2(CO)_9$

Orange-yellow crystals that break down at $100°C$ to yield tetracarbonyl, slightly soluble in alcohol and acetone, almost insoluble in water, ether, and benzene. { 'ī·ərn ,nō·nə'kär·bə ,nil }

iron-ore cement |MATER| A cement in which iron ore is used instead of clay or shale. { 'ī·ərn ⁝ȯr si'ment }

iron oxide |INORG CHEM| Any of the hydrated, synthetic, or natural oxides of iron: ferrous oxide, ferric oxide, ferriferous oxide. { 'ī·ərn 'äk,sīd }

iron pentacarbonyl |INORG CHEM| $Fe(CO)_5$ An oily liquid that decomposes upon exposure to light, soluble in most organic solvents; used as a source of a pure iron catalyst and for magnet cores. Also known as iron carbonyl. { 'ī·ərn ,pen·tə'kär·bə,nil }

iron phosphate *See* ferric phosphate. { 'ī·ərn 'fäs,fāt }

iron red |MATER| Any of the pigments made from red varieties of ferric oxide. { 'ī·ərn 'red }

iron-retention agent |MATER| Complexing agent that ties dissolved iron into complex ions to prevent reprecipitation and wellbore plugging after well acidizing. { 'ī·ərn ri'ten·chən ,ā·jənt }

iron saffron *See* Indian red. { 'ī·ərn 'saf·rən }

iron scurf |MATER| Glazing material used to color blue bricks; a mixture of stone and iron particles from grinding of gun barrels. { 'ī·ərn 'skərf }

iron soldering |MET| Soldering in which a soldering iron provides the required heat. { 'ī·ərn 'säd·ə·riŋ }

iron sulfate *See* ferric sulfate;ferrous sulfate. { 'ī·ərn 'səl,fāt }

iron sulfide *See* ferrous sulfide. { 'ī·ərn 'səl,fīd }

iron tetracarbonyl |INORG CHEM| $Fe_3(CO)_{12}$ Dark-green lustrous crystals that break down at $140–50°C$; soluble in organic solvents. { 'ī·ərn ,te·trə'kär·bə,nil }

iron whiskers |MET| Single-crystal pure iron filaments or fibers. { 'ī·ərn 'wis·kərz }

irregular polymer |POLYM CHEM| A polymer whose molecular structure does not consist of only one species of constitutional unit in a single sequential arrangement. { i⁝reg·yə·lər 'päl·i·mər }

irreversible energy loss |THERMO| Energy transformation process in which the resultant condition lacks the driving potential needed to reverse the process; the measure of this loss is expressed by the entropy increase of the system. { ,i·ri'vər·sə·bəl 'en·ər·je ,lȯs }

irreversible process |THERMO| A process which cannot be reversed by an infinitesimal change in external conditions. { ,i·ri'vər·sə·bəl 'prä·səs }

irreversible thermodynamics *See* nonequilibrium thermodynamics. { ,i·ri'vər·sə·bəl ⁝thər·mə·dī'nam·iks }

isenergic flow |THERMO| Fluid flow in which the sum of the kinetic energy, potential energy, and enthalpy of any part of the fluid does not change as that part is carried along with the fluid. { ⁝ī·sə ,nər·jik 'flō }

isenthalpic expansion |THERMO| Expansion

which takes place without any change in enthalpy. { ¦īs·ən¦thal·mik ik'span·chən }

isenthalpic process |THERMO| A process that is carried out at constant enthalpy. { ˌī·sən ¦thal·pik 'prä¸ses }

isentrope |THERMO| A line of equal or constant entropy. { 'īs·ən¸trōp }

isentropic |THERMO| Having constant entropy; at constant entropy. { ¦īs·ən¦träp·ik }

isentropic compression |THERMO| Compression which occurs without any change in entropy. { ¦īs·ən¦träp·ik kəm'presh·ən }

isentropic expansion |THERMO| Expansion which occurs without any change in entropy. { ¦īs·ən'träp·ik ik'span·chən }

isentropic flow |THERMO| Fluid flow in which the entropy of any part of the fluid does not change as that part is carried along with the fluid. { ¦īs·ən'träp·ik 'flō }

isentropic process |THERMO| A change that takes place without any increase or decrease in entropy, such as a process which is both reversible and adiabatic. { ¦īs·ən'träp·ik 'prä·ses }

Ising coupling |SOLID STATE| A model of coupling between two atoms in a lattice, used to study ferromagnetism, in which the spin component of each atom along some axis is taken to be +1 or −1, and the energy of interaction is proportional to the negative of the product of the spin components along this axis. { 'ī·ziŋ ˌkəp·liŋ }

isinglass |MATER| A gelatin made from the dried swim bladders of sturgeon and other fishes; used in glues, cements, and printing inks. Also known as fish gelatin; ichthyocolla. { 'īz·ən ˌglas }

Ising model |SOLID STATE| A crude model of a ferromagnetic material or an analogous system, used to study phase transitions, in which atoms in a one-, two-, or three-dimensional lattice interact via Ising coupling between nearest neighbors, and the spin components of the atoms are coupled to a uniform magnetic field. { 'ī·ziŋ ˌmäd·əl }

iso- |CHEM| A prefix indicating an isomer of an element in which there is a difference in the nucleus when compared to the most prevalent form of the element. |ORG CHEM| A prefix indicating a single branching at the end of the carbon chain. { 'ī·sō }

isobaric |THERMO| Of equal or constant pressure, with respect to either space or time. { ¦ī·sə ¦bär·ik }

isobaric process |THERMO| A thermodynamic process of a gas in which the heat transfer to or from the gaseous system causes a volume change at constant pressure. { ¦ī·sə¦bär·ik 'prä·səs }

isochromatic |OPTICS| **1.** Pertaining to a variation of certain quantities related to light (such as density of the medium through which the light is passing, index of refraction), in which the color or wavelength of the light is held constant. **2.** Pertaining to lines connecting points of the same color. { ¦ī·sō·krə'mad·ik }

isoclinic line |SOLID STATE| A line joining points

in a plate at which the principal stresses have parallel directions. { ¦ī·sə¦klin·ik 'līn }

isocyanate resin |POLYM CHEM| A linear alkyd resin lengthened by reaction with isocyanates, then treated with a glycol or diamine to crosslink the molecular chain; the product has good abrasion resistance. { ¦ī·sō'sī·ə¸nāt 'rez·ən }

isodesmic structure |SOLID STATE| An ionic crystal structure in which all bonds are of the same strength, so that no distinct groups of atoms are formed. { ¦ī·sə¦dez·mik 'strək·chər }

isodisperse |CHEM| **1.** Having dispersed particles, of colloidal dimensions, that are all of the same size. **2.** Dispersible in solutions with the same pH value. { ˌīs·ə·di'spərs }

isodynamic |MECH| Pertaining to equality of two or more forces or to constancy of a force. { ¦ī·sō·dī'nam·ik }

isoelectric |ELEC| Pertaining to a constant electric potential. { ¦ī·sō·i'lek·trik }

isoelectric point |PHYS CHEM| The pH value of the dispersion medium of a colloidal suspension at which the colloidal particles do not move in an electric field. { ¦ī·sō·i'lek·trik 'pȯint }

isoelectronic principle |CHEM| The concept that molecules having the same number of electrons and the same number of atoms whose atomic masses are greater than that of hydrogen (heavy atoms) tend to have similar electronic structures, similar chemical properties, and heavy-atom geometries. { ¦ī·sō·i¸lek'trän·ik 'prin·sə·pəl }

isolated system See closed system. { 'ī·sə ˌlād·əd 'sis·təm }

isomer |CHEM| One of two or more chemical substances having the same elementary percentage composition and molecular weight but differing in structure, and therefore in properties; there are many ways in which such structural differences occur; one example is provided by the compounds n-butane, $CH_3(CH_2)_2CH_3$, and isobutane, $CH_3CH(CH_3)_2$. { 'ī·sə·mər }

isomerism |CHEM| The phenomenon whereby certain chemical compounds have structures that are different although the compounds possess the same elemental composition. { ī'säm·ə ˌriz·əm }

isomerization |CHEM| A process whereby a compound is changed into an isomer; for example, conversion of butane into isobutane. { ī¸säm·ə·rə'zā·shən }

isometric process |THERMO| A constant-volume, frictionless thermodynamic process in which the system is confined by mechanically rigid boundaries. { ¦ī·sə'me·trik 'prä·səs }

isometric system |CRYSTAL| The crystal system in which the forms are referred to three equal, mutually perpendicular axes. Also known as cubic system. { ¦ī·sə'me·trik 'sis·təm }

isopolymolybdate |INORG CHEM| A class of compounds formed by the acidification of a molybdate solution, or in some cases by heating normal molybdates. { ¦ī·sō¸päl·i·mə'lib¸dāt }

isopolytungstate |INORG CHEM| A compound formed by the condensation of tungstate compounds, usually classified into metatungstates, such as $Na_6W_{12}O_{40}\cdot xH_2O$, and paratungstates, such as $Na_{10}W_{12}O_{41}\cdot xH_2O$. { ¦ī·sō¦päl·i'taŋ, stāt }

isoprene |ORG CHEM| C_5H_8 A conjugated di-olefin; a mobile, colorless liquid having a boiling point of 34.1°C; insoluble in water, soluble in alcohol and ether; polymerizes readily to form dimers and high-molecular-weight elastomer resins. It is abundant in nature and is the building block of many plant-synthesized organic compounds, including natural rubber and steroids. { 'ī·sə,prēn }

isoprenoid See terpene. { ,ī·sə·'prē,nȯid }

isostatic compacting |MET| In powder metallurgy, a process in which pressure from a gas or liquid is applied uniformly to a metal powder contained in a flexible mold. { ,ī·sō¦stad·ik 'käm ,pak·tiŋ }

isostatics |MECH| In photoelasticity studies of stress analyses, those curves, the tangents to which represent the progressive change in principal-plane directions. Also known as stress lines; stress trajectories. { ¦ī·sə¦stad·iks }

isostatic surface |MECH| A surface in a three-dimensional elastic body such that at each point of the surface one of the principal planes of stress at that point is tangent to the surface. { ¦ī·sə'stad·ik 'sər·fəs }

isostructural |CRYSTAL| Pertaining to crystalline materials that have corresponding atomic positions, and have a considerable tendency for ionic substitution. { ¦ī·sō'strək·chə·rəl }

isotactic |POLYM CHEM| Designating crystalline polymers in which substituents in the asymmetric carbon atoms have the same (rather than random or alternating) configuration in relation to the main chain. { ¦ī·sə¦tak·tik }

isotherm |THERMO| A curve or formula showing the relationship between two variables, such as pressure and volume, when the temperature is held constant. Also known as isothermal. { 'ī·sə ,thərm }

isothermal |THERMO| Having constant temperature; at constant temperature. { ¦ī·sə¦thər·məl }

isothermal annealing |MET| Transformation of an austenitic steel to ferrite and pearlite at constant temperature. Also known as isothermal transformation. { ¦ī·sə¦thər·məl ə'nēl·iŋ }

isothermal calorimeter |THERMO| A calorimeter in which the heat received by a reservoir, containing a liquid in equilibrium with its solid at the melting point or with its vapor at the boiling point, is determined by the change in volume of the liquid. { ¦ī·sə¦thər·məl ,kal·ə'rim·əd·ər }

isothermal compression |THERMO| Compression at constant temperature. { ¦ī·sə¦thər·məl kəm'presh·ən }

isothermal equilibrium |THERMO| The condition in which two or more systems are at the same temperature, so that no heat flows between them. { ¦ī·sə¦thər·məl ,ē·kwə'lib·rē·əm }

isothermal expansion |THERMO| Expansion of a substance while its temperature is held constant. { ¦ī·sə¦thər·məl ik'span·chən }

isothermal flow |THERMO| Flow of a gas in which its temperature does not change. { ¦ī·sə ¦thər·məl 'flō }

isothermal layer |THERMO| A layer of fluid, all points of which have the same temperature. { ¦ī·sə¦thər·məl 'lā·ər }

isothermal magnetization |THERMO| Magnetization of a substance held at constant temperature; used in combination with adiabatic demagnetization to produce temperatures close to absolute zero. { ¦ī·sə¦thər·məl ,mag·nə·tə'zā·shən }

isothermal process |THERMO| Any constant-temperature process, such as expansion or compression of a gas, accompanied by heat addition or removal from the system at a rate just adequate to maintain the constant temperature. { ¦ī·sə¦thər·məl 'prä·səs }

isothermal transformation |MET| See isothermal annealing. |THERMO| Any transformation of a substance which takes place at a constant temperature. { ¦ī·sə¦thər·məl ,tranz·fər'mā·shən }

isothermal treatment |MET| Heat treatment of metals at constant temperature. { ¦ī·sə ¦thər·məl 'trēt·mənt }

isotope effect |PHYS CHEM| The effect of difference of mass between isotopes of the same element on nonnuclear physical and chemical properties, such as the rate of reaction or position of equilibrium, of chemical reactions involving the isotopes. |SOLID STATE| Variation of the transition temperatures of the isotopes of a superconducting element in inverse proportion to the square root of the atomic mass. { 'ī·sə ,tōp i,fekt }

isotropic |PHYS| Having identical properties in all directions. { ¦ī·sə¦trä·pik }

isotropic fluid |FL MECH| A fluid whose properties are not dependent on the direction along which they are measured. { ¦ī·sə¦trä·pik 'flü·əd }

isotropic flux |PHYS| Radiation, or a flow of particles or matter, which reaches a location from all directions with equal intensity. { ¦ī·sə¦trä·pik 'fləks }

isotropic material |PHYS| A material whose properties are not dependent on the direction along which they are measured. { ¦ī·sə¦trä·pik mə'tir·ē·əl }

isotropy |PHYS| The quality of a property which does not depend on the direction along which it is measured, or of a medium or entity whose properties do not depend on the direction along which they are measured. { ī'sä·trə·pē }

isotypic |CRYSTAL| Pertaining to a crystalline substance whose chemical formula is analogous to, and whose structure is like, that of another specified compound. { ¦ī·sə¦tip·ik }

ISP See imperial smelting process.

ISTS See impulsive stimulated thermal scattering.

IT calorie See calorie. { ¦ī'tē ,kal·ə·rē }

ITS-90 See international temperature scale.

i-type semiconductor See intrinsic semiconductor. { 'ī ,tīp ¦sem·i·kən¦dək·tər }

ivory |MATER| The ivory-white material composing the tusks and teeth of the elephant; specific gravity is 1.87; takes a high polish and is used for ornamental parts and art objects, and formerly for piano keys.   { 'īv·rē }

ivory black   [MATER] Animal black made by calcining ivory; used as a pigment.   { 'īv·rē 'blak }

ivory board   [MATER] A highly finished cardboard that is clay-coated on both sides; used for art printing and menu cards.   { 'īv·rē ‚bȯrd }

Izod test   [MET] An impact test in which a falling pendulum strikes a fixed, usually notched specimen with 120 foot-pounds (163 joules) of energy at a velocity of 11.5 feet (3.5 meters) per second; the height of the pendulum swing after striking is a measure of the energy absorbed and thus indicates impact strength.   { 'ī‚zäd ‚test }

J

jacaranda *See* carob wood. { ˌjak·əˈran·də }

Jaccarino-Peter effect |SOLID STATE| The production of superconductivity in certain ferromagnetic metals through the application of an external magnetic field that compensates for the polarization of the conduction electrons. Also known as compensation effect. { ˌjak·ə¦rē·nō ˈpēd·ər iˌfekt }

Jaeger-Steinwehr method |THERMO| A refinement of the Griffiths method for determining the mechanical equivalent of heat, in which a large mass of water, efficiently stirred, is used, the temperature rise of the water is small, and the temperature of the surroundings is carefully controlled. { ˈyā·gər ˈshtīn·ver ˌmeth·əd }

jamb brick |MATER| A brick with one rounded corner; used to provide a rounded edge on wall openings. { ˈjam ˌbrik }

japan |MATER| A glossy, black baking paint or varnish that consists primarily of a hard asphalt base. { jəˈpan }

japanning |MET| The finishing of metal objects with japan. { jəˈpan·iŋ }

Japan paper |MATER| A special paper made with an irregular mottled effect on the surface; used for greeting cards and other decorative applications. { jəˈpan ¦pā·pər }

Japan tallow *See* Japan wax. { jəˈpan ¦tal·ō }

Japan wax |MATER| A pale-yellow wax with rancid aroma obtained from the berries of sumac; soluble in benzene and naphtha, insoluble in water; melts at 53°C; used in wax products, polishes, and soaps, and as a beeswax substitute. Also known as Japan tallow; sumac wax. { jəˈpan ¦waks }

jardiniere glaze |MATER| A type of unfritted soft and hard lead glaze; contains oxides of lead, zinc, potassium, calcium, aluminum, and silicon. { ¦¦järd·ənˌir ˌglāz }

jarosite process |MET| A zinc electrometallurgical process in which ferric ions are precipitated from feed solutions in the form of jarosite, a hydrous sulfate of iron and potassium or sodium. { jəˈrō·sīt ˌprä·səs }

Jeans viscosity equation |THERMO| An equation which states that the viscosity of a gas is proportional to the temperature raised to a constant power, which is different for different gases. { ˈjēnz viˈskäs·əd·ē iˌkwā·zhən }

jellium model |PHYS CHEM| A model describing the delocalized valence electrons in a metallic atom cluster in which the positive charge is regarded as being smeared out over the entire volume of the cluster while the valence electrons are free to move within this homogeneously distributed, positively charged background. |SOLID STATE| A model of electron-electron interactions in a metal in which the positive charge associated with the ion cores immersed in the sea of conduction electrons is replaced by a uniform positive background charge terminating along a plane that represents the surface of the metal. { ˈjel·ē·əm ˌmäd·əl }

jeweler's rouge *See* ferric oxide. { ˈjü·lərz ˈrüzh }

jewelry alloy |MET| Any ductile, malleable alloy, usually bronze, of good corrosion resistance, used as a base metal in jewelry. { ˈjül·rē ˌalˌòi }

J factor |THERMO| A dimensionless equation used for the calculation of free convection heat transmission through fluid films. { ˈjā ˌfak·tər }

jog |CRYSTAL| A shift in a dislocation from one crystal plane to another. { jäg }

Johann crystal geometry |CRYSTAL| The focusing shape of a diffracting crystal for x-ray dispersion used in electron-probe microanalysis; less stringent than Johannson crystal geometry. { ˈyō·hän ¦krist·əl jēˈäm·ə·trē }

Johannson crystal geometry |CRYSTAL| The full-focusing shape of a diffracting crystal for x-ray dispersion used in electron-probe microanalyzers; more stringent than Johann crystal geometry. { jōˈhan·sən ¦krist·əl jēˈäm·ə·trē }

Johnson and Lark-Horowitz formula |SOLID STATE| A formula according to which the resistivity of a metal or degenerate semiconductor resulting from impurities which scatter the electrons is proportional to the cube root of the density of impurities. { ˈjän·sən ən ¦lärk ˈhär·əˌwitz ˌfòr·myə·lə }

joint |ELEC| A juncture of two wires or other conductive paths for current. |ENG| The surface at which two or more mechanical or structural components are united. { jòint }

joint buildup sequence |MET| The sequence in which weld beads are deposited relative to the cross section of a multiple-pass joint. { ˈjòint ˈbil·dəp ˌsē·kwəns }

joint compound |MATER| A material used primarily to lubricate the threads of pipe joints and secondarily to prevent joint leakage. { 'jóint ˌkäm‚paúnd }

joint efficiency |MET| A numerical value expressed as the ratio of the strength of a riveted, welded, or brazed joint to the strength of the parent metal. { 'jóint ə‚fish·ən·sē }

joint penetration |MET| The distance extended into a weld joint by the weld metal or fusion zone. { 'jóint ‚pen·ə'trā·shən }

Jominy end quench test *See* Jominy test. { 'jäm·ə·nē 'end ‚kwench ‚test }

Jominy test |MET| A hardenability test in which a steel bar is heated to the desired austenitizing temperature and quench-hardened at one end and then measured for hardness along its length, beginning at the quenched end. Also known as Jominy end quench test. { 'jäm·ə·nē ‚test }

Josephson current |CRYO| The current across a Josephson junction in the absence of voltage across the junction, resulting from the Josephson effect. { 'jō·səf·sən ‚kə·rənt }

Josephson effect |CRYO| The tunneling of electron pairs through a thin insulating barrier between two superconducting materials. Also known as Josephson tunneling. { 'jō·səf·sən i ‚fekt }

Josephson equation |CRYO| An equation according to which the Josephson current is a sinusoidally varying function of the applied magnetic field. { 'jō·səf·sən i‚kwā·zhən }

Josephson junction |CRYO| A thin insulator separating two superconducting materials; it displays the Josephson effect. { 'jō·səf·sən ‚jəŋk·shən }

Josephson penetration depth |CRYO| A measure of the distance that a magnetic field extends into a Josephson junction. { 'jō·səf·sən 'pen·ə'trā·shən ‚depth }

Josephson tunneling *See* Josephson effect. { 'jō·səf·sən ‚tən·əl·iŋ }

Joule and Playfairs' experiment |THERMO| An experiment in which the temperature of the maximum density of water is measured by taking the mean of the temperatures of water in two columns whose densities are determined to be equal from the absence of correction currents in a connecting trough. { ‚jül and 'plā‚färz ik ‚sper·ə·mənt }

Joule effect |PHYS| 1. The heating effect produced by the flow of current through a resistance. 2. A change in the length of a ferromagnetic substance which occurs parallel to an applied magnetic field. Also known as Joule magnetorestriction; longitudinal magnetorestriction. { 'jül i‚fekt }

Joule equivalent |THERMO| The numerical relation between quantities of mechanical energy and heat; the present accepted value is 1 fifteen-degrees calorie equals 4.1855 ± 0.0005 joules. Also known as mechanical equivalent of heat. { 'jül i‚kwiv·ə·lənt }

Joule experiment |THERMO| 1. An experiment to detect intermolecular forces in a gas, in which one measures the heat absorbed when gas in a small vessel is allowed to expand into a second vessel which has been evacuated. 2. An experiment to measure the mechanical equivalent of heat, in which falling weights cause paddles to rotate in a closed container of water whose temperature rise is measured by a thermometer. { 'jül ik‚sper·ə·mənt }

Joule heat |ELEC| The heat which is evolved when current flows through a medium having electrical resistance, as given by Joule's law. { 'jül ‚hēt }

Joule-Kelvin effect *See* Joule-Thomson effect. { 'jül 'kel·vən i‚fekt }

Joule magnetorestriction *See* Joule effect. { 'jül mag‚ned·ō·ri'strik·shən }

Joule's law |ELEC| The law that when electricity flows through a substance, the rate of evolution of heat in watts equals the resistance of the substance in ohms times the square of the current in amperes. |THERMO| The law that at constant temperature the internal energy of a gas tends to a finite limit, independent of volume, as the pressure tends to zero. { 'jülz ‚ló }

Joule-Thomson coefficient |THERMO| The ratio of the temperature change to the pressure change of a gas undergoing isenthalpic expansion. { 'jül 'täm·sən ‚kō·ə‚fish·ənt }

Joule-Thomson effect |THERMO| A change of temperature in a gas undergoing Joule-Thomson expansion. Also known as Joule-Kelvin effect. { 'jül 'täm·sən i‚fekt }

Joule-Thomson expansion |THERMO| The adiabatic, irreversible expansion of a fluid flowing through a porous plug or partially opened valve. Also known as Joule-Thomson process. { 'jül 'täm·sən ik‚span·chən }

Joule-Thomson inversion temperature |THERMO| A temperature at which the Joule-Thomson coefficient of a given gas changes sign. { ‚jül ‚täm·sən in'vər·zhən ‚tem·prə·chər }

Joule-Thomson process *See* Joule-Thomson expansion. { 'jül 'täm·sən ‚prä·səs }

Joule-Thomson valve |CRYO| A valve through which a gas is allowed to expand adiabatically, resulting in lowering of its temperature; used in production of liquid hydrogen and helium. { 'jül 'täm·sən ‚valv }

junction |ELECTR| A region of transition between two different semiconducting regions in a semiconductor device, such as a *pn* junction, or between a metal and a semiconductor. { 'jəŋk·shən }

junction diode |ELECTR| A semiconductor diode in which the rectifying characteristics occur at an alloy, diffused, electrochemical, or grown junction between *n*-type and *p*-type semiconductor materials. Also known as junction rectifier. { 'jəŋk·shən ‚dī‚ōd }

junction laser |OPTICS| A laser in which a junction in a semiconductor serves as the source of the coherent laser beam. { 'jəŋk·shən ‚lā·zər }

junction magnetoresistance *See* tunneling

magnetoresistance. { ¦jəŋk·shən mag₁ned·ō·ri'zis·təns }

junction rectifier See junction diode. { 'jəŋk·shən ¦rek·tə₁fī·ər }

jute board [MATER] A fiberboard made of jute fiber. { 'jüt ₁bȯrd }

jute paper [MATER] A strong paper composed principally of jute fiber. { 'jüt ₁pā·pər }

K

K   *See* potassium.

kadaya gum   *See* karaya gum.   { kə'dī·ə·ə,gəm }

kalium   *See* potassium.   { 'kāl·ēl·əm }

kalsomine   *See* calcimine.   { 'kal·sə,mīn }

Kapitza resistance |CRYO| A thermal resistance to the flow of heat across the interface between liquid helium and a solid.   { 'kä·pit·sə ri¦zis·təns }

karat |MET| A unit for designating the fineness of gold in an alloy; represents a twenty-fourth part; thus, 18-karat gold is 18/24 or 75% pure.   { 'kar·ət }

karaya gum |MATER| The exudation from the Indian tree (*Sterculia urens*); white to dark-colored tears; used in pharmaceuticals, textiles, and foods, and as tragacanth gum substitute. Also known as Indian tragacanth; kadaya gum.   { kə'rī·ə ,gəm }

kauri gum |MATER| Hard copal resins from kauri pine (*Agathis australis*); used in lacquers and varnishes.   { 'kau̇·rē ,gəm }

K band |SOLID STATE| An optical absorption band which appears together with an F-band and has a lower intensity and shorter wavelength than the latter.   { 'kā ,band }

kcal   *See* kilocalorie.

keel block |MET| A simple shape from which a test casting is made in the form of a large head, which is removed and discarded, with a keel on the bottom.   { 'kēl ,bläk }

Keene's cement |MATER| An anhydrous calcined gypsum mixed with an accelerator; used as a hard-finish plaster.   { 'kēnz si,ment }

Kellogg equation |THERMO| An equation of state for a gas, of the form

$$p = RT\rho + \sum_{n=2}^{\infty} \left[b_n T - a_n - \left(c_n / T^2 \right) \right] \rho^n$$

where p is the pressure, T the absolute temperature, ρ the density, R the gas constant, and a_n, b_n, and c_n are constants.   { 'kel,äg i,kwā·zhən }

kelvin |THERMO| A unit of absolute temperature equal to 1/273.16 of the absolute temperature of the triple point of water. Symbolized K. Formerly known as degree Kelvin.   { 'kel·vən }

Kelvin absolute temperature scale |THERMO| A temperature scale in which the ratio of the temperatures of two reservoirs is equal to the ratio of the amount of heat absorbed from one of them by a heat engine operating in a Carnot cycle to the amount of heat rejected by this engine to the other reservoir; the temperature of the triple point of water is defined as 273.16 K. Also known as Kelvin temperature scale.   { 'kel·vən ¦ab·sə ,lūt 'tem·prə·chər ,skāl }

Kelvin equation |THERMO| An equation giving the increase in vapor pressure of a substance which accompanies an increase in curvature of its surface; the equation describes the greater rate of evaporation of a small liquid droplet as compared to that of a larger one, and the greater solubility of small solid particles as compared to that of larger particles.   { 'kel·vən i,kwā·zhən }

Kelvin scale |THERMO| The basic scale used for temperature definition; the triple point of water (comprising ice, liquid, and vapor) is defined as 273.16 K; given two reservoirs, a reversible heat engine is built operating in a cycle between them, and the ratio of their temperatures is defined to be equal to the ratio of the heats transferred.   { 'kel·vən ,skāl }

Kelvin's statement of the second law of thermodynamics |THERMO| The statement that it is not possible that, at the end of a cycle of changes, heat has been extracted from a reservoir and an equal amount of work has been produced without producing some other effect.   { 'kel·vənz 'stāt·mənt əv t͟hə 'sek·ənd ,lȯ əv ,thər·mō·dī'nam·iks }

Kelvin temperature scale |THERMO| An International Temperature Scale which agrees with the Kelvin absolute temperature scale within the limits of experimental determination.   { 'kel·vən 'tem·prə·chər ,skāl }

keto- |ORG CHEM| Organic chemical prefix for the keto or carbonyl group, C:O, as in a ketone.   { 'kēd·ō }

ketone |ORG CHEM| One of a class of chemical compounds of the general formula RR'CO, where R and R' are alkyl, aryl, or heterocyclic radicals; the groups R and R' may be the same or different, or incorporated into a ring; the ketones, acetone, and methyl ethyl ketone are used as solvents, and ketones in general are important intermediates in the synthesis of organic compounds.   { 'kē,tōn }

Keyes equation |THERMO| An equation of state of a gas which is designed to correct the van der Waals equation for the effect of surrounding molecules on the term representing the volume of a molecule. { 'kēz i‚kwä·zhən }

keyhole |MET| A welding method wherein the heat source, because of its concentration, causes a hole through the surface immediately ahead of the molten weld metal in the direction of welding; the hole is filled as the welding progresses, ensuring complete joint penetration. { 'kē‚hōl }

keyhole specimen |MET| A metal specimen containing a keyhole shaped notch and used in certain impact tests. { 'kē‚hōl ‚spes·ə·mən }

keystone |MATER| Small crushed stone used as filler for the large aggregate in bituminous bound roads. { 'kē‚stōn }

kg-cal See kilocalorie.

Kikuchi lines |CRYSTAL| A pattern consisting of pairs of white and dark parallel lines, obtained when an electron beam is scattered (diffracted) by a crystalline solid; the pattern gives information on the structure of the crystal. { kē'kü·chē ‚līnz }

kill |MATER| To treat in such a way as to destroy undesirable properties; for example, neutralization of an acid by the addition of an alkali. |MET| To add a strong deoxidizer, such as silicon or aluminum, to molten steel in order to stop the reaction between carbon and oxygen forming gaseous carbon monoxide and dioxide during solidification. { kil }

killed spirits |MATER| An aqueous solution of zinc(II) chloride used as a flux for solder. { 'kild 'spir·əts }

killed steel |MET| Thoroughly deoxidized steel, for example, by additions of aluminum or silicon, in which the reaction between carbon and oxygen during solidification is suppressed. { 'kild 'stēl }

killer |SOLID STATE| An impurity that inhibits luminescence in a solid. { 'kil·ər }

kilocalorie |THERMO| A unit of heat energy equal to 1000 calories. Abbreviated kcal. Also known as kilogram-calorie (kg-cal); large calorie (Cal). { 'kil·ə‚kal·ə·rē }

kilogram-calorie See kilocalorie. { 'kil·ə‚gram 'kal·ə·rē }

kinematic fluidity |FL MECH| The reciprocal of the kinematic viscosity. { ¦kin·ə¦mad·ik flü'id·əd·ē }

kinematics |MECH| The study of the motion of a system of material particles without reference to the forces which act on the system. { ¦kin·ə ¦mad·iks }

kinematic viscosity |FL MECH| The absolute viscosity of a fluid divided by its density. Also known as coefficient of kinematic viscosity. { ¦kin·ə¦mad·ik vi'skäs·əd·ē }

kinetic energy |MECH| The energy which a body possesses because of its motion; in classical mechanics, equal to one-half of the body's mass times the square of its speed. { kə'ned·ik 'en·ər·jē }

kinetic friction |MECH| The friction between two surfaces which are sliding over each other. { kə'ned·ik 'frik·shən }

kinetic momentum |MECH| The momentum which a particle possesses because of its motion;

in classical mechanics, equal to the particle's mass times its velocity. { kə'ned·ik mə'men·təm }

kinetic reaction |MECH| The negative of the mass of a body multiplied by its acceleration. { kə'ned·ik rē'ak·shən }

kinetics |MECH| The dynamics of material bodies. { kə'ned·iks }

king's blue See cobalt blue. { 'kiŋz 'blü }

Kirchhoff formula |THERMO| A formula for the dependence of vapor pressure p on temperature T, valid over limited temperature ranges; it may be written log $p = A − (B/T) − C$ log T, where A, B, and C are constants. { 'kərk‚hôf ‚fôr·myə·lə }

Kirchhoff's equations |THERMO| Equations which state that the partial derivative of the change of enthalpy (or of internal energy) during a reaction, with respect to temperature, at constant pressure (or volume) equals the change in heat capacity at constant pressure (or volume). { 'kərk‚hôfs i‚kwä·zhənz }

Kirchhoff's law |THERMO| The law that the ratio of the emissivity of a heat radiator to the absorptivity of the same radiator is the same for all bodies, depending on frequency and temperature alone, and is equal to the emissivity of a blackbody. Also known as Kirchhoff's principle. { 'kərk‚hôfs ‚lô }

Kirchhoff vapor pressure formula |THERMO| An approximate formula for the variation of vapor pressure p with temperature T, valid over a limited temperature range; it is ln $p = A − B/T − C$ ln T, where A, B, and C are constants. { ¦kirch‚hôf 'vā·pər ‚pre·shər ‚fôr·myə·lə }

Kirkendall effect |MET| The phenomenon whereby a marker placed at the interface between an alloy and a metal moves toward the alloy region when the temperature of the system is raised to the point where diffusion can occur. { 'kərk·ən‚dôl i‚fekt }

kish |MET| Free graphite that floats to the surface of molten hypereutectic cast iron as it cools. { kish }

K Monel |MET| A nonmagnetic, age-hardenable alloy of nickel (28–34%) and copper and 2.75% aluminum that can be heat-treated after finishing to produce a material that is both corrosion-resistant and extra strong. { 'kā mō‚nel }

knee |MET| The lower supporting structure for an arm in a resistance welding machine. { nē }

knife-line attach |MET| Intergranular corrosion of an alloy adjacent to a weldment after heating the joint above the sensitization temperature. { 'nīf ¦līn ə‚tach }

knock-on atom |SOLID STATE| An atom which is knocked out of its equilibrium position in a crystal lattice by an energetic bombarding particle, and is displaced many atomic distances away into an interstitial position, leaving behind a vacant lattice site. { 'näk‚ön ‚ad·əm }

Knoop hardness |MET| The relative microhardness of a material, such as metal, determined by the Knoop indentation test. { 'nüp ‚härd·nəs }

Knoop indentation test |MET| A diamond pyramid hardness test employing the Knoop indenter; hardness is determined by the depth to which the Knoop indenter penetrates. { 'nüp ‚in‚den'tā·shən ‚test }

Knoop indenter |MET| A diamond indenter which has a rhombic base with diagonals in a 1:7 ratio and included apical angles of 130° and 172°30'; used in the Knoop indentation test. { 'nüp in'den·tər }

knurl |ENG| To provide a surface, usually a metal, with small ridges or knobs to ensure a firm grip or as a decorative feature. { nərl }

Kohn effect |SOLID STATE| A sharp change in the phonon dispersion curve of a material when the wave-number vector of the phonons corresponds to the diameter of the Fermi sphere, because of the production of standing waves. { 'kōn i‚fekt }

Kolosov-Muskhelishvili formulas |MECH| Formulas which express plane strain and plane stress in terms of two holomorphic functions of the complex variable $z = x + iy$, where x and y are plane coordinates. { ¦kōl·ə‚sóf ‚músh'kel·ish ‚vil·ē ‚fór·myə·ləz }

Kondo alloy |MET| A dilute alloy of a magnetic material in a nonmagnetic host which exhibits the Kondo effect. { 'kän·dō ‚al‚ói }

Kondo effect |MET| The large anomalous increase in the resistance of certain dilute alloys of magnetic materials in nonmagnetic hosts as the temperature is lowered. { 'kän·dō i‚fekt }

Kondo temperature |MET| The temperature below which the Kondo effect predominates for a specified magnetic impurity and host material. { 'kän·dō ‚tem·prə·chər }

Kosterlitz-Thouless transition |CRYO| The transition from the superfluid to the normal state in thin films of helium-3, which proceeds through the unbinding of vortices in the phase of the order parameter having opposite directions of rotation. { 'käs·tər‚lits tü'les tran‚zish·ən }

Kourbatoff's reagents |MET| Etching agents used for microanalysis of carbon steels; there are four different formulations, three with nitric acid, one with hydrochloric acid. { 'kür·bə‚tófs rē‚ā·jəns }

kovar |MET| An alloy (54% iron-29% nickel-17% cobalt) that exhibits low thermal expansion; widely used for electronic devices with glass-to-metal seals because its expansion is low enough to match that of glass. { 'kō‚vär }

Kr *See* krypton.

kraft paper |MATER| A strong paper or cardboard made from sulfate-process wood pulp; unbleached varieties are used for wrapping paper and shipping cartons. { 'kraft ‚pā·pər }

Kramer's theorem |SOLID STATE| The theorem that the states of a system consisting of an odd number of electrons in an external electrostatic field are at least twofold degenerate. { 'krä·mərz ‚thir·əm }

Krause rolling mill |MET| A type of rolling mill in which the rolls translate as well as rotate, accomplishing a high reduction of thickness for the single passage of a metal sheet. { 'kraús 'rōl·iŋ ‚mil }

Krigar-Menzel law |MECH| A generalization of the second Young-Helmholtz law which states that when a string is bowed at a point which is at a distance of p/q times the string's length from one of the ends, where p and q are relative primes, then the string moves back and forth with two constant velocities, one of which is $q - 1$ times as large as the other. { ¦krē·gər 'menz·əl ‚lò }

Kroll process |MET| A reduction process for the production principally of titanium metal sponge from titanium tetrachloride. { 'kröl ‚prä·səs }

Kronig-Penney model |SOLID STATE| An idealized one-dimensional model of a crystal in which the potential energy of an electron is an infinite sequence of periodically spaced square wells. { 'krō·nig 'pen·ē ‚mäd·əl }

krypton |CHEM| A colorless, inert gaseous element, symbol Kr, atomic number 36, atomic weight 83.80; it is odorless and tasteless; used to fill luminescent electric tubes. { 'krip·tän }

k-space *See* wave-vector space. { 'kā ‚spās }

Kundt effect |OPTICS| **1.** The occurrence of a very large magnetic rotation when polarized light passes through very thin films of pure ferromagnetic materials. **2.** *See* Faraday effect. { 'künt i ‚fekt }

L

La *See* lanthanum.

lac |MATER| A resinous material secreted by some insects that live on the sap of certain trees, principally in India; used in the manufacture of shellac. { lak }

lacmus *See* litmus. { 'lak·məs }

lacquer |MATER| A homogeneous solution of a film-forming substance, often a polymer, for example, nitrocellulose, used to give a glossy finish, especially on brass and other bright metals. { 'lak·ər }

lacquer diluent |MATER| An organic liquid with no solvent power added to lacquer formulations to reduce viscosity and to adjust flow or other properties. { 'lak·ər 'dil·yə·wənt }

lactam |ORG CHEM| An internal (cyclic) amide formed by heating amino acids; thus γ-aminobutyric acid readily forms γ-butyrolactam (pyrrolidone); many lactams have physiological activity. { 'lak,tam }

lactate |ORG CHEM| A salt or ester of lactic acid. { 'lak,tāt }

lactide |ORG CHEM| A cyclic, intermolecular, double ester formed from α-hydroxy acids; most lactides are relatively low melting solids and are easily hydrolyzed by base to form salts of the parent acid, such as sodium lactate. { 'lak,tīd }

lactim |ORG CHEM| A tautomeric enol form of a lactam with which it forms an equilibrium whenever the lactam nitrogen carries a free hydrogen. { 'lak·təm }

lactone |ORG CHEM| An internal cyclic mono ester formed by hydroxy acids spontaneously; thus γ-hydroxybutyric acid forms γ-butyrolactone. { 'lak,tōn }

lactoprene |MATER| Any of several synthetic rubbers that have good resistance to hydrocarbon oil, ozone, oxygen, and other weather elements excepting cold, and that are polymers or copolymers of an acrylic acid ester. { 'lak·tə,prēn }

ladle |MET| A receptacle for transporting and pouring molten metal. { 'lād·əl }

Lafarge cement |MATER| A cement made of plaster of paris, lime, and marble powder; used in mortar for marble and limestone pieces because it is nonstaining. { lə'färzh si,ment }

lagging |MATER| Asbestos and magnesia plaster that is used as a thermal insulation on process equipment and piping. { 'lag·iŋ }

Lagrange-Helmholtz equation *See* Helmholtz equation. { lə'gränj 'helm,hōlts i,kwā·zhən }

laid paper |MATER| A paper with a pattern of parallel lines spaced so as to give a ribbed effect. { 'lād ,pā·pər }

laitance |MATER| Weak material, consisting principally of lime, that is formed on the surface of concrete, especially when excess water is mixed with the cement. { 'lat·əns }

lake |MATER| Any of a large group of pigments made from dyes that have been combined with or adsorbed by salts of calcium, barium, chromium, aluminum, phosphotungstic acid, or phosphomolybdic acid; used to color foods and pharmaceuticals, and for textile dyeing. Also known as color lake. { lāk }

lake copper |MET| A pure type of copper produced from ores taken from the Lake Superior region; has high conductivity. { 'lāk ,käp·ər }

lambda leak |CRYO| A leak of liquid helium II through small holes where normal liquids cannot pass. Also known as superleak. { 'lam·də ,lēk }

lambda point |CRYO| The temperature (2.1780 K), at atmospheric pressure, at which the transformation between the liquids helium I and helium II takes place; a special case of the thermodynamics definition. |THERMO| A temperature at which the specific heat of a substance has a sharply peaked maximum, observed in many second-order transitions. { 'lam·də ,pȯint }

Lambert surface |THERMO| An ideal, perfectly diffusing surface for which the intensity of reflected radiation is independent of direction. { 'lam·bərt ,sər·fəs }

Lamé constants |MECH| Two constants which relate stress to strain in an isotropic, elastic material. { lä'mā ,kän·stəns }

lamellar crystal |CRYSTAL| A polycrystalline substance whose grains are in the form of thin sheets. { lə'mel·ər 'krist·əl }

lamina |MATER| A flat or curved arrangement of unidirectional or woven fibers in a matrix. { 'lam·ə·nə }

laminar composite |MATER| A composite material that consists of two or more layers of different materials that are bonded together. { 'lam·ə·nər kəm'päz·ət }

laminar flow |FL MECH| Streamline flow of an incompressible, viscous Newtonian fluid; all

laminate

particles of the fluid move in distinct and separate lines. { 'lam·ə·nər 'flō }

laminate [MATER] A sheet of material made of several different bonded layers. { 'lam·ə,nāt }

laminated composite [MATER] A composite material consisting of layers of various materials. { 'lam·ə,nād·əd kəm'päz·ət }

laminated glass See nonshattering glass. { 'lam·ə,nād·əd 'glas }

laminated metal [MET] A sheet or bar of composite metal composed of two or more bonded layers. { 'lam·ə,nād·əd 'med·əl }

laminated plastic [MATER] A thin sheet made of superposed layers of plastic bonded or impregnated with resin or compressed under heat. { 'lam·ə,nād·əd 'plas·tik }

laminated wood [MATER] Board or timber composed of layers of wood glued together with the grains parallel. { 'lam·ə,nād·əd 'wùd }

lampblack [MATER] A grayish-black amorphous, practically pure form of carbon made by burning oil, coal tar, resin, or other carbonaceous substance in an insufficient supply of air; used in making paints, lead pencils, metal polishes, electric brush carbons, crayons, and carbon papers. { 'lamp,blak }

lampshade paper [MATER] Paper that is translucent and either flame-resistant or flame-retardant; often made of wood pulp, vegetable parchment, or laminated glassine. { 'lamp ,shād ,pā·pər }

lance [MET] To cut into but not through the piece of work. { lans }

land [ELECTR] 1. One of the regions between pits on a track on an optical disk. 2. See terminal area. { land }

Landau-Ginzburg theory See Ginzburg-Landau theory. { 'lan,daù 'ginz·bərg ,thē·ə·rē }

Landau levels [SOLID STATE] Energy levels of conduction electrons which occur in a metal subjected to a magnetic field at very low temperatures and which are quantized because of the quantization of the electron motion perpendicular to the field. { 'lan,daù ,lev·əlz }

land plaster [MATER] Finely ground gypsum, used as a fertilizer and as a corrective for soil with excess sodium and potassium carbonates. { 'land ,plas·tər }

Langevin function [ELECTROMAG] A mathematical function, $L(x)$, which occurs in the expressions for the paramagnetic susceptibility of a classical (non-quantum-mechanical) collection of magnetic dipoles, and for the polarizability of molecules having a permanent electric dipole moment; given by $L(x) = \coth x - 1/x$. { länzh·van ,fəŋk·shən }

Langevin theory of diamagnetism [ELECTROMAG] A theory based on the idea that diamagnetism results from electronic currents caused by Larmor precession of electrons inside atoms. { länzh·van ¦thē·ə·rē əv ,dī·ə'mag·nə,tiz·əm }

Langevin theory of paramagnetism [ELECTROMAG] A theory which treats a substance as a classical (non-quantum-mechanical) collection of permanent magnetic dipoles with no in-

teractions between them, having a Boltzmann distribution with respect to energy of interaction with an applied field. { länzh·van ¦thē·ə·rē əv ,par·ə'mag·nə,tiz·əm }

Langmuir-Blodgett film [PHYS CHEM] A highly ordered monomolecular film that results from compressing a surface layer of amphiphilic molecules into a floating monolayer and transferring it to a substrate by dipping. { 'laŋ,myùr 'bläj·ət ,film }

Langmuir effect [SOLID STATE] The ionization of atoms of low ionization potential that come into contact with a hot metal with a high work function. { 'laŋ,myùr i,fekt }

lanthana See lanthanum oxide. { 'lan·thə·nə }

lanthanide series [CHEM] Rare-earth elements of atomic numbers 57 through 71; their chemical properties are similar to those of lanthanum, atomic number 57. { 'lan·thə,nīd ,sir·ēz }

lanthanum [CHEM] A chemical element, symbol La, atomic number 57, atomic weight 138.91; it is the second most abundant element in the rare-earth group. [MET] A white, soft, malleable metal; tarnishes in moist air; a major component of misch metal. { 'lan·thə·nəm }

lanthanum-doped lead zirconate-lead titanate See lead lanthanum zirconate titanate. { 'lan·thə·nəm ¦dōpt 'led 'zərk·ən,āt ¦led 'tīt·ən,āt }

lanthanum nitrate [INORG CHEM] La(NO₃)₃· 6H₂O Hygroscopic white crystals melting at 40°C; soluble in alcohol and water; used as an antiseptic and in gas mantles. { 'lan·thə·nəm 'nī,trāt }

lanthanum oxide [INORG CHEM] La₂O₃ A white powder melting at about 2000°C; soluble in acid, insoluble in water; used to replace lime in calcium lights and in optical glass. Also known as lanthana; lanthanum sesquioxide; lanthanum trioxide. { 'lan·thə·nəm 'äk,sīd }

lanthanum sesquioxide See lanthanum oxide. { 'lan·thə·nəm ,ses·kwē'äk,sīd }

lanthanum sulfate [INORG CHEM] La₂(SO₄)₃· 9H₂O White crystals; slightly soluble in water, soluble in alcohol; used for atomic weight determinations for lanthanum. { 'lan·thə·nəm 'səl,fāt }

lanthanum trioxide See lanthanum oxide. { 'lan·thə·nəm trī'äk,sīd }

lap [MATER] An abrasive material used for lapping. [MET] A defect caused by folding and then rolling or forging a hot metal fin or corner onto a surface without welding. Also known as fold. { lap }

Laplace law See Ampère law. { lə'pläs ,lò }

lapping [ELECTR] Moving a quartz, semiconductor, or other crystal slab over a flat plate on which a liquid abrasive has been poured, to obtain a flat polished surface or to reduce the thickness a carefully controlled amount. [MET] Polishing with a material such as cloth, lead, plastic, wood, iron, or copper having fine abrasive particles incorporated or rubbed into the surface. { 'lap·iŋ }

lap-rivet [MET] To rivet a lap joint. { 'lap ,riv·ət }

lap weld [MET] A welded lap joint. { 'lap ,weld }

Laray viscometer [ENG] An instrument designed to measure viscosity and other properties of ink. { lə'rā vi'skäm·əd·ər }

large calorie See kilocalorie. { 'lärj 'kal·ə·rē }

large polaron [SOLID STATE] An electron in a crystal lattice together with the surrounding lattice deformation, for the case in which the deformation extends over many lattice sites so that the lattice can be treated as a continuum. { 'lärj 'pō·lə,rän }

Larson-Miller parameter [MECH] The effects of time and temperature on creep, being defined empirically as $P = T (C + \log t) \times 10^{-3}$, where $T =$ test temperature in degrees Rankine (degrees Fahrenheit + 460) and $t =$ test time in hours; the constant C depends upon the material but is frequently taken to be 20. { 'lärs·ən 'mil·ər pə'ram·əd·ər }

larvicide [MATER] A pesticide used to kill larvae. { 'lär·və,sīd }

laser [OPTICS] An active electron device that converts input power into a very narrow, intense beam of coherent visible or infrared light; the input power excites the atoms of an optical resonator to a higher energy level, and the resonator forces the excited atoms to radiate in phase. Derived from light amplification by stimulated emission of radiation. { 'lā·zər }

laser beam [OPTICS] A narrow beam of coherent, powerful, and nearly monochromatic electromagnetic radiation emitted by a laser. { 'lā·zər ,bēm }

laser-beam cutting See laser cutting. { 'la·zər ,bēm ,kəd·iŋ }

laser cutting [MET] A process by which a laser beam impinges on the workpiece in order to heat and sever the piece. Also known as laser-beam cutting. { 'lā·zər ,kəd·iŋ }

laser deposition [MET] A vapor deposition process in which films are grown on substrates from a material that is evaporated by laser radiation. { ¦lā·zər dep·ə'zish·ən }

laser diode See semiconductor laser. { 'lā·zər ¦dī,ōd }

laser drill [OPTICS] A drill in which concentrated light from a ruby laser generates intense heat for drilling holes as small as 0.0001 inch (2.5 micrometers) in diameter in tungsten, gemstones, and other hard materials. { 'lā·zər ,dril }

laser glazing [MET] A surface alloying process in which a continuous high-energy carbon dioxide laser traverses the surface of a metal part, creating a thin layer of melt. { 'lā·zər ,glāz·iŋ }

laser-solid interaction [SOLID STATE] Interaction of laser light with a solid, especially the thermal effects of absorption of a high-intensity laser beam. { 'lā·zər 'säl·əd ,in·tər'ak·shən }

laser welding [MET] Micro-spot welding with a laser beam. { 'lā·zər ¦wel·diŋ }

lasing [OPTICS] Generation of visible or infrared light waves having very nearly a single frequency by pumping or exciting electrons into high-energy states in a laser. { 'lāz·iŋ }

latent heat [THERMO] The amount of heat absorbed or evolved by 1 mole, or a unit mass, of a substance during a change of state (such as fusion, sublimation or vaporization) at constant temperature and pressure. { 'lāt·ənt 'hēt }

latent heat of fusion See heat of fusion. { 'lāt·ənt ¦hēt əv 'fyü·zhən }

latent heat of sublimation See heat of sublimation. { 'lāt·ənt ¦hēt əv ,səb·lə'mā·shən }

latent heat of vaporization See heat of vaporization. { 'lāt·ənt ¦hēt əv ,vā·pə·rə'zā·shən }

latex [POLYM CHEM] **1.** Milky colloid in which natural or synthetic rubber or plastic is suspended in water. **2.** An elastomer product made from latex. { 'lā,teks }

latex cement [MATER] A highly adhesive solvent solution of latex. { 'lā,teks si'ment }

latex paint [MATER] A paint consisting of a water suspension or emulsion of latex combined with pigments and additives such as binders and suspending agents. { 'lā,teks 'pānt }

lath brick [MATER] A long, narrow brick. { 'lath ,brik }

latten [MET] A thin metal sheet, particularly of brass or similar alloy, hot-rolled steel, or tin-covered iron, used for ornamental purposes. { 'lat·ən }

lattice [CRYSTAL] A regular periodic arrangement of points in three-dimensional space; it consists of all those points P for which the vector from a given fixed point to P has the form $n_1\mathbf{a} + n_2\mathbf{b} + n_3\mathbf{c}$, where n_1, n_2, and n_3 are integers, and $\mathbf{a}$, $\mathbf{b}$, and $\mathbf{c}$ are fixed, linearly independent vectors. Also known as periodic lattice; space lattice. { 'lad·əs }

lattice constant [CRYSTAL] A parameter defining the unit cell of a crystal lattice, that is, the length of one of the edges of the cell or an angle between edges. Also known as lattice parameter. { 'lad·əs ,kän·stənt }

lattice defect See crystal defect. { 'lad·əs ,dē,fekt }

lattice dynamics [SOLID STATE] The study of the thermal vibrations of a crystal lattice. Also known as crystal dynamics. { 'lad·əs dī,nam·iks }

lattice energy [SOLID STATE] The energy required to separate ions in an ionic crystal an infinite distance from each other. { 'lad·əs ,en·ər·jē }

lattice parameter See lattice constant. { 'lad·əs pə,ram·əd·ər }

lattice polarization [SOLID STATE] Electric polarization of a solid due to displacement of ions from equilibrium positions in the lattice. { 'lad·əs pō·lə·rə'zā·shən }

lattice scattering [SOLID STATE] Scattering of electrons by collisions with vibrating atoms in a crystal lattice, reducing the mobility of charge carriers in the crystal and thereby affecting its conductivity. { 'lad·əs ,skad·ə·riŋ }

lattice vibration [SOLID STATE] A periodic oscillation of the atoms in a crystal lattice about their equilibrium positions. { 'lad·əs vī'brā·shən }

lattice wave [SOLID STATE] A disturbance propagated through a crystal lattice in which atoms oscillate about their equilibrium positions. { 'lad·əs ,wāv }

lauan [MATER] Wood from any of several genera of trees of the Philippines, Malaya, and Sarawak;

resembles mahogany but shrinks and swells more with changes in moisture. { lə'wän }

Laue camera |CRYSTAL| The apparatus used in the Laue method; the x-ray beam usually enters through a hole in the x-ray film, which records beams bent through an angle of nearly 180° by the crystal; less commonly, the film is placed beyond the crystal. { 'laů·ə ,kam·rə }

Laue condition |CRYSTAL| **1.** The condition for a vector to lie in a Laue plane: its scalar product with a specified vector in the reciprocal lattice must be one-half of the scalar product of the latter vector with itself. **2.** See Laue equations. { 'laů·ə kən,dish·ən }

Laue equations |CRYSTAL| Three equations which must be satisfied for an x-ray beam of specified wavelength to be diffracted through a specified angle by a crystal; they state that the scaler products of each of the crystallographic axial vectors with the difference between unit vectors in the directions of the incident and scattered beams, are integral multiples of the wavelength. Also known as Laue condition. { 'laů·ə i,kwā·zhənz }

Laue method |CRYSTAL| A method of studying crystalline structures by x-ray diffraction, in which a finely collimated beam of polychromatic x-rays falls on a single crystal whose orientation can be set as desired, and diffracted beams are recorded on a photographic film. { 'laů·ə ,meth·əd }

Laue pattern |CRYSTAL| The characteristic photographic record obtained in the Laue method. { 'laů·ə ,pad·ərn }

Laue plane |CRYSTAL| A plane which is the perpendicular bisector of a vector in the reciprocal lattice; such planes form the boundaries of Brillouin zones. { 'laů·ə ,plān }

Laue theory |CRYSTAL| A theory of diffraction of x-rays by crystals, based on the Laue equations. { 'laů·ə ,thē·ə·rē }

Laughlin state |CRYO| The simplest type of quantum Hall state, which contains only one component of incompressible fluid and has a filling factor equal to $1/m$, where m is an integer. { 'läk·lin ,stāt }

lautal |MET| A hard aluminum alloy with small percentages of copper and silicon and traces of iron, manganese, or magnesium. { ¦laů¦tal }

Laves phases |MET| Alloy phases which have the general formula AB_2, and the crystal structures of either $MgCu_2$ (cubic) or $MgZn_2$ or $MgNi_2$ (both hexagonal). { 'lä·vəz ,fāz·əz }

law of constant angles |CRYSTAL| The law that the angles between the faces of a crystal remain constant as the crystal grows. { 'lò əv ¦kän·stənt 'aŋ·gəlz }

law of electric charges |ELEC| The law that like charges repel, and unlike charges attract. { 'lò əv i¦lek·trik 'chärj·əz }

law of electromagnetic induction See Faraday's law of electromagnetic induction. { 'lò əv i¦lek·trō·mag¦ned·ik in'dək·shən }

law of electrostatic attraction See Coulomb's law. { 'lò əv i¦lek·trə¦stad·ik ə'trak·shən }

law of magnetism |ELECTROMAG| The law that like poles repel, and unlike poles attract. { 'lò əv 'mag·nə,tiz·əm }

law of partial pressures See Dalton's law. { 'lò əv ¦pär·shəl 'presh·ərz }

law of rational intercepts See Miller law. { 'lò əv ¦rash·ən·əl 'int·ər,seps }

lawrencium |CHEM| A chemical element, symbol Lr, atomic number 103; isotopes with mass numbers 251–263 have been discovered, all unilable; mass number 262 has the longest half-life (3.6 hours). { 'lò'ren·sē·əm }

lay |MET| The direction of the prevailing surface pattern on a piece of metal after grinding, cutting, lapping, or other processing. { lā }

layer |MET| The stratum of weld metal consisting of one or more passes and lying parallel to the welding surface. { 'lā·ər }

layer lattice See layer structure. { 'lā·ər ,lad·əs }

layer structure |CRYSTAL| A crystalline structure found in substances such as graphites and clays, in which the atoms are largely concentrated in a set of parallel planes, with the regions between the planes comparatively vacant. Also known as layer lattice. { 'lā·ər ,strək·chər }

LCD See liquid crystal display.

LDPE See low-density polyethylene.

leachate |CHEM| A solution formed by leaching. { 'lē,chāt }

leach ion-exchange flotation process |MET| A method of extraction developed for treatment of copper ores not amenable to direct flotation; the metal is dissolved by leaching, for example, with sulfuric acid, in the presence of an ion-exchange resin; the resin recaptures the dissolved metal and is then recovered in a mineralized froth by the flotation process. { 'lēch ¦ī,än iks¦chänj flō'tā·shən ,prä·səs }

lead |CHEM| A chemical element, symbol Pb, atomic number 82, atomic weight 207.19. |ELEC| A wire used to connect two points in a circuit. |MET| A soft, heavy metal with a silvery-bluish color; when freshly cut it is malleable and ductile; occurs naturally, mostly in combination; used principally in alloys in pipes, cable sheaths, type metal, and shields against radioactivity. { led }

lead angle |MET| The angle at the point of welding between an electrode and a line perpendicular to the weld axis. { 'lēd ,aŋ·gəl }

lead antimonite |INORG CHEM| $Pb_3(SbO_4)_2$ Poisonous, water-insoluble orange-yellow powder; used as a paint pigment and to stain glass and ceramics. Also known as antimony yellow; Naples yellow. { 'led an'tim·ə,nīt }

lead arsenate |INORG CHEM| $Pb_3(AsO_4)_2$ Poisonous, water-insoluble white crystals; soluble in nitric acid; used as an insecticide. { 'led 'ärs·ən ,āt }

lead azide |INORG CHEM| $Pb(N_3)_2$ Unstable, colorless needles that explode at 350°C; lead azide is shipped submerged in water to reduce sensitivity; used as a detonator for high explosives. { 'led 'ā,zīd }

lead-base babbitt |MET| Alloy of 10–15% antimony, 2–10% tin, up to 0.2% copper, sometimes with arsenic, the remainder being lead; used as a bearing metal; a variation used in diesel engine and railway bearings contains alkaline-earth metals. Also known as white-metal bearing alloy. { 'led ,bās 'bab·ət }

lead-base grease [MATER] A mixture of soap and mineral oil, usually prepared by the reaction of lead oxide and fatty acid; holds up to extreme pressure and is useful as a gear lubricant. { 'led ,bās 'grēs }

lead borate [INORG CHEM] $Pb(BO_2)_2 \cdot H_2O$ Poisonous, water-insoluble white powder; soluble in dilute nitric acid; used as varnish and paint drier, for galvanoplastic work, in lead glass, and in waterproofing paints. { 'led 'bȯr,āt }

lead bromide [INORG CHEM] $PbBr_2$ An alcohol-insoluble white powder melting at 373°C, boiling at 916°C; slightly soluble in hot water. { 'led 'brō,mīd }

lead bronze [MET] An alloy of 60–70% copper, up to 2% nickel, and up to 15% tin with the balance lead; used as a bearing metal. { 'led 'bränz }

lead burning See lead welding. { 'led ,bər·niŋ }

lead carbonate [INORG CHEM] $PbCO_3$ Poisonous, acid-soluble white crystals decomposing at 315°C; insoluble in alcohol and water; used as a paint pigment. { 'led 'kär·bə,nāt }

lead chloride [INORG CHEM] $PbCl_2$ Poisonous white crystals melting at 498°C, boiling at 950°C; slightly soluble in hot water, insoluble in alcohol and cold water; used to make lead salts and lead chromate pigments and as an analytical reagent. { 'led 'klȯr,īd }

lead chromate [INORG CHEM] $PbCrO_4$ Poisonous, water-insoluble yellow crystals melting at 844°C; soluble in acids; used as a paint pigment. { 'led 'krō,māt }

lead cyanide [INORG CHEM] $Pb(CN)_2$ Poisonous white to yellow powder; slightly soluble in water, decomposed by acids; used in metallurgy. { 'led 'sī·ə,nīd }

lead dioxide [INORG CHEM] PbO_2 Poisonous brown crystals that decompose when heated; insoluble in water and alcohol, soluble in glacial acetic acid; used as an oxidizing agent, in electrodes, batteries, matches, and explosives, as a textile mordant, in dye manufacture, and as an analytical reagent. Also known as anhydrous plumbic acid; brown lead oxide; lead peroxide. { 'led dī'äk,sīd }

leaded alloy [MET] An alloy, especially of brass, bronze, or steel, to which lead is added to improve machinability and mechanical properties. { 'led·əd 'al,ȯi }

leaded zinc oxide [MATER] A mixture of zinc oxide and basic lead sulfate; used in paints. { 'led·əd 'ziŋk 'äk,sīd }

lead fluoride [INORG CHEM] PbF_2 A crystalline solid with a melting point of 824°C; used for laser crystals and electronic and optical applications. { 'led 'flȯr,īd }

lead foil [MET] A foil of lead or of lead alloy containing, for example, 10–12% tin and 1% copper. { 'led ;fȯil }

lead glass [MATER] Glass into which lead oxide is incorporated to give high refractive index, optical dispersion, and surface brilliance; used in optical glass. { 'led ;glas }

lead halide [INORG CHEM] PbX_2, where X is a halogen (such as F, Br, Cl, or I). { 'led 'ha,līd }

lead hexafluorosilicate [INORG CHEM] $PbSiF_6 \cdot 2H_2O$ Poisonous, colorless, water-soluble crystals; used in the electrolytic method for refining lead. { 'led ;hek·sə,flür·ə'sil·ə,kāt }

lead-I-lead junction [SOLID STATE] A Josephson junction consisting of two pieces of lead separated by a thin insulating barrier of lead oxide. Abbreviated Pb-I-Pb junction. { 'led ;ī 'led ,jəŋk·shən }

lead iodide [INORG CHEM] PbI_2 Poisonous, water- and alcohol-insoluble golden-yellow crystals melting at 402°C, boiling at 954°C; used in photography, medicine, printing, mosaic gold, and bronzing. { 'led 'ī·ə,dīd }

lead lanthanum zirconate titanate [MATER] A ferroelectric, ceramic, electrooptical material whose optical properties can be changed by an electric field or by being placed in tension or compression; used in optoelectronic storage and display devices. Also known as lanthanum-doped lead zirconate-lead titanate. { 'led 'lan·thə·nəm 'zərk·ən,āt 'tīt·ən,āt }

lead metallurgy [MET] The science and technology of lead. { 'led 'med·əl,ər·jē }

lead metasilicate See lead silicate. { 'led ,med·ə'sil·ə,kāt }

lead molybdate [INORG CHEM] $PbMoO_4$ Poisonous, acid-soluble yellow powder; insoluble in water and alcohol; used in pigments and as an analytical reagent. { 'led mə'lib,dāt }

lead monoxide [INORG CHEM] PbO Yellow, tetragonal crystals that melt at 888°C and are soluble in alkalies and acids; used in storage batteries, ceramics, pigments, and paints. Also known as litharge; plumbous oxide. { 'led mə'näk,sīd }

lead naphthenate [MATER] Soft, combustible, alcohol-soluble, transparent, resinous material, melting about 100°C; made from addition of lead salt to solution of sodium naphthenate; used as a paint and varnish drier, wood preservative, catalyst, insecticide, and lubricating oil additive. { 'led 'naf·thə,nāt }

lead nitrate [INORG CHEM] $Pb(NO_3)_2$ Strongly oxidizing, poisonous, water- and alcohol-soluble white crystals that decompose at 205–223°C; used as a textile mordant, paint pigment, and photographic sensitizer and in medicines, matches, explosives, tanning, and engraving. { 'led 'nī,trāt }

lead orthoplumbate See lead tetroxide. { 'led ,ȯr·thō'pləm,bāt }

lead oxide red See lead tetroxide. { 'led ;äk,sīd 'red }

lead peroxide See lead dioxide. { 'led pə'räk ,sīd }

lead phosphate [INORG CHEM] Pb_3PO_4 A poisonous, white powder that melts at 1014°C; soluble in nitric acid and in fixed alkali hydroxide; used as a stabilizer in plastics. { 'led 'fäs,fāt }

lead pigments [CHEM] Chemical compounds of lead used in paints to give color; examples are white lead; basic lead carbonate; lead carbonate; lead thiosulfate; lead sulfide; basic lead sulfate (sublimed white lead); silicate white lead; basic lead silicate; lead chromate; basic lead chromate; lead oxychloride; and lead oxide (monoxide and dioxide). { 'led 'pig·məns }

lead silicate [INORG CHEM] $PbSiO_3$ Toxic, insoluble white crystals; used in ceramics, paints, and enamels, and to fireproof fabrics. Also known as lead metasilicate. { 'led 'sil·ə,kāt }

lead-silver babbitt [MET] Alloy of lead with 10–15% antimony, 2.5–5.1% silicon, up to 5% tin, and up to 0.2% copper; used as a bearing metal. { 'led 'sil·vər 'bab·ət }

lead-soap lubricant [MATER] Hard, high-melting-point, extreme-pressure lubricant made of lead salts saponified with fats. { 'led ,sōp 'lü·brə·kənt }

lead sodium hyposulfate See lead sodium thiosulfate. { 'led 'sōd·ē·əm ,hī·pō'səl,fāt }

lead sodium thiosulfate [INORG CHEM] $Na_4Pb(S_2O_3)_3$ Poisonous, small, white, heavy crystals that are soluble in thiosulfate solutions; used in the manufacture of matches. Also known as lead sodium hyposulfate; sodium lead hyposulfate; sodium lead thiosulfate. { 'led 'sōd·ē·əm ,thī·ə'səl,fāt }

lead solder [MET] Solder composed of a lead alloy. { 'led 'säd·ər }

lead sulfate [INORG CHEM] $PbSO_4$ Poisonous white crystals melting at 1170°C; slightly soluble in hot water, insoluble in alcohol; used in storage batteries and as a paint pigment. { 'led 'səl,fāt }

lead sulfide [INORG CHEM] PbS Blue, metallic, cubic crystals that melt at 1120°C, derived from the mineral galena or by reacting hydrogen sulfide gas with a solution of lead nitrate; used in semiconductors and ceramics. Also known as plumbous sulfide. { 'led 'səl,fīd }

lead telluride [INORG CHEM] $PbTe$ A crystalline solid that is very toxic if inhaled or ingested; melts at 902°C; used as a semiconductor and photoconductor in the form of single crystals. { 'led 'tel·yə,rīd }

lead tetroxide [INORG CHEM] Pb_3O_4 A poisonous, bright-red powder, soluble in excess glacial acetic acid and dilute hydrochloric acid; used in medicine, in cement for special applications, in manufacture of colorless glass, and in ship paint. Also known as lead orthoplumbate; lead oxide red; red lead. { 'led ,te'träk,sīd }

lead thiocyanate [INORG CHEM] $Pb(SCN)_2$ Yellow, monoclinic crystals, soluble in potassium thiocyanate and slightly soluble in water; used in the powder mixture that primes small arm cartridges, in dyes, and in safety matches. { 'led ,thī·ō'sī·ə,nāt }

lead titanate [INORG CHEM] $PbTiO_3$ A water-insoluble, pale-yellow solid; used as coloring matter in paints. { 'led 'tīt·ən,āt }

lead tungstate [INORG CHEM] $PbWO_4$ A yellowish powder, melting at 1130°C; insoluble in water, soluble in acid; used as a pigment. Also known as lead wolframate. { 'led 'təŋ·stə,nāt }

lead vanadate [INORG CHEM] $Pb(VO_3)_2$ A water-insoluble, yellow powder; used as a pigment and for the preparation of other vanadium compounds. { 'led 'van·ə,dāt }

lead welding [MET] Welding of lead by fusion. Incorrectly known as lead burning. { 'led ,wel·diŋ }

lead wolframate See lead tungstate. { 'led 'wül·frə,māt }

lead wool [MATER] A coarse lead fiber used to caulk pipe joints. { 'led 'wül }

lead zirconate titanate [MATER] A ferroelectric, ceramic, electrooptic material that has lower optical transparency than lead lanthanum zirconate titanate but similar other properties. { 'led 'zərk·ən,āt 'tīt·ən,āt }

leakage conductance [ELEC] The conductance of the path over which leakage current flows; it is normally a low value. { 'lēk·ij, kən¦dək·təns }

leakage current [ELEC] **1.** Undesirable flow of current through or over the surface of an insulating material or insulator. **2.** The flow of direct current through a poor dielectric in a capacitor. [ELECTR] The alternating current that passes through a rectifier without being rectified. { 'lēk·ij ,kə·rənt }

leakage indicator [ELEC] An instrument used to measure or detect current leakage from an electric system to earth. Also known as earth detector. { 'lēk·ij ,in·də,kād·ər }

leakage resistance [ELEC] The resistance of the path over which leakage current flows; it is normally high. { 'lēk·ij ri¦zis·təns }

lean [MATER] **1.** Of concrete or mortar, containing little or insufficient cement. **2.** Of clay, deficient in plasticity. **3.** Of coal, having little or no volatile matter. **4.** Of lime, containing impurities. **5.** Of ore, being low-grade. { lēn }

lecithin [MATER] **1.** A mixture of phosphatides and oil obtained by drying the separate gums from the degumming of soybean oil; consists of the phosphatides (lecithin), cephalin, other fatlike phosphorus-containing compounds, and 30–35% entrained soybean oil; may be treated to produce more refined grades; used in foods, cosmetics, and paints. **2.** A waxy mixture of phosphatides obtained by refining commercial lecithin to remove the soybean oil and other materials; used in pharmaceuticals. { 'les·ə·thən }

LED See light-emitting diode.

ledeburite [MET] The eutectic of the iron-carbon system, the constituents being cementite and austenite at high temperatures; cooling decomposes the austenite to ferrite and cementite. { 'lā·də,bü,rīt }

ledge [MET] In-gate for a foundry mold. { lej }

LEED See low-energy electron diffraction.

Lee's disk [THERMO] A device for determining the thermal conductivity of poor conductors in which a thin, cylindrical slice of the substance under study is sandwiched between two copper disks, a heating coil is placed between one of these disks and a third copper disk, and the temperatures of the three copper disks are measured. { 'lēz ˌdisk }

left-handed |CRYSTAL| Having a crystal structure with a mirror-image relationship to a right-handed structure. { 'left ¦hand·əd }

leg |MET| In a fillet weld, the distance between the root and the toe. { leg }

Leidenfrost point |THERMO| The lowest temperature at which a hot body submerged in a pool of boiling water is completely blanketed by a vapor film; there is a minimum in the heat flux from the body to the water at this temperature. { 'līd·ən ˌfròst ˌpóint }

Leidenfrost's phenomenon |THERMO| A phenomenon in which a liquid dropped on a surface that is above a critical temperature becomes insulated from the surface by a layer of vapor, and does not wet the surface as a result. { 'līd·ən ˌfròsts fə¦nam·ə¦nän }

lens |OPTICS| A curved piece of ground and polished or molded material, usually glass, used for the refraction of light, its two surfaces having the same axis; or two or more such surfaces cemented together. { lenz }

lens coating |OPTICS| A transparent substance coated on an optical surface to derive maximum light transmission. { 'lenz ˌkōd·iŋ }

lens tissue |MATER| Specially prepared paperlike material for cleaning lenses. { 'lenz ˌtish·ü }

Lenz's law |ELECTROMAG| The law that whenever there is an induced electromotive force (emf) in a conductor, it is always in such a direction that the current it would produce would oppose the change which causes the induced emf. { 'lenz·əz ˌlò }

LEPD See low-energy positron diffraction.

Leslie cube |THERMO| A metal box, with faces having different surface finishes, in which water is heated and next to which a thermopile is placed in order to compare the heat emission properties of different surfaces. { 'lez·lē ˌkyüb }

leucitohedron See trapezohedron. { ˌlü·sə· tō'hē·drən }

leveling |MET| Flattening rolled sheet by evening out irregularities, using a roller or tensile straining. { 'lev·ə·liŋ }

leveling action |MET| The property exhibited by a plating solution in making the coating smoother than the base metal. { 'lev·ə·liŋ ˌak·shən }

levigated abrasive |MATER| A fine abrasive powder for final burnishing of metals or for metallographic polishing; the powder is usually processed to make it chemically neutral. { 'lev·ə ˌgād·əd ə'brā·siv }

levitation heating |MET| Providing heat through high-frequency magnetic fields; employed in levitation melting. { ˌlev·ə'tā·shən ¦hēd·iŋ }

levitation melting |MET| Melting metal out of contact with a supporting material by using the induced current provided by a high-frequency surrounding magnetic field to suspend the melt. { ˌlev·ə'tā·shən ¦mel·tiŋ }

Lewis acid |CHEM| A substance that can accept an electron pair from a base; thus, $AlCl_3$, BF_3, and SO_3 are acids. { 'lü·əs ˌas·əd }

Lewis base |CHEM| A substance that can donate an electron pair; examples are the hydroxide ion, OH^-, and ammonia, NH_3. { 'lü·əs ˌbās }

Li See lithium.

lichen blue See litmus. { 'lī·kən 'blü }

ligand |CHEM| The molecule, ion, or group bound to the central atom in a chelate or a coordination compound; an example is the ammonia molecules in $|Co(NH_3)_6|^{3+}$. { 'lī·gənd }

light |OPTICS| **1.** Electromagnetic radiation with wavelengths capable of causing the sensation of vision, ranging approximately from 400 (extreme violet) to 770 nanometers (extreme red). Also known as light radiation; visible radiation. **2.** More generally, electromagnetic radiation of any wavelength; thus, the term is sometimes applied to infrared and ultraviolet radiation. { līt }

light absorption |OPTICS| The process in which energy of light radiation is transferred to a medium through which it is passing. { 'līt əb ˌsorp·shən }

light bulb See incandescent lamp. { 'līt ˌbəlb }

light-drawn |MET| Cold-worked very slightly; for copper or copper alloy tubing, drawing entails between 10 and 25% reduction in area. { 'līt ¦dròn }

light-emitting diode |ELECTR| A rectifying semiconductor device which converts electrical energy into electromagnetic radiation. The wavelength of the emitted radiation ranges from the near-ultraviolet to the near-infrared, that is, from about 400 to over 1500 nanometers. Abbreviated LED. { 'līt iˌmid·iŋ 'dī·ōd }

light-emitting polymer See polymer light-emitting diode. { ¦lītəˌmid·iŋ 'pöl·ə·mər }

light filter See color filter. { 'līt ˌfil·tər }

light intensity See luminous intensity. { 'līt in ˌten·səd·ē }

light metal |MET| A metal or alloy of low density, especially aluminum and magnesium alloys. { 'līt ¦med·əl }

light radiation See light. { 'līt ˌrād·ē·ā·shən }

light ray |OPTICS| A beam of light having a small cross section. { 'līt ˌrā }

light scattering |OPTICS| The process in which energy is removed from a beam of light radiation and reemitted without appreciable change in wavelength. { 'līt ˌskad·ə·riŋ }

light-sensitive |ELECTR| Having photoconductive, photoemissive, or photovoltaic characteristics. Also known as photosensitive. { 'līt 'sensəd·iv }

light transmission |OPTICS| The process in which light travels through a medium without being absorbed or scattered. { 'līt tranzˌmish·ən }

light valve |ELECTR| **1.** A device whose light transmission can be made to vary in accordance with an externally applied electrical quantity, such as voltage, current, electric field, or magnetic field, or an electron beam. **2.** Any direct-view electronic display optimized for reflecting or transmitting an image with an independent collimated light source for projection purposes. { 'lītˌvalv }

lightweight aggregate |MATER| A lightweight inert material, such as foamed slag, vermiculite, clinker, and perlite, used in unreinforced concrete for making structures of low weight and high insulation. { 'līt‚wāt 'ag·rə·gət }

lightweight concrete |MATER| A type of concrete made with lightweight aggregate. { 'līt‚wāt 'kän ‚krēt }

lignaloe oil *See* linaloe wood oil. { lī'na·lō ‚óil }

lignin |MATER| A colorless to brown substance removed from paper-pulp sulfite liquor. { 'lig· nən }

lignin plastic |MATER| A plastic based on resins derived from lignin; used as a binder or extender. { 'lig·nən 'plas·tik }

lignite wax *See* montan wax. { 'lig‚nīt 'waks }

lime-and-cement mortar |MATER| Mortar made of mortar cement, lime putty or hydrated lime, and sand in proportions, by volume, normally of one part cement, one or two lime, and five or six sand; suited for all kinds of masonry. { ¦līm ən si¦ment 'mórd·ər }

lime glass |MATER| A type of glass containing a high proportion of lime; used in many commercial glass products, such as bottles. { 'līm ‚glas }

lime grease |MATER| A type of grease that emulsifies less readily than one made with a soda base and therefore is used in an environment where water may occur. { 'līm ‚grēs }

lime mortar |MATER| A mixture of hydrated lime, sand, and water having a compressive strength up to 400 pounds per square inch (2.8 × 10^6 pascals); used for interior non-load-bearing walls in buildings. { 'līm ¦mórd·ər }

lime oil |MATER| **1.** An edible essential oil squeezed from the rind of lime and other citrus fruit, whose components are limonene and citral; used in flavorings and perfumes. **2.** The distilled essential oil from citron. { 'līm‚óil }

lime putty |MATER| A puttylike cement made from lime slaked in water. { 'līm ¦pəd·ē }

liminal contrast *See* threshold contrast. { 'lim·ə·nəl 'kän‚trast }

limiting current density |PHYS CHEM| The maximum current density to achieve a desired electrode reaction before hydrogen or other extraneous ions are discharged simultaneously. { 'lim·əd·iŋ ¦kə·rənt ‚den·səd·ē }

limiting oxygen index |MATER| For a specific material, the lowest concentration of oxygen in the atmosphere, expressed in percent, that will support sustained combustion of the material. Abbreviated LOI. { ¦lim·əd·iŋ 'äks·ə·jən ‚in ‚deks }

limiting viscosity number *See* intrinsic viscosity. { 'lim·əd·iŋ vi'skäs·əd·ē ‚nəm·bər }

linaloe wood oil |MATER| An essential oil derived from fruit and wood of *Bursera* species; a colorless to yellow liquid, soluble in fixed oils and alcohol; used in perfumery and for flavoring. Also known as lignaloe oil; Mexican linaloe oil. { lē'näl·ō·ā ¦wúd ‚óil }

lindane |INORG CHEM| The gamma isomer of 1,2,3,4,5,6-hexachlorocyclohexane, constituting a persistent, bioaccumulative pesticide and a neurotoxin. { 'lin‚dān }

Lindemann glass |MATER| A lithium borate-beryllium oxide glass having no element higher in atomic number than oxygen; used as window material for low-voltage x-ray tubes because it will pass x-rays of extremely long wavelength, such as Grenz rays. { 'lin·də·mən ‚glas }

Lindemann theory |SOLID STATE| A theory of the melting point of solids according to which solids melt when the amplitude of oscillation of the atoms becomes so great that neighboring atoms collide. { 'lin·də·mən ‚thē·ə·rē }

Linde process |CRYO| A cyclic process for liquefying gases in which compressed gas is cooled by Joule-Thomson expansion through a valve to a pressure of about 40 atmospheres (4 megapascals), further cools the incoming gas in a heat exchanger, and is compressed for the next cycle. { 'lin·də ‚prä‚ses }

Linde's rule |SOLID STATE| The rule that the increase in electrical resistivity of a monovalent metal produced by a substitutional impurity per atomic percent impurity is equal to $a + b\,(v - 1)^2$, where a and b are constants for a given solvent metal and a given row of the periodic table for the impurity, and v is the valence of the impurity. { 'lin·dəz ‚rül }

lineage structure |CRYSTAL| An imperfection structure characterizing a crystal, parts of which have slight differences in orientation. { 'lin·ē·ij ‚strək·chər }

linear collision cascade |SOLID STATE| A sputtering event in which the bombarding projectile collides directly with a small number of target atoms, which collide with others, and the sharing of energy then proceeds through many generations before one or more target atoms are ejected; the density of atoms in motion remains sufficiently small so that collisions between atoms can be ignored. { 'lin·ē·ər kə'lizh·ən ‚kas ‚kād }

linear expansion |PHYS| Expansion of a body in one direction. { 'lin·ē·ər ik'span·chən }

linear expansity *See* coefficient of linear expansion. { 'lin·ē·ər ik'span·səd·ē }

linear polymer |POLYM CHEM| A polymer whose molecule is arranged in a chainlike fashion with few branches or bridges between the chains. { 'lin·ē·ər 'päl·ə·mər }

linear strain |MECH| The ratio of the change in the length of a body to its initial length. Also known as longitudinal strain. { 'lin·ē·ər ¦strān }

line defect *See* dislocation. { 'līn di‚fekt }

liner |MET| **1.** The cylindrical chamber that holds the billet for extrusion. **2.** The slab of coating metal that is placed on the core alloy and is subsequently rolled down to form a clad composite. { 'līn·ər }

lining |MATER| A material used to protect inner surfaces, as of tunnels, pipes, or process equipment. { 'līn·iŋ }

linishing *See* belt grinding. { 'li·nə·shiŋ }

linoleum |MATER| A floor covering made by applying a mixture of gelled linseed oil, pigments, fillers, and other materials to a burlap backing, and curing to produce a hard, resilient sheet. { lə'nō·lē·əm }

linseed cake [MATER] The residue formed during pressing of commercial linseed oil; used for cattle feed and fertilizer. { 'lin₁sēd ₁kāk }

linseed oil [MATER] A product made from the seeds of the flax plant by crushing and pressing either with or without heat; formulated in various grades and with various drying agents and used as a vehicle in oil paints and as a component of oil varnishes. { 'lin₁sēd ₁óil }

lint [MATER] During the first stage of processing cotton, the fiber that is separated from the seeds in a cotton gin. { lint }

linters [MATER] Short residual fibers that adhere to ginned cottonseed; used for making fabrics that do not require long fibers, as plastic fillers, and in the manufacture of cellulosic plastics. { 'lin·tərz }

lipophilic [CHEM] **1.** Having a strong affinity for fats. **2.** Promoting the solubilization of lipids. { ¦lip·ə¦fil·ik }

lipophobic [CHEM] Lacking an affinity for, repelling, or failing to absorb or adsorb fats. { ₁lip·ə'fōb·ik }

liquation [MET] **1.** Separation of fusible metals from less fusible ones by applying heat. **2.** The partial melting of an alloy. { lī'kwā·shən }

liquid [PHYS] A state of matter intermediate between that of crystalline substances and gases in which a substance has the capacity to flow under extremely small shear stresses and conforms to the shape of a confining vessel, but is relatively incompressible, lacks the capacity to expand without limit, and can possess a free surface. { 'lik·wəd }

liquid A [CRYO] A phase of superfluid helium-3 in which the helium-3 pairs only occur in those two of the three possible nuclear spin states in which the nuclear spins are parallel, and these pairs couple coherently to give macroscopic orbital and spin angular momenta and anisotropic superfluid properties. Also known as A phase. { 'lik·wəd 'ā }

liquid A₁ [CRYO] A phase of liquid helium-3 intermediate between liquid A and liquid B that appears only in the presence of a magnetic field and then only in a narrow portion of the pressure-temperature diagram, and in which only pairs of one of the three possible nuclear spin states are superfluid. Also known as A₁ phase. { 'lik·wəd 'ā 'wən }

liquid B [CRYO] A phase of superfluid helium-3 in which pairs of all three possible nuclear spin states are coupled to give superfluid properties that are isotropic except in the more subtle aspects of the spin configuration. Also known as B phase. { 'lik·wəd 'bē }

liquid blast cleaning [MET] Cleaning metal surfaces with a suspension of abrasive in water accelerated to high velocities by compressed air, or by a centrifugal wheel. { 'lik·wəd 'blast ₁klēn·iŋ }

liquid bright gold [MATER] Any of several gold compounds applied to ceramics in the form of varnish which is dried and heated to redness, decomposing the compound and leaving a thin film of gold firmly attached to the underlying ceramic; used in decorating china and for the production of printed electrical circuits on ceramics. { 'lik·wəd 'brīt ¦gōld }

liquid carburizing [MET] Surface hardening of steel by immersion into a molten bath consisting of cyanides and other salts, for example, at 1600–1750°F (850–950°C). { 'lik·wəd 'kär·bə₁rīz·iŋ }

liquid crystal [PHYS CHEM] A liquid which is not isotropic; it is birefringent and exhibits interference patterns in polarized light; this behavior results from the orientation of molecules parallel to each other in large clusters. { 'lik·wəd 'krist·əl }

liquid crystal display [ELECTR] A digital display that consists of two sheets of glass separated by a sealed-in, normally transparent, liquid crystal material; the outer surface of each glass sheet has a polarizer and the inner surface of each glass plate has a transparent conductive coating such as tin oxide or indium oxide, with the viewing-side coating etched into character-forming segments that have leads going to the edges of the display; a voltage applied between front and back electrode coatings disrupts the orderly arrangement of the molecules, darkening the display enough to form visible characters even though no light is generated. Abbreviated LCD. { 'lik·wəd 'krist·əl di'splā }

liquid crystal polymers [POLYM CHEM] Aromatic copolymers that have characteristically high-temperature resistance, yet can be melted and molded. Upon melting, the polymer chains undergo parallel ordering in the direction of the flow, resulting in superior mechanical properties in that direction. { ¦lik·wəd ¦krist·əl 'päl·ə·mərs }

liquid dioxide See nitrogen dioxide. { 'lik·wəd dī'äk₁sīd }

liquid fluorine [CRYO] Cold, liquefied fluorine gas; used as a cryogenic propellant. { 'lik·wəd 'flúr₁ēn }

liquid glass See sodium silicate. { 'lik·wəd 'glas }

liquid grease [MATER] Lubricating oil of light or medium grade that is thickened with calcium soap. { 'lik·wəd 'grēs }

liquid helium [CRYO] The state of helium which exists at atmospheric pressure at temperatures below −268.95°C (4.2 K), and for temperatures near absolute zero at pressures up to about 25 atmospheres (2.53 megapascals); has two phases, helium I and helium II. { 'lik·wəd 'hē·lē·əm }

liquid honing See vapor blasting. { 'lik·wəd 'hōn·iŋ }

liquid hydrogen [CRYO] Hydrogen that exists as a liquid at atmospheric pressure, at −252.7°C (20.46 K); used for high-impulse rocket fuels. { 'lik·wəd 'hī·drə·jən }

liquid insulator [MATER] A liquid with a resistivity greater than about 10^{14} ohm-centimeters, such as a petroleum oil, silicone oil, or halogenated aromatic hydrocarbon. { 'lik·wəd 'in·sə ₁lād·ər }

liquid junction potential See diffusion potential. { 'lik·wəd ¦jəŋk·shən pə'ten·chəl }

liquid-metal embrittlement [MET] The rapid loss of mechanical properties of a metal or an alloy due to contact with certain liquid metals. { 'lik·wəd ¦med·əl em'brid·əl·mənt }

liquid methane [CRYO] Methane that has been cooled to at least −161°C; used for cryogenic applications and for tankship transport of methane. { 'lik·wəd 'meth,ān }

liquid nitrogen [CRYO] Nitrogen that exists as a liquid at atmospheric pressure, at −195°C (77.4 K); used in research work, cryogenics, and cryosurgery. { 'lik·wəd 'nī·trə·jən }

liquid oxygen [CRYO] Oxygen that exists as a liquid at atmospheric pressure, at −182.97°C (90.18 K); a pale-blue, transparent, mobile liquid. { 'lik·wəd 'äk·sə·jən }

liquid penetrant test [ENG] A penetrant method of nondestructive testing used to locate defects open to the surface of nonporous materials; penetrating liquid is applied to the surface, and after 1−30 minutes excess liquid is removed, and a developer is applied to draw the penetrant out of defects, thus showing their location, shape, and size. { 'lik·wəd 'pen·ə·trənt ,test }

liquid-phase epitaxy [SOLID STATE] A process for growing thin epitaxial layers on a crystalline substrate in which the substrate is sequentially brought into contact with solutions that are at the desired composition and may be supersaturated or cooled to achieve growth. Abbreviated LPE. { 'lik·wəd ¦fāz 'ep·ə,tak·sē }

liquid rosin *See* tall oil. { 'lik·wəd 'räz·ən }

liquidus line [THERMO] For a two-component system, a curve on a graph of temperature versus concentration which connects temperatures at which fusion is completed as the temperature is raised. { 'lik·wəd·əs ,līn }

liquor finish [MET] A bright, smooth finish on wet-drawn wire achieved by using fermented-grain mash liquor as a lubricant. { 'lik·ər ,fin·ish }

lithamide *See* lithium amide. { 'lith·ə,mīd }

litharge *See* lead monoxide. { 'li,thärj }

litharge-glycerin cement [MATER] Mixture of glycerin, water, and litharge (lead monoxide) to give, when cured, an acid-resistant cement. { 'li ,thärj ¦glis·ə·rən si'ment }

lithium [CHEM] A chemical element, symbol Li, atomic number 3, atomic weight 6.939; an alkali metal. { 'lith·ē·əm }

lithium aluminum hydride [INORG CHEM] $LiAlH_4$ A compound made by the reaction of lithium hydride and aluminum chloride; a powerful reducing agent for specific linkages in complex molecules; used in organic synthesis. { 'lith·ē·əm ə'lü·mə·nəm 'hī,drīd }

lithium amide [INORG CHEM] $LiNH_2$ A compound crystallizing in the cubic form, and melting at 380–400°C; used in organic synthesis. Also known as lithamide. { 'lith·ē·əm 'am,īd }

lithium bromide [INORG CHEM] $LiBr·H_2O$ A white, deliquescent, granular powder with a bitter taste, melting at 547°C; soluble in alcohol and glycol; used to add moisture to air-conditioning systems and as a sedative and hypnotic in medicine. { 'lith·ē·əm 'brō,mīd }

lithium carbonate [INORG CHEM] Li_2CO_3 A colorless, crystalline compound that melts at 700°C and has slight solubility in water; used in ceramic industries in the manufacture of powdered glass for porcelain enamel formulation. { 'lith·ē·əm 'kär·bə,nāt }

lithium cell [CHEM] An electrolytic cell for the production of metallic lithium. [ELEC] A primary cell for producing electrical energy by using lithium metal for one electrode immersed in usually an organic electrolyte. { 'lith·ē·əm ,sel }

lithium chloride [INORG CHEM] $LiCl·2H_2O$ A colorless, water-soluble compound, forming octahedral crystals and melting at 614°C; used to form concentrated brine in commercial air-conditioning systems and as a pyrotechnic in welding and brazing fluxes. { 'lith·ē·əm 'klòr ,īd }

lithium fluoride [INORG CHEM] LiF Poisonous, white powder melting at 870°C, boiling at 1670°C; insoluble in alcohol, slightly soluble in water, and soluble in acids; used as a heat-exchange medium, as a welding and soldering flux, in ceramics, and as crystals in infrared instruments. { 'lith·ē·əm 'flúr,īd }

lithium grease [MATER] Heat-stable, water-resistant lubricating grease with lithium salts of higher fatty acids (or lithium soaps of fatty glycerides) as a base; used for low-temperature service in aircraft. { 'lith·ē·əm 'grēs }

lithium halide [INORG CHEM] A binary compound of lithium, LiX, where X is a halide; examples are lithium chloride, LiCl, and lithium fluoride, LiF. { 'lith·ē·əm 'hal,īd }

lithium hydride [INORG CHEM] LiH Flammable, brittle, white, translucent crystals; decomposes in water; insoluble in ether, benzene, and toluene; used as a hydrogen source and desiccant, and to prepare lithium amide and double hydrides. { 'lith·ē·əm 'hī,drīd }

lithium hydroxide [INORG CHEM] $LiOH$; $LiOH·H_2O$ Colorless crystals; used as a storage-battery electrolyte, as a carbon dioxide absorbent, and in lubricating greases and ceramics. { 'lith·ē·əm hī'dräk,sīd }

lithium iodide [INORG CHEM] LiI; $LiI·3H_2O$ White, water- and alcohol-soluble crystals; LiI melts at 446°C; $LiI·3H_2O$ loses water at 72°C; used in medicine, photography, and mineral waters. { 'lith·ē·əm 'ī·ə,dīd }

lithium molybdate [INORG CHEM] Li_2MoO_4 Water-soluble white crystals melting at 705°C; used as a catalytic cracking (petroleum) catalyst and as a mill additive for steel. { 'lith·ē·əm mə'lib,dāt }

lithium nitrate [INORG CHEM] $LiNO_3$ Water- and alcohol-soluble colorless powder melting at 261°C; used as a heat-exchange medium and in ceramics, pyrotechnics, salt baths, and refrigeration systems. { 'lith·ē·əm 'nī,trāt }

lithium perchlorate [INORG CHEM] $LiClO_4·3H_2O$ A compound with high oxygen content (60% available oxygen), used as a source of oxygen in rockets and missiles. { 'lith·ē·əm pər'klòr,āt }

lithium tetraborate [INORG CHEM] $Li_2B_4O_7·5H_2O$ White crystals that lose water at 200°C; insoluble in alcohol, soluble in water; used in ceramics. { 'lith·ē·əm ˌte·trə'bȯr,āt }

lithium titanate [INORG CHEM] Li_2TiO_3 A water-insoluble white powder with strong fluxing ability when used in titanium-containing enamels; also used as a mill additive in vitreous and semivitreous glazes. { 'lith·ē·əm 'tī·tən,āt }

lithography [ELECTR] A process for defining a pattern, representing a layer of an integrated circuit, on the surface of a silicon wafer; most commonly, the wafer is coated with a light-sensitive material (resist), and then the resist is exposed and developed to reveal a pattern which is transformed to the underlying surface through etching techniques. { lə'thäg·rə·fē }

lithol red [MATER] Any of various pigments derived from combination of β-naphthol and Tobias acid; available as sodium, barium, and calcium toners and lakes; used in outside, drum, and toy enamels. { 'li,thȯl 'red }

lithophone *See* lithopone. { 'lith·ə,fōn }

lithopone [MATER] A white pigment produced as a filtered, heated, quenched precipitate from reaction of barium sulfide and zinc sulfide; used as a pigment for paint, ink, filled leather, paper, linoleum, oilcloth, and cosmetics. Also known as Charlton white; Griffith's white; lithophone; Orr's white; zinc baryta white; zinc sulfide white. { 'lith·ə,pōn }

litmus [MATER] Blue, water-soluble powder from various lichens, especially *Variolaria lecanora* and *V. rocella*; turns red in solutions at pH 4.5, and blue at pH 8.3; used as an acid-base indicator. Also known as lacmus; lichen blue. { 'lit·məs }

litmus paper [MATER] White, unsized paper saturated by litmus in water; used as a pH indicator. { 'lit·məs ,pā·pər }

live load [MECH] A moving load or a load of variable force acting upon a structure, in addition to its own weight. { 'līv 'lōd }

living polymerization [POLYM CHEM] A polymerization reaction in which there is no termination reaction (that is, terminating species) and chain propagation continues until all the monomer is consumed. { 'liv·iŋ'pə,lim·ə·rə'zā·shən }

load [MECH] **1.** The weight that is supported by a structure. **2.** Mechanical force that is applied to a body. **3.** The burden placed on any machine, measured by units such as horsepower, kilowatts, or tons. { lōd }

load-bearing tile [MATER] A tile with the capacity to support superimposed loads. { 'lōd ¦ber·iŋ ,tīl }

loaded concrete [MATER] Concrete to which elements of high atomic number or capture cross section have been added to increase its effectiveness as a radiation shield in nuclear reactors. { 'lōd·əd kän'krēt }

load factor [MECH] The ratio of load to the maximum rated load. { 'lōd ,fak·tər }

loading [FL MECH] **1.** The relative concentration of particles in a flowing fluid. **2.** In particular, the ratio of particle mass flow to fluid mass flow. [MET] Filling of a die cavity with powdered metal. { 'lōd·iŋ }

load stress [MECH] Stress that results from a pressure or gravitational load. { 'lōd ,stres }

loam [MET] Molding material consisting of sand, silt, and clay used over backup material for producing massive castings, usually of iron or steel. { lōm }

local action [MET] Electrochemical corrosion resulting from the action of local cells. { 'lō·kəl 'ak·shən }

local buckling [MECH] Buckling of thin elements of a column section in a series of waves or wrinkles. { 'lō·kəl 'bək·liŋ }

local coefficient of heat transfer [THERMO] The heat transfer coefficient at a particular point on a surface, equal to the amount of heat transferred to an infinitesimal area of the surface at the point by a fluid passing over it, divided by the product of this area and the difference between the temperatures of the surface and the fluid. { 'lō·kəl ,kō·i'fish·ənt əv 'hēt ,tranz·fər }

local preheating [MET] The heating of a specific portion of a material or structure prior to the performance of a joining or fabrication process. { 'lō·kəl prē'hēd·iŋ }

local structural discontinuity [MECH] The effect of intensified stress on a small portion of a structure. { 'lō·kəl 'strək·chə·rəl dis,känt·ən'ü·əd·ē }

lock [MET] A condition in forging in which the flash line is in more than one plane. { läk }

lockalloy [MET] A beryllium-base alloy composed of 62% beryllium and 38% aluminum; used as a structural aerospace alloy because of low density and high (47,000 pounds per square inch or 3.2×10^8 pascals) yield strength. { 'läk·ə,lȯi }

LOI *See* limiting oxygen index.

London equations [SOLID STATE] Equations for the time derivative and the curl of the current in a superconductor in terms of the electric and magnetic field vectors, respectively, derived in the London superconductivity theory. { 'lən·dən i ¦kwā·zhənz }

London penetration depth [SOLID STATE] A measure of the depth which electric and magnetic fields can penetrate beneath the surface of a superconductor from which they are otherwise excluded, according to the London superconductivity theory. { 'lən·dən ,pen·ə'trā·shən ,depth }

London superconductivity theory [SOLID STATE] An extension of the two-fluid model of superconductivity, in which it is assumed that superfluid electrons behave as if the only force acting on them arises from applied electric fields, and that the curl of the superfluid current vanishes in the absence of a magnetic field. { 'lən·dən ¦sü·pər ,kän,dək'tiv·əd·ē ,thē·ə·rē }

London superfluidity theory [CRYO] A theory, based on the fact that helium-4 obeys Bose-Einstein statistics, in which helium-4 is treated as an ideal Bose-Einstein gas, and its superfluid component is equated with the finite fraction of the atoms of such a gas which are in the ground state at very low temperatures. { 'lən·dən ,sü·pər,flü'id·əd·ē ,thē·ə·rē }

long clay [MATER] A clay used in ceramics that has a high degree of plasticity. { 'lȯŋ ¦klā }

longitudinal direction [MET] The principal direction of flow in a plastically deformed metal. { ˌlän·jə'tüd·ən·əl di'rek·shən }

longitudinal magnetorestriction See Joule effect. { ˌlän·jə'tüd·ən·əl mag¦ned·ō·ri'strik·shən }

longitudinal resistance seam welding [MET] The performance of resistance seam welding parallel to the throat depth of the welding machine. { ˌlän·jə'tüd·ən·əl ri¦zis·təns 'sēm ˌweld·iŋ }

longitudinal sequence [MET] The sequence in which successive welds are deposited along the length of a continuous weld. { ˌlän·jə'tüd·ən·əl 'sē·kwəns }

longitudinal strain See linear strain. { ˌlän·jə'tüd·ən·əl 'strān }

long oil [MATER] Varnish containing a large percentage of oil. { 'lòŋ ¦òil }

long-range order [SOLID STATE] A tendency for some property of atoms in a lattice (such as spin orientation or type of atom) to follow a pattern which is repeated every few unit cells. { 'lòŋ ˌrānj 'òr·dər }

looping mill [MET] An arrangement of mills such that a hot bar discharged from one mill is fed into a second mill in the opposite direction. { 'lüp·iŋ ˌmil }

Lorentz electron [ELECTROMAG] A model of the electron as a damped harmonic oscillator; used to explain the variation of the real and imaginary parts of the index of refraction of a substance with frequency. { 'lòr·ens i'lek·trän }

Lorentz equation [ELECTROMAG] The equation of motion for a charged particle, which sets the rate of change of its momentum equal to the Lorentz force. { 'lòr·ens i'kwā·zhən }

Lorentz force [ELECTROMAG] The force on a charged particle moving in electric and magnetic fields, equal to the particle's charge times the sum of the electric field and the cross product of the particle's velocity with the magnetic flux density. { 'lòr·ens ˌfòrs }

Lorentz local field [ELEC] In a theory of electric polarization, the average electric field due to the polarization at a molecular site that is calculated under the assumption that the field due to polarization by molecules inside a small sphere centered at the site may be neglected. Also known as Mossotti field. { 'lòr·ens ¦lō·kəl 'fēld }

Lorentz number [SOLID STATE] The thermal conductivity of a metal divided by the product of its temperature and its electrical conductivity, according to the Wiedemann-Franz law. { 'lòr ˌens ˌnəm·bər }

Lorentz relation See Wiedemann-Franz law. { 'lòr·ens ri¦lā·shən }

loss current [ELEC] The current which passes through a capacitor as a result of the conductivity of the dielectric and results in power loss in the capacitor. [ELECTROMAG] The component of the current across an inductor which is in phase with the voltage (in phasor notation) and is associated with power losses in the inductor. { 'lòs ˌkə·rənt }

lossless material [PHYS] An ideal material that dissipates none of the energy of electromagnetic or acoustic waves passing through it. { 'lòs·ləs mə'tir·ē·əl }

lossy material [PHYS] A material that dissipates energy of electromagnetic or acoustic energy passing through it. { 'lòs·ē mə'tir·ē·əl }

lost circulation material See bridging material. { 'lòst ˌsər·kyə'lā·shən mə,tir·ē·əl }

lost-wax process [MET] A method used in investment casting in which a wax pattern between a two-layered mold is removed by melting and replaced with molten metal; used for casting bronze statues and in jewelry casting. Also known as investment process. { 'lòst¦waks ˌprä·səs }

low-alloy steel [MET] A hardenable carbon steel generally containing not more than about 1% carbon and one or more of the following alloyed components: < (less than) 2% manganese, <4% nickel, <2% chromium, <0.6% molybdenum, and <0.2% vanadium. { ¦lō ¦al,òi 'stēl }

low boiler [MATER] A fast-evaporating solvent used in lacquer thinner to give a rapid initial set. { 'lō 'bòil·ər }

low brass [MET] Brass containing 20% zinc, 80% copper. { 'lō ¦bras }

low-carbon steel [MET] Steel containing 0.15% or less of carbon. { 'lō ˌkär·bən 'stēl }

low-density polyethylene [POLYM CHEM] A thermoplastic polymer with a density of 0.910–0.940 gram per cubic centimeter (0.526–0.543 ounce per cubic inch). Abbreviated LDPE. { 'lō ˌden·səd·ē ˌpäl·ē'eth·ə,lēn }

low-energy electron diffraction [SOLID STATE] A technique for studying the atomic structure of single crystal surfaces, in which electrons of uniform energy in the approximate range 5–500 electronvolts are scattered from a surface, and those scattered electrons that have lost no energy are selected and accelerated to a fluorescent screen where the diffraction pattern from the surface can be observed. Abbreviated LEED. { 'lō ˌen·ər·jē i,lek,trän di'frak·shən }

low-energy positron diffraction [SOLID STATE] A technique for studying the atomic structure of solid surfaces in which a narrow beam of low-energy monoenergetic positrons is made to strike a solid surface, and the diffracted beams in certain directions that are permitted by the regular array of surface atoms are observed. Abbreviated LEPD. { 'lō ˌen·ər·jē 'päz·ə,trän di ˌfrak·shən }

lower critical field [SOLID STATE] The magnetic field strength below which magnetic flux is completely excluded from type II superconductor and above which it penetrates the superconductor as microscopic filaments called fluxoids. Symbolized H_{c1}. { ¦lō·ər ¦krid·i·kəl 'fēld }

lower heating value See low heat value. { 'lō·ər 'hēd·iŋ ˌval·yü }

lower punch [MET] In powder metallurgy, the portion of the die forming the bottom of the die cavity. { 'lō·ər 'pənch }

lower yield point [MET] In annealed carbon steels, the lowest value of stress after the initial dropoff and before the load begins to rise continuously. { 'lō·ər 'yēld ˌpòint }

low-expansion alloy [MET] An alloy whose dimensions do not vary appreciably with temperature. { 'lō ik,span·chən 'al,ói }

low-frequency current [ELEC] An alternating current having a frequency of less than about 300 kilohertz. { 'lō ,frē·kwən·sē 'kə·rənt }

low-frequency cycle [MET] In resistance welding, one positive- and one negative-current pulse within a heat time at a lower frequency than the electrical power source. { 'lō ,frē·kwən·sē 'sī·kəl }

low-heat cement [MATER] A chemically altered portland cement with a low initial heat liberation. { 'lō ,hēt si'ment }

low heat value [THERMO] The heat value of a combustion process assuming that none of the water vapor resulting from the process is condensed out, so that its latent heat is not available. Also known as lower heating value; net heating value. { 'lō 'hēt ,val·yü }

low-hydrogen electrode [MET] A covered electrode used in arc welding that provides an atmosphere low in hydrogen. { 'lō ,hī·drə·jən i'lek,trōd }

low-loss [ELEC] Having a small dissipation of electric or electromagnetic power. { 'lō ¦lòs }

low-melting glass [MATER] Glass to which selenium, thallium, arsenic, or sulfur is added to give melting points of 260–660°F (127–349°C). { 'lō ,mel·tiŋ 'glas }

low-melting solder See soft solder. { 'lō ,mel·tiŋ 'säd·ər }

low-pressure laminate [MATER] A plastic laminate molded and cured at pressures in general of 400 pounds per square inch (approximately 27 atmospheres or 2.8 × 10⁶ pascals). { 'lō ¦presh·ər 'lam·ə·nət }

low-residual-phosphorus copper [MET] Deoxidized copper with a 0.004–0.012% residual phosphorus content. { 'lō ri¦zij·ə·wəl ¦fäs·fə·rəs ,käp·ər }

low-shaft furnace [MET] A blast furnace having a short shaft; used to produce pig iron, ferroalloys, alumina, and other products from low-grade ores using low-grade fuel. { 'lō ,shaft 'fər·nəs }

low-temperature physics [CRYO] A study of the properties of gross matter at low temperatures, especially at temperatures so low that the quantum character of the substance becomes observable in effects such as superconductivity, superfluid liquid helium, magnetic cooling, and nuclear orientation. { 'lō ,tem·prə·chər 'fiz·iks }

low-temperature production [CRYO] Production of temperatures from about 80 K down to about 10⁻⁶ K by techniques such as isentropic expansion of gases, refrigeration cycles, and adiabatic demagnetization. { 'lō ,tem·prə·chər prə'dək·shən }

low-temperature thermometry [CRYO] The assignment of numbers on the Kelvin absolute temperature scale to achievable and reproducible low-temperature states, and the choice and calibration of suitable instruments for the practical measurement of low temperatures, such as thermocouples, and resistance, vapor-

pressure, gas, and magnetic thermometers. { 'lō ,tem·prə·chər thər'mäm·ə·trē }

LPE See liquid-phase epitaxy.

Lr See lawrencium.

Lu See lutetium.

lubricant [MATER] A substance used to reduce friction between parts or objects in relative motion. { 'lü·brə·kənt }

lubricant additive [MATER] Any material added to lubricants (greases or oils) to give the product special properties, such as resistance to extremes of pressure, cold, or heat, improved viscosity, and detergency. { 'lü·brə·kənt ,ad·əd·iv }

lubricating film [MATER] A thin layer of oil or grease applied between rubbing surfaces. { 'lü·brə,kād·iŋ ,film }

lubricating grease [MATER] A solid or semisolid lubricant consisting of a thickening agent (soap or other additives) in a fluid lubricant (usually petroleum lubricating oil). { 'lü·brə,kād·iŋ ,grēs }

lubricating oil [MATER] Selected fractions of refined petroleum or other oils (with or without additives) used to lessen friction between moving surfaces. { 'lü·brə,kād·iŋ ,öil }

lubrication action [MATER] The ability of the lubricant to maintain a fluid film between solid surfaces and to prevent their physical contact. { ,lü·brə'kā·shən ,ak·shən }

lubricity [MATER] The ability of a material to lubricate. { lü'bris·əd·ē }

Lüders' lines [MET] Surface markings on a metal caused by flow of the material strained beyond its elastic limit. Also known as deformation bands; Hartmann lines; Piobert lines; stretcher strains. { 'lüd·ərz ,līnz }

Ludwig-Soret effect [THERMO] A phenomenon in which a temperature gradient in a mixture of substances gives rise to a concentration gradient. { ¦lùd,vik sə'rā i,fekt }

lug brick [MATER] A brick with lugs for spacing adjacent bricks. { 'ləg ,brik }

lumber [MATER] Logs that have been sawed and prepared for market. { 'ləm·bər }

lumen [OPTICS] The unit of luminous flux, equal to the luminous flux emitted within a unit solid angle (1 steradian) from a point source having a uniform intensity of 1 candela, or to the luminous flux received on a unit surface, all points of which are at a unit distance from such a source. Symbolized lm. { 'lü·mən }

lumen-hour [OPTICS] A unit of quantity of light (luminous energy), equal to the quantity of light radiated or received for a period of 1 hour by a flux of 1 lumen. Abbreviated lm-hr. { 'lü·mən ¦aùr }

lumen per watt [OPTICS] The unit of luminosity factor and of luminous efficacy. Abbreviated lm/w. { 'lü·mən pər 'wät }

lumen-second [OPTICS] A unit of quantity of light (luminous energy), equal to the quantity of light radiated or received for a period of 1 second by a flux of 1 lumen. Abbreviated lm-sec. { 'lü·mən ¦sek·ənd }

luminance [OPTICS] The ratio of the luminous intensity in a given direction of an infinitesimal

luminance factor

element of a surface containing the point under consideration, to the orthogonally projected area of the element on a plane perpendicular to the given direction. Formerly known as brightness. { 'lü·mə·nəns }

luminance factor [OPTICS] The ratio of the luminance of a body when illuminated and observed under certain conditions to that of a perfect diffuser under the same conditions. { 'lü·mə·nəns ,fak·tər }

luminescence [PHYS] Light emission that cannot be attributed merely to the temperature of the emitting body, but results from such causes as chemical reactions at ordinary temperatures, electron bombardment, electromagnetic radiation, and electric fields. { ,lü·mə'nes·əns }

luminescent [PHYS] Capable of exhibiting luminescence. { ,lü·mə'nes·ənt }

luminescent center [SOLID STATE] A point-lattice defect in a transparent crystal that exhibits luminescence. { ,lü·mə'nes·ənt 'sen·tər }

luminescent dye [MATER] A dye that is made luminous by excitation with an outside energy source; used in luminous paint. { ,lü·mə'nes·ənt 'dī }

luminophor [PHYS] A luminescent material that converts part of the absorbed primary energy into emitted luminescent radiation. Also known as fluophor; phosphor. { lü'min·ə,fór }

luminosity See luminosity factor. { ,lü·mə'näs·əd·ē }

luminosity factor [OPTICS] The ratio of luminous flux in lumens emitted by a source at a particular wavelength to the corresponding radiant flux in watts at the same wavelength; thus this is a measure of the visual sensitivity of the eye. Also known as luminosity. { ,lü·mə'näs·əd·ē ,fak·tər }

luminosity function [OPTICS] A standard measure of the response of an eye to monochromatic light at various wavelengths; the function is normalized to unity at its maximum value. Also known as spectral luminous efficiency. { ,lü·mə'näs·əd·ē ,fəŋk·shən }

luminous coefficient [OPTICS] A measure of the fraction of the radiant power of a light source which contributes to its luminous properties, equal to the average of the luminosity function at various wavelengths, weighted according to the spectral intensity of the source. Also known as luminous efficiency. { 'lü·mə·nəs ,kō·i'fish·ənt }

luminous efficacy [OPTICS] **1.** The ratio of the total luminous flux in lumens emitted by a light source over all wavelengths to the total radiant flux in watts. Formerly known as luminous efficiency. **2.** The ratio of the total luminous flux emitted by a light source to the power input of the source; expressed in lumens per watt. { 'lü·mə·nəs ,ef·ə·kə,sē }

luminous efficiency See luminous coefficient; luminous efficacy. { 'lü·mə·nəs i'fish·ən·sē }

luminous emittance [OPTICS] The emittance of visible radiation weighted to take into account the different response of the human eye to different wavelengths of light; in photometry, luminous emittance is always used as a property of a self-luminous source, and therefore should be distinguished from luminance. Also known as luminous exitance. { 'lü·mə·nəs i'mit·əns }

luminous energy [OPTICS] The total radiant energy emitted by a source, evaluated according to its capacity to produce visual sensation; measured in lumen-hours or lumen-seconds. { 'lü·mə·nəs 'en·ər·jē }

luminous exitance See luminous emittance. { 'lü·mə·nəs 'ek·səd·əns }

luminous flux [OPTICS] The time rate of flow of radiant energy, evaluated according to its capacity to produce visual sensations; measured in lumens. { 'lü·mə·nəs 'fləks }

luminous flux density See illuminance. { 'lü·mə·nəs 'fləks 'den·səd·ē }

luminous intensity [OPTICS] The luminous flux incident on a small surface which lies in a specified direction from a light source and is normal to this direction, divided by the solid angle (in steradians) which the surface subtends at the source of light. Also known as light intensity. { 'lü·mə·nəs in'ten·səd·ē }

luminous paint [MATER] A type of paint in which luminous pigments are used. { 'lü·mə·nəs 'pānt }

luminous pigment [MATER] A pigment that absorbs light energy and radiates visible light when exposed to ultraviolet light; made of phosphors such as strontium, zinc, and cadmium sulfides. { 'lü·mə·nəs 'pig·mənt }

luminous quantities [OPTICS] Physical quantities used in photometry, such as luminous intensity and luminance, which are based on the response of the human eye, and are thus weighted to take into account the difference in response at different wavelengths of light. { 'lü·mə·nəs 'kwän·əd·ēz }

lunar caustic [MATER] A form of toughened silver nitrate consisting of 97–98% silver nitrate and 2–3% silver chloride. Also known as fused silver nitrate; molded silver nitrate. { 'lün·ər 'kȯs·tik }

luster [OPTICS] The appearance of a surface dependent on reflected light; types include metallic, vitreous, resinous, adamantine, silky, pearly, greasy, dull, and earthy; applied to minerals, textiles, and many other materials. { 'ləs·tər }

lusterless paint [MATER] Paint which absorbs light rays so that no shine or polish appears on its surface; used extensively on U.S. Army vehicles. { 'ləs·tər·ləs 'pānt }

lute [MATER] A substance, such as cement or clay, for packing a joint or coating a porous surface to produce imperviousness to gas or liquid. { lüt }

lutetium [CHEM] A chemical element, symbol Lu, atomic number 71, atomic weight 174.967; a very rare metal and the heaviest member of the rare-earth group. { lü'tē·shəm }

lycopodium [MATER] A yellow powder prepared from the spores of *Lycopodium clavatum*; used as a desiccant and absorbent. { ,lī·kə'pōd·ē·əm }

Lyddane-Sachs-Teller relation [SOLID STATE] For an infinite ionic crystal, the relation $\epsilon(0)/\epsilon(\infty) = \omega_L^2/\omega_T^2$, where $\epsilon(0)$ is the crystal's static dielectric constant, $\epsilon(\infty)$ is the dielectric constant at a frequency at which electronic polarizability is effective but ionic polarizability is not, ω_L is the frequency of longitudinal optical phonons with zero wave vectors, and ω_T is the frequency of transverse optical phonons with large wave vector.   { lə'dān 'saks 'tel·ər ri,lā·shən }

lye |INORG CHEM| **1.** A solution of potassium hydroxide or sodium hydroxide used as a strong alkaline solution in industry.   **2.** The alkaline solution that is obtained from the leaching of wood ashes.   { lī }

lymphocyte transformation   *See* transformation.   { ¦lim·fə,sīt ,tranz·fər'mā·shən }

lyophilic |CHEM| Referring to a substance which will readily go into colloidal suspension in a liquid.   { ¦lī·ə¦fil·ik }

lyophobic |CHEM| Referring to a substance in a colloidal state that has a tendency to repel liquids.   { ¦lī·ə¦fō·bik }

lyotropic liquid crystal |PHYS CHEM| A liquid crystal prepared by mixing two or more components, one of which is polar in character (for example, water).   { ¦lī·ə¦träp·ik ¦lik·wəd 'krist·əl }

M

M *See* molarity.

M$_f$ |MET| In a carbon steel, the temperature at which martensite formation finishes during cooling of austenite.

M$_s$ |MET| In a carbon steel, the temperature at which martensite formation begins during cooling of austenite.

macadam |CIV ENG| Uniformly graded stones consolidated by rolling to form a road surface; may be bound with water or cement, or coated with tar or bitumen. { mə'kad·əm }

Macarthur and Forest cyanide process |MET| A process for recovering gold by leaching the pulped gold ore with a solution of 0.2–0.8% potassium cyanide and next with water; the gold is then obtained by precipitation on zinc or aluminum or by electrolysis. { mə'kär·thər ən 'fär·əst 'sī·ə,nīd ,präs·əs }

machinability |MET| **1.** The ability of a metal to be machined. **2.** The difficulty or ease with which a metal can be machined. { mə ,shēn·ə'bil·əd·ē }

machinability index |MET| A numerical value that designates the degree of difficulty or ease with which a particular material can be machined; originally based on turning B1112 steel at 180 feet per minute (0.9144 meter per second) with a high-speed tool for an index of 100; with replacement of high-speed steels with carbides in turning operations, it has been found that the machinability index of a given material changes with the type of operation and the tool material. { mə,shēn·ə'bil·əd·ē ,in,deks }

machinable carbide |MET| Titanium carbide in a matrix of Ferro-Tic C tool steel. { mə'shēn·ə·bəl 'kär,bīd }

machine forging |MET| Forging operations performed in and by certain machines. { mə'shēn ,fórj·iŋ }

machine oil |MATER| Medium-density lubricating oil used for machine parts. { mə'shēn ,óil }

machine steel |MET| Plain carbon steel with a 0.2–0.3% carbon content. { mə'shēn ,stēl }

machine welding |MET| Welding with a machine under the control and observation of an operator; may be loaded and unloaded either manually or mechanically. { mə'shēn 'weld·iŋ }

machining stress |MET| Residual stress in the work caused by machining. { mə'shēn·iŋ ,stress }

Mack's cement |MATER| Cement made of dehydrated gypsum with a small amount of calcined sodium sulfate and potassium sulfate. { 'maks si'ment }

macle |CRYSTAL| A twinned crystal. { 'mak·əl }

Macquer's salt *See* potassium arsenate. { mə'kerz ,sólt }

macrocycle *See* macrocyclic compound. { 'mak·rō,sī·kəl }

macrocyclic compound |ORG CHEM| An organic compound containing a large ring, that is, a closed chain of 12 or more carbon atoms; examples include crown ethers, cryptands, spherands, carcerands, cyclodextrins, cyclophanes, and calixarenes. Also known as macrocycle. { ¦mak·rō ,sī·klik 'käm,paúnd }

macrodome |CRYSTAL| Dome of a crystal in which planes are parallel to the longer lateral axis. { 'mak·rə,dōm }

macroencapsulation |BIOMATER| The envelopment of a large mass of xenotransplanted cells or tissue in planar membranes, hollow fibers, or diffusion chambers to isolate the cells from the body, thereby avoiding the immune responses that the foreign cells could initiate, and also to allow the desired metabolites (such as insulin and glucose for pancreatic islet cells) to diffuse in and out of the membrane. { ,mak·rō·in ,kap·sə'lā·shən }

macromolecular |POLYM CHEM| Composed of or characterized by large molecules. { ¦mak· rō·mə'lek·yə·lər }

macromolecule |POLYM CHEM| A large molecule in which there is a large number of one or several relatively simple structural units, each consisting of several atoms bonded together. { ¦mak·rō'mäl·ə,kyül }

macropinacoid *See* front pinacoid. { ¦mak· rō'pin·ə,kòid }

macropore |CHEM| A pore in a catalytic material whose width is greater than 0.05 micrometer. { 'mak·rə,pór }

macroporous resin |POLYM CHEM| A member of a class of very small, highly cross-linked polymer particles penetrated by channels through which solutions can flow; used as ion exchanger. Also known as macroreticular resin. { ¦mak·rə ¦pór·əs 'rez·ən }

macroreticular resin *See* macroporous resin. { ¦mak·rō·rə'tik·yə·lər 'rez·ən }

macrorheology [MECH] A branch of rheology in which materials are treated as homogeneous or quasi-homogeneous, and processes are treated as isothermal. { ¦mak·rō·rē'äl·ə·jē }

macroscopic property See thermodynamic property. { ¦mak·rə¦skäp·ik 'präp·ərd·ē }

macroscopic stress [MET] Residual stress in a metal in a distance comparable to the gage length of strain measurement specimens and therefore detectable by x-ray or dissection techniques. Also known as macrostress. { ¦mak·rə¦skäp·ik 'stres }

macrostress See macroscopic stress. { 'mak·rō ‚stres }

macrostructure [MET] Structure of an etched metal visible to the naked eye or at magnifications up to 10 diameters. { ¦mak·rō'strək·chər }

MAD See multiwavelength anomalous dispersion. { ¦em¦ā'dē or mad }

madder [MATER] The root of the madder plant (*Rubia tinctorium*), pulverized and used as source of glucosides to produce alizarin by fermentation. Also known as gamene. { 'mad·ər }

madder lake [MATER] Bluish-red, transparent pigment produced from alizarin red; used to make stains and inks, and as a component of artists' oil colors. { 'mad·ər 'lāk }

Madelung constant [SOLID STATE] A dimensionless constant which determines the electrostatic energy of a three-dimensional periodic crystal lattice consisting of a large number of positive and negative point charges when the number and magnitude of the charges and the nearest-neighbor distance between them is specified. { 'mä·də‚lůŋ ‚kän·stənt }

magic acid [INORG CHEM] A superacid consisting of equal molar quantities of fluorosulfonic acid (HSO_3F) and antimony pentafluoride (SbF_5). { 'maj·ik ¦as·əd }

magnesia [INORG CHEM] Magnesium oxide that is processed for a particular purpose. { mag'nē·zhə }

magnesia brick [MATER] A type of refractory brick composed of magnesium oxide with about 15% of other oxides. Also known as magnesite brick. { mag'nē·zhə ¦brik }

magnesia cement See magnesium oxychloride cement. { mag'nē·zhə si'ment }

magnesia refractory [MATER] Heat- and corrosion-resistant material made of magnesium oxide; used in cement or brick form to line high-temperature process vessels or furnaces. { mag'nē·zhə ri¦frak·trē }

magnesite brick See magnesia brick. { 'mag·nə ‚sīt ¦brik }

magnesium [CHEM] A metallic element, symbol Mg, atomic number 12, atomic weight 24.305. [MET] A silvery-white, lightweight, malleable, ductile metal, used in metallurgical and chemical processes, photography, pyrotechny, and light alloys. { mag'nē·zē·əm }

magnesium arsenate [INORG CHEM] $Mg_3(AsO_4)_2 \cdot xH_2O$ A white, poisonous, water-insoluble powder used as an insecticide. { mag'nē·zē·əm 'ärs·ən‚āt }

magnesium borate [INORG CHEM] $3MgO \cdot B_2O_3$ Crystals that are white or colorless and transparent; soluble in alcohol and acids, slightly soluble in water; used as a fungicide, antiseptic, and preservative. { mag'nē·zē·əm 'bór‚āt }

magnesium boride See magnesium diboride. { mag'nē·zē·əm 'bór‚īd }

magnesium bromate [INORG CHEM] $Mg(BrO_3)_2 \cdot 6H_2O$ A white crystalline compound, insoluble in alcohol, soluble in water; a fire hazard; used as an analytical reagent. { mag'nē·zē·əm 'brō ‚māt }

magnesium bromide [INORG CHEM] $MgBr_2 \cdot 6H_2O$ Deliquescent, colorless, bitter-tasting crystals, melting at $172°C$; soluble in water, slightly soluble in alcohol; used in medicine and in the synthesis of organic chemicals. { mag'nē·zē·əm 'brō‚mīd }

magnesium carbonate [INORG CHEM] $MgCO_3$ A water-insoluble, white powder, decomposing at about $350°C$; used as a refractory material. { mag'nē·zē·əm 'kär·bə‚nāt }

magnesium chlorate [INORG CHEM] $Mg(ClO_3)_2 \cdot 6H_2O$ A white powder, bitter-tasting and hygroscopic; slightly soluble in alcohols, soluble in water; used in medicine. { mag'nē·zē·əm 'klór ‚āt }

magnesium chloride [INORG CHEM] $MgCl_2 \cdot 6H_2O$ Deliquescent white crystals; soluble in water and alcohol; used in disinfectants and fire extinguishers, and in ceramics, textiles, and paper manufacture. { mag'nē·zē·əm 'klór‚īd }

magnesium diboride [INORG CHEM] MgB_2 A crystalline intermetallic compound, produced as a black powder, that becomes superconducting at the unusually high temperature of 39 K ($-389°F$; $-234°C$); melts at $800°C$. Also known as magnesium boride. { mag'nē·zē·əm dī'bór ‚īd }

magnesium dust [MET] Magnesium metal powder; flammable; used in photographic flash lights and pyrotechnics. { mag'nē·zē·əm 'dəst }

magnesium fluoride [INORG CHEM] MgF_2 White, fluorescent crystals; insoluble in water and alcohol, soluble in nitric acid; melts at $1263°C$; used in ceramics and glass. Also known as magnesium flux. { mag'nē·zē·əm 'flůr‚īd }

magnesium fluosilicate [INORG CHEM] $MgSiF_6 \cdot 6H_2O$ Water-soluble, efflorescent white crystals; used in ceramics, in mothproofing and waterproofing, and as a concrete hardener. Also known as magnesium silicofluoride. { mag'nē·zē·əm 'flů·ə'sil·ə‚kāt }

magnesium flux See magnesium fluoride. { mag'nē·zē·əm 'fləks }

magnesium halide [INORG CHEM] A compound formed from the metal magnesium and any of the halide elements; an example is magnesium bromide. { mag'nē·zē·əm 'ha‚līd }

magnesium hydrate See magnesium hydroxide. { mag'nē·zē·əm 'hī‚drāt }

magnesium hydride [INORG CHEM] MgH_2 A hydride compound formed from the metal magnesium; it decomposes violently in water, and in a vacuum at about $280°C$. { mag'nē·zē·əm 'hī‚drīd }

magnesium hydroxide |INORG CHEM| $Mg(OH)_2$ A white powder, very slightly soluble in water, decomposing at $350°C$; used as an intermediate in extraction of magnesium metal, and as a reagent in the sulfite wood pulp process. Also known as magnesium hydrate. { mag'nē·zē·əm hī'dräk,sīd }

magnesium hyposulfite See magnesium thiosulfate. { mag'nē·zē·əm ,hī·pō'səl,fīt }

magnesium iodide |INORG CHEM| $MgI_2·8H_2O$ Crystalline powder, white and deliquescent, discoloring in air; soluble in water, alcohol, and ether; used in medicine. { mag'nē·zē·əm 'ī·ə ,dīd }

magnesium lime |MATER| Lime containing more than 20% magnesium oxide; slakes more slowly, evolves less heat, expands less, sets more rapidly, and produces higher-strength mortars than does high-calcium quicklime. { mag'nē·zē·əm 'līm }

magnesium nitrate |INORG CHEM| $Mg(NO_3)_2·6H_2O$ Deliquescent white crystals; soluble in alcohol and water; a fire hazard; used as an oxidizing material in pyrotechnics. { mag'nē·zē·əm 'nī,trāt }

magnesium oxide |INORG CHEM| MgO A white powder that (depending on the method of preparation) may be light and fluffy, or dense; melting point $2800°C$; insoluble in acids, slightly soluble in water; used in making refractories, and in cosmetics, pharmaceuticals, insulation, and medicine. { mag'nē·zē·əm 'äk,sīd }

magnesium oxychloride cement |MATER| Cement made by adding a magnesium chloride solution to magnesia; used for interior flooring. Also known as magnesia cement. { mag'nē·zē·əm ,äk·sə'klȯr,īd si'ment }

magnesium perchlorate |INORG CHEM| $Mg(ClO_4)_2·6H_2O$ White, deliquescent crystals; soluble in water and alcohol; explosive when in contact with reducing materials; used as a drying agent for gases. { mag'nē·zē·əm pər'klȯr ,āt }

magnesium peroxide |INORG CHEM| MgO_2 A tasteless, odorless white powder; soluble in dilute acids, insoluble in water; a fire hazard; used as a bleaching and oxidizing agent, and in medicine. { mag'nē·zē·əm pə'räk,sīd }

magnesium phosphate |INORG CHEM| A compound with three forms: monobasic, $MgH_4(PO_4)_2·2H_2O$, used in medicine and wood fireproofing; dibasic, $MgHPO_4·3H_2O$, used in medicine and as a plastics stabilizer; tribasic, $Mg_3(PO_4)_2·8H_2O$, used in dentifrices, as an adsorbent, and in pharmaceuticals. { mag'nē·zē·əm 'fäs,fāt }

magnesium silicate |INORG CHEM| $3MgSiO_3·5H_2O$ White, water-insoluble powder, containing variable proportions of water of hydration; used as a filler for rubber and in medicine. { mag'nē·zē·əm 'sil·ə,kāt }

magnesium silicofluoride See magnesium flusilicate. { mag'nē·zē·əm ,sil·ə·kō'flúr,īd }

magnesium sulfate |INORG CHEM| $MgSO_4$ Colorless crystals with a bitter, saline taste; soluble in glycerol; used in fireproofing, textile processes, ceramics, cosmetics, and fertilizers. { mag'nē·zē·əm 'səl,fāt }

magnesium sulfite |INORG CHEM| $MgSO_3·6H_2O$ A white, crystalline powder; insoluble in alcohol, slightly soluble in water; used in medicine and paper pulp. { mag'nē·zē·əm 'səl,fīt }

magnesium thiosulfate |INORG CHEM| $MgS_2O_3·6H_2O$ Colorless crystals that lose water at $170°C$; used in medicine. Also known as magnesium hyposulfite. { mag'nē·zē·əm ,thī· ə'səl,fāt }

magnesium trisilicate |INORG CHEM| $Mg_2Si_3O_8·5H_2O$ A white, odorless, tasteless powder; insoluble in water and alcohol; used as an industrial odor absorbent and in medicine. { mag'nē·zē·əm ,trī'sil·ə,kāt }

magnesium tungstate |INORG CHEM| $MgWoO_4$ White crystals, insoluble in alcohol and water, soluble in acid; used in luminescent paint and for fluorescent x-ray screens. { mag'nē·zē·əm 'təŋ ,stāt }

magnet |ELECTROMAG| A piece of ferromagnetic or ferrimagnetic material whose domains are sufficiently aligned so that it produces a net magnetic field outside itself and can experience a net torque when placed in an external magnetic field. { 'mag·nət }

magnet alloy |MET| An alloy such as Alnico or Alcomax having strong magnetic properties; used in making permanent magnets. { 'mag·nət ¦al ,ȯi }

magnetic |ELECTROMAG| Pertaining to magnetism or a magnet. { mag'ned·ik }

magnetically hard alloy |MET| A ferromagnetic alloy that can be permanently magnetized after the removal of an externally applied magnetic field. { mag'ned·ə·klē ¦härd 'al,ȯi }

magnetically soft alloy |MET| A ferromagnetic alloy which is capable of being magnetized upon application of an external magnetic field, but which returns to a nonmagnetic condition when the field is removed. { mag'ned·ə·klē ¦sȯft 'al ,ȯi }

magnetic analysis inspection |MET| A nondestructive inspection method to determine the presence of variations in magnetic flux in ferromagnetic materials of constant cross section caused by defects, variations in hardness, discontinuities, or other irregularities. { mag'ned· ikə'nal·ə·səs in,spek·shən }

magnetic anisotropy |ELECTROMAG| The dependence of the magnetic properties of some materials on direction. { mag'ned·ik ,an·ə'sä· trə·pē }

magnetic annealing |MET| Annealing and cooling in a strong magnetic field. { mag'ned· ik ə'nēl·iŋ }

magnetic axis |ELECTROMAG| A line through the center of a magnet such that the torque exerted on the magnet by a magnetic field in the direction of this line equals 0. { mag'ned· ik 'ak·səs }

magnetic bias |ELECTROMAG| A steady magnetic field applied to the magnetic circuit of a relay or other magnetic device. { mag'ned·ik 'bī·əs }

magnetic bubble |SOLID STATE| A cylindrical stable (nonvolatile) region of magnetization produced in a thin-film magnetic material by an external magnetic field; direction of magnetization is perpendicular to the plane of the material. Also known as bubble. { mag'ned·ik'bəb·əl }

magnetic coercive force See coercive force. { mag'ned·ik kō'ər·siv 'fȯrs }

magnetic constant |ELECTROMAG| The absolute permeability of empty space, equal to 1 electromagnetic unit in the centimeter-gram-second system, and to $4\pi \times 10^{-7}$ henry per meter or, numerically, to 1.25664×10^{-6} henry per meter in the International System of units. Symbolized μ_0. { mag'ned·ik 'kän·stənt }

magnetic cooling See adiabatic demagnetization. { mag'ned·ik 'kül·iŋ }

magnetic coupling |ELECTROMAG| For a pair of particles or systems, the effect of the magnetic field created by one system on the magnetic moment or angular momentum of the other. { mag'ned·ik 'kəp·liŋ }

magnetic crack detection See magnetic particle test. { mag'ned·ik 'krak di,tek·shən }

magnetic Curie temperature |SOLID STATE| The temperature below which a magnetic material exhibits ferromagnetism, and above which ferromagnetism is destroyed and the material is paramagnetic. { mag'ned·ik 'kyur·ē ,tem·prə·chər }

magnetic diffusivity |ELECTROMAG| A measure of the tendency of a magnetic field to diffuse through a conducting medium at rest; it is equal to the partial derivative of the magnetic field strength with respect to time divided by the Laplacian of the magnetic field, or to the reciprocal of $4\pi\mu\sigma$, where μ is the magnetic permeability and σ is the conductivity in electromagnetic units. { mag'ned·ik ,di,fyü'siv·əd·ē }

magnetic dipole |ELECTROMAG| An object, such as a permanent magnet, current loop, or particle with angular momentum, which experiences a torque in a magnetic field, and itself gives rise to a magnetic field, as if it consisted of two magnetic poles of opposite sign separated by a small distance. { mag'ned·ik 'dī,pōl }

magnetic dipole density See magnetization. { mag'ned·ik 'dī,pōl ,den·səd·ē }

magnetic dipole moment |ELECTROMAG| A vector associated with a magnet, current loop, particle, or such, whose cross product with the magnetic induction (or alternatively, the magnetic field strength) of a magnetic field is equal to the torque exerted on the system by the field. Also known as dipole moment; magnetic moment. { mag'ned·ik 'dī,pōl ,mō·mənt }

magnetic displacement See magnetic flux density. { mag'ned·ik di'splās·mənt }

magnetic domain See ferromagnetic domain. { mag'ned·ik də'mān }

magnetic double refraction |OPTICS| The double refraction of light passing through certain substances when the substance is placed in a

transverse magnetic field. { mag'ned·ik 'dəbəl ri,frak·shən }

magnetic energy |ELECTROMAG| The energy required to set up a magnetic field. { mag'ned·ik 'en·ər·jē }

magnetic ferroelectric |SOLID STATE| A substance which possesses both magnetic ordering and spontaneous electric polarization. { mag'ned·ik ¦fer·ō·i'lek·trik }

magnetic field |ELECTROMAG| **1.** One of the elementary fields in nature; it is found in the vicinity of a magnetic body or current-carrying medium and, along with electric field, in a light wave; charges moving through a magnetic field experience the Lorentz force. **2.** See magnetic field strength. { mag'ned·ik 'fēld }

magnetic field intensity See magnetic field strength. { mag'ned·ik 'fēld in,ten·səd·ē }

magnetic field strength |ELECTROMAG| An auxiliary vector field, used in describing magnetic phenomena, whose curl, in the case of static charges and currents, equals (in meter-kilogram-second units) the free current density vector, independent of the magnetic permeability of the material. Also known as magnetic field; magnetic field intensity; magnetic force; magnetic intensity; magnetizing force. { mag'ned·ik 'fēld ,streŋkth }

magnetic film See magnetic thin film. { mag 'ned·ik 'film }

magnetic fluid |MATER| A mixture of iron particles in oil or other liquid; viscosity increases sharply in a strong magnetic field. { mag'ned·ik 'flü·əd }

magnetic flux |ELECTROMAG| The integral over a specified surface of the component of magnetic induction perpendicular to the surface. See magnetic lines of force. { mag'ned·ik 'fləks }

magnetic flux density |ELECTROMAG| A vector quantity that is used as a quantitative measure of magnetic field; the force on a charged particle moving in the field is equal to the particle's charge times the cross product of the particle's velocity with the magnetic flux density (SI units). Also known as magnetic displacement; magnetic induction; magnetic vector. { mag'ned·ik 'fləks ,den·səd·ē }

magnetic force See magnetic field strength. { mag'ned·ik 'fȯrs }

magnetic force microscopy |ENG| The use of an atomic force microscope to measure the gradient of a magnetic field acting on a tip made of a magnetic material, by monitoring the shift of the natural frequency of the cantilever due to the magnetic force as the tip is scanned over the sample. { mag¦ned·ik ¦fȯrs mī'krä·skə·pē }

magnetic force welding |MET| A welding process in which the mechanical force is exerted by a magnetic field. { mag'ned·ik ¦fȯrs 'weld·iŋ }

magnetic forming |MET| The forming of metal into desired shapes by using strong magnetic fields, produced by charging a large capacitor bank and then discharging it into an induction coil in less than 10^{-6} second, to push the metal against a forming die. { mag'ned·ik 'fȯr·miŋ }

magnetic gap [ELECTROMAG] The space between a magnet's pole faces. { mag′ned·ik ′gap }

magnetic groups *See* Shubnikov groups. { mag′ned·ik ′grüps }

magnetic hardness comparator [ENG] A device for checking the hardness of steel parts by placing a unit of known proper hardness within an induction coil; the unit to be tested is then placed within a similar induction coil, and the behavior of the induction coils compared; if the standard and test units have the same magnetic properties, the hardness of the two units is considered to be the same. { mag′ned·ik ′härd·nəs kəm ‚par·əd·ər }

magnetic hysteresis [ELECTROMAG] Lagging of changes in the magnetization of a substance behind changes in the magnetic field as the magnetic field is varied. Also known as hysteresis. { mag′ned·ik ‚his·tə′rē·səs }

magnetic induction *See* magnetic flux density. { mag′ned·ik in′dək·shən }

magnetic ink [MATER] Ink containing magnetic particles to permit reading of printed characters by a magnetic character reader as well as by humans. { mag′ned·ik ′iŋk }

magnetic inspection oil [MET] A light petroleum oil, such as kerosine or naphtha, to which has been added fine ferromagnetic particles (usually colored black or red for contrast) to form an inspection penetrant; when the penetrant is applied to a metal surface being inspected, the ferrous particles accumulate in any surface cracks by magnetic attraction, thereby permitting the cracks to be discernible. { mag′ned·ik in′spek·shən ‚öil }

magnetic inspection paste [MET] A paste containing ferromagnetic particles designed to be added to a light distilled petroleum oil, such as kerosine or naphtha, to form an inspection penetrant; when the inspection penetrant is applied to a metal, the ferrous particles accumulate in any surface cracks by magnetic attraction, thereby permitting the cracks to be discerned. { mag′ned·ik in′spek·shən ‚pāst }

magnetic inspection powder [MET] A dry powder containing ferromagnetic particles colored gray, black, or red for contrast, designed to be dusted on metal parts being inspected by a magnetic inspection machine; the ferrous powder accumulates in any surface cracks (flaws) by magnetic attraction, thereby permitting the cracks to be readily discerned; if the ferrous particles are fluorescent, surface cracks will be brilliantly illuminated under black light. { mag′ned·ik in ′spek·shən ‚paüd·ər }

magnetic intensity *See* magnetic field strength. { mag′ned·ik in′ten·səd·ē }

magnetic lines of flux *See* magnetic lines of force. { mag′ned·ik ′līnz əv ′fləks }

magnetic lines of force [ELECTROMAG] Lines used to represent the magnetic induction in a magnetic field, selected so that they are parallel to the magnetic induction at each point, and so that the number of lines per unit area of a surface perpendicular to the induction is equal to the induction. Also known as magnetic flux; magnetic lines of flux. { mag′ned·ik ′līnz əv ′fòrs }

magnetic material [ELECTROMAG] A material exhibiting ferromagnetism. { mag′ned·ikmə′tir· ē·əl }

magnetic moment *See* magnetic dipole moment. { mag′ned·ik ′mō·mənt }

magnetic particle test [MET] A nondestructive test to determine the existence and extent of macrodefects such as cracks in ferromagnetic materials; discontinuities in the material create variations of magnetic field which are outlined by fine magnetic particles. Also known as magnetic crack detection. { mag′ned·ik ¦pärd·ə·kəl ‚test }

magnetic permeability *See* permeability. { mag′ned·ik ‚pər·mē·ə′bil·əd·ē }

magnetic pole [ELECTROMAG] **1.** One of two regions located at the ends of a magnet that generate and respond to magnetic fields in much the same way that electric charges generate and respond to electric fields. **2.** A particle which generates and responds to magnetic fields in exactly the same way that electric charges generate and respond to electric fields; the particle probably does not have physical reality, but it is often convenient to imagine that a magnetic dipole consists of two magnetic poles of opposite sign, separated by a small distance. { mag′ned·ik ′pōl }

magnetic pole strength [ELECTROMAG] The magnitude of a (fictional) magnetic pole, equal to the force exerted on the pole divided by the magnetic induction (or, alternatively, by the magnetic field strength). Also known as pole strength. { mag′ned·ik ′pōl ‚streŋkth }

magnetic potential *See* magnetic scalar potential. { mag′ned·ik pə′ten·chəl }

magnetic recorder [ELECTR] An instrument that records information, generally in the form of audio-frequency or digital signals, on magnetic tape or magnetic wire as magnetic variations in the medium. { mag′ned·ik ri′kòrd·ər }

magnetic recording [ELECTR] Recording by means of a signal-controlled magnetic field. { mag′ned·ik ri′kòrd·iŋ }

magnetic recording paper [MATER] A particle-oriented paper in which both machine-readable and visible traces can be produced by a magnetic recording head; reusable because the trace can be erased by a combination of alternating-current and direct-current magnetic fields. { mag′ned·ik ri′kòrd·iŋ ‚pā·pər }

magnetic refrigerator [CRYO] A device for keeping substances cooled to about 0.2 K, in which a working substance consisting of a paramagnetic salt undergoes a cycle of processes which approximates a Carnot cycle between a high-temperature reservoir consisting of a liquid-helium bath at 1.2 K and a low-temperature reservoir consisting of the substance to be cooled, and isentropic cooling of the working substance is accomplished by demagnetization. { mag′ned·ik ri′frij·ə‚rād·ər }

magnetic relaxation [PHYS] The approach of a magnetic system to an equilibrium or steady-state condition, over a period of time. { mag'ned·ik ‚rē‚lak'sā·shən }

magnetic reluctance See reluctance. { mag 'ned·ik ri'lək·təns }

magnetic reluctivity See reluctivity. { mag 'ned·ik ‚rē‚lək'tiv·əd·ē }

magnetic resonance [PHYS] A phenomenon exhibited by the magnetic spin systems of certain atoms whereby the spin systems absorb energy at specific (resonant) frequencies when subjected to magnetic fields alternating at frequencies which are in synchronism with natural frequencies of the system. Also known as spin resonance. { mag'ned·ik ¦rez·ən·əns }

magnetic rotation [OPTICS] **1.** In a weak magnetic field, the rotation, of the plane of polarization of fluorescent light emitted perpendicular to the field and perpendicular to the propagation direction of the incident light. **2.** See Faraday effect. { mag'ned·ik rō'tā·shən }

magnetic rubber [MATER] Synthetic rubber to which magnetic metal powder is added; produced in sheets or strips. { mag'ned·ik 'rəb·ər }

magnetics [ELECTROMAG] The study of magnetic phenomena, comprising magnetostatics and electromagnetism. { mag'ned·iks }

magnetic saturation [ELECTROMAG] The condition in which, after a magnetic field strength becomes sufficiently large, further increase in the magnetic field strength produces no additional magnetization in a magnetic material. Also known as saturation. { mag'ned·ik ‚sach·ə'rā·shən }

magnetic scalar potential [ELECTROMAG] The work which must be done against a magnetic field to bring a magnetic pole of unit strength from a reference point (usually at infinity) to the point in question. Also known as magnetic potential. { mag'ned·ik ¦skāl·ər pə'ten·chəl }

magnetic scattering [PHYS] Scattering of neutrons as a result of the interaction of the magnetic moment of the neutron with the magnetic moments of atoms or other particles. { mag'ned·ik 'skad·ə·riŋ }

magnetic separatrix [ELECTROMAG] A surface that forms the‚ boundary between an internal region of closed magnetic surfaces and an external region of open field lines. { mag'ned·ik 'sep·rə‚triks }

magnetic shell [ELECTROMAG] Two layers of magnetic charge of opposite sign, separated by an infinitesimal distance. { mag'ned·ik 'shel }

magnetic shunt [ELECTROMAG] Piece of iron, usually adjustable as to position, used to divert a portion of the magnetic lines of force passing through an air gap in an instrument or other device. { mag'ned·ik 'shənt }

magnetic strain energy [SOLID STATE] The potential energy of a magnetic domain, subject to both a tensile stress and a magnetic field, associated with the domain's magnetostriction expansion. { mag'ned·ik 'strān ‚en·ər·jē }

magnetic superconductor [SOLID STATE] A superconductor which is not magnetic in the ordinary sense, but which contains elements with large magnetic moments or large spin. { mag'ned·ik 'sü·pər·kən‚dək·tər }

magnetic susceptibility [ELECTROMAG] The ratio of the magnetization of a material to the magnetic field strength; it is a tensor when these two quantities are not parallel; otherwise it is a simple number. Also known as susceptibility. { mag'ned·ik sə‚sep·tə'bil·əd·ē }

magnetic tape [ELECTR] A plastic, paper, or metal tape that is coated or impregnated with magnetizable iron oxide particles; used in magnetic recording. { mag'ned·ik 'tāp }

magnetic tension effect [ELECTROMAG] The ability of stresses on a ferromagnetic material to alter its remanence. Also known as ·inverse magnetostriction. { mag¦ned·ik 'ten·shən i ‚fekt }

magnetic thermometer [SOLID STATE] A sample of a paramagnetic salt whose magnetic susceptibility is measured and whose temperature is then calculated from the inverse relationship between the two quantities; useful at temperatures below about 1 K. { mag'ned· ik thər'mäm·əd·ər }

magnetic thin film [SOLID STATE] A sheet or cylinder of magnetic material less than 5 micrometers thick, usually possessing uniaxial magnetic anisotropy; used mainly in computer storage and logic elements. Also known as ferromagnetic film; magnetic film. { mag'ned·ik 'thin ¦film }

magnetic transducer [ELECTROMAG] A device for transforming mechanical into electrical energy, which consists of a magnetic field including a variable-reluctance path and a coil surrounding all or a part of this path, so that variation in reluctance leads to a variation in the magnetic flux through the coil and a corresponding induced emf (electromotive force). { mag'ned·ik tranz'dü·sər }

magnetic tunnel junction [ELECTR] A magnetic storage and switching device in which two magnetic layers are separated by an insulating barrier, typically aluminum oxide, that is only 1–2 nanometers thick, allowing an electronic current whose magnitude depends on the orientation of both magnetic layers to tunnel through the barrier when it is subject to a small electric bias. { mag¦ned·ik 'tən·əl ‚jəŋk·shən }

magnetic vector See magnetic flux density. { mag'ned·ik 'vek·tər }

magnetic viscosity [ELECTROMAG] The existence of a time delay between a change in the magnetic field applied to a ferromagnetic material and the resulting change in magnetic induction which is too great to be explained by the existence of eddy currents. { mag'ned·ik vis'käs·əd·ē }

magnetic wave [SOLID STATE] The spread of magnetization from a small portion of a substance where an abrupt change in the magnetic field has taken place. { mag'ned·ik 'wāv }

magnetic wire [MATER] A wire made from magnetic material suitable for magnetic recording. { mag'ned·ik 'wīr }

magnetism [PHYS] Phenomena involving magnetic fields and their effects upon materials. { 'mag·nə,tiz·əm }

magnetization [ELECTROMAG] **1.** The property and in particular, the extent of being magnetized; quantitatively, the magnetic moment per unit volume of a substance. Also known as magnetic dipole density; magnetization intensity. **2.** The process of magnetizing a magnetic material. { ,mag·nəd·ə'zā·shən }

magnetization intensity See magnetization. { ,mag·nəd·ə'zā·shən in'ten·səd·ē }

magnetizing force See magnetic field strength. { 'mag·nə,tīz·iŋ ,fȯrs }

magnetocaloric effect [THERMO] The reversible change of temperature accompanying the change of magnetization of a ferromagnetic material. { mag¦nēd·ō·kə'lȯr·ik i,fekt }

magnetoelastic coupling [SOLID STATE] The interaction between the magnetization and the strain of a magnetic material. { mag¦nēd·ō·i'las·tik 'kəp·liŋ }

magnetoelasticity [SOLID STATE] Phenomenon in which an elastic strain alters the magnetization of a ferromagnetic material. { mag¦nēd·ō,i,las 'tis·əd·ē }

magnetoelectric effect [SOLID STATE] A linear coupling between magnetization and polarization found in certain magnetic ferroelectrics, such as BaMnF$_4$ at low temperatures. { mag ¦nēd·ō·i'lek·trik i,fekt }

magnetoelectricity [SOLID STATE] The appearance of an electric field in certain substances, such as chromic oxide (Cr$_2$O$_3$), when they are subjected to a static magnetic field. { mag ¦nēd·ō·i,lek'tris·əd·ē }

magnetoelectronics [ELECTR] The use of electron spin (as opposed to charge) in electronic devices. Also known as spin electronics; spintronics. { mag,ned·ō·i·lek'trän·iks }

magnetofluid [PHYS CHEM] A Newtonian or shear-thinning fluid whose flow properties become viscoplastic when it is modulated by a magnetic field. { ¦mag·nəd·ō'flü·əd }

magnetomechanics [PHYS] The study of the effects which the magnetization of a material and its strain have on each other. { mag ¦nēd·ō·mi'kan·iks }

magneton [PHYS] A unit of magnetic moment used for atomic, molecular, or nuclear magnets, such as the Bohr magneton, Weiss magneton, or nuclear magneton. { 'mag·nə,tän }

magneton number [PHYS] The ratio of the magnetic moment per atom, ion, or molecule of a paramagnetic or ferromagnetic material to the Bohr magneton. { 'mag·nə,tän ,nəm·bər }

magnetooptic material [OPTICS] A material whose optical properties are changed by an applied magnetic field. { mag¦nēd·ō¦äp·tik mə 'tir·ē·əl }

magnetooptics [OPTICS] The study of the effect of a magnetic field on light passing through a substance in the field. { mag¦nēd·ō¦äp·tiks }

magnetoresistance [ELECTR] The change in the electrical resistance of a material when it is subjected to an applied magnetic field, this

property has widespread application in sensors and magnetic read heads. [ELECTROMAG] The change in electrical resistance produced in a current-carrying conductor or semiconductor on application of a magnetic field. { mag ¦nēd·ō·ri'zis·təns }

magnetoresistivity [ELECTROMAG] The change in resistivity produced in a current-carrying conductor or semiconductor on application of a magnetic field. { mag,nēd·ō·ri,zis'tiv·əd·ē }

magnetoresistor [ELECTR] Magnetic field-controlled variable resistor. { mag¦nēd·ō·ri'zis·tər }

magnetorheological fluid [MATER] A low-viscosity fluid containing a suspension of micrometer-size magnetic particles that increases in viscosity proportionally to the strength of an applied magnetic field; it is used as an adaptive shock absorber for actively controlled viscous damping. { mag,ned·ō,rē·ə,läj·ə·kəl 'flü·əd }

magnetostatic [ELECTROMAG] Pertaining to magnetic properties that do not depend upon the motion of magnetic fields. { mag¦nēd· ə¦stad·ik }

magnetostatic mode [SOLID STATE] A spin wave in a magnetic material whose wavelength is greater than about one-tenth the size of the sample. { mag¦nēd·ə¦stad·ik 'mōd }

magnetostatics [ELECTROMAG] The study of magnetic fields that remain constant with time. Also known as static magnetism. { mag¦nēd·ō ¦stad·iks }

magnetostriction [ELECTROMAG] The dependence of the state of strain (dimensions) of a ferromagnetic sample on the direction and extent of its magnetization. { mag,nēd·ō'strik· shən }

magnetostriction transducer [ELECTROMAG] A transducer used with sonar equipment to change an alternating current to sound energy at the same frequency and to form the sound energy into a beam; its operation depends on the interaction between the magnetization and the deformation of a material having magnetostrictive properties. { mag,nēd·ō'strik·shən tranz'dü·sər }

magnetostrictive resonator [SOLID STATE] Ferromagnetic rod so designed that it can be excited magnetically into resonant vibration at one or more definite and known frequencies. { mag ¦nēd·ō¦strik·tiv 'rez·ən,ād·ər }

magnet power [ELECTROMAG] The electric power supplied to the coils of an electromagnet. { 'mag·nət ,paú·ər }

magno [MET] An alloy of 95.5% nickel and 4.5% manganese, used in the manufacture of incandescent lamps and radio tubes. { 'mag,nō }

magnon [SOLID STATE] A quasi-particle which is introduced to describe small departures from complete ordering of electronic spins in ferro-, ferri-, antiferro-, and helimagnetic arrays. Also known as quantized spin wave. { 'mag ,nän }

maguey [MATER] A fiber obtained from the agave (*Agave cantala*); maguey fibers are white, stiff, brilliant, and light in weight, and are used chiefly for binder twine. { mə'gä }

mahogany |MATER| The hard wood of these trees, especially the red or yellow-brown wood of the West Indies mahogany tree (*Swietenia mahagoni*). { mə'häg·ə·nē }

mahogany soap |MATER| The sodium salt of crude or refined petroleum sulfonic acids, used as flotation agents and to increase the oil absorption of mineral pigments in paint. { mə'häg·ə·nē 'sōp }

majority carrier |ELECTR| The type of carrier, that is, electron or hole, that constitutes more than half the carriers in a semiconductor. { mə'jär·əd·ē 'kar·ē·ər }

malleable |MET| Capable of undergoing plastic deformation without rupture; a property characteristic of metals. { 'mal·yə·bəl }

malleable brass See Muntz metal. { 'mal·yə·bəl 'bras }

malleable iron |MET| White cast iron which has been rendered malleable by heat treatment. { 'mal·yə·bəl 'ī·ərn }

malleableize |MET| To render a material malleable, such as by heat-treating white cast iron. { 'mal·yə·bə,līz }

malleable pig iron |MET| A grade of pig iron suitable for production of white cast iron from which malleable iron is made. { 'mal·yə·bəl 'pig ,ī·ərn }

malm brick |MATER| A brick made of natural malm or an artificial mixture consisting of pulverized chalk, sand, and bits of coke or furnace clinker. { 'mäm ,brik }

malm rubber |MATER| A comparatively soft malm brick that can be worked to a desired shape by rubbing. { 'mäm ,rəb·ər }

malmstone |MATER| A name applied to chert when it is used in building and paving. { 'mäm ,stōn }

malodorant See odorant. { mal'ōd·ə·rənt }

Malotte's metal |MET| A fusible alloy composed of 46% bismuth, 20% lead, and 34% tin; melts at 96–123°C. { mə'läts ,med·əl }

Malter effect |SOLID STATE| A phenomenon in which a metal with a nonconducting surface film has a large coefficient of secondary electron emission; this is particularly notable in aluminum whose surface has been oxidized and then coated with cesium oxide. { 'mäl·tər i,fekt }

mandrel |ENG| The core around which continuous strands of impregnated reinforcement materials are wound to fabricate hollow objects made of composite materials. |MET| A metal bar serving as a core around which other metals are cast, forged, or extruded, forming a true central hole. { 'man·drəl }

mandrel forging See saddling. { 'man·drəl ,fórj·iŋ }

manganate |INORG CHEM| **1.** Salts that have manganese in the anion. **2.** In particular, a salt of manganic acid formed by fusion of manganese dioxide with an alkali. { 'maŋ·gə,nāt }

manganese |CHEM| A metallic element, symbol Mn, atomic weight 54.938, atomic number 25; a transition element whose properties fall between those of chromium and iron. |MET| A hard, brittle, grayish-white metal used chiefly in making steel. { 'maŋ·gə,nēs }

manganese-aluminum |MET| A hardener alloy employed for making additions of manganese to aluminum alloys such as Duralumin; typical composition is 25% manganese, 75% aluminum. { 'maŋ·gə,nēs ə'lü·mə·nəm }

manganese binoxide See manganese dioxide. { 'maŋ·gə,nēs bi'näk,sīd }

manganese black See manganese dioxide. { 'maŋ·gə,nēs 'blak }

manganese borate |INORG CHEM| MnB_4O_7 Water-insoluble, reddish-white powder; used as a varnish and oil drier. { 'maŋ·gə,nēs 'bór,āt }

manganese-boron |MET| Manganese alloyed with boron; used as an ingredient for hardening and deoxidizing bronze. { 'maŋ·gə,nēs 'bór ,än }

manganese brass |MET| A brass containing about 70% copper, 29% zinc, and 1.3% manganese. { 'maŋ·gə,nēs 'bras }

manganese bromide See manganous bromide. { 'maŋ·gə,nēs 'brō,mīd }

manganese bronze |MET| A type of brass or bronze containing about 59% copper, 39% zinc, 1.5% iron, 1% tin, and 0.1% manganese; another composition by the same name contains about 66% copper, 23% zinc, 3% iron, 4.5% aluminum, and 3.7% manganese. { 'maŋ·gə ,nēs 'bränz }

manganese carbonate |INORG CHEM| $MnCO_3$ Rose-colored crystals found in nature as rhodocrosite; soluble in dilute acids, insoluble in water; used in medicine, in fertilizer, and as a paint pigment. { 'maŋ·gə,nēs 'kär·bə,nāt }

manganese dioxide |INORG CHEM| MnO_2 A black, crystalline, water-insoluble compound, decomposing to manganese sesquioxide, Mn_2O_3, and oxygen when heated to 535°C; used as a depolarizer in certain dry-cell batteries, as a catalyst, and in dyeing of textiles. Also known as battery manganese; manganese binoxide; manganese black; manganese peroxide. { 'maŋ·gə,nēs dī'äk,sīd }

manganese fluoride See manganous fluoride. { 'maŋ·gə,nēs 'flúr,īd }

manganese green See barium manganate. { 'maŋ·gə,nēs 'grēn }

manganese halide |INORG CHEM| Compound of manganese with a halide, such as chlorine, bromine, fluorine, or iodine. { 'maŋ·gə,nēs 'ha ,līd }

manganese heptoxide |INORG CHEM| Mn_2O_7 A compound formed as an explosive dark-green oil by the action of concentrated sulfuric acid on permanganate compounds. { 'maŋ·gə,nēs hep'täk,sīd }

manganese hydroxide See manganous hydroxide. { 'maŋ·gə,nēs hī'dräk,sīd }

manganese hypophosphite |INORG CHEM| $Mn(H_2PO_2)_2 \cdot H_2O$ Odorless, tasteless pink crystals which explode if heated with oxidants; used in medicine. { 'maŋ·gə,nēs ,hī·pō'fäs,fīt }

manganese iodide See manganous iodide. { 'maŋ·gə,nēs 'ī·ə,dīd }

manganese monoxide See manganese oxide. { 'maŋ·gə,nēs mə'näk,sīd }

manganese oxide [INORG CHEM] MnO Green powder, soluble in acids, insoluble in water; melts at 1650°C; used in medicine, in textile printing, as a catalyst, in ceramics, and in dry batteries. Also known as manganese monoxide. { 'maŋ·gə,nēs 'äk,sīd }

manganese peroxide See manganese dioxide. { 'maŋ·gə,nēs pə'räk,sīd }

manganese silicate See manganous silicate. { 'maŋ·gə,nēs 'sil·ə,kāt }

manganese-silicon [MET] An alloy that contains 73–78% silicon, 20–25% manganese, a maximum of 1.5% iron, and a maximum of 0.25% carbon; used for adding manganese and silicon to metals. { 'maŋ·gə,nēs 'sil·ə·kən }

manganese steel See Hadfield manganese steel. { 'maŋ·gə,nēs 'stēl }

manganese sulfate See manganese sulfate. { 'maŋ·gə,nēs 'səl,fāt }

manganese sulfide See manganese sulfide. { 'maŋ·gə,nēs 'səl,fīd }

manganese-titanium [MET] An alloy usually composed of 38% manganese, 29% titanium, 8% aluminum, 3% silicon, 22% iron, and no carbon; used as a deoxidizer for high-grade steels and for nonferrous alloys. { 'maŋ·gə,nēs tī'tān·ē·əm }

manganic fluoride [INORG CHEM] MnF_3 Poisonous red crystals, decomposed by heat and water; used as a fluorinating agent. { man'gan·ik 'flür,īd }

manganic hydroxide [INORG CHEM] $Mn(OH)_3$ A brown powder that rapidly loses water to form MnO(OH); used in ceramics and as a fabric pigment. Also known as hydrated manganic hydroxide. { man'gan·ik hī'dräk,sīd }

manganic oxide [INORG CHEM] Mn_2O_3 Hard black powder, insoluble in water, soluble in cold hydrochloric acid, hot nitric acid, and sulfuric acid; occurs in nature as manganite. { man'gan·ik 'äk,sīd }

manganous bromide [INORG CHEM] $MnBr_2 \cdot 4H_2O$ Water-soluble, deliquescent red crystals. Also known as manganese bromide. { 'maŋ·gə·nəs 'brō,mīd }

manganous chloride [INORG CHEM] $MnCl_2 \cdot 4H_2O$ Water-soluble, deliquescent rose-colored crystals melting at 88°C; used as a catalyst and in paints, dyeing, and pharmaceutical preparations. { 'maŋ·gə·nəs 'klór,īd }

manganous fluoride [INORG CHEM] MnF_2 Reddish powder, insoluble in water, soluble in acid. Also known as manganese fluoride. { 'maŋ·gə·nəs 'flür,īd }

manganous hydroxide [INORG CHEM] $Mn(OH)_2$ Heat-decomposable white-pink crystals; insoluble in water and alkali, soluble in acids; occurs in nature as pyrochroite. Also known as manganese hydroxide. { 'maŋ·gə·nəs hī'dräk,sīd }

manganous iodide [INORG CHEM] $MnI_2 \cdot 4H_2O$ Water-soluble, deliquescent yellowish-brown crystals. Also known as manganese iodide. { 'maŋ·gə·nəs 'ī·ə,dīd }

manganous silicate [INORG CHEM] $MnSiO_3$ Water-insoluble red crystals or yellowish-red powder; occurs in nature as rhodonite. Also known as manganese silicate. { 'maŋ·gə·nəs 'sil·ə,kāt }

manganous sulfate [INORG CHEM] $MnSO_4 \cdot 4H_2O$ Water-soluble, translucent, efflorescent rose-red prisms; melts at 30°C; used in medicine, textile printing, and ceramics, as a fungicide and fertilizer, and in paint manufacture. Also known as manganese sulfate. { 'maŋ·gə·nəs 'səl,fāt }

manganous sulfide [INORG CHEM] MnS An almost water-insoluble powder that decomposes on heating; used as a pigment and as an additive in making steel. Also known as manganese sulfide. { 'maŋ·gə·nəs 'səl,fīd }

manganous sulfite [INORG CHEM] $MnSO_3$ Grayish-black or brownish-red powder, soluble in sulfur dioxide, insoluble in water. { 'maŋ·gə·nəs 'səl,fīt }

manifold paper [MATER] An extremely thin paper used for making duplicate copies, such as onionskin paper. { 'man·ə,fōld ,pā·pər }

Manila copal See Manila resin. { mə'nil·ə 'kō·pəl }

Manila paper [MATER] Yellowish paper or Bristol board; the term originally referred to paper manufactured from Manila hemp. { mə'nil·ə ,pā·pər }

Manila resin [MATER] A type of resin extracted from trees of the genus *Agathis* in the Philippines that is soluble in ethyl and methyl alcohol, insoluble in water; used in printing ink, varnishes, paints, and linoleum. Also known as Manila copal. { mə'nil·ə 'rez·ən }

manjak [MATER] A variety of grahamite, manjak is the blackest of the asphalts; used for insulation and varnishes. { 'man,jak }

Mannesmann mill [MET] A mill consisting of two rolls mounted with their axes slightly inclined. { 'män·əs,män ,mil }

Mannesmann process [MET] A process for making seamless tubing by forcing a billet between the rolls of a Mannesmann mill so as to pierce the center, and then forcing the metal over a mandrel to form the central bore. { 'män·əs ,män ,präs·əs }

manocryometer [THERMO] An instrument for measuring the change of a substance's melting point with change in pressure; the height of a mercury column in a U-shaped capillary supported by an equilibrium between liquid and solid in an adjoining bulb is measured, and the whole apparatus is in a thermostat. { ,man·ō ,krī'äm·əd·ər }

mantle [ENG] A lacelike hood or envelope (sack) of refractory material which, when positioned over a flame and heated to incandescence, gives light. [MET] That part of the outer wall and casing of a blast furnace located above the hearth. { 'mant·əl }

manual spinning [MET] A sheet-metal forming process that forms the material over a rotating

215

mandrel with little or no change in the thickness of the original blank. Also known as conventional spinning. { 'man·yə·wəl 'spin·iŋ }

manual welding |MET| A welding method in which the operator manually guides an electrode, clamped in a hand-held electrode holder. { 'man·yə·wəl 'weld·iŋ }

manure salts |INORG CHEM| Potash salts that have a high proportion of chloride and 20–30% potash; used in fertilizers. { mə'núr ,sòlts }

maple |MATER| The hard, light-colored, close-grained wood, especially from sugar maple (A. saccharum). { 'mā·pəl }

maraging steel |MET| High-strength, low-carbon iron-nickel alloy in which a martensitic structure is formed on cooling; contains 6–7% nickel, 0–11% cobalt, 0–5% molybdenum, and small percentages of titanium, aluminum, and columbium; hardening is accomplished by heating the quenched alloy at 400–500°C. { mär'ā·jiŋ ,stēl }

marble dust See marble flour. { 'mär·bəl ,dəst }

marble flour |MATER| Finely divided marble chips; used as a filler or abrasive in hand soaps and for casting. Also known as marble dust. { 'mär·bəl ,flaú·ər }

Margoulis number See Stanton number. { mär 'gü·ləs ,nəm·bər }

marine glue |MATER| An adhesive that is insoluble in water; usually made of rubber and shellac, sometimes with resins. { mə'rēn 'glü }

marquenching See martempering. { 'mär ,kwench·iŋ }

Mars pigments |INORG CHEM| A group of five pigments produced when milk of lime is added to a ferrous sulfate solution, and the precipitate is calcined; color is controlled by calcination temperature to give yellow, orange, brown, red, or violet. { 'märz ,pig·məns }

martempering |MET| Quenching austenitized steel to a temperature just above, or in the upper part of, the martensite range, holding it at this point until the temperature is equalized throughout, and then cooling in air to room temperature. Also known as marquenching. { 'mär,tem·pə·riŋ }

martensite |MET| A metastable transitional structure formed by a shear process during a phase transformation, characterized by an acicular or needlelike pattern; in carbon steel it is a hard, supersaturated solid solution of carbon in a body-centered tetragonal lattice of iron. { 'mär,ten,zīt }

martensite range |MET| The temperature interval between the temperature (M_s) at which formation of martensite initiates and the temperature (M_f) at which the formation is complete. { 'mär ,ten,zīt ,rānj }

martensitic stainless steel |MET| A hard, quenched magnetic martensitic steel containing principally 11–18% chromium and 0.1–1.2% carbon. { ¦mär,ten¦zid·ik 'stān·ləs ¦stēl }

martensitic steel |MET| Quenched carbon steel composed chiefly of martensite. { ¦mär,ten ¦zid·ik 'stēl }

martensitic structure |MET| Of, pertaining to, or having the structure of martensite, that is, an interstitial, supersaturated solid solution of carbon in iron having a body-centered tetragonal lattice; the microstructure is characterized by an acicular or needlelike pattern. { ¦mär,ten¦zid·ik 'strək·chər }

martensitic transformation |MET| A phase transformation which occurs in some metals, resulting in the formation of martensite. Also known as shear transformation. { ¦mär,ten ¦zid·ik ,tranz·fər'mā·shən }

Martin's cement |MATER| A gypsum cement made with potassium carbonate instead of alum. { 'märt·ənz si,ment }

martonite |MATER| A poison gas composed of 20% chloroacetone and 80% bromoacetone; acts as a powerful lacrimator. { 'märt·ən,īt }

Marx effect |SOLID STATE| The effect wherein the energy of photoelectrons emitted from an illuminated surface is decreased when the surface is simultaneously illuminated by light of lower frequency than that causing the emission. { 'märks i,fekt }

mash seam weld |MET| A seam weld at a lap joint in which the overall lap thickness is reduced plastically to the approximate thickness of one of the lapped parts. { 'mash 'sēm ,weld }

mask |ELECTR| A thin sheet of metal or other material containing an open pattern, used to shield selected portions of a semiconductor or other surface during a deposition process. |MET| A protective device in thermal spraying against blasting or coating effects which are reflected from the substrate surface. { mask }

masonry |CIV ENG| A construction of stone or similar materials such as concrete or brick. { 'mās·ən·rē }

masonry cement |MATER| A blended cement, made by combining either natural or portland cements with fattening materials such as hydrated lime and, sometimes, with air-entraining mixtures; used in the mortar of brick and block masonry. { 'mās·ən·rē si,ment }

mason's hydrated lime |MATER| Any hydrated lime suitable for use in mortars, base-coat plasters, and concrete. { 'mās·ənz ¦hī,drād·əd 'līm }

mass |MECH| A quantitative measure of a body's resistance to being accelerated; equal to the inverse of the ratio of the body's acceleration to the acceleration of a standard mass under otherwise identical conditions. { mas }

mass absorption coefficient |PHYS| The linear absorption coefficient divided by the density of the medium. { 'mas əb'sórp·shən ,kō·i,fish·ənt }

mass color See masstone. { ¦mas ¦kəl·ər }

mass concrete |CIV ENG| Concrete set without structural reinforcement. { 'mas ¦kän,krēt }

mass flow |FL MECH| The mass of a fluid in motion which crosses a given area in a unit time. { 'mas 'flō }

Massieu function |THERMO| The negative of the Helmholtz free energy divided by the temperature. { ma'syü ,fəŋk·shən }

mass resistivity |ELEC| The product of the electrical resistance of a conductor and its mass, divided by the square of its length; the product of the electrical resistivity and the density. { 'mas ,rē,zis'tiv·əd·ē }

masstone |MATER| The undiluted color of a pigment or a pigmented paint coating. Also known as mass color. { 'ma,stōn }

mass transport |FL MECH| **1.** Carrying of loose materials in a moving medium such as water or air. **2.** The movement of fluid, especially water, from one place to another. { 'mas 'tranz,pȯrt }

master alloy |MET| An alloy of selected elements that can be added to a charge of molten metal to provide a desired composition or texture or to deoxidize the material. Also known as foundry alloy. { 'mas·tər 'al,ȯi }

masterbatch |MATER| A plastic, rubber, or elastomer mixture in which there is a high additives concentration, such as rubber with carbon black, or plastic with color pigment; used to proportion additives accurately into large bulks of plastic, rubber, or elastomer. { 'mas·tər,bach }

mastic |MATER| **1.** A glasslike, brittle, yellow to greenish yellow resinous exudation of the mastic tree (*Pistacia lentiscus*); used in medicine, condiments, adhesive, incense, and lacquer. Also known as mastiche; mastix. **2.** Mixture of finely powdered rock and asphaltic material used for highway construction. { 'mas·tik }

mastic asphalt *See* asphalt mastic. { 'mas·tik 'as,fȯlt }

mastiche *See* mastic. { 'mas·tik }

mastix *See* mastic. { 'mas·tiks }

mat |MATER| Randomly distributed felt or glass fibers used in reinforced-plastics lay-up molding. { mat }

match plate |MET| A plate on which metal-casting patterns are mounted or formed as an integral part, to facilitate the molding operation. { 'mach ,plāt }

materials science |ENG| The study of the nature, behavior, and use of materials applied to science and technology. { mə'tir·ē·əlz ,sī·əns }

matrix |MATER| A binding agent used to make an agglomerate mass. |MET| **1.** The principal component of an alloy. **2.** The precisely shaped form used as the cathode in electroforming. { 'mā·triks }

matrix fiber *See* biconstituent fiber. { 'mā·triks ,fī·bər }

matte |MATER| Dull, as applied to appearance of a surface. |MET| An impure metallic sulfide mixture produced by smelting the sulfide ores of such metals as copper, lead, or nickel. { mat }

matte dip |MET| An etching solution which reacts with the surface of a metal, giving a dull finish. { 'mat ,dip }

matter |PHYS| The substance composing bodies perceptible to the senses; includes any entity possessing mass when at rest. { 'mad·ər }

matte smelting |MET| Smelting of copper-bearing materials in a reverberatory furnace. { 'mat ,smelt·iŋ }

Matthias' rules |SOLID STATE| Several empirical rules giving the dependence of the transition temperatures of superconducting metals and alloys on the position of the metals in the periodic table and in the composition of the alloys. { mə'thī·əs ,rülz }

Matthiessen sinker method |THERMO| A method of determining the thermal expansion coefficient of a liquid, in which the apparent weight of a sinker when immersed in the liquid is measured for two different temperatures of the liquid. { ¦math·ə·sən 'siŋ·kər ,meth·əd }

Matthiessen's rule |SOLID STATE| An empirical rule which states that the total resistivity of a crystalline metallic specimen is the sum of the resistivity due to thermal agitation of the metal ions of the lattice and the resistivity due to imperfections in the crystal. { 'math·ə·sənz ,rül }

mattress |CIV ENG| A woven mat, often of wire and cement blocks, used to prevent erosion of dikes, jetties, or river banks. { 'ma·trəs }

Maxwell equal-area rule |THERMO| At temperatures for which the theoretical isothermal of a substance, on a graph of pressure against volume, has a portion with positive slope (as occurs in a substance with liquid and gas phases obeying the van der Waals equation), a horizontal line drawn at the equilibrium vapor pressure and connecting two parts of the isothermal with negative slope has the property that the area between the horizontal and the part of the isothermal above it is equal to the area between the horizontal and the part of the isothermal below it. { 'mak,swel ¦ē·kwəl 'er·ē·ə ,rül }

Maxwell relation |ELECTROMAG| According to Maxwell's electromagnetic theory, that relation wherein the dielectric constant of a substance equals the square of its index of refraction. |THERMO| One of four equations for a system in thermal equilibrium, each of which equates two partial derivatives, involving the pressure, volume, temperature, and entropy of the system. { 'mak,swel ri'lā·shən }

Maxwell's electromagnetic theory |ELECTROMAG| A mathematical theory of electric and magnetic fields which predicts the propagation of electromagnetic radiation, and is valid for electromagnetic phenomena where effects on an atomic scale can be neglected. { 'mak,swelz i,lek·trō·mag'ned·ik 'thē·ə·rē }

Maxwell's law |ELECTROMAG| A movable portion of a circuit will always move in such a direction as to give maximum magnetic flux linkages through the circuit. { 'mak,swelz 'lȯ }

Maxwell's theorem |MECH| If a load applied at one point A of an elastic structure results in a given deflection at another point B, then the same load applied at B will result in the same deflection at A. { 'mak,swelz 'thir·əm }

mayer |THERMO| A unit of heat capacity equal to the heat capacity of a substance whose temperature is raised 1° Celsius by 1 joule. { 'mī·ər }

Mayer's formula |THERMO| A formula which states that the difference between the specific heat of a gas at constant pressure and its specific heat at constant volume is equal to the gas constant divided by the molecular weight of the gas. { 'mī·ərz ‚fȯr·myə·lə }

MBE *See* molecular-beam epitaxy.

M-C asphalt *See* medium-curing asphalt. { ¦em ¦sē 'as‚fȯlt }

M center |SOLID STATE| A color center consisting of an F center combined with two ion vacancies. { 'em ‚sen·tər }

McQuaid-Ehn test |MET| A test for determining the grain-size characteristics of a steel in which a sample is carburized for 8 hours at 1700°F (927°C) and cooled slowly; the high-carbon case on slow cooling will reject cementite at the austenite grain boundaries and, by polishing and etching, the grains will clearly be seen under a microscope. { mə'kwäd 'än ‚test }

Md *See* mendelevium.

mean British thermal unit *See* British thermal unit. { 'mēn ¦brid·ish 'thər·məl ‚yü·nət }

mean calorie |THERMO| One-hundredth of the heat needed to raise 1 gram of water from 0 to 100°C. { 'mēn 'kal·ə·rē }

mean normal stress |MECH| In a system stressed multiaxially, the algebraic mean of the three principal stresses. { 'mēn ¦nȯrm·əl 'stres }

mean specific heat |THERMO| The average over a specified range of temperature of the specific heat of a substance. { 'mēn spə'sif·ik 'hēt }

mean stress |MECH| The algebraic mean of the maximum and minimum values of a periodically varying stress. { 'mēn 'stres }

mechanical alloying |MET| A materials processing method for assembling metal constituents with a controlled microstructure by repeated welding, fracturing, and rewelding of a mixture of powder particles, generally in a high-energy ball mill. { mi'kan·i·kəl ə'lȯi·iŋ }

mechanical equation of state |MET| An equation that expresses the relation of stress, strain, strain rate, and temperature for a metal. { mi'kan·ə·kəl i'kwā·zhən əv 'stāt }

mechanical equivalent of heat |THERMO| The amount of mechanical energy equivalent to a unit of heat. { mi'kan·ə·kəl i'kwiv·ə·lənt əv 'hēt }

mechanical hysteresis |MECH| The dependence of the strain of a material not only on the instantaneous value of the stress but also on the previous history of the stress; for example, the elongation is less at a given value of tension when the tension is increasing than when it is decreasing. { mi'kan·ə·kəl ‚his·tə'rē·səs }

mechanical impedance |MECH| The complex ratio of a phasor representing a sinusoidally varying force applied to a system to a phasor representing the velocity of a point in the system. { mi'kan·ə·kəl im'pēd·əns }

mechanically foamed plastic |MATER| A foamed plastic having its cellular structure produced by gases that are physically incorporated. { mi'kan·ə·klē ¦fōmd 'plas·tik }

mechanical metallurgy |MET| The science and technology of the behavior of metals relating to

mechanical forces imposed on them; includes rolling, extruding, deep drawing, bending, and other processes. { mi'kan·ə·kəl 'med·əl‚ər·jē }

mechanical ohm |MECH| A unit of mechanical resistance, reactance, and impedance, equal to a force of 1 dyne divided by a velocity of 1 centimeter per second. { mi'kan·ə·kəl 'ōm }

mechanical plating |MET| Deposition of one metal on another by a cold-peening process, such as tumbling. { mi'kan·ə·kəl 'plād·iŋ }

mechanical property |MECH| A property that involves a relationship between stress and strain or a reaction to an applied force. { mi'kan·ə·kəl 'präp·ərd·ē }

mechanical pulp |MATER| Wood pulp produced by grinding and soaking the wood fibers. Also known as groundwood pulp. { mi'kan·ə·kəl 'pəlp }

mechanical reactance |MECH| The imaginary part of mechanical impedance. { mi'kan·ə·kəl rē'ak·təns }

mechanical resistance *See* resistance. { mi'kan·ə·kəl ri'zis·təns }

mechanical twin |MET| A twin formed in a metal crystal by plastic deformation, involving shear of the lattice. { mi'kan·ə·kəl 'twin }

mechanical vapor diffusion |MET| A process in which a metal and a metal surface to which the first metal is to be added are vaporized by a pulsating arc and become an integral part of the total mass. Abbreviated MVD. { mi'kan·ə·kəl 'vāp·ər də'fyü·shen }

mechanical vibration |MECH| The continuing motion, often repetitive and periodic, of parts of machines and structures. { mi'kan·ə·kəl vī'brā·shən }

mechanical working |MET| Formation of a desired shape and physical properties of a metal by subjecting it to pressure by rolls, presses, or hammers. { mi'kan·ə·kəl 'wərk·iŋ }

mechanocaloric effect |CRYO| An effect resulting from the fact that a temperature gradient in helium II is invariably accompanied by a pressure gradient, and conversely; examples are the fountain effect, and the heating of liquid helium left behind in a container when part of it leaks out through a small orifice. { ¦mek·ə·nō·kə'lȯr·ik i‚fekt }

mechanochemical effect |PHYS CHEM| Changes in the dimensions of certain polymers, particularly photoelectrolytic gels and crystalline polymers, in response to changes in their chemical environment. { ¦mek·ə·nō'kem·ə·kəl i‚fekt }

mechanochemistry |PHYS CHEM| The study of the conversion of mechanical energy into chemical energy in polymers. { ¦mek·ə·nō'kem·ə·strē }

mechanomotive force |MECH| The root-mean-square value of a periodically varying force. { ¦mek·ə·nō¦mōd·iv ‚fȯrs }

mechanophotochemistry |PHYS CHEM| The study of changes in the dimensions of certain photoresponsive polymers upon exposure to light. { ¦mek·ə·nō‚fō·dō'kem·ə·strē }

medium |CHEM ENG| **1.** The carrier in which a

chemical reaction takes place. **2.** Material of controlled pore size used to remove foreign particles or liquid droplets from fluid carriers. |PHYS| That entity in which objects exist and phenomena take place; examples are free space and various fluids and solids. { 'mē·dē·əm }

medium boiler |MATER| A solvent, intermediate in volatility between high and low boilers, which is added to lacquer thinner and influences flow. { 'mē·dē·əm 'bȯil·ər }

medium carbon steel |MET| Steel containing 0.15-0.30% carbon. { 'mē·dē·əm ¦kär·bən ‚stēl }

medium-curing asphalt |MATER| Liquid product composed of asphalt cement and a kerosine-type diluent. Also known as M-C asphalt. { 'mē·dē·əm ¦kyúr·iŋ 'as‚fȯlt }

megohm |ELEC| A unit of resistance, equal to 1,000,000 ohms. { 'me‚gōm }

Meissner effect |SOLID STATE| The expulsion of magnetic flux from the interior of a piece of superconducting material as the material undergoes the transition to the superconducting phase. Also known as flux jumping; Meissner-Ochsenfeld effect. { 'mīs·nər i‚fekt }

Meissner-Ochsenfeld effect See Meissner effect. { 'mīs·nər 'äk·sən‚feld i‚fekt }

meitnerium |CHEM| A chemical element, symbolized Mt, atomic number 109, a synthetic element; the seventeenth transuranium element. { mīt'nir·ē·əm }

melamine resin |MATER| An amino resin made from formaldehyde and melamine; it is used as a molding compound with fillers added to it; it may also be used for laminating. { 'mel·ə‚mēn 'rez·ən }

melt |CHEM| **1.** To change a solid to a liquid by the application of heat. **2.** A melted material. |MET| A charge of molten metal. { melt }

melt extraction |MET| A rapid quenching process in which the molten metal is brought into contact with the periphery of a rotating heat-extracting disk; quench rates exceed 1,000,000 K per second. { 'melt ik‚strak·shən }

melt-fabricable |MATER| Referring to a plastic material that can be shaped as a melt without decomposing, and is capable of being extruded. { 'melt ¦fab·rə·kə·bəl }

melt fracture |MECH| Melt flow instability through a die during plastics molding, leading to helicular, rippled surface irregularities on the finished product. { 'melt ‚frak·chər }

melt index |ENG| Number of grams of thermoplastic resin at 190°C that can be forced through a 0.0825-inch (2.0955-millimeter) orifice in 10 minutes by a 2160-gram force. { 'melt ‚in‚deks }

melting loss |MET| Weight loss due to volatilization or oxidation during metal melting in a foundry. { 'melt·iŋ ‚lȯs }

melting point |THERMO| **1.** The temperature at which a solid of a pure substance changes to a liquid. Abbreviated mp. **2.** For a solution of two or more components, the temperature at which the first trace of liquid appears as the solution is heated. { 'melt·iŋ ‚pȯint }

melting rate |MET| In electric arc welding, the

weight or length of electrode melted in a specified unit of time. Also known as melt-off rate. { 'melt·iŋ ‚rāt }

melting ratio |MET| The ratio of metal weight to fuel weight in a melting process. { 'melt·iŋ ‚rā·shō }

melt instability |MECH| Instability of the plastic melt flow through a die. { 'melt ‚in·stə'bil·əd·ē }

melt-off rate See melting rate. { 'melt‚ȯf ‚rāt }

melt strength |MECH| Strength of a molten plastic. { 'melt ‚streŋkth }

membrane mimetic chemistry |ORG CHEM| The study of processes and reactions that have been developed by using information obtained from biological membrane systems. { ¦mem‚brän mi ¦med·ik 'kem·ə·strē }

MEMS See micro-electro-mechanical system. { memz or ¦em¦ē¦em'es }

mendelevium |CHEM| Synthetic radioactive element, symbol Md, with atomic number 101; made by bombarding lighter elements with light nuclei accelerated in cyclotrons. { ‚men·də'lē·vē·əm }

meniscus |FL MECH| The free surface of a liquid which is near the walls of a vessel and which is curved because of surface tension. |MET| In reference to a solder joint, the minimum angle at which the solder tapers from the joint to the flat area. { mə'nis·kəs }

mercapt-, mercapto- |CHEM| A combining form denoting the presence of the thiol (SH) group. { mər'kap·tō }

mercapto compound See sulfhydryl compound. { mər'kap·tō ‚käm‚pau̇nd }

mercerizing assistant |MATER| A wetting agent, such as cresylic acid and derivatives or oils, used to increase the penetration of textile mercerization baths. { 'mər·sə‚riz·iŋ ə·‚sis·tənt }

merchant mill |MET| A mill, consisting of a group of stands of three rolls each, used to roll rounds, flats, or squares of smaller dimensions than could be rolled on a bar mill. { 'mər·chənt ‚mil }

mercuric |INORG CHEM| The mercury ion with a 2+ oxidation state, for example $Hg(NO_3)_2$. { mər'kyúr·ik }

mercuric arsenate |INORG CHEM| $HgHAsO_4$ A poisonous yellow powder; soluble in hydrochloric acid, insoluble in water; used in antifouling and waterproof paints and in medicine. Also known as mercury arsenate; mercury arseniate. { mər'kyúr·ik 'ärs·ən‚āt }

mercuric barium iodide |INORG CHEM| $HgI_2 \cdot BaI_2 \cdot 5H_2O$ Crystals that are yellow or reddish and deliquescent; soluble in alcohol and water; used in aqueous solution as Rohrbach's solution for mineral separation on the basis of density. Also known as barium mercury iodide; mercury barium iodide. { mər'kyúr·ik 'bar·ē·əm 'ī·ə‚dīd }

mercuric bromide |INORG CHEM| $HgBr_2$ Poisonous white crystals, sensitive to light, melting at 235°C; soluble in alcohol and ether; used in medicine. Also known as mercury bromide. { mər'kyúr·ik 'brō‚mīd }

mercuric chloride |INORG CHEM| $HgCl_2$ An extremely toxic compound that forms white, rhombic crystals which sublime at 300°C and are soluble in alcohol or benzene; used for the manufacture of other mercuric compounds, as a fungicide, and in medicine and photography. Also known as bichloride of mercury; corrosive sublimate. { mər'kyùr·ik 'klór,īd }

mercuric cyanide [INORG CHEM] $Hg(CN)_2$ Poisonous, colorless, transparent crystals that darken in light, decompose when heated; soluble in water and alcohol; used in photography, medicine, and germicidal soaps. Also known as mercury cyanide. { mər'kyùr·ik 'sī·ə,nīd }

mercuric fluoride [INORG CHEM] HgF_2 Poisonous, transparent crystals that decompose when heated; moderately soluble in alcohol and water; used to synthesize organic fluorides. { mər'kyùr·ik 'flùr,īd }

mercuric iodide [INORG CHEM] HgI_2 Poisonous red crystals that turn yellow when heated to $150°C$; soluble in boiling alcohol; used in medicine and in Nessler's and Mayer's reagents. { mər'kyùr·ik 'ī·ə,dīd }

mercuric nitrate [INORG CHEM] $Hg(NO_3)_2·H_2O$ Poisonous, colorless crystals that decompose when heated; soluble in water and nitric acid, insoluble in alcohol; a fire hazard; used in medicine, in nitrating organic aromatics, and in felt manufacture. Also known as mercury nitrate; mercury pernitrate. { mər'kyùr·ik 'nī,trāt }

mercuric oxide [INORG CHEM] HgO A compound of mercury that exists in two forms, red mercuric oxide and yellow mercuric oxide; the red form decomposes upon heating, is insoluble in water, and is used in pigments and paints, and in ceramics; the yellow form is insoluble in water, decomposes upon heating, and is used in medicine. Also known as mercury oxide; red precipitate; yellow precipitate. { mər'kyùr·ik 'äk ,sīd }

mercuric phosphate [INORG CHEM] $Hg_3(PO_4)_2$ Poisonous yellowish or white powder; insoluble in alcohol and water, soluble in acids; used in medicine. Also known as mercury phosphate; trimercuric orthophosphate. { mər'kyùr·ik 'fäs ,fāt }

mercuric sulfate [INORG CHEM] $HgSO_4$ A toxic, white, crystalline powder, soluble in acids; used in medicine, as a catalyst, and for galvanic batteries. Also known as mercury persulfate; mercury sulfate. { mər'kyùr·ik 'səl,fāt }

mercuric sulfide [INORG CHEM] HgS **1.** The black variety is a poisonous powder; insoluble in water, alcohol, and nitric acid, soluble in sodium sulfide solution; sublimes at $583°C$; used as a pigment. **2.** The red variety is a poisonous powder; insoluble in water and alcohol; sublimes at $446°C$; used as a medicine and pigment. Also known as Chinese vermilion; quicksilver vermilion; red mercury sulfide; vermilion. { mər'kyùr·ik 'səl,fīd }

mercuric thiocyanate [INORG CHEM] $Hg(SCN)_2$ Poisonous white powder; soluble in alcohol, slightly soluble in water; decomposes when heated; used in photography. Also known as mercury thiocyanate. { mər'kyùr·ik ,thī·ə'sī·ə ,nāt }

mercurous [INORG CHEM] Referring to mercury with a valence of 1; for example, mercurous chloride, Hg_2Cl_2, where the mercury is covalently bonded, as Cl–Hg–Hg–Cl. { mər'kyùr·əs }

mercurous bromide [INORG CHEM] HgBr Poisonous white powder, crystals, or fibrous mass; odorless and tasteless; darkens in light; soluble in hot sulfuric acid and fuming nitric acid, insoluble in alcohol and ether; used in medicine. Also known as mercury bromide. { mər'kyùr·əs 'brō,mīd }

mercurous chlorate [INORG CHEM] $Hg_2(ClO_3)_2$ Poisonous white crystals that decompose at $250°C$; soluble in alcohol and water; explodes in contact with combustible substances. Also known as mercury chlorate. { mər'kyùr·əs 'klór ,āt }

mercurous chloride [INORG CHEM] Hg_2Cl_2 Odorless, nonpoisonous white crystals that darken in light; insoluble in water, alcohol, and ether; melts at $302°C$; used in medicine and pyrotechnics. Also known as mercury monochloride; mercury protochloride; mild mercury chloride. { mər'kyùr·əs 'klór,īd }

mercurous chromate [INORG CHEM] Hg_2CrO_4 Red powder with variable composition; decomposes when heated; soluble in nitric acid, insoluble in water and alcohol; used to color ceramics green. Also known as mercury chromate. { mər'kyùr·əs 'krō,māt }

mercurous iodide [INORG CHEM] Hg_2I_2 Odorless, tasteless, poisonous yellow powder; darkens when heated; insoluble in water, alcohol, and ether; sublimes at $140°C$; used as external medicine. Also known as mercury protoiodide. { mər'kyùr·əs 'ī·ə,dīd }

mercurous oxide [INORG CHEM] Hg_2O A poisonous black powder; insoluble in water, soluble in acids; decomposes at $100°C$. { mər'kyùr·əs 'äk,sīd }

mercurous phosphate [INORG CHEM] Hg_3PO_4 Light-sensitive white powder with variable composition; insoluble in alcohol and water, soluble in nitric acids; used in medicine. Also known as mercury phosphate; trimercurous orthophosphate. { mər'kyùr·əs 'fäs,fāt }

mercurous sulfate [INORG CHEM] Hg_2SO_4 Poisonous yellow-to-white powder; soluble in hot sulfuric acid or dilute nitric acid, insoluble in water; used as a catalyst and in laboratory batteries. { mər'kyùr·əs 'səl,fāt }

mercury [CHEM] A metallic element, symbol Hg, atomic number 80, atomic weight 200.59, existing at room temperature as a silvery, heavy liquid. Also known as quicksilver. { 'mər·kyə·rē }

mercury arsenate See mercuric arsenate. { 'mər·kyə·rē 'ärs·ən,āt }

mercury arseniate See mercuric arsenate. { 'mər·kyə·rē är'sē·nē,āt }

mercury barium iodide See mercuric barium iodide. { 'mər·kyə·rē ba·rē·əm 'ī·ō·dīd }

mercury bromide See mercuric bromide; mercurous bromide. { 'mər·kyə·rē 'brō,mīd }

mercury chlorate See mercurous chlorate. { 'mər·kyə·rē 'klór,āt }

mercury chromate See mercurous chromate. { 'mər·kyə·rē 'krō,māt }

mercury cyanide See mercuric cyanide. { 'mər·kyə·rē 'sī·ə,nīd }

mercury monochloride See mercurous chloride. { 'mər·kyə·rē ,män·ə'klòr,īd }

mercury nitrate See mercuric nitrate. { 'mər·kyə·rē 'nī,trāt }

mercury oxide See mercuric oxide. { 'mər·kyə·rē 'äk,sīd }

mercury pernitrate See mercuric nitrate. { 'mər·kyə·rē pər'nī,trāt }

mercury persulfate See mercuric sulfate. { 'mər·kyə·rē pər'səl,fāt }

mercury phosphate See mercuric phosphate; mercurous phosphate. { 'mər·kyə·rē 'fäs,fāt }

mercury protochloride See mercurous chloride. { 'mər·kyə·rē ,prō·dō'klòr,īd }

mercury protoiodide See mercurous iodide. { 'mər·kyə·rē ,prō·dō'ī·ə,dīd }

mercury sulfate See mercuric sulfate. { 'mər·kyə·rē 'səl,fāt }

mercury thiocyanate See mercuric thiocyanate. { 'mər·kyə·rē ,thī·ə'sī·ə,nāt }

mercury-vapor lamp [ELECTR] A lamp in which light is produced by an electric arc between two electrodes in an ionized mercury-vapor atmosphere; it gives off a bluish-green light rich in ultraviolet radiation. { 'mər·kyə·rē ¦vā·pər ,lamp }

merit [ELECTR] A performance rating that governs the choice of a device for a particular application; it must be qualified to indicate type of rating, as in gain-bandwidth merit or signal-to-noise merit. { 'mer·ət }

merohedral [CRYSTAL] Of a crystal class in a system, having a general form with only one-half, one-fourth, or one-eighth the number of equivalent faces of the corresponding form in the holohedral class of the same system. Also known as merosymmetric. { ¦mer·ə¦hē·drəl }

merosymmetric See merohedral. { ¦mer·ə·si ¦me·trik }

Merrill-Crowe process [MET] Removal of gold from cyanide solution by deoxygenation followed by precipitation on zinc dust, the work being completed by filtration to give the resultant auriferous gold slimes. { 'mer·əl 'krō ,prä·səs }

MESFET See metal semiconductor field-effect transistor. { 'mes,fet }

mesh weld [MET] A seam weld in which the finished weld is only slightly thicker than the sheets, and the lap disappears. { 'mesh ,weld }

meso- [CHEM] A prefix meaning intermediate or middle, as in denoting inactive optical isomers, the form of intermediate inorganic acid, the middle position in cyclic organic compounds, or a ring system with middle ring positions. { 'me·zō }

mesomorphism [PHYS CHEM] A state of matter intermediate between a crystalline solid and a normal isotropic liquid, in which long rod-shaped organic molecules contain dipolar and polarizable groups. { ¦mez·ə¦mòr,fiz·əm }

mesopore [CHEM] A pore in a catalytic material whose width ranges from 2 nanometers to 0.05 micrometer. { 'mez·ə,pòr }

mesoscopic [PHYS] Pertaining to a size regime, intermediate between the microscopic and the macroscopic, that is characteristic of a region where a large number of particles can interact in a quantum-mechanically correlated fashion. { ¦mez·ə¦skäp·ik }

mesoscopic physics [PHYS] A subdiscipline of condensed-matter physics that focuses on the properties of solids in a size range intermediate between bulk matter and individual atoms or molecules. { ,mez·ə,skäp·ik 'fiz·iks }

meta- [ORG CHEM] A prefix for benzene-ring compounds when two side chains are connected to carbon atoms with an unsubstituted carbon atom between them. { 'med·ə }

metahydrate sodium carbonate [INORG CHEM] $Na_2CO_3 \cdot H_2O$ Water-soluble, white crystals with an alkaline taste, loses water at 109°C, melts at 851°C; used in medicine, photography, and water pH control, and as a food additive. Also known as soda crystals. { ,med·ə'hī,drāt ¦sōd·ē·əm 'kär·bə,nāt }

metal [MATER] An opaque crystalline material usually of high strength with good electrical and thermal conductivities, ductility, and reflectivity; properties are related to the structure, in which the positively charged ions are bonded through a field of free electrons which surrounds them forming a close-packed structure. { 'med·əl }

metal arc cutting [MET] Cutting metal with the heat of an arc between a metal electrode and the base metal. { 'med·əl 'ärk ,kəd·iŋ }

metal arc welding [MET] Arc welding using covered metal electrodes. { 'med·əl 'ärk ,weld·iŋ }

metal casting [MET] A metal-forming process whereby molten metal is poured into a cavity or mold and, when cooled, solidifies, taking on the characteristic shape of the mold. { 'med·əl ¦kast·iŋ }

metal ceramic See cermet. { 'med·əl sə'ram·ik }

metal cluster compound [CHEM] A compound in which two or more metal atoms aggregate so as to be within bonding distance of one another and each metal atom is bonded to at least two other metal atoms; some nonmetal atoms may be associated with the cluster. { 'med·əl ¦kləs·tər 'käm,paùnd }

metal coating [MET] A thin film of metal bonded to a base material in order to add specific surface properties, such as corrosion or oxidation resistance, color, wear resistance, or optical characteristics. { 'med·əl ¦kōd·iŋ }

metal distribution ratio [MET] In plating operations, the ratio of the thickness of metal deposited on a near portion of a cathode to that deposited on a far portion of the cathode. { 'med·əl ,dis·trə'byü·shən ,rā·shō }

metal dye [MATER] Any of the special dyes, such as alizarin cyanin RR or alizarin green S, used to color oxided surfaces of aluminum or steel. { 'med·əl ,dī }

metal-foil paper [MATER] Paper backed with metal foils, manufactured in a number of vivid colors. { 'med·əl ¦fȯil 'pā·pər }

metal forming [MET] Any manufacturing process by which parts of components are fabricated by shaping or molding a piece of metal stock. { 'med·əl ˌförm·iŋ }

metal halide lamp [ELECTR] A discharge lamp in which metal halide salts are added to the contents of a discharge tube in which there is a high-pressure arc in mercury vapor; the added metals generate different wavelengths, to give substantially white light at an efficiency approximating that of high-pressure sodium lamps. { 'med·əl 'ha,līd ,lamp }

metaliding [MET] A process of depositing a metal as an alloy on a substrate from a fused complex metal salt. { 'med·əl,īd·iŋ }

metal inert-gas welding [MET] A welding procedure in which an electric current heats one metal which then joins to a second metal, with an inert gas preventing oxidation. { 'med·əl i¦nərt ¦gas 'weld·iŋ }

metal-insulator semiconductor [SOLID STATE] Semiconductor construction in which an insulating layer, generally a fraction of a micrometer thick, is deposited on the semiconducting substrate before the pattern of metal contacts is applied. Abbreviated MIS. { 'med·əl ¦in·sə ,lād·ər 'sem·i·kən'dək·tər }

metal-insulator transition [SOLID STATE] The change of certain low-dimensional conductors from metals to insulators as the temperature is lowered through a certain value, due to the lattice distortion and band gap accompanying the onset of a charge-density wave. { 'med·əl ¦in·sə,lād·ər tran'zish·ən }

metal leaf [MET] Metal sheet, thinner than foil, formed by beating. { 'med·əl 'lef }

metallic [OPTICS] Having a brilliant mineral luster characteristic of metals. { mə'tal·ik }

metallic bond [PHYS CHEM] The type of chemical bond that is present in all metals, and may be thought of as resulting from a sea of valence electrons which are free to move throughout the metal lattice. { mə'tal·ik 'bänd }

metallic corrosion [MET] Destruction of a metal by dissolution, oxidation, or other chemical reaction of the metal with its environment. { mə'tal·ik kə'rō·zhən }

metallic element [CHEM] An element generally distinguished (from a nonmetallic one) by its luster, electrical conductivity, malleability, and ability to form positive ions. { mə'tal·ik 'el·ə·mənt }

metallic glass See glassy alloy. { mə'tal·ik 'glas }

metallic hydrogen [PHYS CHEM] **1.** A phase of hydrogen believed to occur at extremely high pressures, in which the material transforms to a conducting molecular solid. **2.** A phase of hydrogen believed to occur at still higher pressures, in which the molecular bonds that exist at lower pressures are broken and an atomic solid with the structure of an alkali metal is formed. { mə'tal·ik 'hī·drə·jən }

metallic mortar [MATER] Mortar made with ceramic oxide binders and containing a high percentage of lead powder; mixed with water to form plasters or for casting sections and blocks; used for x-ray and nuclear installation shielding. { mə'tal·ik 'mȯrd·ər }

metallic paint [MATER] **1.** Paint used for covering metal surfaces; the pigment is commonly iron oxide. **2.** Paint with a metal pigment. { mə'tal·ik 'pānt }

metallic paper [MATER] **1.** A paper coated with zinc white, or with clay and other materials; the surface can be marked on with metal points (of silver, aluminum, or gold, for example), but it cannot be erased. **2.** Paper coated with finely flaked metal. { mə'tal·ik 'pā·pər }

metallic pigment [MATER] Thin, opaque aluminum or copper alloy flakes that are incorporated into plastic masses to produce metallike effects. { mə'tal·ik 'pig·mənt }

metallize [ENG] To coat or impregnate a metal or nonmetal surface with a metal, as by metal spraying or by vacuum evaporation. { 'med·əl ,īz }

metallized wood [MATER] Wood impregnated with molten metal, filling the cells in the wood to increase hardness, compressive strength, and flexural strength; the wood becomes an electrical conductor lengthwise of the grain. { 'med·əl,īzd 'wud }

metallocene [ORG CHEM] Organometallic coordination compound which is obtained as a cyclopentadienyl derivative of a transition metal or a metal halide. { mə'tal·ə,sēn }

metallocene catalyst [POLYM CHEM] A molecular structure with a well-defined single catalytic site, consisting of an organometallic coordination compound in which one or two cyclopentadienyl rings (with or without substituents) are bonded to a central transition-metal atom; used to produce uniform polyolefins with unique structures and physical properties. { mə¦tal·ə ,sēn 'kad·ə,list }

metallographic test [MET] A test to determine the structural composition of a metal as shown at low and high magnification and by x-ray diffraction methods; tests include macroexamination, microexamination, and x-ray diffraction studies. { mə'tal·ə,graf·ik 'test }

metallography [MET] The study of the structure of metals and alloys by various methods, especially by the optical and the electron microscope, and by x-ray diffraction. { ,med·əl'äg·rə·fē }

metalloid [CHEM] An element whose properties are intermediate between those of metals and nonmetals. Also known as semimetal. { 'med·ə ,lȯid }

metallostatic pressure [MET] Pressure developed within a volume of molten metal. { mə ¦tal·ə,stad·ik 'presh·ər }

metallurgical balance sheet [MET] Material balance of a metallurgical process. { ,med·əl'ər· jə·kəl 'bal·əns ,shēt }

metallurgical coke [MATER] Coke resulting from high-temperature retorting of suitable coal; a dense, crush-resistant fuel for use in shaft furnaces. { ‚med·əl'ər·jə·kəl 'kōk }

metallurgical dust [MET] A mixture of particles of elements and nonmetallic and metallic compounds. { ‚med·əl'ər·jə·kəl 'dəst }

metallurgical fume [MET] A mixture of fine particles of elements and metallic and nonmetallic compounds either sublimed or condensed from the vapor state. { ‚med·əl'ər·jə·kəl 'fyüm }

metallurgical microscope [ENG] A microscope used in the study of metals, usually optical. { ‚med·əl'ər·jə·kəl 'mī·krə‚skōp }

metal matrix composite [MATER] A material in which a continuous metallic phase (the matrix) is combined with another phase (the reinforcement) to strengthen the metal and increase high-temperature stability. The reinforcement is typically a ceramic in the form of particulates, platelets, whiskers, or fibers. The metals are typically alloys of aluminum, magnesium, or titanium. { ¦med·əl ¦mā·triks kəm'päz·ət }

metal-nitride-oxide semiconductor [SOLID STATE] A semiconductor structure that has a double insulating layer; typically, a layer of silicon dioxide (SiO_2) is nearest the silicon substrate, with a layer of silicon nitride (Si_3N_4) over it. Abbreviated MNOS. { 'med·əl ¦nī‚trīd ¦äk‚sīd 'sem·i·kən‚dək·tər }

metal-organic chemical vapor deposition [SOLID STATE] A technique for growing thin layers of compound semiconductors in which metal organic compounds, having the formula MR_x, where M is a group III metal and R is an organic radical, are decomposed near the surface of a heated substrate wafer, in the presence of a hydride of a group V element. Abbreviated MOCVD. { 'med·əl òr'gan·ik 'kem·ə·kəl 'vā·pər ‚dep·ə'zish·ən }

metal oxide semiconductor [SOLID STATE] A metal insulator semiconductor structure in which the insulating layer is an oxide of the substrate material; for a silicon substrate, the insulating layer is silicon dioxide (SiO_2). Abbreviated MOS. { 'med·əl ¦äk‚sīd 'sem·i·kən‚dək·tər }

metal oxide semiconductor field-effect transistor [ELECTR] A field-effect transistor having a gate that is insulated from the semiconductor substrate by a thin layer of silicon dioxide. Abbreviated MOSFET; MOST; MOS transistor. Formerly known as insulated-gate field-effect transistor (IGFET). { 'med·əl ¦äk‚sīd 'sem·i·kən‚dək·tər 'fēld i‚fekt tran'zis·tər }

metal oxide semiconductor integrated circuit [ELECTR] An integrated circuit using metal oxide semiconductor transistors; it can have a higher density of equivalent parts than a bipolar integrated circuit. { 'med·əl ¦äk‚sīd 'sem·i·kən ‚dək·tər 'int·ə‚grād·əd 'sər·kət }

metal plating See plating. { 'med·əl 'plād·iŋ }

metal pointing See pointing. { 'med·əl ‚pòint· iŋ }

metal powder [MET] A finely divided metal or alloy. { 'med·əl ‚paúd·ər }

metal replacement See immersion plating. { 'med·əl ri'plās·mənt }

metal rolling See rolling. { 'med·əl ‚rōl·iŋ }

metal semiconductor field-effect transistor [ELECTR] A field-effect transistor that uses a thin film of gallium arsenide, with a Schottky barrier gate formed by depositing a layer of metal directly onto the surface of the film. Abbreviated MESFET. { 'med·əl 'sem·i·kən‚dək·tər 'fēld i‚fekt tran'zis·tər }

metal spraying [ENG] Coating a surface with droplets of molten metal or alloy by using a compressed gas stream. { 'med·əl 'sprā·iŋ }

metal vapor laser [OPTICS] An ion laser based on vaporization of a solid or liquid metal, such as cadmium, calcium, copper, lead, manganese, selenium, strontium, and tin, vaporized with a buffer gas such as helium. { 'med·əl ‚vā·pər 'lā·zər }

metarheology [MECH] A branch of rheology whose approach is intermediate between those of macrorheology and microrheology; certain processes that are not isothermal are taken into consideration, such as kinetic elasticity, surface tension, and rate processes. { ‚med·ə·rē'äl·ə·jē }

metastable phase [PHYS CHEM] Existence of a substance as either a liquid, solid, or vapor under conditions in which it is normally unstable in that state. { ¦med·ə'stā·bəl ¦fāz }

metatitanic acid See titanic acid. { ¦med·ə· tī'tan·ik 'as·əd }

meter-kilogram-second system [MECH] A metric system of units in which length, mass, and time are fundamental quantities, and the units of these quantities are the meter, the kilogram, and the second respectively. Abbreviated mks system. { 'mēd·ər 'kil·ə‚gram 'sek·ənd ‚sis· təm }

methacrylate ester [ORG CHEM] CH_2:$C(CH_3)$· COOR Methacrylic acid ester used to make thermoplastic polymers or copolymers. R is often methyl, ethyl, butyl, and so on. { meth'ak·rə ‚lāt 'es·tər }

methacrylic acid [ORG CHEM] CH_2:$C(CH_3)$· COOH Easily polymerized, colorless liquid melting at 15–16°C; soluble in water and most organic solvents; used to make water-soluble polymers and as a chemical intermediate. { ¦meth·ə¦kril·ik 'as·əd }

methacrylic polymer [POLYM CHEM] A polymer whose monomer is a methacrylic ester with the general formula H_2C=$C(CH_3)COOR$. { 'meth·ə ‚kril·ik 'päl·ə·mər }

methanal See formaldehyde. { 'meth·ə‚nal }

methane hydrate [CHEM] Methane gas trapped or dissolved in ice formed in deep-sea sediments. { 'mēth‚ān 'hī‚drāt }

method of mixtures [THERMO] A method of determining the heat of fusion of a substance whose specific heat is known, in which a known amount of the solid is combined with a known amount of the liquid in a calorimeter, and the decrease in the liquid temperature during melting of the solid is measured. { 'meth·əd əv 'miks·chərz }

methoxy- [ORG CHEM] OCH_3—A combining form

indicating the oxygen-containing methane radical, found in many organic solvents, insecticides, and plasticizer intermediates. { mə'thäk·sē }

methyl [ORG CHEM] The alkyl group derived from methane and usually written CH_3-. { 'meth·əl }

methyl acetone [MATER] Flammable, water-white liquid, a mixture of acetone, methanol, and methyl acetate in various proportions; miscible with water, oils, and hydrocarbons; used as a solvent. { 'meth·əl 'as·ə,tōn }

methyl acrylate [ORG CHEM] $CH_2{:}CHCOOCH_3$ A readily polymerized, volatile, colorless liquid boiling at 80°C; slightly soluble in water; used as a chemical intermediate and in making polymers. { 'meth·əl 'ak·rə,lāt }

methylated spirits [MATER] Denatured ethanol produced by addition of about 9.5% methanol, 0.5% pyridine, and a blue dye. { 'meth·ə,lād·əd 'spir·əts }

methylcellulose [POLYM CHEM] A grayish-white powder derived from cellulose; swells in water to a colloidal solution; soluble in glacial acetic acid; used in water-based paints and ceramic glazes; for leather tanning, and as a thickening and sizing agent, adhesive, and food additive. Also known as cellulose methyl ether. { 'meth·əl'sel·yə,lōs }

methylene oxide See formaldehyde. { 'meth·ə ,lēn 'äk,sīd }

methylethylcellulose [POLYM CHEM] A combustible, white to cream-colored, fibrous solid or powder; disperses in cold water, forming solutions which undergo reversible transformation from sol to gel; used as an emulsifier and foaming agent. { ¦meth·əl¦eth·əl'sel·yə,lōs }

methyl ethylene See propylene. { 'meth·əl 'eth·ə,lēn }

metric system [MECH] A system of units used in scientific work throughout the world and employed in general commercial transactions and engineering applications; its units of length, time, and mass are the meter, second, and kilogram respectively, or decimal multiples and submultiples thereof. { 'me·trik ,sis·təm }

metrology [PHYS] The science of measurement. { mə'träl·ə·jē }

Mexican linaloe oil See linaloe wood oil. { 'mek·si·kən lē'nä·lō·a ,òil }

Mg See magnesium.

M glass [MATER] A glass with a high content of beryllium oxide and a high modulus of elasticity. { 'em ,glas }

mho See siemens. { mō }

micelle [PHYS CHEM] A colloidal aggregate of a unique number (between 50 and 100) of amphipathic molecules, which occurs at a well-defined concentration known as the critical micelle concentration. { mī'sel }

microalloy diffused transistor [ELECTR] A microalloy transistor in which the semiconductor wafer is first subjected to gaseous diffusion to produce a nonuniform base region. Abbreviated MADT. { ¦mī·krō'al,òi də'fyüzd tran'zis·tər }

microalloy transistor [ELECTR] A transistor in which the emitter and collector electrodes are formed by etching depressions, then electroplating and alloying a thin film of the impurity metal to the semiconductor wafer, somewhat as in a surface-barrier transistor. { ¦mī·krō'al,òi tran'zis·tər }

microbicide [MATER] An agent that kills microbes. { mī'krō·bə,sīd }

microbridge [CRYO] A Josephson junction formed by configuration of thin superconducting films. { 'mī·krō,brij }

microchip See chip. { 'mī·krō,chip }

microcrack See microfissure. { 'mī·krō,krak }

microcrystalline [CRYSTAL] Composed of or containing crystals that are visible only under the microscope. { ¦mī·krō'krist·əl·ən }

microcrystalline wax [MATER] A petroleum wax containing small, indistinct crystals, and having a higher molecular weight, melting point, and viscosity than paraffin wax; used in laminated paper and electrical coil coating. { ¦mī·krō'krist·əl·ən 'waks }

micro-electro-mechanical system [ENG] A system in which micromechanisms are coupled with microelectronics, most commonly fabricated as microsensors or microactuators. Abbreviated MEMS. Also known as microsystem. { ¦mī·krō̇i ,lek·trə·mə'kan·ə·kəl ,sis·təm }

microemulsion [MATER] A thermodynamically stable dispersion of two immiscible liquids, stabilized by surfactants; it is typically clear because the dispersed droplets are less than 100 nanometers in diameter. { ¦mī·krō·i'məl·shən }

microencapsulation [CHEM ENG] Enclosing of materials in capsules from well below 1 micrometer to over 2000 micrometers in diameter. { ¦mī·krō·in,kap·sə'lā·shən }

microfabrication [ENG] The technology of fabricating microsystems from silicon wafers, using standard semiconductor process technologies in combination with specially developed processes. { ¦mī·krō,fab·rə'kā·shən }

microfissure [MET] A crack of microscopic dimensions. Also known as microcrack. { ¦mī·krō'fish·ər }

microfluid [FL MECH] A fluid in which the effects of local motion of contained material particles on properties and behavior of the fluid are not disregarded. { 'mī·krō,flü·əd }

microhardness [MET] Hardness of microscopic areas of a metal or alloy. { ¦mī·krō'härd·nəs }

microhysteresis effect [SOLID STATE] Hysteresis that results from the motion of domain walls lagging behind an applied magnetic or elastic stress when these walls are held up by dislocations and other imperfections in the material. { ¦mī·krō,his·tə'rē·səs i,fekt }

microlite [CRYSTAL] A microscopic crystal which polarizes light. Also known as microlith. { 'mī·krə,līt }

microlith See microlite. { 'mī·krə,lith }

microlithography [MATER] The transfer of a pattern or image from one medium to another, as from a mask to a wafer, with image features in the micrometer range or smaller. { ˌmī·krō·liˈthäg·rə·fē }

micromachine [MATER] A micrometer-size mechanical device; compared with an integrated circuit, it has some mechanical parts that stand above the substrate or move freely over it. { ˈmī·krō·mə‚shēn }

micromachining [ENG] The use of standard semiconductor process technologies in combination with specially developed processes to fabricate miniature mechanical devices and components on silicon and other materials. { ˈmī·krō·mə‚shēn·iŋ }

micromechanics [ENG] **1.** The design and fabrication of micromechanisms. **2.** *See* composite micromechanics. { ¦mī·krō·məˈkan·iks }

micromechanism [ENG] A mechanical component with submillimeter dimensions and corresponding tolerances of the order of 1 micrometer or less. { ¦mī·krō ˈmek·ə‚niz·əm }

micromechatronics [ENG] The branch of engineering concerned with micro-electro-mechanical systems. { ¦mī·kro‚mek·əˈträn·iks }

micrometer [ENG] **1.** An instrument attached to a telescope or microscope for measuring small distances or angles. **2.** A caliper for making precise measurements; a spindle is moved by a screw thread so that it touches the object to be measured; the dimension can then be read on a scale. Also known as micrometer caliper. [MECH] A unit of length equal to one-millionth of a meter. Abbreviated μm. Also known as micron (μ). { mīˈkräm·əd·ər }

micrometer caliper *See* micrometer. { mī ˈkräm·əd·ər ˈkal·ə·pər }

micromolding [ENG] An alternative technique to micromachining for fabricating microsystems, in which a sacrificial material serves as a mold to which a deposited material conforms. { ˈmī·krō ‚mōld·iŋ }

micromotor [MATER] A micromachine and forerunner of micro-electro-mechanical systems. { ˈmī·krə‚mōd·ər }

micron *See* micrometer. { ˈmī‚krän }

micronized clay [MATER] A pure kaolin pulverized to a fineness of 400 to 800 mesh; used as a filler material in rubber. { ˈmī·krə‚nīzd ˈklā }

micronized mica [MATER] Powdered mica of a fineness of 400 to 1000 mesh; used as a filler. { ˈmī·krə‚nīzd ˈmī·kə }

micro-opto-electro-mechanical system [ENG] A microsystem that combines the functions of optical, mechanical, and electronic components in a single, very small package or assembly. Abbreviated MOEMS. { ¦mī·kro¦äp·tō i¦lek·trō mə¦kan·ə·kəl ˈsis·təm }

micro-opto-mechanical system [ENG] A microsystem that combines optical and mechanical functions without the use of electronic devices or signals. Abbreviated MOMS. { ¦mī·krō ¦op·to·mə¦kan·ə·kəl ‚sis·təm }

micropore [CHEM] A pore in a catalytic material whose diameter is less than 2 nanometers. { ˈmī·krə‚pór }

microporosity [MET] Extremely fine porosity,

visible only with the aid of a microscope. { ¦mī·krō·pəˈräs·əd·ē }

micro-reciprocal-degree *See* mired. { ¦mī·krō riˈsip·rə·kəl diˈgrē }

microrheology [MECH] A branch of rheology in which the heterogeneous nature of dispersed systems is taken into account. { ¦mī·krō·rēˈäl·ə·jē }

microscopic stress [MET] Residual stress ranging from compression to tension in a metal within a distance often comparable to the grain size. Also known as microstress. { ¦mī·krə¦skäp·ik ˈstres }

microsegregation [MET] Segregation within a grain, crystal, or particle of microscopic size. { ¦mī·krō‚seg·rəˈgā·shən }

microsensor [ENG] A submicrometer- to millimeter-size device that converts a nonelectrical physical or chemical quantity, such as pressure, acceleration, temperature, or gas concentration, into an electrical signal; it is generally able to offer better sensitivity, accuracy, dynamic range, and reliability, as well as lower power consumption, compared to larger counterparts. { ˈmī·krō‚sen·sər }

microshrinkage [MET] A casting defect consisting of interdendritic voids, visible only at magnifications over 10 diameters. { ¦mī·krōˈshriŋ·kij }

microsilica *See* silica fume. { ˌmī·krōˈsil·ə·kə }

microsphere [MATER] A sphere sized from about 0.5 to 100 micrometers and made of any material. { ˈmī·krə‚sfir }

microstress *See* microscopic stress. { ˈmī·krə ‚stres }

microsystem *See* micro-electro-mechanical system. { ˈmī·krō‚sis·təm }

Mie-Grüneisen equation [THERMO] An equation of state particularly useful at high pressure, which states that the volume of a system times the difference between the pressure and the pressure at absolute zero equals the product of a number which depends only on the volume times the difference between the internal energy and the internal energy at absolute zero. { ˈmē ˈgrü ‚nīz·ən i‚kwā·zhən }

migration [MET] The uncontrolled movement of certain metals, particularly silver, from one location to another, usually with associated undesirable effects such as oxidation or corrosion. [SOLID STATE] **1.** The movement of charges through a semiconductor material by diffusion or drift of charge carriers or ionized atoms. **2.** The movement of crystal defects through a semiconductor crystal under the influence of high temperature, strain, or a continuously applied electric field. { mīˈgrā·shən }

migration current [PHYS CHEM] Additional current produced by electrostatic attraction of cations to the surface of a dropping electrode; an unpredictable and undesirable effect to be avoided during analytical voltammetry. { mīˈgrā·shən ‚kə·rənt }

mil [MECH] A unit of length, equal to 0.001 inch, or to 2.54×10^{-5} meter. Also known as milli-inch; thou. { mil }

mild abrasive [MATER] An abrasive material,

such as chalk or talc, having a hardness of 1–2 on Mohs scale; used in silver polishes and window cleaners. { 'mīld ə'brā·siv }

mild mercury chloride *See* mercurous chloride. { 'mīld 'mər·kyə·rē 'klòr‚īd }

mild steel [MET] Carbon steel containing 0.05–0.25% carbon. { 'mīld 'stēl }

milk glass [MATER] A white, and sometimes colored, opaque glass made by adding calcium fluoride and alumina to soda-lime glass. { 'milk ‚glas }

millboard [MATER] Hard, strong paperboard; used for furniture panels. { 'mil‚bòrd }

Miller indices [CRYSTAL] Three integers identifying a type of crystal plane; the intercepts of a plane on the three crystallographic axes are expressed as fractions of the crystal parameters; the reciprocals of these fractions, reduced to integral proportions, are the Miller indices. Also known as crystal indices. { 'mil·ər 'in·də‚sēz }

Miller law [CRYSTAL] If the edges formed by the intersections of three faces of a crystal are taken as the three reference axes, then the three quantities formed by dividing the intercept of a fourth face with one of these axes by the intercept of a fifth face with the same axis are proportional to small whole numbers, rarely exceeding 6. Also known as law of rational intercepts. { 'mil·ər ‚lò }

mill finish [MET] The characteristic surface finish on a rolled metal product. { 'mil ‚fin·ish }

milli-inch *See* mil. { 'mil·ē‚inch }

mill scale [MET] A surface layer of ferric oxide (Fe_3O_4) that forms on steel or iron during hot rolling. { 'mil ‚skāl }

mimetic [CRYSTAL] Pertaining to a crystal that is twinned or malformed but whose crystal symmetry appears to be of a higher grade than it actually is. { mə'med·ik }

mineral acid [INORG CHEM] Any one of the major inorganic acids, such as sulfuric, nitric, or hydrochloric acids. { 'min·rəl ‚as·əd }

mineral additive [MATER] A mineral-derived substance added during grease manufacture, particularly to heavy-duty grease. { 'min·rəl 'ad·ə‚tiv }

mineral black [MATER] Black pigment made from ground slate, shale, coke, coal, or slaty coal; used in inks, plastics, and coatings. { 'min·rəl ‚blak }

mineral cotton *See* mineral wool. { 'min·rəl 'kät·ən }

mineral dressing *See* beneficiation. { 'min·rəl ‚dres·iŋ }

mineral dye [MATER] A natural dyestuff made from minerals; examples are ochre, chrome yellow, and Prussian blue. { 'min·rəl ‚dī }

mineral green *See* copper carbonate. { 'min·rəl ‚grēn }

mineral lard oil [MATER] A mixture of refined mineral oil with lard oil, having a fatty content of 25–30, and a flash point about 300°F (149°C). { 'min·rəl 'lärd ‚òil }

mineral rubber [MATER] Asphaltine minerals (such as gilsonite and grahamite) and blown

asphalts; used in compounding of rubber, coatings, and paints. { 'min·rəl 'rəb·ər }

minerals beneficiation *See* beneficiation. { 'min·rəlz ‚ben·ə‚fish·ē'ā·shən }

mineral seal oil [MATER] A petroleum distillate with a boiling point higher than that of kerosine; used as a solvent oil and illuminant. { 'min·rəl 'sēl ‚òil }

mineral wool [MATER] A fibrous substance, technically a glass, made from molten slag, rock, glass, or a selected combination of these ingredients; produced by blowing, drawing, or other means of fabricating into fine fibers; used for insulation, fireproofing, and as a filter medium. Also known as mineral cotton; rock wool; silicate cotton; slag wool. { 'min·rəl ‚wúl }

minimum bend radius [MET] The minimum radius through which a piece of metal can be bent to form a given angle without fracturing. { 'min·ə·məm ‚bend 'rād·ē·əs }

minimum resolvable temperature difference [THERMO] The change in equivalent blackbody temperature that corresponds to a change in radiance which will produce a just barely resolvable change in the output of an infrared imaging device, taking into account the characteristics of the device, the display, and the observer. Abbreviated MRTD. { 'min·ə·məm ri'zäl·və·bəl 'tem·prə·chər ‚dif·rəns }

minority carrier [SOLID STATE] The type of carrier, electron, or hole that constitutes less than half the total number of carriers in a semiconductor. { mə'när·əd·ē 'kar·ē·ər }

minus sieve [MET] In powder metallurgy, the portion of a powder sample that passes through a specified standard sieve. { 'mī·nəs ‚siv }

mired [THERMO] A unit used to measure the reciprocal of color temperature, equal to the reciprocal of a color temperature of 10^6 kelvins. Derived from micro-reciprocal-degree. { mīrd }

mirror plane of symmetry *See* plane of mirror symmetry. { 'mir·ər 'plān əv 'sim·ə·trē }

MIR technique *See* multiple isomorphous replacement technique. { ‚em‚ī'är tek‚nēk }

MIS *See* metal-insulator semiconductor.

misch metal [MET] An alloy consisting of a crude mixture of cerium, lanthanum, and other rare-earth metals obtained by electrolysis of the mixed chlorides of the metals dissolved in fused sodium chloride; used in making aluminum alloys, in some steels and irons, and in coating the cathodes of glow-type voltage regulator tubes. { 'mish ‚med·əl }

miscibility [CHEM] The tendency or capacity of two or more liquids to form a uniform blend, that is, to dissolve in each other; degrees are total miscibility, partial miscibility, and immiscibility. { ‚mis·ə'bil·əd·ē }

mismatch [MET] Failure to match forged surfaces formed in opposite dies. { 'mis‚mach }

misrun [MET] Incompletely formed casting due to premature solidification of metal before the mold is filled. { mis'rən }

mix crystal *See* mixed crystal. { 'miks ‚krist·əl }

mixed acid *See* nitrating acid. { 'mikst 'as·əd }

mixed crystal |CRYSTAL| A crystal whose lattice sites are occupied at random by different ions or molecules of two different compounds. Also known as mix crystal. { 'mikst 'krist·əl }

mixed oil See nitrating acid. { 'mikst 'oil }

mks system See meter-kilogram-second system. { ¦em¦kā'es ‚sis·təm }

Mn See manganese.

MNOS See metal-nitride-oxide semiconductor. { 'em‚nȯs }

Mo See molybdenum.

mobile electron |PHYS CHEM| An electron that can move readily from one atom to another within a chemical structure in response to changes in the external chemical environment. { ‚mō·bəl ə'lek‚trän }

mobility |FL MECH| The reciprocal of the plastic viscosity of a Bingham plastic. |PHYS| Freedom of particles to move, either in random motion or under the influence of fields or forces. |SOLID STATE| See drift mobility. { mō'bil·əd·ē }

mock silver |MET| **1.** An aluminum alloy containing 5% copper and 10% tin, or 5% copper and 5% silver. **2.** A white brass containing 55% zinc and 45% copper. { 'mäk 'sil·vər }

MOCVD See metal-organic chemical vapor deposition.

modified asphalt |MATER| Asphalt modified by addition of a rosin ester or synthetic resin. { 'mäd·ə‚fīd 'as‚fȯlt }

modified gunmetal |MET| Gunmetal containing about 2.5% lead; used for gears and bearings. { 'mäd·ə‚fīd 'gən‚med·əl }

modifier |MATER| In flotation, any of the chemicals which increase the specific attraction between collector agents and particle surfaces or which increase the wettability of those surfaces. { 'mäd·ə‚fī·ər }

modulation-doped structure |SOLID STATE| An epitaxially grown crystal structure in which successive semiconductor layers contain different types of electrical dopants. { ‚mäj·ə'lā·shən ¦dōpt 'strak·chər }

modulus of compression See bulk modulus of elasticity. { 'mäj·ə·ləs əv kəm'presh·ən }

modulus of deformation |MECH| The modulus of elasticity of a material that deforms other than according to Hooke's law. { 'mäj·ə·ləs əv ‚dē ‚fȯr'mā·shən }

modulus of elasticity |MECH| The ratio of the increment of some specified form of stress to the increment of some specified form of strain, such as Young's modulus, the bulk modulus, or the shear modulus. Also known as coefficient of elasticity; elasticity modulus; elastic modulus. { 'mäj·ə·ləs əv i‚las'tis·əd·ē }

modulus of elasticity in shear |MECH| A measure of a material's resistance to shearing stress, equal to the shearing stress divided by the resultant angle of deformation expressed in radians. Also known as coefficient of rigidity; modulus of rigidity; rigidity modulus; shear modulus. { 'mäj·ə·ləs əv i‚las'tis·əd·ē in 'shir }

modulus of resilience |MECH| The maximum mechanical energy stored per unit volume of material when it is stressed to its elastic limit. { 'mäj·ə·ləs əv ri'zil·yəns }

modulus of rigidity See modulus of elasticity in shear. { 'mäj·ə·ləs əv ri'jid·əd·ē }

modulus of rupture in bending |MECH| The maximum stress per unit area that a specimen can withstand without breaking when it is bent, as calculated from the breaking load under the assumption that the specimen is elastic until rupture takes place. { 'mäj·ə·ləs əv 'rəp·chər in 'bend·iŋ }

modulus of rupture in torsion |MECH| The maximum stress per unit area that a specimen can withstand without breaking when its ends are twisted, as calculated from the breaking load under the assumption that the specimen is elastic until rupture takes place. { 'mäj·ə·ləs əv 'rəp·chər in 'tȯr·shən }

modulus of simple longitudinal extension See axial modulus. { ¦mäj·ə·ləs əv ¦sim·pəl ‚län·jə ¦tüd·ən·əl ik'sten·chən }

modulus of strain hardening See rate of strain hardening. { 'mäj·ə·ləs əv 'strān ‚härd·ən·iŋ }

modulus of torsion See torsional modulus. { 'mäj·ə·ləs əv 'tȯr·shən }

modulus of volume elasticity See bulk modulus of elasticity. { 'mäj·ə·ləs əv 'väl·yəm i‚las 'tis·əd·ē }

MOEMS See micro-opto-electro-mechanical system. { 'mō‚emz }

Mohr's circle |MECH| A graphical construction making it possible to determine the stresses in a cross section if the principal stresses are known. { 'mȯrz 'sər·kəl }

Mohr's salt See ferrous ammonium sulfate. { 'mȯrz ‚sȯlt }

moiety |CHEM| A part or portion of a molecule, generally complex, having a characteristic chemical or pharmacological property. { 'mȯi·əd·ē }

moisture barrier |MATER| A material that retards the passage of moisture into walls. { 'mȯis·chər ‚bar·ē·ər }

moisture content |MECH| The quantity of water in a mass of soil, sewage, sludge, or screenings; expressed in percentage by weight of water in the mass. { 'mȯis·chər ‚kän·tent }

mol See mole. { mōl }

molality |CHEM| Concentration given as moles per 1000 grams of solvent. { mō'lal·əd·ē }

molar |PHYS CHEM| Denoting a physical quantity divided by the amount of substance expressed in moles. { 'mō·lər }

molarity |CHEM| Measure of the number of gram-molecular weights of a compound present (dissolved) in 1 liter of solution; it is indicated by M, preceded by a number to show solute concentration. { mō'lar·əd·ē }

moldability |MATER| The capability of being molded. { ‚mōl·də'bil·əd·ē }

molded silver nitrate See lunar caustic. { 'mōl· dəd 'sil·vər 'nī‚trāt }

molding board |MET| In sand-mold metal casting, a component that fits the lower half of the form or flask used for the molding operation to hold the casting pattern. { 'mōld·iŋ ,bȯrd }

molding machine [MET] A machine that compacts sand around a pattern to form a mold. { 'mōl·diŋ mə,shēn }

molding powder [MATER] Powdered plastic-material ingredients (such as resin, filler, pigments, and plasticizers) ready for compression in molding. { 'mōl·diŋ ,paud·ər }

molding press [MET] A press used to form compacts in powder metallurgy. { 'mōl·diŋ ,pres }

molding sand See foundry sand. { 'mōld·iŋ ,sand }

molding time See curing time. { 'mōl·diŋ ,tīm }

mold release See release agent. { 'mōld ri,lēs }

mold shift [MET] The mismatch of mold halves at the parting line, resulting in a casting defect. { 'mōld ,shift }

mold steel [MET] Steel of tool-steel quality used to make molds for plastics; properties include uniform texture, good machinability with die-sinking tools, and lack of microscopic porosity. { 'mōld ,stēl }

mold wash [MET] An aqueous or alcoholic suspension or emulsion used to coat the surfaces of a mold cavity. { 'mōld ,wȧsh }

mole [CHEM] An amount of substance of a system which contains as many elementary units as there are atoms of carbon in 0.012 kilogram of the pure nuclide carbon-12; the elementary unit must be specified and may be an atom, molecule, ion, electron, photon, or even a specified group of such units. Symbolized mol. { mōl }

molecular adhesion [PHYS CHEM] A particular manifestation of intermolecular forces which causes solids or liquids to adhere to each other; usually used with reference to adhesion of two different materials, in contrast to cohesion. { mə'lek·yə·lər ad'hē·zhən }

molecular association [PHYS CHEM] The formation of double molecules or polymolecules from a single species as a result of specific and moderately strong intermolecular forces. { mə'lek·yə·lər ə,sō·sē'ā·shən }

molecular asymmetry See asymmetry. { mə 'lek·yə·lər ,ā'sim·ə·trē }

molecular beam [PHYS] A beam of neutral molecules whose directions of motion lie within a very small solid angle. { mə'lek·yə·lər 'bēm }

molecular-beam epitaxy [SOLID STATE] A technique of growing single crystals in which beams of atoms or molecules are made to strike a single-crystalline substrate in a vacuum, giving rise to crystals whose crystallographic orientation is related to that of the substrate. Abbreviated MBE. { mə'lek·yə·lər¦bēm 'ep·ə,tak·sē }

molecular binding [SOLID STATE] The force which holds a molecule at some site on the surface of a crystal. { mə'lek·yə·lər 'bind·iŋ }

molecular circuit [ELECTR] A circuit in which the individual components are physically indistinguishable from each other. { mə'lek·yə·lər 'sər·kət }

molecular cluster [PHYS CHEM] An assembly of molecules that are weakly bound together and

display properties intermediate between those of isolated gas-phase molecules and bulk condensed media. { mə'lek·yə·lər 'kläs·tər }

molecular crystal [CRYSTAL] A solid consisting of a lattice array of molecules such as hydrogen, methane, or more complex organic compounds, bound by weak van der Waals forces, and therefore retaining much of their individuality. { mə'lek·yə·lər 'krist·əl }

molecular device [CHEM] An assemblage of a discrete number of molecular components (that is, a supramolecular structure) designed to achieve a specific function. { mə'lek·yə·lər di'vīs }

molecular diamagnetism [PHYS CHEM] Diamagnetism of compounds, especially organic compounds whose susceptibilities can often be calculated from the atoms and chemical bonds of which they are composed. { mə'lek·yə·lər ,dī·ə'mag·nə,tiz·əm }

molecular dipole [PHYS CHEM] A molecule having an electric dipole moment, whether it is permanent or produced by an external field. { mə'lek·yə·lər 'dī,pōl }

molecular engineering [ELECTR] The use of solid-state techniques to build, in extremely small volumes, the components necessary to provide the functional requirements of overall equipments, which when handled in more conventional ways are vastly bulkier. { mə'lek·yə·lər ,en·jə'nir·iŋ }

molecular field theory See Weiss theory. { mə'lek·yə·lər 'fēld ,thē·ə·rē }

molecular formula [CHEM] A chemical formula that indicates the actual numbers and kinds of atoms in a molecule, but not the chemical structure. { mə'lek·yə·lər 'fȯr·myə·lə }

molecular gas [CHEM] A gas composed of a single species, such as oxygen, chlorine, or neon. { mə'lek·yə·lər 'gas }

molecular heat [THERMO] The heat capacity per mole of a substance. { mə'lek·yə·lər 'hēt }

molecular heat diffusion [THERMO] Transfer of heat through the motion of molecules. { mə 'lek·yə·lər ¦hēt di,fyü·shən }

molecular imprinting [POLYM CHEM] A technique for creating receptor structures on a polymer surface that can selectively bind to molecules of interest, molecularly imprinted polymers are used for separations, as catalysts, and in biosensors. { mə¦lek·yə·lər 'im,print·iŋ }

molecular ion [ORG CHEM] An ion that results from the loss of an electron by an organic molecule following bombardment with high-energy electrons during mass spectrometry. { mə'lek·yə·lər 'ī,än }

molecular machine [CHEM] A molecular device in which the component parts can display changes (reversible movement) in their relative positions as a result of some external stimulus (such as light, electrical energy, or chemical energy), resulting in a signal (a change in a chemical or physical property of the supramolecular system) that can be used to monitor the operation of the device. { mə¦lek·yə·lər mə'shēn }

molecular magnet |PHYS CHEM| A molecule having a nonvanishing magnetic dipole moment, whether it is permanent or produced by an external field. { mə'lek·yə·lər 'mag·nət }

molecular optics |OPTICS| The study of the propagation of light and associated phenomena, such as refraction, absorption, and scattering, through collections of molecules in gases, liquids, and solids. { mə'lek·yə·lər 'äp·tiks }

molecular orbital |PHYS CHEM| A wave function describing an electron in a molecule. { mə'lek·yə·lər 'ȯr·bəd·əl }

molecular physics |PHYS| The study of the behavior and structure of molecules, including the quantum-mechanical explanation of several kinds of chemical binding between atoms in a molecule, directed valence, the polarizability of molecules, the quantization of vibrational, rotational, and electronic motions of molecules, and the phenomena arising from intermolecular forces. { mə'lek·yə·lər 'fiz·iks }

molecular receptor |ORG CHEM| A species that can select one of many possible binding partners and form a complex that is stabilized by interactions such as hydrogen bonding or changes in solvation. { mə'lek·yər·lər ri'sep·tər }

molecular recognition |CHEM| The (molecular) storage and the (supramolecular) retrieval and processing of molecular structural information and interactions. { mə¦lek·yə·lər ‚rek·ig'nish·ən }

molecular rotation |OPTICS| In a solution of an optically active compound, the specific rotation (angular rotation of polarized light) multiplied by the compound's molecular weight. { mə'lek·yə·lər rō'tā·shən }

molecular self-assembly |ORG CHEM| The spontaneous aggregation of molecules into well-defined, stable, noncovalently bonded assemblies that are held together by intermolecular forces. { mə'lek·yə·lər ‚self ə'sem·blē }

molecular sieve |CHEM| A naturally occurring or synthetic zeolite characterized by the ability to undergo dehydration with little or no change in crystal structure, thereby offering a very high surface area for adsorption of foreign molecules. { mə'lek·yə·lər 'siv }

molecular simulation |CHEM| Computational techniques for predicting many useful functional properties of chemicals and materials, including thermodynamic properties, thermochemical properties, spectroscopic properties, mechanical properties, transport properties, and morphological information. { mə¦lek·yə·lər ‚sim·yə'lā·shən }

molecular structure |PHYS CHEM| The manner in which electrons and nuclei interact to form a molecule, as elucidated by quantum mechanics and a study of molecular spectra. { mə'lek·yə·lər 'strək·chər }

molecular weight |CHEM| The sum of the atomic weights of all the atoms in a molecule. Also known as relative molecular mass. { mə'lek·yə·lər 'wāt }

molecular-weight distribution |POLYM CHEM| Frequency of occurrence of the different molecular-weight chains in a homologous polymeric system. { mə'lek·yə·lər 'wāt ‚di·strə'byü·shən }

molecule |CHEM| A group of atoms held together by chemical forces; the atoms in the molecule may be identical as in H_2, S_2, and S_8, or different as in H_2O and CO_2; a molecule is the smallest unit of matter which can exist by itself and retain all its chemical properties. { 'mäl·ə ‚kyül }

mole fraction |CHEM| The ratio of the number of moles of a substance in a mixture or solution to the total number of moles of all the components in the mixture or solution. { 'mōl ‚frak·shən }

Mollier diagram |THERMO| Graph of enthalpy versus entropy of a vapor on which isobars, isothermals, and lines of equal dryness are plotted. { mól'yā ‚dī·ə‚gram }

molluscicide |MATER| An agent that kills mollusks. { 'mäl·əs·ə‚sīd }

molybdate |INORG CHEM| A salt derived from a molybdic acid. { mə'lib‚dät }

molybdate orange |MATER| Pigment that is a solid solution of lead chromate, molybdate, and sulfate; used in paints, inks, and plastics. { mə'lib‚dät 'är·ənj }

molybdenum |CHEM| A chemical element, symbol Mo, atomic number 42, and atomic weight 95.94. |MET| A silvery-gray metal used in iron-based alloys. { mə'lib·de·nəm }

molybdenum cast iron |MET| Cast iron containing small amounts of molybdenum, added as ferromolybdenum or calcium molybdenum; increases strength, toughness, and wear resistance. { mə'lib·də·nəm 'kast ‚ī·ərn }

molybdenum dioxide |INORG CHEM| MoO_2 Lead-gray powder; insoluble in hydrochloric and hydrofluoric acids; used in pigment for textiles. { mə'lib·də·nəm dī'äk‚sīd }

molybdenum disilicide |INORG CHEM| $MoSi_2$ A dark gray, crystalline powder with a melting range of 1870–2030°C; soluble in hydrofluoric and nitric acids; used in electrical resistors and for protective coatings for high-temperature conditions. { mə'lib·də·nəm dī'sil·ə‚sīd }

molybdenum disulfide |INORG CHEM| MoS_2 A black lustrous powder, melting at 1185°C, insoluble in water, soluble in aqua regia and concentrated sulfuric acid; used as a dry lubricant and an additive for greases and oils. Also known as molybdenum sulfide; molybdic sulfide. { mə'lib·də·nəm dī'səl‚fīd }

molybdenum pentachloride |INORG CHEM| $MoCl_5$ Hygroscopic gray-black needles melting at 194°C; reacts with water and air; soluble in anhydrous organic solvents; used as a catalyst and as raw material to make molybdenum hexacarbonyl. { mə'lib·də·nəm ‚pen·tə'klȯr‚īd }

molybdenum sesquioxide |INORG CHEM| MoO_3 Water-insoluble, gray-black powder with slight solubility in acids; used as a catalyst and as a coating for metal articles. { mə'lib·də·nəm ‚ses·kwē'äk‚sīd }

molybdenum silicide [MET] A mixture of molybdenum, silicon, and iron in the proportion 60:30:10; used to introduce molybdenum into steel melts. { mə'lib·də·nəm 'sil·ə,sīd }

molybdenum steel [MET] **1.** A carbon steel containing usually less than 0.5% molybdenum to aid hardenability. **2.** A tool steel containing up to 10% molybdenum, up to 1.5% carbon, and varying amounts of chromium, vanadium tungsten, and sometimes cobalt. { mə'lib·də·nəm 'stel }

molybdenum sulfide *See* molybdenum disulfide. { mə'lib·də·nəm 'səl,fīd }

molybdenum trioxide [INORG CHEM] MoO₃ A white solid at room temperature, with a melting point of 795°C; soluble in concentrated mixtures of nitric and sulfuric acids and nitric and hydrochloric acids; used as a corrosion inhibitor, in enamels and ceramic glazes, in medicine and agriculture, and as a catalyst in the petroleum industry. { mə'lib·də·nəm trī'äk,sīd }

molybdic acid [INORG CHEM] Any acid derived from molybdenum trioxide, especially the simplest acid H_2MoO_4, obtained as white crystals. { mə'lib·dik 'as·əd }

molybdic sulfide *See* molybdenum disulfide. { mə'lib·dik 'səl,fīd }

moly-blacks [MATER] Lustrous, black, molybdenum-containing decorative coatings used to blacken zinc or zinc-base alloys. { 'mäl·ē,blaks }

MOMS *See* micro-opto-mechanical system. { mämz *or* ¦em¦ō¦em'es }

monatomic gas [CHEM] A gas whose molecules have only one atom; the inert gases are examples. { ¦män·ə¦täm·ik 'gas }

Mond process [MET] A process for extracting and purifying nickel whereby nickel carbonyl is first formed by reaction of the reduced metal with carbon monoxide, and then the nickel carbonyl is decomposed thermally, resulting in deposition of nickel. { 'mänd ,prä·səs }

monoacid [CHEM] **1.** An acid that has only one replaceable hydrogen. **2.** A base or an alcohol that has a single hydroxyl (−OH) group which can be replaced by an atom or a functional group to form a salt or ester. { ¦män·ō'as·əd }

monobasic [CHEM] Pertaining to an acid with one displaceable hydrogen atom, such as hydrochloric acid, HCl. { ¦män·ō'bās·ik }

monobasic calcium phosphate *See* calcium phosphate. { ¦män·ō'bās·ik 'kal·sē·əm 'fäs,fāt }

monobasic sodium phosphate [INORG CHEM] NaH_2PO_4 White crystals that are slightly hygroscopic, soluble in water, insoluble in alcohol; used in baking powders and acid cleansers, and as a cattle-food supplement. { ¦män·ō'bās·ik 'sōd·ē·əm 'fäs,fāt }

monocalcium phosphate *See* calcium phosphate. { ¦män·ō'kal·sē·əm 'fäs,fāt }

monochromatic [OPTICS] Pertaining to the color of a surface which radiates light having an extremely small range of wavelengths. [PHYS] Consisting of electromagnetic radiation having an extremely small range of wavelengths, or particles having an extremely small range of energies. { ¦män·ə·krə'mad·ik }

monochromatic emissivity [THERMO] The ratio of the energy radiated by a body in a very narrow band of wavelengths to the energy radiated by a blackbody in the same band at the same temperature. Also known as color emissivity. { ,män·ə·krə'mad·ik ,ē·mi'siv·əd·ē }

monochromatic interference [OPTICS] Interference between beams coming from a source of monochromatic light. { män·ə·krə'mad·ik ,in·tər'fir·əns }

monochromatic light [OPTICS] Light of one color, having wavelengths confined to an extremely narrow range. { män·ə·krə'mad·ik 'līt }

monochromatic temperature scale [THERMO] A temperature scale based upon the amount of power radiated from a blackbody at a single wavelength. { män·ə·krə'mad·ik 'tem·prə·chər ,skāl }

monochrome [OPTICS] Having only one chromaticity. { 'män·ə,krōm }

monoclinic system [CRYSTAL] One of the six crystal systems characterized by a single, twofold symmetry axis or a single symmetry plane. { ¦män·ə'klin·ik ,sis·təm }

monodispersity [POLYM CHEM] Polymer system that is homogeneous in molecular weight, that is, it does not have a distribution of different molecular-weight chains within the total mass. { ¦män·ō·di'spər·səd·ē }

monofilament [TEXT] A single, large, continuous filament (single-strand thread) of a natural or synthetic fiber. { ¦män·ə'fil·ə·mənt }

monofunctional compound [ORG CHEM] An organic compound whose chemical structure possesses a single highly reactive site. { ,män·ō ¦fəŋk·shən·əl 'käm,paúnd }

monolayer *See* monomolecular film. { 'män·ō ,lā·ər }

monolith [MATER] A large concrete block. { 'män·ə,lith }

monolithic integrated circuit [ELECTR] An integrated circuit having elements formed in place on or within a semiconductor substrate, with at least one element being formed within the substrate. { ,män·ə'lith·ik 'int·ə,grād·əd 'sər·kət }

monomer [POLYM CHEM] A molecule which is capable of combining with like or unlike molecules to form a polymer; it is a repeating structure unit within a polymer. Also known as repeating unit. { 'män·ə·mər }

monomeric unit *See* repeating unit. { ,män·ə ¦mer·ik 'yü·nət }

monomolecular film [PHYS CHEM] A film one molecule thick. Also known as monolayer. { ¦män·ō·mə¦lek·yə·lər 'film }

monotron [MET] A machine for determining indentation hardness by measuring the load required to dent a specimen to a constant depth with a diamond 0.625 millimeter in diameter. { 'män·ə,trän }

monotrophic [CRYSTAL] Of crystal pairs, having one of the pair always metastable with respect to the other. { ¦män·ə¦träf·ik }

monotype metal [MET] A type metal typically composed of 76% lead, 16% antimony, and 8%

tin, with good wear resistance and compressive strength. { 'män·ə,tīp ,med·əl }

monovalent [CHEM] A radical or atom whose valency is 1. { ¦män·ō'vā·lənt }

montan wax [MATER] A hard mineral wax with a melting point of 80–90°C; white after purification and brown in the crude form; soluble in carbon tetrachloride, benzene, and chloroform; used for shoe and furniture polishes, adhesive pastes, roofing paints, and phonograph records. Also known as lignite wax. { 'män,tan ,waks }

mordant [CHEM] An agent, such as alum, phenol, or aniline, that fixes dyes to tissues, cells, textiles, and other materials by combining with the dye to form an insoluble compound. { 'mȯrd·ənt }

mordant dye [MATER] Textile dye that requires a mordant (third substance) to bind the dye onto the fiber. { 'mȯrd·ənt 'dī }

mordanting assistant [MATER] A chemical used with textile dye mordants to cause decomposition of the mordant and uniform deposition on the fibers; examples are sulfuric, oxalic, and lactic acids. { 'mȯrd·ənt·iŋ ə,sis·tənt }

mordant rouge [MATER] Aluminum acetate-acetic acid solution used in dyeing and calico printing. Also known as red acetate; red liquor. { 'mȯrd·ənt 'rüzh }

Morera's stress functions [MECH] Three functions of position, ψ_1, ψ_2, and ψ_3, in terms of which the elements of the stress tensor σ a body may be expressed, if the body is in equilibrium and is not subjected to body forces; the elements of the stress tensor are given by $\sigma_{11} = -2\partial^2\psi_1/\partial x_2\partial x_3$, $\sigma_{23} = \partial^2\psi_2/\partial x_1\partial x_2 + \partial^2\psi_3/\partial x_1\partial x_3$, and cyclic permutations of these equations. { mȯ'rer·əz 'stres ,fəŋk·shənz }

Morgan equation [THERMO] A modification of the Ramsey-Shields equation, in which the expression for the molar surface energy is set equal to a quadratic function of the temperature rather than to a linear one. { 'mȯr·gən i,kwā·zhən }

morphotropism [CRYSTAL] Similarity of structure, axial ratios, and angles between faces of one or more zones in crystalline substances whose formulas can be derived one from another by substitution. { ¦mȯr·fō'trō,piz·əm }

mortar [MATER] A mixture of cement, lime, and sand used for laying bricks or masonry. { 'mȯrd·ər }

MOS See metal oxide semiconductor.

mosaic gold See stannic sulfide. { mō'zā·ik ,gōld }

mosaic structure [CRYSTAL] In crystals, a substructure in which neighboring regions are oriented slightly differently. { mō'zā·ik ¦strək·chər }

MOSFET See metal oxide semiconductor field-effect transistor. { 'mȯs,fet }

Mosotti field See Lorentz local field. { mȯ'säd·ē ,fēld }

mossy zinc [MET] Zinc granules made by pouring molten zinc into water. { 'mȯs·ē 'ziŋk }

MOST See metal oxide semiconductor field-effect transistor.

MOS transistor See metal oxide semiconductor field-effect transistor. { ¦em¦ō'es tran'zis·tər }

mottled iron [MET] A cast iron showing gray areas that contain graphite, perlite, and sometimes ferrite, and white areas containing primarily cementite. { 'mäd·əld 'ī·ərn }

mountain blue [INORG CHEM] $2CuCO_3\cdot Cu(OH)_2$ Ground azurite used as a paint pigment. Also known as copper blue. { 'maúnt·ən 'blü }

mp See melting point.

MRTD See minimum resolvable temperature difference.

mucilage [MATER] **1.** A sticky material employed as an adhesive. **2.** A gummy material derived from plants. { 'myü·sə·lij }

muffin-tin potential [SOLID STATE] A potential function used in the augmented plane-wave method and related methods of approximating the energy states of electrons in a crystal lattice, which is spherically symmetric within spheres centered at each atomic nucleus and constant in the region between these spheres. { 'məf·ən ,tin pə,ten·chəl }

multimeter See volt-ohm-milliammeter. { 'məl·tə,mēd·ər or məl'tim·əd·ər }

multiphonon emission [SOLID STATE] A process of nonradiative recombination of electrons and holes in which an electron is captured into a deep level near the middle of an energy gap associated with a lattice defect, exciting lattice vibrations, and the trapped electron state captures a hole from the valence band. { ¦məl·tə'fō ,nän i'mish·ən }

multiple [MET] A piece of stock cut from bar for use in a forging which provides the exact length needed for a single workpiece. { 'məl·tə·pəl }

multiple-impulse welding [MET] Spot, upset, or projection welding in which more than one current impulse is generated during a single machine cycle. Also known as pulsation welding. { 'məl·tə·pəl ¦im,pəls ,weld·iŋ }

multiple isomorphous replacement technique [CRYSTAL] A technique for overcoming the phase problem by growing crystals in three different isomorphic chemical forms and comparing x-ray diffraction data obtained from all three. Abbreviated MIR technique. { ,məl·tə·pəl ,ī·sə ,mȯr·fəs ri'plās·mənt tek,nēk }

multiple-pass weld [MET] A weld made by depositing filler metal with two or more passes in succession. { 'məl·tə·pəl ¦pas ,weld }

multiple-purpose tester See volt-ohm-milliammeter. { 'məl·tə·pəl ¦pər·pəs 'tes·tər }

multiple spot welding [MET] Spot welding in which several spots are welded during a single machine cycle. { 'məl·tə·pəl 'spät ,weld·iŋ }

multiwavelength anomalous dispersion [CRYSTAL] A technique for overcoming the phase problem by collecting x-ray diffraction data at several wavelengths around the absorption edge of a strongly absorbing atom. Abbreviated MAD. { ,məl·tē¦wāv,leŋkth ə,näm·ə·ləs di'spərzh·ən }

Mumetal [MET] An alloy of high magnetic permeability, containing 14% iron, 5% copper,

1.5% chromium, and the balance nickel. { 'myü ,med·əl }

Munsell chroma *See* chroma. { mən'sel ,krō·mə }

Munsell color system [OPTICS] A system for designating colors which employs three perceptually uniform scales (Munsell hue, Munsell value, Munsell chroma) defined in terms of daylight reflectance. { mən'sel 'kəl·ər ,sis·təm }

Munsell hue [OPTICS] The dimension of the Munsell system of color that determines whether a color is blue, green, yellow, red, purple, or the like, without regard to its lightness or saturation. { mən'sel ,hyü }

Munsell value [OPTICS] The dimension, in the Munsell system of object-color specificiation, that indicates the apparent luminous transmittance or reflectance of the object on a scale having approximately equal perceptual steps under the usual conditions of observation. { mən'sel ,val,yü }

Muntz metal [MET] A 60/40 type of brass composed of 58-61% copper, up to 1% lead, and remainder zinc. Also known as malleable brass; yellow metal. { 'məns ,med·əl }

muriatic acid *See* hydrochloric acid. { myür·ē'ad·ik 'as·əd }

music wire [MET] High-quality, high-carbon steel wire used for making mechanical springs. { 'myü·zik ,wīr }

MVD *See* mechanical vapor diffusion.

myrrh [MATER] A gum resin of species of myrrh (*Commiphora*); partially soluble in water, alcohol, and ether; used in dentifrices, perfumery, and pharmaceuticals. { mər }

N

N *See* newton; nitrogen; normality.

Na *See* sodium.

nanochemistry [CHEM] The study of the synthesis and analysis of materials in the nanoscale range (1–10 nanometers), including large organic molecules, inorganic cluster compounds, and metallic or semiconductor particles. { ˌnan·ō'kem·ə·strē }

nanocomposite [MATER] A material that results from the intimate mixture of two or more nanophase materials. { ˌnan·ō·kəm'päz·ət }

nanocomposite material *See* nanostructured material. { ˌnan·ō·kəm'päz·ət məˌtir·ē·əl }

nanoelectronics [ELECTR] The technology of electronic devices whose dimensions range from atoms up to 100 nanometers. { ˌnan·ō·i ˌlek'trän·iks }

nanophase material [MATER] **1.** A material made up of phases that have dimensions of the order of nanometers. **2.** An ultrafine single solid phase where at least one dimension is in the nanometer range, and typically dimensions are in the 1–20-nanometer range. { 'nan·ō,fāz məˌtir·ē·əl }

nanostructure [SOLID STATE] Something that has a physical dimension smaller than 100 nanometers, ranging from clusters of atoms to dimensional layers. { 'nan·ō,strək·chər }

nanostructured material [MATER] A structure assembled from a layer or cluster of atoms with size the order or nanometers. { 'nan·ōstrək·chərd mətir·ē·əl }

nanotechnology [ENG] **1.** Systems for transforming matter, energy, and information that are based on nanometer-scale components with precisely defined molecular features. **2.** Techniques that produce or measure features less than 100 nanometers in size. { ¦nan·ō·tek'näl·ə·jē }

nantokite *See* cuprous chloride. { 'nan·tə,kīt }

naphtha [MATER] **1.** Petroleum fraction with volatility between gasoline and kerosine; used as a gasoline ingredient, solvent for paints and rubber, and cleaning solvent. **2.** Aromatic solvent from coal tar, either solvent naphtha or heavy naphtha. { 'naf·thə }

Naples yellow *See* lead antimonite. { 'nā·pəlz 'yel·ō }

natrium [CHEM] Latin name for sodium; source of the symbol Na. { 'nā·trē·əm }

natural aging [MET] Spontaneous aging at room temperature of a supersaturated metallic solid solution. { 'nach·rəl 'āj·iŋ }

natural cement [MATER] Hydraulic cement made from pulverized and heated limestone containing clay, magnesia, and iron. { 'nach·rəl si'ment }

natural convection [THERMO] Convection in which fluid motion results entirely from the presence of a hot body in the fluid, causing temperature and hence density gradients to develop, so that the fluid moves under the influence of gravity. Also known as free convection. { 'nach·rəl kən'vek·shən }

natural pressure cycle [MET] A cycle in which pressure buildup conforms proportionately to the buildup of stresses due to forming. { 'nach·rəl 'presh·ər ,sī·kəl }

natural resource [MATER] A deposit of minerals, water, or other materials furnished by nature. { 'nach·rəl 're,sórs }

natural-seasoned lumber *See* air-dried lumber. { 'nach·rəl ¦sēz·ənd 'ləm·bər }

natural steel [MET] **1.** Steel made directly from cast iron. **2.** Steel, such as wootz, made directly from the ore. { 'nach·rəl 'stēl }

naval brass [MET] Brass composed of 60–62% copper, 37–39% zinc, and 0.75–1% tin; relatively resistant to corrosion by seawater. Also known as naval bronze. { 'nā·vəl 'bras }

naval bronze *See* naval brass. { 'nā·vəl 'bränz }

naval stores [MATER] **1.** Pitch and rosin formerly used in the construction of wooden ships. **2.** All pine wood products, including rosin, turpentine, and pine oils. { 'nā·vəl ,storz }

Navier's equation [MECH] A vector partial differential equation for the displacement vector of an elastic solid in equilibrium and subjected to a body force. { nä'vyāz i,kwā·zhən }

Navier-Stokes equations [FL MECH] The equations of motion for a viscous fluid which may be written $dV/dt = -(1/\rho)\nabla p + F + \nu\nabla^2 V + (\frac{1}{3})\nu\nabla(\nabla \cdot V)$, where p is the pressure, ρ the density, F the total external force per unit mass, V the fluid velocity, and ν the kinematic viscosity; for an incompressible fluid, the term in $\nabla \cdot V$ (divergence) vanishes, and the effects of viscosity then play a role analogous to that of temperature in thermal conduction and to that of density in simple diffusion. { nä'vyā 'stōks i,kwā·zhənz }

Nb *See* niobium.

NBR *See* nitrile rubber.

N center |SOLID STATE| A color center which arises from continued exposure to light in the F band or to x-rays and which produces a faint absorption band on the long-wavelength side of the M band. Also known as G center. { 'en ‚sen·tər }

n-channel |ELECTR| A conduction channel formed by electrons in an *n*-type semiconductor, as in an *n*-type field-effect transistor. { 'en ‚chan·əl }

Nd *See* neodymium.

Ne *See* neon.

nearest neighbors |CRYSTAL| Any pair of atoms in a crystal lattice which are as close to each other, or closer to each other, than any other pair. { 'nir·əst 'nā·bərz }

nearly free electron method |SOLID STATE| A method of approximating the energy levels of electrons in a crystal lattice by considering the potential energy resulting from atomic nuclei and from other electrons in the lattice as a perturbation on free electron states. Abbreviated NFE method. { ¦nir·lē ¦frē i¦lek,trän ,meth·əd }

neat cement grout |MATER| Grout made from a mixture of cement and water. { 'nēt si'ment ‚graút }

neat plaster |MATER| A base-coat plaster, having sand added at the job location. { 'nēt ,plas·tər }

neck |ENG| The part of a furnace where the flame is contracted before reaching the stack. |MET| In a tensile test, that portion of the metal at which fracture is imminent during the later stages of plastic deformation in a tensile test. { nek }

neck-down |MET| **1.** A thin core used for restricting the riser neck; facilitates cutting off the riser from the casting. **2.** Localized area reduction of a test piece during plastic deformation. { 'nek ‚daún }

necking |MET| Reducing the diameter or cross-sectional area of a tube or other piece of metal by stretching. { 'nek·iŋ }

necking down |MET| Localized reduction in cross-sectional area of a specimen during tensile deformation. { 'nek·iŋ ¦daún }

Néel ferromagnetism *See* ferrimagnetism. { 'nā·el ,fer·ō'mag·nə,tiz·əm }

Néel point *See* Néel temperature. { 'nā·el ‚póint }

Néel's theory |SOLID STATE| A theory of the behavior of antiferromagnetic and other ferrimagnetic materials in which the crystal lattice is divided into two or more sublattices; each atom in one sublattice responds to the magnetic field generated by nearest neighbors in other sublattices, with the result that magnetic moments of all the atoms in any sublattice are parallel, but magnetic moments of two different sublattices can be different. { 'nā·elz ‚thē·ə·rē }

Néel temperature |SOLID STATE| A temperature, characteristic of certain metals, alloys, and salts, below which spontaneous nonparalleled magnetic ordering takes place so that they become antiferromagnetic, and above which they are paramagnetic. Also known as Néel point. { 'nā·el 'tem·prə·chər }

Néel wall |SOLID STATE| The boundary between two magnetic domains in a thin film in which the magnetization vector remains parallel to the faces of the film in passing through the wall. { 'nā·el ‚wól }

negative |ELEC| Having a negative charge. |GRAPHICS| The image on film in which the dark tones of the original appear transparent, and the light tones appear black and opaque. Also known as reversed image. { 'neg·əd·iv }

negative crystal |CRYSTAL| A crystal containing a cavity, where the form of the cavity is one of the characteristic crystal forms of the mineral in question. |OPTICS| A uniaxial crystal in which the extraordinary wave travels faster than the ordinary wave, such as calcite. { 'neg·əd·iv 'krist·əl }

negative electrode *See* cathode. { 'neg·əd·iv i'lek,trōd }

negative elongation |CRYSTAL| In a section of an anisotropic crystal, a sign of elongation that is parallel to the faster of the two plane-polarized rays. { 'neg·əd·iv ‚ē,lóŋ'gā·shən }

negative ion |CHEM| An atom or group of atoms which by gain of one or more electrons has acquired a negative electric charge. |PHYS| An electron or negatively charged subatomic particle. { 'neg·əd·iv 'ī,än }

negative-ion vacancy |CRYSTAL| A point defect in an ionic crystal in which a negative ion is missing from its lattice site. { 'neg·əd·iv ¦ī,än 'vā·kən·sē }

negative pole *See* south pole. { 'neg·əd·iv 'pōl }

negative temperature |THERMO| The property of a thermally isolated thermodynamic system whose elements are in thermodynamic equilibrium among themselves, whose allowed states have an upper limit on their possible energies, and whose high-energy states are more occupied than the low-energy ones. { 'neg·əd·iv 'tem·prə·chər }

neighbor |CRYSTAL| One of a pair of atoms or ions in a crystal which are close enough to each other for their interaction to be of significance in the physical problem being studied. { 'nā·bər }

nematicide |MATER| A chemical used to kill plant-parasitic nematodes. Also spelled nematocide. { nə'mad·ə,sīd }

nematic phase |PHYS CHEM| A phase of a liquid crystal in the mesomorphic state, in which the liquid has a single optical axis in the direction of the applied magnetic field, appears to be turbid and to have mobile threadlike structures, can flow readily, has low viscosity, and lacks a diffraction pattern. { nə'mad·ik ,fāz }

nematocide *See* nematicide. { nə'mad·ə,sīd }

nematogenic solid |PHYS CHEM| A solid which will form a nematic liquid crystal when heated. { nə'mad·ə,jen·ik 'säl·əd }

neodymium |CHEM| A metallic element, symbol Nd, with atomic weight 144.24, atomic number

60; a member of the rare-earth group of elements. { ‚nē·ō'dim·ē·əm }

neodymium chloride [INORG CHEM] $NdCl_3 \cdot xH_2O$ Water-and acid-soluble, pink lumps; used to prepare metallic neodymium. { ‚nē·ō'dim·ē·əm 'klȯr‚īd }

neodymium glass [MATER] A glass containing small amounts of neodymium oxide; used for color television filter plates since it transmits 90% of the blue, green, and red light rays and no more than 10% of the yellow. { ‚nē·ō'dim·ē·əm 'glas }

neodymium-iron-boron magnet [MET] A permanent-magnet material that has the highest energy product known, and consists of an intermetallic phase, $Nd_2Fe_{14}B$, in a tetragonal crystal structure. { ‚nē·ō'dim·ē·əm 'ī·ərn 'bȯ ‚rän ‚mag·nət }

neodymium oxide [INORG CHEM] Nd_2O_3 A hygroscopic, blue-gray powder; insoluble in water, soluble in acids; used to color glass and in ceramic capacitors. { ‚nē·ō'dim·ē·əm 'äk‚sīd }

neon [CHEM] A gaseous element, symbol Ne, atomic number 10, atomic weight 20.179; a member of the family of noble gases in the zero group of the periodic table. { 'nē‚än }

neophane glass [MATER] A glass containing neodymium oxide to reduce glare; used for yellow sunglasses or for windshield glass. { 'nē·ə‚fän ‚glas }

neoprene [MATER] A synthetic rubber with outstanding resistance to ozone, weathering, various chemicals, oil, and flame, made by polymerization of chloroprene (2-chlorobutadiene-1,3); varies from amber to silver to cream in color; used in paints, putties, adhesives, shoe soles, tank linings, and rubber products. { 'nē·ə‚prēn }

neptunium [CHEM] A chemical element, symbol Np, atomic number 93, atomic weight 237.0482; a member of the actinide series of elements. { nep'tü·nē·əm }

Nernst approximation formula [THERMO] An equation for the equilibrium constant of a gas reaction based on the Nernst heat theorem and certain simplifying assumptions. { 'nernst ə ‚präk·sə'mā·shən ‚fȯr·myə·lə }

Nernst effect [PHYS] The phenomenon that, when a conductor is placed in a magnetic field and an electric current flows through the conductor perpendicular to the field, a temperature gradient arises in the direction of the current. { 'nernst i‚fekt }

Nernst heat theorem [THERMO] The theorem expressing that the rate of change of free energy of a homogeneous system with temperature, and also the rate of change of enthalpy with temperature, approaches zero as the temperature approaches absolute zero. { 'nernst 'hēt ‚thir·əm }

Nernst-Simon statement of the third law of thermodynamics [THERMO] The statement that the change in entropy which occurs when a homogeneous system undergoes an isothermal reversible process approaches zero as the temperature approaches absolute zero. { 'nernst 'sī·mən 'stāt·mənt əv thə 'thərd 'lȯ əv ‚thər·mō·dī'nam·iks }

NETD See noise equivalent temperature difference.

net density See density scale. { 'net ‚den·səd·ē }

net heating value See low heat value. { 'net 'hēd·iŋ ‚val·yü }

network polymer [POLYM CHEM] A three-dimensional material made by crosslinking. { ¦net‚wərk 'päl·ə·mər }

network structure [MET] A crystal structure in a metal in which one constituent occurs primarily at grain boundaries enveloping the grains made up of other constituents. { 'net‚wərk ‚strək·chər }

Neuberg blue [MATER] Pigment made up of a mixture of copper blue and iron blue. { 'nȯi ‚bərk 'blü }

Neugebauer effect [ELEC] A small change in the polarization of an optically isotropic medium in an external electric field, related to the electrooptical Kerr effect. { 'nȯi·gə‚baů·ər i‚fekt }

Neumann bands See Neumann lines. { 'nȯi ‚män ‚banz }

Neumann-Kopp rule [THERMO] The rule that the heat capacity of 1 mole of a solid substance is approximately equal to the sum over the elements forming the substance of the heat capacity of a gram atom of the element times the number of atoms of the element in a molecule of the substance. { 'nȯi‚män 'kȯp ‚rül }

Neumann lines [MET] Mechanical deformation twins seen as straight, serrated narrow bands parallel to preferred planes in the crystals of an etched metal which has been strained, usually by sudden impact; most often observed along the 112 planes of body-centered-cubic ferrite. Also known as Neumann bands. { 'nȯi‚män ‚līnz }

Neumann's principle [CRYSTAL] The principle that the symmetry elements of the point group of a crystal are included among the symmetry elements of any property of the crystal. { 'nȯi ‚mänz ‚prin·sə·pəl }

neuromorphic engineering [ENG] Use of the functional principles of biological nervous systems to inspire the design and fabrication of artificial nervous systems, such as vision chips and roving robots. { ¦nů·rō‚mȯr·fik ‚en·jə'nir·iŋ }

neuronal interface [ENG] An artificial synapse capable of reversible chemical-to-electrical transduction processes between neural tissue and conventional solid-state electronic devices for applications such as aural, visual, and mechanical prostheses, as well as expanding human memory and intelligence. { nů¦rōn·əl 'in·tər‚fās }

neurotechnology [ENG] The application of microfabricated devices to achieve direct contact with the electrically active cells of the nervous system (neurons). { ‚nů·rō·tek'näl·ə·jē }

neutral [CHEM] Property of a solution which is neither acidic nor basic; if acqueous, having the same concentration of hydrogen ions as water. [ELEC] Referring to the absence of a net electric charge. { 'nü·trəl }

neutral axis [MECH] In a beam bent downward, the line of zero stress below which all fibers are in tension and above which they are in compression. { 'nü·trəl 'ak·səs }

neutral fiber [MECH] A line of zero stress in cross section of a bent beam, separating the region of compressive stress from that of tensile stress. { 'nü·trəl 'fī·bər }

neutralize [CHEM] To make a solution neutral (neither acidic nor basic; if acqueous, pH of 7) by adding a base to an acidic solution, or an acid to a basic solution. { 'nü·trə,līz }

neutral point [MET] In rolling mills, the point at which the speed of the work is equal to the peripheral speed of the rolls. [PHYS] A point where two fields are equal in magnitude and opposite in direction so that the net field is zero. { 'nü·trəl ,pȯint }

neutral potassium phosphate See potassium phosphate. { 'nü·trəl pə'tas·ē·əm 'fäs,fāt }

neutral surface [MECH] A surface in a bent beam along which material is neither compressed nor extended. { 'nü·trəl 'sər·fəs }

neutron [PHYS] An elementary particle which has approximately the same mass as the proton but lacks electric charge, and is a constituent of all nuclei having mass number greater than 1. { 'nü,trän }

nevyanskite [MET] A tin-white variety of iridosmine containing 35–50% osmium or more than 50% iridium; occurs in flat scales. { nev'yan ,skīt }

new blue [MATER] Any of several iron blue types of pigments; varieties are called mendola blue and prussian blue. { 'nü 'blü }

New Jersey retort process See vertical retort process. { ¦nü 'jər·zē 'rē,tȯrt ,prä·səs }

newsprint [MATER] The paper used in the publication of newspapers; an impermanent material made from mechanical wood pulp, with some chemical wood pulp. { 'nüz,print }

newton [MECH] The unit of force in the meter-kilogram-second system, equal to the force which will impart an acceleration of 1 meter per second squared to the International Prototype Kilogram mass. Symbolized N. Formerly known as large dyne. { 'nüt·ən }

Newton formula for the stress See Newtonian friction law. { 'nüt·ən 'fȯr·myə·lə fȯr thə 'stres }

Newtonian flow [FL MECH] Flow system in which the fluid performs as a Newtonian fluid, that is, shear stress is proportional to shear rate. { 'nü̇'tō·nē·ən 'flō }

Newtonian fluid [FL MECH] A simple fluid in which the state of stress at any point is proportional to the time rate of strain at that point; the proportionality factor is the viscosity coefficient. { 'nü̇'tō·nē·ən 'flü·əd }

Newtonian friction law [FL MECH] The law that shear stress in a fluid is proportional to the shear rate; it holds only for some fluids, which are then called Newtonian. Also known as Newton formula for the stress. { 'nü̇'tō·nē·ən 'frik·shən ,lȯ }

Newtonian viscosity [FL MECH] The viscosity of a Newtonian fluid. { 'nü̇'tō·nē·ən vi'skäs·əd·ē }

Newton's alloy [MET] A fusible alloy made up of

50% bismuth, 31% lead, and 19% tin; melts at 95°C; used in applications where it is required to fall away at predetermined temperatures, as for automatic sprinkler links. Also known as Newton's metal. { 'nüt·ənz 'al,ȯi }

Newton's law of cooling [THERMO] The law that the rate of heat flow out of an object by both natural convection and radiation is proportional to the temperature difference between the object and its environment, and to the surface area of the object. { 'nüt·ənz 'lȯ əv 'kül·iŋ }

Newton's law of resistance [FL MECH] The law that the force opposing the motion of an object through a fluid at moderate velocities is proportional to the square of the velocity. { 'nüt·ənz 'lȯ əv ri'zis·təns }

Newton's metal See Newton's alloy. { 'nüt·ənz ,med·əl }

Newton's theory of light See corpuscular theory of light. { 'nüt·ənz ,thē·ə·rē əv 'līt }

NFE method See nearly free electron method. { ¦en¦ef'ē ,meth·əd }

Ni See nickel.

nicarbing See carbonitriding. { 'nī,kär·biŋ }

nickel [CHEM] A chemical element, symbol Ni, atomic number 28, atomic weight 58.69. [MET] A silver-gray, ductile, malleable, tough metal; used in alloys, plating, coins (to replace silver), ceramics, and electronic circuits. { 'nik·əl }

nickel-aluminum bronze [MET] An alloy that is composed of an 8–10% aluminum bronze with nickel added to increase strength, corrosion resistance, and heat resistance; used for dies, molds, cast propellers, and valve seats. { 'nik·əl ə'lü·mə·nəm 'bränz }

nickel ammonium sulfate [INORG CHEM] $NiSO_4 \cdot (NH_4)_2SO_4 \cdot 6H_2O$ A green, crystalline compound, soluble in water; used as a nickel electrolyte for electroplating. Also known as ammonium nickel sulfate; double nickel salt. { 'nik·əl ə'mō·nē·əm 'səl,fāt }

nickel arsenate [INORG CHEM] $Ni_3(AsO_4)_2 \cdot H_2O$ Poisonous yellow-green powder; soluble in acids, insoluble in water; used as a fat-hardening catalyst in soapmaking. { 'nik·əl 'ars·ən,āt }

nickel brass See nickel silver. { 'nik·əl 'bras }

nickel bronze [MET] Bronze containing nickel; a common type contains 88% copper, 5% tin, 5% nickel, and 2% zinc. { 'nik·əl 'bränz }

nickel carbonate [INORG CHEM] $NiCO_3$ Light-green crystals that decompose upon heating; soluble in acid, insoluble in water; used in electroplating. { 'nik·əl 'kär·bə,nāt }

nickel carbonyl [INORG CHEM] $Ni(CO)_4$ Colorless, flammable, poisonous liquid boiling at 43°C; soluble in alcohol and concentrated nitric acid, insoluble in water; used in gas plating (vapor decomposes at 60°C) and to produce metallic nickel. { 'nik·əl 'kär·bə,nil }

nickel cast iron [MET] An improved-strength alloy cast iron containing a small percentage of nickel (2–5%); in larger amounts (15–36%) nickel primarily imparts corrosion resistance. { 'nik·əl 'kast 'ī·ərn }

nickel-chromium steel [MET] Steel containing

nickel (0.2–3.75%) and chromium (0.3–1.5%) as alloying elements. { 'nik·əl 'krō·mē·əm 'stēl }

nickel cyanide [INORG CHEM] Ni(CN)$_2$·4H$_2$O Poisonous, water-insoluble apple-green powder; melts and loses water at 200°C, decomposes at higher temperatures; used for electroplating and metallurgy. { 'nik·əl 'sī·ə,nīd }

nickel iodide [INORG CHEM] NiI$_2$ or NiI$_2$·6H$_2$O Hygroscopic black or blue-green solid; soluble in water and alcohol; sublimes when heated. { 'nik·əl 'ī·ə,dīd }

nickel-molybdenum iron [MET] An alloy containing 20–40% molybdenum and up to 60% nickel with some carbon added; has high acid resistance. { 'nik·əl mə'lib·də·nəm 'ī·ərn }

nickel-molybdenum steel [MET] Steel containing 0.2–0.3% molybdenum and 1.65–3.75% nickel as alloying elements. { 'nik·əl mə'lib·də·nəm 'stēl }

nickel nitrate [INORG CHEM] Ni(NO$_3$)$_2$·6H$_2$O Fire-hazardous oxidant; deliquescent, green, water- and alcohol-soluble crystals; used for nickel plating and brown ceramic colors, and in nickel catalysts. { 'nik·əl 'nī,trāt }

nickel oxide [INORG CHEM] NiO Green powder; soluble in acids and ammonium hydroxide; insoluble in water; used to make nickel salts and for porcelain paints. Also known as green nickel oxide. { 'nik·əl 'äk,sīd }

nickel phosphate [INORG CHEM] Ni$_3$(PO$_4$)$_2$·7H$_2$O A light-green powder; soluble in acids and ammonium hydroxide, insoluble in water; used for electroplating and production of yellow nickel. { 'nik·əl 'fäs,fāt }

nickel plating [MET] Electrolytic deposition of a metallic nickel coating. { 'nik·əl ¦plād·iŋ }

nickel rhodium [MET] Nickel alloy with 25–80% rhodium; can also contain other metals, such as platinum or molybdenum; used for pen points, reflectors, electrodes, and chemical equipment. { 'nik·əl 'rō·dē·əm }

nickel silver [MET] A silver-white alloy composed of 52–80% copper, 10–35% zinc, and 5–35% nickel; sometimes also contains a few percent of lead and tin. Also known as German silver; nickel brass. { 'nik·əl 'sil·vər }

nickel steel [MET] Carbon steel containing up to 9% nickel as a major alloying element. { 'nik·əl 'stēl }

nickel-vanadium steel [MET] A nickel steel containing about 1.5% nickel, 1% manganese, 0.28% carbon, and 0.10% vanadium; used for high-strength cast parts. { 'nik·əl və'nā·dē·əm 'stēl }

niello [MATER] Mixture of sulfides of copper, silver, and lead, with black metallic appearance; used in ornamental inlays engraved on metals such as silver. { nē'el·ō }

nigrite [MATER] A mixture of rubber with ozocerite distillation residue; used as a substitute for gutta-percha. { 'nī,grīt }

nigrosine [MATER] Any of a group of blue or black azine dyes used for coloring inks, shoe polish, leather, and wood; can be water-, alcohol-, or oil-soluble. { 'nī·grə,sēn }

niobic acid [INORG CHEM] Nb$_2$O$_5$·nH$_2$O Family of hydrates; white precipitate, soluble in inorganic acids and bases, insoluble in water; its formation is part of the analytical determination of niobium. { nī'ō·bik 'as·əd }

niobium [CHEM] A chemical element, symbol Nb, atomic number 41, atomic weight 92.904. [MET] A platinum-gray, ductile metal with brilliant luster; used in alloys, especially stainless steels. Also known as columbium. { nī'ō·bē·əm }

niobium carbide [INORG CHEM] NbC A lavender gray powder with a melting point of 3500°C; used for carbide-tipped tools and special steels. { nī'ō·bē·əm 'kär,bīd }

niter See potassium nitrate. { 'nīd·ər }

niter cake See sodium bisulfate. { 'nīd·ər ,kāk }

nitrate [CHEM] **1.** A salt or ester of nitric acid. **2.** Any compound containing the ion NO$_3^-$. { 'nī,trāt }

nitrating acid [INORG CHEM] Sulfuric-nitric acid mix used to nitrate cellulosics and aromatic chemicals. Also known as mixed acid. { 'nī ,trād·iŋ 'as·əd }

nitric acid [INORG CHEM] HNO$_3$ Strong oxidant that is fire-hazardous; colorless or yellowish liquid, miscible with water; boils at 86°C; used for chemical synthesis, explosives, and fertilizer manufacture, and in metallurgy, etching, engraving, and ore flotation. Also known as aquafortis. { 'nī·trik 'as·əd }

nitride [INORG CHEM] Compound of nitrogen and a metal, such as Mg$_3$N$_2$. { 'nī,trīd }

nitriding [MET] Surface hardening of steel by formation of nitrides; nitrogen is introduced into the steel usually by heating in gaseous ammonia. { 'nī,trīd·iŋ }

nitrile [ORG CHEM] RC≡N Cyanide derived by removal of water from an acid amide. { 'nī ,trīl }

nitrile-butadiene rubber See nitrile rubber. { 'nī ,trīl ,byüd·ə'dī,ēn ,rəb·ər }

nitrile rubber [MATER] A synthetic rubber formed by polymerization of acrylonitrile with butadiene; the structure of the polymer is —CH$_2$CH=CHCH$_2$CH$_2$CH(CN)—. Also known as acrylonitrile-butadiene rubber; acrylonitrile rubber; NBR; nitrile-butadiene rubber; NR. { 'nī,trīl ,rəb·ər }

nitrite [CHEM] **1.** A salt or ester of nitrous acid, HNO$_2$. **2.** A compound containing the radical NO$_2^-$; can be organic or inorganic. { 'nī,trīt }

nitro- [CHEM] Chemical prefix showing the presence of the NO$_2^-$ radical. { 'nī·trō }

nitrobarite See barium nitrate. { ¦nī·trō'ba,rīt }

nitrocalcite See calcium nitrate. { ¦nī·trō'kal,sīt }

nitrocellulose See cellulose nitrate. { ¦nī·trō'sel· yə,lōs }

nitrocotton See cellulose nitrate. { ¦nī·trō'kät· ən }

nitrogen [CHEM] A chemical element, symbol N, atomic number 7, atomic weight 14.0067; it is a gas, diatomic (N$_2$) under normal conditions; about 78% of the atmosphere is N$_2$; in the combined form the element is a constituent of all proteins. { 'nī·trə·jən }

nitrogen acid anhydride *See* nitrogen pentoxide. { 'nī·trə·jən ¦as·əd an'hī,drīd }

nitrogen dioxide [INORG CHEM] NO_2 A reddish-brown gas; it exists in varying degrees of concentration in equilibrium with other nitrogen oxides; used to produce nitric acid. Also known as dinitrogen tetroxide; liquid dioxide; nitrogen peroxide; nitrogen tetroxide. { 'nī·trə·jən dī'äk,sīd }

nitrogen oxides [INORG CHEM] NO_x Chemical compounds of nitrogen and oxygen; produced primarily from the combustion of fossil fuels, they contribute to the formation of ground-level ozone. { 'nī·trə·jən 'äk,sīdz }

nitrogen pentoxide [INORG CHEM] N_2O_5 Colorless crystals, soluble in water (forms HNO_3); decomposes at 46°C. Also known as nitrogen acid anhydride. { 'nī·trə·jən pen 'täk,sīd }

nitrogen peroxide *See* nitrogen dioxide. { 'nī·trə·jən pə'räk,sīd }

nitrogen solution [INORG CHEM] Mixture used to neutralize super-phosphate in fertilizer manufacture; consists of 60% ammonium nitrate, and the balance a 50% aqua ammonia solution. { 'nī·trə·jən sə,lü·shən }

nitrogen tetroxide *See* nitrogen dioxide. { 'nī·trə·jən te'träk,sīd }

nitrogen trifluoride [INORG CHEM] NF_3 A colorless gas that has a melting point of −206.6°C and a boiling point of −128.8°C; used as an oxidizer for high-energy fuels. { 'nī·trə·jən trī'flùr,īd }

nitrogen trioxide [INORG CHEM] N_2O_3 Green, water-soluble liquid; boils at 3.5°C. { 'nī·trə·jən trī'äk,sīd }

nitroso [CHEM] The radical NO^- with trivalent nitrogen. Also known as hydroximino; oximido. { nī'trō·sō }

nitrous acid [INORG CHEM] HNO_2 Aqueous solution of nitrogen trioxide, N_2O_3. { 'nī·trəs 'as·əd }

nitryl halide [INORG CHEM] NO_2X Compound containing a halide (X) and a nitro group (NO_2). { 'nī,tril 'ha,līd }

nn junction [ELECTR] In a semiconductor, a region of transition between two regions having different properties in *n*-type semiconducting material. { ¦en¦en ,jəŋk·shən }

No *See* nobelium.

nobelium [CHEM] A chemical element, symbol No, atomic number 102; a synthetic element, in the actinium series; isotopes with mass numbers 250–260 and 262 have been produced in the laboratory, with mass number 259 having the longest known half-life, 58 minutes. { nō'bel·ē·əm }

noble gas [CHEM] A gas in group 0 of the periodic table of the elements; it is monatomic and, with limited exceptions, chemically inert. Also known as inert gas; rare gas. { 'nō·bəl 'gas }

noble metal [MET] A metal, or alloy, such as gold, silver, or platinum having high resistance to corrosion and oxidation; used in the construction of thin-film circuits, metal-film resistors, and other metal-film devices. { 'nō·bəl 'med·əl }

noble potential [PHYS CHEM] A potential equaling or approaching that of the noble elements, such as gold, silver, or copper, of the electromotive series. { 'nō·bəl pə'ten·chəl }

no-draft forging [MET] A forging designed with little or no taper for removal from dies, and with extremely fine tolerances for closer control of grain flow during production of the final part. { 'nō ,draft ,fòrj·iŋ }

nodular cast iron [MET] Cast iron treated in the molten state with a master alloy containing an element such as magnesium which favors formation of spheroidal graphite. Also known as ductile iron; spheroidal graphite cast iron. { 'näj·ə·lər ¦kast 'ī·ərn }

nodular powder [MET] Irregularly shaped metal powder particles. { 'näj·ə·lər 'paúd·ər }

no-fines concrete [MATER] Concrete made without sand and therefore containing a high proportion of communicating pores which provide thermal insulation and drainage. { 'nō ¦fīnz kän'krēt }

noise equivalent temperature difference [THERMO] The change in equivalent blackbody temperature that corresponds to a change in radiance which will produce a signal-to-noise ratio of 1 in an infrared imaging device. Abbreviated NETD. { 'nóiz i¦kwiv·ə·lənt 'tem·prə·chər ,dif·rəns }

nominal stress [MET] The stress calculated by simple elasticity theory, ignoring stress raisers and plastic flow; in tensile testing of a notched specimen, the load applied at the notch divided by the initial cross-sectional area at the notch. { 'näm·ə·nəl 'stres }

nonabelian quantum Hall state A quantum Hall state that is not amenable to a description in terms of electron dancing steps and that cannot be characterized by a symmetric matrix and a charge vector. { ,nän·ə¦bēl·yən ¦kwän·təm 'hòl ,stāt }

nonadiabatic *See* diabatic. { ¦nän·ad·ē·ə¦bad·ik }

nonaqueous [CHEM] Pertaining to a liquid or solution containing no water. { ¦nän'ā·kwē·əs }

nonasphaltic road oil [MATER] Any of the nonhardening petroleum distillates or residual oils used to lay road dust; they have viscosities low enough to be applied without prior heating. { ¦nän·as'fòl·tik 'rōd ,òil }

nonblackbody [THERMO] A body that reflects some fraction of the radiation incident upon it; all real bodies are of this nature. { ¦nän'blak ,bäd·ē }

nonconsumable electrode [MET] An electrode, such as of carbon or tungsten, that is not consumed during a welding or melting operation. { ¦nän·kən'sü·mə·bəl i'lek,trōd }

noncorrosive flux [MET] A soldering flux composed of rosin or of rosin in a volatile solvent; the residue is nonhygroscopic, noncorrosive, and nonconducting; suitable for soldering electronic components. Also known as activated rosin flux. { ¦nän·kə¦rō·siv 'fləks }

noncovalent bonds [CHEM] Weak chemical bonds that are electrostatic and hydrophobic in nature, for example, hydrogen bonds; important in determining complex biological structures. { ¦nän·kō¦vāl·ənt 'bänd }

nondeforming steel [MET] A group of alloy steels which do not easily deform when heat-treated. Also known as nonshrinking steel. { ¦nän·di'fȯr·miŋ 'stēl }

nondestructive testing [ENG] A technique for revealing flaws and defects in a material or device without damaging or destroying the test sample; includes use of x-rays, ultrasonics, radiography, and magnetic flux. { ¦nän·di'strək·div 'test·iŋ }

nonequilibrium thermodynamics [THERMO] A quantitative treatment of irreversible processes and of rates at which they occur. Also known as irreversible thermodynamics. { ¦nän,ē·kwə'lib·rē·əm ‚thər·mō·dī'nam·iks }

nonfaradaic path [PHYS CHEM] One of the two available paths for transfer of energy across an electrolyte-metal interface, in which energy is carried by capacitive transfer, that is, by charging and discharging the double-layer capacitance. { ¦nän,far·ə'dā·ik 'path }

nonferrous metal [MET] Any metal other than iron and its alloys. { ¦nän'fer·əs 'med·əl }

nonferrous metallurgy [MET] A branch of metallurgy that deals with metals other than iron and iron-base alloys. { ¦nän'fer·əs 'med·əl‚ər·jē }

nonionic detergent [MATER] A detergent with molecules that do not ionize in aqueous solution, for example, detergents derived from condensation products of long-chain glycols and octyl or nonyl phenols. { ¦nän·ī'än·ik di'tər·jənt }

nonlinear [PHYS] Pertaining to a response which is other than directly or inversely proportional to a given variable. { 'nän‚lin·ē·ər }

nonlinear crystal [SOLID STATE] A crystal in which some influence (such as stress, electric field, or magnetic field) produces a response (such as strain, electric polarization, or magnetization) which is not proportional to the influence. { 'nän‚lin·ē·ər 'krist·əl }

nonlinear damping [PHYS] Damping that is not proportional to velocity. { 'nän‚lin·ē·ər 'damp·iŋ }

nonlinear material [PHYS] A material in which some specified influence (such as stress, electric field, or magnetic field) produces a response (such as strain, electric polarization, or magnetization) which is not proportional to the influence. { 'nän‚lin·ē·ər mə'tir·ē·əl }

nonlinear optical device [OPTICS] A device based on one of a class of optical effects that result from the interaction of electromagnetic radiation from lasers with nonlinear materials. { 'nän‚lin·ē·ər 'äp·tə·kəl di‚vīs }

nonlinear optics [OPTICS] The study of the interaction of radiation with matter in which certain variables describing the response of the matter (such as electric polarization or power absorption) are not proportional to variables describing the radiation (such as electric field strength or energy flux). { 'nän‚lin·ē·ər 'äp‚tiks }

nonlinear physics [PHYS] The study of situations where the measure of an effect is not proportional to the measure of what is considered to be its cause. { ‚nän‚lin·ē·ər 'fiz·iks }

nonlinear refraction [OPTICS] The phenomenon whereby the refractive index of certain substances varies with light intensity. { 'nän‚lin·ē·ər ri'frak·shən }

nonlinear Schrödinger equation [OPTICS] A special form into which the Maxwell equations can be transformed in a medium with an optical nonlinearity that gives rise to self-action effects; this equation resembles the Schrödinger equation of quantum mechanics with the potential term in the latter equation replaced by a nonlinear term proportional to the local intensity of the light field, and it possesses soliton solutions. { ¦nän‚lin·ē·ər 'shräd·iŋ·ər i‚kwä·zhən }

nonlinear vibration [MECH] A vibration whose amplitude is large enough so that the elastic restoring force on the vibrating object is not proportional to its displacement. { 'nän‚lin·ē·ər vī'brā·shən }

nonlinear viscoelasticity [FL MECH] The behavior of a fluid which does not obey a first-order differential equation in stress and strain. { 'nän‚lin·ē·ər ¦vis·gō·i‚las'tis·əd·ē }

non-load-bearing tile [MATER] Tile unable to carry superimposed loads. { ¦nän 'lōd ‚ber·iŋ 'tīl }

nonlocalized bond See delocalized bond. { ¦nän'lō·kə‚līzd 'bänd }

nonmagnetic [ELECTROMAG] Not magnetizable, and therefore not affected by magnetic fields. { ¦nän·mag'ned·ik }

nonmagnetic steel [MET] A steel alloy containing about 12% manganese and sometimes a small quantity of nickel; it is practically nonmagnetic at ordinary temperatures. { ¦nän·mag'ned·ik 'stēl }

non-Newtonian fluid [FL MECH] A fluid whose flow behavior departs from that of a Newtonian fluid, so that the rate of shear is not proportional to the corresponding stress. Also known as non-Newtonian system. { ‚nän·nü'tō·nē·ən 'flü·əd }

non-Newtonian fluid flow [FL MECH] The flow behavior of non-Newtonian fluids, whose study has applications in many important problems of practical significance such as flow in tubes, extrusion, flow through dies, coating operations, rolling operations, and mixing of fluids. { ‚nän·nü'tō·nē·ən 'flü·əd ‚flō }

non-Newtonian system See non-Newtonian fluid. { ‚nän·nü'tō·nē·ən 'sis·təm }

non-Newtonian viscosity [FL MECH] The behavior of a fluid which, when subjected to a constant rate of shear, develops a stress which is not proportional to the shear. Also known as anomalous viscosity. { ‚nän·nü'tō·nē·ən vi'skäs·əd·ē }

nonpolar [CHEM] Pertaining to an element or compound which has no permanent electric dipole moment. { ¦nän'pō·lər }

nonpolar solvent [MATER] A solvent that does not have a permanent electric dipole moment and therefore has no tendency for intramolecular association with polar species. { ¦nän'pō·lər 'säl·vənt }

nonrigid plastic [MATER] A plastic with modulus of elasticity not greater than 50,000 pounds per square inch (3.45×10^8 pascals) at 25°C, according to standard American Society for Testing and Materials test procedures. { ¦nän ¦rij·əd 'plas·tik }

nonselective radiator See graybody. { ¦nän· si'lek·tiv 'rād·ē,ād·ər }

nonshattering glass [MATER] Two sheets of plate glass with a sheet of transparent resinoid between, the whole molded together under heat and pressure; it will crack without shattering. Also known as laminated glass; shatterproof glass. { 'nän,shad·ə·riŋ 'glas }

nonshrinking steel See nondeforming steel. { ¦nän,shriŋk·iŋ 'stēl }

nonslip concrete [MATER] Rough-surface concrete made by applying oxide grains to the mixture before it hardens; used for steps. { 'nän ¦slip 'kan,krēt }

nonsoap grease [MATER] Mineral oil thickened with solid lubricants such as graphite, mica, talc, molybdenum sulfide, asbestos fiber, uncombined fats, or rosin oils; used as a lubricant. { 'nän,sōp 'grēs }

nonsynchronous initiation [MET] In resistance welding, random starting and stopping of transformer primary current relative to the voltage wave. { ¦nän'siŋ·krə·nəs i,nish·ē'ā·shən }

nontransferred arc [MET] In arc welding and cutting, an arc made between the electrode and constricting nozzle, excluding the workpiece from the circuit. { ¦nän'tranz·fərd 'ärk }

nonviscous fluid See inviscid fluid. { 'nän,vis· kəs 'flü·əd }

nonwetting [MET] Of a metal or alloy, when it is molten, not adhering to or wetting the surface of a diamond which is to be set. { 'nän,wed·iŋ }

nor- [CHEM] Chemical formula prefix for normal; indicates a parent for another compound to be formed by removal of one or more carbons and associated hydrogens. { nȯr }

Nordheim's rule [SOLID STATE] The rule that the residual resistivity of a binary alloy that contains mole fraction x of one element and $1 - x$ of the other is proportional to $x(1 - x)$. { 'nȯrd,hīmz ,rül }

norm See fineness. { nȯrm }

normal fluid [CRYO] The component of liquid helium II, postulated in the two-fluid theory, that has viscosity and behaves like an ordinary fluid. { 'nȯr·məl 'flü·əd }

normal frequencies [MECH] The frequencies of the normal modes of vibration of a system. { 'nȯr·məl 'frē·kwən,sēz }

normal Hall effect [ELECTROMAG] The development, in a current-carrying conductor in a magnetic field, of a transverse voltage in the direction of the deflection of negative charge carriers (electrons) by the Lorentz force. { 'nȯrm·əl 'hȯl i,fekt }

normal induction [ELECTROMAG] Limiting induction, either positive or negative, in a magnetic material that is under the influence of a magnetizing force which varies between two specific limits. { 'nȯr·məl in'dək·shən }

normality [CHEM] Measure of the number of

gram-equivalent weights of a compound per liter of solution. Abbreviated N. { nȯr'mal· əd·ē }

normalize [MET] To heat a ferrous alloy to some temperature above the transformation range, followed by air cooling. { 'nȯr·mə,līz }

normal mode of vibration [MECH] Vibration of a coupled system in which the value of one of the normal coordinates oscillates and the values of all the other coordinates remain stationary. { 'nȯrmǝl ¦mōd əv vī'brā·shən }

normal potassium pyrophosphate See potassium pyrophosphate. { 'nȯr·məl pə'tas·ē·əm ,pī·rō'fäs,fāt }

normal reaction [MECH] The force exerted by a surface on an object in contact with it which prevents the object from passing through the surface; the force is perpendicular to the surface, and is the only force that the surface exerts on the object in the absence of frictional forces. { 'nȯr·məl rē'ak·shən }

normal segregation [MET] Segregation of the lower-melting-point constituents of an alloy primarily near the center of a casting (last portion to solidify). { 'nȯr·məl ,seg·rə'gā·shən }

normal silver sulfate See silver sulfate. { 'nȯr· məl 'sil·vər 'səl,fāt }

normal stress [MECH] The stress component at a point in a structure which is perpendicular to the reference plane. { 'nȯr·məl 'stres }

normal temperature and pressure See standard conditions. { 'nȯr·məl 'tem·prə·chər ən 'presh·ər }

normal thorium sulfate See thorium sulfate. { 'nȯr·məl 'thȯr·ē·əm 'səl,fāt }

normal twin [CRYSTAL] A twin crystal whose twin axis is perpendicular to the composition surface. { 'nȯr·məl 'twin }

north pole [ELECTROMAG] The pole of a magnet at which magnetic lines of force are considered as leaving the magnet; the lines enter the south pole; if the magnet is freely suspended, its north pole points toward the north geomagnetic pole. Also known as positive pole. { 'nȯrth 'pōl }

Notarys-Mercereau microbridge See proximity-effect microbridge. { nō'tar·əs 'mer·sə,rō 'mī· krō,brij }

notch acuity [MET] The severity of the stress concentration produced by a given notch in a structure; it is expressed as the ratio of the notch depth to the notch radius (depth is small compared to width, or diameter, or the narrowest cross section). { 'näch ə'kyü·əd·ē }

notch brittleness [MET] Susceptibility of a material to brittle fracture at areas of stress concentration; in notch tensile testing, a material has notch brittleness if the notch strength lies below the tensile strength. { 'näch 'brid·əl·nəs }

notch depth [MET] The distance from the surface of a metal specimen to the bottom of the notch. { 'näch ,depth }

notch ductility [MET] Percentage reduction in area of the specimen after failure in a notched tensile test. { 'näch dək'til·əd·ē }

notched-bar test [MET] Test in which a notched

metal specimen is bent with the notch in tension. { 'nächt ¦bär ‚test }

notch sensitivity |MET| A measure of the reduction in strength of a metal caused by the presence of a notch. { 'näch ‚sen·sə'tiv·əd·ē }

notch strength |MET| The ratio of maximum tensional load required to fracture a notched specimen to the original minimum cross-sectional area. { 'näch ‚streŋkth }

notch test |MET| A tensile or creep test of a metal to determine the effect of a surface notch. { 'näch ‚test }

novolac resin |POLYM CHEM| Any of the thermoplastic phenol-formaldehyde resins made with an excess of phenol in the reaction; used in varnishes. { 'nō·və‚lak 'rez·ən }

Np See neptunium.

npin transistor |ELECTR| An npn transistor which has a layer of high-purity germanium between the base and collector to extend the frequency range. { 'en‚pin tran'zis·tər }

npnp diode See pnpn diode. { ¦en‚pē¦en‚pē 'dī ‚ōd }

npnp transistor |ELECTR| An npn-junction transistor having a transition or floating layer between p and n regions, to which no ohmic connection is made. Also known as pnpn transistor. { ¦en‚pē¦en‚pē tran'zis·tər }

npn semiconductor |ELECTR| Double junction formed by sandwiching a thin layer of p-type material between two layers of n-type material of a semiconductor. { 'en‚pē'en 'sem·i·kən‚dək·tər }

npn transistor |ELECTR| A junction transistor having a p-type base between an n-type emitter and an n-type collector; the emitter should then be negative with respect to the base, and the collector should be positive with respect to the base. { 'en‚pē'en tran'zis·tər }

np semiconductor |ELECTR| Region of transition between n- and p-type material. { ¦en¦pē 'sem·i·kən‚dək·tər }

NR See nitrile rubber.

n-type germanium |ELECTR| Germanium to which more impurity atoms of donor type (with valence 5, such as antimony) than of acceptor type (with valence 3, such as indium) have been added, with the result that the conduction electron density exceeds the hole density. { 'en ‚tīp jər'mā·nē·əm }

n-type semiconductor |ELECTR| An extrinsic semiconductor in which the conduction electron density exceeds the hole density. { 'en ‚tīp 'sem·i·kən‚dak·tər }

nuclear |CHEM| Pertaining to a group of atoms joined directly to the central group of atoms or central ring of a molecule. { 'nü·klē·ər }

nuclear adiabatic demagnetization |CRYO| A technique for cooling substances, in which the sample is first cooled to temperatures on the order of 10^{-2} K in an extremely intense magnetic field and is then thermally isolated and removed from the field to reach temperatures on the order of 10^{-6} K. { 'nü·klē·ər ‚ad·ē·ə'bad·ik dē ‚mag·nə·tə'zā·shən }

nucleated glass |MATER| Glass treated with a nucleating agent to transform it into a crystalline material. { 'nü·klē·ād·əd 'glas }

nucleic acid |BIOMATER| A large, acidic, chainlike molecule containing phosphoric acid, sugar, and purine and pyrimidine bases; two types are ribonucleic acid and deoxyribonucleic acid. { nü'klē·ik 'as·əd }

nucleophile |PHYS CHEM| A species possessing one or more electron-rich sites, such as an unshared pair of electrons, the negative end of a polar bond, or pi electrons. Also known as electron donor. { 'nü·klē·ə‚fīl }

nugget |MET| A weld bead. { 'nəg·ət }

Nusselt equation |THERMO| Dimensionless equation used to calculate convection heat transfer for heating or cooling of fluids outside a bank of 10 or more rows of tubes to which the fluid flow is normal. { 'nús·əlt i‚kwā·zhən }

Nusselt number |PHYS| A dimensionless number used in the study of mass transfer, equal to the mass-transfer coefficient times the thickness of a layer through which mass transfer is taking place divided by the molecular diffusivity. Symbolized Nu_m; N_{Nu_m}. Also known as Sherwood number (N_{Sh}). |THERMO| A dimensionless number used in the study of forced convection which gives a measure of the ratio of the total heat transfer to conductive heat transfer, and is equal to the heat-transfer coefficient times a characteristic length divided by the thermal conductivity. Symbolized N_{Nu}. { 'nús·əlt ‚nəm·bər }

nylon |POLYM CHEM| Generic name for longchain polymeric amide molecules in which recurring amide groups are part of the main polymer chain; used to make fibers, fabrics, sheeting, and extruded forms. { 'nī‚län }

O

O See oxygen.

oakum |MATER| Old hemp or jute fiber, loosely twisted and impregnated with tar or a tar derivative, used to caulk sides and decks of ships and to pack joints of pipes and caissons. { 'ōk·əm }

oatmeal paper |MATER| A paper in which fine sawdust is added to produce a sheet with a coarse texture; it can be used as an inexpensive sketching paper for work in pastels and charcoal, or as wallpaper. { 'ōt,mēl ,pā·pər }

obscure glass |MATER| Translucent glass. { əb'skyūr 'glas }

obtuse bisectrix |CRYSTAL| The bisectrix of the obtuse angle between the axes of a biaxial crystal. { äb'tüs bī'sek·triks }

occlusion |ENG| The retention of undissolved gas in a solid during solidification. |PHYS| Adhesion of gas or liquid on a solid mass, or the trapping of a gas or liquid within a mass. { ə'klü·zhən }

octahedral cleavage |CRYSTAL| Crystal cleavage in the four planes parallel to the face of the octahedron. { ¦äk·tə¦hē·drəl 'klē·vij }

octahedral plane |CRYSTAL| The plane in a cubic lattice having three numerically equal Miller indices. { ¦äk·tə¦hē·drəl 'plān }

octet rule |CHEM| A concept of chemical bonding theory based on the assumption that in the formation of compounds, atoms exhibit a tendency for their valence shells either to be empty or to have a full complement of eight electrons (octet); for some elements there are more than the usual eight valence electrons in some of their compounds. { äk,tet ,rül }

odorant |MATER| Material added to odorless fuel gases to give them a distinctive odor for safety purposes; usually a sulfur- or mercaptan-containing compound. Also known as malodorant; stench; warning agent. { 'ō·də·rənt }

offset |MECH| The value of strain between the initial linear portion of the stress-strain curve and a parallel line that intersects the stress-strain curve of an arbitrary value of strain; used as an index of yield stress; a value of 0.2% is common. { 'óf,set }

offset paper |MATER| Paper with a certain degree of porosity as a result of coating with an alkali-swelling resin; used for offset printing. { 'óf,set ,pā·pər }

offset yield strength |MECH| That stress at which the strain surpasses by a specific amount (called the offset) an extension of the initial proportional portion of the stress-strain curve; usually expressed in pounds per square inch. { 'óf,set 'yēld ,streŋkth }

off time |MET| In resistance welding, usually in repetitive cycles, the time that the electrode is not in contact with the work. { 'óf ,tīm }

ogdosymmetric class |CRYSTAL| A merohedral crystal class whose general form has one-eighth the number of equivalent faces of the corresponding holohedral form. { ¦äg·dō·si'me·trik 'klas }

ohm |ELEC| The unit of electrical resistance in the rationalized meter-kilogram-second system of units, equal to the resistance through which a current of 1 ampere will flow when there is a potential difference of 1 volt across it. Symbolized Ω. { ōm }

ohmic |ELEC| Pertaining to a substance or circuit component that obeys Ohm's law. { 'ō·mik }

ohmic contact |ELEC| A region where two materials are in contact, which has the property that the current flowing through it is proportional to the potential difference across it. { 'ō·mik 'kän ,takt }

ohmic dissipation |ELECTR| Loss of electric energy when a current flows through a resistance due to conversion into heat. Also known as ohmic loss. { 'ō·mik ,dis·ə'pā·shən }

ohmic loss See ohmic dissipation. { 'ō·mik 'lós }

ohmic resistance |ELEC| Property of a substance, circuit, or device for which the current flowing through it is proportional to the potential difference across it. { 'ō·mik ri'zis·təns }

Ohm's law |ELEC| The law that the direct current flowing in an electric circuit is directly proportional to the voltage applied to the circuit; it is valid for metallic circuits and many circuits containing an electrolytic resistance. { 'ōmz ,ló }

oil |MATER| Any of various viscous, combustible, water-immiscible liquids that are soluble in certain organic solvents, as ether and naphtha; may be of animal, vegetable, mineral, or synthetic origin; examples are fixed oils, volatile or essential oils, and mineral oils. { óil }

oil asphalt |MATER| Water-insoluble, heavy black

residue left after removing the tar tailings during the distillation of petroleum; used in roofing, paints, and coatings. { 'òil 'as‚fólt }

oil bath |ENG| **1.** Oil, in a container, within which a mechanism works or into which it dips. **2.** Oil in which a piece of apparatus is submerged. **3.** Oil that is poured on a cutting tool. |MET| Oil used in tempering. { 'òil ‚bath }

oil blue |INORG CHEM| Violet-blue copper sulfide pigment used in varnishes. { 'òil 'blü }

oil core |MET| A core in which sand is held together by an oil binder. { 'òil ‚kòr }

oil emulsion |MATER| Suspension of oil droplets in another liquid in which the oil is insoluble. { 'òil i‚məl·shən }

oil-extended rubber |MATER| Synthetic rubber into which 25–50% of a petroleum oil emulsion has been incorporated to decrease cost and increase low-temperature flexibility and resilience. { 'òil ik¦stend·əd 'rəb·ər }

oil hardening |MET| Quenching of carbon steel in an oil bath; the steel cools slowly, and a more uniform and desirable hardness is attained. { 'òil ‚härd·ən·iŋ }

oil length |MATER| The ratio of oil to resin in varnish; expressed as gallons of oil per 100 pounds (45.3 kilograms) of resin. { 'òil ‚leŋkth }

oil of vitriol See sulfuric acid. { 'òil əv 'vit·rē‚òl }

oil paint |MATER| A paint made with a vegetable oil as the filmogen. { 'òil ‚pānt }

oil-soluble resin |MATER| A resin that, at moderate temperatures, dissolves in, disperses in, or reacts with drying oils to produce a homogeneous film of modified characteristics. { 'òil ¦säl·yə·bəl 'rez·ən }

oil stain |MATER| Thin oil paint with very little pigment, used to stain wood surfaces. { 'òil ‚stān }

oilstone |MATER| A whetstone used with oil. { 'òil‚stōn }

oil varnish |MATER| A varnish composed of resins dissolved in oil. { 'òil ‚vär·nish }

oil-well cement |MATER| A type of hydraulic cement which has a slow setting rate under the high temperatures obtained in oil wells; uses include support of tubing and bypassing of unwanted zones. { 'òil ‚wel si‚ment }

oiticica oil |MATER| A light-yellowish oil obtained from seeds of the Brazilian oiticica tree (*Licania rigida*); raw oil becomes buttery unless heat-treated (semipolymerized); used principally in paint and varnish as a drying oil as a substitute for tung oil or with tung oil. { ‚oid·ə'sē·kə ‚òil }

okonite |MATER| Insulating material made from the vulcanization of ozokerite and resin with rubber and sulfur. { ō·kə‚nīt }

olefiant gas See ethylene. { ¦ō·lə¦fī·ənt 'gas }

olefin |ORG CHEM| C_nH_{2n} A family of unsaturated, chemically active hydrocarbons with one carbon-carbon double bond; includes ethylene and propylene. |TEXT| A manufactured fiber in which the fiber-forming substance is any long-chain synthetic polymer composed of at least 85% by weight of ethylene, propylene, or other

olefin units except amorphous (noncrystalline) polyolefins qualifying as rubber. { 'ō·lə‚fən }

olefin copolymer |POLYM CHEM| Polymer made by the interreaction of two or more kinds of olefin monomers, such as butylene and propylene. { 'ō·lə·fən kō'päl·ə·mər }

olefin resin |POLYM CHEM| Long-chain polymeric material produced by the chain reaction of olefinic monomers, such as polyethylene from ethylene, or polypropylene from propylene. { 'ō·lə·fən 'rez·ən }

oleoresin |MATER| A resin-essential oil mixture with pungent taste; extracted from various plants; used in pharmaceutical preparations; examples are Peru, tulu, and styrax balsams. { ¦ō·ō'rez·ən }

oleoresinous varnish |MATER| A varnish made by compounding the resin with oxidizable oil, such as linseed oil. { ¦ō·lē·ō'rez·ən·əs 'vär·nish }

oleostearin |MATER| Edible solid fat from tissues of cattle (genus *Bos*); the solid remaining after oleo oil or tallow oil is removed from tallow. Also known as beef stearin. { ¦ō·lē·ō'stir·ən }

oleum |CHEM| Latin name for oil. |INORG CHEM| See fuming sulfuric acid. { 'ō·lē·əm }

oleum gossypii seminis See cotton oil. { 'ō·lē·əm gä'sip·ē‚ī 'sem·ə·nəs }

oligomer |POLYM CHEM| A molecule made up of a relatively small number of monomer units. { ə'lig·ə·mər }

oligopeptide |ORG CHEM| A peptide composed of no more than 10 amino acids. { ‚äl·ə·gō'pep‚tīd }

Olsen ductility test |MET| A cupping test in which a piece of sheet metal is deformed at the center by a steel ball until fracture occurs; ductility is measured by the height of the cup at the time of failure. { 'ōl·sən dək'til·ə·dē ‚test }

omission solid solution |CRYSTAL| A crystal with certain atomic sites incompletely filled. { ō'mish·ən ¦säl·əd sə'lü·shən }

on composition See on grade. { ¦òn ‚käm·pə'zish·ən }

one-dimensional lattice |CRYSTAL| A simplified model of a crystal lattice consisting of particles lying along a straight line at either equal or periodically repeating distances. { 'wən di‚men·chən·əl 'lad·əs }

one-face-centered lattice |CRYSTAL| A crystal lattice in which there are lattice points at the centers of one pair of faces in each unit cell as well as at the corners. { 'wən ¦fās ¦sen·tərd 'lad·əs }

on grade |MET| A classification for an alloy indicating that it has a chemical composition within the range of the specified composition limits for that particular alloy. Also known as on composition. { 'òn 'grād }

onionskin paper |MATER| A lightweight, durable bond paper; usually quite translucent, resembling the dry outer skin of an onion; used for duplicate typewriter copies and in interleaving order books. { 'ən‚yən‚skin 'pā·pər }

Onsager reciprocal relations |THERMO| A set of conditions which state that the matrix, whose elements express various fluxes of a system (such as diffusion and heat conduction) as linear functions of the various conjugate affinities (such as mass and temperature gradients) for systems close to equilibrium, is symmetric when certain definitions are chosen for these fluxes and affinities. { 'ȯn₁säg·ər ri'sip·rə·kəl ri'lā·shənz }

Onsager theory of dielectrics |ELEC| A theory for calculating the dielectric constant of a material with polar molecules in which the local field at a molecule is calculated for an actual spherical cavity of molecular size in the dielectric using Laplace's equation, and the polarization catastrophe of the Lorentz field theory is thereby avoided. { 'ȯn₁säg·ər ₁thē·ə·rē əv ₁dī·ə'lek·triks }

opacifier |MATER| A material used in ceramic glazes and vitreous enamels to render them nontransparent and to improve other properties. { ō'pas·ə₁fī·ər }

opacity |OPTICS| The light flux incident upon a medium divided by the light flux transmitted by it. { ō'pas·əd·ē }

opalescence |OPTICS| The milky, iridescent appearance of a dense, transparent medium or colloidal system when it is illuminated by polychromatic radiation in the visible range, such as sunlight. { ₁ō·pə'les·əns }

opal glass |MATER| Translucent or opaque glass, often milky white, made by adding impurities such as fluorine compounds to the melt; it appears white by reflected light but shows color images through thin sections; used for ornamental glass and as an efficient light diffuser. { 'ō·pəl ₁glas }

opaque medium |OPTICS| A medium impervious to rays of light, that is, not transparent to the human eye. |PHYS| **1.** A medium which does not transmit electromagnetic radiation of a specified type, such as that in the infrared, x-ray, ultraviolet, and microwave regions. **2.** A medium which prevents the passage of particles of a specified type. { ō'pāk ₁mēd·ē·əm }

open |ELEC| **1.** Condition in which conductors are separated so that current cannot pass. **2.** Break or discontinuity in a circuit which can normally pass a current. { 'ō·pən }

open-arc furnace |MET| An electrosmelting furnace in which the arc is generated above the level of the furnace feed. { 'ō·pən ¦ärk ₁fər·nəs }

open-cell foam |MATER| Foamed material, natural or synthetic, rigid or flexible, organic or metallic, in which there is interconnection between the cells. { 'ō·pən ₁sel 'fōm }

open-circuit potential |PHYS CHEM| Steady-state or equilibrium potential of an electrode in absence of external current flow to or from the electrode. { 'ō·pən ¦sər·kət pə'ten·chəl }

open cycle |THERMO| A thermodynamic cycle in which new mass enters the boundaries of the system and spent exhaust leaves it; the automotive engine and the gas turbine illustrate this process. { ¦ō·pən ¦sī·kəl }

open die |MET| A forming or forging die in which there is little or no restriction to the lateral flow of metal within the die set. { 'ō·pən 'dī }

open-die forging |MET| Forging performed with open dies. { 'ō·pən ¦dī 'fȯrj·iŋ }

open form |CRYSTAL| A crystal form in which the crystal faces do not entirely enclose a space. { 'ō·pən ₁fȯrm }

open-hearth furnace |MET| A reverberatory smelting furnace with a shallow hearth and a low roof, in which the charge is heated both by direct flame and by radiation from the roof and walls of the furnace. { 'ōpən ¦härth 'fər·nəs }

open-hearth process |MET| A steel-making process carried out in an open-hearth furnace in which selected pig iron and malleable scrap iron are melted, with the addition of pure iron ore. { ¦ō·pən ¦härth ₁prä·səs }

opening material |MATER| A material added to plastic clay in ceramic making to speed drying and reduce shrinking. { 'ōp·ə·niŋ mə₁tir·ē·əl }

open-packed structure |CRYSTAL| A crystal structure corresponding to the stacking of spheres in an orthogonal arrangement so that each sphere is in contact with six others. { ¦ō·pən ¦pakt 'strək·chər }

open system |THERMO| A system across whose boundaries both matter and energy may pass. { 'ō·pən 'sis·təm }

operating stress |MECH| The stress to which a structural unit is subjected in service. { 'äp·ə ₁rād·iŋ ₁stres }

optic |OPTICS| Pertaining to the lenses, prisms, and mirrors of a camera, microscope, or other conventional optical instrument. { 'äp·tik }

optical |OPTICS| Pertaining to or utilizing visible or near-visible light; the extreme limits of the optical spectrum are about 100 nanometers (0.1 micrometer or 3×10^{15} hertz) in the far ultraviolet and 30,000 nanometers (30 micrometers or 10^{13} hertz) in the far infrared. { 'äp·tə·kəl }

optical activity |OPTICS| The behavior of substances which rotate the plane of polarization of plane-polarized light, as it passes through them. Also known as rotary polarization. { 'äp·tə·kəl ak'tiv·əd·ē }

optical anisotropy |OPTICS| The behavior of a medium, or of a single molecule, whose effect on electromagnetic radiation depends on the direction of propagation of the radiation. { 'äp·tə·kəl ₁an·ə'sä·trə·pē }

optical antipode See enantiomorph. { 'äp·tə· kəl 'ant·i₁pōd }

optical branch |SOLID STATE| The vibrations of an optical mode plotted on a graph of frequency versus wave number; it is separated from, and has higher frequencies than, the acoustic branch. { 'äp·tə·kəl 'branch }

optical coating |OPTICS| Either a mirror coating, or a film of the proper thickness and refractive index applied to the air-glass surface of a lens to reduce reflection. { 'äp·tə·kəl 'kōd·iŋ }

optical crystal |CRYSTAL| Any natural or synthetic crystal, such as sodium chloride, calcium fluoride, silver chloride, potassium iodide, or stilbene, that is used in infrared and ultraviolet optics and for its piezoelectric effects. { 'äp·tə·kəl 'krist·əl }

optical density

optical density |OPTICS| The degree of opacity of a translucent medium expressed by $\log I_0/I$, where I_0 is the intensity of the incident ray, and I is the intensity of the transmitted ray. Abbreviated OD. { 'äp·tə·kəl 'den·səd·ē }

optical dispersion |OPTICS| Separation of different colors of light such as occurs when it passes from one medium to another or is reflected from a diffraction grating. { 'äp·tə·kəl di'sper·shən }

optical fiber |OPTICS| A long, thin thread of fused silica, or other transparent substance, used to transmit light. { 'äp·tə·kəl 'fī·bər }

optical glass |MATER| A type of glass which is free from imperfections, such as unmelted particles, bubbles, and chemical inhomogeneities, which would affect its transmission of light. { 'äp·tə·kəl 'glas }

optical harmonic |SOLID STATE| Light, generated by passing a laser beam with a power density on the order of 10^{10} watts per square centimeter or more through certain transparent materials, which has a frequency which is an integral multiple of that of the incident laser light. { 'äp·tə·kəl här'män·ik }

optical isomer See enantiomorph. { 'äp·tə·kəl 'ī·sə·mər }

optical lithography |ELECTR| Lithography in which an integrated circuit pattern is first created on a glass plate or mask and is then transferred to the resist by one of a number of optical techniques by using visible or ultraviolet light. { 'äp·tə·kəl li'thäg·rə·fē }

optically pumped laser |OPTICS| A laser that uses absorption of light from an auxiliary light source to excite electrons into an upper energy state. { 'äp·tə·klē ¦pəmpt 'lā·zər }

optical mask |ELECTR| A thin sheet of metal or other substance containing an open pattern, used to suitably expose to light a photoresistive substance overlaid on a semiconductor or other surface to form an integrated circuit. { 'äp·tə·kəl 'mask }

optical material |MATER| A material which is transparent to light or to infrared, ultraviolet, or x-ray radiation, such as glass and certain single crystals, polycrystalline materials (chiefly for the infrared), and plastics. { 'äp·tə·kəl mə'tir·ē·əl }

optical mode |SOLID STATE| A type of vibration of a crystal lattice whose frequency varies with wave number only over a limited range, and in which neighboring atoms or molecules in different sublattices move in opposition to each other. { 'äp·tə·kəl ˌmōd }

optical phonon |SOLID STATE| A quantum of an optical mode of vibration of a crystal lattice. { 'äp·tə·kəl 'fō,nän }

optical plastic |MATER| A plastic which is transparent to light, occasionally used in optical systems for reasons of economy, special index-dispersion relation, light weight, and non-brittleness. { 'äp·tə·kəl 'plas·tik }

optical prism See prism. { 'äp·tə·kəl 'priz·əm }

optical pyrometer |ENG| An instrument which determines the temperature of a very hot surface from its incandescent brightness; the image of the surface is focused in the plane of an electrically heated wire, and current through the wire is adjusted until the wire blends into the image of the surface. { 'äp·tə·kəl pī'räm·əd·ər }

optical rotation |OPTICS| Rotation of the plane of polarization of plane-polarized light, or of the major axis of the polarization ellipse of elliptically polarized light by transmission through a substance or medium. { 'äp·tə·kəl rō'tā·shən }

optical rotatory dispersion |OPTICS| Specific rotation, considered as a function of wavelength. Abbreviated ORD. { 'äp·tə·kəl 'rōd·ə ˌtör·ē di'spər·zhən }

optical thickness |OPTICS| The thickness of an optical material times its index of refraction. { 'äp·tə·kəl 'thik·nəs }

optical twinning |CRYSTAL| Growing together of two crystals which are the same except that the structure of one is the mirror image of the structure of the other. Also known as chiral twinning. { 'äp·tə·kəl 'twin·iŋ }

optic angle See axial angle. { 'äp·tik ,aŋ·gəl }

optic-axial angle See axial angle. { 'äp·tik ¦ak·sē·əl ,aŋ·gəl }

optics |PHYS| **1.** Narrowly, the science of light and vision. **2.** Broadly, the study of the phenomena associated with the generation, transmission, and detection of electromagnetic radiation in the spectral range extending from the longwave edge of the x-ray region to the shortwave edge of the radio region, or in wavelength from about 1 nanometer to about 1 millimeter. { 'äp·tiks }

optoelectronics |ELECTR| **1.** The branch of electronics that deals with solid-state and other electronic devices for generating, modulating, transmitting, and sensing electromagnetic radiation in the ultraviolet, visible-light, and infrared portions of the spectrum. **2.** See photonics. { ¦äp·tō·i,lek'trän·iks }

OPW method See orthogonalized plane-wave method. { ¦ō¦pē'dəb·əl,yü ,meth·əd }

orange |OPTICS| The hue evoked in an average observer by monochromatic radiation having a wavelength in the approximate range from 597 to 622 nanometers; however, the same sensation can be produced in a variety of other ways. { 'är·inj }

orange lake |MATER| Any of various transparent orange pigments from the precipitation of an orange dyestuff on aluminum hydrate or other base; used to produce transparent coatings for metal cans and bottle caps. { 'är·inj 'lāk }

orange mineral |MATER| A bright orange-red lead oxide pigment used in printing inks and primers. { 'är·inj 'min·rəl }

orange oxide See uranium trioxide. { 'är·inj 'äk ,sīd }

orange peel |MATER| A pebbled film surface, resembling an orange skin, on lacquer or enamel as a result of too rapid drying after spraying, or failure to exhibit the desired leveling effects. |MET| A rough, pebble-grained metal surface resulting from either plastic deformation or electropolishing. Also known as alligator effect; pebbling. { 'är·inj ¸pēl }

orchil [MATER] Dark-brownish-red coloring matter derived from lichens as paste or aqueous extract; main components are orcin and orcein; used as carpet-yarn dye. Also known as orseille. { 'ôr·chəl }

order [PHYS] A range of magnitudes of a quantity (and of all other quantities having the same physical dimensions) extending from some value of the quantity to some small multiple of the quantity (usually 10). Also known as order of magnitude. { 'ôrd·ər }

order-disorder transition [SOLID STATE] The transition of an alloy or other solid solution between a state in which atoms of one element occupy certain regular positions in the lattice of another element, and a state in which this regularity is not present. { 'ôrd·ər 'dis¸ôrd·ər tran'zish·ən }

ordering [SOLID STATE] A solid-state transformation in certain solid solutions, in which a random arrangement in the lattice is transformed into a regular ordered arrangement of the atoms with respect to one another; a so-called superlattice is formed. { 'ôrd·ə·riŋ }

order of magnitude See order. { 'ôrd·ər əv 'mag·nə¸tüd }

order of phase transition [THERMO] A phase transition in which there is a latent heat and an abrupt change in properties, such as in density, is a first-order transition; if there is not such a change, the order of the transition is one greater than the lowest derivative of such properties with respect to temperature which has a discontinuity. { 'ôrd·ər əv 'fāz tran¸zish·ən }

ore flotation promoter [MATER] Material that gives a water-repellent surface to mineral particles so that air bubbles will adhere and cause selective flotation. { 'ôr flō'tā·shən prə¸mōd·ər }

oreide bronze [MET] A series of brass compositions containing 68–87% copper, 10–32% zinc, and sometimes small amounts of tin; used for hardware. { 'ō·rē¸īd 'bränz }

organic [ORG CHEM] Of chemical compounds, based on carbon chains or rings and also containing hydrogen with or without oxygen, nitrogen, or other elements. { ôr'gan·ik }

organic acid [ORG CHEM] A chemical compound with one or more carboxyl radicals (COOH) in its structure; examples are butyric acid, $CH_3(CH_2)_2COOH$, maleic acid, HOOC-CHCHCOOH, and benzoic acid, C_6H_5COOH. { ôr'gan·ik 'as·əd }

organic coating [MATER] Material used to protect metal surfaces from chemical or atmospheric attack; includes latex paints, plastics, asphaltic materials, rubbers, and elastomers. { ôr'gan·ik 'kōd·iŋ }

organic conductor [MATER] A two-component material containing anion and cation charged species originating from a charge transfer between two inorganic molecules or between one organic molecule and one inorganic ion. { ôr ¦gan·ik kən'dək·tər }

organic glass [MATER] An amorphous, solid, glasslike material made of transparent plastic. { ôr'gan·ik 'glas }

organic pigment [ORG CHEM] Any of the materials with organic-chemical bases used to add color to dyes, plastics, linoleum, tones, and lakes. { ôr'gan·ik 'pig·mənt }

organic salt [ORG CHEM] The reaction product of an organic acid and an inorganic base, for example, sodium acetate (CH_3COONa) from the reaction of acetic acid (CH_3COOH) and sodium hydroxide (NaOH). { ôr'gan·ik 'sôlt }

organic semiconductor [MATER] An organic material having unusually high conductivity, often enhanced by the presence of certain gases, and other properties commonly associated with semiconductors; an example is anthracene. { ôr'gan·ik 'sem·i·kən¸dək·tər }

organic solvent [ORG CHEM] Liquid organic compound with the power to dissolve solids, gases, or liquids (miscibility); examples are methanol (methyl alcohol), CH_3OH, and benzene, C_6H_6. { ôr'gan·ik 'säl·vənt }

organometallic compound [ORG CHEM] Molecules containing carbon-metal linkage; a compound containing an alkyl or aryl radical bonded to a metal, such as tetraethyllead, $Pb(C_2H_5)_4$. { ôr¦gan·ə·mə'tal·ik 'käm¸paúnd }

organometallic polymer [POLYM CHEM] A special class of inorganic polymer in which the metal, metalloid, or other element is covalently bonded to carbon. { ôr¦gan·ə·mə'tal·ik 'päl·ə·mər }

organosol [MATER] **1.** Finely divided or colloidal suspension of insoluble material in a suspending organic liquid; known as plastisol when the solid is a synthetic resin suspended in an organic liquid; used for coatings, moldings, and casting of films. **2.** A dispersion of very finely divided resin particles that are suspended in an organic-liquid mixture which cannot dissolve the resin at normal temperatures. { ôr'gan·ə¸sòl }

oriental linaloe [MATER] A rosewood oil distilled from highly perfumed parts of *Aquilaria agollocha* trees of Burma, eastern India, and Java. Also known as agar attar; aloe wood oil. { ¸ôr·ē'ent·əl lə'nal·ō }

orientation [CRYSTAL] The directions of the axes of a crystal lattice relative to the surfaces of the crystal, to applied fields, or to some other planes or directions of interest. { ¸ôr·ē·ən'tā·shən }

orientation effect [ELEC] Those bulk properties of a material which result from orientation polarization. { ¸ôr·ē·ən'tā·shən i¸fekt }

orientation polarization [ELEC] Polarization arising from the orientation of molecules which have permanent dipole moments arising from an asymmetric charge distribution. Also known as dipole polarization. { ¸ôr·ē·ən'tā·shən ¸pō·lə·rə¸zā·shən }

orifice gas [MET] The torch gas in a plasma arc welding or cutting process which becomes ionized in the arc to form plasma and is ejected from the orifice in a jet stream. { 'ôr·ə·fəs 'gas }

Orr's white See lithopone. { 'ôrz 'wīt }

orseille See orchil. { ôr'sā }

orthoarsenic acid See arsenic acid. { ¦ȯr·thō·är¦sen·ik 'as·əd }

orthoaxis |CRYSTAL| The diagonal or lateral axis perpendicular to the vertical axis in the monoclinic system. { ¦ȯr·thō'ak·səs }

orthoboric acid See boric acid. { ¦ȯr·thə¦bȯr·ik 'as·əd }

orthogonal crystal |CRYSTAL| A crystal whose axes are mutually perpendicular. { ȯr'thäg·ən·əl 'krist·əl }

orthogonalized plane-wave method |SOLID STATE| A method of approximating the energy states of electrons in a crystal lattice: trial wave functions (the orthogonalized plane waves) are constructed which are linear combinations of plane waves and Bloch functions based on core states, and which are orthogonal to the Bloch functions, and linear combinations of these trial functions are then determined by the variational method. Abbreviated OPW method. { ȯr¦thäg·ən·əl,īzd 'plān ,wāv ,meth·əd }

orthographic projection |CRYSTAL| A projection for displaying the poles of a crystal in which the poles are projected from a reference sphere onto an equatorial plane by dropping perpendiculars from the poles to the plane. { ¦ȯr·thə¦graf·ik prə'jek·shən }

orthohexagonal axes |CRYSTAL| A set of crystallographic axes, two of which have a fixed ratio, as in hexagonal or trigonal crystals. { ¦ȯr·thō·hek'säg·ən·əl 'ak,sēz }

Orthonik |MET| A magnetic alloy composed of 45–50% nickel, with the remainder iron, having a grainy structure, high permeability, and a rectangular hysteresis loop; used in magnetic cores. { 'ȯr·thə,nik }

orthophosphate |INORG CHEM| One of the possible salts of orthophosphoric acid; the general formula as in M_3PO_4, where M may be potassium as in potassium orthophosphate, K_3PO_4. { ¦ȯr·thə'fäs,fāt }

orthophosphoric acid See phosphoric acid. { ¦ȯr·thə·fäs'fȯr·ik 'as·əd }

orthopinacoid See front pinacoid. { ¦ȯr·thə 'pin·ə,kȯid }

orthorhombic lattice |CRYSTAL| A crystal lattice in which the three axes of a unit cell are mutually perpendicular, and no two have the same length. Also known as rhombic lattice. { ¦ȯr·thə¦räm·bik 'lad·əs }

orthorhombic system |CRYSTAL| A crystal system characterized by three axes of symmetry that are mutually perpendicular and of unequal length. Also known as rhombic system. { ¦ȯr·thə ¦räm·bik 'sis·təm }

orthosymmetric crystal |CRYSTAL| A crystal that has orthorhombic symmetry. { ¦ȯr·thō· si'me·trik 'krist·əl }

orthotropic |MECH| Having elastic properties such as those of timber, that is, with considerable variations of strength in two or more directions perpendicular to one another. { ¦ȯr·thə¦träp·ik }

orthotungstic acid See tungstic acid. { ¦ȯr· thō'təŋ·stik 'as·əd }

Os See osmium.

oscillating magnetic field |ELECTROMAG| A magnetic field which varies periodically in time. { 'äs·ə,lād·iŋ mag'ned·ik 'fēld }

oscillation |PHYS| Any effect that varies periodically back and forth between two values. { ,äs·ə'lā·shən }

oscillation photography |SOLID STATE| A method of x-ray diffraction analysis in which a single crystal is made to oscillate through a small angle about an axis perpendicular to a beam of monochromatic x-rays or particles. { ,äs·ə ¦lā·shən fə¦täg·rə·fē }

oscillator |ELECTR| **1.** An electronic circuit that converts energy from a direct-current source to a periodically varying electric output. **2.** The stage of a superheterodyne receiver that generates a radio-frequency signal of the correct frequency to mix with the incoming signal and produce the intermediate-frequency value of the receiver. **3.** The stage of a transmitter that generates the carrier frequency of the station or some fraction of the carrier frequency. |PHYS| Any device (mechanical or electrical) which, in the absence of external forces, can have a periodic back-and-forth motion, the frequency determined by the properties of the oscillator. { 'äs·ə,lād· ər }

oscillatory twinning |CRYSTAL| Repeated, parallel twinning. { 'äs·ə·lə,tȯr·ē 'twin·iŋ }

osmate |INORG CHEM| A salt or ester of osmic acid, containing the osmate radical, $OsO_4{}^{2-}$; for example, potassium osmate (K_2OsO_4). { 'äz·māt }

osmic acid anhydride |INORG CHEM| OsO_4 Poisonous yellow crystals with disagreeable odor; melts at 40°C; soluble in water, alcohol, and ether; used in medicine, photography, and catalysis. Also known as osmium oxide; osmium tetroxide. { 'äz·mik 'as·əd an'hī,drīd }

osmium |CHEM| A chemical element, symbol Os, atomic number 76, atomic weight 190.2. |MET| A hard white metal of rare natural occurrence. { 'äz·mē·əm }

osmium oxide See osmic acid anhydride. { 'äz·mē·əm 'äk,sīd }

osmium tetroxide See osmic acid anhydride. { 'äz·mē·əm te'träk,sīd }

osmosis |PHYS CHEM| The transport of a solvent through a semipermeable membrane separating two solutions of different solute concentration, from the solution that is dilute in solute to the solution that is concentrated. { ä'smō·səs }

osmotic gradient See osmotic pressure. { äz'mäd·ik 'grād·ē·ənt }

osmotic pressure |PHYS CHEM| **1.** The applied pressure required to prevent the flow of a solvent across a membrane which offers no obstruction to passage of the solvent, but does not allow passage of the solute, and which separates a solution from the pure solvent. **2.** The applied pressure required to prevent passage of a solvent across a membrane which separates solutions of different concentration, and which allows passage of the solute, but may also allow limited passage of the solvent. Also known as osmotic gradient. { äz'mäd·ik 'presh·ər }

osseine [MATER] The organic residue formed when bone is dissolved in hydrochloric acid; used to make glue and gelatin. { 'äs·ē·ən }

Ostwald's adsorption isotherm [THERMO] An equation stating that at a constant temperature the weight of material adsorbed on an adsorbent dispersed through a gas or solution, per unit weight of adsorbent, is proportional to the concentration of the adsorbent raised to some constant power. { 'ost,välts ad'sörp·shən 'ī·sə ,thərm }

Otto cycle [THERMO] A thermodynamic cycle for the conversion of heat into work, consisting of two isentropic phases interspersed between two constant-volume phases. Also known as spark-ignition combustion cycle. { 'äd·ō ,sī·kəl }

ounce metal [MET] An alloy composed of 1 ounce each of lead, tin, and zinc to 1 pound of copper. Also known as composition metal. { 'aúns ,med·əl }

ouricury wax [MATER] A hard brown wax obtained from leaves of the ouricury palm (*Cocos coronapa*); similar to carnauba wax in use and properties. { ¦úr·ə·kə¦rē ,waks }

outer-shell electron See conduction electron. { 'aúd·ər ¦shel i'lek,trän }

outgassing [ENG] The release of adsorbed or occluded gases or water vapor, usually by heating, as from a vacuum tube or other vacuum system. { 'aút,gas·iŋ }

overaging [MET] Aging at a higher temperature or for a longer time than is required to produce maximum or optimum properties. { ¦ō·vər ¦āj·iŋ }

overbending [MET] Compensation for springback in metalforming by bending the material through a greater arc than that required for the finished part. { ¦ō·vər¦bend·iŋ }

overcuring [CHEM ENG] A condition resulting from vulcanizing longer than necessary to achieve full development of physical strength; causes softness or brittleness and impaired age-resisting quality of the material. { 'ō·vər ,kyúr·iŋ }

overdraft [MET] Upward curving of a piece of metal after leaving the rolls during forming, due to higher speed of the lower roll. { 'ō·vər,draft }

overglaze color [MATER] Any of the mixtures of ground pigment and low-melting glass melting at 704–816°C; used for decorative designs fired onto china and ceramics. { 'ō·vər,glāz ,kəl·ər }

overgrowth [CRYSTAL] A crystal growth in optical and crystallographic continuity around another crystal of different composition. { 'ō·vər,grōth }

overhauling [MET] Removing scale and surface defects from metal castings or slabs by cutting away surface layers. { 'ō·vər,hól·iŋ }

overhead position [MET] In welding, the position by which the deposit is made from the underside of the joint. { 'ō·vər,hed pə,zish·ən }

overheat [MET] To heat a metal or alloy to such high temperatures that its physical properties are impaired. { ¦ō·vər¦hēt }

overheating effect [SOLID STATE] The effect whereby, under certain conditions, a superconductor can be heated above its critical temperature without losing superconductivity. { ,ō·vər'hēd·iŋ i,fekt }

overlap [MET] **1.** Projection of the weld metal beyond the bond at the toe of the weld. **2.** Extension of one sheet over another in spot, seam, or projection welding. { 'ō·vər,lap }

overlay plywood [MATER] Plywood with a resin-treated fiber surface. { 'ō·vər,lā 'plī,wúd }

overpotential See overvoltage. { ¦ō·vər·pə'ten·chəl }

oversize powder [MET] A metal powder having coarser particles than the maximum permitted. { 'ō·vər,sīz ,paúd·ər }

overvoltage [PHYS CHEM] The difference between electrode potential under electrolysis conditions and the thermodynamic value of the electrode potential in the absence of electrolysis for the same experimental conditions. Also known as overpotential. { ¦ō·vər¦vōl·tij }

oxalyl chloride [INORG CHEM] (COCl)$_2$ Toxic, colorless liquid boiling at 64°C; soluble in ether, benzene, and chloroform; used as a chlorinating agent and for military poison gas. { 'äk·sə,lil 'klór,īd }

oxidant See oxidizing agent. { 'äk·səd·ənt }

oxidation [CHEM] **1.** A chemical reaction that increases the oxygen content of a compound. **2.** A chemical reaction in which a compound or radical loses electrons, that is in which the positive valence is increased. { ,äk·sə'dā·shən }

oxidation number [CHEM] **1.** Numerical charge on the ions of an element. **2.** See oxidation state. { ,äk·sə'dā·shən ,nəm·bər }

oxidation-reduction reaction [CHEM] An oxidizing chemical change, where an element's positive valence is increased (electron loss), accompanied by a simultaneous reduction of an associated element (electron gain). { ,äk·sə'dā·shən ri'dək·shən rē,ak·shən }

oxidation state [CHEM] The number of electrons to be added (or subtracted) from an atom in a combined state to convert it to elemental form. Also known as oxidation number. { ,äk·sə'dā·shən ,stāt }

oxide [CHEM] Binary chemical compound in which oxygen is combined with a metal (such as Na$_2$O; basic) or nonmetal (such as NO$_2$; acidic). { 'äk,sīd }

oxidized cellulose See oxycellulose. { 'äk·sə ,dīzd 'sel·yə,lōs }

oxidized microcrystalline wax [MATER] Refined, oxidized wax from bottoms of storage tanks for solvent-extracted petroleum; used in floor polishes. { 'äk·sə,dīzd ¦mī·krō¦krist·əl·ən 'waks }

oxidizing agent [CHEM] Compound that gives up oxygen easily, removes hydrogen from another compound, or attracts negative electrons. Also known as oxidant. { 'äk·sə,dīz·iŋ ,ā·jənt }

oxidizing atmosphere [CHEM] Gaseous atmosphere in which an oxidation reaction occurs;

usually refers to the oxidation of solids. { 'äk·sə ,dīz·iŋ 'at·mə,sfir }

oxidizing flame |CHEM| A flame, or the portion of it, that contains an excess of oxygen. { 'äk·sə ,dīz·iŋ ,flām }

oximido See nitroso. { äk'sim·ə·dō }

oxirane See epoxide;ethylene oxide. { 'äk·sə·rān }

oxyacetylene welding |MET| A welding process in which the heat is supplied by an oxyacetylene torch. Also known as acetylene welding. { ¦äk·sē·ə'sed·əl,ēn ¦weld·iŋ }

oxycellulose |MATER| Cellulose mixed with reaction products from oxidation of cellulose in the presence of steam or alkalies or by strong sunlight. Also known as oxidized cellulose. { ¦äk·sē'sel·yə,lōs }

oxychloride cement |MATER| A strong, hard cement composed of magnesium chloride and calcined magnesia; used for floors and stucco. Also known as Sorel cement. { ¦äk·sē'klȯr,īd si'ment }

oxygen |CHEM| A gaseous chemical element, symbol O, atomic number 8, and atomic weight 15.9994; an essential element in cellular respiration and in combustion processes; the most abundant element in the earth's crust, and about 20% of the air by volume. { 'äk·sə·jən }

oxygen corrosion |MET| The reaction of oxygen with metallic surfaces to form an oxide of the metal or alloy. { 'äk·sə·jən kə,rōzh·ən }

oxygen-free copper |MET| Pure copper having a conductivity greater than that of copper containing impurities such as cuprous oxide; used for the construction of high-power electron tubes because it does not release appreciable gas when hot. { 'äk·sə·jən ¦frē 'käp·ər }

oxygen furnace steel |MET| Steel made by a process in which oxygen under pressure is directed onto or into the molten metal. { 'äk·sə·jən ¦fər·nəs ,stēl }

oxygen gouging See flame gouging. { 'äk·sə·jən ,gau̇j·iŋ }

oxygen lance |MET| A pipe used to direct oxygen under pressure into a bath of molten steel. { 'äk·sə·jən ,lans }

oxygen point |THERMO| The temperature at which liquid oxygen and its vapor are in equilibrium, that is, the boiling point of oxygen, at standard atmospheric pressure; it is taken as a fixed point on the International Practical Temperature Scale of 1968, at $-182.962°C$. { 'äk·sə·jən ,pȯint }

oxygen steelmaking |MET| The manufacture of steel from molten pig iron and steel scrap by methods which employ pure oxygen gas (99+%) and suitable fluxes to remove carbon and phosphorus (and in part, sulfur) without introducing nitrogen or hydrogen. { 'äk·sə·jən 'stēl,māk·iŋ }

oxyhydrogen flame |CHEM| A flame obtained from the combustion of a mixture of oxygen and hydrogen. { ¦äk·sē'hī·drə·jən ¦flām }

oxyhydrogen welding |MET| Welding with an oxyhydrogen flame. { ¦äk·sē'hī·drə·jən ¦weld·iŋ }

ozocerite See ceresin. { ō'zäs·ə,rīt }

ozone |CHEM| O_3 Unstable blue gas with pungent odor; an allotropic form of oxygen; a powerful oxidant boiling at $-112°C$; used as an oxidant, bleach, and water purifier, and to treat industrial wastes. { 'ō,zōn }

ozonization |CHEM| The process of treating, impregnating, or combining with ozone. { ,ō ,zō·nə'zā·shən }

P

P See phosphorus; poise.

Pa See pascal; protactinium.

pachimeter |ENG| An instrument for measuring the limit beyond which shear of a solid ceases to be elastic. { pə'kim·əd·ər }

pachymeter |ENG| An instrument used to measure the thickness of a material, for example, a sheet of paper. { pə'kim·əd·ər }

packaging |ELEC| The process of physically locating, connecting, and protecting devices or components. { 'pak·ə·jiŋ }

pack carburizing |MET| A method of surface hardening of steel in which parts are packed in a steel box with the carburizing compound and heated to elevated temperatures. { 'pak 'kär·bə ,rīz·iŋ }

pack hardening |MET| A process of heat treating in which the workpiece is packed in a metal box together with carbonaceous material; carbon penetration is proportional to the length of heating; after treatment the workpiece is reheated and quenched. { 'pak ,hard·ən·iŋ }

packing |CRYSTAL| Arrangement of atoms or ions in a crystal lattice. |MET| In powder metallurgy, a material in which compacts are embedded during presintering or sintering operations. { 'pak·iŋ }

packing house pitch |MATER| Dark-brown to black by-product residue from manufacturing soap and candle stock or from refining vegetable oils, refuse, or wool grease; soluble in naphtha and carbon disulfide; used to make paints, varnishes, and tar paper, and in marine caulking and waterproofing. Also known as fatty-acid pitch. { 'pak·iŋ ,haùs ,pich }

packing index |CRYSTAL| The volume of ion divided by the volume of the unit cell in a crystal. { 'pak·iŋ ,in,deks }

packing radius |CRYSTAL| One-half the smallest approach distance of atoms or ions. { 'pak·iŋ ,rād·ē·əs }

pack rolling |MET| Hot rolling of two or more sheets of metal packed together; a thin surface oxide film prevents their welding. { 'pak ,rōl·iŋ }

pad See terminal area. |MET| The brickwork that is beneath the molten iron at the base of a blast furnace. { 'pad }

paint |MATER| A mixture of a pigment and a vehicle, such as oil or water, that together form a liquid or paste that can be applied to a surface to provide an adherent coating that imparts color to and often protects the surface. { pānt }

paint base |MATER| A vehicle into which pigment is mixed to form a paint. { 'pānt ,bās }

paint clay |MATER| A light-yellow to dark-reddish-brown iron- or manganese-bearing clay that mixes easily with linseed oil. { 'pānt ,klā }

paint remover |MATER| Liquid or paste formulation used to remove dried paint, varnish, enamel, or lacquer; contains solvents such as methanol, ethyl alcohol, acetone, toluene, benzene, and ethyl acetate. { 'pānt ri,müv·ər }

paint vehicle |MATER| The liquid constituent of paint; consists of volatile solvent or thinner and a film-forming component. { 'pānt ,vē·ə·kəl }

palau |MET| A palladium-gold alloy; used as a platinum substitute in analytical chemistry. { pə'laù }

palba wax |MATER| A grayish-yellow wax from older green leaves of the palm tree *Copernicia cerifera*. { 'päl·bə ,waks }

pale oil |MATER| A petroleum lubricating or process oil refined until its color (measured by transmitted light) is straw to pale yellow. { 'pāl ,òil }

palladium |CHEM| A chemical element, symbol Pd, atomic number 46, atomic weight 106.4. |MET| A white, ductile malleable metal that resembles platinum and follows it in abundance and importance of applications; does not tarnish at normal temperatures. { pə'lād·ē·əm }

palladium chloride |INORG CHEM| PdCl₂ or PdCl₂·2H₂O Dark-brown, deliquescent powder that decomposes at 501°C; soluble in water, alcohol, acetone, and hydrochloric acid; used in medicine, analytical chemisty, photographic chemicals, and indelible inks. { pə'lād·ē·əm 'klòr,īd }

palladium iodide |INORG CHEM| PdI₂ Black powder that decomposes above 100°C; soluble in potassium iodide solution, insoluble in water and alcohol. { pə'lād·ē·əm 'ī·ə,dīd }

palladium nitrate |INORG CHEM| Pd(NO₃)₂ Brown, water-soluble, deliquescent salt; used as an analytical reagent. { pə'lād·ē·əm 'nī,trāt }

palladium oxide |INORG CHEM| PdO Amber or black-green powder that decomposes at 750°C; soluble in dilute acids; used in chemical synthesis as a reduction catalyst. { pə'lād·ē·əm 'äk ,sīd }

pancake forging [MET] A rough, flat, forged shape made quickly with a minimum of tooling. { 'pan₁kāk ¦fórj·iŋ }

panel board [MATER] A rigid paperboard used for paneling in buildings and automobile bodies. { 'pan·əl ₁bȯrd }

paper [MATER] Felted or matted sheets of cellulose fibers, formed on a fine-wire screen from a dilute water suspension, and bonded together as the water is removed and the sheet is dried. { 'pā·pər }

paperboard [MATER] A composition board available in varying thicknesses and degrees of rigidity. { 'pā·pər₁bȯrd }

paper clay [MATER] A special-grade clay that is mixed with paper pulp to add body, weight, and finish to paper products. { 'pā·pər ₁klā }

paper coating [MATER] Surface coating for paper; made from suspension of clays, starches, casein, rosin, polymers, wax, or various combinations; used to give strength and special surface qualities. { 'pā·pər ₁kȯd·iŋ }

paper insulation [MATER] Electrical insulation made of paper, chiefly from coniferous woods but also from rags, rope, and other materials, which are chemically treated, beaten into a dispersed pulp, formed into a loose sheet by filtering on a moving wire screen, and compacted into paper by calendering with heated rolls. { 'pā·pər ₁in·sə'lā·shən }

papier maché [MATER] A lightweight molding material made from paper pulped with glue and other additives; dries to a hard finish that can be drilled, sanded, or painted. { ₁pā·pə·mə'shā }

papyrus [MATER] A paperlike material made by pressing the pith of the papyrus plant in water. { pə'pī·rəs }

paracyanogen [INORG CHEM] (CN)$_x$ A white solid produced by polymerization of cyanogen gas when heated to 400°C. { ¦par·ə·sī'an·ə·jən }

paraffin See alkane;paraffin wax. { 'par·ə·fən }

paraffinic hydrocarbon See alkane. { ₁par·ə¦fin·ik 'hī·drə₁kär·bən }

paraffin scale [MATER] Unrefined paraffin wax remaining in the chamber after oil has been removed from a mixture of oil and paraffin by sweating. { 'par·ə·fən ¦skāl }

paraffin wax [MATER] A solid, crystalline hydrocarbon mixture derived ₁from the paraffin distillate portion of crude petroleum; used in paper coating, candles, creams, emollients, and lipsticks. Also known as paraffin. { 'par·ə·fən ₁waks }

paraformaldehyde [ORG CHEM] (HCHO)$_n$ Polymer of formaldehyde where n is greater than 6; white, alkali-soluble solid, insoluble in alcohol, ether, and water; used as a disinfectant, fumigant, and fungicide, and to make resins. { ¦par·ə·fȯr'mal·də₁hīd }

parallel growth See parallel intergrowth. { 'par·ə₁lel 'grōth }

parallel intergrowth [CRYSTAL] Intergrowth of two or more crystals in such a way that one or more axes in each crystal are approximately parallel. Also known as parallel growth. { 'par·ə ₁lel 'in·tər₁grōth }

parallel laminate [MATER] A laminate in which all the layers of material are set approximately parallel with respect to a particular characteristic, such as the grain or the direction of tension. { 'par·ə₁lel 'lam·ə·nət }

parallel twin [CRYSTAL] A twinned crystal whose twin axis is parallel to the composition surface. { 'par·ə₁lel 'twin }

paramagnetic [ELECTROMAG] Exhibiting paramagnetism. { ¦par·ə·mag'ned·ik }

paramagnetic alloy [MET] An alloy whose permeability is slightly greater than that of vacuum and is independent of the magnetic field strength, such as intermetallic compounds of nickel and titanium. { ¦par·ə·mag'ned·ik 'al₁ȯi }

paramagnetic cooling See adiabatic demagnetization. { ¦par·ə·mag'ned·ik 'kül·iŋ }

paramagnetic crystal [ELECTROMAG] A crystal whose permeability is slightly greater than that of vacuum and is independent of the magnetic field strength. { ¦par·ə·mag'ned·ik 'krist·əl }

paramagnetic iron [MET] Iron which has been transformed from a ferromagnetic to a paramagnetic substance by application of a pressure somewhat greater than 10^5 bars (10^{10} pascals). { ¦par·ə·mag'ned·ik 'ī·ərn }

paramagnetic material [ELECTROMAG] A material within which an applied magnetic field is increased by the alignment of electron orbits. { ¦par·ə·mag'ned·ik mə'tir·ē·əl }

paramagnetic relaxation [ELECTROMAG] The approach of a system, which displays paramagnetism because of electronic magnetic moments of atoms or ions, to an equilibrium or steady-state condition over a period of time, following a change in the magnetic field. { ¦par·ə·mag'ned·ik ₁rē₁lak'sā·shən }

paramagnetic resonance See electron paramagnetic resonance. { ¦par·ə·mag'ned·ik 'rez·ən·əns }

paramagnetic susceptibility [ELECTROMAG] The susceptibility of a paramagnetic substance, which is a positive number and is, in general, much smaller than unity. { ¦par·ə·mag'ned·ik sə₁sep·tə'bil·əd·ē }

paramagnetism [ELECTROMAG] A property exhibited by substances which, when placed in a magnetic field, are magnetized parallel to the field to an extent proportional to the field (except at very low temperatures or in extremely large magnetic fields). { ¦par·ə'mag·nə₁tiz·əm }

parameter [CRYSTAL] Any of the axial lengths or interaxial angles that define a unit cell. { pə'ram·əd·ər }

para toner [MATER] Water-insoluble red pigment made from β-naphthol and paranitroaniline; used in paints, in printing, and to make para lakes. { 'par·ə₁tōn·ər }

parchment [MATER] **1.** The skin of a goat or sheep that has been treated so that it can be used to write upon. **2.** A drawing or written text on this material. { 'parch·mənt }

parchment paper [MATER] Paper that has been

manufactured so that its appearance resembles parchment.   { 'parch·mənt ˌpä·pər }

parent metal   See base metal.   { 'per·ənt ˌmed·əl }

Parian cement |MATER| Gypsum plaster containing borax, which dries to a hard finish.   { 'par·ē·ən siˌment }

parkerizing |MET| Trade name for a process for the production of phosphate coating on steel articles by immersion in an aqueous solution of manganese or zinc acid with phosphate.   { 'pär·kəˌrīz·iŋ }

Parker-Washburn boundary |SOLID STATE| A surface which separates two regions in a solid in which the crystal axes point in different directions, and which is made up of a single array of dislocations.   { 'pär·kər 'wäsh·bərn ˌbaún·drē }

Parkes process |MET| A process for recovering precious metals from lead by stirring about 2% zinc into the melt to form zinc compounds with gold and silver which can then be skimmed off the surface.   { 'pärks ˌprä·səs }

partial dislocation |CRYSTAL| The line at the edge of an extended dislocation where a slip through a fraction of a lattice constant has occurred.   { 'pär·shəl ˌdis·lō'kā·shən }

partial wetting |FL MECH| The situation in which the contact angle between a solid and a liquid is greater than zero but less than 90°.   { 'pär·shəl 'wed·iŋ }

particle |PHYS| 1. Any very small part of matter, such as a molecule, atom, or electron. Also known as fundamental particle.   2. Any relatively small subdivision of matter, ranging in diameter from a few angstroms (as with gas molecules) to a few millimeters (as with large raindrops).   { 'pärd·ə·kəl }

particle board |MATER| Construction board made with wood particles impregnated with low-molecular-weight resin and then cured.   { 'pärd·ə·kəl ˌbórd }

particle flow |FL MECH| The transport of particles in fluids and gases.   { 'pärd·i·kəl ˌflō }

particle-oriented paper |MATER| A chart paper that has a magnetic coating which is produced by combining microscopic magnetic flakes with oil to form droplets and then forming these particles into an emulsion that can be applied to the surface of ordinary bond paper or to a clear plastic substrate; the magnetic field of a small-diameter recording head rotates the magnetic flakes so that they absorb or scatter incident light to give a visible dark trace that can also be read magnetically.   { 'pärd·ə·kəl ˌór·ē¦ent·əd 'pā·pər }

particle size |MET| The average and controlling lineal dimension of an individual particle of metal powder as determined by suitable screens or other methods of analysis.   { 'pärd·ə·kəl ˌsīz }

particulate composite |MATER| A composite material composed of particles embedded in a matrix.   { pär'tik·yə·lət kəm'päz·ət }

particulates |MATER| Fine solid particles which remain individually dispersed in gases and stack emissions.   { pär'tik·yə·ləts }

particulate solid   See bulk solid.   { pär'tik·yə·lət ˌsäl·əd }

parting |MET| 1. Recovery of gold (or occasionally another metal) from its alloys by a corrosion process.   2. Zone of separation between cope and drag portions of mold or flask in sand casting.   3. In sand molding, a composition to facilitate removal of the pattern.   4. A shearing operation to produce two or more parts from a stamping.   { 'pärd·iŋ }

parting agent   See release agent.   { 'pärd·iŋ ˌā·jənt }

parting compound |MET| A material, such as silica or graphite, used to facilitate the separation of the cope and drag parting surfaces.   { 'pärd·iŋ ˌkäm,paúnd }

parting line |MET| 1. The line along which a mold is separated.   2. A line or seam on a casting corresponding to the joint of mold parts.   { 'pärd·iŋ ˌlīn }

parting sand |MET| Fine, dry sand applied to the faces of a sand mold to allow disassembly.   { 'pärd·iŋ ˌsand }

partridgewood   See acapau.   { 'pär·trij,wúd }

parylene |POLYM CHEM| Polyparaxylylene, used in ultrathin plastic films for capacitor dielectrics, and as a pore-free coating.   { 'par·ə,lēn }

pascal |MECH| A unit of pressure equal to the pressure resulting from a force of 1 newton acting uniformly over an area of 1 square meter. Symbolized Pa.   { pa'skal }

pass |MET| 1. Passage of a metal bar between rolls.   2. Open space between two grooved rolls through which metal is processed.   3. Weld metal deposited in one trip along the axis of a weld.   { pas }

passivation |ELECTR| Growth of an oxide layer on the surface of a semiconductor to provide electrical stability by isolating the transistor surface from electrical and chemical conditions in the environment; this reduces reverse-current leakage, increases breakdown voltage, and raises power dissipation rating. |MET| The process of rendering passive; reducing the reactivity of a chemically active metal surface by electrochemical polarization or by immersion in a passivating solution.   { ˌpas·ə'vā·shən }

passive-active cell |MET| An electrochemical corrosion cell established between passive and active areas on a metal surface.   { 'pas·iv 'ak·tiv ˌsel }

passive metal |MET| A metal on which a surface film forms by natural process or by immersion in a passivating solution, making the metal resistant to corrosion.   { 'pas·iv 'med·əl }

passivity |CHEM| A state of chemical inactivity, especially of a metal that is relatively resistant to corrosion due to loss of chemical activity. |MET| The property of a metal that has been made passive.   { pə'siv·əd·ē }

paste |ELEC| In batteries, the medium in the form of a paste or jelly, containing an electrolyte; it is positioned adjacent to the negative electrode of a dry cell; in an electrolytic cell, the paste serves as one of the conducting plates. |MATER|

An adhesive mixture with a characteristic plastic consistency, a high order of yield value, and a low bond strength; for example, a paste prepared by heating a starch and water mixture, then cooling the hydrolyzed product. |MET| Finely divided particles of ferromagnetic material in paste form used in the wet method of magnetic particle inspection. { pāst }

pasteboard |MATER| A type of thin cardboard made from gluing together two or more sheets of paper. { 'pāst,bórd }

paste ink |MATER| A pastelike mixture of pigment or dye with oil and other additives, such as resins, driers, tackifiers, and adhesives; used in paper and textile printing and ballpoint pens. { 'pāst ,iŋk }

pastel |MATER| A chalk or crayon made of a finely ground pigment and a minimum of nongreasy binder, such as gum tragacanth or methylcellulose; since pastels are blended on the painting itself, a larger assortment of shades and tints is required than with oil or other color mediums. { pa'stel }

pastel fixative |MATER| A thin varnish material for the simple and even application to drawings. { pa'stel 'fik·səd·iv }

paste resin |MATER| Solventless, fluid or semisolid mixture of powdered resin and plasticizer. { 'pāst ,rez·ən }

patenting |MET| A process used in the production of high-strength steel wire containing 0.35–0.85% carbon, in which the wire is heated to above the transformation temperature, then quenched in molten lead or molten salt, or cooled in air. { 'pat·ənt·iŋ }

patent leveling See stretcher leveling. { 'pat·ənt 'lev·ə·liŋ }

Patera process |MET| A method used in the 19th century for extracting silver from ore: the ore was roasted with sodium chloride, a solution of sodium thiosulfate was then added to leach out the silver chloride, and sodium sulfide precipitated the silver as silver sulfide. { pə'ter·ə ,prä·səs }

patina |MET| The greenish product, usually basic copper sulfate, formed on copper and copper-rich alloys as a result of prolonged atmospheric corrosion. { 'pat·ən·ə or pə'tē·nə }

patio process |MET| A crude chemical method of reducing silver from its ores, followed by amalgamation in low heaps with the aid of salt and copper sulfate. { 'pad·ē·ō 'prä·səs }

Patterson function |SOLID STATE| A function of three spatial coordinates, constructed in the Patterson-Harker method, which has peaks at all vectors between two atoms in a crystal, the heights of the peaks being approximately proportional to the product of the atomic numbers of the corresponding atoms. { 'pad·ər·sən ,fəŋk·shən }

Patterson-Harker method |SOLID STATE| A method of analyzing the structure of a crystal from x-ray diffraction results; a Fourier series involving squares of the absolute values of the structure factors, which are directly observable, is used to construct a vectorial representation of interatomic distances in the crystal (Patterson map). { 'pad·ər·sən 'här·kər ,meth·əd }

Patterson map |SOLID STATE| A contour chart of the Patterson function. { 'pad·ər·sən ,map }

Patterson projection |SOLID STATE| A projection of the Patterson function on a section through a crystal. { 'pad·ər·sən prə,jek·shən }

Patterson vectors |SOLID STATE| In analysis of crystal structure, the vectors of peaks relative to the origin in a Patterson function or Patterson projection. { 'pad·ər·sən ,vek·tərz }

Pattinson process |MET| A method for separating silver from its alloys rich in lead by slow cooling of the melt so that silver-poor lead crystals separate out and are removed. { 'pat·ən·sən ,prä·səs }

paulin |MATER| A fabricated textile item generally used as a weather protection cover for various items or materials during storage or transit. { 'pól·ən }

Pauling rule |SOLID STATE| A rule governing the number of ions of opposite charge in the neighborhood of a given ion in an ionic crystal, in accordance with the requirement of local electrical neutrality of the structure. { 'pól·iŋ ,rül }

Pauli spin susceptibility |SOLID STATE| The susceptibility of free electrons in a metal due to the tendency of their spins to align with a magnetic field. { 'pól·ē ¦spin sə,sep·tə'bil·əd·ē }

paving brick |MATER| A vitrified clay brick used in the construction of pavements. { 'pāv·iŋ ,brik }

paving-brick clay |MATER| Impure shale or fire clay with good tensile strength and plasticity; used to make paving bricks. { 'pāv·iŋ ,brik ,klā }

Pb See lead.

Pb-I-Pb junction See lead-I-lead junction.

p-block elements |CHEM| Elements of the main groups 13 through 18, minus helium, in the periodic table whose outer electronic configurations have occupied p levels. { ¦pē 'bläk 'el·ə,mənts }

Pd See palladium.

peak effect |SOLID STATE| In certain hard superconductors, the occurrence of a maximum in the value of the critical current as the external magnetic field is varied, near the critical magnetic field. { 'pēk i,fekt }

pearl |MATER| A dense, more or less round, white or light-colored concretion having various degrees of luster formed within or beneath the mantle of various mollusks by deposition of thin concentric layers of nacre about a foreign particle. { pərl }

pearl ash |MATER| An impure substance derived from potash following partial purification from wood ash. { 'pərl ,ash }

pearl essence |MATER| A brilliant, translucent, lustrous material obtained from fish scales; used in pearl lacquers and to make artificial pearls. Also known as pearl white. { 'pərl ,es·əns }

pearl hardening |INORG CHEM| Commerical name for a crystallized grade of calcium sulfate; used as a paper filler. { 'pərl ¦härd·ən·iŋ }

pearlite |MET| A lamellar aggregate of ferrite

(almost pure iron) and cementite (Fe_3C) often occurring in carbon steels and in cast iron. { 'pər₁līt }

pearl white See pearl essence. { 'pərl ₁wīt }

pebbles |MATER| Grinding media for pebble mills, usually balls of hard flint or hard burned white porcelain. { 'peb·əlz }

pebbling See orange peel. { 'peb·liŋ }

pedial class |CRYSTAL| That class in the triclinic system which has no symmetry. { 'ped·ē·əl ₁klas }

pedion |CRYSTAL| A crystal form with only one face; member of the asymmetric class of the triclinic system. { 'ped· ē·ən }

peeling |MATER| **1.** Stripping or detaching a rubber coating from a metal, cloth, or other material. **2.** Pulling a layer of material away from another layer, breaking one row of bonds at a time. { 'pēl·iŋ }

peening |MET| Surface-hardening a piece of metal by hammering or by bombarding with hard shot. { 'pēn·iŋ }

peg See plug. { peg }

pegging rammer |MET| A rod with an oblong piece of iron at its end; used to compact sand in a mold. { 'peg·iŋ ₁ram·ər }

Peierls-Nabarro force |SOLID STATE| The force required to displace a dislocation along its slip plane. { 'pā·ərlz nə'bär·ō ₁fórs }

pendant-drop melt extraction |MET| Melt extraction in which the molten metal is produced by heating the end of a rod above a disk. { 'pen·dənt ₁dräp 'melt ik₁strak·shən }

penetrant |MATER| A liquid with low surface tension, usually containing a dye or fluorescent chemical; when flowed over a metal surface, it is used to determine the existence and extent of cracks and other discontinuities. { 'pen·ə·trənt }

penetrating oil |MATER| Low-viscosity oil that can penetrate between closely fitted parts, such as the leaves of springs and screw threads; used to loosen rusted parts. { 'pen·ə₁träd·iŋ ₁óil }

penetration |MET| **1.** The distance from the original surface of the base metal to that point at which weld fusion ends. **2.** A surface defect on a casting caused by molten metal filling voids in the sand mold. { ₁pen·ə'trā·shən }

penetration depth |CRYO| The depth beneath the surface of superconductor in a magnetic field at which the magnetic field strength has fallen to $1/e$ of its value at the surface. |ELEC| In induction heating, the thickness of a layer, extending inward from a conductor's surface, whose resistance to direct current equals the resistance of the whole conductor to alternating current of a given frequency. { ₁pen·ə'trā·shən ₁depth }

penetration hardness See indentation hardness. { ₁pen·ə'trā·shən ₁härd·nəs }

penetration macadam |MATER| A paving material consisting of crushed stone in two sizes bound together by asphalt or tar. { ₁pen·ə'trā·shən mə'kad·əm }

penetration twin See interpenetration twin. { ₁pen·ə'trā·shən ₁twin }

pentaborane |INORG CHEM| B_5H_9 Flammable liquid boiling at 48°C; ignites spontaneously in air; proposed as high-energy fuel for aircraft and missiles. { ₁pen·tə'bór₁ān }

pentacite |MATER| Alkyd resin in which pentaerythritol is the polyhydric alcohol; used in coatings and printing inks. { 'pen·tə₁sīt }

pentadentate ligand |INORG CHEM| A chelating agent having five groups capable of attachment to a metal ion. { ₁pen·tə¦den₁tāt 'lī·gənd }

pentagonal dodecahedron See pyritohedron. { pen'tag·ən·əl dō·dek·ə'hē·drən }

pentoxide |INORG CHEM| A compound that is binary and has five atoms of oxygen; for example, phosphorus pentoxide, P_2O_5. { pen 'täk₁sīd }

pentyl See amyl. { 'pent·əl }

peptide |BIOMATER| A compound of two or more amino acids joined by peptide bonds. { 'pep ₁tīd }

peptide bond |ORG CHEM| A bond in which the carboxyl group of one amino acid is condensed with the amino group of another to form a —CO·NH—linkage. Also known as peptide linkage. { 'pep₁tīd ₁bänd }

peptide linkage See peptide bond. { 'pep₁tīd ₁liŋ·kij }

per- |CHEM| Prefix meaning: **1.** Complete, as in hydrogen peroxide. **2.** Extreme, or the presence of the peroxy (—O—O—) group. **3.** Exhaustive (complete) substitution, as in perchloroethylene. { pər or per }

peracid |CHEM| Acid containing the peroxy (—O—O—) group, such as peracetic acid or perchloric acid. { ¦pər'as·əd }

perchlorate |INORG CHEM| A salt of perchloric acid containing the ClO_4^- radical; for example, potassium perchlorate, $KClO_4$. { pər'klór₁āt }

perchloric acid |INORG CHEM| $HClO_4$ Strongly oxidizing, corrosive, colorless, hygroscopic liquid, boiling at 16°C (8 mmHg, or 1067 pascals); soluble in water; unstable in pure form, but stable when diluted in water; used in medicine, electrolytic baths, electropolishing, explosives, and analytical chemistry, and as a chemical intermediate. Also known as Fraude's reagent. { pər'klór·ik 'as·əd }

perchloryl fluoride |INORG CHEM| $ClFO_3$ A colorless gas with a sweet odor; boiling point is −46.8°C and melting point is −146°C; used as an oxidant in rocket fuels. { pər'klór·əl 'flùr₁īd }

percolation limit |SOLID STATE| In a disordered crystalline alloy having one constituent with a magnetic moment, the concentration of the magnetic element above which the spin-glass phase is replaced by the ferromagnetic state. { pər·kə'lā·shən ₁lim·ət }

percussion figure |CRYSTAL| Radiating lines on a crystal section produced by a sharp blow. { pər'kəsh·ən ₁fig·yər }

percussion welding |MET| Resistance welding with arc heat and simultaneously applied pressure from a hammerlike blow. { pər'kəsh·ən ,weld·iŋ }

perfect crystal [CRYSTAL] A crystal without lattice defects; it is an unattained ideal or standard. { 'pər·fikt 'krist·əl }

perfect dielectric See ideal dielectric. { 'pər·fikt ,dī·ə'lek·trik }

perfect fluid See inviscid fluid. { 'pər·fikt 'flü·əd }

perfect gas See ideal gas. { 'pər·fikt 'gas }

perforated metal [MATER] Sheet metal with round, square, diamond, or rectangular perforations; used for screens and for construction. { 'pər·fə,rād·əd 'med·əl }

pericline twin law [CRYSTAL] A parallel twin law in triclinic feldspars, in which the b axis is the twinning axis and the composition surface is a rhombic section. { 'per·ə,klīn 'twin ,lö }

period [CHEM] A series of elements with consecutive atomic numbers in the periodic table. [PHYS] The duration of a single repetition of a cyclic phenomenon. { 'pir·ē·əd }

periodate [INORG CHEM] A salt of periodic acid, HIO_4, for example, potassium periodate, KIO_4. { 'pər'ī·ə,dāt }

periodic acid [INORG CHEM] $HIO_4 \cdot 2H_2O$ Water- and alcohol-soluble white crystals; loses water at 100°C; used as an oxidant. { ¦pir·ē¦äd·ik 'as·əd }

periodic lattice See lattice. { ¦pir·ē¦äd·ik 'lad·əs }

periodic law [CHEM] The law that the properties of the chemical elements and their compounds are a periodic function of their atomic weights. Periodicity is, however, more closely a function of atomic number. { ¦pir·ē¦äd·ik ¦lö }

periodic table [CHEM] A table of the elements, written in sequence in the order of atomic number or atomic weight and arranged in horizontal rows (periods) and vertical columns (groups) to illustrate the occurrence of similarities in the properties of the elements as a periodic function of the sequence. { ¦pir·ē¦äd·ik 'tā·bəl }

peripheral milling [MET] Removing metal from a surface parallel to the axis of a milling cutter. { pə'rif·ə·rəl 'mil·iŋ }

permafil [MATER] Polymerizable mixture that cures without any evaporation. { 'pər,mə,fil }

permalloy [MET] A trade name for any of several highly magnetically permeable iron-base alloys containing about 45–80% nickel. { 'pər,mə,lói }

permanent anode [MET] A very-corrosion-resistant anode, made of a material such as a carbon, aluminum, or lead alloy, or 14.5% silicon iron; used in cathodic protection against corrosion. { 'pər·mə·nənt 'an,ōd }

permanent gas [THERMO] A gas at a pressure and temperature far from its liquid state. { 'pər·mə·nənt 'gas }

permanent hardness [CHEM] The hardness of water persisting after boiling. { 'pər·mə·nənt 'härd·nəs }

permanent ink [MATER] Ink that contains up to 1% dissolved iron to prevent fading or washing away when dried. { 'pər·mə·nənt 'iŋk }

permanent magnet [ELECTROMAG] A piece of hardened steel or other magnetic material that has been strongly magnetized and retains its magnetism indefinitely. Abbreviated PM. { 'pər·mə·nənt 'mag·nət }

permanent mold [MET] A reusable metal mold for the production of many castings of the same kind. { 'pər·mə·nənt 'mōld }

permanent set [MECH] Permanent plastic deformation of a structure or a test piece after removal of the applied load. Also known as set. { 'pər·mə·nənt 'set }

permanent starch [MATER] An emulsion of polyvinyl acetate used for starching clothing and textiles; it is not removed by washing. { 'pər·mə·nənt 'stärch }

permanganate [INORG CHEM] A purple salt of permanganic acid containing the MnO_4^- radical; used as an oxidizing agent and a disinfectant. { ¦pər'maŋ·gə,nāt }

permanganic acid [INORG CHEM] $HMnO_4$ An unstable acid that exists only in dilute solutions; decomposes to manganese dioxide and oxygen. { ¦pər·man'gan·ik 'as·əd }

permeability [ELECTROMAG] A factor, characteristic of a material, that is proportional to the magnetic induction produced in a material divided by the magnetic field strength; it is a tensor when these quantities are not parallel. Also known as magnetic permeability. [FL MECH] **1.** The ability of a membrane or other material to permit a substance to pass through it. **2.** Quantitatively, the amount of substance which passes through the material under given conditions. { ,pər·mē·ə'bil·əd·ē }

permeability alloy [MET] An iron-nickel alloy having greater magnetic susceptibility than iron. { ,pər·mē·ə'bil·əd·ē 'al,ói }

permeable membrane [CHEM] A thin sheet or membrane of material through which selected liquid or gas molecules or ions will pass, either through capillary pores in the membrane or by ion exchange; used in dialysis, electrodialysis, and reverse osmosis. { 'pər·mē·ə·bəl 'mem ,brān }

permeance [ELECTROMAG] A characteristic of a portion of a magnetic circuit, equal to magnetic flux divided by magnetomotive force; the reciprocal of reluctance. Symbolized P. { 'pər·mē·əns }

Permendur [MET] A magnetic alloy which is composed of equal parts of iron and cobalt and has an extremely high permeability when saturated. { 'pər·mən,dúr }

permenorm alloy [MET] An alloy containing 50% nickel and 50% iron; used as magnet core material and in magnetic amplifiers. { 'pər·mə,nórm 'al ,ói }

permittivity [ELEC] The dielectric constant multiplied by the permittivity of empty space, where the permittivity of empty space (ϵ_0) is a constant appearing in Coulomb's law, having the value of 1 in centimeter-gram-second electrostatic units, and of 8.854 × 10^{-12} farad/meter in rationalized meter-kilogram-second units. Symbolized ϵ. { ,pər·mə'tiv·əd·ē }

permselective membrane [PHYS CHEM] An ion-exchange material that allows ions of one electrical sign to enter and pass through. { ¦pərm·si¦lek·tiv 'mem₁brān }

peroxide [CHEM] A compound containing the peroxy (−O−O−) group, as in hydrogen peroxide. [INORG CHEM] See hydrogen peroxide. { pə'räk₁sīd }

peroxydol See sodium perborate. { pə'räk·sə₁dȯl }

peroxynitrite [INORG CHEM] A nitrogen oxyanion containing an O-O peroxo bond that is a structural isomer of the nitrate ion. Species are generally distinguished by writing the chemical formula for peroxynitrite as $ONOO^-$ and nitrate as NO^{3-}. Other names that have been given to peroxynitrite include pernitrite and peroxonitrite; its recommended IUPAC name is oxoperoxonitrate(1-). { pə₁räk·sē'nī₁trīt }

Persian red [INORG CHEM] Red pigment made from basic lead chromate or ferric oxide. { 'pər·zhən 'red }

persistent current [CRYO] **1.** A magnetically induced electric current that flows undiminished in a superconducting material or circuit. **2.** A superfluid current that flows undiminished around a closed path. { pər'sis·tənt 'kə·rənt }

Persoz's reagent [MATER] Chemical reagent used to detect the presence of silk with wool (only silk dissolves); consists of zinc chloride and zinc oxide in water. { pər'sōz·əz rē₁ā·jənt }

persulfate [INORG CHEM] Salt derived from persulfuric acid and containing the radical $S_2O_8{}^{2-}$; made by electrolysis of sulfate solutions. { ¦pər'səl₁fāt }

persulfuric acid [INORG CHEM] $H_2S_2O_8$ Acid formed in lead-cell batteries by electrolyzing sulfuric acid; strong oxidizing agent. { ¦pər·səl'fyur·ik 'as·əd }

pesticide [MATER] A chemical agent that destroys pests. Also known as biocide. { 'pes·tə₁sīd }

PET See polyethylene terephthalate. { ¦pē¦ē'tē or pet }

petrolatum [MATER] A smooth, semisolid blend of mineral oil with waxes crystallized from the residual type of petroleum lubricating oil; the wax molecules contain 30–70 carbon atoms and are straight chains with a few branches or naphthene rings; used as a lubricant, as a carrier in polishes and cosmetics, and as a rust preventive. { ₁pe·trə'lād·əm }

petroleum asphalt [MATER] Asphalt recovered or made from petroleum. { pə'trō·lē·əm 'as₁fȯlt }

petroleum resin [POLYM CHEM] Any one of a family of polymers produced from mixed unsaturated monomers recovered from petroleum processing streams. { pə'trō·lē·əm ₁rez·ən }

petroleum sulfonates [MATER] Sulfonated petroleum products made as by-products of SO_3 treatment of white oil or lube-stock; used as lube-oil additives, textile-processing emulsifiers, and rust preventives. { pə'trō·lē·əm 'səl·fə₁nāts }

petroleum tar [MATER] A viscous, black or dark-brown product of petroleum refining; yields substantial quantity of solid residue when partly evaporated or fractionally distilled. { pə'trō·lē·əm ₁tär }

petroleum wax [MATER] A wax occurring naturally in various fractions of crude petroleum; there are two groups: paraffin wax and microcrystalline wax. { pə'trō·lē·əm ₁waks }

petrous [MATER] Referring to a material whose hardness resembles that of stone. { 'pe·trəs }

pewter [MET] An alloy that typically contained tin as the principal component and some antimony and copper; older produced pewter typically contains lead along with the other components. { 'pyüd·ər }

PFE See photoferroelectric effect.

pH [CHEM] A term used to describe the hydrogen-ion activity of a system; it is equal to $-\log a_H{}^+$; here $a_H{}^+$ is the activity of the hydrogen ion; in dilute solution, activity is essentially equal to concentration and pH is defined as $-\log_{10}[H^+]$, where H^+ is hydrogen-ion concentration in moles per liter; a solution of pH 0 to 7 is acid, pH of 7 is neutral, pH over 7 to 14 is alkaline. { pē'āch }

phantom crystal [CRYSTAL] A crystal containing an earlier stage of crystallization outlined by dust, minute inclusions, or bubbles. Also known as ghost crystal. { 'fan·təm 'krist·əl }

phase [CHEM] Portion of a physical system (liquid, gas, solid) that is homogeneous throughout, has definable boundaries, and can be separated physically from other phases. [MET] A constituent of an alloy that is physically distinct and is homogeneous in chemical composition. [PHYS] **1.** The fractional part of a period through which the time variable of a periodic quantity (alternating electric current, vibration) has moved, as measured at any point in time from an arbitrary time origin; usually expressed in terms of angular measure, with one period being equal to $360°$ or 2π radians. **2.** For a sinusoidally varying quantity, the phase (first definition) with the time origin located at the last point at which the quantity passed through a zero position from a negative to a positive direction. **3.** The argument of the trigonometric function describing the space and time variation of a sinusoidal disturbance, $y = A \cos [(2\pi/\lambda)(x - vt)]$, where x and t are the space and time coordinates, v is the velocity of propagation, and λ is the wavelength. [THERMO] The type of state of a system, such as solid, liquid, or gas. { fāz }

phase change [PHYS] The metamorphosis of a material or mixture from one phase to another, such as gas to liquid, solid to gas. { 'fāz ₁chānj }

phase-change material [ENG] A material which is used to store the latent heat absorbed in the material during a phase transition. { 'fāz ₁chānj mə₁tir·ē·əl }

phase diagram [MET] See constitution diagram. [PHYS CHEM] A graphical representation of the equilibrium relationships between phases (such as vapor-liquid, liquid-solid) of a chemical compound, mixture of compounds, or solution. |THERMO| **1.** A graph showing the pressures at which phase transitions between different states of a pure compound occur, as a function of temperature. **2.** A graph showing the temperatures at which transitions between different phases of a binary system occur, as a function of the relative concentrations of its components. { 'fāz ,dī·ə ,gram }

phase equilibria |PHYS CHEM| The equilibrium relationships between phases (such as vapor, liquid, solid) of a chemical compound or mixture under various conditions of temperature, pressure, and composition. { 'fāz ,ē·kwə'lib·rē·ə }

phase factor |SOLID STATE| The argument (phase) of a structure factor; it cannot be directly observed. { 'fāz ,fak·tər }

phase problem |CRYSTAL| The problem that arises in determining the electron density function of a crystal from x-ray diffraction data, namely that a complete determination requires knowledge of both the magnitudes and phases of the structure factors, but experimental measurements yield only the magnitudes. { 'fāz ,präb·ləm }

phase transformation See phase transition. { 'fāz ,tranz·fər,mā·shən }

phase transition |PHYS| A change of a substance from one phase to another. Also known as phase transformation. { 'fāz tran,zish·ən }

phasor |SOLID STATE| A low-energy collective excitation of the conduction electrons in a metal, corresponding to a slowly varying phase modulation of a charge-density wave. { 'fāz·ər }

phenol |ORG CHEM| **1.** C_6H_5OH White, poisonous, corrosive crystals with sharp, burning taste; melts at $43°C$, boils at $182°C$; soluble in alcohol, water, ether, carbon disulfide, and other solvents; used to make resins and weed killers, and as a solvent and chemical intermediate. Also known as carbolic acid. **2.** A chemical compound based on the substitution product of phenol, for example, ethylphenol ($C_2H_4C_4H_5OH$), the ethyl substitute of phenol. { 'fē,nȯl }

phenol-formaldehyde resin |POLYM CHEM| Thermosetting resin made by the reaction of phenol and formaldehyde; has good strength and chemical resistance and low cost; used as a molding material for mechanical and electrical parts. Originally known as Bakelite. { 'fē,nȯl fər'mal·də,hīd ,rez·ən }

phenol-furfural resin |POLYM CHEM| A phenolic resin characterized by the ability to be fabricated by injection molding since it hardens after curing conditions are reached. { 'fē,nȯl 'fər·fə,ral ,rez·ən }

phenolic laminate |MATER| Canvas, linen, kraft paper, glass fiber, or other substrate impregnated with 30% or more of thermosetting phenolic resin and cured; used for structural, mechanical, and electrical purposes. { fə'näl·ik 'lam·ə·nət }

phenolic plastic |POLYM CHEM| A thermosetting plastic material available in many combinations of phenol and formaldehyde, often with

added fillers to provide a broad range of physical, electrical, chemical, and molding properties. { fə'näl·ik 'plas·tik }

phenoxy resin |POLYM CHEM| A high-molecular-weight thermoplastic polyether resin based on bisphenol-A and epichlorohydrin with bisphenol-A terminal groups; used for injection molding, extrusion, coatings, and adhesives. { fə'näk·sē 'rez·ən }

phenyl |ORG CHEM| $C_6H_5−$ A functional group consisting of a benzene ring from which a hydrogen has been removed. { 'fen·əl }

phenylbenzene See biphenyl. { ¦fen·əl'ben ,zēn }

phenylethylene See styrene. { ¦fen·əl'eth·ə,lēn }

philosopher's wool See zinc oxide. { fə'läs·ə· fərz 'wul̇ }

phonon |SOLID STATE| A quantum of an acoustic mode of thermal vibration in a crystal lattice. { 'fō,nän }

phonon-drag effect See Gurevich effect. { 'fō ,nän ,drag i,fekt }

phonon-electron interaction |SOLID STATE| An interaction between an electron and a vibration of a lattice, resulting in a change in both the momentum of the particle and the wave vector of the vibration. { 'fō,nän i'lek,trän ,in·tər,ak·shən }

phonon emission |SOLID STATE| The production of a phonon in a crystal lattice, which may result from the interaction of other phonons via anharmonic lattice forces, from scattering of electrons in the lattice, or from scattering of x-rays or particles which bombard the crystal. { 'fō,nän i,mish·ən }

phonon wind |SOLID STATE| A stream of nonthermal phonons that is effective in propelling electron-hole droplets through a crystal. { 'fō ,nän ,wind }

phosphate |CHEM| **1.** Generic term for any compound containing a phosphate group ($PO_4{}^{3−}$), such as potassium phosphate, K_3PO_4. **2.** Generic term for a phosphate-containing fertilizer material. { 'fä,sfāt }

phosphate anion |INORG CHEM| $PO_4{}^{3−}$ The negative ion of phosphoric acid. { 'fä,sfāt 'an ,ī·ən }

phosphate coating |MET| A conversion coating on metal, usually steel, produced by dipping in an aqueous solution of zinc or manganese acid phosphate; used to furnish a black finish to small arms, artillery, or automotive components to provide resistance to corrosion. { 'fä,sfāt ,kōd·iŋ }

phosphate glass |MATER| Glass in which phosphorus pentoxide is a major component; resistant to hydrofluoric acid. { 'fä,sfāt ,glas }

phosphate rock |MATER| **1.** A rock that is naturally high enough in phosphorus to be used directly in fertilizer manufacturing. **2.** The beneficiated concentrate of a phosphate deposit. { 'fä,sfāt 'räk }

phosphating |MET| Forming a phosphate coating on a metal. Also known as phosphatizing. { 'fäs,fād·iŋ }

phosphatizing *See* phosphating. { 'fäs·fə ,tīz·iŋ }

phosphide [INORG CHEM] Binary compound of trivalent phosphorus, as in Na₃P. { 'fä,sfīd }

phosphine [INORG CHEM] PH₃ Poisonous, colorless, spontaneously flammable gas with garlic aroma; soluble in alcohol, slightly soluble in cold water; boils at −85°C; used in organic reactions. Also known as hydrogen phosphide; phosphoretted hydrogen. { 'fä,sfēn }

phosphite [INORG CHEM] Salt of phosphorous acid; contains the radical PO₃³⁻; an example is normal sodium phosphite, Na₃PO₃. { 'fä,sfīt }

phosphomolybdic acid [INORG CHEM] H₃PO₄· 12MoO₃·xH₂O Yellow crystals; soluble in alcohol, ether, and water; used as an alkaloid reagent and a pigment. Abbreviated PMA. { ¦fä·sfō·mə'lib·dik 'as·əd }

phosphor *See* luminophor. { 'fäs·fər }

phosphor bronze [MET] A hard copper-base alloy containing several percent tin, and sometimes smaller percentages of lead, deoxidized with phosphorus. { 'fäs·fər 'bränz }

phosphor dot [ELECTR] One of the tiny dots of phosphor material that are used in groups of three, one group for each primary color, on the screen of a color television picture tube. { 'fäs·fər ,dät }

phosphorescent paint [MATER] A luminous paint containing phosphors or phosphorogens which requires activation from an outside source of light, depending upon the ability of the chemical to absorb light energy, and to emit it in the form of photons of light. { ,fäs·fə'res·ənt pānt }

phosphoretted hydrogen *See* phosphine. { 'fäs·fə,red·əd 'hī·drə·jən }

phosphoric acid [INORG CHEM] H₃PO₄ Water-soluble, transparent crystals, melting at 42°C; used as a fertilizer, in soft drinks and flavor syrups, pharmaceuticals, water treatment, and animal feeds and to pickle and rust-proof metals. Also known as orthophosphoric acid. { fä'sfȯr·ik 'as·əd }

phosphoric anhydride [INORG CHEM] P₂O₅ A flammable, dangerous, soft-white deliquescent powder; used as a dehydrating agent, in medicine and sugar refining, and as a chemical intermediate and analytical reagent. Also known as anhydrous phosphoric acid; phosphoric oxide; phosphorus pentoxide. { fä'sfȯr·ik an'hī,drīd }

phosphoric oxide *See* phosphoric anhydride. { fä'sfȯr·ik 'äk,sīd }

phosphorized copper [MET] A phosphorus deoxidized copper. { 'fäs·fə,rīzd 'käp·ər }

phosphorogen [PHYS] A substance that promotes phosphorescence in another substance, as manganese does in zinc sulfide. { fä'sfȯr·ə·jən }

phosphorous acid [INORG CHEM] H₃PO₃ Alcohol- and water-soluble deliquescent white or yellowish crystals; decomposes at 200°C; used as an analytical reagent and reducing agent. { 'fäs·fə·rəs 'as·əd }

phosphor tin [MET] A master alloy of tin and phosphorus, usually containing up to 5%

phosphorus; used to make phosphor bronze. { 'fäs·fər 'tin }

phosphorus [CHEM] A nonmetallic element, symbol P, atomic number 15, atomic weight 30.97376; used to manufacture phosphoric acid, in phosphor bronzes, incendiaries, pyrotechnics, matches, and rat poisons; the white (or yellow) allotrope is a soft waxy solid melting at 44.5°C, is soluble in carbon disulfide, insoluble in water and alcohol, and is poisonous and self-igniting in air; the red allotrope is an amorphous powder subliming at 416°C, igniting at 260°C, is insoluble in all solvents, and is nonpoisonous; the black allotrope comprises lustrous crystals similar to graphite, and is insoluble in most solvents. { 'fäs·fə·rəs }

phosphorus nitride [INORG CHEM] P₃N₅ Amorphous white solid that decomposes in hot water; insoluble in cold water, soluble in organic solvents; used to dope semiconductors. { 'fäs·fə·rəs 'nī,trīd }

phosphorus oxide [INORG CHEM] An oxygen compound of phosphorus; examples are phosphorus monoxide (P₂O), phosphorus trioxide (P₂O₃), phosphorus suboxide (P₄O). { 'fäs·fə·rəs 'äk,sīd }

phosphorus oxychloride [INORG CHEM] POCl₃ Toxic, colorless, fuming liquid with pungent aroma; boils at 107°C; decomposes in water or alcohol; causes skin burns; used as a catalyst, chlorinating agent, and in manufacture of various anhydrides. Also known as phosphoryl chloride. { 'fäs·fə·rəs ¦äk·sē'klȯr,īd }

phosphorus pentabromide [INORG CHEM] PBr₅ Yellow crystals, decomposing at 106°C and in water; used in organic synthesis. { 'fäs·fə·rəs ¦pen·tə'brō,mīd }

phosphorus pentachloride [INORG CHEM] PCl₅ Toxic, yellowish crystals with irritating aroma; an eye irritant; sublimes on heating, but will melt at 148°C under pressure; soluble in carbon disulfide; decomposes in water; used as a catalyst and chlorinating agent. { 'fäs·fə·rəs ¦pen·tə'klȯr ,īd }

phosphorus pentasulfide [INORG CHEM] P₂S₅ Flammable, hygroscopic, yellow crystals, melting at 281°C; decomposes in moist air; soluble in alkali hydroxides; used to make lube-oil additives, rubber additives, and flotation agents. { 'fäs·fə·rəs ¦pen·tə'səl,fīd }

phosphorus pentoxide *See* phosphoric anhydride. { 'fäs·fə·rəs pen'täk,sīd }

phosphorus sesquisulfide [INORG CHEM] P₄S₃ Flammable, yellow crystals, melting at 172°C; decomposed by hot water, insoluble in water, soluble in carbon disulfide; used as chemical intermediate and to make matches. Also known as tetraphosphorus trisulfide. { 'fäs·fə·rəs ¦ses·kwē'səl,fīd }

phosphorus thiochloride [INORG CHEM] PSCl₃ Yellow liquid, boiling at 125°C; used to make insecticides and oil additives. { 'fäs·fə·rəs ,thī·ə'klȯr,īd }

phosphorus tribromide [INORG CHEM] PBr₃ A corrosive, fuming, colorless liquid with penetrating aroma; soluble in acetone, alcohol, carbon disulfide, and hydrogen sulfide; decomposes in water; used as an analytical reagent to test for sugar and oxygen. { 'fäs·fə·rəs trī'brō,mīd }

phosphorus trichloride [INORG CHEM] PCl_3 A colorless, fuming liquid that decomposes rapidly in moist air and water; soluble in ether, benzene, carbon disulfide, and carbon tetrachloride; boils at 76°C; used as a chlorinating agent, phosphorus solvent, and in saccharin manufacture. { 'fäs·fə·rəs trī'klȯr,īd }

phosphorus triiodide [INORG CHEM] PI_3 Hygroscopic, red crystals, melting at 61°C; soluble in alcohol and carbon disulfide; decomposes in water; used in organic syntheses. { 'fäs·fə·rəs trī'ī·ə,dīd }

phosphorus trisulfide [INORG CHEM] P_2S_3 or P_4S_6 Grayish-yellow, tasteless, odorless solid that burns in air; soluble in alcohol, carbon disulfide, and ether; melts at 290°C; used as an analytical reagent. { 'fäs·fə·rəs trī'səl,fīd }

phosphoryl chloride *See* phosphorus oxychloride. { 'fäs·fə·rəl 'klȯr,īd }

phosphotungstic acid [INORG CHEM] $H_3PO_4 \cdot 12WO_3 \cdot xH_2O$ Heavy-greenish, water- and alcohol-soluble crystals; used as an analytical reagent and in the manufacture of organic pigments. Also known as heavy acid; phosphowolframic acid; PTA. { ¦fä·sfō¦təŋ·stik 'as·əd }

phosphowolframic acid *See* phosphotungstic acid. { ¦fä·sfō·wu̇l'fram·ik 'as·əd }

photoabsorption [PHYS] A process in which a photon transfers all its energy to an atom, molecule, or nucleus. { ,fōd·ō·əb'sȯrp·shən }

photobleach [PHYS CHEM] Upon exposure to light, to decrease in absorbance intensity or, for fluorescent compounds, to decrease in emission intensity. { ¦fōd·ō,blēch }

photocapacitative effect [ELEC] A change in the capacitance of a bulk semiconductor or semiconductor surface film upon exposure to light. { ,fōd·ō·kə'pas·ə,tā·tiv i,fekt }

photocatalysis [PHYS CHEM] The phenomenon by which a relatively small amount of light-absorbing material, called a photocatalyst, changes the rate of chemical reaction without itself being consumed. { ,fōd·ō·kə'tal·ə·səs }

photocatalyst [PHYS CHEM] A light-absorbing substance which, when added to a reaction, facilitates the reaction, while remaining unchanged at the end of the reaction. { ,fōd·ō'kad·əl·ist }

photochemical reaction [PHYS CHEM] A chemical reaction influenced or initiated by light. { ¦fōd·ō¦kem·ə·kəl rē'ak·shən }

photochemical reduction *See* photoreduction. { ,fōd·ō¦kem·ə·kəl ri'dək·shən }

photochromic compound [CHEM] A chemical compound that changes in color when exposed to visible or near-visible radiant energy; the effect is reversible; used to produce very-high-density microimages. { ¦fōd·ō¦krō·mik 'käm,paůnd }

photochromic glass [MATER] A glass that darkens when exposed to light but regains its original transparency a few minutes after light is removed; the rate of clearing increases with temperature. { ¦fōd·ō¦krō·mik 'glas }

photochromic reaction [CHEM] A chemical reaction that produces a color change. { ¦fōd·ō ¦krō·mik rē'ak·shən }

photochromism [CHEM] The ability of a chemically treated plastic or other transparent material to darken reversibly in strong light. { ¦fōd·ō'krō ,miz·əm }

photoconduction [SOLID STATE] An increase in conduction of electricity resulting from absorption of electromagnetic radiation. { ¦fōd·ō·kən'dək·shən }

photoconductive film [ELECTR] A film of material whose current-carrying ability is enhanced when illuminated. { ¦fōd·ō·kən'dək·tiv 'film }

photoconductivity [SOLID STATE] The increase in electrical conductivity displayed by many nonmetallic solids when they absorb electromagnetic radiation. { ¦fōd·ō,kän,dək'tiv·əd·ē }

photoconductor [SOLID STATE] A nonmetallic solid whose conductivity increases when it is exposed to electromagnetic radiation. { ¦fōd·ō·kən'dək·tər }

photodegradation [POLYM CHEM] Chemical changes resulting from the absorption of light that reduce the useful properties of materials, particularly polymers. The chemical changes can include bond scission (especially of the molecular backbone), color formation, crosslinking, and chemical rearrangements. { ¦fōd·ō,deg·rə'dā·shən }

photodichroic material [OPTICS] A material which exhibits photoinduced dichroism and birefringence. { ¦fōd·ō·dī'krō·ik mə'tir·ē·əl }

photoelastic effect [OPTICS] Changes in optical properties of a transparent dielectric when it is subjected to mechanical stress, such as mechanical birefringence. Also known as photoelasticity. { ¦fōd·ō·i¦las·tik i'fekt }

photoelasticity [OPTICS] An experimental technique for the measurement of stresses and strains in material objects by means of the phenomenon of mechanical birefringence. *See* photoelastic effect. { ¦fōd·ō,i,las'tis·əd·ē }

photoelectret [SOLID STATE] An electret produced by the removal of light from an illuminated photoconductor in an electric field. { ¦fōd·ō·i'lek·trət }

photoelectric effect *See* photoelectricity. { ¦fōd·ō·i'lek·trik i,fekt }

photoelectricity [ELECTR] The liberation of an electric charge by electromagnetic radiation incident on a substance; includes photoemission, photoionization, photoconduction, the photovoltaic effect, and the Auger effect (an internal photoelectric process). Also known as photoelectric effect; photoelectric process. { ¦fōd·ō,i ,lek'tris·əd·ē }

photoelectric process *See* photoelectricity. { ¦fōd·ō·i'lek·trik 'prä·səs }

photoelectromagnetic effect |ELECTR| The effect whereby, when light falls on a flat surface of an intermetallic semiconductor located in a magnetic field that is parallel to the surface, excess hole-electron pairs are created, and these carriers diffuse in the direction of the light but are deflected by the magnetic field to give a current flow through the semiconductor that is at right angles to both the light rays and the magnetic field. { ¦fōd·ō·i¦lek·trō·mag'nedik i'fekt }

photoemission |ELECTR| The ejection of electrons from a solid (or less commonly, a liquid) by incident electromagnetic radiation. Also known as external photoelectric effect. { ¦fōd·ō·i'mish·ən }

photoemissivity |ELECTR| The property of a substance that emits electrons when struck by light. { ¦fōd·ō,ē·ma'siv·əd·ē }

photoemitter |SOLID STATE| A material that emits electrons when sufficiently illuminated. { ¦fōd·ō·i¦mid·ər }

photoengraving zinc |MET| Pure zinc mixed with a small amount of iron to reduce grain size, and alloyed with a maximum of 0.2% each of cadmium, manganese, and magnesium; used for printing plates. { ¦fōd·ō·in'grāv·iŋ ,ziŋk }

photoferroelectric effect |SOLID STATE| An effect observed in ferroelectric ceramics such as PLZT materials, in which light at or near the band-gap energy of the material has an effect on the electric field in the material created by an applied voltage, and, at a certain value of the voltage, also influences the degree of ferroelectric remanent polarization. Abbreviated PFE. { ¦fōd·ō¦fer·ō·i'lek·trik i,fekt }

photoflash composition |MATER| A pyrotechnic material which, when loaded in a suitable casing and ignited, produces a flash of sufficient intensity and duration for photographic purposes; used as the filler in photoflash bombs and cartridges. { 'fōd·ə,flash ,käm·pə'zish·ən }

photographic emulsion |GRAPHICS| Microscopic grains of light-sensitive silver halide suspended in a gelatin surface on paper, plastic, metal, or glass; used to coat photographic film. { ¦fōd·ə¦graf·ik i'məl·shən }

photographic film |GRAPHICS| Sensitized material (emulsion) coated on a flexible support, usually a transparent plastic material. { ¦fōd·ə¦graf·ik 'film }

photographic fixing |GRAPHICS| The process by which the unexposed and unreduced silver halide in a negative is removed in an exposed film; sodium thiosulfate (hypo) is the chemical usually used. { ¦fōd·ə¦graf·ik 'fiks·iŋ }

photo-Hall effect |PHYS| An effect in which the illumination of a semiconductor in a magnetic field produces a change in its Hall resistance. { ¦fōd·ō 'hol i,fekt }

photoinitiated polymerization |POLYM CHEM| A chain reaction of monomer to polymer initiated by a photogenerated radical or ion. { ,fōd·ō·ə¦nish·ē,ād·əd pə,lim·ə·rə'zā·shən }

photoinitiator |PHYS CHEM| A substance (other than reactant) which, on absorption of light,

generates a reactive species (ion or radical), initiates a chemical reaction or transformation, and is consumed. { ,fōd·ō·ə'nish·ē,ād·ər }

photolysis |PHYS CHEM| The use of radiant energy to produce chemical changes. { fō'täl·ə·səs }

photomagnetic effect |PHYS| 1. The direct effect of light on the magnetic susceptibility of certain substances. 2. Paramagnetism displayed by certain substances when they are in a phosphorescent state. { ¦fōd·ō·mag¦ned·ik i'fekt }

photomagnetoelectric effect |ELECTROMAG| The generation of a voltage when a semiconductor material is positioned in a magnetic field and one face is illuminated. { ¦fōd·ō·mag¦ned·ō·i'lek·trik i,fekt }

photomask |ELECTR| A film or glass negative that has many high-resolution images, used in the production of semiconductor devices and integrated circuits. { 'fōd·ō,mask }

photomechanochemistry |PHYS CHEM| A branch of polymer sciences that deals with photochemical conversion of chemical energy into mechanical energy. { ¦fōd·ō·mə,kan·ō'kem·ə·strē }

photon |QUANT MECH| A massless particle, the quantum of the electromagnetic field, carrying energy, momentum, and angular momentum. { 'fō,tän }

photon antibunching |OPTICS| A quantum phenomenon that occurs in certain types of light emission such as resonance fluorescence, in which the emission of one photon reduces the probability that another photon will be emitted immediately afterward. { 'fō,tän 'an·ti,bənch·iŋ }

photon bunching |OPTICS| The tendency of photoelectric pulses from an illuminated photodetector to occur in bunches rather than at random. { 'fō,tän ,bənch·iŋ }

photon coupling |ELECTR| Coupling of two circuits by means of photons passing through a light pipe. { 'fō,tän ,kəp·liŋ }

photonegative |ELECTR| Having negative photoconductivity, hence decreasing in conductivity (increasing in resistance) under the action of light; selenium sometimes exhibits photonegativity. { ¦fōd·ō'neg·ə·tiv }

photon flux |OPTICS| The number of photons in a light beam reaching a surface, such as the surface of the photocathode of a photomultiplier tube, in a unit of time. { 'fō,tän ,fləks }

photon-gated material |MATER| A material in which persistent spectral holeburning with a narrow-band laser occurs only in the presence of a second enabling light beam. { 'fō,tän ¦gād·əd mə'tir·ē·əl }

photonic band-gap material *See* photonic crystal. { fə,täm·ik 'band,gap mə,tir·ē·əl }

photonic crystal |OPTICS| A macroscopic, periodic dielectric structure that possesses spectral gaps (stop bands) for electromagnetic waves, in analogy with the energy bands and gaps in regular semiconductors. Also known as photonic band-gap material. { fə,tän·ik 'krist·əl }

photonics |ELECTR| The electronic technology involved with the practical generation, manipulation, analysis, transmission, and reception of electromagnetic energy in the visible, infrared, and ultraviolet portions of the light spectrum. It contributes to many fields, including astronomy, biomedicine, data communications and storage, fiber optics, imaging, optical computing, optoelectronics, sensing, and telecommunications. Also known as optoelectronics.   { fō'tän·iks }

photooxidation |PHYS CHEM| **1.** The loss of one or more electrons from a photoexcited chemical species. **2.** The reaction of a substance with oxygen and light. When oxygen remains in the product, the reaction is also known as photooxygenation.   { ˌfōd·ō‚äk·sə'dā·shən }

photooxygenation See photooxidation.   { ˌfōd·ō‚äk·sə·jə'nā·shən }

photopolymer |POLYM CHEM| Any polymer which, on exposure to light, undergoes a spontaneous and permanent change in physical properties, such as crosslinking or depolymerization.   { ˌfōd·ō'päl·ə·mər }

photopositive |ELECTR| Having positive photoconductivity, hence increasing in conductivity (decreasing in resistance) under the action of light; selenium ordinarily has photopositivity.   { ¦fōd·ō'päz·əd·iv }

photoreduction |CHEM| A chemical reduction that is produced by electromagnetic radiation. Also known as photochemical reduction. |PHYS CHEM| **1.** Addition of one or more electrons to a photoexcited chemical species. **2.** Photochemical hydrogenation of a substance.   { ˌfōd·ō·ri'dək·shən }

photorefractive material |MATER| A material whose index of refraction changes when it is exposed to light.   { ˌfōd·ō·ri'frak·tiv mə'tir·ē·əl }

photoresist |GRAPHICS| A light-sensitive coating that is applied to a substrate, exposed, and developed prior to chemical etching; the exposed areas serve as a mask for selective etching.   { 'fōd·ō·ri‚zist }

photosensitive See light-sensitive.   { ¦fōd·ō'sen·səd·iv }

photosensitive glass |GRAPHICS| Glass containing submicroscopic metallic particles; when ultraviolet light passes through a negative on the glass, it precipitates the particles, with shadowed areas of the negative permitting deeper penetration into the glass than highlight areas, giving the picture three dimensions and color; photograph is developed by heating the glass to 1000°F (538°C).   { ¦fōd·ō'sen·səd·iv 'glas }

photosensitizer |PHYS CHEM| A light-absorbing substance that initiates a photochemical or photophysical reaction in another substance (molecule), and is not consumed in the reaction.   { ¦fōd·ō'sen·sə‚tīz·ər }

photostabilize |POLYM CHEM| To incorporate stabilizers in polymers, such as ultraviolet absorbers, to prevent photodegradation.   { ˌfōd·ō'stā·bə‚līz }

photostriction |PHYS| The changes in the dimensions of piezoelectric materials that also exhibit one of the photoelectric effects when they are illuminated by light.   { ˌfōd·ə'strik·shən }

phototropism |SOLID STATE| A reversible change in the structure of a solid exposed to light or other radiant energy, accompanied by a change in color.   { fō'tä·trə‚piz·əm }

photoviscoelasticity |OPTICS| Changes in optical properties of a transparent, viscoelastic substance when it is subjected to stress.   { ¦fōd·ō‚vis·gō‚i‚las'tis·əd·ē }

photovoltaic |ELECTR| Capable of generating a voltage as a result of exposure to visible or other radiation.   { ¦fōd·ō·vōl'tā·ik }

photovoltaic effect |ELECTR| The production of a voltage in a nonhomogeneous semiconductor, such as silicon, or at a junction between two types of material, by the absorption of light or other electromagnetic radiation.   { ¦fōd·ō·vōl'tā·ik i‚fekt }

phthalate |ORG CHEM| A salt of phthalic acid; contains the radical $C_6H_4(COO)_2{}^{2-}$; an example is dibutylphthalate, $C_{16}H_{22}O_4$; used as a plasticizer in plastics, and as a buffer in standard laboratory solutions.   { 'tha‚lāt }

phthalate ester |ORG CHEM| Any of a group of plastics plasticizers made by the direct action of alcohol on phthalic anhydride; generally characterized by moderate cost, good stability, and good general properties.   { 'tha‚lāt 'es·tər }

phthalic acid |ORG CHEM| $C_6H_4(CO_2H)_2$ Any of three isomeric benzene dicarboxylic acids; the ortho form is usually called phthalic acid, comprises alcohol-soluble, colorless crystals decomposing at 191°C, slightly soluble in water and ether, is used to make dyes, medicine, and synthetic perfumes, and as a chemical intermediate, and is also known as benzene orthodicarboxylic acid; the para form, known as terephthalic acid, is used to make polyester resins (Dacron) and as poultry feed additives; the meta form is isophthalic acid.   { 'thal·ik 'as·əd }

physical law |PHYS| A property of a physical phenomenon, or a relationship between the various quantities or qualities which may be used to describe the phenomenon, that applies to all members of a broad class of such phenomena, without exception.   { 'fiz·ə·kəl 'lö }

physical metallurgy |MET| The branch of metallurgy concerned with physical and mechanical properties of metals as affected by composition, mechanical working, and heat treatment.   { 'fiz·ə·kəl 'med·əl·ər‚jē }

physical property |CHEM| Property of a compound that can change without involving a change in chemical composition; examples are the melting point and boiling point.   { 'fiz·ə·kəl 'präp·ərd·ē }

physical testing |ENG| Determination of physical properties of materials based on observation and measurement.   { 'fiz·ə·kəl 'test·iŋ }

physical vapor deposition |MATER| A thin-film deposition process in which a material (metal, alloy, compound, cermet, or composite) is either evaporated or sputtered onto a substrate in a vacuum. Abbreviated PVD. { ˌfiz·i·kəl 'vā·pər ˌdep·ə·zish·ən }

piano wire [MET] High-tensile-strength, 0.75 to 0.85% carbon steel wire cold-drawn to uniform thickness. { pe'an·ō ˌwīr }

pickle liquor [MET] A spent pickling solution. { 'pik·əl ˌlik·ər }

pickle patch [MET] A coating of oxide or scale that remains adherent after pickling. { 'pik·əl ˌpach }

pickle stain [MET] Discoloration of a metal surface due to chemical cleaning without adequate washing and drying. { 'pik·əl ˌstān }

pickling [MET] Preferential removal of oxide or mill scale from the surface of a metal by immersion usually in an acidic or alkaline solution. { 'pik·liŋ }

pickling acid [CHEM] Any of the acids used in pickling solutions, such as hydrochloric, sulfuric, nitric, phosphoric, or hydrofluoric acid. { 'pik·liŋ ˌas·əd }

pickup [MET] Transfer of metal from the work to the tool, or from the tool to the work, during a forming operation. { 'pik·əp }

Pictet's liquid [MATER] Liquid mixture of carbon dioxide and sulfur dioxide; used to produce low temperatures. { pik'tāz ˌlik·wəd }

picture element [ELECTR] **1.** That portion, in facsimile, of the subject copy which is seen by the scanner at any instant; it can be considered a square area having dimensions equal to the width of the scanning line. **2.** In television, any segment of a scanning line, the dimension of which along the line is exactly equal to the nominal line width; the area which is being explored at any instant in the scanning process. Also known as critical area; elemental area; pixel; recording spot; scanning spot. { 'pik·chər ˌel·ə·mənt }

picture tube [ELECTR] A cathode-ray tube used in television receivers to produce an image by varying the electron-beam intensity as the beam is deflected from side to side and up and down to scan a raster on the fluorescent screen at the large end of the tube. { 'pik·chər ˌtüb }

Pidgeon process [MET] A method for producing magnesium from calcined dolomite by reduction with ferrosilicon. Also known as ferrosilicon process; silicothermic process. { 'pij·ən ˌpräs·əs }

piezocaloric effect [SOLID STATE] The production of entropy in a crystal that is subjected to mechanical stress. { pē,ā·zō·kə'lór·ik iˌfekt }

piezoceramic [MATER] A ceramic, such as lead zirconate titanate, that converts an electrical field to a mechanical strain or a mechanical strain to an electrical charge. In smart structures, piezoceramics are used as sensors, actuators, or both, for vibration suppression applications. Also known as piezoelectric ceramic. { pē ¦ā·zō·sə'ram·ik }

piezochemistry [CHEM] The field of chemical reactions under high pressures. { pē ¦ā·zō'kem·ə·strē }

piezocomposite [MATER] A composite material

that has piezoelectric properties and is fabricated by interleaving a cut or preshaped piezoceramic with a passive polymer or epoxy host matrix compound. { pē¦ā·zō·kəm'päz·ət }

piezoelectric [SOLID STATE] Having the ability to generate a voltage when mechanical force is applied, or to produce a mechanical force when a voltage is applied, as in a piezoelectric crystal. { pē¦ā·zō·ə'lek·trik }

piezoelectric ceramic See piezoceramic. { pē ¦ā·zō·ə'lek·trik sə'ram·ik }

piezoelectric crystal [SOLID STATE] A crystal which exhibits the piezoelectric effect; used in crystal loudspeakers, crystal microphones, and crystal cartridges for phono pickups. { pē ¦ā·zō·ə'lek·trik 'krist·əl }

piezoelectric effect [SOLID STATE] **1.** The generation of electric polarization in certain dielectric crystals as a result of the application of mechanical stress. **2.** The reverse effect, in which application of a voltage between certain faces of the crystal produces a mechanical distortion of the material. { pē¦ā·zō·ə'lek·trik i'fekt }

piezoelectric element [ELECTR] A piezoelectric crystal used in an electric circuit, for example, as a transducer to convert mechanical or acoustical signals to electric signals, or to control the frequency of a crystal oscillator. { pē¦ā·zō·ə'lek·trik 'el·ə·mənt }

piezoelectric hysteresis [SOLID STATE] Behavior of a piezoelectric crystal whose electric polarization depends not only on the mechanical stress to which the crystal is subjected, but also on the previous history of this stress. { pē ¦ā·zō·ə'lek·trik ˌhis·tə'rē·səs }

piezoelectricity [SOLID STATE] Electricity or electric polarization resulting from the piezoelectric effect. { pē¦ā·zō·ə,lek'tris·əd·ē }

piezoelectric polymer See piezopolymer. { pē ¦ā·zō·ə'lek·trik 'päl·ə·mər }

piezoelectric semiconductor [SOLID STATE] A semiconductor exhibiting the piezoelectric effect, such as quartz, Rochelle salt, and barium titanate. { pē¦ā·zō·ə'lek·trik 'sem·i·kən,dək·tər }

piezoelectric transducer [ELECTR] A piezoelectric crystal used as a transducer, either to convert mechanical or acoustical signals to electric signals, as in a microphone, or vice versa, as in ultrasonic metal inspection. { pē¦ā·zō·ə'lek·trik tranz'dü·sər }

piezoelectric vibrator [SOLID STATE] An element cut from piezoelectric material, usually in the form of a plate, bar, or ring, with electrodes attached to or supported near the element to excite one of its resonant frequencies. { pē ¦ā·zō·ə'lek·trik 'vī,brād·ər }

piezojunction effect [ELECTR] A change in the current-voltage characteristic of a pn junction that is produced by a mechanical stress. { pē,ā·zō'jəŋk·shən i,fekt }

piezomagnetism [SOLID STATE] Stress dependence of magnetic properties. { pē¦ā·zō'mag·nə ˌtiz·əm }

piezooptical effect [OPTICS] The change

produced in the index of refraction of a light-transmitting material by externally applied stress. { pe¦ā·zō′äp·tə·kəl i¸fekt }

piezopolymer [POLYM CHEM] A polymeric film that has the ability to reversibly convert heat and pressure to electricity. Also known as piezoelectric polymer. { pē¦ā·zō′päl·ə·mər }

piezoresistance effect [SOLID STATE] The change in the electrical resistance of a metal or semiconductor that is produced by mechanical stress. { pē¸ā·zō·ri′zis·təns i¸fekt }

piezoresistive material [MATER] A metal or semiconductor in which a change in electrical resistance occurs in response to changes in the applied stress. { pē¦ā·zō·ri¦zis·tiv mə′tir·ē·əl }

pig [MET] A crude metal casting prepared for storage, transportation, or remelting. { pig }

pig iron [MET] **1.** Crude, high-carbon iron produced by reduction of iron ore in a blast furnace. **2.** Cast iron in the form of pigs. { ′pig ¸ī·ərn }

pigment [MATER] A solid that reflects light of certain wavelengths while absorbing light of other wavelengths, without producing appreciable luminescence; used to impart color to other materials. { ′pig·mənt }

Pilger tube-reducing process [MET] A tube-reducing process in which pierced billets are forced over a mandrel between two rolls with inclined axes and given a rotary forging treatment; the tube is advanced during the gap in each revolution. { ′pil·gər ′tüb ri¸düs·iŋ ¸prä·səs }

pimple [MATER] A small, conical elevation on the surface of a plastic. { ′pim·pəl }

pinacoid [CRYSTAL] An open crystal form that comprises two parallel faces. { ′pin·ə¸kóid }

pinacoidal class [CRYSTAL] That crystal class in the triclinic system having only a center of symmetry. { ¦pin·ə¦kóid·əl ¦klas }

pinacoidal cleavage [CRYSTAL] A type of crystal cleavage that is parallel to one of the crystal's pinacoidal surfaces. { ¦pin·ə¦kóid·əl ′klē·vij }

pin bar [MET] A small-diameter, case-hardened steel rod used for making dowel pins. { ′pin ¸bär }

pinch pass [MET] A cold rolling of sheet metal to effect a very small reduction in thickness and to produce a piece of accurate dimensions. { ′pinch ¸pas }

pinch trimming [MET] Trimming a tubular or hollow part by pinching the flange or lip over the cutting edge of a punch. { ′pinch ¸trim·iŋ }

pine oleoresin [MATER] Fused solid blend of turpentine and rosin. { ′pīn ¦ō·lē·ō′rez·ən }

pine tar [MATER] A viscous black mass obtained as a by-product in the distillation of pine wood; used for roofing. { ′pīn ¸tär }

pine tar pitch [MATER] Viscous residue resulting from distillation of volatile oils from pine tar. { ′pīn ¸tär ¸pich }

pine terpene [MATER] Any terpene in the essential oils obtained from various *Pinus* species. { ¦pīn ′tər¸pēn }

pin expansion test [MET] A test for determining tube expandability or for revealing longitudinal weaknesses by forcing a tapered pin into the open end of the tube. { ′pin ik¦span·shən ¸test }

pinhole [MET] A material fault resulting from small blisters that have burst in a casting or that have formed during electroplating. { ′pin¸hōl }

pin junction [ELECTR] A semiconductor device having three regions: p-type impurity, intrinsic (electrically pure), and n-type impurity. { ′pin ¸jəŋk·shən }

pin metal [MET] Brass with a composition of 63% copper and 37% zinc, used as cold-drawn wire for making ordinary dressmaking pins. { ′pin ¸med·əl }

pinning [SOLID STATE] The hindering of motion of dislocations in a solid, and the consequent hardening of the solid, by impurities which collect near the dislocations, resulting in a large energy barrier being imposed against the motion of the dislocations. { ′pin·iŋ }

Piobert lines See Lüders' lines. { pyō′ber ¸līnz }

pipe [MET] **1.** The central cavity in an ingot or casting formed by contraction of the metal during solidification. **2.** An extrusion defect caused by the oxidized surface of the billet flowing toward the center of the rod at the back end. { pīp }

pit [MET] A small hole in the surface of a metal; usually caused by corrosion or formed during electroplating operations. { pit }

pitch [MATER] A dark heavy liquid or solid substance obtained as a residue after distillation of tar, oil, and such materials; occurs naturally as asphalt. { pich }

pitch coke [MATER] Coke made from coal tar pitch, characterized by high carbon and low ash content, and used mainly for production of electrode carbon. { ′pich ¸kōk }

pit furnace [MET] A low-temperature furnace in which steel is tempered. { ′pit ¸fər·nəs }

pitting [MET] Selective localized formation of rounded cavities in a metal surface due to corrosion or to nonuniform electroplating. { ′pid·iŋ }

pitting potential [MET] The electrochemical potential in a given environment above which, but not below, a corrosion pit initiates in a metal surface. { ′pid·iŋ pə¸ten·chəl }

pixel [ELECTR] The smallest addressable element in an electronic display; a short form for picture element. { pik′sel }

pK [CHEM] The logarithm (to base 10) of the reciprocal of the equilibrium constant for a specified reaction under specified conditions.

Pl See poiseuille.

plagiohedral [CRYSTAL] Pertaining to obliquely arranged spiral faces; in particular, to a member of a group in the isometric system with 13 axes but no center or planes. { ¦plā·jē·ō¦hē·drəl }

plait point [CHEM] Composition conditions in which the three coexisting phases of partially soluble components of a three-phase liquid system approach each other in composition. { ′plāt ¸póint }

planchet [MATER] A milled metal disk ready for coining. { ′plan·chət }

Planck function [THERMO] The negative of the

Gibbs free energy divided by the absolute temperature. { 'plänk ,fəŋk·shən }

plane defect |CRYSTAL| A type of crystal defect that occurs along the boundary plane of two regions of a crystal, or between two grains. { 'plān di,fekt }

plane dendrite See plane-dendritic crystal. { 'plān ¦den,drīt }

plane-dendritic crystal |CRYSTAL| An ice crystal exhibiting an elaborately branched (dendritic) structure of hexagonal symmetry, with its much larger dimension lying perpendicular to the principal (c-axis) of the crystal. Also known as plane dendrite; stellar crystal. { 'plān den ¦drid·ik 'krist·əl }

plane group |CRYSTAL| The group of operations (rotations, reflections, translations, and combinations of these) which leave a regular, periodic structure in a plane unchanged. { 'plān ,grüp }

plane lattice |CRYSTAL| A regular, periodic array of points in a plane. { 'plān 'lad·əs }

plane of maximum shear stress |MECH| Either of two planes that lie on opposite sides of and at angels of 45° to the maximum principal stress axis and that are parallel to the intermediate principal stress axis. { 'plān əv ¦mak·si·məm 'shir ,stres }

plane of mirror symmetry |CRYSTAL| In certain crystals, a symmetry element whereby reflection of the crystal through a certain plane leaves the crystal unchanged. Also known as mirror plane of symmetry; plane of symmetry; reflection plane; symmetry plane. { 'plān əv 'mir·ər ,sim·ə·trē }

plane of reflection |MATH| See plane of mirror symmetry. |OPTICS| A plane containing the direction of propagation of radiation reflected from a surface, and the normal to the surface. Also known as reflection plane. { 'plān əv ri'flek·shən }

plane of symmetry See plane of mirror symmetry. { 'plān əv 'sim·ə·trē }

plane strain |MECH| A deformation of a body in which the displacements of all points in the body are parallel to a given plane, and the values of these displacements do not depend on the distance perpendicular to the plane. { 'plān ,strān }

plane stress |MECH| A state of stress in which two of the principal stresses are always parallel to a given plane and are constant in the normal direction. { 'plān ,stres }

plane symmetry group See plane group. { 'plān 'sim·ə·trē ,grüp }

planimetric method |MET| A method of measuring grain size by counting the number of grains in a given area. { ¦plan·ə¦me·trik 'meth·əd }

plasma-arc welding |MET| Welding metal in a gas stream heated by a tungsten arc to temperatures approaching 60,000°F (33,315°C). { 'plaz·mə ¦ärk 'weld·iŋ }

plasma etching |ELECTR| A method of forming integrated-circuit patterns on a surface, in which charged species in a plasma formed above a masked surface are directed to impact the nonmasked regions of the surface and knock out substrate atoms. Also known as dry plasma etching. { 'plaz·mə 'ech·iŋ }

plasma generator |ELECTR| Any device that produces a high-velocity plasma jet, such as a plasma accelerator, engine, oscillator, or torch. { 'plaz·mə ¦jen·ə,rād·ər }

plasma-source ion implantation |ENG| A method of ion implantation in which the workpiece is placed in a plasma containing the appropriate ion species and is repetitively pulse-biased to a high negative potential so that positive plasma ions are accelerated to the surface and implant in the bulk material. Abbreviated PSII. { ¦plaz·mə ,sȯrs 'ī·ən ,im·plan ,tā·shən }

plasma spraying |MET| In thermal spraying, melting and transference of a metal coating to a workpiece by use of a nontransferred arc. { 'plaz·mə 'sprā·iŋ }

plasmon |SOLID STATE| A quantum of a collective longitudinal wave in the electron gas of a solid. { 'plaz,män }

plaster |MATER| A plastic mixture of various materials, such as lime or gypsum, and water which sets to a hard, coherent solid. { 'plas·tər }

plasterboard |MATER| A large, thin sheet of pulpboard, paper, or felt bonded to a hardened gypsum plaster core and used as a wall backing or as a substitute for plaster. { 'plas·tər,bȯrd }

plaster of paris |INORG CHEM| White powder consisting essentially of the hemihydrate of calcium sulfate ($CaSO_4 \cdot \frac{1}{2} H_2O$ or $2CaSO_4 \cdot H_2O$), produced by calcining gypsum until it is partially dehydrated; forms with water a paste that quickly sets; used for casts and molds, building materials, and surgical bandages. Also known as calcined gypsum. { 'plas·tər əv 'par·əs }

plastic |POLYM CHEM| A polymeric material (usually organic) of large molecular weight which can be shaped by flow; usually refers to the final product with fillers, plasticizers, pigments, and stabilizers included (versus the resin, the homogeneous polymeric starting material); examples are polyvinyl chloride, polyethylene, and urea-formaldehyde. |MECH| Displaying, or associated with, plasticity. { 'plas·tik }

plastic bonding |ENG| The joining of plastics by heat, solvents, adhesives, pressure, or radio frequency. { 'plas·tik 'bänd·iŋ }

plastic bronze |MET| A copper alloy containing lead, usually on the order of 30%, of sufficient plasticity to make a good bearing. { 'plas·tik 'bränz }

plastic cement |MATER| A plastic material used to seal narrow openings in buildings. { 'plas·tik si'ment }

plastic clay |MATER| Fireclay which forms a moldable mass when mixed with water. { 'plas·tik 'klā }

plastic collision |MECH| A collision in which one or both of the colliding bodies suffers plastic deformation and mechanical energy is dissipated. { ¦plas·tik kə'lizh·ən }

plastic deformation |MECH| Permanent change

in shape or size of a solid body without fracture resulting from the application of sustained stress beyond the elastic limit. { 'plas·tik ‚dē ‚for'mā·shən }

plastic dielectric |MATER| A plastic used in an application in which its high resistance, dielectric strength, or other electrical properties are important, such as for electrical insulation or in a capacitor. { 'plas·tik ‚dī·ə'lek·trik }

plastic film |MATER| Film with thickness from 0.0015 to 0.006 inch (0.0038 to 0.015 centimeter); made from polyvinyl chloride, polyethylene, polypropylene, polystyrene, Mylar, and other resins; used for wrapping, sealing, garment waterproofing, and coating wood, paper, or fabric. { 'plas·tik 'film }

plastic flow |PHYS| Rheological phenomenon in which flowing behavior of the material occurs after the applied stress reaches a critical (yield) value, such as with putty. { 'plas·tik 'flō }

plastic foam See expanded plastic. { 'plas·tik 'fōm }

plasticity |MECH| The property of a solid body whereby it undergoes a permanent change in shape or size when subjected to a stress exceeding a particular value, called the yield value. { plas'tis·əd·ē }

plasticize |ENG| To soften a material to make it plastic or moldable by adding a plasticizer or by using heat. { 'plas·tə‚sīz }

plasticizer See flexibilizer. { 'plas·tə‚sīz·ər }

plasticizing oil |MATER| Coal tar distillate or solvent naphthas distilling in a wide range above 300°C; used with plastics as a plasticizer. { 'plas·tə‚sīz·iŋ ‚oil }

plasticoviscosity |MECH| Plasticity in which the rate of deformation of a body subjected to stresses greater than the yield stress is a linear function of the stress. { ‚plas·tə‚kō·vi'skäs·əd·ē }

plastic paint |MATER| Paint composed of a plastic (such as vinyl or nitrocellulose) in a solvent. { 'plas·tik ‚pānt }

plastic semiconductor |MATER| An organic plastic resin with a conjugated double-bond structure, such as polyacetylene; the material is a semiconductor due to resistance of electrons to transfer from one molecule to another. { 'plas·tik 'sem·i·kən‚dək·tər }

plastic viscosity |FL MECH| A measure of the internal resistance to fluid flow of a Bingham plastic, expressed as the tangential shear stress in excess of the yield stress divided by the resulting rate of shear. { 'plas·tik vi'skäs·əd·ē }

plastic wood |MATER| Wood flour or wood cellulose compounded with a synthetic resin of high molecular weight; it is adhesive but does not penetrate wood, and is used to fill cavities or seams in wood products. { 'plas·tik 'wud }

plastigel |MATER| A plastisol with gellike flow properties achieved by adding a thixotropic agent (such as bentonite) to the plastisol. { 'plas·tə ‚jel }

plastisol |MATER| A vinyl resin dissolved in a

plasticizer to make a pourable liquid. { 'plas·tə ‚sól }

plastizymes |POLYM CHEM| Artificial enzymes (artificial polymeric materials with molecule-shaped pores) that possess catalytic properties. { 'plas·ti‚zīmz }

plate |ELEC| 1. One of the conducting surfaces in a capacitor. 2. One of the electrodes in a storage battery. |ELECTR| See anode. |MET| A thick flat particle of metal powder. { plāt }

plate amalgamation |MET| Use of copper-alloy plates or copper coated with mercury in order to trap gold from crushed ore pulp as it flows over the plates. { 'plāt ə‚mal·gə'mā·shən }

plate glass |MATER| Flat, high-quality glass with plane, parallel surfaces. { 'plāt 'glas }

plate modulus |MECH| The ratio of the stress component T_{xx} in an isotropic, elastic body obeying a generalized Hooke's law to the corresponding strain component S_{xx}, when the strain components S_{yy} and S_{zz} are 0; the sum of the Poisson ratio and twice the rigidity modulus. { 'plāt ‚mäj·ə·ləs }

platina |MET| A white brittle brass containing 75% zinc and 25% copper; used for jewelry. { plə'tēn·ə }

plating |MET| Forming a thin, adherent layer of metal on an object. Also known as metal plating. { 'plād·iŋ }

plating rack |MET| A fixture that holds, and conducts current to, a piece of work during electrodeposition. { 'plād·iŋ ‚rak }

platinic chloride See chloroplatinic acid. { plə'tin·ik 'klòr‚īd }

platinic sodium chloride See sodium chloroplatinate. { plə'tin·ik 'sōd·ē·əm 'klòr‚īd }

platinic sulfate See platinum sulfate. { plə'tin·ik 'səl‚fāt }

platinochloride See chloroplatinate. { ‚plat·ən·ō'klòr‚īd }

platinocyanide |INORG CHEM| A double salt of platinous cyanide and another cyanide, such as $K_2Pt(CN)_4$; used in photography and fluorescent x-ray screens. Also known as cyanoplatinate. { ‚plat·ən·ō'sī·ə‚nīd }

platinoid |MET| 1. Resembling or related to platinum. 2. A copper-nickel-zinc alloy used for electrical resistance wire. { 'plat·ən‚oid }

platinous chloride See platinum dichloride. { 'plat·ən·əs 'klòr‚īd }

platinous iodide See platinum iodide. { 'plat·ən·əs 'ī·ə‚dīd }

platinum |CHEM| A chemical element, symbol Pt, atomic number 78, atomic weight 195.08. |MET| A soft, ductile, malleable, grayish white noble metal with relatively high electric resistance; used in alloys, in electrical and electronic devices, and in jewelry. { 'plat·ən·əm }

platinum bichloride See platinum dichloride. { 'plat·ən·əm bī'klòr‚īd }

platinum black |MET| Black-colored, finely divided metallic platinum; soluble in aqua regia; used as a catalyst, as an absorbent for gases (hydrogen, oxygen), and for gas ignition. Also known as platinum Mohr. { 'plat·ən·əm 'blak }

platinum chloride [INORG CHEM] $PtCl_4$ or $PtCl_4 \cdot 5H_2O$ A brown solid or red crystals; soluble in alcohol and water; decomposes when heated (loses $4H_2O$ at $100°C$); used as an analytical reagent. { 'plat·ən·əm 'klȯr,īd }

platinum dichloride [INORG CHEM] $PtCl_2$ Water-insoluble, green-gray powder; decomposes to platinum at red heat; used to make platinum salts. Also known as platinous chloride; platinum bichloride. { 'plat·ən·əm dī'klȯr,īd }

platinum diiodide See platinum iodide. { 'plat·ən·əm dī'ī·ə,dīd }

platinum electrode [PHYS CHEM] A solid platinum wire electrode used during voltammetric analyses of electrolytes. { 'plat·ən·əm i'lek,trōd }

platinum iodide [INORG CHEM] PtI_2 Water- and alkali-insoluble black powder; slightly soluble in hydrochloric acid; decomposes at $300–350°C$. Also known as platinous iodide; platinum diiodide. { 'plat·ən·əm 'ī·ə,dīd }

platinum-iridium alloy [MET] An alloy with 1–30% iridium; as concentration of iridium increases, so do hardness, chemical resistance, and melting point; used in jewelry, electrical contacts, and hypodermic needles. { 'plat·ən·əm ə'rid·ē·əm 'al,ȯi }

platinum metal [CHEM] A group of transition metals that includes ruthenium, osmium, rhodium, iridium, palladium, and platinum. { 'plat·ən·əm 'med·əl }

platinum Mohr See platinum black. { 'plat·ən·əm 'mȯr }

platinum oxide [INORG CHEM] An oxide of platinum; examples are platinum monoxide (or platinous oxide), PtO, and platinum dioxide (or platinic oxide), PtO_2. { 'plat·ən·əm 'äk,sīd }

platinum potassium chloride See potassium chloroplatinate. { 'plat·ən·əm pə'tas·ē·əm 'klȯr,īd }

platinum-rhodium alloy [MET] An alloy with up to 40% rhodium; as concentration of rhodium increases, so do chemical resistance and hardness (although less hard than for platinum-iridium alloys); used as a catalyst to make nitric acid, in thermocouples, and in rayon spinnerets. { 'plat·ən·əm 'rō·dē·əm 'al,ȯi }

platinum sodium chloride See sodium chloroplatinate. { 'plat·ən·əm 'sōd·ē·əm 'klȯr,īd }

platinum sponge [MET] Porous, grayish-black mass of finely divided platinum; soluble in aqua regia; used as a catalyst, and for ignition of combustible gases. { 'plat·ən·əm 'spənj }

platinum sulfate [INORG CHEM] $Pt(SO_4)_2$ A hygroscopic, dark mass; soluble in alcohol, ether, water, and dilute acids; used in microanalysis for halogens. Also known as platinic sulfate. { 'plat·ən·əm 'səl,fāt }

Plattner's process [MET] A process for extracting gold in which a charge of gold-bearing pulp is placed in a revolving iron drum lined with lead, and a stream of chlorine gas is conducted through the pulp, producing chloride of gold, which is soluble in water. { 'plat·nərz ,prä·səs }

pleochroism [OPTICS] Phenomenon exhibited by certain transparent crystals in which light viewed through the crystal has different colors when it passes through the crystal in different directions. Also known as polychroism. { plē'äk·rə,wiz·əm }

pleomorphism See polymorphism. { ,plē·ō 'mȯr,fiz·əm }

Plessy's green [INORG CHEM] $CrPO_4 \cdot xH_2O$ Deep-green pigment made of chromium phosphate mixed with chromium oxide and calcium phosphate. { ple'sēz 'grēn }

plow steel [MET] High-quality, high-strength steel with 0.5 to 0.95% carbon content, used for wire rope. { plaù ,stēl }

plug [MET] **1.** A rod or mandrel over which a pierced tube is forced, or that fills a tube as it is drawn through a die. **2.** A punch or mandrel over which a cup is drawn. **3.** A protruding portion of a die impression for forming a corresponding recess in the forging. **4.** A false bottom in a die. Also known as peg. { pləg }

plug die See floating plug. { 'pləg ,dī }

plug drawing [MET] Drawing tubing over a plug or mandrel and through a die simultaneously to reduce diameter and thickness and to produce a smooth symmetrical bore surface. { 'pləg ,drȯ·iŋ }

plug weld [MET] A circular fusion weld made in the hole of a slotted lap or tee joint. { 'pləg ,weld }

plumbous oxide See lead monoxide. { 'pləm·bəs 'äk,sīd }

plumbous sulfide See lead sulfide. { 'pləm·bəs 'səl,fīd }

plumbum [CHEM] Latin name for lead; source of the element symbol Pb. { 'pləm·bəm }

plus sieve [MET] That portion of a powder sample that is retained by a standard sieve of a specified number. { 'pləs ,siv }

plutonium [CHEM] A reactive metallic element, symbol Pu, atomic number 94, in the transuranium series of elements; the first isotope to be identified was plutonium-239; used as a nuclear fuel, to produce radioactive isotopes for research, and as the fissile agent in nuclear weapons. { plü'tō·nē·əm }

plutonium oxide [INORG CHEM] PuO_2 A radioactively poisonous pyrophoric oxide of plutonium; particles may be easily airborne. { plü'tō·nē·əm 'äk,sīd }

ply [MATER] A thin sheet of wood or other material bonded to one or more additional thin sheets, as in plywood. [TEXT] A strand of yarn made by twisting together two or more strands. { plī }

plymetal [MATER] **1.** A material consisting of layers of dissimilar metals bonded together. **2.** Plywood faced with aluminum on both sides. { 'plī,med·əl }

plywood [MATER] A material composed of thin sheets of wood glued together, with the grains of adjacent sheets oriented at right angles to each other. { 'plī,wùd }

Pm See promethium.

PMA See phosphomolybdic acid.

pneumatic steelmaking [MET] Any steelmaking process which employs air or oxygen and for which all heat is derived from the initial heat content of the charge materials and from the thermal energy of the refining reactions. { nü'mad·ik 'stēl,māk·iŋ }

pn junction [ELECTR] The interface between two regions in a semiconductor crystal which have been treated so that one is a p-type semiconductor and the other is an n-type semiconductor; it contains a permanent dipole charge layer. { ¦pē ¦en ,jəŋk·shən }

pnpn diode [ELECTR] A semiconductor device consisting of four alternate layers of p-type and n-type semiconductor material, with terminal connections to the two outer layers. Also known as *npnp* diode. { ¦pē¦en¦pē¦en ,dī,ōd }

pnpn transistor See *npnp* transistor. { ¦pē¦en¦pē ¦en tran,zis·tər }

Po See polonium.

point defect [CRYSTAL] A departure from crystal symmetry which affects only one, or, in some cases, two lattice sites. { 'pȯint di,fekt }

point group [CRYSTAL] A group consisting of the symmetry elements of an object having a single fixed point; 32 such groups are possible. { 'pȯint ,grüp }

pointing [MET] Reducing the diameter and tapering a short length at the end of a rod, wire, or tube. Also known as metal pointing. { 'pȯint·iŋ }

point of contraflexure [MECH] A point at which the direction of bending changes. Also known as point of inflection. { 'pȯint əv ,kän·trə'flek·shər }

point of inflection See point of contraflexure. { 'pȯint əv in'flek·shən }

point of operation [MET] That portion of a metal-forming press in which the material is positioned and the work is performed. { ¦pȯint əv ,äp·ə'rā·shən }

poise [FL MECH] A unit of dynamic viscosity equal to the dynamic viscosity of a fluid in which there is a tangential force 1 dyne per square centimeter resisting the flow of two parallel fluid layers past each other when their differential velocity is 1 centimeter per second per centimeter of separation. Abbreviated P. { pȯiz }

poiseuille [FL MECH] A unit of dynamic viscosity of a fluid in which there is a tangential force of 1 newton per square meter resisting the flow of two parallel layers past each other when their differential velocity is 1 meter per second per meter of separation; equal to 10 poise; used chiefly in France. Abbreviated Pl. { pwä'zə·ē }

poison [CHEM] A substance that exerts inhibitive effects on catalysts, even when present only in small amounts; for example, traces of sulfur or lead will poison platinum-based catalysts. [ELECTR] A material which reduces the emission of electrons from the surface of a cathode. [MATER] A substance that in relatively small doses has an action that either destroys life or impairs seriously the functions of organs or tissues. { 'pȯiz·ən }

Poisson number [MECH] The reciprocal of the Poisson ratio. { pwä'sōn ,nəm·bər }

Poisson ratio [MECH] The ratio of the transverse contracting strain to the elongation strain when a rod is stretched by forces which are applied at its ends and which are parallel to the rod's axis. { pwä'sōn ,rā·shō }

poke welding See push welding. { 'pōk ,weld·iŋ }

polar axis [CRYSTAL] An axis of crystal symmetry which does not have a plane of symmetry perpendicular to it. { 'pō·lər 'ak·səs }

polar compound [CHEM] Molecules which contain polar covalent bonds; they can ionize when dissolved or fused; polar compounds include inorganic acids, bases, and salts. { 'pō·lər 'käm ,paúnd }

polar crystal See ferroelectric crystal. { 'pō·lər 'krist·əl }

polariton [SOLID STATE] A coupled mode of motion in an ionic crystal due to the coupling between the electromagnetic field and transverse optical phonons of long wavelength. { pə'lar·ə ,tän }

polarity [PHYS] Property of a physical system which has two points with different (usually opposite) characteristics, such as one which has opposite charges or electric potentials, or opposite magnetic poles. { pə'lar·əd·ē }

polarizability [ELEC] The electric dipole moment induced in a system, such as an atom or molecule, by an electric field of unit strength. { ,pō·lə,rīz·ə'bil·əd·ē }

polarization [ELEC] **1.** The process of producing a relative displacement of positive and negative bound charges in a body by applying an electric field. **2.** A vector quantity equal to the electric dipole moment per unit volume of a material. Also known as dielectric polarization; electric polarization. **3.** A chemical change occurring in dry cells during use, increasing the internal resistance of the cell and shortening its useful life. [PHYS] **1.** Phenomenon exhibited by certain electromagnetic waves and other transverse waves in which the direction of the electric field or the displacement direction of the vibrations is constant or varies in some definite way. Also known as wave polarization. **2.** The direction of the electric field or the displacement vector of a wave exhibiting polarization (first definition). **3.** The process of bringing about polarization (first definition) in a transverse wave. **4.** Property of a collection of particles with spin, in which the majority have spin components pointing in one direction, rather than at random. { ,pō·lə·rə'zā·shən }

polarization charge See bound charge. { ,pō·lə·rə'zā·shən ,chärj }

polarized ceramics [MATER] A substance, such as lead zirconate and barium titanate, having high electromechanical conversion efficiency and used as a transducer element in an ultrasonic system. { 'pō·lə,rīzd sə'ram·iks }

polarized light [OPTICS] Polarized electromagnetic radiation whose frequency is in the optical region. { 'pō·lə,rīzd 'līt }

polaron [SOLID STATE] An electron in a crystal

268

lattice together with a cloud of phonons that result from the deformation of the lattice produced by the interaction of the electron with ions or atoms in the lattice.   { 'pō·lə,rän }

polar solvent |MATER| A solvent in whose molecules there is either a permanent separation of positive and negative charges, or the centers of positive and negative charges do not coincide; these solvents have high dielectric constants, are chemically active, and form coordinate covalent bonds; examples are alcohols and ketones.   { 'pō·lər 'säl·vənt }

polar symmetry |CRYSTAL| A type of crystal symmetry in which the two ends of the central crystallographic axis are not symmetrical.   { 'pō·lər 'sim·ə·trē }

pole |CRYSTAL| **1.** A direction perpendicular to one of the faces of a crystal.   **2.** One of the points at which normals to crystal faces or planes intersect a reference sphere at whose center the crystal is located. |ELEC| **1.** One of the electrodes in an electric cell.   **2.** An output terminal on a switch; a double-pole switch has two output terminals.   { pōl }

pole strength   See magnetic pole strength.   { 'pōl ,strenkth }

poling |MET| A technique used in the refining of copper that consists of the thrusting of greenwood poles into the molten metal in order to generate the reducing gases that react with the oxides in the metal.   { 'pōl·iŋ }

polish |MATER| A powder, liquid, or semiliquid used to give smoothness, surface protection, or decoration to finishes; for example, finely ground red oxide (rouge) is used to polish plate glass, mirror backs, and optical glass; solvent-wax liquids and pastes are used to protect and enhance leather and wood surfaces; nitrocellulose lacquers are used to paint finger- and toenails.   { 'päl·ish }

polonium |CHEM| A chemical element, symbol Po, atomic number 84; all polonium isotopes are radioactive; polonium-210 is the naturally occurring isotope found in pitchblende.   { pə'lō·nē·əm }

poly- |POLYM CHEM| A chemical prefix meaning many; for example, a polymer is made of a number of single molecules known as monomers, as polyethylene is made from ethylene.   { 'päl·ē, 'päl·ə, 'päl·i }

polyacetals   See acetal resins.   { |päl·ē'as·ə,talz }

polyacrylamide |POLYM CHEM| $(CH_2CHCONH_2)_x$ A white, water-soluble high polymer based on acrylamide; used as a thickening or suspending agent in water-base formulations.   { |päl·ē·ə'kril·ə·məd }

polyacrylate |POLYM CHEM| A polymer of an ester or salt of acrylic acid.   { |päl·ē'ak·rə,lāt }

polyacrylic acid |POLYM CHEM| $(CH_2CHCO-OH)_x$ An acrylic or acrylate resin formed by the polymerization of acrylic acid; water-soluble; used as a suspending and textile-sizing agent, and in adhesives, paints, and hydraulic fluids.   { |päl·ē·ə|kril·ik 'as·əd }

polyacrylic fiber |POLYM CHEM| Continuous-strand fiber extruded from an acrylate resin.   { |päl·ē·ə|kril·ik 'fī·bər }

polyacrylonitrile |POLYM CHEM| Polymer of acrylonitrile; semiconductive; used like an inorganic oxide catalyst to dehydrogenate *tert*-butyl alcohol to produce isobutylene and water.   { |päl·ē|ak·rə·lō'nī·trəl }

polyalcohol   See polyhydric alcohol.   { |päl·ē'al·kə,hȯl }

polyallomer |ORG CHEM| A copolymer of propylene with other olefins.   { |päl·ē'al·ə·mər }

polyamide |POLYM CHEM| Any member of a class of polymers in which individual structural units are joined by amide bonds.   { |päl·ē'am·əd }

polyamide resin |POLYM CHEM| Product of polymerization of amino acid or the condensation of a polyamine with a polycarboxylic acid; an example is the nylons.   { |päl·ē'am·əd 'rez·ən }

polybasic |CHEM| A chemical compound in solution that yields two or more H^- ions per molecule, such as sulfuric acid, H_2SO_4.   { |päl·i ,bā·sik }

polyblend |MATER| A mechanical mixture of two or more polymers, such as polystyrene and rubber.   { 'päl·i,blend }

polybutadiene |POLYM CHEM| Oil-extendable synthetic elastomer polymer made from butadiene; resilience is similar to natural rubber; it is blended with natural rubber for use in tire and other rubber products. Also known as butadiene rubber.   { |päl·i,byüd·ə'dī,ēn }

polybutene |POLYM CHEM| A polymer of isobutene, $(CH_3)_2CCH_2$; made in varying chain lengths to give a wide range of properties from oily to solid; used as a lube-oil additive, in adhesives, and in rubber products.   { |päl·i'byü ,tēn }

polybutylene |POLYM CHEM| A polymer of one or more butylenes whose consistency ranges from a viscous liquid to a rubbery solid.   { |päl·i'byüd·ə,lēn }

polycarbonate |POLYM CHEM| $|OC_6H_4C(CH_3)_2-C_6H_4OCO|_x$ A linear polymer of carbonic acid which is a thermoplastic synthetic resin made from bisphenol A and phosgene; used as high-strength glazing (for example, airplane windows) and to manufacture compact disks.   { |päl·i'kär·bə·nət }

polycarboxylic |POLYM CHEM| Prefix for a compound containing two or more carboxyl (−COOH) groups.   { |päl·i|kär,bäk|sil·ik }

polychroism   See pleochroism.   { ,päl·i'krō,iz·əm }

polycondensation |POLYM CHEM| A chemical condensation leading to the formation of a polymer by the linking together of molecules of a monomer and the releasing of water or a similar simple substance.   { |päl·i,kän·dən'sā·shən }

polycrystal |MATER| A polycrystalline solid.   { |päl·i'krist·əl }

polycrystalline |MATER| **1.** Pertaining to a material composed of aggregates of individual crystals. **2.** Characterized by variously oriented crystals. { ¦päl·i'krist·əl·ən }

polydispersity [POLYM CHEM] Molecular-weight nonhomogeneity in a polymer system; that is, there is some molecular-weight distribution throughout the body of the polymer. { ¦päl·i·di'spər·səd·ē }

polyelectrolyte [POLYM CHEM] A natural or synthetic electrolyte with high molecular weight, such as proteins, polysaccharides, and alkyl addition products of polyvinyl pyridine; can be a weak or strong electrolyte; when dissociated in solution, it does not give uniform distribution of positive and negative ions (the ions of one sign are bound to the polymer chain while the ions of the other sign diffuse through the solution). { ¦päl·ē·ə'lek·trə,līt }

polyene [ORG CHEM] Compound containing many double bonds, such as the carotenoids. { 'päl·ē,ēn }

polyester fiber [TEXT] A fiber filament made from a material that is 85% or more thermoplastic polyester resin. { 'päl·ē,es·tər 'fī·bər }

polyester film [MATER] Thin film made of polyester resin; used for packaging food and other products. { 'päl·ē,es·tər 'film }

polyester laminate [MATER] Glass fabric or fiber mat impregnated with a polyester resin slurry, and cured; used to make sheets, bars, and structural shapes. { 'päl·ē,es·tər 'lam·ə·nət }

polyester-reinforced urethane [MATER] A poromeric material which may have a urethane impregnation or a silicone coating; used for shoe uppers and as a substitute for industrial leathers. { 'päl·ē,es·tər ,rē·in,först 'yür·ə,thān }

polyester resin [POLYM CHEM] A thermosetting or thermoplastic synthetic resin made by esterification of polybasic organic acids with polyhydric acids; examples are Dacron and Mylar; the resin has high strength and excellent resistance to moisture and chemicals when cured. { 'päl·ē ,es·tər 'rez·ən }

polyether [POLYM CHEM] Any compound whose molecular structure contains linked ethers, R—O—R', where R and R' represent functional groups. { 'päl·ē,ē·thər }

polyether resin [POLYM CHEM] Any member of a large group of thermoplastic or thermosetting polymers that contain the typical polyether linkages in the polymer chain. { 'päl·ē,ē·thər 'rez·ən }

polyethylene See ethylene resin. { ¦päl·ē'eth·ə ,lēn }

polyethylene glycol [POLYM CHEM] Any of a family of colorless, water-soluble liquids with molecular weights from 200 to 6000; soluble also in aromatic hydrocarbons (not aliphatics) and many organic solvents; used to make emulsifying agents and detergents, and as plasticizers, humectants, and water-soluble textile lubricants. { ¦päl·ē'eth·ə,lēn 'glī,kòl }

polyethylene resin See ethylene resin. { ¦päl·ē'eth·ə,lēn 'rez·ən }

polyethylene terephthalate [POLYM CHEM] A thermoplastic polyester resin made from ethylene glycol and terephthalic acid; melts at 265°C; used to make films or fibers. Abbreviated PET. { ¦päl·ē'eth·ə,lēn ,ter·ə'tha,lāt }

polyformaldehyde See polyoxymethylene. { ,päl· ē·fór'mal·də,hīd }

polyglycol [ORG CHEM] A dihydroxy ether derived from the dehydration (removal of a water molecule) of two or more glycol molecules; an example is diethylene glycol, $CH_2OHCH_2OCH_2CH_2OH$. { 'päl·i,glī,kòl }

polyglycolic acid polymer [BIOMATER] A synthetic biodegradable polymer material used as dissolvable sutures and tissue engineering scaffolds. { ,päl·ē·glī¦käl·ik ¦as·əd 'päl·i·mər }

polygonization [SOLID STATE] A phenomenon observed during the annealing of plastically bent crystals in which the edge dislocations created by cold working organize themselves vertically above each other so that polygonal domains are formed. { pə,lig·ə·nə'zā·shən }

polygon wall See tilt boundary. { 'päl·i,gän ,wòl }

polyhydric alcohol [ORG CHEM] An alcohol with many hydroxyl (—OH) radicals, such as glycerol, $C_3H_5(OH_3)$. Also known as polyalcohol; polyol. { ¦päl·i¦hī·drək 'al·kə,hòl }

polyimide [POLY ENG] A group of polymers that contain a repeating imide group (—CONHCO—). Aromatic polyimides are noted for their resistance to high temperatures, wear, and corrosion. { ¦päl·ē'ī,mīd }

polyimide resin [POLYM CHEM] An aromatic polyimide made by reacting pyromellitic dianhydride with an aromatic diamine; has high resistance to thermal stresses; used to make components of internal combustion engines. { ¦päl·ē'i,mīd ,rez·ən }

polyisoprene [POLYM CHEM] $(C_5H_8)_x$ The basis of natural rubber, balata, gutta-percha, and other rubberlike materials; can also be made synthetically; the stereospecific forms are *cis*-1,4- and *trans*-1,4-polyisoprene; the polymer is thermoplastic. { ¦päl·ē'īs·ə,prēn }

polylactic resin [POLYM CHEM] A soft, elastic resin made by the heat reaction of lactic acid with castor oil or other fatty oils; used to produce tough, water-resistant coatings. { päl·i¦lak·tik ,rez·ən }

polyLED See polymer light-emitting diode. { ,päl·ē¦el¦ē'dē }

polymer [ORG CHEM] High-molecular-weight materials composed of repeating subunits; these materials may be organic, inorganic, as well as synthetic or natural in origin. { 'päl·ə·mər }

polymer blend [POLYM CHEM] A homogeneous mixture of two or more different polymers. { 'päl·ə·mər ,blend }

polymeric [POLYM CHEM] Made of repeating subunits. { ,päl·ə'mer·ik }

polymerization [POLYM CHEM] **1.** The bonding of two or more monomers to produce a polymer. **2.** Any chemical reaction that produces such a bonding. { pə,lim·ə·rə'zā·shən }

polymer light-emitting diode [POLYM CHEM] An organic polymeric material that emits light

polyterpene resin

in response to the application of an electric field. It may be an organic semiconductor sandwiched between metals of high and low work functions or a heterostructure made of two polymers, which increases the likelihood of radiative electron-hole recombination because of the energy-band structure. Also known as light-emitting polymer; polyLED. { ¦päl·ə·mər ¸līt·i¸mid·iŋ 'dī¸ōd }

polymer paint [MATER] A paint made of acrylic resin or vinyl resin, or a combination of both resins, in a liquid form with water as the base; it spreads out in a layer, and the water evaporates to leave a continuous, flexible, and waterproof film of plastic. { 'päl·ə·mər ¦pānt }

polymer plastic [MATER] The product of a high polymer with or without additives, such as plasticizers, autooxidants, colorants, or fillers; can be sprayed, shaped, molded, extruded, cast or foamed, depending on whether it is thermoplastic or thermosetting. { 'päl·ə·mər ¦plas·tik }

polymethyl methacrylate [POLYM CHEM] A thermoplastic polymer that is derived from methyl methacrylate, $CH_2=C(CH_3)COOCH_3$; transparent solid with excellent optical qualities and water resistance; used for aircraft domes, lighting fixtures, optical instruments, and surgical appliances. { 'päl·i¸meth·əl mə'thak·rə¸lāt }

polymolecular assembly [CHEM] The spontaneous association of a large number of components into a specific phase (films, layers, membranes, vesicles, micelles, mesophases, surfaces, solids, and so on). { ¸päl·ē·mə¸lek·yə·lər ə'sem·blē }

polymorph [CRYSTAL] One of the crystal forms of a substance displaying polymorphism. Also known as polymorphic modification. { 'päl·i ¸mórf }

polymorphic modification See polymorph. { ¸päl·i¦mór·fik ¸mäd·ə·fə'kā·shən }

polymorphism [CRYSTAL] The property of a chemical substance crystallizing into two or more forms having different structures, such as diamond and graphite. Also known as pleomorphism. { ¸päl·i·mór¸fiz·əm }

polyol See polyhydric alcohol. { 'päl·ē¸ól }

polyolefin [POLYM CHEM] A resinous material made by the polymerization of olefins, such as polyethylene from ethylene, polypropylene from propylene, or polybutene from butylene. { ¸päl·ē'ól·ə·fən }

polyolefin fiber [MATER] Continuous-strand fiber made from a polyolefin. { ¸päl·ē'ól·ə·fən 'fī·bər }

polyoxyalkylene resin [POLYM CHEM] Condensation polymer produced from an oxyalkene, such as polyethylene glycol from oxyethylene or ethylene glycol. { ¦päl·ē¦äk·sē'al·kə¸lēn 'rez·ən }

polyoxymethylene [POLYM CHEM] $(OCH_2)_n$ A polymer of formaldehyde that has excellent mechanical and high-temperature properties. Also know as polyacetal; polyformaldehyde. { ¸päl·ē¸äk·sē'meth·ə¸lēn }

polyphenylene oxide [POLYM CHEM] A polyether resin of 2,6-dimethylphenol, $(CH_3)_2$-

C_6H_3OH; useful temperature range is −275 to 375°F (−168 to 191°C), with intermittent use possible up to 400°F (204°C). { ¸päl·i'fen·əl¸ēn 'äk¸sīd }

polyphosphazene [POLYM CHEM] A high-molecular-weight, essentially linear polymer with alternating phosphorus and nitrogen atoms in the skeleton and two side groups attached to each phosphorus. { ¸päl·i'fä·sfə¸zēn }

polyphosphoric acid [INORG CHEM] $H_6P_4O_{13}$ Viscous, water-soluble, hygroscopic, water-white liquid; used wherever concentrated phosphoric acid is needed. { ¸päl·i·fä'sfór·ik 'as·əd }

polypropylene [POLYM CHEM] $(C_3H_6)_x$ A crystalline, thermoplastic resin made by the polymerization of propylene, C_3H_6; the product is hard and tough, resists moisture, oils, and solvents, and withstands temperatures up to 170°C; used to make molded articles, fibers, film, rope, printing plates, and toys. { ¸päl·ə'prō·pə ¸lēn }

polypropylene glycol [POLYM CHEM] CH_3CH-$OH(CH_2OCH$-$CH_3)_xCH_2OH$ Polymeric material similar to polyethylene glycol, but with greater oil solubility and less water solubility; used as a solvent for vegetable oils, waxes, and resins, in hydraulic fluids and as a chemical intermediate. { ¸päl·ə'prō·pə¸lēn 'glī¸kól }

polysiloxane [POLYM CHEM] $(R_2SiO)_n$ A polymer in which the chain contains alternate silicon and oxygen atoms; in the formula, R can be H or an alkyl or aryl group; commercially, the R is usually CH_3 (the methylsiloxanes); properties vary with molecular weight, from oils to greases to rubbers to plastics. { ¸päl·i·si'läk¸sän }

polystyrene [POLYM CHEM] $(C_6H_5CHCH_2)_x$ A water-white, tough synthetic resin made by polymerization of styrene; soluble in aromatic and chlorinated hydrocarbon solvents; used for injection molding, extrusion or casting for electrical insulation, fabric lamination, and molding of plastic objects. { ¸päl·i'stī¸rēn }

polysulfide rubber [POLYM CHEM] A synthetic polymer made by the reaction of sodium polysulfide with an organic dichloride; resistant to light, oxygen, oils, and solvents; impermeable to gases; poor tensile strength and abrasion resistance. { ¸päl·i'səl¸fīd 'rəb·ər }

polysulfone resin [POLYM CHEM] A thermoplastic polymer containing the sulfone linkage (O=S=O); these resins have exceptional high-temperature, low-creep, and arc-resistance properties, and are self-extinguishing. { ¸päl· i'səl¸fōn ¸rez·ən }

polysynthetic twinning [CRYSTAL] Repeated twinning that involves three or more individual crystals according to the same twin law and on parallel twin planes. { ¸päl·i·sin'thed·ik 'twin·iŋ }

polyterpene resin [POLYM CHEM] A thermoplastic resin or viscous liquid from polymerization of turpentine; used in paints, polishes, and rubber plasticizers; used to cure concrete and impregnate paper. { ¦päl·i'tər ¸pēn ¸rez·ən }

271

polytetrafluoroethylene [POLYM CHEM] $(CF_2CF_2)_n$ A highly crystalline perfluorinated polymer that is characteristically resistant to heat and chemicals. { ¦päl·ē¸te·trə¸flür·ō'eth·ə¸lēn }

polythene [POLYM CHEM] Common name for polyethlylene in Great Britain. { 'päl·i¸thēn }

polytrifluorochloroethylene resin See chlorotrifluoroethylene polymer. { ¦päl·i·trī¦flür·ō ¦klôr·ō'eth mbə¸lēn ¸rez·ən }

polytropic process [THERMO] An expansion or compression of a gas in which the quantity pV^n is held constant, where p and V are the pressure and volume of the gas, and n is some constant. { ¦päl·i¦träp·ik 'prä·səs }

polytype [CRYSTAL] A type of polymorph whose different forms are due to more than one possible mode of atomic packing. { 'päl·i¸tīp }

polytypism [CRYSTAL] The ability of a mineral to crystallize into more than one form, because of more than one possible mode of atomic packing. { ¦päl·i'ti¸piz·əm }

polyunsaturated acid [ORG CHEM] A fatty acid with two or more double bonds per molecule, such as linoleic or linolenic acid. { ¦päl·ē ¸ən'sach·ə¸rād·əd 'as·əd }

polyunsaturated fat [MATER] A fat or oil based on fatty acids such as linoleic or linolenic acids which have two or more double bonds in each molecule; corn oil and safflower oil are examples. { ¦päl·ē¸ən'sach·ə¸rād·əd 'fat }

polyurethane foam [MATER] A solid or spongy cellular material produced by the reaction of a polyester (such as glycerin) with a diisocyanate (such as toluene diisocyanate) while carbon dioxide is liberated by the reaction of a carboxyl with the isocyanate; used for thermal insulation, soundproofing, and padding. { ¦päl·ē'yür·ə¸thän 'fōm }

polyurethane resin [POLYM CHEM] Any resin resulting from the reaction of diisocyanates (such as toluene diisocyanate) with a phenol, amine, or hydroxylic or carboxylic compound to produce a polymer with free isocyanate groups; used as protective coatings, potting or casting resins, adhesives, rubbers, and foams, and in paints, varnishes, and adhesives. { ¦päl·ē'yür·ə¸thän 'rez·ən }

polyurethane rubber [POLYM CHEM] A synthetic polyurethane-resin elastomer made by the reaction of a diisocyanate with a polyol, such as a polyester (for example, the glycol-adipic acid ester); has high resistance to abrasion, oil, ozone, and high temperatures. { ¦päl·ē'yür·ə ¸thän 'rəb·ər }

polyvinyl acetal resin See vinyl acetal resin. { ¦päl·i'vīn·əl 'as·ə¸tal 'rez·ən }

polyvinyl acetate [POLYM CHEM] $(H_2CCHOOCCH_3)_x$ A thermoplastic polymer; insoluble in water, gasoline, oils, and fats, soluble in ketones, alcohols, benzene, esters, and chlorinated hydrocarbons; used in adhesives, films, lacquers, inks, latex paints, and paper sizes. Abbreviated PVA; PVAc. { ¦päl·i'vīn·əl 'as·ə¸tāt }

polyvinyl alcohol [POLYM CHEM] Water-soluble polymer made by hydrolysis of a polyvinyl ester (such as polyvinyl acetate); used in adhesives, as textile and paper sizes, and for emulsifying, suspending, and thickening of solutions. Abbreviated PVA. { ¦päl·i'vīn·əl 'al·kə¸hôl }

polyvinyl carbazole [POLYM CHEM] Thermoplastic resin made by reaction of acetylene with carbazole; softens at 150°C; has good electrical properties and heat and chemical stabilities; used as a paper-capacitor impregnant and as a substitute for electrical mica. { ¦päl·i'vīn·əl 'kär·bə¸zōl }

polyvinyl chloride [POLYM CHEM] $(H_2CCHCl)_x$ Polymer of vinyl chloride; tasteless, odorless; insoluble in most organic solvents; a member of the family of vinyl resins; used in soft flexible films for food packaging and in molded rigid products such as pipes, fibers, upholstery, and bristles. Abbreviated PVC. { ¦päl·i'vīn·əl 'klôr ¸īd }

polyvinyl chloride acetate [POLYM CHEM] Thermoplastic copolymer of vinyl chloride, CH_2CHCl, and vinyl acetate, $CH_3COOCH=CH_2$; colorless solid with good resistance to water, concentrated acids, and alkalies; compounded with plasticizers, it yields a flexible material superior to rubber in aging properties; used for cable and wire coverings and protective garments. { ¦päl·i'vīn·əl 'klôr¸īd 'as·ə¸tāt }

polyvinyl dichloride [POLYM CHEM] A high-strength polymer of chlorinated polyvinyl chloride; it is self-extinguishing and has superior chemical resistance; used for pipes carrying hot, corrosive materials. Abbreviated PVDC. { ¦päl·i'vīn·əl dī'klôr¸īd }

polyvinyl ether See polyvinyl ethyl ether. { ¦päl·i'vīn·əl 'ē·thər }

polyvinyl ethyl ether [POLYM CHEM] $[-CH(OC_2H_5)CH_2-]_x$ A viscous gum to rubbery solid, soluble in organic solvents; used for pressure-sensitive tape. Also known as polyvinyl ether. { ¦päl·i'vīn·əl 'eth·əl 'ē·thər }

polyvinyl fluoride [POLYM CHEM] $(-H_2CCHF-)_x$ Vinyl fluoride polymer; has superior resistance to weather, chemicals, oils, and stains, and has high strength; used for packaging (but not of food) and electrical equipment. { ¦päl·i'vīn·əl 'flür¸īd }

polyvinyl formate resin [POLYM CHEM] $(CH_2=CHOOCH)_x$ Clear-colored resin that is hard and solvent-resistant; used to make clear, hard plastics. { ¦päl·i'vīn·əl 'fôr¸māt ¸rez·ən }

polyvinylidene chloride [POLYM CHEM] Thermoplastic polymer of vinylidene chloride, $H_2C=CCl_2$; white powder softening at 185–200°C; used to make soft-flexible to rigid products. { ¦päl·i·vī'nil·ə¸dēn 'klôr¸īd }

polyvinylidene fluoride [POLYM CHEM] Fluorocarbon polymer made from vinylidene fluoride, $(H_2C=CF_2)$; has good tensile and compressive strength and high impact strength; used in chemical equipment for gaskets, impellers, and other pump parts, and for drum linings and protective coatings. { ¦päl·i·vī'nil·ə¸dēn 'flür¸īd }

polyvinylidene resin See vinylidene resin. { ¦päl·i·vī'nil·ə¸dēn 'rez·ən }

polyvinyl isobutyl ether |POLYM CHEM| |−CH₂-CHOCH₂CH(CH₃)₂−|ₓ An odorless synthetic resin; elastomer to viscous liquid depending on molecular weight; soluble in hydrocarbons, esters, ethers, and ketones, insoluble in water; used in adhesives, waxes, plasticizers, lubricating oils, and surface coatings. Abbreviated PVI. { ¦päl·i'vīn·əl ¦ī·sə¦byüd·əl 'e·thər }

polyvinyl methyl ether |POLYM CHEM| (−CH₂-CHOCH₃−)ₓ A colorless, tacky liquid, soluble in organic solvents, except aliphatic hydrocarbons, and in water below 32°C; used for pressure-sensitive adhesives, as a heat sensitizer for rubber latex, and as a pigment binder in inks and textile finishing. Abbreviated PVM. { ¦päl·i'vīn·əl 'meth·əl 'ē·thər }

polyvinyl pyrrolidone |POLYM CHEM| (C₆H₉-NO)ₓ A water-soluble, white, resinous solid; used in pharmaceuticals, cosmetics, detergents, and foods, and as a synthetic blood plasma. Abbreviated PVP. { ¦päl·i'vīn·əl pə'räl·ə ‚dōn }

polyvinyl resin |POLYM CHEM| Any resin or polymer derived from vinyl monomers. Also known as vinyl plastic. { ¦päl·i'vīn·əl 'rez·ən }

Pomeranchuk cooling |CRYO| A method of attaining temperatures as low as 1 millikelvin in which helium-3 is cooled by adiabatic compression at temperatures below 0.3 K. { ¦päm·ə ¦ran·chək 'kül·iŋ }

Poole-Frenkel effect |ELEC| An increase in the electrical conductivity of insulators and semiconductors in strong electric fields. { ¦pül 'freŋ·kəl i‚fekt }

porcelain |MATER| A high-grade ceramic ware characterized by high strength, a white color, very low absorption, good translucency, and a hard glaze. Also known as European porcelain; hard paste porcelain; true porcelain. { 'pòrs·lən }

porcelain cement |MATER| A cement for bonding porcelain to porcelain, such as a mixture of gutta-percha and shellac. { 'pòrs·lən si ‚ment }

porcelain clay |MATER| A clay suitable for use in the manufacture of porcelain, specifically kaolin. Also known as porcelain earth. { 'pòrs·lən ‚klā }

porcelain earth See porcelain clay. { 'pòrs·lən ‚ərth }

porcelain enamel See vitreous enamel. { 'pòrs·lən i'nam·əl }

porcelain insulator |MATER| An electrical insulator made from porcelain; the porcelain is often made in a one-fire process, the glaze being applied to the green or unfired ware, in contrast to the two-fire process used in making ordinary porcelain. { 'pòrs·lən 'in·sə‚lād·ər }

pore |MET| A minute cavity in a powder compact, metal casting, or electroplated coating. { pòr }

porosity |PHYS| 1. Property of a solid which contains many minute channels or open spaces. 2. The fraction as a percent of the total volume occupied by these channels or spaces; for example, in petroleum engineering the ratio (expressed in percent) of the void space in a rock to the bulk volume of that rock. { pə'räs·əd·ē }

porous |MATER| 1. Filled with pores. 2. Capable of absorbing liquids. { 'pòr·əs }

porous alum See aluminum sodium sulfate. { 'pòr·əs 'al·əm }

porous carbon |MATER| Plates, tubes, or disks of uniform carbon particles pressed together without a binder; used for the filtration of corrosive liquids and gases. { 'pòr·əs 'kär·bən }

porous graphite |MATER| Plates, tubes, or disks of uniform graphite particles pressed together without a binder; more resistant to oxidation but lower in strength than porous carbon. { 'pòr·əs 'gra‚fīt }

porous metals |MET| Metals, made by powder metallurgy, having uniformly distributed controlled pore sizes, in the form of sheets, tubes, and shapes; used for filtering liquids and gases at elevated temperatures. { 'pòr·əs 'med·əlz }

Portevin-Le Chatelier effect |SOLID STATE| The effect of foreign atoms on the deformation curve of a material, in which steps appear in what was initially a smooth curve. { ¦port‚van lə‚shat·lē'ā i‚fekt }

porthole die |MET| An extrusion die having two or more sections in which metal is extruded separately in each section and welded before leaving the die to form intricate hollow shapes. { 'pòrt‚hōl ‚dī }

portland cement |MATER| A hydraulic cement made of pulverized, calcined argillaceous and calcareous materials; the proper name for ordinary cement. { 'pòrt·lənd si'ment }

portland-pozzolana cement |MATER| Portland cement to which pozzolana has been added, in the amount of about 20%, to reduce the liability of leaching. { 'pòrt·lənd ‚pät·sə'län·ə si'ment }

portland stone |MATER| A type of oolitic limestone that consists of fossils cemented together with lime; found only on the Isle of Portland, Dorset, England. { 'pòrt·lənd 'stōn }

positioned weld |MET| A weld made in a joint in which members have been positioned to facilitate welding. { pə'zish·ənd 'weld }

positive |ELEC| Having fewer electrons than normal, and hence having ability to attract electrons. |GRAPHICS| Having the same rendition of light and shade as in the original scene. { 'päz·əd·iv }

positive charge |ELEC| The type of charge which is possessed by protons in ordinary matter, and which may be produced in a glass object by rubbing with silk. { 'päz·əd·iv 'chärj }

positive crystal |OPTICS| 1. Uniaxial anisotropic crystal having the ordinary index of refraction greater than the extraordinary index. 2. Biaxial anisotropic crystal having the intermediate index of refraction beta closer in value to alpha, and with Z the acute bisectrix. { 'päz·əd·iv 'krist·əl }

positive electrode See anode. { 'päz·əd·iv i'lek ‚trōd }

positive ion |CHEM| An atom or group of atoms which by loss of one or more electrons has acquired a positive electric charge; occurs in ionization of chemical compounds as H⁺ from ionization of hydrochloric acid, HCl. { 'päz·əd·iv 'ī‚än }

positive pole See north pole. { 'päz·əd·iv 'pōl }

positive temperature coefficient [THERMO]
The condition wherein the resistance, length,
or some other characteristic of a substance
increases when temperature increases.
{ 'päz·əd·iv 'tem·prə·chər ‚kō·i‚fish·ənt }

positron depth profiling [SOLID STATE] A technique in which the spread in stopping depths
from a low-energy monoenergetic positron beam
is measured and used to obtain information
on the presence and depth of various crystal
defects below the surface. { 'päz·ə‚trän 'depth
‚prō‚fīl·iŋ }

postcard paper [MATER] Lightweight Bristol
board, made from soda and sulfite pulps, that
has a smooth, firm surface for writing with pen
or pencil, or for ordinary printing. { 'pōs‚kärd
‚pā·pər }

postcure bonding [ENG] A method of postcur-
ing at elevated temperatures of parts previously
subjected to autoclave or press in order to obtain
higher heat-resistant properties of the adhesive
bond. { 'pōst‚kyür 'bänd·iŋ }

poster board [MATER] A stiff cardboard used for
show cards, posters, display advertising, and
signs; it may be white or colored on one side.
{ 'pōs·tər ‚bórd }

poster paint [MATER] A water paint with a gum
binder, which is brilliant, opaque, and fast-
drying, and usually sold in jars. Also known as
show-card color. { 'pōs·tər ‚pānt }

poster paper [MATER] A strong, waterproofed
paper used for billboard poster work; it is white
or colored with nonfading pigments, and does
not curl when paste is applied to it. { 'pōs·tər
‚pā·pər }

postheating [MET] Application of heat after thermal spraying, brazing, or welding to control
cooling rate. { ¦pōst'hēd·iŋ }

postweld interval [MET] The elapsed time between the end of a resistance welding current and
the start of hold time. { 'pōst‚weld 'in·tər·vəl }

potash See potassium carbonate. { 'päd‚ash }

potash blue [INORG CHEM] A pigment made by
oxidizing ferrous ferrocyanide; used in making
carbon paper. { 'päd‚ash 'blü }

potassium [CHEM] A chemical element, symbol K, atomic number 19, atomic weight
39.0983; an alkali metal. Also known as kalium.
{ pə'tas·ē·əm }

potassium acid carbonate See potassium bicarbonate. { pə'tas·ē·əm 'as·əd 'kär·bə‚nāt }

potassium acid fluoride See potassium bifluoride. { pə'tas·ē·əm 'as·əd 'flür‚īd }

potassium acid phosphate See potassium phosphate. { pə'tas·ē·əm 'as·əd 'fäs‚fāt }

potassium acid sulfate See potassium bisulfate.
{ pə'tas·ē·əm 'as·əd 'səl‚fāt }

potassium acid sulfite See potassium bisulfite.
{ pə'tas·ē·əm 'as·əd 'səl‚fīt }

potassium alum See potassium aluminum sulfate. { pə'tas·ē·əm 'al·əm }

potassium aluminate [INORG CHEM] $K_2Al_2O_4 \cdot$
$3H_2O$ Water-soluble, alcohol-insoluble, lustrous
crystals; used as a dyeing and printing mordant,

and as a paper sizing. { pə'tas·ē·əm ə'lüm·ə
‚nāt }

potassium aluminum fluoride [INORG CHEM]
K_3AlF_6 A toxic, white powder used as an insecticide. { pə'tas·ē·əm ə'lüm·ə·nəm 'flür‚īd }

potassium aluminum sulfate [INORG CHEM]
$KAl(SO_4)_2 \cdot 12H_2O$ White, odorless crystals that
are soluble in water; used in medicines and
baking powder, in dyeing, papermaking, and
tanning. Also known as alum; aluminum potassium sulfate; potassium alum. { pə'tas·ē·əm
ə'lüm·ə·nəm 'səl‚fāt }

potassium antimonate [INORG CHEM] $KSbO_3$
White, water-soluble crystals. Also known as
potassium stibnate. { pə'tas·ē·əm 'ant·ə·mə
‚nāt }

potassium arsenate [INORG CHEM] K_3AsO_4
Poisonous, colorless crystals; soluble in water,
insoluble in alcohol; used as an insecticide,
analytical reagent, and in hide preservation and
textile printing. Also known as Macquer's salt.
{ pə'tas·ē·əm 'ärs·ən‚āt }

potassium arsenite [INORG CHEM] $KH(AsO_2)_2$
Poisonous, hygroscopic, white powder; soluble
in alcohol; decomposes slowly in air; used
in medicine, on mirrors, and as an analytical
reagent. Also known as potassium metarsenite.
{ pə'tas·ē·əm 'ärs·ən‚īt }

potassium aurichloride See potassium gold
chloride. { pə'tas·ē·əm ¦ór·ə'klór‚īd }

potassium bicarbonate [INORG CHEM] $KHCO_3$
A white powder or granules, or transparent
colorless crystals; used in baking powder and
in medicine as an antacid. Also known as
potassium acid carbonate. { pə'tas·ē·əm
bī'kär·bə‚nāt }

potassium bichromate See potassium dichromate. { pə'tas·ē·əm bī'krō‚māt }

potassium bifluoride [INORG CHEM] KHF_2 Colorless, corrosive, poisonous crystals; soluble in
water and dilute alcohol; used to etch glass and
as a metallurgy flux. Also known as potassium
acid fluoride. { pə'tas·ē·əm bī'flür‚īd }

potassium bisulfate [INORG CHEM] $KHSO_4$
Water-soluble, colorless crystals, melting
at 214°C; used in winemaking, fertilizer
manufacture, and as a flux and food preservative.
Also known as acid potassium sulfate; potassium
acid sulfate. { pə'tas·ē·əm bī'səl‚fāt }

potassium bisulfite [INORG CHEM] $KHSO_3$
White, water-soluble powder with sulfur dioxide
aroma; insoluble in alcohol; decomposes when
heated; used as an antiseptic and reducing
chemical, and in analytical chemistry, tanning,
and bleaching. Also known as potassium acid
sulfite. { pə'tas·ē·əm bī'səl‚fīt }

potassium borohydride [INORG CHEM] KBH_4 A
white, crystalline powder, soluble in water, alcohol, and ammonia; used as a hydrogen source
and a reducing agent for aldehydes and ketones.
{ pə'tas·ē·əm ¦bór·ō'hī‚drīd }

potassium bromate [INORG CHEM] $KBrO_3$ Water-
soluble, white crystals, melting at 434°C; insoluble in alcohol; strong oxidizer and a fire hazard;
used in analytical chemistry and as an additive

for permanent-wave compounds. { pə'tas·ē·əm 'brō₁māt }

potassium bromide |INORG CHEM| KBr White, hygroscopic crystals with bitter taste; soluble in water and glycerin, slightly soluble in alcohol and ether; melts at 730°C; used in medicine, soaps, photography, and lithography. { pə'tas·ē·əm 'brō₁mīd }

potassium cadmium iodide See potassium tetraiodocadmate. { pə'tas·ē·əm 'kad·mē·əm 'ī·ə₁dīd }

potassium carbonate |INORG CHEM| K₂CO₃ White, water-soluble, deliquescent powder, melting at 891°C; insoluble in alcohol; used in brewing, ceramics, explosives, fertilizers, and as a chemical intermediate. Also known as potash; salt of tartar. { pə'tas·ē·əm 'kär·bə₁nāt }

potassium chlorate |INORG CHEM| KClO₃ Transparent, colorless crystals or a white powder with a melting point of 356°C; soluble in water, alcohol, and alkalies; used as an oxidizing agent, for explosives and matches, and in textile printing and paper manufacture. { pə'tas·ē·əm 'klȯr₁āt }

potassium chloride |INORG CHEM| KCl Colorless crystals with saline taste; soluble in water, insoluble in alcohol; melts at 776°C; used as a fertilizer and in photography and pharmaceutical preparations. Also known as potassium muriate. { pə'tas·ē·əm 'klȯr₁īd }

potassium chloroaurate See potassium gold chloride. { pə'tas·ē·əm ₁klȯr·ō'ȯr₁āt }

potassium chloroplatinate |INORG CHEM| K₂PtCl₆ Orange-yellow crystals or powder which decomposes when heated (250°C); used in photography. Also known as platinum potassium chloride; potassium platinichloride. { pə'tas·ē·əm ₁klȯr·ō'plat·ən₁āt }

potassium chromate |INORG CHEM| K₂CrO₄ Yellow crystals, melting at 971°C; soluble in water, insoluble in alcohol; used as an analytical reagent and textile mordant, in enamels, inks, and medicines, and as a chemical intermediate. { pə'tas·ē·əm 'krō₁māt }

potassium chromium sulfate See chrome alum. { pə'tas·ē·əm 'krō·mē·əm 'səl₁fāt }

potassium cobaltinitrite See cobalt potassium nitrite. { pə'tas·ē·əm kō₁bȯl·tə'nī₁trāt }

potassium cyanate |INORG CHEM| KOCN Colorless, water-soluble crystals; used as an herbicide and for the manufacture of drugs and organic chemicals. { pə'tas·ē·əm 'sī·ə₁nāt }

potassium cyanide |INORG CHEM| KCN Poisonous, white, deliquescent crystals with bitter almond taste; soluble in water, alcohol, and glycerol; used for metal extraction, for electroplating, for heat-treating steel, and as an analytical reagent and insecticide. { pə'tas·ē·əm 'sī·ə₁nīd }

potassium cyanoaurite See potassium gold cyanide. { pə'tas·ē·əm ₁sī·ə·nō'ȯr₁īt }

potassium dichloroisocyanurate |INORG CHEM| White, crystalline powder or granules; strong oxidant used in dry household bleaches,

detergents, and scouring powders. { pə'tas·ē·əm dī₁klȯr·ō₁ī'sō₁sī'an·yür₁āt }

potassium dichromate |INORG CHEM| K₂Cr₂O₇ Poisonous, yellowish-red crystals with metallic taste; soluble in water, insoluble in alcohol; melts at 396°C, decomposes at 500°C; used as an oxidizing agent and analytical reagent, and in explosives, matches, and electroplating. Also known as potassium bichromate; red potassium chromate. { pə'tas·ē·əm dī'krō₁māt }

potassium dihydrogen phosphate See potassium phosphate. { pə'tas·ē·əm dī'hī·drə·jən 'fäs₁fāt }

potassium diphosphate See potassium phosphate. { pə'tas·ē·əm dī'fäs₁fāt }

potassium ferric oxalate |INORG CHEM| K₃Fe(C₂O₄)₃·3H₂O Green crystals decomposing at 230°C, soluble in water and acetic acid; used in photography and blueprinting. { pə'tas·ē·əm 'fer·ik 'äk·sə₁lāt }

potassium ferricyanide |INORG CHEM| K₃Fe(CN)₆ Poisonous, water-soluble, bright-red crystals; decomposes when heated; used in calico printing and wool dyeing. Also known as red potassium prussiate; red prussiate of potash. { pə'tas·ē·əm ₁fer·ə'sī·ə₁nīd }

potassium ferrocyanide |INORG CHEM| K₄Fe(CN)₆·3H₂O Yellow crystals with saline taste; soluble in water, insoluble in alcohol; loses water at 60°C; used in medicine, dry colors, explosives, and as an analytical reagent. Also known as yellow prussiate of potash. { pə'tas·ē·əm ₁fer·ō'sī·ə₁nīd }

potassium fluoborate |INORG CHEM| KBF₄ White powder or gelatinous crystals that decompose at high temperatures; slightly soluble in water and hot alcohol; used as a sand agent to cast magnesium and aluminum, and in electrochemical processes. { pə'tas·ē·əm ₁flü·ə'bȯr₁āt }

potassium fluoride |INORG CHEM| KF or KF·2H₂O Poisonous, white, deliquescent crystals with saline taste; soluble in water and hydrofluoric acid, insoluble in alcohol; melts at 846°C; used to etch glass and as a preservative and insecticide. { pə'tas·ē·əm 'flür₁īd }

potassium fluosilicate |INORG CHEM| K₂SiF₆ An odorless, white crystalline compound; slightly soluble in water; used in vitreous frits, synthetic mica, metallurgy, and ceramics. Also known as potassium silicofluoride. { pə'tas·ē·əm ₁flü·ə'sil·ə₁kət }

potassium gold chloride |INORG CHEM| KAuCl₄·2H₂O Yellow crystals, soluble in water, ether, and alcohol; used in photography and medicine. Also known as gold potassium chloride; potassium aurichloride; potassium chloroaurate. { pə'tas·ē·əm 'gōld 'klȯr₁īd }

potassium gold cyanide |INORG CHEM| KAu(CN)₂ A white, water-soluble, crystalline powder; used in medicine and for gold plating. Also known as gold potassium cyanide; potassium cyanoaurite. { pə'tas·ē·əm 'gōld 'sī·ə₁nīd }

potassium hydrate See potassium hydroxide. { pə'tas·ē·əm 'hī₁drāt }

potassium hydrogen phosphate See potassium phosphate. { pə'tas·ē·əm 'hī·drə·jən 'fäs,fāt }

potassium hydroxide [INORG CHEM] KOH Toxic, corrosive, water-soluble, white solid, melting at 360°C; used to make soap and matches, and as an analytical reagent and chemical intermediate. Also known as caustic potash; potassium hydrate. { pə'tas·ē·əm hī'dräk,sīd }

potassium hyperchlorate See potassium perchlorate. { pə'tas·ē·əm ,hī·pər'klȯr,āt }

potassium hypophosphite [INORG CHEM] KH₂PO₂ White, opaque crystals or powder, soluble in water and alcohol; used in medicine. { pə'tas·ē·əm ,hī·pō'fäs,fīt }

potassium iodate [INORG CHEM] KIO₃ Odorless, white crystals; soluble in water, insoluble in alcohol; melts at 560°C; used as an analytical reagent and in medicine. { pə'tas·ē·əm 'ī·ə,dāt }

potassium iodide [INORG CHEM] KI Water- and alcohol-soluble, white crystals with saline taste; melts at 686°C; used in medicine and photography, and as an analytical reagent. { pə'tas·ē·əm 'ī·ə,dīd }

potassium manganate [INORG CHEM] K₂MnO₄ Water-soluble dark-green crystals, decomposing at 190°C; used as an analytical reagent, bleach, oxidizing agent, disinfectant, mordant for dyeing wool and in photography, printing, and water purification. { pə'tas·ē·əm 'maŋ·gə,nāt }

potassium metabisulfite [INORG CHEM] K₂S₂O₅ White granules or powder, decomposing at 150–190°C; used as an antiseptic, for winemaking, food preservation, and process engraving, and as a source for sulfurous acid. Also known as potassium pyrosulfite. { pə'tas·ē·əm ,med·ə·bī'səl,fīt }

potassium metarsenite See potassium arsenite. { pə'tas·ē·əm ¦med·ə'ärs·ən,īt }

potassium monophosphate See potassium phosphate. { pə'tas·ē·əm ,män·ō'fäs,fāt }

potassium muriate See potassium chloride. { pə'tas·ē·əm 'myůr·ē,āt }

potassium nitrate [INORG CHEM] KNO₃ Flammable, water-soluble, white crystals with saline taste; melts at 337°C; used in pyrotechnics, explosives, and matches, as a fertilizer, and as an analytical reagent. Also known as niter. { pə'tas·ē·əm 'nī,trāt }

potassium nitrite [INORG CHEM] KNO₂ White, deliquescent prisms, melting at 297–450°C; soluble in water, insoluble in alcohol; strong oxidizer, exploding at over 550°C; used as an analytical reagent, in medicine, organic synthesis, pyrotechnics, and explosives. { pə'tas·ē·əm 'nī,trīt }

potassium oxide [INORG CHEM] K₂O Gray, water-soluble crystals; melts at red heat; forms potassium hydroxide in water. { pə'tas·ē·əm 'äk,sīd }

potassium percarbonate [INORG CHEM] K₂C₂O₆·H₂O White, granular, water-soluble mass with a melting point of 200–300°C; used in microscopy, photography, and textile printing. { pə'tas·ē·əm pər'kär·bə,nāt }

potassium perchlorate [INORG CHEM] KClO₄ Explosive, oxidative, colorless crystals; soluble in water, insoluble in alcohol; decomposes at 400°C; used in explosives, medicine, pyrotechnics, analysis, and as a reagent and oxidizing agent. Also known as potassium hyperchlorate. { pə'tas·ē·əm pər'klȯr,āt }

potassium permanganate [INORG CHEM] KMnO₄ Highly oxidative, water-soluble, purple crystals with sweet taste; decomposes at 240°C; and explodes in contact with oxidizable materials; used as a disinfectant and analytical reagent, in dyes, bleaches, and medicines, and as a chemical intermediate. Also known as purple salt. { pə'tas·ē·əm pər'maŋ·gə,nāt }

potassium peroxide [INORG CHEM] K₂O₂ Yellow mass with a melting point of 490°C; decomposes with oxygen evolution in water; used as an oxidizing and bleaching agent. { pə'tas·ē·əm pə'räk,sīd }

potassium peroxydisulfate See potassium persulfate. { pə'tas·ē·əm pə¦räk·sē·dī'səl,fāt }

potassium persulfate [INORG CHEM] K₂S₂O₈ White, water-soluble crystals, decomposing below 100°C; used for bleaching and textile desizing, as an oxidizing agent and antiseptic, and in the manufacture of soap and pharmaceuticals. Also known as potassium peroxydisulfate. { pə'tas·ē·əm pər'səl,fāt }

potassium phosphate [INORG CHEM] Any one of three orthophosphates of potassium. The monobasic form, KH₂PO₄, consists of colorless, water-soluble crystals melting at 253°C; used in sonar transducers, optical modulation, medicine, baking powders, and nutrient solutions; also known as potassium acid phosphate, potassium dihydrogen phosphate (KDP), potassium diphosphate, potassium orthophosphate. The dibasic form, K₂HOP₄, consists of white, water-soluble crystals; used in medicine, fermentation, and nutrient solutions; also known as potassium hydrogen phosphate, potassium monophosphate. The tribasic form, K₃PO₄, is a water-soluble, hygroscopic white powder, melting at 1340°C; used to purify gasoline, to soften water, and to make liquid soaps and fertilizers; also known as neutral potassium phosphate, tripotassium orthophosphate. { pə'tas·ē·əm 'fäs,fāt }

potassium platinichloride See potassium chloroplatinate. { pə'tas·ē·əm ¦plat·ən·ə'klȯr,īd }

potassium polymetaphosphate [INORG CHEM] (KPO₃)ₙ White powder with a molecular weight up to 500,000; used in foods as a fat emulsifier and moisture-retaining agent. { pə'tas·ē·əm päl·e,med·ə'fäs,fāt }

potassium pyrophosphate [INORG CHEM] K₄P₂O₇·3H₂O Water-soluble, colorless crystals; dehydrates below 300°C, melts at 1090°C; used in tin plating, china-clay purification, dyeing, oil-drilling muds, and synthetic rubber production. Also known as normal potassium pyrophosphate; tetrapotassium pyrophosphate. { pə'tas·ē·əm ,pī·rō'fäs,fāt }

potassium pyrosulfite See potassium metabisulfite. { pə'tas·ē·əm ,pī·rō'səl,fīt }

potassium silicate [INORG CHEM] SiO_2=K_2O A compound existing in two forms, solution and solid (glass); as a solution, it is colorless to turgid in water, and is used in paints and coatings, as an arc-electrode binder and catalyst and in detergents; as a solid, it is colorless and water-soluble solid, and is used in glass manufacture and for dyeing and bleaching. { pə'tas·ē·əm 'sil·ə,kāt }

potassium silicofluoride See potassium fluosilicate. { pə'tas·ē·əm ¦sil·ə·kō'flur,īd }

potassium sodium ferricyanide [INORG CHEM] $K_2NaFe(CN)_6$ Red, water-soluble crystals; used for blueprint paper and in photography. { pə'tas·ē·əm 'sōd·ē·əm ,fer·ə'sī·ə,nīd }

potassium sodium tartrate [INORG CHEM] $KNaC_4H_4O_6·4H_2O$ Colorless, water-soluble, efflorescent crystals or white powder with a melting point of 70–80°C; used in medicine and as a buffer and sequestrant in foods. Also known as Rochelle salt; Seignette salt. { pə'tas·ē·əm 'sōd·ē·əm 'tär,trāt }

potassium stannate [INORG CHEM] $K_2SnO_3·3H_2O$ White crystals; soluble in water, insoluble in alcohol; used in textile printing and dyeing, and in tin-plating baths. { pə'tas·ē·əm 'stan ,āt }

potassium stibnate See potassium antimonate. { pə'tas·ē·əm 'stib,nīt }

potassium sulfate [INORG CHEM] K_2SO_4 Colorless crystals with bitter taste; soluble in water, insoluble in alcohol; melts at 1072°C; used as an analytical reagent, medicine, and fertilizer, and in aluminum and glass manufacture. Also known as salt of Lemery. { pə'tas·ē·əm 'səl ,fāt }

potassium sulfide [INORG CHEM] K_2S Moderately flammable, water-soluble, deliquescent red crystals; melts at 840°C; used in analytical chemistry, medicine, and depilatories. Also known as fused potassium sulfide; hepar sulfuris; potassium sulfuret. { pə'tas·ē·əm 'səl,fīd }

potassium sulfite [INORG CHEM] $K_2SO_3·2H_2O$ Water-soluble, white crystals; used in medicine and photography. { pə'tas·ē·əm 'səl,fīt }

potassium sulfuret See potassium sulfide. { pə'tas·ē·əm 'səl·fə,ret }

potassium tetraiodocadmate [INORG CHEM] $K_2(CdI_4)·2H_2O$ A crystalline compound; used in analytical chemistry for alkaloids, amines, and other compounds. Also known as cadmium potassium iodide; potassium cadmium iodide. { pə'tas·ē·əm ¦te·trə¦ī·ə·dō'kad,māt }

potassium thiocyanate [INORG CHEM] KCNS Water- and alcohol-soluble, colorless, odorless hygroscopic crystals with saline taste; decomposes at 500°C; used as an analytical reagent and in freezing mixtures, chemicals manufacture, textile printing and dyeing, and photographic chemicals. { pə'tas·ē·əm ¦thī·ō'sī·ə,nāt }

pot clay [MATER] Refractory clay used to make the pots in which glass is produced. { 'pät ,klā }

pot earth See potter's clay. { 'pät ,ərth }

potential See electric potential. { pə'ten·chəl }

potential difference [ELEC] Between any two points, the work which must be done against electric forces to move a unit charge from one point to the other. Abbreviated PD. { pə'ten·chəl ¦dif·rəns }

potential drop [ELEC] The potential difference between two points in an electric circuit. { pə'ten·chəl ¦dräp }

potential gradient [ELEC] Difference in the values of the voltage per unit length along a conductor or through a dielectric. { pə'ten·chəl 'grād·ē·ənt }

potential sputtering [ELECTR] The ejection of mainly neutral atoms from the surface of a solid insulator due to the impact of slow, multiply charged ions whose kinetic energy alone is incapable of initiating sputtering. { pə'ten·chəl ,spəd·ə·riŋ }

potential temperature [THERMO] The temperature that would be reached by a compressible fluid if it were adiabatically compressed or expanded to a standard pressure, usually 1 bar. { pə'ten·chəl 'tem·prə·chər }

pot life [POLYM CHEM] The period of time during which paint remains useful after its original package has been opened or after a catalyst or other additive has been incorporated. Also known as spreadable life; usable life. { 'pät ,līf }

potter's clay [MATER] A plastic clay, free from iron and devoid of fissility, suitable for modeling or pottery making or adapted for use on a potter's wheel. Also known as argil; pot earth; potter's earth. { 'päd·ərz 'klā }

potter's earth See potter's clay. { 'päd·ərz 'ərth }

pottery [MATER] Objects made of clay which may be nonvitreous, porous, opaque, and glazed or unglazed; also included is earthenware such as stoneware. { 'päd·ə·rē }

potting [ELECTR] Process of filling a complete electronic assembly with a thermosetting compound for resistance to shock and vibration, and for exclusion of moisture and corrosive agents. { 'päd·iŋ }

pounce [MATER] Pumice in the form of a very fine powder, used for preparing parchment and tracing cloth. { pauns }

pouncing paper [MATER] Paper coated with pumice for polishing felt hats. { 'pauns·iŋ ,pā·pər }

Pourbaix diagram [MET] A plot of potential versus pH used to predict the thermodynamic tendency of a metal to corrode. { ¦pur,bā 'dī·ə ,gram }

pour depressant See pour-point depressant. { 'pòr di,pres·ənt }

pouring [MET] Transferring molten metal to a mold or ladle. { 'pòr·iŋ }

pour point [FL MECH] Lowest test temperature at which a liquid will flow. [MET] Temperature at which a molten alloy is cast. { 'pòr ,pòint }

pour-point depressant [MATER] An additive that lowers the pour point of a wax-containing petroleum-base lubricating oil by reducing the tendency of the wax to solidify. Also known as pour depressant; pour-point inhibitor. { 'pór ‚póint di‚pres·ənt }

pour-point inhibitor *See* pour-point depressant. { 'pór ‚póint in'hib·əd·ər }

pour reversion |MATER| The difference between the original American Society for Testing and Materials (ASTM) pour point of a lubricating oil and the relatively high solidification temperature observed in the field. { 'pór ri‚vər·zhən }

pour stability |MATER| Ability of a pour-depressant-treated petroleum lubricating oil to maintain its original American Society for Testing and Materials (ASTM) pour point at low temperatures approximating winter conditions. { 'pór stə‚bil·əd·ē }

powder |MATER| A loose grouping or aggregation of solid particles, usually smaller than 1000 micrometers. *See* bulk solid. { 'paúd·ər }

powder coating |MATER| A dry coating method in which fine clear or pigmented powder particles, containing resin, modifiers, and possibly a curing agent, are electrostatically sprayed onto a substrate and heated (melted) in a oven to form a continuous film. { 'paúd·ər ‚kōt·iŋ }

powder diffraction camera |CRYSTAL| A metal cylinder having a window through which an x-ray beam of known wavelength is sent by an x-ray tube to strike a finely ground powder sample mounted in the center of the cylinder; crystal planes in this powder sample diffract the x-ray beam at different angles to expose a photographic film that lines the inside of the cylinder; used to study crystal structure. Also known as x-ray powder diffractometer. { 'paúd·ər di'frak·shən ‚kam·rə }

powder insulation |MATER| Thermal insulation material made up of a finely divided solid held between two surfaces (one hot, one cold); the powder reduces both convection and radiation heat flow between the surfaces. { 'paúd·ər ‚in·sə ‚lā·shən }

powder lubricant |MET| An agent mixed with a powder metal to facilitate formation and ejection of a compact. { 'paúd·ər 'lü·brə·kənt }

powder metal |MET| Finely divided particles of a metal. { 'paúd·ər ‚med·əl }

powder metallurgy |MET| A metalworking process used to fabricate parts of simple or complex shape from a wide variety of metals and alloys in the form of powders. The process involves shaping of the powder and subsequent bonding of its individual particles by heating or mechanical working. { 'paúd·ər 'med·əl ‚ər·jē }

powder method |SOLID STATE| A method of x-ray diffraction analysis in which a collimated, monochromatic beam of x-rays is directed at a sample consisting of an enormous number of tiny crystals having random orientation, producing a diffraction pattern that is recorded on film or with a counter tube. Also known as x-ray powder method. { 'paúd·ər ‚meth·əd }

powder pattern |CRYSTAL| In the powder method of x-ray diffraction analysis, the display of lines made on film by the Debye-Scherrer method or on paper by a recording diffractometer. { 'paúd·ər ‚pad·ərn }

power |OPTICS| *See* focal power. |PHYS| The time rate of doing work. { 'paú·ər }

Poynting effect |MECH| The effect of torsion of a very long cylindrical rod on its length. { 'póin·tiŋ i‚fekt }

Poynting's law |THERMO| A special case of the Clapeyron equation, in which the fluid is removed as fast as it forms, so that its volume may be ignored. { 'póint·iŋz ‚ló }

pozzolan |MATER| Cement made by mixing and grinding together slaked lime and pozzolan without burning; sometimes used for concrete not exposed to the air. { 'pät·sə·lən }

pp junction |ELECTR| A region of transition between two regions having different properties in *p*-type semiconducting material. { ¦pē¦pē ‚jəŋk·shən }

Pr *See* praseodymium.

practical entropy *See* virtual entropy. { 'prak·ti·kəl 'en·trə·pē }

Prandtl number |FL MECH| A dimensionless number used in the study of diffusion in flowing systems, equal to the kinematic viscosity divided by the molecular diffusivity. Symbolized Pr_m. Also known as Schmidt number 1 (N_{Sc}). |THERMO| A dimensionless number used in the study of forced and free convection, equal to the dynamic viscosity times the specific heat at constant pressure divided by the thermal conductivity. Symbolized N_{Pr}. { 'pränt·əl ‚nəm·bər }

praseodymium |CHEM| A chemical element, symbol Pr, atomic number 59, atomic weight 140.9077; a metallic element of the rare-earth group. { ‚prā·zē·ó'dim·ē·əm }

precession camera |CRYSTAL| An x-ray diffraction camera used in the Buerger precession method for recording the diffractions of an individual crystal. { prē'sesh·ən ‚kam·rə }

precharge |MET| The pressure introduced into the cavity of a mold before forming a part. { prē'chärj }

precious metal |MET| A relatively scarce, valuable metal, such as gold, silver, and members of the platinum group. { 'presh·əs 'med·əl }

precipitate |CHEM| **1.** A substance separating, in solid particles, from a liquid as the result of a chemical or physical change; **2.** To form a precipitate. { prə'sip·ə‚tāt }

precipitation |CHEM| The process of producing a separable solid phase within a liquid medium; represents the formation of a new condensed phase, such as a vapor or gas condensing to liquid droplets; a new solid phase gradually precipitates within a solid alloy as a result of slow, inner chemical reaction; in analytical chemistry, precipitation is used to separate a solid phase in an aqueous solution. { prə‚sip·ə'tā·shən }

precipitation hardening *See* age hardening. { prə‚sip·ə'tā·shən ‚härd·ən·iŋ }

precision casting |MET| A metal casting of accurately reproducible dimensions. { prə'sizh·ən ¦kast·iŋ }

precoat |MET| In casting, thin coating of refractory slurry applied to an expendable wax or plastic pattern as a base for the application of the main slurry. { ¦prē'kōt }

preforming |MET| **1.** Initial pressing of a powder metal to form a compact. **2.** Preliminary shaping of a refractory metal compact after presintering. **3.** In wire rope manufacturing, a process that sets a crimp in the strands, providing a permanent set and controlling the flexibility of the rope. { prē'fȯrm·iŋ }

preheat current |MET| Current impulses which occur prior to and apart from the electric current in a resistance welding process. { 'prē¸hēt ¸kə·rənt }

preparing salt *See* sodium stannate. { prə 'per·iŋ ¸sȯlt }

prepolymer |POLYM CHEM| A reactive low-molecular-weight macromolecule or an oligomer, capable of further polymerization. { prē'päl·i·mər }

prepolymer molding |ENG| A urethane-foam-producing system in which a portion of the polyol is prereacted with the isocyanate to form a liquid prepolymer with a pumpable viscosity; when combined with a second blend containing more polyol, catalyst, or blowing agent, the two components react and a foamed plastic results. { prē'päl·i·mər 'mōld·iŋ }

prepreg |ENG| A reinforced-plastics term for the reinforcing material that contains or is combined with the full complement of resin before the molding operation. { 'prē¸preg }

preservative |MATER| A chemical added to foodstuffs to prevent oxidation, fermentation, or other deterioration, usually by inhibiting the growth of bacteria. { pri'zər·vəd·iv }

presintering |MET| Heating a compact to a temperature lower than the final sintering temperature to facilitate handling or to remove a binder or lubricant. { prē'sint·ə·riŋ }

press drip |MATER| Oil that drips from the wax press after pressed petroleum distillate has been removed. { 'pres ¸drip }

pressed brick |MATER| Brick subjected to pressure before burning to eliminate imperfections of shape and texture. { 'prest 'brik }

pressed density |MET| The density of a metal powder compact before sintering. { 'prest 'den·səd·ē }

pressed distillate |MATER| Oil recovered when refinery paraffin distillate is pressed to separate the liquid from the solid wax. { 'prest 'dis·tə·lət }

pressed glass |MATER| Glass shaped by being poured into a mold under pressure or pressed into a mold in a plastic state. { 'prest 'glas }

press forging |MET| Forging hot metal between dies in a press. { 'pres ¸fȯrj·iŋ }

press forming |MET| A metal-forming operation performed with a mechanical or hydraulic press. { 'pres ¸fȯrm·iŋ }

pressing |MET| **1.** Shallow-drawing metal sheet or plate. **2.** Using compressive force to form a metal powder compact. { 'pres·iŋ }

pressure |MECH| A type of stress which is exerted uniformly in all directions; its measure is the force exerted per unit area. { 'presh·ər }

pressure casting |MET| Making castings of molten or plastic metal in metal molds under applied pressures. { 'presh·ər ¸kast·iŋ }

pressure coefficient |THERMO| The ratio of the fractional change in pressure to the change in temperature under specified conditions, usually constant volume. { 'presh·ər ¸kō·i¸fish·ənt }

pressure-sensitive adhesive |MATER| An adhesive that develops maximum bonding power when applied by a light pressure only. { 'presh·ər ¦sen·səd·iv ad'hē·siv }

pressure-sensitive paint |FL MECH| A flow visualization technique in which ultraviolet light is used to excite specific molecules in a special paint affixed to a test surface positioned in a wind tunnel flow. The resulting phosphorescence of these molecules indicates the amount of oxygen in contact with the paint and, thereby, the spatial distribution of surface pressure. { ¸presh·ər ¸sen·səd·iv 'pānt }

pressure welding |MET| Welding of metal surfaces by the application of pressure; examples are percussion welding, resistance welding, seam welding, and spot welding. { 'presh·ər ¸weld·iŋ }

prestress |ENG| To apply a force to a structure to condition it to withstand its working load more effectively or with less deflection. { ¦prē'stres }

prestressed concrete |MATER| Concrete compressed with heavily loaded wires or bars to reduce or eliminate cracking and tensile forces. { ¦prē'strest 'kän¸krēt }

Prevost's theory |THERMO| A theory according to which a body is constantly exchanging heat with its surroundings, radiating an amount of energy which is independent of its surroundings, and increasing or decreasing its temperature depending on whether it absorbs more radiation than it emits, or vice versa. { 'prä·vōz ¸thē·ə·rē }

preweld interval |MET| In resistance spot welding, elapsed time between the end of squeeze time and the beginning of welding current. { prē'weld 'in·tər·vəl }

prill |MATER| Spherical particles about the size of buckshot. { pril }

primary |CHEM| A term used to distinguish basic compounds from similar or isomeric forms; in organic compounds, for example, RCH_2OH is a primary alcohol, R_1R_2CHOH is a secondary alcohol, and $R_1R_2R_3COH$ is a tertiary alcohol; in inorganic compounds, for example, NaH_2PO_4 is primary sodium phosphate, Na_2HPO_4 is the secondary form, and Na_3PO_4 is the tertiary form. |MET| Of a metal, obtained directly from ore. { 'prī ¸mer·ē }

primary colors |OPTICS| **1.** Three colors, red, yellow, and blue, which can be combined in various proportions to produce any other color. **2.** Any three colors that can be mixed in proper proportions to specify other colors; they need not be physically realizable. { 'prī¸mer·ē 'kəl·ərz }

primary creep |MECH| The initial high strain-rate region in a material subjected to sustained stress. { 'prī,mer·ē 'krēp }

primary extinction |SOLID STATE| A weakening of the stronger beams produced in x-ray diffraction by a very perfect crystal, as compared with the weaker. { 'prī,mer·ē ik'stiŋk·shən }

primary lead |MET| Lead recovered from ore, as contrasted with recycled scrap (secondary) lead. { 'prī,mer·ē 'led }

primary phase |THERMO| The only crystalline phase capable of existing in equilibrium with a given liquid. { 'prī,mer·ē 'fāz }

primary phase region |THERMO| On a phase diagram, the locus of all compositions having a common primary phase. { 'prī,mer·ē 'fāz ,rē·jən }

primary plasticizer |MATER| A plasticizer material for plastics formulations that has sufficient affinity to a polymer or resin so that it is considered compatible and therefore may be used as the sole plasticizer. { 'prī,mer·ē 'plas·tə ,sīz·ər }

primary stress |MECH| A normal or shear stress component in a solid material which results from an imposed loading and which is under a condition of equilibrium and is not self-limiting. { 'prī,mer·ē 'stres }

primary structure |POLYM CHEM| The chemical structure of a polymer chain. { 'prī,mer·ē 'strək·chər }

prime coat See primer. { 'prīm 'kōt }

primer |MATER| A prefinishing coat applied to surfaces that are to be painted or otherwise finished. Also known as prime coat. { 'prīm·ər }

primer mixture |MATER| An explosive mixture containing a sensitive explosive and other ingredients, used in a primer. { 'prīm·ər ,miks·chər }

primes |MET| High-quality metal products, particularly sheet and plate, that are free from visible defects. { 'prīmz }

priming composition |MATER| A physical mixture of materials that is very sensitive to impact or percussion and, when so exploded, undergoes very rapid autocombustion, producing hot gases and incandescent solid particles; priming compositions are used for the ignition of primary high explosives, black powder igniter charges, and propellants in small arms ammunition. { 'prīm·iŋ ,kəm·pə,zish·ən }

primitive cell |CRYSTAL| A parallelepiped whose edges are defined by the primitive translations of a crystal lattice; it is a unit cell of minimum volume. { 'prim·əd·iv 'sel }

primitive lattice |CRYSTAL| A crystal lattice in which there are lattice points only at its corners. Also known as simple lattice. { 'prim·əd·iv 'lad·əs }

primitive translation |CRYSTAL| For a space lattice, one of three translations which can be repeatedly applied to generate any translation which leaves the lattice unchanged. { 'prim·əd·iv tranz'lā·shən }

principal axis |CRYSTAL| The longest axis in a crystal. { 'prin·sə·pəl 'ak·səs }

principle of inaccessibility See Carathéodory's principle. { 'prin·sə·pəl əv ,in·ak,ses·ə'bil·əd·ē }

principle of superposition |MECH| The principle that when two or more forces act on a particle at the same time, the resultant force is the vector sum of the two. |PHYS| **1.** Also known as superposition principle. **2.** A general principle applying to many physical systems which states that if a number of independent influences act on the system, the resultant influence is the sum of the individual influences acting separately. **3.** In all theories characterized by linear homogeneous differential equations, such as optics, acoustics, and quantum theory, the principle that the sum of any number of solutions to the equations is another solution. { 'prin·sə·pəl əv ,sü·pər·pə'zish·ən }

printing ink |MATER| Ink generally made from carbon black, lampblack, or other pigment suspended in an oil vehicle, with a resin, solvent, adhesive, and drier; available in many variations. { 'print·iŋ ,iŋk }

prism |CRYSTAL| A crystal which has three, four, six, eight, or twelve faces, with the face intersection edges parallel, and which is open only at the two ends of the axis parallel to the intersection edges. |OPTICS| An optical system consisting of two or more usually plane surfaces of a transparent solid or embedded liquid at an angle with each other. { 'priz·əm }

prismatic cleavage |CRYSTAL| A type of crystal cleavage that occurs parallel to the faces of a prism. { priz'mad·ik 'klē·vij }

prismatic plane |MET| Any plane parallel to the principal c axis in noncubic metals. { priz'mad·ik 'plān }

process annealing |MET| Softening a ferrous alloy by heating to a temperature close to but below the lower limit of the transformation range and then cooling. { 'prä,sas ə,nēl·iŋ }

process metallurgy |MET| The branch of metallurgy concerned with the extraction of metals from ore, and with the refining of metals; usually synonymous with extractive metallurgy. { 'prä ,sas ,med·əl,ər·jē }

prochirality |ORG CHEM| The property displayed by a molecule or atom, which contains, or is bonded to, two constitutionally identical ligands (atoms or groups), whereby replacement of one of which by a different ligand makes the molecule or atom chiral. { ¦prō·kī'ral·əd·ē }

product |CHEM| A substance formed as a result of a chemical reaction. { 'präd·əkt }

progenitor cell |BIOMATER| A precursor cell that completes a series of cell divisions to produce a distinct cell lineage. { prə'jen·əd·ər ,sel }

progression |MET| The fixed dimension between adjacent stations in a progressive die and thus the precise distance the strip must advance between successive cycles of the press. { prə'gresh·ən }

progressive aging |MET| Aging of metals achieved by increasing the temperature in stages or by continuous elevation of the temperature. { prə'gres·iv 'āj·iŋ }

progressive block sequence |MET| A welding sequence in which the joint is completed in sections from one end to the other or from the center alternately to both ends. { prə'gres·iv 'bläk 'sē·kwəns }

progressive die |MET| A die in which two or more operations are performed sequentially at different positions. { prə'gres·iv 'dī }

progressive forming |MET| Sequential forming at consecutive stations either with a single die or with separate dies. { prə'gres·iv 'fȯrm·iŋ }

projection welding |MET| Resistance welding in which the welds are localized at projections, intersections, and overlaps on the parts. { prə'jek·shən ‚weld·iŋ }

promethium |CHEM| A chemical element, symbol Pm, atomic number 61, produced artificially in nuclear reactors; atomic weight of the most abundant separated isotope is 147; a member of the rare-earth group of metals. { prə'mē·thē·əm }

promoter |CHEM| A chemical which itself is a feeble catalyst, but greatly increases the activity of a given catalyst. { prə'mōd·ər }

proof resilience |MECH| The tensile strength necessary to stretch an elastomer from zero elongation to the breaking point, expressed in foot-pounds per cubic inch of original dimension. { 'prüf ri‚zil·yəns }

proof stress |MECH| **1.** The stress that causes a specified amount of permanent deformation in a material. **2.** A specified stress to be applied to a member or structure in order to assess its ability to support service loads. { 'prüf ‚stres }

propadiene See allene. { ‚präp·ə'dī‚ēn }

propagation rate |CHEM| The speed at which a flame front progresses through the body of a flammable fuel-oxidizer mixture, such as gas and air. { ‚präp·ə'gā·shən ‚rāt }

propagation step |CHEM| In a chain reaction, one of the fundamental steps that take place repeatedly until the reaction is complete. { ‚präp·ə'gā·shən ‚step }

proplatinum |MET| A nickel-silver-bismuth alloy used as a substitute for platinum. { prō'plat·ən·əm }

proportional elastic limit |MECH| The greatest stress intensity for which stress is still proportional to strain. { prə'pȯr·shən·əl i'las·tik ‚lim·ət }

proportional limit |MECH| The greatest stress a material can sustain without departure from linear proportionality of stress and strain. { prə'pȯr·shən·əl 'lim·ət }

propylene |ORG CHEM| $CH_3CH=CH_2$ Colorless unsaturated hydrocarbon gas, with boiling point of $-47°C$; used to manufacture plastics and as a chemical intermediate. Also known as methyl ethylene. { 'prō·pə‚lēn }

prosthesis |BIOMATER| An artificial substitute for a missing part of the body, such as a substitute hand, leg, eye, or denture. { präs'thē·səs }

protactinium |CHEM| A chemical element, symbol Pa, atomic number 91; the third member of the actinide group of elements; all the isotopes

are radioactive; the longest-lived isotope is protactinium-231. { ¦prōd‚ak'tin·ē·əm }

protective atmosphere |MET| A substance such as inert or combusted fuel gas which surrounds a workpiece to be heat-treated, welded, brazed, or thermally sprayed under controlled conditions. { prə'tek·tiv 'at·mə‚sfir }

protein |BIOMATER| Any of a class of high-molecular-weight polymer compounds composed of a variety of α-amino acids joined by peptide linkages. { 'prō‚tēn }

proteinaceous |MATER| **1.** Pertaining to any material having a protein base. **2.** Pertaining to adhesive materials having a protein base such as animal glue, casein, and soya. { ¦prōt·ən'ā·shəs }

proton |PHYS| An elementary particle that is the positively charged constituent of ordinary matter and, together with the neutron, is a building block of all atomic nuclei; its mass is approximately 938 megaelectronvolts and spin $1/2$. { 'prō‚tän }

proton acid See Brønsted acid. { 'prō‚tän 'as·əd }

protonic acid See Brønsted acid. { prō'tän·ik 'as·əd }

proton scattering microscope |SOLID STATE| A microscope in which protons produced in a cold-cathode discharge are accelerated and focused on a crystal in a vacuum chamber; protons reflected from the crystal strike a fluorescent screen to give a visual and photographable display that is related to the structure of the target crystal. { 'prō‚tän ‚skad·ə·riŋ 'mī·krə‚skōp }

proximity effect |ELEC| Redistribution of current in a conductor brought about by the presence of another conductor. { präk'sim·əd·ē i‚fekt }

proximity-effect microbridge |CRYO| A Josephson junction formed by overcoating a few micrometers of thin superconducting film with normal metal, thereby weakening the superconductivity in the film beneath the metal. Also known as Notarys-Mercereau microbridge. { präk'sim·əd·ē i¦fekt 'mī·krō‚brij }

Prussian blue |INORG CHEM| $Fe_4[Fe(CN)_6]_3$ Ferric ferrocyanide, used as a blue pigment and in the removal of hydrogen sulfide from gases. { 'prəsh·ən 'blü }

prussic acid See hydrocyanic acid. { 'prəs·ik 'as·əd }

pseudocarburizing See blank carburizing. { ¦sü·dō'kär·bə‚rīz·iŋ }

pseudocrystal |CRYSTAL| A substance that appears to be crystalline but does not have a true crystalline diffraction pattern. { ¦sü·dō'krist·əl }

pseudoplastic fluid |FL MECH| A fluid whose apparent viscosity or consistency decreases instantaneously with an increase in shear rate. { ¦sü·dō'plas·tik 'flü·əd }

pseudopotential |SOLID STATE| The common effective potential for electrons in a crystal lattice that is calculated in the orthogonalized plane-wave method and in the pseudopotential method, and that is relatively weak (except for diffracted electrons) because the electrons are moving rapidly past the atoms in the lattice. { ,sü·də·pə'ten·chəl }

pseudopotential method [SOLID STATE] A method of approximating the energy states of electrons in a crystal lattice in which the electrons are assumed to move in a common effective potential that is calculated from the experimentally determined energy levels and the effective masses of the electrons. { ,sü·də·pə,ten·chəl 'meth·əd }

pseudorotaxane [CHEM] A supramolecular species consisting of a linear molecular component (without bulky end groups) encircled by a macrocyclic component. { ,süd·ō·rō'tak ,sān }

pseudosymmetry [CRYSTAL] Apparent symmetry of a crystal, resembling that of another system; generally due to twinning. { ¦sü·dō'sim·ə·trē }

psychromatic ratio [THERMO] Ratio of the heat-transfer coefficient to the product of the mass-transfer coefficient and humid heat for a gas-vapor system; used in calculation of humidity or saturation relationships. { ,sī·krə'mad·ik 'rā·shō }

psychrometric chart [THERMO] A graph each point of which represents a specific condition of a gas-vapor system (such as air and water vapor) with regard to temperature (horizontal scale) and absolute humidity (vertical scale); other characteristics of the system, such as relative humidity, wet-bulb temperature, and latent heat of vaporization, are indicated by lines on the chart. { ¦sī·krə¦me·trik 'chärt }

psychrometric formula [THERMO] The semiempirical relation giving the vapor pressure in terms of the barometer and psychrometer readings. { ¦sī·krə¦me·trik 'fȯr·myə·lə }

psychrometric tables [THERMO] Tables prepared from the psychrometric formula and used to obtain vapor pressure, relative humidity, and dew point from values of wet-bulb and dry-bulb temperature. { ¦sī·krə¦me·trik 'tā·bəlz }

Pt See platinum.

PTA See phosphotungstic acid.

p-type conductivity [ELECTR] The conductivity associated with holes in a semiconductor, which are equivalent to positive charges. { 'pē ¦tīp ,kän,dək'tiv·əd·ē }

p-type semiconductor [ELECTR] An extrinsic semiconductor in which the hole density exeeds the conduction electron density. { 'pē ¦tīp 'sem·i·kən,dək·tər }

p⁺-type semiconductor [ELECTR] A p-type semiconductor in which the excess mobile hole concentration is very large. { 'pē¦pläs ,tīp 'sem·i·kən,dək·tər }

p-type silicon [ELECTR] Silicon to which more impurity atoms of acceptor type (with valence of 3, such as boron) than of donor type (with valence of 5, such as phosphorus) have been added, with the result that the hole density exceeds the conduction electron density. { 'pē ¦tīp 'sil·ə ,kän }

Pu See plutonium.

puckering [MET] Corrugations in metal parts resulting from pressing or drawing. { 'pək·ə· riŋ }

puddle [MET] A batch of molten iron within the puddling furnace. { pəd·əl }

puddling [MET] A process for the production of wrought iron by agitation of a bath of molten pig iron with iron oxide in order to reduce the carbon, silicon, phosphorus, and manganese content. { 'pəd·liŋ }

puddling furnace [MET] A coal-fired reverberatory furnace for puddling pig iron. { 'pəd·liŋ ,fər·nəs }

pull crack [MET] A crack in a casting caused by contraction strains during cooling and resulting from the shape of the casting. { 'pủl ,krak }

pull strength [MECH] A unit in tensile testing; the bond strength in pounds per square inch. { 'pủl ,streŋkth }

pulp [MATER] The cellulosic material produced by reducing wood mechanically or chemically and used in making paper and cellulose products. Also known as wood pulp. { pəlp }

pulpboard [MATER] Chipboard to which is added a percentage of mechanical wood pulp. { 'pəlp ,bȯrd }

pulpstone [MATER] A block of sandstone cut into wheels for grinding, especially wood pulp in paper manufacture. { 'pəlp,stōn }

pulpwood [MATER] Any wood that can be reduced to pulp. { 'pəlp,wủd }

pulsation welding See multiple-impulse welding. { pəl'sā·shən ,weld·iŋ }

pulsed laser [OPTICS] A laser in which a pulse of coherent light is produced at fixed time intervals, as required for ranging and tracking applications or to permit higher output power than can be obtained with continuous operation. { 'pəlst 'lā·zər }

pulsed ruby laser [OPTICS] A laser in which ruby is used as the active material; the extremely high pumping power required is obtained by discharging a bank of capacitors through a special high-intensity flash tube, giving a coherent beam that lasts for about 0.5 millisecond. { 'pəlst 'rü·bē 'lā·zər }

pulse-echo method [MET] A nondestructive test in which pulses of energy are directed into a part, and the time for the echo to return from one or more surfaces is measured. { 'pəls ¦ek·ō ,meth·əd }

pulse hardening [MET] A surface-hardening process performed by heating to the required temperature in a span of several milliseconds by using an energy and time-controlled pulse of very high power at a very high frequency, about 27 megahertz. { 'pəls ¦härd·ən·iŋ }

pumpability [MATER] **1.** The property of a lubricating grease that causes it to flow under pressure through lines, nozzles, and fittings. **2.** The ability of any liquid, slurry, or suspension to be moved through a flow conduit by pressure from a pump. { ,pəm·pə'bil·əd·ē }

pure shear [MECH] A particular example of irrotational strain or flattening in which a body is elongated in one direction and shortened at right angles to it as a consequence of differential displacements on two sets of intersecting planes. { 'pyür 'shir }

pure substance [CHEM] A sample of matter, either an element or a compound, that consists of only one component with definite physical and chemical properties and a definite composition. { 'pyür 'səb·stəns }

purity [CHEM] The degree to which the content of impurity can be detected by an analytical procedure in a sample of matter that is classified as a pure substance; the grade of purity is in inverse proportion to the amount of impurity present. [OPTICS] The degree to which a primary color is pure and not mixed with the other two primary colors. { 'pyür·əd·ē }

purple lakes [MATER] A class of lake (pigment) used in printing inks; derived from a combination of such compounds as β-hydroxynaphthoic acid and 2-diazonaphthalene-1-sulfonic acid. { 'pər·pəl 'lāks }

purple of Cassius See gold tin purple. { 'pər·pəl əv 'kash·əs }

purple salt See potassium permanganate. { 'pər·pəl 'sȯlt }

pusher furnace [MET] A type of continuous furnace in which the stock to be heated is charged at one end, carried through a sequence of one or more heating zones, and discharged at the opposite end. { 'pùsh·ər ,fər·nəs }

push welding [MET] Spot or projection welding in which the force is applied manually by one electrode; the work takes the place of the other electrode. Also known as poke welding. { 'pùsh ,weld·iŋ }

putty [MATER] A cement of dough consistency made of whiting and boiled linseed oil and used in fastening glass in sashes and sealing crevices in woodwork. { 'pəd·ē }

putty oil [MATER] Petroleum oil that is added to putty; serves to lubricate the putty and keep it soft after the linseed oil dries. { 'pəd·ē ,ȯil }

PVA See polyvinyl acetate; polyvinyl alcohol.

PVAc See polyvinyl acetate.

PVC See polyvinyl chloride.

PVD See physical vapor deposition.

PVDC See polyvinyl dichloride.

PVI See polyvinyl isobutyl ether.

PVM See polyvinyl methyl ether.

PVP See polyvinyl pyrrolidone.

pyramid [CRYSTAL] An open crystal having three, four, six, eight, or twelve nonparallel faces that meet at a point. { 'pir·ə,mid }

pyramidal cleavage [CRYSTAL] A type of crystal cleavage that occurs parallel to the faces of a pyramid. { ¦pir·ə¦mid·əl 'klē·vij }

pyritohedron [CRYSTAL] A dodecahedral crystal with 12 irregular pentagonal faces; it is characteristic of pyrite. Also known as pentagonal dodecahedron; pyritoid; regular dodecahedron. { pə¦rīd·ō'hē·drən }

pyritoid See pyritohedron. { 'pī,rīd,ȯid }

pyro- [CHEM] A chemical prefix for compounds formed by heat, such as pyrophosphoric acid, an inorganic acid formed by the loss of one water molecule from two molecules of an ortho acid. { 'pī·rō, 'pī·rə }

pyroconductivity [SOLID STATE] Electrical conductivity that develops in a material only at high temperature, chiefly at fusion, in solids that are practically nonconductive at atmospheric temperatures. { ¦pī·rō,kän·dək'tiv·əd·ē }

pyroelectric crystal [SOLID STATE] A crystal exhibiting pyroelectricity, such as tourmaline, lithium sulfate monohydrate, cane sugar, and ferroelectric barium titanate. { ¦pī·rō·i¦lek·trik 'krist·əl }

pyroelectricity [SOLID STATE] **1.** The property of certain crystals to produce a state of electrical polarity by a change of temperature. **2.** An electric charge released as the result of a temperature change. { ¦pī·rō,i,lek'tris·əd·ē }

pyrolysis [CHEM] The breaking apart of complex molecules into simpler units by the use of heat, as in the pyrolysis of heavy oil to make gasoline. Also known as thermolysis. { pə'räl·ə·səs }

pyrometallurgy [MET] Processes that use chemical reactions at elevated temperatures for the extraction of metals from raw materials such as ores and concentrates, or for the treatment of recycled scrap. { ¦pī·rō'med·əl,ər·jē }

pyrometer [ENG] Any of a broad class of temperature-measuring devices; they were originally designed to measure high temperatures, but some are now used in any temperature range; includes radiation pyrometers, thermocouples, resistance pyrometers, and thermistors. { pī'räm·əd·ər }

pyrometric cone See Seger cone. { ¦pī·rə¦me·trik 'kōn }

pyrometry [THERMO] The science and technology of measuring high temperatures. { pī'räm·ə·trē }

pyrophoric alloy [MET] **1.** An alloy such as ferrocerium that produces a spark when struck with metal (steel) at an angle; used for automatic cigarette lighters. **2.** An alloy in powder form that spontaneously oxidizes in air, reaching high temperatures. { ¦pī·rə¦fȯr·ik 'al,ȯi }

pyrophoric material [CHEM] A material that spontaneously ignites in air below 113°C (45°C), such as fine metal powder, alkali metal, partially or fully alkylated metal or nonmetal hydride, and metal carbonyl. { ,pī·rə,fȯr·ik mə'tir·ē·əl }

pyrophosphoric acid [INORG CHEM] $H_4P_2O_7$ Water-soluble, syrupy liquid melting at 61°C; used as a catalyst and to make organic phosphate esters. { ¦pī·rō·fä'sfȯr·ik 'as·əd }

pyroxylin cement [MATER] A solution of nitrocellulose in a chemical solvent, compounded with a resin, or plasticized with a gum or synthetic; dries by evaporation of the solvent. { pə'räk·sə·lən si'ment }

Q

Q |PHYS| A measure of the ability of a system with periodic behavior to store energy equal to 2π times the average energy stored in the system divided by the energy dissipated per cycle. Also known as Q factor; quality factor; storage factor. |THERMO| A unit of heat energy, equal to 10^{18} British thermal units, or approximately 1.055×10^{21} joules. **Q factor** See Q. { 'kyü ,fak·tər }

Q unit |THERMO| A unit of energy, used in measuring the heat energy of fuel reserves, equal to 10^{18} British thermal units, or approximately 1.055×10^{21} joules. { 'kyü ,yü·nət }

quadridentate ligand |INORG CHEM| A group which forms a chelate and has four points of attachment. { ¦kwä·drə¦den,tāt 'līg·ənd }

quadrupole |ELECTROMAG| A distribution of charge or magnetization which produces an electric or magnetic field equivalent to that produced by two electric or magnetic dipoles whose dipole moments have the same magnitude but point in opposite directions, and which are separated from each other by a small distance. { 'kwä·drə,pōl }

quadrupole field |ELECTROMAG| **1.** An electric or magnetic field equivalent to that produced by two electric or magnetic dipoles whose dipole moments have the same magnitude but point in opposite directions, and which are separated from each other by a small distance. **2.** The field produced by a quadrupole lens. { 'kwä·drə,pōl ,fēld }

quadrupole moment |ELECTROMAG| A quantity characterizing a distribution of charge or magnetization; it is given by integrating the product of the charge density or divergence of magnetization density, the second power of the distance from the origin, and a spherical harmonic Y^*_{2m} over the charge or magnetization distribution. { 'kwä·drə,pōl ,mō·mənt }

quality factor See Q. { 'kwäl·əd·ē ,fak·tər }

quantity of electricity See charge. { 'kwän·əd·ē əv ,i,lek'tris·əd·ē }

quantized electronic structure |ELECTR| A material that confines electrons in such a small space that their wave-like behavior becomes important and their properties are strongly modified by quantum-mechanical effects. { ¦kwän ,tīzd i·lek·¦trän·ik 'strək·chər }

quantized spin wave See magnon. { 'kwän,tīzd 'spin ,wāv }

quantized vortex |CRYO| A circular flow pattern observed in superfluid helium and type II superconductors, in which a superfluid flows about a normal (nonsuperfluid) cylindrical region or core which has the form of a thin line, and either the circulation or the magnetic flux is quantized. { 'kwän,tīzd 'vór,teks }

quantum dot |ELECTR| A quantized electronic structure in which electrons are confined with respect to motion in all three dimensions. { ,kwänt·əm 'dät }

quantum Hall liquid See quantum Hall state. { 'kwän·təm 'hól 'lik·wəd }

quantum Hall state |CRYO| A kind of incompressible liquid state obtained by placing a two-dimensional electron gas, confined on the interface of two different semiconductors, in a strong magnetic field at low temperature. Also known as quantum Hall liquid. { ¦kwän·təm 'hól ,stāt }

quantum hydrodynamics |CRYO| The mechanics of a superfluid, such as helium II, investigating phenomena such as the fountain effect and second sound. { 'kwän·təm ,hī·drō·dī'nam·iks }

quantum size effects |SOLID STATE| Unusual properties of extremely small crystals that arise from confinement of electrons to small regions of space in one, two, or three dimensions. { ¦kwänt·əm 'sīz i,feks }

quantum solid |SOLID STATE| A solid whose atoms or molecules undergo large zero-point motion even in the quantum ground state (at absolute zero temperature) as a result of their small mass and the weak attractive part of their interaction potential. { 'kwän·təm ,säl·əd }

quantum turbulence |CRYO| A phenomenon observed in a channel filled with superfluid and subjected to a heat flux which exceeds a certain critical value, in which the superfluid becomes filled with a tangled mass of quantized vortex lines. { 'kwän·təm 'tər·byə·ləns }

quantum well |ELECTR| A thin layer of material (typically between 1 and 10 nanometers thick) within which the potential energy of an electron is less than outside the layer, so that the motion of the electron perpendicular to the layer is quantized. { 'kwän·təm 'wel }

quantum wire |ELECTR| A strip of conducting material about 10 nanometers or less in width and thickness that displays quantum-mechanical effects such as the Aharanov-Bohm effect and universal conductance fluctuations. { 'kwän·təm 'wīr }

quartation *See* inquartation. { kwȯr'tā·shən }

quartz crystal [ELECTR] A natural or artificially grown piezoelectric crystal composed of silicon dioxide, from which thin slabs or plates are carefully cut and ground to serve as a crystal plate. { 'kwȯrts ¦krist·əl }

quartz lamp [ELECTR] A mercury-vapor lamp having a transparent envelope made from quartz instead of glass; quartz resists heat, permitting higher currents, and passes ultraviolet rays that are absorbed by ordinary glass. { 'kwȯrts ¦lamp }

quasi-crystal [CRYSTAL] A phase of solid matter that, like a crystal, exhibits long-range orientational order and translational order but whose atoms and clusters repeat in a sequence defined by a sum of periodic functions whose periods are in an irrational ratio. { ¦kwä·zē 'krist·əl }

quasi-free-electron theory [SOLID STATE] A modification of the free-electron theory of metals to take into account the periodic variation of the potential acting on a conduction electron, in which these electrons are assigned an effective scalar mass which differs from their real mass. { ¦kwä·zē ¦frē i'lek,tran ,thē·ə·rē }

quasi-static process *See* reversible process. { ¦kwä·zē 'stad·ik 'prä·səs }

quaternary alloy [MET] An alloy containing four principal elements apart from accidental impurities. { 'kwät·ən,er·ē 'al,ȯi }

quebracho [MATER] A drilling-fluid additive used for thinning or dispersing in order to control viscosity and thixotropy; made from an extract of the quebracho tree and consisting essentially of tannic acid. { kā'bra·chō }

queen's metal [MET] An alloy consisting principally of tin to which antimony, zinc, and lead or copper are added. { 'kwēnz ,med·əl }

quench aging [MET] Aging of metal induced by rapid cooling from solution heat-treatment temperatures. { 'kwench ,āj·iŋ }

quench annealing [MET] Annealing an austenitic ferrous alloy by heating followed by quenching from solution temperatures. { 'kwench ə,nēl·iŋ }

quench bath [ENG] A liquid medium, such as oil, fused salt, or water, into which a material is plunged for heat-treatment purposes. { 'kwench ,bath }

quench hardening [MET] The hardening of a ferrous alloy by quenching from a temperature above the transformation range. { 'kwench ,hard·ən·iŋ }

quenching [ENG] Shock cooling by immersing liquid or molten material into a cooling medium (liquid or gas); used in metallurgy, plastics forming, and petroleum refining. [SOLID STATE] Reduction in the intensity of sensitized luminescence radiation when energy migrating through a crystal by resonant transfer is dissipated in crystal defects or impurities rather than being reemitted as radiation. { 'kwench·iŋ }

quenching oil [MET] Animal, vegetable, or mineral oil, such as fish oil, cottonseed oil, or lard, used in quenching baths for carbon and alloy steels; removes heat from the steel more slowly and uniformly than water. { 'kwench·iŋ ,ȯil }

quenching stress [MET] Internal stresses set up in a metal as a result of quenching. { 'kwench·iŋ ,stres }

quicklime *See* calcium oxide. { 'kwik,līm }

quick malleable iron [MET] Malleable iron containing 2.2% carbon, 1.5% silicon, 0.30–0.60% manganese, and 0.75–1% copper. { 'kwik 'mal·yə·bəl 'ī·ərn }

quicksilver *See* mercury. { 'kwik,sil·vər }

quicksilver vermilion *See* mercuric sulfide. { 'kwik,sil·vər vər'mil·yən }

R

Ra *See* radium.

rabbit ear [MET] A recess in the corner of a die allowing wrinkling or folding of the blank. { 'rab·ət ,ir }

rabble [MET] An iron bar for skimming the bath in a smelting or refining furnace or for stirring the ore in a roasting furnace either manually or mechanically. { 'rab·əl }

racemate [ORG CHEM] An equimolar mixture of the two enantiomers (+ and −, or R and S) of a substance; it is optically inactive. { 'ras·ə ,māt }

racemic mixture [ORG CHEM] According to the IUPAC, this usage is strongly discouraged, racemate is preferred. { rə'sēm·ik 'miks·chər }

racemization [ORG CHEM] A process by which an optically active form of a substance is converted into a racemic mixture. { ,rā·sə· mə'zā·shən }

racking [MET] Suspending work from a frame that holds and conducts current to one or more electrodes for electroplating and other electrochemical operations. { 'rak·iŋ }

radar-absorbing material [MATER] A material that is designed to reduce the reflection of electromagnetic radiation by a conducting surface in the frequency range from approximately 100 megahertz to 100 gigahertz. { 'rā,där əb ,sórb·iŋ mə,tir·ē·əl }

radar paint [MATER] A coating that absorbs radar waves. { 'rā,där ,pānt }

radial heat flow [THERMO] Flow of heat between two coaxial cylinders maintained at different temperatures; used to measure thermal conductivities of gases. { 'rād·ē·əl 'hēt ,flō }

radiance [OPTICS] The radiant flux per unit solid angle per unit of projected area of the source; the usual unit is the watt per steradian per square meter. { 'rād·ē·əns }

radiating power *See* emittance. { 'rād·ē,ād·iŋ 'paú·ər }

radiation [PHYS] **1.** The emission and propagation of waves transmitting energy through space or through some medium; for example, the emission and propagation of electromagnetic, sound, or elastic waves. **2.** The energy transmitted by waves through space or some medium; when unqualified, usually refers to electromagnetic radiation. Also known as radiant energy. **3.** A stream of particles, such as electrons, neutrons, protons, α-particles, or high-energy photons, or a mixture of these. { ,rād·ē'ā·shən }

radiation correction *See* cooling correction. { ,rād·ē'ā·shən kə,rek·shən }

radiation corrosion [MET] Accelerated corrosion of a metal caused by radiation. { ,rād·ē'ā· shən kə,rō·zhən }

radiation noise *See* electromagnetic noise. { ,rād·ē'ā·shən ,nóiz }

radical *See* free radical. { 'rad·ə·kəl }

radical ion [CHEM] A charged compound that has an unpaired electron; it may be either a radical cation (positively charged) or radical anion (negatively charged). { ,rad·ə·kəl 'ī·ən }

radioactive heat [THERMO] Heat produced within a medium as a result of absorption of radiation from decay of radioisotopes in the medium, such as thorium-232, potassium-40, uranium-238, and uranium-235. { ¦rād·ē·ō'ak· tiv 'hēt }

radioactive paint [MATER] A luminous paint that gives off light without being activated. { ¦rād·ē·ō'ak·tiv 'pānt }

radiochemistry [CHEM] That area of chemistry concerned with the study of radioactive substances. { ¦rad·ē·ō'kem·ə·strē }

radio-frequency resistance *See* high-frequency resistance. { 'rād·ē·ō ¦frē·kwən·sē ri'zis·təns }

radio-frequency welding *See* high-frequency welding. { 'rād·ē·ō ¦frē·kwən·sē 'weld·iŋ }

radiosensitive [MATER] Sensitive to damage by radiant energy. { ¦rād·ē·ō 'sen·səd·iv }

radium [CHEM] **1.** A radioactive member of group II, symbol Ra, atomic number 88; the most abundant naturally occurring isotope has mass number 226 and a half-life of 1620 years. **2.** A highly toxic solid that forms water-soluble compounds; decays by emission of α, β, and γ-radiation; melts at 700°C; boils at 1140°C; turns black in air; used in medicine, in industrial radiography, and as a source of neutrons and radon. { 'rād·ē·əm }

radium bromide [INORG CHEM] RaBr₂ Water-soluble, poisonous, radioactive white powder, corrosive to skin or flesh; melts at 728°C; used in medicine, physical research, and luminous paint. { 'rād·ē·əm 'brō,mīd }

radium carbonate [INORG CHEM] RaCO₃ Water-insoluble, poisonous, radioactive, white powder; used in medicine. { 'rād·ē·əm 'kär·bə,nāt }

radium chloride |INORG CHEM| $RaCl_2$ Water- and alcohol-soluble, poisonous, radioactive, yellow-white crystals; corrosive effect on skin and flesh; melts at 1000°C; used in medicine, physical research, and luminous paint. { 'rād·ē·əm 'klòr ,īd }

radium sulfate |INORG CHEM| $RaSO_4$ Water-insoluble, radioactive, poisonous, white crystals; used in medicine. { 'rād·ē·əm 'səl,fāt }

radon |CHEM| A chemical element, symbol Rn, atomic number 86; all isotopes are radioactive, the longest half-life being 3.82 days for mass number 222; it is the heaviest element of the noble-gas group, produced as a gaseous emanation from the radioactive decay of radium. { 'rā,dän }

rag papers |MATER| The most expensive papers, made wholly or partly from cotton or linen rags; rag content is expressed as 25, 50, 75, or 100%; pure rag papers are the strongest and most resistant to the discoloration and deterioration due to age. { 'rag ,pāpərs }

railroad thermit |MET| Red thermit to which is added up to 16% nickel, manganese, and steel. { 'rāl,rōd 'thər·mət }

rail steel |MET| Steel used to make rail track. { 'rāl ,stēl }

ramoff |MET| A defect in a casting due to improper ramming of the sand. { 'ram,óf }

Ramsay-Shields-Eötvös equation |THERMO| An elaboration of the Eötvös rule which states that at temperatures not too near the critical temperature, the molar surface energy of a liquid is proportional to t_c-t-6 K, where t is the temperature and t_c is the critical temperature. { 'ram·zē 'shēlz 'öt·vósh i,kwā·zhən }

Ramsay-Young method |THERMO| A method of measuring the vapor pressure of a liquid, in which a thermometer bulb is surrounded by cotton wool soaked in the liquid, and the pressure, measured by a manometer, is reduced until the thermometer reading is steady. { ¦ram·zē 'yəŋ ,meth·əd }

Ramsay-Young rule |THERMO| An empirical relationship which states that the ratio of the absolute temperatures at which two chemically similar liquids have the same vapor pressure is independent of this vapor pressure. { 'ram·zē 'yəŋ ,rül }

random copolymer |POLYM CHEM| Resin copolymer in which the molecules of each monomer are randomly arranged in the polymer backbone. { 'ran·dəm kō'päl·i·mər }

random interstratification |SOLID STATE| A crystalline structure in which two or more types of layers alternate in a random fashion. { 'ran·dəm ,in·tər,strad·i·fə'kā·shən }

random sequence |MET| A longitudinal sequence of weld beads deposited in random increments. { 'ran·dəm 'sē·kwəns }

random structure |CRYSTAL| A crystal structure in which different types of atoms are associated with the various points in a crystal lattice in a random fashion. { 'ran·dəm 'strək·chər }

Raney nickel |MET| A nickel powder prepared from an alloy of nickel and aluminum in equal parts by preferentially dissolving the aluminum in a warm solution of sodium hydroxide. { 'rā·nē ,nik·əl }

Rankine cycle |THERMO| An ideal thermodynamic cycle consisting of heat addition at constant pressure, isentropic expansion, heat rejection at constant pressure, and isentropic compression; used as an ideal standard for the performance of heat-engine and heat-pump installations operating with a condensable vapor as the working fluid, such as a steam power plant. Also known as steam cycle. { 'raŋ·kən ,sī·kəl }

Rankine-Hugoniot equations |THERMO| Equations, derived from the laws of conservation of mass, momentum, and energy, which relate the velocity of a shock wave and the pressure, density, and enthalpy of the transmitting fluid before and after the shock wave passes. { 'raŋ·kən yü'gō·nē·ō i,kwā·zhənz }

Rankine temperature scale |THERMO| A scale of absolute temperature; the temperature in degrees Rankine (°R) is equal to $\%$ of the temperature in kelvins and to the temperature in degrees Fahrenheit plus 459.67. { 'raŋ·kən 'tem·prə·chər ,skāl }

rapid-curing asphalt |MATER| A liquid asphalt composed of asphalt cement and a gasoline- or naphtha-type diluent. Abbreviated RC asphalt. { 'rap·əd ¦kyúr·iŋ 'as,fólt }

rapid quenching |MET| Superfast cooling (1–5 × 10^6 K per second) of a molten metal to produce new and amorphous alloys and new crystalling material with improved properties. { 'rap·əd 'kwench·iŋ }

rare-earth alloy |MET| An alloy containing rare-earth materials. { 'rer ,ərth 'al,ói }

rare-earth element |CHEM| The name given to any of the group of chemical elements with atomic numbers 58 to 71; the name is a misnomer since they are neither rare nor earths; examples are cerium, erbium, and gadolinium. { 'rer ,ərth 'el·ə·mənt }

rare-earth garnet |MATER| A synthetic garnet having the general structure of grossularite, but with calcium replaced by a rare-earth metal, and aluminum and silicon replaced by iron; used for electronic applications. { 'rer ,ərth 'gär·nət }

rare-earth magnet |ELECTROMAG| Any of several types of magnets made with rare-earth elements, such as rare-earth-cobalt magnets, which have coercive forces up to ten times that of ordinary magnets; used for computers and signaling devices. { 'rer ,ərth 'mag·nət }

rare-earth salts |INORG CHEM| Salts derived from monazite, and with rare earths in similar proportions as in monazite; contains La, Ce, Pr, Nd, Sm, Gd, and Y as acetates, carbonates, chlorides, fluorides, nitrates, sulfates, and so on. { 'rer ,ərth 'sóls }

rare gas See noble gas. { 'rer 'gas }

rare metal |MET| Any metal that is difficult to extract from ore and is rare and expensive commercially; includes masurium, alabamine, and virginium. { 'rer 'med·əl }

raster |ELECTR| A predetermined pattern of scanning lines that provides substantially uniform coverage of an area; in television the raster is seen as closely spaced parallel lines, most evident when there is no picture. { 'ras·tər }

rate-determining step |CHEM| In a multistep chemical reaction, the step with the lowest velocity, which determines the rate of the overall reaction. { 'rāt di¦tər·mən·iŋ ,step }

rate of reaction |CHEM| A measurement based on the mass of reactant consumed in a chemical reaction during a given period of time. { 'rāt əv rē'ak·shən }

rate of strain hardening |MET| Rate of change of true stress with respect to true strain in the plastic range. Also known as modulus of strain hardening. { 'rāt əv 'strān 'härd·ən·iŋ }

rattail |MET| A small irregular line marking a minor buckle on the surface of a casting. { 'rat ,tāl }

Rayleigh balance |ELECTROMAG| An apparatus for assigning the value of the ampere in which the force exerted on a movable circular coil by larger circular coils above and below, but coaxial with, the movable coil is compared with the gravitational force on a known mass. { 'rā·lē ,bal·əns }

Rayleigh cycle |ELECTROMAG| A cycle of magnetization that does not extend beyond the initial portion of the magnetization curve, between zero and the upward bend. { 'rā·lē ,sī·kəl }

Rayleigh law |ELECTROMAG| **1.** For small values of the magnetic field strength H, the normal permeability of a material is approximately by $a + bH$, where a is the initial permeability and b is a constant. **2.** In a magnetic material subject to cyclic magnetization, with maximum magnetic field strength small compared with the coercive force, the hysteresis loss per cycle is proportional to the cube of the maximum value of the magnetic induction. |OPTICS| In Rayleigh scattering, the intensity of light scattered in a direction making an angle θ with the incident direction is proportional to $1 + \cos^2 \theta$ and inversely proportional to the fourth power of the wavelength of the incident radiation. { 'rā·lē ,lò }

Rayleigh loop |ELECTROMAG| A parabolic approximation to a magnetic hysteresis loop. { 'rā·lē ,lüp }

Rayleigh number 2 |THERMO| A dimensionless number used in studying free convection, equal to the product of the Grashof number and the Prandtl number. Symbolized R'_2. { 'rā·lē ¦nəm·bər 'tü }

Rayleigh number 3 |THERMO| A dimensionless number used in the study of combined free and forced convection in vertical tubes, equal to Rayleigh number 2 times the Nusselt number times the tube diameter divided by its entry length. Symbolized Ra_3. { 'rā·lē ¦nəm·bər 'thrē }

rayon |TEXT| A fiber made from regenerated cellulose by the viscose or cuprammonium process. { 'rā,än }

rayon coning oil |MATER| Lubricant oil used to reduce static in yarns being wound on cones; composed of low-viscosity mineral oils emulsifiable in water. { 'rā,än 'kōn·iŋ ,òil }

Rb See rubidium.

RC asphalt See rapid-curing asphalt. { ¦är¦sē 'as ,fòlt }

R center |SOLID STATE| A color center whose absorption band lies between the F band and M band, and which is produced by prolonged irradiation with light in the F band or prolonged x-ray exposure at room temperature. Also known as D center; E center. { 'är ,sen·tər }

Re See rhenium.

reactance |ELEC| The imaginary part of the impedance of an alternating-current circuit. { rē'ak·təns }

reactant |CHEM| A substance that reacts with another one to produce a new set of substances (products). { rē'ak·tənt }

reaction mechanism |CHEM| The sequence of steps during which a chemical reaction occurs, including the transition state during which the reactants are converted into products. { rē'ak·shən ,mek·ə,niz·əm }

reactive bond |CHEM| A bond between atoms that is easily invaded (reacted to) by another atom or radical; for example, the double bond in $CH_2{=}CH_2$ (ethylene) is highly reactive to other ethylene molecules in the reaction known as polymerization to form polyethylene. { rē'ak·tiv 'bänd }

reactive dye |MATER| Dye that reacts with the textile fiber to produce both a hydroxyl and an oxygen linkage, the chlorine combining with the hydroxyl to form a strong ether linkage; gives fast, brilliant colors. { rē'ak·tiv 'dī }

reactive intermediate |CHEM| An unstable compound formed as an intermediate during a chemical reaction. { rē'ak·tiv ,in·tər'mē·dē·ət }

reactive ion etching |ELECTR| A directed chemical etching process used in integrated circuit fabrication in which chemically active ions are accelerated along electric field lines to meet a substrate perpendicular to its surface. { rē'ak·tiv 'ī ,än ,ech·iŋ }

reactivity |CHEM| The relative capacity of an atom, molecule, or radical to combine chemically with another atom, molecule, or radical. { ,rē·ak'tiv·əd·ē }

ready-mixed concrete |MATER| Concrete mixed away from the construction site and delivered in readiness for placing. { 'red·ē ¦mikst kän 'krēt }

reagent |CHEM| The compound that supplies the molecule, ion, or free radical which is arbitrarily considered as the attacking species in a chemical reaction. { rē'ā·jənt }

real crystal |CRYSTAL| A crystal for which the finite extent of the crystal and its various imperfections and defects are taken into account. { 'rēl 'krist·əl }

real fluid flow |FL MECH| The flow in which

effects of tangential or shearing forces are taken into account; these forces give rise to fluid friction, because they oppose the sliding of one particle past another. { 'rēl 'flü·əd ,flō }

real gas [THERMO] A gas, as considered from the viewpoint in which deviations from the ideal gas law, resulting from interactions of gas molecules, are taken into account. Also known as imperfect gas. { 'rēl 'gas }

ream [MATER] **1.** A layer of nonhomogeneous material in flat glass. **2.** Five hundred sheets of paper; a printer's ream consists of 516 sheets. { rēm }

reamed extrusion ingot [MET] A hollow extrusion ingot whose original inside surface has been removed by reaming. { 'rēmd ik'strü·zhən ,iŋ·gət }

Réaumur temperature scale [THERMO] Temperature scale where water freezes at 0°R and boils at 80°R. { ¦rā·ō¦myür 'tem·prə·chər ,skāl }

rebar [CIV ENG] A steel bar or rod used to reinforce concrete. { 'rē,bär }

recalescence [MET] Brightening (reglowing) of iron on cooling through the gamma- to alpha-phase transformation temperature caused by liberation of the latent heat of transformation. { ,rē·kə'les·əns }

recarburize [MET] **1.** To increase the carbon content of molten steel or cast iron. **2.** To carburize a metal part, making up for any loss of carbon during processing. { rē'kär·bə,rīz }

rechargeable battery See storage battery. { rē'chär·jə·bəl'bad·ə·rē }

reciprocal impedance [ELEC] Two impedances Z_1 and Z_2 are said to be reciprocal impedances with respect to an impedance Z (invariably a resistance) if they are so related as to satisfy the equation $Z_1 Z_2 = Z^2$. { ri'sip·rə·kəl im'pēd·əns }

reciprocal lattice [CRYSTAL] A lattice array of points formed by drawing perpendiculars to each plane (hkl) in a crystal lattice through a common point as origin; the distance from each point to the origin is inversely proportional to spacing of the specific lattice planes; the axes of the reciprocal lattice are perpendicular to those of the crystal lattice. { ri'sip·rə·kəl 'lad·əs }

reciprocal ohm See siemens. { ri'sip·rə·kəl 'ōm }

reciprocal space See wave-vector space. { ri'sip·rə·kəl ,spās }

reciprocal vectors [CRYSTAL] For a set of three vectors forming the primitive translations of a lattice, the vectors that form the primitive translations of the reciprocal lattice. { ri'sip·rə·kəl 'vek·tərz }

reclaimed rubber [MATER] Scrap rubber (natural or synthetic) prepared for reuse; fragmented scrap is digested in hot caustic solution to which reclaiming agents have been added; reclaimed rubber is used to blend with virgin rubber, or for low-grade rubber products. { rē'klāmd 'rəb·ər }

reclaim rinse [MET] A nonflowing rinse used to recover dragout in electroplating operations. { 'rē,klām ,rins }

recoil implantation [ELECTR] A mechanism for ion-beam mixing of a film and a substrate in which atoms are driven from the film into the substrate as a result of direct collisions with incident ions. { ¦rē,kȯil ,im·plan'tā·shən }

recombination radiation [SOLID STATE] The radiation emitted in semiconductors when electrons in the conduction band recombine with holes in the valence band. { ,rē,käm·bə'nā·shən ,rād·ē,ā·shən }

reconstituted mica [MATER] Mica sheets or shaped objects made by breaking up scrap natural mica, combining with a binder, and pressing into forms suitable for use as electrical insulating material. { rē'kän·stə,tüd·əd 'mī·kə }

reconstruction [SOLID STATE] A process in which atoms at the surface of a solid displace and form bands different from those existing in the bulk solid. { ,rē·kən'strək·shən }

reconstructive processing [INORG CHEM] The spinning of an inorganic compound of an organic support or binder subsequently removed by oxidation or volatilization to form an inorganic polymer. { ,rē·kən'strək·tiv 'prä,ses·iŋ }

reconstructive transformation [CRYSTAL] A type of crystal transformation that involves the breaking of either first- or second-order coordination bonds. { ,rē·kən'strək·tiv ,tranz·fər'mā·shən }

recording spot See picture element. { ri'kȯrd·iŋ ,spät }

recoverable shear [FL MECH] Measure of the elastic content of a fluid, related to elastic recovery (mechanicallike property of elastic recoil); found in unvulcanized, unfilled natural rubber, and certain polymer solutions, soap gels, and biological fluids. { ri'kəv·rə·bəl 'shir }

recovery [MECH] The return of a body to its original dimensions after it has been stressed, possibly over a considerable period of time. [MET] **1.** The percentage of valuable material obtained from a processed ore. **2.** Reduction or elimination of work-hardening effects, usually by heat treatment. { ri'kəv·ə·rē }

recrystallization [CHEM] Repeated crystallization of a material from fresh solvent to obtain an increasingly pure product. [CRYSTAL] A change in the structure of a crystal without a chemical alteration. [MET] A process which takes place in metals and alloys following distortion and fragmentation of constituent crystals by severe mechanical deformation, in which some fragments grow at the expense of others, so that larger, strain-free grains are formed; it progresses slowly at room temperature, but is greatly speeded by annealing. { rē,krist·əl·ə'zā·shən }

recrystallization annealing [MET] Producing a new grain structure without phase change by annealing cold-worked metal. { rē,krist·əl·ə'zā·shən ə,nēl·iŋ }

recrystallization temperature [MET] The minimum temperature at which complete recrystallization occurs in an annealed cold-worked metal within a specified time. { rē,krist·əl·ə'zā·shən ,tem·prə·chər }

red [OPTICS] The hue evoked in an average

observer by monochromatic radiation having a wavelength in the approximate range from 622 to 770 nanometers; however, the same sensation can be produced in a variety of other ways. { red }

red acetate   See mordant rouge. { 'red 'as·ə,tāt }

red brass |MET| Brass containing 85% copper, 5% zinc, 5% tin, and 5% lead. { 'red 'bras }

red charcoal |MATER| An impure charcoal made by heating wood to 300°C; much of the oxygen and hydrogen is retained. { 'red 'chär,kōl }

red glass |MATER| Soda-zinc glass to which small amounts of cadmium and selenium are added. { 'red 'glas }

red-hardness |MET| In reference to high-speed steel and other cutting tool materials, the property of being hard enough to cut metals even when heated to a dull-red color. { 'red ¦härd·nəs }

red lake C pigment |MATER| An organic azo pigment; made by coupling the diazonium salt (barium or sodium) of *ortho*-chloro-*meta*-toluidine-*para*-sulfonic acid with β-naphthol; used to color inks, plastics, and rubbers. { 'red ¦lāk 'sē ,pig·mənt }

red lead   See lead tetroxide. { 'red 'led }

red liquor   See mordant rouge. { 'red 'lik·ər }

red mercury sulfide   See mercuric sulfide. { 'red 'mər·kyə·rē 'səl,fīd }

red metal |MET| A copper matte having a copper content of about 48%. { 'red 'med·əl }

red mud |MET| An iron oxide-rich residue obtained in purifying bauxite in the Bayer process. { 'red 'məd }

red ocher   See ferric oxide. { 'red 'ō·kər }

redox polymer |POLYM CHEM| A polymer whose structure contains functional groups that can be reversibly reduced or oxidized. Also known as electron exchanger. { 'rē,däks ,päl·ə·mər }

red phosphorus |CHEM| An allotropic form of the element phosphorus; violet-red, amorphous powder subliming at 416°C, igniting at 260°; insoluble in all solvents,; nonpoisonous. { 'red 'fä·sfə·rəs }

red potassium chromate   See potassium dichromate. { 'red pə'tas·ē·əm 'krō,māt }

red potassium prussiate   See potassium ferricyanide. { 'red pə'tas·ē·əm 'prəs·ē,āt }

red precipitate   See mercuric oxide. { 'red prə 'sip·ə,tāt }

red prussiate of potash   See potassium ferricyanide. { 'red 'prəs·ē,āt əv 'päd,ash }

red prussiate of soda   See sodium ferricyanide. { 'red 'prəs·ē,āt əv 'sōd·ə }

red thermit |MET| Thermit made with red iron oxide. { 'red 'thər·mət }

reduced nickel |MET| Nickel obtained by the precipitation of nickel hydroxide or nickel carbonate onto kieselguhr, then reducing the precipitate by heating it with hydrogen. { ri'düst 'nik·əl }

reduced pressure |THERMO| The ratio of the pressure of a substance to its critical pressure. { ri'düst 'presh·ər }

reduced property   See reduced value. { ri'düst 'präp·ərd·ē }

reduced temperature |THERMO| The ratio of the temperature of a substance to its critical temperature. { ri'düst 'tem·prə·chər }

reduced value |THERMO| The actual value of a quantity divided by the value of that quantity at the critical point. Also known as reduced property. { ri'düst 'val·yü }

reduced volume |THERMO| The ratio of the specific volume of a substance to its critical volume. { ri'düst 'väl·yəm }

reducer   See reducing agent. { ri'dü·sər }

reducing agent |CHEM| **1.** Also known as reducer. **2.** A material that adds hydrogen to an element or compound. **3.** A material that adds an electron to an element or compound, that is, decreases the positiveness of its valence. { ri'düs·iŋ ,ā·jənt }

reducing atmosphere |CHEM| An atmosphere of hydrogen (or other substance that readily provides electrons) surrounding a chemical reaction or physical device; the effect is the opposite to that of an oxidizing atmosphere. { ri'düs·iŋ 'at·mə,sfir }

reducing flame |CHEM| A flame having excess fuel and being capable of chemical reduction, such as extracting oxygen from a metallic oxide. { ri'düs·iŋ ,flām }

reductant |MET| Coal or other reducing materials introduced in a smelting process to remove oxygen from ores or concentrates. { ri'dək·tənt }

reduction |CHEM| **1.** Reaction of hydrogen with another substance. **2.** Chemical reaction in which an element gains an electron (has a decrease in positive valence). { ri'dək·shən }

reduction of area |MET| In tensile testing, the percentage of decrease in cross-sectional area of a specimen at the point of rupture. { ri'dək·shən əv 'er·ē·ə }

reduction roll |MET| A roller used to reduce the thickness of a piece of metal. { ri'dək·shən ,rōl }

reentrant angle |CRYSTAL| The angle between two plane surfaces on a crystalline solid, in which the external angle is less than 180°. { rē'en·trənt ,aŋ·gəl }

refined paraffin wax |MATER| A grade of paraffin wax; a hard, crystalline hydrocarbon wax derived from mixed-base or paraffin-base crude oils. { ri'fīnd 'par·ə·fən ¦waks }

refined tar |MATER| A tar from which water has been extracted by evaporation or distillation. { ri'fīnd 'tär }

refinery |POLY ENG| System of process units used to convert crude petroleum into fuels, lubricants, and other petroleum-derived products. |MET| System of process units used to convert nonferrous-metal ores into pure metals, such as copper or zinc. { ri'fīn·rē }

refining temperature |MET| The temperature just above the transformation range employed in the heat treatment of steel in order to refine grain size. { ri'fīn·iŋ ,tem·prə·chər }

reflectance   See reflectivity. { ri'flek·təns }

reflection angle   See angle of reflection. { ri'flek·shən ,aŋ·gəl }

reflection coefficient |PHYS| The ratio of the

amplitude of a wave reflected from a surface to the amplitude of the incident wave. Also known as coefficient of reflection. { ri'flek·shən ˌkō·i ˌfish·ənt }

reflection diffraction [PHYS] Type of electron diffraction analysis in which the electron beam grazes the sample surface. { ri'flek·shən di ˌfrak·shən }

reflection plane See plane of mirror symmetry;plane of reflection. { ri'flek·shən ˌplān }

reflection twin [CRYSTAL] A crystal twin whose symmetry is formed by an apparent mirror image across a plane. { ri'flek·shən ˌtwin }

reflective insulation [MATER] An insulating material used to retard the flow of heat by reflecting heat radiation; usually made of aluminum foil or sheets, although coated steel sheets, aluminized paper, gold and silver surfaces, and refractory metals at higher temperatures are also used. { ri'flek·tiv ˌin·sə'lā·shən }

reflectivity [PHYS] The ratio of the energy carried by a wave which is reflected from a surface to the energy carried by the wave which is incident on the surface. Also known as reflectance. { ˌrē ˌflek'tiv·əd·ē }

reflectometer [ENG] A photoelectric instrument for measuring the optical reflectance of a reflecting surface. { ˌrē ̣flek'täm·əd·ər }

reflowing [ENG] Melting and resolidifying an electrodeposited or other type coating. { rē'flō· iŋ }

refraction [ELECTROMAG] The change in direction of lines of force of an electric or magnetic field at a boundary between media with different permittivities or permeabilities. [PHYS] The change of direction of propagation of any wave, such as an electromagnetic or sound wave, when it passes from one medium to another in which the wave velocity is different, or when there is a spatial variation in a medium's wave velocity. { ri'frak·shən }

refractive constant See index of refraction. { ri'frak·tiv 'kän·stənt }

refractive index See index of refraction. { ri 'frak·tiv ˌin,deks }

refractivity [ELECTROMAG] 1. Some quantitative measure of refraction, usually a measure of the index of refraction. 2. The index of refraction minus 1. { ˌrē ̣frak'tiv·əd·ē }

refractometer [ENG] An instrument used to measure the index of refraction of a substance in any one of several ways, such as measurement of the refraction produced by a prism, measurement of the critical angle, observation of an interference pattern produced by passing light through the substance, and measurement of the substance's dielectric constant. { ˌrē ˌfrak'täm·əd·ər }

refractometry [OPTICS] The measurement of the index of refraction of a substance; it is an important tool in analytical chemistry. { ˌrē ̣frak'täm·ə·trē }

refractory [MATER] 1. A material of high melting point. 2. The property of resisting heat. { ri'frak·trē }

refractory cement [MATER] Any of a variety of mixtures, such as fireclay-silica-ganister mixture, or fireclay mixed with crushed brick, or fireclay and silica sand, with a refractory range of 2600–2800°F (1412–1523°C); used for furnace and oven linings and for fillers. { ri'frak·trē si'ment }

refractory clay [MATER] Clay with a melting point above 1600°C; used to make firebrick and linings for furnaces and reactors. { ri'frak·trē ̣klā }

refractory coating [MATER] A coating composed of a refractory material. { ri'frak·trē ̣kōd·iŋ }

refractory concrete [MATER] Heat-resistant concrete made with high-alumina or calcium-aluminate cement and a refractory aggregate. { ri'frak·trē 'kän,krēt }

refractory enamel [MATER] An enamel for coating and protecting metals against attack by hot gases. { ri'frak·trē i'nam·əl }

refractory hard metals [CHEM] True chemical compounds composed of two or more metals in the crystalline form, and having a very high melting point and high hardness. { ri'frak·trē 'härd 'med·əlz }

refractory metal [MET] A metal or alloy that is heat-resistant, having a high melting point. { ri'frak·trē ̣med·əl }

refractory sand [MATER] Sand used for refractory which is capable of resisting high temperatures. { ri'frak·trē ̣sand }

refrangible [PHYS] Capable of being refracted. { ri'fran·jə·bəl }

refrigerant [MATER] A substance that by undergoing a change in phase (liquid to gas, gas to liquid) releases or absorbs a large latent heat in relation to its volume, and thus effects a considerable cooling effect; examples are ammonia, sulfur dioxide, ethyl or methyl chloride (these are no longer widely used), and the fluorocarbons, such as Freon, Ucon, and Genetron. { ri'frij·ə·rənt }

refrigeration cycle [THERMO] A sequence of thermodynamic processes whereby heat is withdrawn from a cold body and expelled to a hot body. { ri,frij·ə'rā·shən ,sī·kəl }

refrigeration oil [MATER] A mineral oil with all moisture and wax removed; used for lubricating refrigerating machinery. { ri,frij·ə'rā·shən ,óil }

regelation [THERMO] Phenomenon in which ice (or any substance which expands upon freezing) melts under intense pressure and freezes again when this pressure is removed; accounts for phenomena such as the slippery nature of ice and the motion of glaciers. { ̣rē·jə'lā·shən }

regenerated cellulose [MATER] 1. Rayon in which the raw cellulose is changed physically but not chemically, such as viscose, cuprammonium, and nitrocellulose rayons. 2. A transparent cellulose plastic material made by mixing cellulose xanthate with a dilute sodium hydroxide solution to form a viscose. { rē'jen·ə,rād·əd 'sel·yə,lōs }

regenerative cycle [THERMO] An engine cycle in which low-grade heat that would ordinarily be lost is used to improve the cyclic efficiency. { rē'jen·rəd·iv ,sī·kəl }

regular dodecahedron See pyritohedron. { 'reg·yə·lər dō,dek·ə'hē·drən }

regular polymer [POLYM CHEM] A polymer whose molecules possess only one kind of constitutional unit in a single sequential structure. { 'reg·yə·lər 'päl·ə·mər }

regulus [MET] Impure metal formed beneath the slag during smelting or reduction of ores. { 'reg·yə·ləs }

reheating [THERMO] A process in which the gas or steam is reheated after a partial isentropic expansion to reduce moisture content. Also known as resuperheating. { rē'hēd·iŋ }

reinforced beam [CIV ENG] A concrete beam provided with steel bars for longitudinal tension reinforcement and sometimes compression reinforcement and reinforcement against diagonal tension. { ¦rē·ən'fȯrst 'bēm }

reinforced brickwork [CIV ENG] Brickwork strengthened by expanded metal, steel-wire mesh, hoop iron, or thin rods embedded in the bed joints. { ¦rē·ən'fȯrst 'brik,wərk }

reinforced column [CIV ENG] **1.** A long concrete column reinforced with longitudinal bars with ties or circular spirals. **2.** A composite column. **3.** A combination column. { ¦rē·ən'fȯrst 'käl·əm }

reinforced concrete [CIV ENG] Concrete containing reinforcing steel rods or wire mesh. { ¦rē·ən'fȯrst 'kän,krēt }

reinforced molding compound [MATER] A compound containing polymer or resin and a reinforcing filler, supplied in the form of ready-to-use material as distinguished from premix. { ¦rē·ən'fȯrst 'mōld·iŋ ,käm,paủnd }

reinforced plastic [MATER] High-strength filled plastic product used for mechanical, construction, and electrical products, automotive components, and ablative coatings; filling can be whiskers of glass, metal, boron, or other materials. { ¦rē·ən'fȯrst 'plas·tik }

reinforcement [CIV ENG] Strengthening concrete, plaster, or mortar by embedding steel rods or wire mesh in it. [MATER] A strong inert material bonded to a plastic to enhance its strength, stiffness, and resistance to impact. { ,rē·ən'fȯrs·mənt }

reinforcement of weld [MET] Weld metal that extends beyond the surface or plane of the weld joint. { ,rē·ən'fȯrs·mənt əv 'weld }

reinforcing bars [CIV ENG] Steel rods that are embedded in building materials such as concrete for reinforcement. { ¦rē·ən'fȯrs·iŋ ,bärz }

rel [ELECTROMAG] Unit of reluctance equal to 1 ampere-turn per magnetic line of force. { rel }

relative atomic mass See atomic weight. { 'rel·əd·iv ə'täm·ik 'mas }

relative dielectric constant See dielectric constant. { 'rel·əd·iv ¦dī·i'lek·trik 'kän·stənt }

relative index of refraction [OPTICS] The ratio of the velocity of light in one medium to that in another medium. { 'rel·əd·iv 'in,deks əv ri'frak·shən }

relative molecular mass See molecular weight. { 'rel·əd·iv mə'lek·yə·lər 'mas }

relative permittivity See dielectric constant. { 'rel·əd·iv ,pər·mə'tiv·əd·ē }

relative resistance [ELEC] The ratio of the resistance of a piece of a material to the resistance of a piece of specified material, such as annealed copper, having the same dimensions and temperature. { 'rel·əd·iv ri'zis·təns }

relaxation [MECH] **1.** Relief of stress in a strained material due to creep. **2.** The lessening of elastic resistance in an elastic medium under·an applied stress resulting in permanent deformation. { ,rē,lak'sā·shən }

relaxation test [ENG] A creep test in which the decrease of stress with time is measured while the total strain (elastic and plastic) is maintained constant. { ,rē,lak'sā·shən ,test }

release agent [MATER] A lubricant, such as wax or silicone oil, used to coat a mold cavity to prevent the molded piece from sticking when removed. Also known as mold release; parting agent. { ri'lēs ,ā·jənt }

relief [CRYSTAL] The apparent topography exhibited by minerals in thin section as a consequence of refractive index. { ri'lēf }

reluctance [ELECTROMAG] A measure of the opposition presented to magnetic flux in a magnetic circuit, analogous to resistance in an electric circuit; it is equal to magnetomotive force divided by magnetic flux. Also known as magnetic reluctance. { ri'lək·təns }

reluctivity [PHYS] The reciprocal of magnetic permeability; the reluctivity of empty space is unity. Also known as magnetic reluctivity; specific reluctance. { ,rē,lək'tiv·əd·ē }

remanence [ELECTROMAG] The magnetic flux density that remains in a magnetic circuit after the removal of an applied magnetomotive force; if the magnetic circuit has an air gap, the remanence will be less than the residual flux density. { 'rem·ə·nəns }

remoistening adhesive [MATER] Any adhesive material, such as dextrin, animal glue, or gum arabic, which is reactivated with the application of water upon the adhesive surface. { rē'mȯis·ən·iŋ ad,hē·ziv }

Renn-Walz process [MET] A method of reclaiming iron and other metals from the waste materials produced in the smelting of zinc and lead ores; this material is brought up to 1000°C in the preheating zone of the kiln by the countercurrent gases, and the oxidized metal vapors are carried off in the flue gases, from which they are subsequently filtered. { 'ren 'välts ,prä·səs }

repeated load [MECH] A force applied repeatedly, causing variation in the magnitude and sometimes in the sense, of the internal forces. { ri'pēd·əd 'lōd }

repeated twinning [CRYSTAL] Crystal twinning that involves more than two simple crystals. { ri'pēd·əd 'twin·iŋ }

repeating unit [POLYM CHEM] The group of atoms that is derived from a monomer and repeats throughout a polymer. Also known as monomeric unit. { ri¦pēd·iŋ ,yü·nət }

repellency [CHEM] Ability to repel water, or

being hydrophobic; opposite to water wettability. { ri'pel·ən·sē }

repressing |MET| Applying pressure to a pressed and sintered compact to improve some physical property. { ri 'pres·iŋ }

repulsion |MECH| A force which tends to increase the distance between two bodies having like electric charges, or the force between atoms or molecules at very short distances which keeps them apart. Also known as repulsive force. { ri'pəl·shən }

repulsive force See repulsion. { ri'pəl·siv 'fórs }

residual elements |MET| Elements present in small amounts in a metal or alloy, not added intentionally. { rə'zij·ə·wəl 'el·ə·məns }

residual field |ELECTROMAG| The magnetic field left in an iron core after excitation has been removed. { rə'zij·ə·wəl ¦fēld }

residual flux density |ELECTROMAG| The magnetic flux density at which the magnetizing force is zero when the material is in a symmetrically and cyclically magnetized condition. Also known as residual induction; residual magnetic induction; residual magnetism. { rə'zij·ə·wəl 'fləks ‚den·səd·ē }

residual induction See residual flux density. { rə'zij·ə·wəl in'dək·shən }

residual magnetic induction See residual flux density. { rə'zij·ə·wəl mag'ned·ik in'dək·shən }

residual magnetism See residual flux density. { rə'zij·ə·wəl 'mag·nə,tiz·əm }

residual method |MET| Magnetic particle inspection in which particles are supplied to a specimen after the magnetizing force has been removed. { rə'zij·ə·wəl ¦meth·əd }

residual resistance |SOLID STATE| The value to which the electrical resistance of a metal drops as the temperature is lowered to near absolute zero, caused by imperfections and impurities in the metal rather than by lattice vibrations. { rə'zij·ə·wəl ri'zis·təns }

residual stress See internal stress. { rə'zij·ə·wəl 'stres }

residual tack See aftertack. { rə'zij·ə·wəl 'tak }

resilience |MECH| 1. Ability of a strained body, by virtue of high yield strength and low elastic modulus, to recover its size and form following deformation. 2. The work done in deforming a body to some predetermined limit, such as its elastic limit or breaking point, divided by the body's volume. { rə'zil·yəns }

resin |POLYM CHEM| Any of a class of solid or semisolid organic products of natural or synthetic origin with no definite melting point, generally of high molecular weight; most resins are polymers. { 'rez·ən }

resin emulsion |MATER| Stable emulsion of a resin in a solvent carrier, such as the latex emulsions used in water-based latex paints. { 'rez·ən i,məl·shən }

resin of copper See cuprous chloride. { 'rez·ən əv 'käp·ər }

resinoid |POLYM CHEM| A thermosetting synthetic resin either in its initial (temporarily fusible) or in its final (infusible) state. { 'rez·ən ‚óid }

resinous cement |MATER| An acid-proof cement with a base of synthetic resin. { 'rez·ən·əs si'ment }

resist |GRAPHICS| A protective layer applied to the image, or other parts of a plate, to protect that portion of the metal from the action of an etching bath or a sandblasting operation. |MATER| An acid-resistant nonconducting coating used to protect desired portions of a wiring pattern from the action of the etchant during manufacture of printed wiring boards. |MET| An insulating material, for example lacquer, applied to the surface of work to prevent electroplating or electrolytic action at the coated area. Also known as stopoff. { ri'zist }

resistance |ELEC| 1. The opposition that a device or material offers to the flow of direct current, equal to the voltage drop across the element divided by the current through the element. Also known as electrical resistance. 2. In an alternating-current circuit, the real part of the complex impedance. |MECH| In damped harmonic motion, the ratio of the frictional resistive force to the speed. Also known as damping coefficient; damping constant; mechanical resistance. { ri'zis·təns }

resistance brazing |MET| Brazing employing the heat developed by an electric current, the joint being part of the electric circuit. { ri'zis·təns ‚brāz·iŋ }

resistance coefficient 2 See Darcy number 1. { ri'zis·təns,kō·i,fish·ənt 'tü }

resistance drop |ELEC| The voltage drop occurring between two points on a conductor due to the flow of current through the resistance of the conductor; multiplying the resistance in ohms by the current in amperes gives the voltage drop in volts. Also known as IR drop. { ri'zis·təns ‚dräp }

resistance loss |ELEC| Power loss due to current flowing through resistance; its value in watts is equal to the resistance in ohms multiplied by the square of the current in amperes. { ri'zis·təns ‚lós }

resistance material |ELEC| Material having sufficiently high resistance per unit length or volume to permit its use in the construction of resistors. { ri'zis·təns mə'tir·ē·əl }

resistance measurement |ELEC| The quantitative determination of that property of an electrically conductive material, component, or circuit called electrical resistance. { ri'zis·təns ‚mezh·ər·mənt }

resistance seam welding |MET| Resistance welding process which produces a series of individual spot welds, overlapping spot welds, or a continuous nugget weld made by circular or wheel-type electrodes. { ri'zis·təns 'sēm ‚weld·iŋ }

resistance spot welding |MET| Resistance welding process in which the parts are lapped and held in place under pressure; the size and shape of the electrodes (usually circular) control the size and shape of the welds. { ri'zis·təns 'spät ‚weld·iŋ }

resistance welding |MET| Joining metals together under pressure by making use of heat developed by an electric current, the work being part of the electrical circuit. { ri'zis·təns ˌweld·iŋ }

resistance wire |MET| Wire made from a metal or alloy having high resistance per unit length, such as Nichrome; used in wire-wound resistors and heating elements. { ri'zis·təns ˌwīr }

resisting moment |MECH| A moment produced by internal tensile and compressive forces that balances the external bending moment on a beam. { ri'zist·iŋ ˌmō·mənt }

resistivity See electrical resistivity. { ˌrē,zis'tiv·əd·ē }

resistor |ELEC| A device designed to have a definite amount of resistance; used in circuits to limit current flow or to provide a voltage drop. { ri'zis·tər }

resite See C stage. { 're,zīt }

resonance |PHYS| **1.** A phenomenon exhibited by a physical system acted upon by an external periodic driving force, in which the resulting amplitude of oscillation of the system becomes large when the frequency of the driving force approaches a natural free oscillation frequency of the system. **2.** In general, any phenomenon which is greatly enhanced at frequencies or energies that are at or very close to a given characteristic value. { 'rez·ən·əns }

resonance vibration |MECH| Forced vibration in which the frequency of the disturbing force is very close to the natural frequency of the system, so that the amplitude of vibration is very large. { 'rez·ən·əns vī,brā·shən }

restitution coefficient See coefficient of restitution. { ˌres·tə'tü·shən ˌkō·i,fish·ənt }

restricted adhesive |MATER| An adhesive which for any reason cannot satisfactorily pass its evaluation test; as a result, the maximum time required for curing, that is, its usable life, cannot be assigned; it cannot be used for structural bonding. { ri'strik·təd ad'hē·ziv }

resultant of forces |MECH| A system of at most a single force and a single couple whose external effects on a rigid body are identical with the effects of the several actual forces that act on that body. { ri'zəlt·ənt əv 'fȯrs·əz }

resuperheating See reheating. { rē¦sü·pər'hēd·iŋ }

retardant See retarder. { ri'tärd·ənt }

retarder |MATER| A material that inhibits the action of another substance, such as flameproofing agents or substances added to cement to retard setting time. Also known as retardant. { ri 'tärd·ər }

retentivity |ELECTROMAG| The residual flux density corresponding to the saturation induction of a magnetic material. { ˌrē,ten'tiv·əd·ē }

Retgers' law |SOLID STATE| The law that the properties of crystalline mixtures of isomorphous substances are continuous functions of the percentage composition. { 'ret·gərz,lȯ }

reticulated glass |MATER| Ornamental glass-ware containing interlacing sets of lines. { rə'tik·yə,lād·əd 'glas }

reverse bias |ELECTR| A bias voltage applied to a diode or a semiconductor junction with polarity such that little or no current flows; the opposite of forward bias. { ri'vərs 'bī·əs }

reverse Brayton cycle |THERMO| A refrigeration cycle using air as the refrigerant but with all system pressures above the ambient. Also known as dense-air refrigeration cycle. { ri'vərs 'brāt·ən ˌsī·kəl }

reverse Carnot cycle |THERMO| An ideal thermodynamic cycle consisting of the processes of the Carnot cycle reversed and in reverse order, namely, isentropic expansion, isothermal expansion, isentropic compression, and isothermal compression. { ri'vərs kär'nō ˌsī·kəl }

reverse-current cleaning See anodic cleaning. { ri'vərs ¦kə·rənt ˌklēn·iŋ }

reversed image |GRAPHICS| **1.** A mirror image in which the right and left sides of the picture are interchanged. **2.** See negative. { ri'vərst 'im·ij }

reverse drawing |MET| Drawing for a second time, in a direction opposite to the original drawing. { ri'vərs 'drȯ·iŋ }

reverse polarity |MET| An arc-welding circuit in which the electrode is connected to the positive terminal. { ri'vərs pə'lar·əd·ē }

reversible path |THERMO| A path followed by a thermodynamic system such that its direction of motion can be reversed at any point by an infinitesimal change in external conditions; thus the system can be considered to be at equilibrium at all points along the path. { ri'vər·sə·bəl 'path }

reversible process |THERMO| An ideal thermodynamic process which can be exactly reversed by making an indefinitely small change in the external conditions. Also known as quasi-static process. { ri'vər·sə·bəl 'prä·səs }

reversing mill |MET| A rolling mill in which the workpiece is passed forward and backward through a given pair of rolls. { ri'vərs·iŋ ˌmil }

Reynolds equation |FL MECH| A form of the Navier-Stokes equation, which is $\rho \partial u/\partial t = (\partial/\partial x)(p_{xx} - \rho u^2) + (\partial/\partial y) \cdot (p_{xy} - \rho u v) + (\partial/\partial z)(p_{xz} - \rho u w)$, where ρ is the fluid density, u, v, and w are the components of the fluid velocity, and p_{xx}, p_{xy}, and p_{xz} are normal and shearing stresses. { 'ren·əlz i,kwā·zhən }

Reynolds number |FL MECH| A dimensionless number which is significant in the design of a model of any system in which the effect of viscosity is important in controlling the velocities or the flow pattern of a fluid; equal to the density of a fluid, times its velocity, times a characteristic length, divided by the fluid viscosity. Symbolized N_{Re}. Also known as Damköhler number V (DaV). { 'ren·əlz ˌnəm·bər }

Reynolds stress |FL MECH| The net transfer of momentum across a surface in a turbulent fluid because of fluctuations in fluid velocity. Also known as eddy stress. { 'ren·əlz ˌstres }

Reynolds stress tensor |FL MECH| A tensor whose components are the components of the

Reynolds stress across three mutually perpendicular surfaces. { 'ren·əlz 'stres ˌten·sər }

Rf See rutherfordium.

Rh See rhodium.

rhe [FL MECH] **1.** A unit of dynamic fluidity, equal to the dynamic fluidity of a fluid whose dynamic viscosity is 1 centipoise. **2.** A unit of kinematic fluidity, equal to the kinematic fluidity of a fluid whose kinematic viscosity is 1 centistoke. { rē }

rhenium [CHEM] A metallic element, symbol Re, atomic number 75, atomic weight 186.207; a transition element. { 'rē·nē·əm }

rhenium halide [INORG CHEM] Halogen compound of rhenium; examples are $ReCl_3$, $ReCl_4$, ReF_4, and ReF_6. { 'rē·nē·əm 'ha,līd }

rheocasting [MET] A process in which a liquid metal is vigorously agitated during initial stages of solidification to produce a globular semisolid structure which remains highly fluid when more than 60% solidification has occurred. { 'rē·ō ˌkast·iŋ }

rheogoniometry [MECH] Rheological tests to determine the various stress and shear actions on Newtonian and non-Newtonian fluids. { ¦rē·ə·gō·nē'äm·ə·trē }

rheology [MECH] The study of the deformation and flow of matter, especially non-Newtonian flow of liquids and plastic flow of solids. { rē'äl·ə·jē }

rheometer [ENG] An instrument for determining flow properties of solids by measuring relationships between stress, strain, and time. { rē'äm·əd·ər }

rheopectic fluid [FL MECH] A fluid for which the structure builds up on shearing; this phenomenon is regarded as the reverse of thixotropy. { ¦rē·ə¦pek·tik 'flü·əd }

rheopexy [PHYS CHEM] A property of certain sols, having particles shaped like rods or plates, which set to gel form more quickly when mechanical means are used to hasten the orientation of the particles. { 'rē·ə,pek·sē }

rheotropic brittleness [MET] A low-temperature or high-strain-rate brittleness that may be eliminated by prestraining under milder conditions. { ¦rē·ə¦träp·ik 'brid·əl·nəs }

rhinestone [MATER] A clear, colorless imitation of diamond, made of glass, paste, or gem quartz, backed with metallic foil. { 'rīn,stōn }

rhodamine toner [MATER] Rhodamine dye and phosphotungstic or phosphomolybdic acid; red to maroon; used in printing inks. { 'rōd·ə,mēn 'tōn·ər }

rhodanic acid See thiocyanic acid. { rō'dan·ik 'as·əd }

rhodium [CHEM] A chemical element, symbol Rh, atomic number 45, atomic weight 102.9055. [MET] A silver-white metal in the platinum family; sometimes alloyed with platinum for thermocouples or used as a tarnish-resistant electrode posit. { 'rōd·ē·əm }

rhodium chloride [INORG CHEM] $RhCl_3$ Water-insoluble, brown-red powder, soluble in cyanides and alkalies; decomposes at 450–500°C. Also known as rhodium trichloride. { 'rōd·ē·əm 'klór ,īd }

rhodium trichloride See rhodium chloride. { 'rōd·ē·əm trī'klór,īd }

rhomb See rhombohedron. { räm }

rhombic dodecahedron [CRYSTAL] A crystal form in the cubic system that is a dodecahedron whose faces are equal rhombuses. { 'räm·bik ,dō·dek·ə'hē·drən }

rhombic lattice See orthorhombic lattice. { 'räm·bik 'lad·əs }

rhombic system See orthorhombic system. { 'räm·bik 'sis·təm }

rhombohedral [CRYSTAL] **1.** Of or pertaining to the rhombohedral system. **2.** Of or pertaining to crystal cleavage in or a centered lattice of the hexagonal system. { ¦räm·bō¦hē·drəl }

rhombohedral close packing See rhombohedral packing. { ¦räm·bō¦hē·drəl 'klōs 'pak·iŋ }

rhombohedral lattice [CRYSTAL] A crystal lattice in which the three axes of a unit cell are of equal length, and the three angles between axes are the same, and are not right angles. Also known as trigonal lattice. { ¦räm·bō¦hē·drəl ¦lad·əs }

rhombohedral packing [CRYSTAL] The tightest manner of systematic arrangement of uniform solid spheres in a clastic sediment or crystal lattice, characterized by a unit cell of six planes passed through eight sphere centers situated at the corners of a regular rhombohedron. Also known as rhombohedral close packing. { ¦räm·bō¦hē·drəl ¦pak·iŋ }

rhombohedral system [CRYSTAL] A division of the trigonal crystal system in which the rhombohedron is the basic unit cell. { ¦räm·bō¦hē·drəl ¦sis·təm }

rhombohedron [CRYSTAL] A trigonal crystal form that is a parallelepiped, the six identical faces being rhombs. Also known as rhomb. { ¦räm·bō¦hē·drən }

rice glue [MATER] A paste made from ground rice boiled in soft water; used in molded objects such as statuary. { 'rīs ,glü }

rice paper [MATER] **1.** A product, not a true paper and not made from rice, but manufactured from the pith of a tree grown in Taiwan; tissue-thin sheets of the pith are peeled away as a cylindrical section of the wood rotates against a knife. **2.** Any of various oriental papers used in block printing. { 'rīs ,pā·pər }

Richard's solder [MET] A yellow brass containing 3% aluminum and 3% phosphor tin. { 'rich·ərdz ,säd·ər }

rich concrete [MATER] Concrete with a high cement content. { 'rich 'kän,krēt }

right-handed [CRYSTAL] Having a crystal structure with a mirror-image relationship to a left-handed structure. { 'rīt ¦han·dəd }

rigid body [MECH] An idealized extended solid whose size and shape are definitely fixed and remain unaltered when forces are applied. { 'rij·id 'bäd·ē }

rigidity modulus See modulus of elasticity in shear. { ri'jid·əd·ē ,mäj·ə·ləs }

rigid pavement [CIV ENG] A thick portland cement pavement on a gravel base and subbase, with steel reinforcement and often with transverse joints. { 'rij·əd 'pāv·mənt }

rigid resin [MATER] A resin with a modulus of 10,000 pounds per square inch (6.895 × 10⁷ pascals) or greater. { 'rij·əd 'rez·ən }

rimmed steel [MET] Low-carbon steel, partially deoxidized, which on cooling continuously, evolves sufficient carbon monoxide to form a case or rim of metal virtually free of voids. { 'rimd 'stēl }

ring [ORG CHEM] A closed loop of bonded atoms in a chemical structure, for example, benzene or cyclohexane. { riŋ }

ring-rolling [MET] Producing a thin, large-diameter ring from a thicker, smaller-diameter ring by placing the ring between two rotating rolls. { 'riŋ ,rōl·iŋ }

ring whizzer [INORG CHEM] A fluxional molecule frequently encountered in organometallic chemistry in which rapid rearrangements occur by migrations about unsaturated organic rings. { 'riŋ ,wiz·ər }

riser See feedhead. { 'rīz·ər }

Ritchie's experiment [THERMO] An experiment that uses a Leslie cube and a differential air thermometer to demonstrate that the emissivity of a surface is proportional to its absorptivity. { 'rich·ēz ik,sper·ə·mənt }

riveling See wrinkling. { 'riv·əl·iŋ }

rivet weld [MET] A weld shaped like a countersunk rivet. { 'riv·ət ,weld }

Rn See radon.

road oil [MATER] A heavy residual petroleum oil, usually one of the slow-curing grades of liquid asphalt. { 'rōd ,óil }

roast [MET] To heat ore to effect some chemical change that will facilitate smelting. { rōst }

robber [MET] An extra cathode that reduces current density at local areas of the work being electroplated for the purpose of producing a more uniform thickness coating. { 'räb·ər }

Rochelle-electric See ferroelectric. { rō'shel·i ,lek·trik }

Rochelle salt See potassium sodium tartrate. { rō'shel,sòlt }

rocklath [MATER] A sheet of gypsum used as a base for plaster. { 'räk,lath }

Rockwell hardness [ENG] A measure of hardness of a material as determined by the Rockwell hardness test. { 'räk,wel 'härd·nəs }

Rockwell hardness test [ENG] One of the arbitrarily defined measures of resistance of a material to indentation under static or dynamic load; depth of indentation of either a steel ball or a 120° conical diamond with rounded point, 1/16, 1/8, 1/4, or 1/2 inch (1.5875, 3.175, 6.35, 12.7 millimeters) in diameter, called a brale, under prescribed load is the basis for Rockwell hardness; 60, 100, 150 kilogram load is applied with a special machine, and depth of impression under initial minor load is indicated on a dial whose graduations represent hardness number. { 'räk,wel 'härd·nəs ,test }

rock wool See mineral wool. { 'räk ,wúl }

rod mill [MET] A mill for making metal rods. { 'räd ,mil }

roentgen diffractometry See x-ray crystallography. { 'rent·gən ,dē,frak'täm·ə·trē }

roll compacting [MET] Compacting a metal powder by using a rolling mill. { 'rōl kəm,pak·tiŋ }

rolled glass [MATER] Thick flat glass made by passing a roller over the molten glass. { 'rōld 'glas }

rolled gold [MET] A metal or alloy of low value covered by a layer of gold alloy where the proportion of gold alloy to total weight of the article may be less than 1:20 and fineness of the gold alloy may not be less than 10 karats. { 'rōld 'gōld }

roller coating [ENG] The application of paints, lacquers, or other coatings onto raised designs or letters by means of a roller. { 'rōl·lər ,kōd·iŋ }

roll flattening See flattening. { 'rōl ,flat·ən·iŋ }

roll forging [MET] Forging metal by using grooved rotating dies. { 'rōl ,fòrj·iŋ }

roll forming [MET] Metal forming by using contoured rolls. { 'rōl ,fórm·iŋ }

Rollin film See helium film. { 'räl·ən ,film }

rolling [MET] Reducing or changing the cross-sectional area of a workpiece by the compressive forces exerted by rotating rolls. Also known as metal rolling. { 'rōl·iŋ }

roll resistance spot welding [MET] Resistance spot welding using rotating circular electrodes. { 'rōl ri,zis·təns 'spät ,weld·iŋ }

roll roofing [MATER] Composition sheet roofing supplied in rolls from which it is laid in overlapping strips. { 'rōl ,rüf·iŋ }

roll welding [MET] Forge welding by heating in a furnace and applying pressure with rolls. { 'rōl ,weld·iŋ }

roofing [MATER] Material used in roof construction, such as tar, tar paper, shingles, slate, and tin. { 'rüf·iŋ }

roofing copper [MET] Copper that has been hot-rolled to sheets in 14- to 32-ounce (400- to 900-gram) weights. { 'rüf·iŋ ,käp·ər }

roofing felt [MATER] Thick asphalt-impregnated paper used for roofing. { 'rüf·iŋ ,felt }

roofing granules [MATER] Graded particles of crushed rock, slate, slag, porcelain, or tile, used as surfacing on asphalt roofing and shingles. { 'rüf·iŋ ,gran·yülz }

roofing putty [MATER] Heavy consistency asphalt solution with asbestos fibers; used for caulking metal roofs. { 'rüf·iŋ ,pəd·ē }

roofing slate [MATER] Hard varieties of slate varying in size from 12 × 6 inches (30 × 15 centimeters) to 24 × 14 inches (60 × 35 centimeters), and from 1/8 to 3/4 inch (3 to 19 millimeters) in thickness. { 'rüf·iŋ ,slāt }

root See root of weld. { rüt }

root crack [MET] A crack in the weld or in the heat-affected zone at the root of the weld. { 'rüt ,krak }

root face [MET] The part of a fusion face that

is not beveled in a welding operation. { 'rüt
‚fās }

root of joint |MET| The area of closest proximity between members of a joint to be welded. { 'rüt əv 'jȯint }

root of weld |MET| The points at which the bottom of the weld and the base metal surfaces intersect. Also known as root. { 'rüt əv 'weld }

root opening |MET| In welding, the distance between members at the root of the joint. { 'rüt 'ō·pən·iŋ }

root pass |MET| The first weld bead deposited in a multiple pass weld. Also known as root sealer bead. { 'rüt ‚pas }

root penetration |MET| The depth of penetration of the weld metal into the root of a joint. { 'rüt ‚pen·ə‚trā·shən }

root sealer bead See root pass. { 'rüt ‚sēl·ər ‚bēd }

rope |MATER| A long, flexible object which consists of many strands of wire, plastic, or vegetable fiber such as manila. { rōp }

Rose's metal |MET| An alloy of bismuth tin and lead; melts at 94°C. { 'rōz·əz ‚med·əl }

rosette |MET| **1.** Rounded constituents in a microstructure arranged in whorls. **2.** Strain gages arranged to indicate at a single position the strains in three different directions. { rō'zet }

rosin |MATER| A translucent yellow, umber, or reddish resinous residue from the distillation of crude turpentine from the sap of pine trees (gum rosin) or from an extract of the stumps and other parts of the tree (wood rosin); used in varnishes, lacquers, printing inks, adhesives, and soldering fluxes, in medical ointments, and as a preservative. { 'räz·ən }

rosin-core solder |MATER| Solder made up in tubular or other hollow form, with the inner space filled with noncorrosive rosin flux. { 'räz·ən ¦kȯr 'säd·ər }

rosin essence |MATER| That part of rosin that can be distilled off at a temperature below 360°C. { 'räz·ən 'es·əns }

rosin-extended rubber |MATER| Cold rubber with up to 50% rosin. { 'räz·ən ik¦sten·dəd 'rəb·ər }

rosin oil |MATER| Viscous, water-insoluble, white-to-brown liquid; soluble in ether, chloroform, carbon disulfide, and fatty oils; distilled from rosin; used as a lubricant and in adhesives, inks, and linoleum. { 'räz·ən ‚ȯil }

rosin size |MATER| An alkali-treated rosin used as a dry powder or emulsion to surface-size paper products. { 'räz·ən ‚sīz }

Rossby diagram |THERMO| A thermodynamic diagram, named after its designer, with mixing ratio as abscissa and potential temperature as ordinate; lines of constant equivalent potential temperature are added. { 'rȯs·bē ‚dī·ə‚gram }

rot See curl. { rät }

rotary dispersion |OPTICS| The change in the angle through which an optically active substance rotates the plane of polarization of plane polarized light as the wavelength of the light is varied. Also known as rotatory dispersion. { 'rōd·ə·rē di'spər·zhən }

rotary polarization See optical activity. { 'rōd·ə·rē ‚pō·lə·rə'zā·shən }

rotary reflection axis See rotoreflection axis. { 'rōd·ə·rē ri'flek·shən ‚ak·səs }

rotating crystal method |SOLID STATE| Any method of studying crystalline structures by x-ray or neutron diffraction in which a monochromatic, collimated beam of x-rays or neutrons falls on a single crystal that is rotated about an axis perpendicular to the beam. { 'rō ‚tād·iŋ ¦krist·əl ‚meth·əd }

rotating-cylinder method |FL MECH| A method of measuring the viscosity of a fluid in which the fluid fills the space between two concentric cylinders, and the torque on the stationary inner cylinder is measured when the outer cylinder is rotated at constant speed. { 'rō ‚tād·iŋ ¦sil·ən·dər ‚meth·əd }

rotating Reynolds number |FL MECH| A nondimensional number arising in problems of a rotating viscous fluid and, in particular, in problems involving the agitation of such a fluid by an impeller, equal to the product of the square of the impeller's diameter and its angular velocity divided by the kinematic viscosity of the fluid. Symbolized Re_r. { 'rō‚tād·iŋ 'ren·əlz ‚nəm·bər }

rotational transformation |CRYSTAL| A type of crystal transformation that is a change from an ordered phase to a partially disordered phase by rotation of groups of atoms. { rō'tā·shən·əl ‚tranz·fər'mā·shən }

rotation axis |CRYSTAL| A symmetry element of certain crystals in which the crystal can be brought into a position physically indistinguishable from its original position by a rotation through an angle of 360°/n about the axis, where n is the multiplicity of the axis, equal to 2, 3, 4, or 6. Also known as symmetry axis. { rō'tā·shən ‚ak·səs }

rotation camera |SOLID STATE| An instrument for studying crystalline structure by x-ray or neutron diffraction, in which a monochromatic, collimated beam of x-rays or neutrons falls on a single crystal which is rotated about an axis perpendicular to the beam and parallel to one of the crystal axes, and the various diffracted beams are registered on a cylindrical film concentric with the axis of rotation. { rō'tā·shən ‚kam·rə }

rotation-inversion axis |CRYSTAL| A symmetry element of certain crystals in which a crystal can be brought into a position physically indistinguishable from its original position by a rotation through an angle of 360°/n about the axis followed by an inversion, where n is the multiplicity of the axis, equal to 1, 2, 3, 4, or 6. Also known as inversion axis. { rō'tā·shən in'vər·zhən ‚ak·səs }

rotation-reflection axis |CRYSTAL| A symmetry element of certain crystals in which a crystal can be brought into a position physically indistinguishable from its original position by a rotation through an angle of 360°/n about the axis followed by a reflection in the plane perpendicular to the axis, where n is the multiplicity of the axis, equal to 1, 2, 3, 4, or 6. { rō'tā·shən ri'flek·shən ˌak·səs }

rotation twin [CRYSTAL] A twin crystal in which the parts will coincide if one part is rotated 180° (sometimes 30, 60, or 120°). { rō'tā·shən ˌtwin }

rotatory dispersion *See* rotary dispersion. { 'rōd·ə,tȯr·ē di'spər·zhən }

rotaxane [ORG CHEM] A compound with two or more independent portions not bonded to each other but linked by a linear portion threaded through a ring and maintained in this position by bulky end groups. { rō'tak,sān }

rotoinversion axis [CRYSTAL] A type of crystal symmetry element that combines a rotation of 60, 90, 120, or 180° with inversion across the center. Also known as symmetry axis of rotary inversion; symmetry axis of rotoinversion. { ˌrōd·ō·in'vər·zhən ˌak·səs }

rotoreflection axis [CRYSTAL] A type of symmetry element that combines a rotation of 60, 90, 120, or 180° with reflection across the plane perpendicular to the axis. Also known as rotoreflection axis. { ˌrōd·ō·ri'flek·shən ˌak·səs }

rottenstone [MATER] A soft, decomposed limestone, light gray to olive in color; used in powder form as a polishing material for metal and wood. { 'rät·ən,stōn }

rouge [MATER] Finely divided, hydrated iron oxide, used in polishing glass, metal, or gems, and as a pigment. { 'rüzh }

roughening transition [PHYS] A change in the behavior of the interface between the solid and liquid phases of a substance at a certain critical temperature, below which the surface is flat or sharp and displays distinct terraces and ledges on an atomic level, while above the critical temperature the surface is rough or rounded. { 'rəf·ən·iŋ tran'zish·ən }

roughing stand [MET] The first stand of rolls, or the last stand before the finishing rolls, through which a preheated billet is passed. { 'rəf·iŋ ˌstand }

roving [MATER] Fibrous glass in which spun strands are woven into a tubular rope. [TEXT] Natural fiber yarns that have been drawn out and slightly twisted in preparation for spinning. { 'rōv·iŋ }

Ru *See* ruthenium.

rubber [POLYM CHEM] A natural, synthetic, or modified high polymer with elastic properties and, after vulcanization, elastic recovery. { 'rəb·ər }

rubber accelerator [POLYM CHEM] A substance that increases the speed of curing of rubber, such as thiocarbanilide. { 'rəb·ər ak'sel·ə,rād·ər }

rubber adhesive [MATER] An adhesive made with a rubber base by using natural or synthetic rubber in an evaporative solvent; a tacky mixture of rubber and filler material, as used on pressure-sensitive tapes; or rubber-solvent-catalyst mixtures (usually two-part) that cure in place. { 'rəb·ər ad,hē·ziv }

rubber-base paint [MATER] A paint in which chlorinated rubber or synthetic latex is the nonvolatile vehicle. { 'rəb·ər ¦bās 'pānt }

rubber cement [MATER] An adhesive composed of unvulcanized rubber in an organic solvent. { 'rəb·ər si,ment }

rubber fiber [MATER] A fiber composed of natural or synthetic rubber; used to make elastic yarn for clothing. { 'rəb·ər 'fī·bər }

rubber foam *See* rubber sponge. { 'rəb·ər 'fōm }

rubber solvent [MATER] Fast-evaporating petroleum distillate used as a solvent for tackifying rubber during plying (laminating) operations, and in compounding rubber cements. { 'rəb·ər ˌsäl·vənt }

rubber sponge [MATER] Foamed, flexible rubber; produced by beating air into unvulcanized latex, or by incorporating a gas-producing ingredient (such as sodium bicarbonate) into a strongly masticated rubber stock; used for comfort cushioning, packaging, and shock insulation. Also known as cellular rubber; foam rubber; rubber foam; sponge rubber. { 'rəb·ər ¦spənj }

rubbing oil [MATER] **1.** A low-viscosity petroleum oil used either with or without an abrasive to polish dried surfaces, such as paint. **2.** A nonviscous oil used for polishing wood furniture. { 'rəb·iŋ ˌȯil }

rubidium [CHEM] A chemical element, symbol Rb, atomic number 37, atomic weight 85.4678; a reactive alkali metal; salts of the metal may be used in glass and ceramic manufacture. { rü'bid·ē·əm }

rubidium bromide [INORG CHEM] RbBr Colorless, regular crystals, melting at 683°C; soluble in water; used as a nerve sedative. { rü'bid·ē·əm 'brō,mīd }

rubidium chloride [INORG CHEM] RbCl A water-soluble, white, lustrous powder melting at 715°C; used as a source for rubidium metal, and as a laboratory reagent. { rü'bid·ē·əm 'klȯr,īd }

rubidium halide [INORG CHEM] Any of the halogen compounds of rubidium; examples are RbBr, RbCl, RbF, RbIBrCl, RbBr₂Cl, and RbIBr₂. { rü'bid·ē·əm 'ha,līd }

rubidium halometallate [INORG CHEM] Halogen-metal-containing compounds of rubidium; examples are Rb₂GeF₆ (rubidium hexafluorogermanate), Rb₂PtCl₆ (rubidium chloroplatinate), and Rb₂PdCl₅ (rubidium palladium chloride). { rü'bid·ē·əm ,ha·lō'med·əl,āt }

rubidium sulfate [INORG CHEM] Rb₂SO₄ Colorless, water-soluble rhomboid crystals, melting at 1060°C; used as a cathartic. { rü'bid·ē·əm 'səl ,fāt }

ruby glass [MATER] Glass of a rich red color produced by adding selenium or cadmium sulfide, or copper oxide to the glass. { 'rü·bē 'glas }

runaround scrap *See* in-house scrap. { 'rən·ə ,raùnd ,skrap }

runner [ENG] In a plastics injection or transfer mold, the channel (usually circular) that connects the sprue with the gate to the mold cavity. [MET] **1.** The part of a casting between itself and the gate assembly of the mold. **2.** A channel through which molten metal flows from one receptacle to another. { 'rən·ər }

runner box |MET| A box that divides the molten metal into several streams before it enters the cavity of the mold. { 'rən·ər ,bäks }

running gate |MET| A gate through which molten metal enters a mold. { 'rən·iŋ ,gāt }

run-of-bank gravel *See* bank-run gravel. { ¦rən əv 'baŋk ¦grav·əl }

runout |MET| **1.** Escape of molten metal from a casting mold, crucible, or furnace. **2.** Defect in a casting caused by escape of metal from a mold. { 'rən,aút }

runout table |MET| A roll table used to receive a rolled or extruded member. { 'rən,aút ,tā·bəl }

rust |MET| The iron oxides formed on corroded ferrous metals and alloys. { rəst }

rusting |MET| The formation of rust on ferrous metals and alloys. { 'rəst·iŋ }

rust preventive |MATER| One of a group of products, often with petroleum thinners, used to prevent corrosion to metal surfaces. { 'rəst pri ,ven·tiv }

rusty gold |MET| Native gold that has a thin coat of iron oxide or silica that prevents it from amalgamating readily. { 'rəs·tē 'gōld }

ruthenic chloride *See* ruthenium chloride. { rü'then·ik 'klȯr,īd }

ruthenium |CHEM| A chemical element, symbol Ru, atomic number 44, atomic weight 101.07. |MET| A hard, brittle, grayish-white metal used as a catalyst; workable only at high temperatures. { rü'thē·nē·əm }

ruthenium chloride |INORG CHEM| $RuCl_3$ Black, deliquescent, water-insoluble solid that decomposes in hot water and above 500°C; used as a laboratory reagent. Also known as ruthenic chloride; ruthenium sesquichloride. { rü'thē·nē·əm 'klȯr,īd }

ruthenium halide |INORG CHEM| Halogen compound of ruthenium; examples are $RuCl_2$, $RuCl_3$, $RuCl_4$, $RuBr_3$, and RuF_5. { rü'thē·nē·əm 'ha,līd }

ruthenium red |INORG CHEM| $Ru_2(OH)_2Cl_4 \cdot 7NH_3 \cdot 3H_2O$ A water-soluble, brownish-red powder; used as an analytical reagent and stain. { rü'thē·nē·əm 'red }

ruthenium sesquichloride *See* ruthenium chloride. { rü'thē·nē·əm ,ses·kwi'klȯr,īd }

ruthenium tetroxide |INORG CHEM| RuO_4 A yellow, toxic solid, melting at 25°C; used as an oxidizing agent. { rü'thē·nē·əm te'träk,sīd }

rutherfordium |CHEM| A chemical element, symbolized Rf, atomic number 104, a synthetic element; the first element beyond the actinide series, and the twelfth transuranium element. { ,rəth·ər'fȯr·dē·əm }

S

S *See* siemens; sulfur.

sacrificial anode |PHYS CHEM| A protective coating applied to a metal surface to act as an anode and be consumed in an electrochemical reaction, thereby preventing electrolytic corrosion of the metal. { ¦sak·rə,fish·əl 'an,ōd }

sacrificial compliant substrate *See* compliant substrate. { ,sak·rə¦fish·əl kəm¦plī·ənt 'səb ,strāt }

sacrificial metal |PHYS CHEM| A metal that can be used for a sacrificial anode. { ,sak·rə'fish·əl 'med·əl }

saddling |MET| Forming a seamless ring by forging a pierced disk over a mandrel (or saddle). Also known as mandrel forging. { 'sad·liŋ }

safety factor *See* factor of safety. { 'sāf·tē ,fak·tər }

safety glass |MATER| **1.** A glass that resists shattering (such as a glass containing a net of wire or constructed of sheets separated by plastic film). **2.** A glass that has been tempered so that when it shatters, it breaks up into grains instead of jagged fragments. { 'sāf·tē ,glas }

sag |MET| Decrease in the section thickness of a casting caused by weakness of the sand mold. { sag }

saggar clay |MATER| A fire clay of which the case is made that is used for the firing of porcelain and pottery. Also spelled sagger clay. { 'sag·ər ,klā }

sagger clay *See* saggar clay. { 'sag·ər ,klā }

sago |MATER| A starch obtained from the trunks of certain tropical palms, such as the sago; used as a thickening agent in food and as textile stiffening. { 'sā·gō }

Saint Joseph retort process |MET| An electrothermic retort process for processing zinc ore and zinc from secondary sources into zinc; heat of reaction between the sintered zinc concentrate and the coke mixture is supplied by passage of heavy electric current through the resistance of the charge. { sānt 'jō·səf ri'tȯrt ,prä·səs }

sal soda |INORG CHEM| $Na_2CO_3 \cdot 10H_2O$ White, water-soluble crystals, insoluble in alcohol; melts and loses water at about 33°C; mild irritant to mucous membrane; used in cleansers and for washing textiles and bleaching linen and cotton. Also known as sodium carbonate decahydrate; washing soda. { 'sal ,sō·də }

salt |CHEM| The reaction product when a metal displaces the hydrogen of an acid; for example, $H_2SO_4 + 2NaOH \rightarrow Na_2SO_4$ (a salt) $+ 2H_2O$. { sȯlt }

salt bath |MET| Molten salts in which steel is heated for hardening and tempering. { 'sȯlt ,bath }

salt cake |INORG CHEM| Impure sodium sulfate; used in soaps, paper pulping, detergents, glass, ceramic glaze, and dyes. { 'sȯlt ,kāk }

salt-fog test |MET| An accelerated corrosion test in which a piece of metal is subjected to a fine spray of sodium chloride solution. Also known as salt-spray test. { 'sȯlt ¦fäg ,test }

salt of Lemery *See* potassium sulfate. { 'sȯlt əv lem'rē }

salt of tartar *See* potassium carbonate. { 'sȯlt əv 'tär·tər }

saltpeter *See* potassium nitrate. { sȯlt'pēd·ər }

salt-spray test *See* salt-fog test. { 'sȯlt ¦sprā ,test }

samarium |CHEM| A rare-earth metal, atomic number 62, symbol Sm; melts at 1350°C, tarnishes in air, ignites at 200–400°C. { sə'mar·ē·əm }

samarium-cobalt magnet |ELECTROMAG| A rare-earth permanent magnet that is more efficient, has lower leakage and greater resistance to demagnetization, and can be magnetized to higher levels than conventional permanent magnets. { sə'mar·ē·əm 'kō,bȯlt 'mag·nət }

samarium oxide |INORG CHEM| Sm_2O_3 A cream-colored powder with a melting point of 2300°C; soluble in acids; used for infrared-absorbing glass and as a neutron absorber. { sə'mar·ē·əm 'äk,sīd }

SAN *See* styrene-acrylonitrile resin.

sandarac gum |MATER| Yellow, brittle, water-insoluble, natural resin obtained from the African sandarac tree of Morocco; used in varnishes and lacquers. { 'san·də,rak ,gəm }

sand-cast |MET| Made by pouring molten metal into a mold made of sand. { 'san,kast }

sand control |MET| A process to regulate the properties of foundry sand to produce defect-free castings. { 'san kən,trōl }

sand-lime brick |MATER| **1.** Decorative brick made of sand and lime pressed in an atmosphere of steam. **2.** A firebrick made of refractory silica sand with lime as a bonding agent. { 'san 'līm 'brik }

sandpaper |MATER| Paper with abrasive glued to the surface. { 'san,pā·pər }

sandwich braze |MET| A technique by which a shim is placed between materials to be brazed as a transition layer to decrease thermal stress. { 'san,wich ,brāz }

sandwich rolling |MET| Rolling strips of metal together to form a metallurgically bonded composite sheet. { 'san,wich ,rōl·iŋ }

Sapele mahogany |MATER| A figured wood from E*ntandrophragma cylindricum*, a big tree growing on the Ivory Coast, Ghana, and Nigeria. Also known as aboundikro; scented mahogany; West African cedar. { sə'pē·lē mə'häg·ə·nē }

saponification |CHEM| The process of converting chemicals into soap; involves the alkaline hydrolysis of a fat or oil, or the neutralization of a fatty acid. { sə,pän·ə·fə'kā·shən }

sapphire whiskers See alumina fibers. { 'sa,fīr ,wis·kərz }

Sargent cycle |THERMO| An ideal thermodynamic cycle consisting of four reversible processes: adiabatic compression, heating at constant volume, adiabatic expansion, and isobaric cooling. { 'sär·jənt ,sī·kəl }

SAS See aluminum sodium sulfate.

satin finish |MET| A finish involving soft scratch-brushing of polished metal surfaces to produce a soft sheen. Also known as Butler finish; scratch-brush finish. { 'sat·ən 'fin·ish }

saturated color |OPTICS| A pure color not contaminated by white. { 'sach·ə,rād·əd 'kəl·ər }

saturated compound |ORG CHEM| An organic compound with all carbon bonds satisfied; it does not contain double or triple bonds and thus cannot add elements or compounds. { 'sach·ə ,rād·əd 'käm,pau̇nd }

saturated hydrocarbon |ORG CHEM| A saturated carbon-hydrogen compound with all carbon bonds filled; that is, there are no double or triple bonds as in olefins and acetylenics. { 'sach·ə ,rād·əd ¦hī·drə'kär·bən }

saturated vapor |THERMO| A vapor whose temperature equals the temperature of boiling at the pressure existing on it. { 'sach·ə,rād·əd 'vā·pər }

saturation |OPTICS| See color saturation. |PHYS| **1.** The condition in which a further increase in some cause produces no further increase in the resultant effect. **2.** The property exhibited by certain forces between particles wherein each particle can interact strongly with only a limited number of other particles, as in the forces between atoms in a molecule, and between nucleons in a nucleus. { ,sach·ə'rā·shən }

saturation magnetization |ELECTROMAG| The maximum possible magnetization of a material. { ,sach·ə'rā·shən ,mag·nəd·ə'zā·shən }

saturation scale |OPTICS| A series of colors which appear to have equal differences in color saturation. { ,sach·ə'rā·shən ¦skāl }

saturation specific humidity |THERMO| A thermodynamic function of state; the value of the

specific humidity of saturated air at the given temperature and pressure. { ,sach·ə'rā·shən spə'sif·ik hyü'mid·əd·ē }

saturation vapor pressure |THERMO| The vapor pressure of a thermodynamic system, at a given temperature, wherein the vapor of a substance is in equilibrium with a plane surface of that substance's pure liquid or solid phase. { ,sach·ə'rā·shən 'vā·pər ,presh·ər }

sawdust |MATER| Wood fragments made by a saw in cutting. { 'sȯ,dəst }

sawdust concrete |MATER| Concrete containing sawdust as the principal aggregate. { 'sȯ,dəst 'kän,krēt }

Saybolt Furol viscosity |FL MECH| The time in seconds for 60 milliliters of fluid to flow through a capillary tube in a Saybolt Furol viscosimeter at specified temperatures between 70 and 210°F (21 and 99°C); used for high-viscosity petroleum oils, such as transmission and gear oils, and heavy fuel oils. { 'sā,bōlt 'fyu̇,ról vi'skäs·əd·ē }

Saybolt Seconds Universal |FL MECH| A unit of measurement for Saybolt Universal viscosity. Abbreviated SSU. { 'sā,bōlt 'sek·ənz ,yü·nə'vər·səl }

Saybolt Universal viscosity |FL MECH| The time in seconds for 60 milliliters of fluid to flow through a capillary tube in a Saybolt Universal viscosimeter at a given temperature. { 'sā,bōlt ,yü·nə'vər·səl vi'skäs·əd·ē }

Sb See antimony.

s-block element |CHEM| A chemical element whose valence shell contains s-electrons only; found in groups I and 2 of the periodic table. { 'es ,bläk ,el·ə·mənt }

SBR See styrene-butadiene rubber.

Sc See scandium.

scab |MET| A defect consisting of a flat, partially detached piece of metal joined to the surface of a casting or piece of rolled metal. { skab }

scalar |PHYS| **1.** A quantity which has magnitude only and no direction, in contrast to a vector. **2.** A quantity which has magnitude only, and has the same value in every coordinate system. { 'skā·lər }

scale |MET| A thick metallic oxide coating formed usually by heating metals in air. |PHYS| **1.** A one-to-one correspondence between numbers and the value of some physical quantity, such as the centigrade or Kelvin temperature scales on the API (American Petroleum Institute) or Baumé scales of specific gravity. **2.** To determine a quantity at some order of magnitude by using data or relationships which are known to be valid at other (usually lower) orders of magnitude. { skāl }

scaleboard |MATER| Thin sheets of wood used as veneer. { 'skāl,bȯrd }

scalenohedron |CRYSTAL| A closed crystal form whose faces are scalene triangles. { skə'lē·nō'hē·drən }

scale wax |MATER| The paraffin wax derived by sweating the greater part of the oil from slack wax; contains up to 6% oil. Also known as crude scale; paraffin scale. { 'skāl ,waks }

scaling |ENG| Removing scale (rust or salt) from a metal or other surface. |MET| **1.** Forming of a thick layer of metallic oxide on metals at high temperatures. **2.** Depositing of solid inorganic solutes from water on a metal surface, such as a cooling tube or boiler. { 'skāl·iŋ }

scalp |MET| To remove surface layers, and thereby defects, from ingots, billets, or slabs by machining. { skalp }

scalped extrusion ingot |MET| A cast, solid, or hollow extrusion ingot which has been machined to remove outside surface layers. { 'skalpt ik'strü·zhən ,iŋ·gət }

scalping chips |MET| Material removed from the surface of cast ingots to reduce surface roughness and to provide a smooth, clean surface for the rolling mill. { 'skalp·iŋ ,chips }

scandia See scandium oxide. { 'skan·dē·ə }

scandium |CHEM| A transition element, symbol Sc, atomic number 21; melts at 1200°C; found associated with rare-earth elements. { 'skan·dē·əm }

scandium halide |INORG CHEM| A compound of scandium and a halogen; for example, scandium chloride, $ScCl_3$. { 'skan·dē·əm 'ha,līd }

scandium oxide |INORG CHEM| Sc_2O_3 White powder, soluble in hot acids; used to prepare scandium. Also known as scandia. { 'skan·dē·əm 'äk,sīd }

scandium sulfate |INORG CHEM| $Sc_2(SO_4)_3$ Water-soluble, colorless crystals. { 'skan·dē·əm 'səl ,fāt }

scandium sulfide |INORG CHEM| Sc_3S_3 Yellowish powder; decomposes in dilute acids and boiling water to give off hydrogen sulfide. { 'skan·dē·əm 'səl,fīd }

scanning electron microscope |ELECTR| A type of electron microscope in which a beam of electrons, a few hundred angstroms in diameter, systematically sweeps over the specimen; the intensity of secondary electrons generated at the point of impact of the beam on the specimen is measured, and the resulting signal is fed into a cathode-ray-tube display which is scanned in synchronism with the scanning of the specimen. Abbreviated SEM. { 'skan·iŋ i'lek,trän 'mī·krə ,skōp }

scanning proton microprobe |ENG| An instrument used for determining the spatial distribution of trace elements in samples, in which a beam of energetic protons is focused on a narrow spot which is swept over the sample, and the characteristic x-rays emitted from the target are measured. { 'skan·iŋ 'prō,tän 'mī·krə,skōp }

scanning spot See picture element. { 'skan·iŋ ,spät }

scanning transmission electron microscope |ELECTR| A type of electron microscope which scans with an extremely narrow beam that is transmitted through the sample; the detection apparatus produces an image whose brightness depends on atomic number of the sample. Abbreviated STEM. { 'skan·iŋ tranz'mish·ən i'lek ,trän 'mī·krə,skōp }

scanning tunneling microscope |ELECTR| An instrument for producing surface images with atomic-scale lateral resolution, in which a fine probe tip is raster-scanned over the surface at a distance of 0.5–1 nanometer, and the resulting tunneling current, or the position of the tip required to maintain a constant tunneling current, is monitored. Also known as tunneling microscope. { 'skan·iŋ |tən·əl·iŋ 'mī·krə,skōp }

scarfing |MET| **1.** Cutting away of surface defects on metals by use of a gas torch. **2.** A forging process in which the ends of two pieces to be joined are tapered to avoid an enlarged joint. { 'skärf·iŋ }

SC asphalt See slow-curing liquid asphaltic material. { |es|sē 'as,fólt }

scavenger |CHEM| A substance added to a mixture or other system to remove or inactivate impurities. Also known as getter. |MET| A reactive metal added to a molten metal to combine with and remove dissolved gases. { 'skav·ən·jər }

scavenging |MET| Removal of dissolved gases from molten metal. { 'skav·ən·jiŋ }

scene paint |MATER| A paint used in theatrical scene painting; it is a dry pigment mixed with a glue-water mixture called size water. { 'sēn ,pānt }

scented mahogany See Sapele mahogany. { 'sent·əd mə'häg·ə·nē }

Scheele's green See copper arsenite. { 'shā·ləz 'grēn }

schlanite |MATER| The soluble resin extracted from anthracoxene by ether. { 'shlä,nīt }

Schleiermacher's method |THERMO| A method of determining the thermal conductivity of a gas, in which the gas is placed in a cylinder with an electrically heated wire along its axis, and the electric energy supplied to the wire and the temperatures of wire and cylinder are measured. { 'shlī·ər,mäk·ərz ,meth·əd }

Schönflies crystal symbols |CRYSTAL| Symbols denoting the 32 crystal point groups or symmetry classes; capital letters indicate the general type of class, and subscripts the multiplicity of rotation axes and the existence of additional symmetries. { 'shən,flēs 'krist·əl ,sim·bəlz }

Schottky anomaly |SOLID STATE| A contribution to the heat capacity of a solid arising from the thermal population of discrete energy levels as the temperature is raised; the effect is particularly prominent at low temperatures. { 'shät·kē ə ,näm·ə·lē }

Schottky barrier |ELECTR| A transition region formed within a semiconductor surface to serve as a rectifying barrier at a junction with a layer of metal. { 'shät·kē ,bar·ē·ər }

Schottky defect |SOLID STATE| **1.** A defect in an ionic crystal in which a single ion is removed from its interior lattice site and relocated in a lattice site at the surface of the crystal. **2.** A defect in an ionic crystal consisting of the smallest number of positive-ion vacancies and negative-ion vacancies which leave the crystal electrically neutral. { 'shät·kē di,fekt }

Schottky effect |SOLID STATE| The enhancement of the thermionic emission of a conductor resulting from an electric field at the conductor surface. { 'shät·kē i,fekt }

Schottky line |SOLID STATE| A graph of the logarithm of the saturation current from a thermionic cathode as a function of the square root of anode voltage; it is a straight line according to the Schottky theory. { 'shät·kē ,līn }

Schottky theory |SOLID STATE| A theory describing the rectification properties of the junction between a semiconductor and a metal that result from formation of a depletion layer at the surface of contact. { 'shät·kē ,thē·ə·rē }

sclerometer |ENG| An instrument used to determine the hardness of a material by measuring the pressure needed to scratch or indent a surface with a diamond point. { sklə'räm·əd·ər }

scleroscope |ENG| An instrument used to determine the hardness of a material by measuring the height to which a standard ball rebounds from its surface when dropped from a standard height. { 'skler·ə,skōp }

scleroscope hardness test   See Shore scleroscope hardness test. { 'skler·ə,skōp 'härd·nəs ,test }

scoria |MATER| Refuse after melting metals or reducing ore. { 'skòr·ē·ə }

scorification |MET| Concentration of precious metals, such as gold and silver, in molten lead by oxidation employing appropriate fluxes. { ,skòr·ə·fə'kā·shən }

scoring |ENG| Scratching the surface of a material. |MATER| See attrition. { 'skòr·iŋ }

scotophor |MATER| A solid that exhibits reversible darkening and bleaching actions of tenebrescence under suitable irradiation. { 'skäd·ə ,fòr }

scouring |MATER| See attrition. |TEXT| **1.** Removal of grease and dirt from wool. **2.** The cleaning of fabric before the dyeing step. { 'skaùr·iŋ }

scrap mica |MATER| Mica whose size, color, or quality is below specifications for sheet mica. { 'skrap ,mī·kə }

scratch-brush finish   See satin finish. { 'skrach ,brəsh ,fin·ish }

scratch hardness |MATER| A measure of the resistance of minerals or metals to scratching; for minerals it is defined by comparison with 10 selected minerals which are numbered in order of increasing hardness according to the Mohs scale. { 'skrach ,härd·nəs }

scratch hardness test |MET| A hardness test in which a cutting point under given pressure is drawn across the surface of a metal, and the width of the scratch is measured. { 'skrach ¦härd·nəs ,test }

screen |ENG| **1.** A large sieve of suitably mounted wire cloth, grate bars, or perforated sheet iron used to sort rock, ore, or aggregate according to size. **2.** A covering to give physical protection from light, noise, heat, or flying particles. **3.** A filter medium for liquid-solid separation. { skrēn }

screen analysis |ENG| A method for finding the particle-size distribution of any loose, flowing, conglomerate material by measuring the percentage of particles that pass through a series of standard screens with holes of various sizes. { 'skrēn ə,nal·ə·səs }

screening |ENG| **1.** The separation of a mixture of grains of various sizes into two or more size-range portions by means of a porous or woven-mesh screening media. **2.** The removal of solid particles from a liquid-solid mixture by means of a screen. **3.** The material that has passed through a screen. { 'skrēn·iŋ }

screen mesh |ENG| A wire network or cloth mounted in a frame for separating and classifying materials. { 'skrēn ,mesh }

screw axis |CRYSTAL| A symmetry element of some crystal lattices, in which the lattice is unaltered by a rotation about the axis combined with a translation parallel to the axis and equal to a fraction of the unit lattice distance in this direction. { 'skrü ,ak·səs }

screw dislocation |CRYSTAL| A dislocation in which atomic planes form a spiral ramp winding around the line of the dislocation. { 'skrü ,dis·lō ,kā·shən }

scrim |MATER| A coarse mesh made of wire, fiberglass, or other heavy fibers and used to bridge and reinforce a joint or used as a base for painting or plastering. { skrim }

scruff |MET| A mixture of tin oxide and iron-tin alloy formed as dross on a molten tin-coating bath. { skrəf }

scum |MATER| **1.** A film of impurities that rises to or is formed on the surface of a liquid. **2.** A slimy film formed on the surface of a solid object. { skəm }

Se   See selenium.

seaborgium |CHEM| A chemical element, symbolized Sg, atomic number 106, a synthetic element; the fourteenth transuranium element. { sē'bòrg·ē·əm }

seal coat |MATER| A layer of bituminous material applied to bituminous macadam or concrete to seal the surface. { 'sēl ,kōt }

sealer |MATER| A preliminary coating applied to seal the pores in a porous, uncoated surface, such as wood. { 'sēl·ər }

sealing |MET| **1.** Impregnation of porous castings with resins to overcome porosity. **2.** Reducing porosity of an anodic oxide film on aluminum and aluminum alloys by immersion in boiling water. { 'sēl·iŋ }

sealing compound |ELEC| A compound used in dry batteries, capacitor blocks, transformers, and other components to keep out air and moisture. { 'sēl·iŋ ,käm,paùnd }

sealing tape |MATER| Gummed tape for sealing packages. { 'sēl·iŋ ,tāp }

sealing wax |MATER| A colored, scented mixture of resins and shellac; used for sealing containers and documents. { 'sēl·iŋ ,waks }

seal weld |MET| A weld designed primarily for preventing leakage. { 'sēl ,weld }

seam |ENG| **1.** A mechanical or welded joint. **2.** A mark on ceramic or glassware where matching mold parts join. **3.** A line occurring on a molded or laminated piece of plastic material that differs in appearance from the rest of the surface and is caused by a parting of the mold. |MET| An unwelded fold or lap which appears as a crack on the surface of a casting or wrought product. { sēm }

seaming |MET| The joining of the edges of sheet-metal parts by interlocking folds. { 'sēm·iŋ }

seamless ring rolling |MET| The hot-rolling of a circular blank, with a hole in the center, to form a seamless ring. { 'sēm·ləs ¦riŋ ,rōl·iŋ }

seamless tubing |MET| A tubing made by extrusion or by piercing and rolling a billet. { 'sēm·ləs 'tüb·iŋ }

seam weld |MET| **1.** A longitudinal weld joining of sheet-metal parts or in making tubing. **2.** Arc or resistance welding in which a series of overlapping spot welds is produced. { 'sēm ,weld }

season crack |MET| A stress-corrosion crack produced in a copper-base alloy subject to a residual or applied tensile stress and exposed to a specific environment such as moist air containing traces of ammonia. { 'sēz·ən ¦krak }

seasoned lumber |MATER| Lumber which has been cured by drying to ensure a uniform moisture content. { 'sēz·ənd 'ləm·bər }

secondary amine |ORG CHEM| An organic compound that may be written R_1R_2NH, where R_1 and R_2 designate either identical or different groups. { 'sek·ən,der·ē 'am,ēn }

secondary battery See storage battery. { 'sek·ən,der·ē 'bad·ə·rē }

secondary cell See storage cell. { 'sek·ən,der·ē 'sel }

secondary circuit |MET| The part of a welding machine which conducts secondary current between transformer and electrodes or between electrodes and workpiece. { 'sek·ən ,der·ē 'sər·kət }

secondary creep |MECH| The change in shape of a substance under a minimum and almost constant differential stress, with the strain-time relationship a constant. Also known as steady-state creep. { 'sek·ən,der·ē 'krēp }

secondary hardening |MET| The hardening of certain alloy steels at moderate temperatures (250–650°C) by the precipitation of carbides; the resultant hardness is greater than that obtained by tempering the steel at some lower temperature for the same time. { 'sek·ən,der·ē 'härd·ən·iŋ }

secondary metal |MET| Metal recovered from scrap by remelting and refining. { 'sek·ən,der·ē 'med·əl }

secondary plasticizer |MATER| A plastics plasticizer that has insufficient affinity for a resin for it to be the sole plasticizer, and must be blended with a primary plasticizer. Also known as extender plasticizer. { 'sek·ən,der·ē 'plas·tə,sīz·ər }

secondary twinning |CRYSTAL| Twinning of a crystal caused by an external influence, such as pressure in rock. { 'sek·ən,der·ē 'twin·iŋ }

second law of thermodynamics |THERMO| A general statement of the idea that there is a preferred direction for any process; there are

many equivalent statements of the law, the best known being those of Clausius and of Kelvin. { 'sek·ənd 'lȯ əv ,thər·mə·dī'nam·iks }

second-order transition |THERMO| A change of state through which the free energy of a substance and its first derivatives are continuous functions of temperature and pressure, or other corresponding variables. { 'sek·ənd ¦ȯr·dər tran'zish·ən }

second sound |CRYO| A type of wave propagated in the superfluid phase of liquid helium (helium II), in which temperature and entropy variations propagate with no appreciable variation in density or pressure. { 'sek·ənd 'saùnd }

sedimentation |CHEM| The settling of suspended particles within a liquid under the action of gravity or a centrifuge. |MET| Classification of metal powders by the rate of settling in a fluid. { ,sed·ə·mən'tā·shən }

Seebeck coefficient |ELECTR| The ratio of the open-circuit voltage to the temperature difference between the hot and cold junctions of a circuit exhibiting the Seebeck effect. { 'zā,bek ,kō·i'fish·ənt }

Seebeck effect |ELECTR| The development of a voltage due to differences in temperature between two junctions of dissimilar metals in the same circuit. { 'zā,bek i,fekt }

seed |CHEM| A small, single crystal of a desired substance added to a solution to induce crystallization. |SOLID STATE| A small, single crystal of semiconductor material used to start the growth of a large, single crystal for use in cutting semiconductor wafers. { sēd }

Seger cone |MATER| Any of a series of conical shaped thermometric devices made of materials that deform at specified temperatures; consists of mixtures of clay, salt, and other materials in such proportions that their softening temperatures vary progressively through the series; used to indicate temperatures of furnaces, particularly in ceramic industries. Also known as pyrometric cone. { 'zā·gər ,kōn }

segment die |MET| A die made of parts that can be disassembled to facilitate removal of the workpiece. Also known as split die. { 'seg·mənt ,dī }

segregation |MET| The nonuniform distribution of alloying elements, impurities, or microphases, resulting in localized concentrations. { ,seg·rə'gā·shən }

Seignette-electric See ferroelectric. { sen¦yet i'lek·trik }

Seignette salt See potassium sodium tartrate. { sen'yet ,sȯlt }

seizing |MET| Welding a workpiece to a die member under the combined forces of pressure and sliding friction. { 'sēz·iŋ }

Sejournet process |MET| During hot extrusion, the lubrication and insulation of a metal billet with molten glass. Also known as Ugine Sejournet process. { sə·zhȯr'nā ,prä·səs }

selective heating |MET| Heating only certain portions of a workpiece to impart desired properties. { si'lek·tiv 'hēd·iŋ }

selective plating [MET] An electrochemical process in which the base metal is masked, except the area to receive the plate, with a nonconductive material; the masked metal, with an electric current running through it, is then sprayed with a solution of plating metal which adheres only to the unmasked section. { si'lek·tiv 'plād·iŋ }

selective polymerization [POLYM CHEM] The polymerization of a single type of molecule in a mixture of monomers; for example, the production of diisobutylene from a mixture of butylenes. { si'lek·tiv pə,lim·ə·rə'zā·shən }

selective quenching [MET] Quenching only certain portions of a piece of metal. { si'lek·tiv 'kwench·iŋ }

selector [MET] A converter that separates purified copper from residue in a single operation. { si'lek·tər }

selenic acid [INORG CHEM] H_2SeO_4 A highly toxic, water-soluble, white solid, melting point 58°C, decomposing at 260°C. { sə'len·ik 'as·əd }

selenide [INORG CHEM] M_2Se A binary compound of divalent selenium, such as Ag_2Se, silver selenide. [ORG CHEM] An organic compound containing divalent selenium, such as $(C_2H_5)_2Se$, ethyl selenide. { 'sel·ə,nīd }

selenious acid See selenous acid. { sə'lē·nē·əs 'as·əd }

selenium [CHEM] A highly toxic, nonmetallic element in group 16, symbol Se, atomic number 34; steel-gray color; soluble in carbon disulfide, insoluble in water and alcohol; melts at 217°C; and boils at 690°C; used in analytical chemistry, metallurgy, and photoelectric cells, and as a lube-oil stabilizer and chemicals intermediate. { sə'lē·nē·əm }

selenium bromide [INORG CHEM] Any of three compounds of selenium and bromine: Se_2Br_2, a red liquid that melts at −46°C, also known as selenium monobromide; $SeBr_2$, a brown liquid, also known as selenium dibromide; and $SeBr_4$, orange, carbon-disulfide-soluble crystals, also known as selenium tetrabromide. { sə'lē·nē·əm 'brō,mīd }

selenium dibromide See selenium bromide. { sə'lē·nē·əm dī'brō,mīd }

selenium dioxide [INORG CHEM] SeO_2 Water- and alcohol-soluble, white to reddish, lustrous crystals; melts at 340°C; used in medicine, and as an oxidizing agent and catalyst. Also known as selenous acid anhydride; selenous anhydride; selenium oxide. { sə'lē·nē·əm dī'äk,sīd }

selenium halide [INORG CHEM] A compound of selenium and a halogen, for example, Se_2Br_2, $SeBr_2$, $SeBr_4$; Se_2Cl_2, $SeCl_2$, $SeCl_4$; Se_2I_2, SeI_4. { sə'lē·nē·əm 'ha,līd }

selenium monobromide See selenium bromide. { sə'lē·nē·əm ,män·ō'brō,mīd }

selenium nitride [INORG CHEM] Se_2N_2 A water-insoluble, yellow solid that explodes at 200°C. { sə'lē·nē·əm 'nī,trīd }

selenium oxide See selenium dioxide. { sə'lē·nē·əm 'äk,sīd }

selenium stainless steel [MET] Stainless steel to which about 0.1 percent or more selenium is added to improve machinability. { sə'lē·nē·əm 'stān·ləs 'stēl }

selenium tetrabromide See selenium bromide. { sə'lē·nē·əm ,te·trə'brō,mīd }

selenous acid [INORG CHEM] H_2SeO_3 Colorless, transparent crystals; soluble in water and alcohol, insoluble in ammonia; decomposes when heated; used as an analytical reagent. Also spelled selenious acid. { sə'lē·nəs 'as·əd }

selenous acid anhydride See selenium dioxide. { sə'lē·nəs 'as·əd an'hī,drīd }

selenous anhydride See selenium dioxide. { sə'lē·nəs an'hī,drīd }

self-action effect [OPTICS] In a medium with a third-order optical nonlinearity, the modification of the refractive index and absorption coefficient of a light field present in the medium by the strength of the light intensity, so that the light field effectively acts on itself. { ,self 'ak·shən i ,fekt }

self-assembling monolayer [MATER] Monomolecular films formed by immersing an appropriate substrate into a solution of an active surfactant. { ,self ə,sem·bliŋ 'män·ō,lā·ər }

self-diffusion [SOLID STATE] The spontaneous movement of an atom to a new site in a crystal of its own species. { ,self di¦fyü·zhən }

self-extinguishing [MATER] The ability of a material to cease burning once the source of the flame has been removed. { ,self ik¦stiŋ·gwə·shiŋ }

self-fluxing alloy [MET] Any alloy used in thermal spraying which does not require the addition of a flux in order to wet the substrate and coalesce when heated. { ,self ¦flək·siŋ 'al,ȯi }

self-focusing fiber [OPTICS] A type of optical fiber in which the refractive index decreases continuously along the radius, but progressively more rapidly with distance from the radius, so that light rays which travel longer distances are speeded up, and nearly all light rays travel with the same net axial velocity. { ,self 'fō·kə·siŋ 'fī·bər }

self-hardening steel See air-hardening steel. { ,self ¦härd·ən·iŋ 'stēl }

self-healing dielectric breakdown [ELECTR] A dielectric breakdown in which the breakdown process itself causes the material to become insulating again. { ,self ¦hēl·iŋ ,dī·ə¦lek·trik 'brāk,daun }

self-organization [CHEM] The capability of a system to spontaneously generate a well-defined supramolecular entity by self-assembling from components in a given set of conditions. { ,self ,ȯr·gə·nə'zā·shən }

sellite [INORG CHEM] A solution of sodium sulfite (Na_2SO_3) used in the purification of 2,4,6-trinitrotoluene to remove unsymmetrical isomers. { 'se,līt }

SEM See scanning electron microscope.

semiautomatic welding [MET] An arc-welding method in which the electrode, a long length of small-diameter bare wire, usually in coil form, is positioned and advanced by the operator from a hand-held welding gun which feeds the

electrode through the nozzle. { ¦sem·ē‚ȯd·
ə'mad·ik 'weld·iŋ }

semicompreg |MATER| Resin-impregnated wood compressed to a density not exceeding 1.25. { ¦sem·i'käm‚preg }

semiconducting compound |SOLID STATE| A compound which is a semiconductor, such as copper oxide, mercury indium telluride, zinc sulfide, cadmium selenide, and magnesium iodide. { ¦sem·i·kən¦dək·tiŋ 'käm‚paünd }

semiconducting crystal |SOLID STATE| A crystal of a semiconductor, such as silicon, germanium, or gray tin. { ¦sem·i·kən¦dək·tiŋ ¦krist·əl }

semiconductor |SOLID STATE| A solid crystalline material whose electrical conductivity is intermediate between that of a conductor and an insulator, ranging from about 10^5 mhos to 10^{-7} mho per meter, and is usually strongly temperature-dependent. { ¦sem·i·kən¦dək·tər }

semiconductor diode |ELECTR| **1.** Also known as crystal diode; crystal rectifier; diode. **2.** A two-electrode semiconductor device that utilizes the rectifying properties of a *pn* junction or a point contact. **3.** More generally, any two-terminal electronic device that utilizes the properties of the semiconductor from which it is constructed. { ¦sem·i·kən¦dək·tər 'dī‚ōd }

semiconductor doping See doping. { ¦sem·i·kən¦dək·tər 'dōp·iŋ }

semiconductor heterostructure |ELECTR| A structure of two different semiconductors in junction contact having useful electrical or electrooptical characteristics not achievable in either conductor separately; used in certain types of lasers and solar cells. { ¦sem·i·kən ¦dək·tər 'hed·ə·rō‚strək·chər }

semiconductor intrinsic properties |SOLID STATE| Properties of a semiconductor that are characteristic of the ideal crystal. { ¦sem·i·kən¦dək·tər in'trin·sik 'präp·ərd·ēz }

semiconductor junction |ELECTR| Region of transition between semiconducting regions of different electrical properties, usually between *p*-type and *n*-type material. { ¦sem·i·kən¦dək·tər ‚jəŋk·shən }

semiconductor laser |OPTICS| A laser in which stimulated emission of coherent light occurs at a *pn* junction when electrons and holes are driven into the junction by carrier injection, electron-beam excitation, impact ionization, optical excitation, or other means. Also known as diode laser; laser diode. { ¦sem·i·kən¦dək·tər 'lā·zər }

semiconductor thermocouple |ELECTR| A thermocouple made of a semiconductor, which offers the prospect of operation with high-temperature gradients, because semiconductors are good electrical conductors but poor heat conductors. { ¦sem·i·kən¦dək·tər 'thər·mə‚kəp·əl }

semiconductor trap See trap. { ¦sem·i·kən¦dək·tər ‚trap }

semifinishing |MET| The preliminary finishing operations. { ¦sem·i'fin·ish·iŋ }

semigloss |MATER| Pertaining to a surface finish intermediate between flat and glossy; used especially of paint and varnish. { 'sem·i‚gläs }

semikilled steel |MET| Incompletely deoxidized steel containing enough dissolved oxygen to react with the carbon it contains to form carbon monoxide, the latter offsetting solidification shrinkage. { 'sem·i‚kild 'stēl }

semimat |MATER| Intermediate between glossy and mat, as photographic paper. { ¦sem·i'mat }

semimetal See metalloid. { ¦sem·ē'med·əl }

semipermanent mold |MET| A reusable metal mold with expendable sand cores. { ¦sem·i'pər·mə·nənt 'mōld }

semipermeable membrane |PHYS| A membrane which allows a solvent to pass through it, but not certain dissolved or colloidal substances. { ¦sem·i'pər·mē·ə·bəl 'mem‚brān }

semirefined wax |MATER| Commercial grades of petroleum wax which are inferior to fully refined grades but which meet specified requirements as to color and oil content. { ¦sem·i·ri'fīnd 'waks }

semirigid plastic |MATER| A plastic that has a stiffness or apparent modulus of elasticity of between 10,000 and 100,000 pounds per square inch (6.895 × 10^7 and 6.895 × 10^8 pascals) under prescribed test conditions. { ¦sem·i'rij·əd 'plas·tik }

semisilica refractory |MATER| A silica refractory made from clay with a high silica (sand) content (over 70% total silica); characterized by dimensional stability when heated or fired. { ¦sem·i'sil·ə·kə ri'frak·trē }

semisteel |MET| Low-carbon steel made by replacing about one-fourth of the pig iron in the cupola with steel scrap. { ¦sem·i'stēl }

semivitreous |MATER| Pertaining to ceramics whose glassy content is not sufficient to reduce porosity below 0.2%. { ¦sem·i'vi·trē·əs }

Sendust |MATER| The trade name for an alloy consisting of approximately 85% iron, 6% aluminum, and 9% silicon; its powdered form is compacted to manufacture low-loss magnetic cores. { 'sen‚dəst }

Sendzimir mill |MET| A mill having small-diameter working rolls, each backed by a pair of supporting rolls, and each pair of these supported by a cluster of three rolls; used for cold-rolling wide sheets of metal to close tolerance. { 'sen·zə‚mir ‚mil }

sensible heat |THERMO| The heat absorbed or evolved by a substance during a change of temperature that is not accompanied by a change of state. See enthalpy. { 'sen·sə·bəl 'hēt }

sensible-heat factor |THERMO| The ratio of space sensible heat to space total heat; used for air-conditioning calculations. Abbreviated SHF. { 'sen·sə·bəl ¦hēt ‚fak·tər }

sensible-heat flow |THERMO| The heat given up or absorbed by a body upon being cooled or heated, as the result of the body's ability to hold heat; excludes latent heats of fusion and vaporization. { 'sen·sə·bəl ¦hēt 'flō }

separated aggregate |MATER| Aggregate for concrete that has been separated into fine and coarse constituents. { 'sep·ə‚rād·əd 'ag·rə·gət }

sepia |MATER| A brown pigment prepared from

the dried, inky exudation of a cuttlefish; used as a dye and in watercolors and ink. { 'sē·pē·ə }

sequence timer |MET| A device used in resistance welding to control the sequence and duration of all elements of the weld cycle, except weld time or heat time. { 'sē·kwəns ,tīm·ər }

sequence weld timer |MET| A sequence timer which also controls weld time or heat time. { 'sē·kwəns 'weld ,tīm·ər }

sequestering agent |CHEM| A substance that removes a metal ion from a solution system by forming a complex ion that does not have the chemical reactions of the ion that is removed; can be a chelating or a complexing agent. { si'kwes·tə·riŋ ,ā·jənt }

series welding |MET| Making two or more resistance welds simultaneously by using a single welding transformer with three or more electrodes forming a series circuit. { 'sir·ēz ,weld·iŋ }

service rating |MATER| A classification for an engine-lubricating oil that indicates the type of service for which the oil is most appropriate. { 'sər·vəs ,rād·iŋ }

sessile dislocation |MET| A dislocation in a metal lattice that is relatively immobile, offering an obstacle to the movement of other dislocations. { 'ses·əl ,dis·lō'kā·shən }

set |CHEM| The hardening or solidifying of a plastic or liquid substance. |MATER| **1.** The hardening or firmness displayed by some materials when left undisturbed. **2.** Permanent deformation of a material, such as metal or plastic, when stressed beyond the elastic limit. |MECH| See permanent set. { set }

set copper |MET| An intermediate copper product obtained at the end of the oxidizing portion of the fire-refining cycle and containing about 3–4% cuprous oxide. { 'set 'käp·ər }

setup time |MATER| The time required for a cement or a gelatin to harden. { 'sed,əp ,tīm }

S glass |MATER| A glass containing magnesia, alumina, and a silicate. { 'es ,glas }

shade |OPTICS| The color of a mixture of pigments or dyes which has some black pigment or dye in it. { shād }

shadow |OPTICS| A region of darkness caused by the presence of an opaque object interposed between such a region and a source of light. |PHYS| A region which some type of radiation, such as sound or x-rays, does not reach because of the presence of an object, which the radiation cannot penetrate, interposed between the region and the source of radiation. { 'shad·ō }

shadow mask |ELECTR| A thin, perforated metal mask mounted just back of the phosphor-dot faceplate in a three-gun color picture tube; the holes in the mask are positioned to ensure that each of the three electron beams strikes only its intended color phosphor dot. Also known as aperture mask. { 'shad·ō ,mask }

shake |MATER| **1.** Separation between adjoining layers of wood, due to causes other than drying. **2.** A thick hand-cut shingle. { shāk }

shakeout |MET| Removing a casting from a sand mold. { 'shāk,aut }

shale clay |MATER| A clay made from ground shale. { 'shāl ,klā }

shape memory alloy |MET| An alloy that, after being deformed, can recover its original shape when it is heated. { 'shāp ¦mem·rē 'al,ói }

shattercrack See flake. { 'shad·ər,krak }

shattering |MECH| The breaking up into highly irregular, angular blocks of a very hard material that has been subjected to severe stresses. { 'shad·ə·riŋ }

shatterproof glass See nonshattering glass. { ¦shad·ər¦prüf 'glas }

Shaw process |MET| A foundry molding process which makes use of wood or metal patterns and a refractory mold bonded with an ethyl silicate base material. { 'shó ,prä·səs }

shear See shear strain. { shir }

shear fracture |MECH| A fracture resulting from shear stress. { 'shir ,frak·chər }

shearing forces |MECH| Two forces that are equal in magnitude, opposite in direction, and act along two distinct parallel lines. { 'shēr·iŋ ,fórs·əz }

shearing strain |MECH| The distortion that results from motion of material on opposite sides of a plane in opposite directions parallel to the plane. { 'shir·iŋ ,strān }

shearing stress |MECH| A stress in which the material on one side of a surface pushes on the material on the other side of the surface with a force which is parallel to the surface. Also known as shear stress; tangential stress. { 'shir·iŋ ,stres }

shear lip |MET| An area or ridge at the edge of a shear fracture surface. { 'shir ,lip }

shear modulus See modulus of elasticity in shear. { 'shir ,mäj·ə·ləs }

shear plane |MECH| A confined zone along which fracture occurs in metal cutting. { 'shir ,plān }

shear rate |FL MECH| The relative velocities in laminar flow of parallel adjacent layers of a fluid body under shear force. { 'shir ,rāt }

shear resistance |FL MECH| A tangential stress caused by fluid viscosity and taking place along a boundary of a flow in the tangential direction of local motion. { 'shir ri,zis·təns }

shear steel |MET| A cutlery steel made from short sheared lengths of blister steel; the lengths are heated, joined by rolling or hammering, and finished by hammering. { 'shir ,stēl }

shear strain |MECH| Also known as shear. **1.** A deformation of a solid body in which a plane in the body is displaced parallel to itself relative to parallel planes in the body; quantitatively, it is the displacement of any plane relative to a second plane, divided by the perpendicular distance between planes. **2.** The force causing such deformation. { 'shir ,strān }

shear strength |MECH| **1.** The maximum shear stress which a material can withstand without rupture. **2.** The ability of a material to withstand shear stress. { 'shir ,streŋkth }

shear stress See shearing stress. { 'shir ,stres }

shear thickening [FL MECH] Viscosity increase of non-Newtonian fluids (for example, complex polymers, proteins, protoplasm) that undergo viscosity increases under conditions of shear stress (that is, viscometric flow). { 'shir ,thik·ən·iŋ }

shear thinning [FL MECH] Viscosity reduction of non-Newtonian fluids (for example, polymers and their solutions, most slurries and suspensions, lube oils with viscosity-index improvers) that undergo viscosity reductions under conditions of shear stress (that is, viscometric flow). { 'shir ,thin·iŋ }

shear transformation See martensitic transformation. { 'shir tranz·fər'mā·shən }

shear-viscosity function [FL MECH] The expression of the viscometric flow of a purely viscous, non-Newtonian fluid in terms of velocity gradient and shear stress of the flowing fluid. { 'shir vi'skäs·əd·ē ,faŋk·shən }

sheathing board [MATER] A composition board (for example, of fiber or gypsum cement) used instead of wood sheathing. { 'shēth·iŋ ,bȯrd }

sheathing paper [MATER] A paper that is heavier and of better quality than the usual building paper. { 'shēth·iŋ ,pā·pər }

sheen [OPTICS] A subdued and often iridescent or metallic glitter which approaches, but is just short of, optical reflection and which modifies the surface luster of a mineral. { shēn }

sheet [MATER] A material in a configuration similar to a film except that its thickness is greater than 0.25 millimeter. { shēt }

sheet asphalt [MATER] Asphalt which provides a smooth surface and is used for continuous road surfacing. { 'shēt 'as,fȯlt }

sheet copper [MET] Copper rolled into sheets; for roofing sometimes used as it leaves the rolls, but for other purposes it is commonly employed after it has been cold-rolled to increase hardness and strength. { 'shēt ,käp·ər }

sheet glass [MATER] Flat sections of glass made by drawing a continuous thin film of glass from a molten bath, then cooling and cutting it; used for common glazing. { 'shēt ¦glas }

sheeting [MATER] **1.** A continuous film of a material such as plastic. **2.** Steel or wood members used to face the walls of an excavation such as a basement or a trench. { 'shēd·iŋ }

sheet metal [MET] Thin sections of metal formed by rolling hot metal and usually less than 0.25 inch (6.35 millimeters) thick; when thicker than 0.25 inch, called plate. { 'shēt ,med·əl }

sheet-metal gage [MET] A standard for expressing the thickness of metal sheets; some manufacturers, for example, Brown & Sharpe (B&S), Birmingham (BG), and Imperial, use code numbers with actual thickness in inches or millimeters. { 'shēt ¦med·əl ,gāj }

sheet mica [MATER] Mica that is relatively flat and sufficiently free from structural defects to enable it to be punched or stamped into specified shapes for use by the electronic and electrical industries. { 'shēt ,mī·kə }

sheet plastic [MATER] Flat sections of extruded,

molded, or cast plastic, with a thickness greater than that for film, that is, greater than 0.05 inch (1.3 millimeters). { 'shēt ,plas·tik }

sheet rubber [MATER] Latex that has been rolled into sheets, either smooth or ribbed. { 'shēt ¦rəb·ər }

sheet separation [MET] The gap between faying surfaces surrounding the weld in spot, seam, or projection welding. { 'shēt ,sep·ə,rā·shən }

sheet steel [MET] Steel rolled in the form of sheet, usually used for deep-drawing applications. { 'shēt ¦stēl }

Sheffield plate [MET] A cladding of silver rolled and fused on both sides of a copper sheet. { 'she,fēld 'plāt }

shelf life [ENG] The time that elapses before stored food, chemicals, batteries, and other materials or devices become inoperative or unusable due to age or deterioration. { 'shelf ,līf }

shell [MET] **1.** The outer wall of a metal mold. **2.** The hard layer of sand and thermosetting plastic formed over a pattern and used as a mold wall in shell molding. **3.** The metal sleeve remaining when a billet is extruded with a dummy block at smaller diameter. **4.** A tubular casting used in preparing seamless drawn tubes. **5.** A pierced forging. { shel }

shellac [MATER] A natural, alcohol-soluble, water-insoluble, flammable resin; made from lac resin deposited on tree twigs in India by the lac insect (*Laccifer lecca*) used as an ingredient of wood coatings. { shə'lak }

shellac varnish [MATER] A solution of shellac in denatured alcohol; used in wood finishing where a fast-drying, light-colored, hard finish is desired. { shə'lak ,vär·nish }

shellac wax [MATER] A hard wax with 3% shellac, from which it is extracted, and used in polishes and insulating materials. { shə'lak ,waks }

shell core [MET] A sand core formed by shell molding. { 'shel ,kȯr }

shell molding [MET] Forming a rigid, porous, self-supporting refractory mold by sprinkling molding sand blended with thermosetting plastic or resin over a preheated metal pattern and then curing in an oven. { 'shel ,mōld·iŋ }

Shenstone effect [ELECTR] An increase in photoelectric emission of certain metals following passage of an electric current. { 'shen,stōn i ,fekt }

sherardizing [MET] Coating iron with zinc by tumbling the article in powdered zinc at about 250–375°C. { shə'rär,dīz·iŋ }

Sherwood number See Nusselt number. { 'shər ,wu̇d ,nəm·bər }

SHF See sensible-heat factor.

shielded arc welding [MET] Arc welding in which the electric arc and the weld metal are protected by gas, decomposition products of the electrode covering, or a blanket of fusible flux. { 'shēl·dəd 'ärk ,weld·iŋ }

shielded metal-arc welding [MET] Arc welding in which heating with an electric arc between the electrode and the work produces fusion of the electrode covering which shields the work. { 'shēl·dəd ¦med·əl 'ärk 'weld·iŋ }

shielding |MET| Placing a nonconducting object in an electrolytic bath during plating to alter the current distribution. { 'shēld·iŋ }

shielding gas [MET] Gas, such as nitrogen, oxygen, and carbon dioxide, used in shielded arc welding to protect molten weld from contamination and damage by the atmosphere. { 'shēld·iŋ ¸gas }

shift [MET] A casting defect caused by malalignment of the mold parts. { shift }

shingle [MATER] A rectangular piece of wood, metal, or other material that is used like a tile and arranged in overlapping rows for covering roofs and walls. { 'shiŋ·gəl }

shiplap [MATER] Lumber whose edges are rabbeted in order to make a close overlapping joint. Also known as shiplap board; shiplap siding. { 'ship¸lap }

shiplap board See shiplap. { 'ship¸lap ¸bórd }

shiplap siding See shiplap. { 'ship¸lap ¸sīd·iŋ }

shivering [MATER] Cracks and scales on a pottery glaze caused by unequal contraction during cooling. { 'shiv·ə·riŋ }

Shockley partial dislocation [SOLID STATE] A partial dislocation in which the Burger's vector lies in the fault plane, so that it is able to glide, in contrast to a Frank partial dislocation. Also known as glissile dislocation. { 'shäk·lē 'pär·shəl ¸dis·lō'kā·shən }

shock resistance [ENG] The property which prevents cracking or general rupture when impacted. { 'shäk ri¸zis·təns }

shop lumber [MATER] Softwood lumber graded and used in the factory for general cut-up purposes; similar to factory lumber but of a lower grade. { 'shäp ¸ləm·bər }

Shore hardness [ENG] A method of rating the hardness of a metal or of a plastic or rubber material. { 'shór ¸härd·nəs }

Shore scleroscope [ENG] A device used in rebound hardness testing of rubber, metal, and plastic; consists of a small, conical hammer fitted with a diamond point and acting in a glass tube. { 'shór 'skler·ə¸skōp }

Shore scleroscope hardness test [MET] A rebound hardness test in which a metal body is dropped vertically down a glass tube onto the surface of the material being tested; the height of the rebound is a measure of the hardness. Also known as scleroscope hardness test. { 'shór 'skler·ə¸skōp 'härd·nəs ¸test }

shortness [MET] A form of brittleness in metal, designated as hot, cold, or red to indicate the temperature range in which brittleness occurs. { 'shórt·nəs }

short oil [MATER] Varnish containing a small percentage of oil. { 'shórt ¸óil }

short-pulse laser [OPTICS] A laser designed to generate a pulse of light lasting on the order of nanoseconds or less, and having very high power, such as by Q switching or mode-locking. { 'shórt ¦pəls 'lā·zər }

short run [MET] Pertaining to a mold or casting

filled only partially with molten metal. { 'shórt 'rən }

shot blasting [MET] Cleaning and descaling metal by shot peening or by means of a stream of abrasive powder blown through a nozzle under air pressure in the range 30–150 pounds per square inch (200–1000 kilopascals). { 'shät ¸blast·iŋ }

shot peening [MET] Shot blasting with small steel balls driven by a blast of air. { 'shät ¸pēn·iŋ }

shotting [MET] Making shot by pouring molten metal in finely divided streams; the particles solidify during descent and are cooled in a tank of water. { 'shäd·iŋ }

show-card color See poster paint. { 'shō ¸kärd ¸kəl·ər }

shrinkage cavity [MET] A cavity resulting from shrinkage during casting. { 'shriŋ·kij ¸kav·əd·ē }

shrinkage crack [MET] An irregular interdendritic crack in a casting caused by unequal contraction or inadequate feeding. { 'shriŋ·kij ¸krak }

shrinkage rule See contraction rule. { 'shriŋ·kij ¸rül }

shrink-mixed concrete [MATER] Concrete that is partially mixed before being put in a truck mixer. { 'shriŋk ¦mikst 'kän¸krēt }

shrink rule See contraction rule. { 'shriŋk ¸rül }

Shubnikov-de Haas effect [SOLID STATE] Oscillations of the resistance or Hall coefficient of a metal or semiconductor as a function of a strong magnetic field, due to the quantization of the electron's energy. { 'shüb·nə¸kóf də'häs i ¸fekt }

Shubnikov groups [SOLID STATE] The point groups and space groups of crystals having magnetic moments. Also known as black-and-white groups; magnetic groups. { 'shüb·nə¸kóf ¸grüps }

Si See silicon.

SIC See dielectric constant.

side-centered lattice [CRYSTAL] A type of centered lattice that is centered on the side faces only. { 'sīd ¦sen·tərd 'lad·əs }

side chain [ORG CHEM] A grouping of similar atoms (two or more, generally carbons, as in the ethyl radical, C_2H_5-) that branches off from a straight-chain or cyclic (for example, benzene) molecule. Also known as branch; branched chain. { 'sīd ¸chān }

side-construction tile [MATER] A type of structural clay tile designed to receive its principal stress at right angles to the axis of the cells. { 'sīd kən¦strək·shən 'tīl }

side-cut brick [MATER] Brick cut by taut wire along the long side, as opposed to the edge. { 'sīd ¦kət ¸brik }

side pinacoid [CRYSTAL] A pinacoid with Miller indices (010) in an orthorhombic, monoclinic, or triclinic crystal. { 'sīd 'pin·ə¸kóid }

siding [MATER] Any wall cladding, except masonry or brick. { 'sīd·iŋ }

siemens [ELEC] A unit of conductance, admittance, and susceptance, equal to the conductance between two points of a conductor such that a potential difference of 1 volt between these points produces a current of 1 ampere; the conductance of a conductor in siemens is the reciprocal of its resistance in ohms. Formerly known as mho (Ω); reciprocal ohm. Symbolized S. { 'sē·mənz }

sienna |MATER| Any of various yellowish-brown earthy substances consisting of hydrated iron oxide occurring in limonite; becomes orange-brown when burnt and is generally darker and more transparent in oils than is ocher; used as pigment for oil paints and stains. { sī'en·ə }

sieve |ENG| **1.** A meshed or perforated device or sheet through which dry loose material is refined, liquid is strained, and soft solids are comminuted. **2.** A meshed sheet with apertures of uniform size used for sizing granular materials. { siv }

sieve analysis |ENG| The size distribution of solid particles on a series of standard sieves of decreasing size, expressed as a weight percent. Also known as sieve classification; sieving. { 'siv ə,nal·ə·səs }

sieve classification See sieve analysis. { 'siv ,klas·ə·fə,kā·shən }

sieve diameter |ENG| The size of a sieve opening through which a given particle will just pass. { 'siv dī,am·əd·ər }

sieve fraction |ENG| That portion of solid particles which pass through a standard sieve of given number and is retained by a finer sieve of a different number. { 'siv ,frak·shən }

sieving See sieve analysis. { 'siv·iŋ }

sigma function |THERMO| A property of a mixture of air and water vapor, equal to the difference between the enthalpy and the product of the specific humidity and the enthalpy of water (liquid) at the thermodynamic wet-bulb temperature; it is constant for constant barometric pressure and thermodynamic wet-bulb temperature. { 'sig·mə ,fəŋk·shən }

sigma phase |MET| A brittle, nonmagnetic phase of tetragonal structure occurring in many transition-metal alloys; frequently encountered in high chromium stainless steels. { 'sig·mə ,fāz }

silane |INORG CHEM| Si_nH_{2n+2} A class of silicon-based compounds analogous to alkanes, that is, straight-chain, saturated paraffin hydrocarbons; they can be gaseous or liquid. Also known as silicon hydride. { 'si,lān }

silanol |CHEM| A member of the family of compounds whose structure contains a silicon atom that is bound directly to one or more hydroxyl groups. { 'sī·lə,nȯl }

silex |MATER| Heat- and shock-resistant glass containing about 98% quartz. { 'sī,leks }

silica aerogel |MATER| A colloidal silica powder whose grains have small pores; used as a low-temperature insulator. { 'sil·ə·kə 'er·ə,jel }

silica brick |MATER| A type of refractory brick formed of at least 90% silica cemented with, for example, slurried lime; used to line furnace roofs. { 'sil·ə·kə ¦brik }

silica cement |MATER| A mortar used with silica cement; it is a refractory material. { 'sil·ə·kə si'ment }

silica flour |MET| A sand additive for casting produced by pulverizing quartz sand. { 'sil·ə·kə ¦flaȯ·ər }

silica fume |MATER| A fine-particulate waste product of electric-arc furnaces, consisting primarily of amorphous (noncrystalline) silicon dioxide, its most important use is in the production of high-strength concrete. Also know as microsilica. { 'sil·ə·kə ,fyüm }

silica gel |INORG CHEM| A colloidal, highly absorbent silica used as a dehumidifying and dehydrating agent, as a catalyst carrier, and sometimes as a catalyst. { 'sil·ə·kə ¦jel }

silica glass |MATER| A translucent or transparent vitreous material consisting almost entirely of silica. Also known as fused silica; vitreous silica. { 'sil·ə·kə ¦glas }

silicate |INORG CHEM| The generic term for a compound that contains silicon, oxygen, and one or more metals, and may contain hydrogen. { 'sil·ə·kət }

silicate cement |MATER| The silicate of soda glue, used as an adhesive in cardboard and plywood boxes. { 'sil·ə·kət si,ment }

silicate cotton See mineral wool. { 'sil·ə·kət ¦kät·ən }

silicate of soda See sodium silicate. { 'sil·ə·kət əv 'sōd·ə }

silicate paint |MATER| A paint in which the vehicle is water-soluble sodium silicate; used for painting mortar. { 'sil·ə·kət ¦pānt }

silicic acid |INORG CHEM| $SiO_2 \cdot nH_2O$ A white, amorphous precipitate; used to bleach fats, waxes, and oils. Also known as hydrated silica. { sə'lis·ik 'as·əd }

silicide |CHEM| A binary compound in which silicon is bonded with a more electropositive element. { 'sil·ə,sīd }

silicomanganese |MET| A crude alloy made up of 65–70% manganese, 16–25% silicon, and 1–2.5% carbon; used in the manufacture of low-carbon steel. { ¦sil·ə·kō'maŋ·gə,nēs }

silicon |CHEM| A group 14 nonmetallic element, symbol Si, with atomic number 14, atomic weight 28.086; dark-brown crystals that burn in air when ignited; soluble in hydrofluoric acid and alkalies; melts at 1410°C; used to make silicon-containing alloys, as an intermediate for silicon-containing compounds, and in rectifiers and transistors. { 'sil·ə·kən }

silicon bromide See silicon tetrabromide. { 'sil·ə·kən 'brō,mīd }

silicon bronze |MET| An alloy of copper with 1–5% silicon; it is corrosion-resistant and has good mechanical properties. { 'sil·ə·kən 'bränz }

silicon carbide |INORG CHEM| SiC Water-insoluble, bluish-black crystals, very hard and iridescent; soluble in fused alkalies; sublimes at 2210°C; used as an abrasive and a heat refractory, and in light-emitting diodes to produce green or yellow light. { 'sil·ə·kən 'kär,bīd }

silicon chloride See silicon tetrachloride. { 'sil·ə·kən 'klȯr,īd }

silicon copper |MET| An alloy containing 70–80% copper and 20–30% silicon, used as an addition to molten copper or brass. { 'sil·ə·kən 'käp·ər }

silicon dioxide |INORG CHEM| SiO_2 Colorless, transparent crystals, soluble in molten alkalies and hydrofluoric acid; melts at 1710°C; used to make glass, ceramic products, abrasives, foundry molds, and concrete. { 'sil·ə·kən dī'äk,sīd }

silicone |MATER| A fluid, resin, or elastomer; can be a grease, a rubber, or a foamable powder; the group name for heat-stable, water-repellent, semiorganic polymers of organic radicals attached to the silicones, for example, dimethyl silicone; used in adhesives, cosmetics, and elastomers. { 'sil·ə,kōn }

silicon fluoride See silicon tetrafluoride. { 'sil·ə·kən 'flür,īd }

silicon halide |INORG CHEM| A compound of silicon and a halogen; for example, $SiBr_4$, Si_2Br_6, $SiCl_4$, Si_2Cl_6, Si_3Cl_8, SiF_4, Si_2F_6, SiI_4, and Si_2F_6. { 'sil·ə·kən 'ha,līd }

silicon hydride See silane. { 'sil·ə·kən 'hī,drīd }

siliconized graphite |MATER| A graphite material whose surface has been chemically converted to silicon carbide. { ¦sil·ə·kə,nīzd 'gra,fīt }

siliconizing |MET| Diffusing silicon into solid metal at an elevated temperature. { 'sil·ə·kə ,nīz·iŋ }

silicon monoxide |INORG CHEM| SiO A hard, abrasive, amorphous solid used as thin surface films to protect optical parts, mirrors, and aluminum coatings. { 'sil·ə·kən mə'näk,sīd }

silicon nitride |INORG CHEM| Si_3N_4 A white, water-insoluble powder, resistant to thermal shock and to chemical reagents; used as a catalyst support and for stator blades of high-temperature gas turbines. { 'sil·ə·kən 'nī,trīd }

silicon-on-insulator |ELECTR| A semiconductor manufacturing technology in which thin films of single-crystalline silicon are grown over an electrically insulating substrate. { 'sil·ə·kən ȯn 'in·sə,lād·ər }

silicon solar cell |ELECTR| A solar cell consisting of p and n silicon layers placed one above the other to form a pn junction at which radiant energy is converted into electricity. { 'sil·ə·kən 'sō·lər 'sel }

silicon steel |MET| A steel that contains 0.5–4.5% silicon, used in electric transformer coils. { 'sil·ə·kən 'stēl }

silicon tetrabromide |INORG CHEM| $SiBr_4$ A fuming, colorless liquid that yellows in air; disagreeable aroma; boils at 153°C. Also known as silicon bromide. { 'sil·ə·kən ¦te·trə'brō,mīd }

silicon tetrachloride |INORG CHEM| $SiCl_4$ A clear, corrosive, fuming liquid with suffocating aroma; decomposes in water and alcohol; boils at 57.6°C; used in warfare smoke screens, to make ethyl silicate and silicones, and as a source of pure silicon and silica. Also known as silicon chloride; tetrachlorosilane. { 'sil·ə·kən ¦te·trə'klȯr,īd }

silicon tetrafluoride |INORG CHEM| SiF_4 A colorless, suffocating gas absorbed readily by water, in which it decomposes; boiling point, −86°C;

used in chemical analysis and to make fluosilicic acid. Also known as silicon fluoride. { 'sil·ə·kən ¦te·trə'flür,īd }

silicospiegel |MET| A spiegeleisen pig iron containing 15–20% manganese and 8–15% silicon and up to 4% carbon with the balance iron; used in making steel. { ¦sil·ə·kō'spē·gəl }

silicothermic process See Pidgeon process. { ¦sil·ə·kō'thər·mik 'prä·səs }

silk paper |MATER| **1.** A paper containing a small amount of silk fibers which give a mottled appearance. **2.** A safety paper sometimes used for postage and revenue stamps. { silk ,pā·pər }

silky fracture |MET| A metal fracture in which the broken surface is fine in texture and dull in appearance; characteristic of tough, strong metals. { 'sil·kē 'frak·chər }

Silsbee effect |CRYO| The ability of an electric current to destroy superconductivity by means of the magnetic field that it generates, without raising the cryogenic temperature. { 'silz·bē i,fekt }

silver |CHEM| A white metallic transition element, symbol Ag, with atomic number 47; soluble in acids and alkalies, insoluble in water; melts at 961°C, boils at 2212°C; used in photographic chemicals, alloys, conductors, and plating. |MET| A sonorous, ductile, malleable metal that is capable of a high degree of polish and that has high thermal and electric conductivity. { 'sil·vər }

silver acetylide |INORG CHEM| Ag_2C_2 A white explosive powder used in detonators. { 'sil·vər ə'sed·əl,īd }

silver alloy |MET| A metal consisting of silver and one or more additional metallic components. { 'sil·vər 'al,ȯi }

silver arsenite |INORG CHEM| Ag_3AsO_3 A poisonous, light-sensitive, yellow powder; soluble in acids and alkalies, insoluble in water and alcohol; decomposes at 150°C; used in medicine. { 'sil·vər 'ärs·ən,īt }

silver brazing |MET| Brazing in which silver-base alloys are used as the filler metal. { 'sil·vər 'brāz·iŋ }

silver brazing alloy See silver solder. { 'sil·vər ¦brāz·iŋ 'al,ȯi }

silver bromate |INORG CHEM| $AgBrO_3$ A poisonous, light- and heat-sensitive, white powder; soluble in ammonium hydroxide, slightly soluble in hot water; decomposed by heat. { 'sil·vər 'brō ,māt }

silver bromide |INORG CHEM| AgBr Yellowish, light-sensitive crystals; soluble in potassium bromide and potassium cyanide, insoluble in water; melts at 432°C; used in photographic films and plates. { 'sil·vər 'brō,mīd }

silver carbonate |INORG CHEM| Ag_2CO_3 Yellowish, light-sensitive crystals; insoluble in water and alcohol, soluble in alkalies and acids; decomposes at 220°C; used as a reagent. { 'sil·vər 'kär·bə,nət }

silver chloride |INORG CHEM| AgCl A white, poisonous, light-sensitive powder; slightly soluble in water, soluble in alkalies and acids; melts at

445°C; used in photography, photometry, silver plating, and medicine. { 'sil·vər 'klȯr,īd }

silver chromate [INORG CHEM] Ag_2CrO_4 Dark-colored crystals insoluble in water, soluble in acids and in solutions of alkali chromates; used as an analytical reagent. { 'sil·vər 'krō,māt }

silver coating *See* silver plating. { 'sil·vər 'kōd·iŋ }

silver cyanide [INORG CHEM] AgCN A poisonous, white, light-sensitive powder; insoluble in water, soluble in alkalies and acids; decomposes at 320°C; used in medicine and in silver plating. { 'sil·vər 'sī·ə,nīd }

silver fluoride [INORG CHEM] $AgF·H_2O$ A light-sensitive, yellow or brownish solid, soluble in water; dehydrated form melts at 435°C; used in medicine. Also known as tachiol. { 'sil·vər 'flur ,īd }

silver foil [MET] Silver or a silver-colored metal in very thin sheets. { 'sil·vər ,fȯil }

silver halide [INORG CHEM] A compound of silver and a halogen; for example, silver bromide (AgBr), silver chloride (AgCl), silver fluoride (AgF), and silver iodide (AgI). { 'sil·vər 'ha,līd }

silver iodate [INORG CHEM] $AgIO_3$ A white powder, soluble in ammonium hydroxide and nitric acid, slightly soluble in water; melts above 200°C; used in medicine. { 'sil·vər 'ī·ə,dāt }

silver iodide [INORG CHEM] AgI A pale-yellow powder, insoluble in water, soluble in potassium iodide-sodium chloride solutions and ammonium hydroxide; melts at 556°C; used in medicine, photography, and artificial rainmaking. { 'sil·vər 'ī·ə,dīd }

silver metallurgy [MET] The art and science of extracting silver metal economically from ores, and the reclamation of silver from industrial processes or from scrap metal. { 'sil·vər 'med·əl ,ər·jē }

silver migration [ELEC] A process, causing reduction in insulation resistance and dielectric failure; silver, in contact with an insulator, at high humidity, and subjected to an electrical potential, is transported ionically from one location to another. { 'sil·vər mī'grā·shən }

silver nitrate [INORG CHEM] $AgNO_3$ Poisonous, corrosive, colorless crystals; soluble in glycerol, water, and hot alcohol; melts at 212°C; used in external medicine, photography, hair dyeing, silver plating, ink manufacture, and mirror silvering, and as a chemical reagent. { 'sil·vər 'nī,trāt }

silver nitrite [INORG CHEM] $AgNO_2$ Yellow or grayish-yellow needles which decompose at 140°C; soluble in hot water; used in organic synthesis and in testing for alcohols. { 'sil·vər 'nī,trīt }

silver orthophosphate *See* silver phosphate. { 'sil·vər ,ȯr·thō'fäs,fāt }

silver oxide [INORG CHEM] Ag_2O An odorless, dark-brown powder with a metallic taste; soluble in nitric acid and ammonium hydroxide, insoluble in alcohol; decomposes above 300°C; used in medicine and in glass polishing and coloring, as a catalyst, and to purify drinking water. { 'sil·vər 'äk,sīd }

silver permanganate [INORG CHEM] $AgMnO_4$ Water-soluble, violet crystals that decompose in alcohol; used in medicine and in gas masks. { 'sil·vər pər'maŋ·gə,nāt }

silver phosphate [INORG CHEM] Ag_3PO_4 A poisonous, yellow powder; darkens when heated or exposed to light; soluble in acids and in ammonium carbonate, very slightly soluble in water; melts at 849°C; used in photographic emulsions and in pharmaceuticals, and as a catalyst. Also known as silver orthophosphate. { 'sil·vər 'fä,sfāt }

silver plating [MET] Electrolytically depositing a coating of metallic silver on a base metal. Also known as silver coating. { 'sil·vər ,plād·iŋ }

silver selenide [INORG CHEM] Ag_2Se Gray powder, insoluble in water, soluble in ammonium hydroxide; melts at 880°C. { 'sil·vər 'sel·ə,nīd }

silver solder [MET] A solder composed of silver, copper, and zinc, having a melting point lower than silver but higher than lead-tin solder. Also known as silver brazing alloy. { 'sil·vər ,säd·ər }

silver suboxide [INORG CHEM] AgO A charcoal-gray powder that crystallizes in the cubic or orthorhombic system, and has diamagnetic properties; used in making silver oxide-zinc alkali batteries. Also known as argentic oxide. { 'sil·vər səb'äk,sīd }

silver sulfate [INORG CHEM] Ag_2SO_4 Light-sensitive, colorless, lustrous crystals; soluble in alkalies and acids, insoluble in alcohol; melts at 652°C; used as an analytical reagent. Also known as normal silver sulfate. { 'sil·vər 'səl,fāt }

silver sulfide [INORG CHEM] Ag_2S A dark, heavy powder, insoluble in water, soluble in concentrated sulfuric and nitric acids; melts at 825°C; used in ceramics and in inlay metalwork. { 'sil·vər 'səl,fīd }

silvery iron [MET] A variety of cast iron with a high silicon content, a light-gray color, and a fine grain. { 'sil·və·rē 'ī·ərn }

silylene [CHEM] A divalent silicon species (R_2Si, with two nonbonding electrons, where R = alkyl, aryl, or hydrogen); analogous to a carbene in carbon chemistry. { sə'lī,lēn }

Simon liquefier [CRYO] A device for liquefying helium in which helium is first cooled at high pressure by liquid or solid hydrogen and is then liquefied by a single adiabatic expansion. { ¦sī·mən 'lik·wə,fī·ər }

simple cubic lattice [CRYSTAL] A crystal lattice whose unit cell is a cube, and whose lattice points are located at the vertices of the cube. { 'sim·pəl 'kyü·bik 'lad·əs }

simple lattice *See* primitive lattice. { 'sim·pəl 'lad·əs }

simple metal [SOLID STATE] A metal in which the electrons are basically free to move throughout the volume. { 'sim·pəl 'med·əl }

simple twin [CRYSTAL] A twinned crystal composed of only two individuals in twin relation. { 'sim·pəl 'twin }

single-bevel groove weld [MET] A groove weld in which one member has a joint edge beveled from one side. { 'siŋ·gəl ¦bev·əl 'grüv ,weld }

single-carrier theory |SOLID STATE| A theory of the behavior of a rectifying barrier which assumes that conduction is due to the motion of carriers of only one type; it can be applied to the contact between a metal and a semiconductor. { 'siŋ·gəl ¦kar·ē·ər 'thē·ə·rē }

single crystal |CRYSTAL| A crystal, usually grown artificially, in which all parts have the same crystallographic orientation. { 'siŋ·gəl 'krist·əl }

single-impulse welding |MET| Spot, projection, or upset welding by means of a single current impulse. { 'siŋ·gəl 'im,puls 'weld·iŋ }

single-J groove weld |MET| A groove weld in which one member has a joint edge in the form of a J from one side. { 'siŋ·gəl ¦jā ¦grüv ,weld }

single knock-on |SOLID STATE| A sputtering event in which target atoms are ejected either directly by the bombarding projectiles or after a small number of collisions. { 'siŋ·gəl 'näk·ȯn }

single-pass weld |MET| A weld made by depositing the filler metal with a single pass. { 'siŋ·gəl ¦pas weld }

single-sized aggregate |MATER| Aggregate in which most of the particles lie between narrow size limits. { 'siŋ·gəl ¦sīzd 'ag·rə·gət }

single-stand mill |MET| A rolling mill in which the product is in contact with only two rolls at a time. { 'siŋ·gəl ¦stand 'mil }

single-U groove weld |MET| A groove weld in which the joint edge of both members is prepared in the form of a J from one side, giving a final U form to the completed weld. { 'siŋ·gəl ¦yü ¦grüv ,weld }

single-V groove weld |MET| A groove weld in which the joint edge of each member is beveled from the same side. { 'siŋ·gəl ¦vē ¦grüv ,weld }

single welded joint |MET| A joint welded from one side only. { 'siŋ·gəl 'wel·dəd 'jȯint }

sinkhead See feedhead. { 'siŋk,hed }

sinking tubing |MET| Drawing tubing through a die or passing it through rolls without the use of a tool in the bore to control the inside diameter. { 'siŋk·iŋ ,tüb·iŋ }

sinter |MET| **1.** The product of a sintering operation. **2.** A shaped body composed of metal powders and produced by sintering with or without previous compacting. { 'sin·tər }

sintered copper |MET| Copper prepared by heating a compressed powder of the metal to form a solid mass. { 'sin·tərd 'käp·ər }

sintered steel |MET| Steel prepared by heating compressed iron powder and graphite to form a solid. { 'sin·tərd 'stēl }

sintering |MET| Forming a coherent bonded mass by heating metal powders without melting; used mostly in powder metallurgy. { 'sin·tə·riŋ }

sintering furnace |MET| A furnace in which presintering and sintering operations are carried out. { 'sin·tə·riŋ ,fər·nəs }

siporex |MATER| A building material composed of sand, lime or cement, and aluminum powder which are mixed and cast into molds to be made into roof slabs, door lintels, and wall blocks which give excellent heat and sound insulation. { 'sī·pə,reks }

sisal-hemp wax |MATER| A hard wax derived from sisal waste; melts at 63°C, decomposes at 95°C. Also known as sisal wax. { 'sī·səl ¦hemp ,waks }

sisal wax See sisal-hemp wax. { 'sī·səl ,waks }

Sitka cypress See Alaska cedar. { 'sit·kə 'sī·prəs }

sixty degrees Fahrenheit British thermal unit See British thermal unit. { 'siks·tē di¦grēz 'far·ən,hīt 'brid·ish 'thər·məl ,yü·nət }

size |MATER| Materials used to surface-treat textiles, papers, and leathers; examples are starch, gelatins, casein, water-soluble gums, and waxes. Also known as sizing. { sīz }

size classification See sizing. { 'sīz ,klas·ə·fə ,kā·shən }

size effect |MET| The effect of the size of a piece of metal on its properties and manufacturing variables; in general, mechanical properties are lower for a larger size. { 'sīz i,fekt }

size of weld |MET| **1.** The joint penetration in a groove weld. **2.** The lengths of the nominal legs of a fillet weld. { 'sīz əv 'weld }

sizing |ENG| Separating an aggregate of mixed particles into groups according to size, using a series of screens. Also known as size classification. |MATER| See size. |MET| **1.** Final pressing of a metal powder compact after sintering. **2.** A cold-working operation in which a part is re-pressed in a die to improve surface hardness, smoothness, and dimensional accuracy. { 'sīz·iŋ }

Sk See Stefan number.

skeleton crystal |CRYSTAL| A crystal formed in microscopic outline with incomplete filling in of the faces. { 'skel·ət·ən ,krist·əl }

skelp |MET| A strip or sheet of steel which will be rolled and welded to form a tube. { skelp }

skim gate |MET| A gate used to prevent slag and other undesirable materials from passing into the casting. { 'skim ,gāt }

skin |MET| A thin outside layer of metal differing in composition, structure, or other characteristics from the main mass of metal but not formed by bonding or electroplating. { skin }

skin effect |ELEC| The tendency of alternating currents to flow near the surface of a conductor thus being restricted to a small part of the total sectional area and producing the effect of increasing the resistance. { 'skin i,fekt }

skin lamination |MET| Surface rupture in flat-rolled metals due to exposure of a subsurface lamination. { 'skin ,lam·ə,nā·shən }

skin resistance |ELEC| For alternating current of a given frequency, the direct-current resistance of a layer at the surface of a conductor whose thickness equals the skin depth. { 'skin ri ,zis·təns }

skull |MET| A layer of solidified metal or dross left in the pouring vessel after the molten metal has been poured. { skəl }

skull crucible |MET| A consumable-electrode vacuum arc melting and casting furnace; used in the production of turbine buckets for aircraft jet engines using a nickel-base high-temperature alloy. { 'skəl ,krü·sə·bəl }

slab |CIV ENG| That part of a reinforced concrete floor, roof, or platform which spans beams, columns, walls, or piers. |ELECTR| A relatively thick-cut crystal from which blanks are obtained by subsequent transverse cutting. |MATER| A thin piece of concrete or stone. |MET| A piece of metal, intermediate between ingot and plate, with the width at least twice the thickness. { slab }

slabbing mill |MET| A steel rolling mill for making slabs. { 'slab·iŋ ,mil }

slack quenching |MET| Formation of transformation products other than martensite as a result of quenching at a rate slower than the critical cooling rate. { 'slak ,kwench·iŋ }

slack wax |MATER| A soft, oily, crude wax obtained from the pressing of petroleum paraffin distillate or wax distillate. { 'slak 'waks }

slag |MET| A nonmetallic product resulting from the interaction of flux and impurities in the smelting and refining of metals. Also known as bloom. { slag }

slag cement |MATER| Cement produced by grinding blast-furnace slag and mixing it with lime, portland cement, or dehydrated gypsum. { 'slag si,ment }

slagging |MET| Freeing from or converting into slag. { 'slag·iŋ }

slag inclusion |MET| Slag entrapped in solidified metal. { 'slag in'klü·zhən }

slag sand |MATER| Slag that has been finely crushed for use in mortar and concrete. { 'slag ,sand }

slag wool See mineral wool. { 'slag ,wùl }

slaked lime See calcium hydroxide. { 'slākt 'līm }

sleeve brick |MATER| Tube-shaped firebrick for lining slag vents. { 'slēv ,brik }

slip |CRYSTAL| The movement of one atomic plane over another in a crystal; it is one of the ways that plastic deformation occurs in a solid. Also known as glide. |MATER| A suspension of fine clay in water with a creamy consistency, used in the casting process and in decorating ceramic ware. Also known as slurry. { slip }

slip additive |MATER| A plastics modifier that acts as an internal lubricant by exuding to the surface of the plastic during and immediately after processing to reduce friction and improve slip. { 'slip ,ad·əd·iv }

slipband |CRYSTAL| One of the microscopic parallel lines (Lüders' lines) on the surface of a crystalline material stretched beyond its elastic limit, located at the intersection of the surface with intracrystalline slip planes in the grains of the material. Also known as slip line. { 'slip ,band }

slip clay |MATER| Clay containing a high percentage of fluxing impurities; easily fusible and used in clayware to produce a natural glaze. { 'slip ,klā }

slip direction |CRYSTAL| The crystallographic direction in which the translation of slip occurs. { 'slip di,rek·shən }

slip line See slipband. { 'slip ,līn }

slip plane See glide plane. { 'slip ,plān }

sliver |MATER| A piece of propellant grain of triangular cross section which remains unburned when the web of multiperforated grains has been burned through. |MET| A thin, elongated fragment of metal that has been rolled onto the surface of the parent metal and is attached by only one end. { 'sliv·ər }

slope control |MET| Electronic production of a change in the welding current within set limits and a selected interval of time. { 'slōp kən,trōl }

slot weld |MET| Similar to plug weld, but the hole is elongated and may extend to the edge of a member without closing. { 'slät ,weld }

slow-curing liquid asphaltic material |MATER| An asphalt cement blended with slow-volatilizing gas oil. Also known as SC asphalt. { 'slō ¦kyur·iŋ 'lik·wəd as'fól·tik mə'tir·ē·əl }

slug |MET| **1.** A small, roughly shaped piece of metal for subsequent processing, as by forging or extruding. **2.** The piece of material produced by piercing a hole in a sheet. { sləg }

slugging |MET| Adding a separate piece of material to a weld joint, resulting in a joint which does not meet specifications. { 'sləg·iŋ }

SLUG junction |CRYO| A Josephson junction consisting of a drop of lead-tin solder solidified around a niobium wire. { 'sləg ,jəŋk·shən }

slump test |ENG| Determining the consistency of concrete by filling a conical mold with a sample of concrete, then inverting it over a flat plate and removing the mold; the amount by which the concrete drops below the mold height is measured and this represents the slump. { 'sləmp ,test }

slurry |MATER| **1.** A semiliquid refractory material, such as clay, used to repair furnace refractories. **2.** A free-flowing, pumpable suspension of fine solid material in liquid. **3.** An emulsion of a sulfonated soluble oil in water used to cool and lubricate metal during cutting operations. **4.** A plastic mixture of portland cement and water pumped into an oil well; after hardening, it provides support for the casing and a seal for the well bore. **5.** See slip. { 'slər·ē }

slurry blasting agent |MATER| A dense, insensitive, high-velocity explosive of great power and very high water resistance used principally for blasting hard rock or where blastholes are wet. { 'slər·ē 'blast·iŋ ,ā·jənt }

slush casting |MET| Producing a hollow casting without a core in the mold by rotating a liquid alloy in a hollow metal mold until a solid layer chills onto the mold, and then pouring off the remaining liquid. { 'sləsh ,kast·iŋ }

slushing compound |MATER| A temporary, corrosion-protective coating for metals; made of nondrying oil, grease, or other similar material. { 'sləsh·iŋ ,käm,pund }

slushing grease |MATER| A special grade of grease used as a metal coating to prevent corrosion. { 'sləsh·iŋ ,grēs }

slushing oil |MATER| A nondrying oil which is strongly adhesive to metal and is applied to metal surfaces to minimize corrosion. { 'sləsh·iŋ ,óil }

Sm   *See* samarium.
small calorie   *See* calorie.   { 'smȯl 'kal·ə·rē }
small polaron |SOLID STATE| A quasiparticle comprising a self-trapped electronic charge localized within a small region of a solid of spatial extent comparable to an interatomic dimension, and the atomic displacement pattern which produces the potential well within which the charge is bound.   { 'smȯl 'pō·lə,rän }
smalt |MATER| A blue glass made by fusing silica and potash with cobalt oxides; used as a pigment for glass, ceramics, paints, and dyes.   { smȯlt }
smart materials |MATER| **1.** Materials that can significantly change their mechanical properties (such as shape, stiffness, and viscosity), or their thermal, optical, or electromagnetic properties, in a predictable or controllable manner in response to their environment. **2.** Materials that perform sensing and actuating functions, including piezoelectrics, electrostrictors, magnetostrictors, and shape-memory alloys.   { ,smärt mə'tir·ē·əlz }
smart sensor |ENG| A microsensor integrated with signal-conditioning electronics such as analog-to-digital converters on a single silicon chip to form an integrated microelectromechanical component that can process information itself or communicate with an embedded microprocessor.   { ,smärt 'sen·sər }
smart structures |ENG| Structures that are capable of sensing and reacting to their environment in a predictable and desired manner, through the integration of various elements, such as sensors, actuators, power sources, signal processors, and communications network. In addition to carrying mechanical loads, smart structures may alleviate vibration, reduce acoustic noise, monitor their own condition and environment, automatically perform precision alignments, or change their shape or mechanical properties on command.   { ,smärt 'strək·chərz }
smectic-A |PHYS CHEM| A subclass of smectic liquid crystals in which molecules are free to move within layers and are oriented perpendicular to the layers.   { 'smek·dik 'ā }
smectic-B |PHYS CHEM| A subclass of smectic liquid crystals in which molecules in each layer are arranged in a close-packed lattice and are oriented perpendicular to the layers.   { 'smek·dik 'bē }
smectic-C |PHYS CHEM| A subclass of smectic liquid crystals in which molecules are free to move within layers and are oriented with their axes tilted with respect to the normal to the layers.   { 'smek·dik 'sē }
smectic phase |PHYS CHEM| A form of the liquid crystal (mesomorphic) state in which molecules are arranged in layers that are free to glide over each other with relatively small viscosity.   { 'smek·dik ,fāz }
smectogenic solid |PHYS CHEM| A solid which will form a smectic liquid crystal when heated.   { |smek·tə|jen·ik 'säl·əd }
smelter |MET| A furnace used for smelting.   { 'smel·tər }

smelting |MET| The heating of ore mixtures accompanied by a chemical change resulting in liquid metal.   { 'smelt·iŋ }
smith forging |MET| Manual forging of small, hot metal parts with flat or simple-shaped dies, as with a hammer and anvil.   { 'smith ,fȯrj·iŋ }
S Monel |MET| An alloy similar to H Monel but containing 4% silicon.   { 'es mō'nel }
smut |MET| A reaction product left on the surface of a metal after pickling.   { smət }
Sn   *See* tin.
snake |MET| **1.** A twisted and bent hod rod formed before subsequent rolling operations. **2.** A flexible mandrel used to prevent collapse of a shaped piece during bending operations.   { snāk }
snap flask |MET| A foundry flask having its sides latched on one corner to allow removal of the flask from around the sand mold.   { 'snap ,flask }
Snell laws of refraction |OPTICS| When light travels from one medium into another the incident and refracted rays lie in one plane with the normal to the surface; are on opposite sides of the normal; and make angles with the normal whose sines have a constant ratio to one another.   { 'snel 'lȯz əv ri'frak·shən }
Snoek effect |SOLID STATE| The preferential occupation by carbon impurity atoms of sites on one of the three faces in the cubic lattice of iron.   { 'snük i,fekt }
snowflake   *See* flake.   { 'snō,flāk }
soak cleaning |MET| Cleaning the surface of a metal by immersion in a cleaning solution without electrolysis.   { 'sōk ,klēn·iŋ }
soaking |MET| Heating an alloy, usually an ingot, to a temperature not far below its melting temperature and holding it there for a long time to eliminate segregation that occurred on solidification.   { 'sōk·iŋ }
soaking pit |MET| A high-temperature, gas-fired, tightly covered, refractory-lined hole or pit into which a hot metal ingot (with liquid interior) is held at a fixed temperature until needed for rolling into sheet or other forms.   { 'sōk·iŋ ,pit }
soap |MATER| **1.** A particular type of detergent, in which the water-solubilizing group is a carboxylate, $COO-$, and the positive ion is usually sodium, Na^+, or potassium, K^+. **2.** A soap compound mixed with a fragrance and other ingredients and then cast into soap bars of different shapes.   { sōp }
soap builder |MATER| A substance mixed with soap to modify the alkali content, to add water-softening ability, or to improve otherwise the cleaning properties; examples are rosin and sodium phosphate.   { 'sōp ,bil·dər }
soda   *See* sodium carbonate.   { 'sōd·ə }
soda alum   *See* aluminum sodium sulfate.   { 'sōd·ə 'al·əm }
soda ash |INORG CHEM| Na_2CO_3 The commercial grade of sodium carbonate; a powder soluble in water, insoluble in alcohol; used in glass manufacture and petroleum refining, and for soaps and detergents. Also known as anhydrous sodium carbonate; calcined soda.   { 'sōd·ə 'ash }

soda crystals See metahydrate sodium carbonate. { 'sōd·ə ˌkrist·əlz }

soda lime [MATER] A mixture of sodium or potassium hydroxide with calcium oxide; granules are used to absorb water vapor and carbon dioxide gas. { 'sōd·ə ˌlīm }

soda-lime glass [MATER] Glass made by fusion of sand with sodium carbonate, or sodium sulfate and lime, or limestone; used for window glass. { 'sōd·ə ¦līm ˌglas }

sodamide See sodium amide. { 'sōd·əˌmīd }

sodide [INORG CHEM] A member of a class of alkalides in which the metal anion is sodium (Na⁻). { 'sä,dīd }

sodium 12-molybdophosphate [INORG CHEM] Na₃PMo₁₂O₄₀ Yellow, water-soluble crystals; used in neuromicroscopy and photography, and as a water-resisting agent in plastic adhesives and cements. { 'sōd·ē·əm ¦twelv məˌlib·dō'fä ˌsfāt }

sodium [CHEM] A metallic element of group I, symbol Na, with atomic number 11, atomic weight 22.98977; silver-white, soft, and malleable; oxidizes in air; melts at 97.6°C; used as a chemical intermediate and in pharmaceuticals, petroleum refining, and metallurgy; the source of the symbol Na is natrium. { 'sōd·ē·əm }

sodium acid carbonate See sodium bicarbonate. { 'sōd·ē·əm 'as·əd 'kär·bə·nət }

sodium acid chromate See sodium dichromate. { 'sōd·ē·əm 'as·əd 'krō,māt }

sodium acid fluoride See sodium bifluoride. { 'sōd·ē·əm 'as·əd 'flúr,īd }

sodium acid sulfate See sodium bisulfate. { 'sōd·ē·əm 'as·əd 'səl,fāt }

sodium acid sulfite See sodium bisulfite. { 'sōd·ē·əm 'as·əd 'səl,fīt }

sodium alginate [ORG CHEM] C₆H₇O₆Na Colorless or light yellow filaments, granules, or powder which forms a viscous colloid in water; used in food thickeners and stabilizers, in medicine and textile printing, and for paper coating and water-base paint. Also known as algin; alginic acid sodium salt; sodium polymannuronate. { 'sōd·ē·əm 'al·jəˌnāt }

sodium aluminate [INORG CHEM] Na₂Al₂O₄ A white powder soluble in water, insoluble in alcohol; melts at 1800°C; used as a zeolite-type of material and a mordant, and in water purification, milkglass manufacture, and cleaning compounds. { 'sōd·ē·əm ə'lü·mə,nāt }

sodium aluminosilicate [INORG CHEM] White, amorphous powder or beads of variable stoichiometry, partially soluble in strong acids and alkali hydroxide solutions between 80 and 100°C; used in food as an anticaking agent. Also known as sodium silicoaluminate. { 'sōd·ē·əm ə¦lü·mə·nō'sil·ə,kāt }

sodium aluminum phosphate [INORG CHEM] NaAl₃H₁₄(PO₄)₈·4H₂O or Na₃Al₂H₁₅(PO₄)₈ White powder, soluble in hydrochloric acid; used as a food additive for baked products. { 'sōd·ē·əm ə'lü·mə·nəm 'fä,sfāt }

sodium aluminum silicofluoride [INORG CHEM]

Na₅Al(SiF₆)₄ A toxic, white powder, used for mothproofing and in insecticides. { 'sōd·ē·əm ə'lü·mə·nəm ˌsil·ə·kō'flúr,īd }

sodium aluminum sulfate See aluminum sodium sulfate. { 'sōd·ē·əm ə'lü·mə·nəm 'səl,fāt }

sodium amalgam [INORG CHEM] NaₓHgₓ A fire-hazardous, silver-white crystal mass that decomposes in water; used to make hydrogen and as an analytical reagent. { 'sōd·ē·əm ə'mal·gəm }

sodium amide [INORG CHEM] NaNH₂ White crystals that decompose in water; melts at 210°C; a fire hazard; used to make sodium cyanide. Also known as sodamide. { 'sōd·ē·əm 'am,īd }

sodium antimonate [INORG CHEM] NaSbO₃ A white, granular powder, used as an enamel opacifier and high-temperature oxidizing agent. Also known as antimony sodiate. { 'sōd·ē·əm an'tim·ə,nāt }

sodium arsenate [INORG CHEM] Na₃AsO₄·12H₂O Water-soluble, poisonous, clear, colorless crystals with a mild alkaline taste; melts at 86°C; used in medicine, insecticides, dry colors, and textiles, and as a germicide and a chemical intermediate. { 'sōd·ē·əm 'ärs·ən,āt }

sodium arsenite [INORG CHEM] NaAsO₂ A poisonous, water-soluble, grayish powder; used in antiseptics, dyeing, insecticides, and soaps for taxidermy. { 'sōd·ē·əm 'ärs·ən,īt }

sodium azide [INORG CHEM] NaN₃ Poisonous, colorless crystals; soluble in water and liquid ammonia; decomposes at 300°C; used in medicine and to make lead azide explosives. { 'sōd·ē·əm 'ā,zīd }

sodium bicarbonate [INORG CHEM] NaHCO₃ White, water-soluble crystals with an alkaline taste; loses carbon dioxide at 270°C; used as a medicine and a butter preservative; in food preparation, in effervescent salts and beverages, in ceramics, and to prevent timber mold. Also known as baking soda; bicarbonate of soda; sodium acid carbonate. { 'sōd·ē·əm bī'kär·bə ˌnət }

sodium bichromate See sodium dichromate. { 'sōd·ē·əm bī'krō,māt }

sodium bifluoride [INORG CHEM] NaHF₂ Poisonous, water-soluble, white crystals; decomposes when heated; used as a laundry-rinse neutralizer, preservative, and antiseptic, and in glass etching and tinplating. Also known as sodium acid fluoride. { 'sōd·ē·əm bi'flúr·īd }

sodium bismuthate [INORG CHEM] NaBiO₃ A yellow to brown amorphous powder; used as an analytical reagent and in pharmaceuticals. { 'sōd·ē·əm 'biz·mə,thāt }

sodium bisulfate [INORG CHEM] NaHSO₄ Colorless crystals, soluble in water; the aqueous solution is strongly acidic; decomposes at 315°C; used for flux to decompose minerals, as a disinfectant, and in dyeing and manufacture of magnesia, cements, perfumes, brick, and glue. Also known as niter cake; sodium acid sulfate. { 'sōd·ē·əm bī'səl,fāt }

sodium bisulfide See sodium hydrosulfide. { 'sōd·ē·əm bī'səl,fīd }

sodium bisulfite [INORG CHEM] $NaHSO_3$ A colorless, water-soluble solid; decomposes when heated. Also known as sodium acid sulfite. { 'sōd·ē·əm bī'səl‚fīt }

sodium borate [INORG CHEM] $Na_2B_4O_7 \cdot 10H_2O$ A water-soluble, odorless, white powder; melts between 75 and 200°C; used in glass, ceramics, starch and adhesives, detergents, agricultural chemicals, pharmaceuticals, and photography; the impure form is known as borax. Also known as sodium pyroborate; sodium tetraborate. { 'sōd·ē·əm 'bȯ‚rāt }

sodium borohydride [INORG CHEM] $NaBH_4$ A flammable, hygroscopic, white to gray powder; soluble in water, insoluble in ether and hydrocarbons; decomposes in damp air; used as a hydrogen source, a chemical reagent, and a rubber foaming agent. { 'sōd·ē·əm ‚bȯr·ō'hī‚drīd }

sodium bromate [INORG CHEM] $NaBrO_3$ Odorless, white crystals; soluble in water, insoluble in alcohol; decomposes at 381°C; a fire hazard, used as an analytical reagent. { 'sōd·ē·əm 'brō‚māt }

sodium bromide [INORG CHEM] $NaBr$ White, water-soluble, crystals with a bitter, saline taste; absorbs moisture from air; melts at 758°C; used in photography and medicine, as a chemical intermediate, and to make bromides. { 'sōd·ē·əm 'brō‚mīd }

sodium carbonate [INORG CHEM] Na_2CO_3 A white, water-soluble powder that decomposes when heated to about 852°C; used as a reagent; forms a monohydrate compound, $Na_2CO_3 \cdot H_2O$, and a decahydrate compound, $Na_2CO_3 \cdot 10H_2O$. Also known as soda. { 'sōd·ē·əm 'kär·bə·nət }

sodium carbonate decahydrate See sal soda. { 'sōd·ē·əm 'kär·bə·nət ‚dek·ə'hī‚drāt }

sodium carbonate peroxide [INORG CHEM] $2Na_2CO_3 \cdot 3H_2O$ A white, crystalline powder; used in household detergents, in dental cleansers, and for bleaching and dyeing. { 'sōd·ē·əm 'kär·bə·nət pə'räk‚sīd }

sodium chlorate [INORG CHEM] $NaClO_3$ Water- and alcohol-soluble, colorless crystals with a saline taste; melts at 255°C; used as a medicine, weed killer, defoliant, and oxidizing agent, and in matches, explosives, and bleaching. { 'sōd·ē·əm 'klȯr‚āt }

sodium chloride [INORG CHEM] $NaCl$ Colorless or white crystals; soluble in water and glycerol, slightly soluble in alcohol; melts at 804°C; used in foods and as a chemical intermediate and an analytical reagent. Also known as common salt; table salt. { 'sōd·ē·əm 'klȯr‚īd }

sodium chlorite [INORG CHEM] $NaClO_2$ An explosive, white, mildly hygroscopic, water-soluble powder; decomposes at 175°C; used as an analytical reagent and oxidizing agent. { 'sōd·ē·əm 'klȯr‚īt }

sodium chloroplatinate [INORG CHEM] $Na_2PtCl_6 \cdot 4H_2O$ A yellow powder, soluble in alcohol and water; used for zinc etching, indelible ink, plating, and mirrors, and in photography and medicine. Also known as platinic sodium chloride; plat-

inum sodium chloride; sodium platinichloride. { 'sōd·ē·əm ‚klȯr·ō'plat·ən‚āt }

sodium chromate [INORG CHEM] $Na_2CrO_4 \cdot 10H_2O$ Water-soluble, translucent, yellow, efflorescent crystals that melt at 20°C; used as a rust preventive and in inks, dyeing, and leather tanning. { 'sōd·ē·əm 'krō‚māt }

sodium cobaltinitrite [INORG CHEM] $Na_3Co(NO_2)_6 \cdot \frac{1}{2}H_2O$ Purple, water-soluble, hygroscopic crystals; used as a reagent for analysis of potassium. { 'sōd·ē·əm ‚kō‚bȯl·tə'nī‚trīt }

sodium cyanate [INORG CHEM] $NaOCN$ A poisonous, white powder; soluble in water, insoluble in alcohol and ether; used as a chemical intermediate and for the manufacture of medicine and the heat-treating of steels. { 'sōd·ē·əm 'sī·ə‚nāt }

sodium cyanide [INORG CHEM] $NaCN$ A poisonous, water-soluble, white powder melting at 563°C; decomposes rapidly when standing; used to manufacture pigments, in heat treatment of metals, and as a silver- and gold-ore extractant. { 'sōd·ē·əm 'sī·ə‚nīd }

sodium dichromate [INORG CHEM] $Na_2Cr_2O_7 \cdot 2H_2O$ Poisonous, red to orange deliquescent crystals; soluble in water, insoluble in alcohol; melts at 320°C; loses water of hydration upon prolonged heating at 105°C; used as a chemical intermediate and corrosion inhibitor and in the manufacture of pigments, leather tanning, and electroplating. Also known as sodium acid chromate; sodium bichromate. { 'sōd·ē·əm dī'krō‚māt }

sodium dithionite See sodium hydrosulfite. { 'sōd·ē·əm dī'thī·ə‚nīt }

sodium ferricyanide [INORG CHEM] $Na_3Fe(CN)_6 \cdot H_2O$ A poisonous, deliquescent, red powder; soluble in water, insoluble in alcohol; used in printing and for the manufacture of pigments. Also known as red prussiate of soda. { 'sōd·ē·əm ‚fer·ə'sī·ə‚nīd }

sodium ferrocyanide [INORG CHEM] $Na_4Fe(CN)_6 \cdot 10H_2O$ Semitransparent crystals, soluble in water; insoluble in alcohol; used in photography, dyes, tanning, and blueprint paper. Also known as yellow prussiate of soda. { 'sōd·ē·əm ‚fer·ə'sī·ə ‚nīd }

sodium fluoborate [INORG CHEM] $NaBF_4$ A white powder with a bitter taste; soluble in water, slightly soluble in alcohol; decomposes when heated, fuses below 500°C; used in electrochemical processes, as flux for nonferrous metals refining, and as an oxidation inhibitor. { 'sōd·ē·əm ‚flü·ə'bȯr‚āt }

sodium fluoride [INORG CHEM] NaF A poisonous, water-soluble, white powder, melting at 988°C; used as an insecticide and a wood and adhesive preservative, and in fungicides, vitreous enamels, and dentistry. { 'sōd·ē·əm 'flur‚īd }

sodium fluosilicate [INORG CHEM] Na_2SiF_6 A poisonous, white, amorphous powder; slightly soluble in water; decomposes at red heat; used to fluoridate drinking water and to kill rodents and insects. Also known as sodium silicofluoride. { 'sōd·ē·əm ‚flü·ə'sil·ə‚kət }

318

sodium gold chloride [INORG CHEM] $NaAuCl_4 \cdot 2H_2O$ Yellow crystals, soluble in water and alcohol; used in photography, fine glass staining, porcelain decorating, and medicine. Also known as gold salt; gold sodium chloride. { 'sōd·ē·əm 'gōld 'klȯr,īd }

sodium halide [INORG CHEM] A compound of sodium with a halogen; for example, sodium bromide (NaBr), sodium chloride (NaCl), sodium iodide (NaI), and sodium fluoride (NaF). { 'sōd·ē·əm 'ha,līd }

sodium halometallate [INORG CHEM] A compound of sodium with halogen and a metal; for example, sodium platinichloride, $Na_2PtCl_6 \cdot 6H_2O$. { 'sōd·ē·əm ,hal·ō'med·əl,āt }

sodium hydrate See sodium hydroxide. { 'sōd·ē·əm 'hī,drāt }

sodium hydride [INORG CHEM] NaH A white powder, decomposed by water, and igniting in moist air; used to make sodium borohydride and as a drying agent and a reagent. { 'sōd·ē·əm 'hī ,drīd }

sodium hydrogen phosphate [INORG CHEM] $NaH_2PO_4 \cdot H_2O$ Hygroscopic, transparent, water-soluble crystals; used as a purgative, reagent, and buffer. { 'sōd·ē·əm 'hī·drə·jən 'fä,sfāt }

sodium hydrogen sulfide See sodium hydrosulfide. { 'sōd·ē·əm 'hī·drə·jən 'səl,fīd }

sodium hydrosulfide [INORG CHEM] $NaSH \cdot 2H_2O$ Toxic, colorless, water-soluble needles, melting at $55°C$; used in pulping of paper, processing dyestuffs, hide dehairing, and bleaching. Also known as sodium bisulfide; sodium hydrogen sulfide; sodium sulfhydrate. { 'sōd·ē·əm ,hī·drə'səl,fīd }

sodium hydrosulfite [INORG CHEM] $Na_2S_2O_4$ A fire-hazardous, lemon to whitish-gray powder; soluble in water, insoluble in alcohol; melts at $55°C$; used as a chemical intermediate and catalyst and in ore flotation. Also known as sodium dithionite. { 'sōd·ē·əm ,hī·drə'səl,fīt }

sodium hydroxide [INORG CHEM] NaOH White, deliquescent crystals; absorbs carbon dioxide and water from air; soluble in water, alcohol, and glycerol; melts at $318°C$; used as an analytical reagent and chemical intermediate, in rubber reclaiming and petroleum refining, and in detergents. Also known as sodium hydrate. { 'sōd·ē·əm hī'dräk,sīd }

sodium hypochlorite [INORG CHEM] NaOCl Air-unstable, pale-green crystals with sweet aroma; soluble in cold water, decomposes in hot water; used as a bleaching agent for paper pulp and textiles, as a chemical intermediate, and in medicine. { 'sōd·ē·əm ¦hī·pō'klȯr,īt }

sodium hypophosphite [INORG CHEM] $NaH_2PO_2 \cdot H_2O$ Colorless, pearly, water-soluble crystalline plates or a white, granular powder; used in medicine and electroless nickel plating of plastic and metal. { 'sōd·ē·əm ¦hī·pō'fä ,sfīt }

sodium hyposulfite See sodium thiosulfate. { 'sōd·ē·əm ¦hī·pō'səl,fīt }

sodium iodate [INORG CHEM] $NaIO_3$ A white, water- and acetone-soluble powder; used as a

disinfectant and in medicine. { 'sōd·ē·əm 'ī·ə ,dāt }

sodium iodide [INORG CHEM] NaI A white, air-sensitive powder, deliquescent, with bitter taste; soluble in water, alcohol, and glycerin; melts at $653°C$; used in photography and in medicine and as an analytical reagent. { 'sōd·ē·əm 'ī·ə,dīd }

sodium lead alloy [MATER] A highly toxic, explosion-prone alloy of lead and sodium; contains 10% sodium and 90% lead when used to make tetraethyllead, and 2% sodium and 98% lead when used as a deoxidizer and homogenizer in lead-containing nonferrous alloys; reacts with moisture, acids, and oxidizing agents. { 'sōd·ē·əm 'led 'al,ȯi }

sodium lead hyposulfate See lead sodium thiosulfate. { 'sōd·ē·əm 'led ¦hī·pō'səl,fāt }

sodium lead thiosulfate See lead sodium thiosulfate. { 'sōd·ē·əm 'led ¦thī·ə'səl,fāt }

sodium metaborate [INORG CHEM] $NaBO_2$ Water-soluble, white crystals, melting at $966°C$; the aqueous solution is alkaline; made by fusing sodium carbonate with borax; used as an herbicide. { 'sōd·ē·əm ¦med·ə'bȯr,āt }

sodium metaphosphate [INORG CHEM] $(NaPO_3)_x$ Sodium phosphate groupings; cyclic forms range from $x = 3$ for the trimetaphosphate, to $x = 10$ for the decametaphosphate; sodium hexametaphosphate with $x = 10$ to 20 is probably a polymer; used for dental polishing, building detergents, and water softening, and as a sequestrant, emulsifier, and food additive. { 'sōd·ē·əm ¦med·ə'fä,sfāt }

sodium metasilicate See sodium silicate. { 'sōd·ē·əm ¦med·ə'sil·ə,kāt }

sodium metavanadate [INORG CHEM] $NaVO_3$ Colorless crystals or a pale green, crystalline powder with a melting point of $630°C$; soluble in water; used in inks, fur dyeing, and photography, and as a corrosion inhibitor in gas scrubbers. { 'sōd·ē·əm ¦med·ə'van·ə,dāt }

sodium molybdate [INORG CHEM] Na_2MoO_4 Water-soluble crystals, melting at $687°C$; used as an analytical reagent, corrosion inhibitor, catalyst, and zinc-plating brightening agent, and in medicine. { 'sōd·ē·əm mə'lib,dāt }

sodium monoxide [INORG CHEM] Na_2O A strong basic white powder soluble in molten caustic soda; forms sodium hydroxide in water; used as a dehydrating and polymerization agent. Also known as sodium oxide. { 'sōd·ē·əm mə'näk,sīd }

sodium nitrate [INORG CHEM] $NaNO_3$ Fire-hazardous, transparent, colorless crystals with bitter taste; soluble in glycerol and water; melts at $308°C$; decomposes when heated; used in manufacture of glass and pottery enamel and as a fertilizer and food preservative. { 'sōd·ē·əm 'nī,trāt }

sodium nitrite [INORG CHEM] $NaNO_2$ A fire-hazardous, air-sensitive, yellowish powder, soluble in water; decomposes above $320°C$; used as an intermediate for dyestuffs and for pickling meat, textiles dyeing, and rust-proofing, and in medicine. { 'sōd·ē·əm 'nī,trīt }

sodium nitroferricyanide

sodium nitroferricyanide [INORG CHEM] $Na_2Fe(CN)_5NO \cdot 2H_2O$ Water-soluble, transparent, reddish crystals; slowly decomposes in water; used as an analytical reagent. { 'sōd·ē·əm ¦nī·trō,fer·ə'sī·ə,nīd }

sodium oxide See sodium monoxide. { 'sōd·ē·əm 'äk,sīd }

sodium paraperiodate [INORG CHEM] $Na_3H_2IO_6$ White, crystalline solid, soluble in concentrated sodium hydroxide solutions; used to wet-strengthen paper and to aid in tobacco combustion. { 'sōd·ē·əm ¦par·ə·pər'ī·ə,dāt }

sodium pentaborate [INORG CHEM] $Na_2B_{10}O_{16} \cdot 10H_2O$ A white, water-soluble powder; used in glassmaking, weed killers, and fireproofing compositions. { 'sōd·ē·əm ¦pen·tə'bór,āt }

sodium perborate [INORG CHEM] $NaBO_2 \cdot H_2O_2 \cdot 3H_2O$ A white powder with a saline taste; slightly soluble in water, decomposes in moist air; used in deodorants, in dental compositions, and as a germicide. Also known as peroxydol. { 'sōd·ē·əm pər'bór,āt }

sodium perchlorate [INORG CHEM] $NaClO_4$ Fire-hazardous, white, deliquescent crystals; soluble in water and alcohol; melts at 482°C; explosive when in contact with concentrated sulfuric acid; used in jet fuel, as an analytical reagent, and for explosives. { 'sōd·ē·əm pər'klór,āt }

sodium permanganate [INORG CHEM] $NaMnO_4 \cdot 3H_2O$ A fire-hazardous, water-soluble, purple powder; decomposes when heated; used to make saccharin, as a disinfectant, and as an oxidizing agent. { 'sōd·ē·əm pər'maŋ·gə,nāt }

sodium peroxide [INORG CHEM] Na_2O_2 A fire-hazardous, white powder that yellows with heating; decomposes when heated; causes ignition when in contact with water; used as an oxidizing agent and a bleach, and in medicinal soap. { 'sōd·ē·əm pə'räk,sīd }

sodium persulfate [INORG CHEM] $Na_2S_2O_8$ A white, water-soluble, crystalline powder; used as a bleaching agent and in medicine. { 'sōd·ē·əm pər'səl,fāt }

sodium phosphate [INORG CHEM] A general term encompassing the following compounds: sodium hexametaphosphate, sodium metaphosphate, dibasic sodium phosphate, hemibasic sodium phosphate, monobasic sodium phosphate, tribasic sodium phosphate, sodium pyrophosphate, and acid sodium pyrophosphate. { 'sōd·ē·əm 'fä,sfāt }

sodium phosphite [INORG CHEM] $Na_2HPO_3 \cdot 5H_2O$ White, hygroscopic crystals, melting at 53°C; soluble in water, insoluble in alcohol; used in medicine. { 'sōd·ē·əm 'fä,sfīt }

sodium phosphotungstate See sodium tungstophosphate. { 'sōd·ē·əm ,fä·sfō'təŋ,stāt }

sodium platinichloride See sodium chloroplatinate. { 'sōd·ē·əm ¦plat·ən·ə'klór,īd }

sodium plumbite [INORG CHEM] $Na_2PbO_2 \cdot 3H_2O$ A toxic, corrosive solution of lead oxide (litharge) in sodium hydroxide; used (as doctor solution) to sweeten gasoline. { 'sōd·ē·əm 'pləm,bīt }

sodium polymannuronate See sodium alginate. { 'sōd·ē·əm ¦päl·i·mə'nyůr·ə,nāt }

sodium polysulfide [INORG CHEM] Na_2S_x Yellow-brown granules, used to make dyes and colors, and insecticides, as a petroleum additive, and in electroplating. { 'sōd·ē·əm ¦päl·i'səl,fīd }

sodium pyroborate See sodium borate. { 'sōd·ē·əm ¦pī·rō'bór,āt }

sodium pyrophosphate [INORG CHEM] $Na_4P_2O_7$ A white powder; soluble in water, insoluble in alcohol and ammonia; melts at 880°C; used as a water softener and newsprint deinker, and to control drilling-mud viscosity. Also known as tetrasodium pyrophosphate (TSPP). { 'sōd·ē·əm ¦pī·rō'fä,sfāt }

sodium selenate [INORG CHEM] $Na_2SeO_4 \cdot 10H_2O$ White, poisonous, water-soluble crystals; used as an insecticide. { 'sōd·ē·əm 'sel·ə,nāt }

sodium selenite [INORG CHEM] $Na_2SeO_3 \cdot 5H_2O$ White, water-soluble crystals; used in glass manufacture, as a bacteriological reagent, and for decorating porcelain. { 'sōd·ē·əm 'sel·ə,nīt }

sodium sesquicarbonate [INORG CHEM] $Na_2CO_3 \cdot NaHCO_3 \cdot 2H_2O$ White, water-soluble, needle-shaped crystals; used as a detergent, an alkaline agent for water softening and leather tanning, and a food additive. { 'sōd·ē·əm ¦ses·kwē'kär·bə,nāt }

sodium sesquisilicate [INORG CHEM] $Na_6Si_2O_7$ A white, water-soluble powder; used for metals cleaning and textile processing. { 'sōd·ē·əm ¦ses·kwē'sil·ə,kāt }

sodium silicate [INORG CHEM] Na_2SiO_3 A gray-white powder; soluble in alkalies and water, insoluble in alcohol and acids; used to fireproof textiles, in petroleum refining and corrugated paperboard manufacture, and as an egg preservative. Also known as liquid glass; silicate of soda; sodium metasilicate; soluble glass; water glass. { 'sōd·ē·əm 'sil·ə,kāt }

sodium silicoaluminate See sodium aluminosilicate. { 'sōd·ē·əm ¦sil·ə·kō·ə'lü·mə,nāt }

sodium silicofluoride See sodium fluosilicate. { 'sōd·ē·əm ¦sil·ə·kō'flůr,īd }

sodium stannate [INORG CHEM] $Na_2SnO_3 \cdot 3H_2O$ Water- and alcohol-insoluble, whitish crystals; used in ceramics, dyeing, and textile fireproofing, and as a mordant. Also known as preparing salt. { 'sōd·ē·əm 'sta,nāt }

sodium subsulfite See sodium thiosulfate. { 'sōd·ē·əm ¦səb'səl,fīt }

sodium sulfate [INORG CHEM] Na_2SO_4 Crystalline compound, melts at 888°C, soluble in water; used to make paperboard, kraft paper, glass, and freezing mixtures. { 'sōd·ē·əm 'səl,fāt }

sodium sulfhydrate See sodium hydrosulfide. { 'sōd·ē·əm ¦səlf'hī,drāt }

sodium sulfide [INORG CHEM] Na_2S An irritating, water-soluble, yellow to red, deliquescent powder; melts at 1180°C; used as a chemical intermediate, solvent, photographic reagent, and analytical reagent. Also known as sodium sulfuret. { 'sōd·ē·əm 'səl,fīd }

sodium sulfite [INORG CHEM] Na_2SO_3 White, water-soluble, crystals with a sulfurous, salty taste; decomposes when heated; used as a chemical intermediate and food preservative, in medicine and paper manufacturing, and for dyes and photographic developing.   { 'sōd·ē·əm 'səl ˌfīt }

sodium sulfocyanate   See sodium thiocyanate. { 'sōd·ē·əm ¦səl·fō'sī·əˌnāt }

sodium sulfuret   See sodium sulfide.   { 'sōd·ē· əm 'səl·fyəˌret }

sodium tetraborate   See sodium borate.   { 'sōd· ē·əm ¦te·trə'bȯrˌāt }

sodium tetrasulfide   [INORG CHEM] Na₂S₄ Hygroscopic, yellow or dark-red crystals, melting at 275°C; used for insecticides and fungicides, ore flotation, and dye manufacture, and as a reducing agent.   { 'sōd·ē·əm ¦tet·rə'səlˌfīd }

sodium thiocyanate   [INORG CHEM] NaSCN A poisonous, water- and alcohol-soluble, deliquescent, white powder; melts at 287°C; used as an analytical reagent, solvent, and chemical intermediate, and for rubber treatment and textile dyeing and printing. Also known as sodium sulfocyanate.   { 'sōd·ē·əm ˌthī·ə'sī·ə ˌnāt }

sodium   thiosulfate   [INORG CHEM] Na₂S₂O₃· 5H₂O White, translucent crystals or powder with a melting point of 48°C; soluble in water and oil of turpentine; used as a fixing agent in photography, for extracting silver from ore, in medicine, and as a sequestrant in food. Also known as sodium hyposulfite; sodium subsulfite.   { 'sōd·ē·əm ˌthī·ə'səlˌfāt }

sodium tripolyphosphate   [INORG CHEM] Na₅-P₃O₁₀ A white powder with a melting point of 622°C; used for water softening and as a food additive and texturizer. Abbreviated STPP. { 'sōd·ē·əm trī¸päl·i'fäˌsfāt }

sodium tungstate   [INORG CHEM] Na₂WO₄·2H₂O Water-soluble, colorless crystals; lose water at 100°C, melts at 692°C; used as a chemical intermediate analytical reagent, and for fireproofing. Also known as sodium wolframate.   { 'sōd·ē·əm 'təŋˌstāt }

sodium   tungstophosphate   Approximately 2Na₂O·P₂O₅·12WO₃·18H₂O A yellowish-white powder, soluble in water and alcohols; used to manufacture organic pigments, as an antistatic agent for textiles, in leather tanning, and as a water-resistant agent in plastic films, adhesives, and cements. Also known as sodium phosphotungstate.   { 'sōd·ē·əm ¦twelv ¦təŋ·stō'fäˌsfät }

sodium wolframate   See   sodium   tungstate. { 'sōd·ē·əm 'wùl·frəˌmīt }

softening agent   [MATER] **1.** A substance that is added to another substance to increase softness; for example, stearic acid added to plastics, fatliquoring agents to leather, and fatty alcohol to fabrics.   **2.** A chemical that softens hard water by removing or trapping calcium and magnesium ions.   { 'sȯf·ən·iŋ ˌā·jənt }

softening point   [PHYS] For a substance which does not have a definite melting point, the temperature at which viscous flow changes to plastic flow.   { 'sȯf·ən·iŋ ˌpȯint }

softening range   [PHYS] The temperature range

in which material without a melting point goes from a rigid to a soft condition.   { 'sȯf·ən·iŋ ˌrānj }

soft magnetic material   [ELECTROMAG] A magnetic material which is relatively easily magnetized or demagnetized.   { 'sȯft mag'ned·ik mə'tir·ē·əl }

soft phosphate   [MATER] Powdery, impure tricalcium phosphate separated in fertilizer manufacture from rock and pebble phosphates.   { 'sȯft 'fäˌsfät }

soft rubber   [MATER] A type of rubber that has been cured by adding 0.5 to 8% sulfur, without prolonged vulcanization.   { 'sȯft 'rəb·ər }

soft solder   [MET] Solder composed of an alloy of lead and tin. Also known as low-melting solder. { 'sȯft 'säd·ər }

soft soldering   [MET] Soldering with a soft solder.   { 'sȯft 'säd·ə·riŋ }

soft wood   [MATER] Wood from a coniferous tree. { 'sȯft 'wùd }

soil-cement   [MATER] A compacted mixture of soil, cement, and water used as a base course or surface for roads and airport paving.   { ¦sȯil si¦ment }

sol   [CHEM] A colloidal solution consisting of a suitable dispersion medium, which may be gas, liquid, or solid, and the colloidal substance, the disperse phase, which is distributed throughout the dispersion medium.   { säl }

solar cell   [ELECTR] A pn-junction device which converts the radiant energy of sunlight directly and efficiently into electrical energy.   { 'sō·lər 'sel }

solder   [MET] **1.** To join by means of solder. **2.** An alloy, such as of zinc and copper, or of tin and lead, used when melted to join metallic surfaces.   { 'säd·ər }

solder glass   [MATER] A special glass having a relatively low softening point (below 500°C); used to join two pieces of higher-melting glass without softening and deforming them. { 'säd·ər ˌglas }

soldering embrittlement   [MET] The reduction in mechanical properties of a metal due to the local penetration of solder along grain boundaries. { 'säd·ə·riŋ em'brid·əl·mənt }

soldering   flux   [MET] A chemical substance which aids the flow of solder and serves to remove and prevent the formation of oxides on the pieces to be united.   { 'säd·ə·riŋ ˌfləks }

solder paste   [MET] Finely powdered solder metal combined with a flux.   { 'säd·ər'pāst }

sol-gel glass   [PHYS CHEM] An optically transparent amorphous silica or silicate material produced by forming interconnections in a network of colloidal, submicrometer particles under increasing viscosity until the network becomes completely rigid, with about one-half the density of glass.   { 'säl 'jel 'glas }

sol-gel monolith process   [MATER] A method for fabricating advanced ceramics in which a suspension of colloidal ceramic particles (sol) is converted to a gel by chemical treatment, and then dried and sintered to form a ceramic product. Many ceramic fibers are manufactured by the sol-gel monolith process. { ¦säl ¦jel 'män·ə¦lith ˌprä¦ses }

solid |PHYS| **1.** A substance that has a definite volume and shape and resists forces that tend to alter its volume or shape. **2.** A crystalline material, that is, one in which the constituent atoms are arranged in a three-dimensional lattice, periodic in three independent directions. { 'säl·əd }

solid asphalt |MATER| Asphalt with a penetration number of less than 10 under specified test conditions. { 'säl·əd 'as‚fòlt }

solid helium |CRYO| A certain state which is not attained by helium under its own vapor pressure down to absolute zero, but which requires an external pressure of 25 atmospheres at absolute zero. { 'säl·əd 'hē·lē·əm }

solidification |PHYS| The change of a fluid (liquid or gas) into the solid state. { sə'lid·ə·fə'kā·shən }

solidification shrinkage |MET| Volume contraction of a metal during solidification. { sə'lid·ə·fə'kā·shən ‚shriŋ·kij }

solid insulator |ELEC| An electric insulator made of a solid substance, such as sulfur, polystyrene, rubber, or porcelain. { 'säl·əd 'in·sə‚lād·ər }

solid-phase welding |MET| A welding method in which the weld is consummated by pressure or by heat and pressure without fusion. { 'säl·əd ¦fāz 'weld·iŋ }

solid-piled |MATER| Pertaining to plywood which is fresh from clamps or a hot press, and which is piled onto a solid flat base without stickers and weighted down until it reaches its normal temperature and moisture content. Also known as bulked-down. { 'säl·əd 'pīld }

solid solution |PHYS| A homogeneous crystalline phase composed of several distinct chemical species, occupying the lattice points at random and existing in a range of concentrations. { 'säl·əd sə'lü·shən }

solid state |ENG| Pertaining to a circuit, device, or system that depends on some combination of electrical, magnetic, and optical phenomena within a solid that is usually a crystalline semiconductor material. { 'säl·əd 'stāt }

solid-state amorphizing reaction |MET| An interdiffusion reaction that takes place at constant temperature over long periods of time at a clean, oxide-free boundary between two crystalline metals that have a large chemical affinity, and results in the formation of an amorphous alloy of the two metals (a metallic glass). { 'säl·əd ¦stāt ə'mòr‚fīz·iŋ rē'ak·shən }

solid-state battery |ELEC| A battery in which both the electrodes and the electrolyte are solid-state materials. { 'säl·əd ¦stāt 'bad·ə·rē }

solid-state image sensor *See* charge-coupled image sensor. { 'säl·əd ¦stāt 'im·ij ‚sen·sər }

solid-state laser |OPTICS| A laser in which a semiconductor material produces the coherent output beam. { 'säl·əd ¦stāt 'lā·zər }

solid-state welding |MET| Welding processes which coalesce materials at temperatures below the melting point of the base metal by methods such as welding or diffusion welding without the use of filler metal. { 'säl·əd ¦stāt 'weld·iŋ }

solidus |PHYS CHEM| In a constitution or equilibrium diagram, the locus of points representing the temperature below which the various compositions finish freezing on cooling, or begin to melt on heating. { 'säl·əd·əs }

solidus curve |PHYS CHEM| A curve on the phase diagram of a system with two components which represents the equilibrium between the liquid phase and the solid phase. { 'säl·əd·əs ‚kərv }

solubility |PHYS CHEM| The ability of a substance to form a solution with another substance. { ‚säl·yə'bil·əd·ē }

soluble |CHEM| Capable of being dissolved. { 'säl·yə·bəl }

soluble castor oil *See* Turkey red oil. { 'säl·yə·bəl 'kas·tər ‚òil }

soluble cutting oil |MATER| A petroleum oil containing an emulsifying agent to make it mix easily with water; used as a coolant for metal-cutting tools. { 'säl·yə·bəl 'kəd·iŋ ‚òil }

soluble glass *See* sodium silicate. { 'säl·yə·bəl 'glas }

soluble starch |MATER| A group of water-soluble polymers formed from starch, such as the starches derived from corn or potato, by acetylation, acid hydrolysis, chlorination, or by action of enzymes to form starch acetates, ethers, and esters; used as textile sizing agents, emulsifying agents, and paper coatings. { 'säl·yə·bəl 'stärch }

solute |CHEM| The substance dissolved in a solvent. { 'säl·yüt }

solution |CHEM| A single, homogeneous liquid, solid, or gas phase that is a mixture in which the components (liquid, gas, solid, or combinations thereof) are uniformly distributed throughout the mixture. { sə'lü·shən }

solution ceramic |ELEC| A nonbrittle, inorganic ceramic insulating coating that can be applied to wires at a low temperature; examples include ceria, chromia, titania, and zirconia. { sə'lü·shən sə'ram·ik }

solution heat treatment |MET| Heating and holding an alloy at a temperature at which one (or more) constituent enters into solid solution, then cooling the alloy rapidly to prevent the constituent from precipitating. { sə'lü·shən 'hēt ‚trēt·mənt }

solvation |CHEM| The process of swelling, gelling, or dissolving of a material by a solvent; for resins, the solvent can be a plasticizer. { säl'vā·shən }

solvent |CHEM| That part of a solution that is present in the largest amount, or the compound that is normally liquid in the pure state (as for solutions of solids or gases in liquids). { 'säl·vənt }

solvent naphtha |MATER| Refined petroleum naphtha of restricted boiling range; used as a solvent and paint thinner and in dry cleaning; Stoddard solvent is a special grade of solvent naphtha. { 'säl·vənt ,naf·thə }

Sommerfeld model See free-electron theory of metals. { 'zóm·ər,felt ,mäd·əl }

Sommerfeld theory See free-electron theory of metals. { 'zóm·ər,felt ,thē·ə·rē }

sonochemistry |CHEM| Any chemical change, such as in reaction type or rate, that occurs in response to sound or ultrasound. { ¦sän·ə'kem·ə·strē }

sorbate |CHEM| A substance that has been either adsorbed or absorbed. [ORG CHEM] A salt or an ester of sorbic acid. { 'sór,bāt }

sorbent |MATER| A material, compound, or system that can provide a sorption function, such as adsorption, absorption, or desorption. { 'sór·bənt }

Sorel cement See oxychloride cement. { só'rel si,ment }

source |ELECTR| The terminal in a field-effect transistor from which majority carriers flow into the conducting channel in the semiconductor material. |PHYS| **1.** In general, a device that supplies some extensive entity, such as energy, matter, particles, or electric charge. **2.** A point, line, or area at which mass or energy is added to a system, either instantaneously or continuously. **3.** A point at which lines of force in a vector field originate, such as a point in an electrostatic field where there is positive charge. { sòrs }

south pole |ELECTROMAG| The pole of a magnet at which magnetic lines of force are assumed to enter. Also known as negative pole. { 'saùth 'pōl }

sow |MET| **1.** A mold of larger size than a pig. **2.** A channel that conducts molten metal to molds in a pig bed. { saù }

sow block |MET| In forging, a removable block set into the hammer anvil to reduce wear on the anvil. { 'saù ,bläk }

space charge |ELEC| The net electric charge within a given volume. { 'spās ,chärj }

space-charge layer See depletion layer. { 'spās ¦charj ,lā·ər }

space-charge polarization |ELEC| Polarization of a dielectric which occurs when charge carriers are present which can migrate an appreciable distance through the dielectric but which become trapped or cannot discharge at an electrode. Also known as interfacial polarization. { 'spās ¦chärj ,pō·lə·rə'zā·shən }

space group |CRYSTAL| A group of operations which leave the infinitely extended, regularly repeating pattern of a crystal unchanged; there are 230 such groups. { 'spās ,grüp }

space-group extinction |CRYSTAL| The absence of certain classes of reflections in the x-ray diffraction pattern of a crystal due to the existence of symmetry elements in the space group of the crystal which are not present in its point group. { 'spās ¦grüp ik'stiŋk·shən }

space lattice See lattice. { 'spās ,lad·əs }

spacer strip |MET| A strip or bar of metal placed in the root of a weld joint, prepared for a groove weld, to serve as backing and maintain root opening during welding. { 'spās·ər ,strip }

Spanish white See bismuth subnitrate. { 'span·ish 'wīt }

spark-ignition combustion cycle See Otto cycle. { 'spärk ig¦nish·ən kəm'bəs·chən ,sī·kəl }

sparkle metal |MET| A crude mixture of sulfides containing 74% copper produced by the smelting of copper ore. { 'spär·kəl ,med·əl }

spark machining |MET| Cutting metal by repetitive sparking between a tool (the cathode) and the workpiece (the anode). { 'spärk mə ,shēn·iŋ }

spark test |MET| A technique for estimating the composition of a steel by observing the sparks it produces on a grinding wheel. { 'spärk ,test }

spar varnish |MATER| A flammable varnish made of drying oils, resins, thinners, and driers to provide a durable, water-resistant coating for outside or other severe service. { 'spär ,vär·nish }

spatter |MET| Particles of metal expelled during arc or gas welding. { 'spad·ər }

species |CHEM| A chemical entity or molecular particle, such as a radical, ion, molecule, or atom. Also known as chemical species. { 'spē·shēz }

specific conductance See conductivity. { spə 'sif·ik kən'dək·təns }

specific energy |THERMO| The internal energy of a substance per unit mass. { spə'sif·ik 'en·ər·jē }

specific gravity |MECH| The ratio of the density of a material to the density of some standard material, such as water at a specified temperature, for example, 4°C or 60°F, or (for gases) air at standard conditions of pressure and temperature. Abbreviated sp gr. { spə'sif·ik 'grav·əd·ē }

specific heat |THERMO| **1.** The ratio of the amount of heat required to raise a mass of material 1 degree in temperature to the amount of heat required to raise an equal mass of a reference substance, usually water, 1 degree in temperature; both measurements are made at a reference temperature, usually at constant pressure or constant volume. **2.** The quantity of heat required to raise a unit mass of homogeneous material one degree in temperature in a specified way; it is assumed that during the process no phase or chemical change occurs. { spə'sif·ik 'hēt }

specific inductive capacity See dielectric constant. { spə'sif·ik in'dək·tiv kə'pas·əd·ē }

specific insulation resistance See volume resistivity. { spə'sif·ik ,in·sə'lā·shən ri,zis·təns }

specific reluctance See reluctivity. { spə'sif·ik ri'lək·təns }

specific resistance See electrical resistivity. { spə'sif·ik ri'zis·təns }

specific viscosity |FL MECH| The specific viscosity of a polymer is the relative viscosity of a polymer solution of known concentration minus 1; usually determined at low concentration of the polymer; for example, 0.5 gram per 100 milliliters of solution, or less. { spə'sif·ik vi'skäs·əd·ē }

specific volume [MECH] The volume of a substance per unit mass; it is the reciprocal of the density. Abbreviated sp vol. { spə'sif·ik 'väl·yəm }

specific weight [MECH] The weight per unit volume of a substance. { spə'sif·ik 'wāt }

speckle [OPTICS] A phenomenon in which the scattering of light from a highly coherent source, such as a laser, by a rough surface or inhomogeneous medium generates a random-intensity distribution of light that gives the surface or medium a granular appearance. { 'spek·əl }

spectral color [OPTICS] **1.** A color corresponding to light of a pure frequency; the basic spectral colors are violet, blue-green, yellow, orange, and red. **2.** A color that is represented by a point on the chromaticity diagram that lies on a straight line between some point on the spectral color (first definition) locus and the achromatic points; purple, for example, is not a spectral color. { 'spek·trəl 'kəl·ər }

spectral emissivity [THERMO] The ratio of the radiation emitted by a surface at a specified wavelength to the radiation emitted by a perfect blackbody radiator at the same wavelength and temperature. { 'spek·trəl ,ē,mi'siv·əd·ē }

spectral extinction [OPTICS] The selective absorption of different wavelengths of light as a function of depth in water. { 'spek·trəl ik'stiŋk·shən }

spectral irradiance [OPTICS] The density of the radiant flux that is incident on a surface per unit of wavelength. { 'spek·trəl i'rād·ē·əns }

spectral luminous efficacy [OPTICS] The ratio of the luminous flux emitted by a monochromatic light source in lumens to its radiant flux in watts, as a function of the wavelength of the emitted light. { 'spek·trəl 'lü·mə·nəs 'ef·i·kə·sē }

spectral radiance [OPTICS] The radiant flux per unit wavelength or frequency interval per unit solid angle per unit of projected area of the source; the usual unit is watt per nanometer per steradian per square meter. { 'spek·trəl 'rād·ē·əns }

spectral response See spectral sensitivity. { 'spek·trəl ri'späns }

spectral sensitivity [PHYS] The response of a device or material to monochromatic light as a function of wavelength. Also known as spectral response. { 'spek·trəl ,sen·sə'tiv·əd·ē }

spectral temperature [OPTICS] The temperature of a blackbody that produces the same spectral radiance as a given radiation field at a given wavelength or frequency and in a given direction. { 'spek·trəl 'tem·prə·chər }

spectral transmission [OPTICS] The radiant flux which passes through a filter divided by the radiant flux incident upon it, for monochromatic light of a specified wavelength. { 'spek·trəl tranz'mish·ən }

spectroscopy [PHYS] The branch of physics concerned with the production, measurement, and interpretation of electromagnetic spectra arising from either emission or absorption of radiant energy by various substances. { spek'träs·kə·pē }

spectrum [PHYS] **1.** A display or plot of intensity of radiation (particles, photons, or acoustic radiation) as a function of mass, momentum, wavelength, frequency, or some related quantity. **2.** The set of frequencies, wavelengths, or related quantities, involved in some process; for example, each element has a characteristic discrete spectrum for emission and absorption of light. **3.** A range of frequencies within which radiation has some specified characteristic, such as audio-frequency spectrum, ultraviolet spectrum, or radio spectrum. { 'spek·trəm }

speculum alloy [MET] A brilliant white, hard, brittle alloy composed of copper and tin in a 2:1 proportion and sometimes with additions of other elements. { 'spek·yə·ləm 'al,ói }

speed of travel [MET] The speed at which a weld is made along its longitudinal axis; measured in inches or spots per minute. { 'spēd əv 'trav·əl }

speiss [MET] A mixture of impure metal arsenides and antimonides resulting from the smelting of certain ores such as cobalt and lead. { spīs }

spelter [MET] A commercially pure grade of zinc used in galvanizing; contains lead or iron as impurities. { 'spel·tər }

spelter solder [MET] Brass composed of equal parts of copper and zinc; used in brazing as a filler metal. Also known as brazing brass. { 'spel·tər ,säd·ər }

spent iron sponge [MATER] Iron sponge saturated with sulfur; prone to spontaneous heating. Also known as spent oxide. { 'spent 'ī·ərn ,spänj }

spent oxide See spent iron sponge. { 'spent 'äk ,sīd }

speromagnetic state [SOLID STATE] The condition of a rare-earth glass in which the spins are oriented in fixed directions which are more or less random because of electric fields which exist in the glass. { ¦spir·ə·mag¦ned·ik 'stāt }

sp gr See specific gravity.

sphenoid [CRYSTAL] An open crystal, occurring in monoclinic crystals of the sphenoidal class, and characterized by two nonparallel faces symmetrical with an axis of twofold symmetry. { 'sfē ,nóid }

spherand [ORG CHEM] A macrocyclic compound capable of completely enveloping a cation, having donor atoms (O, N, S) arranged such that they provide a solvation sphere for the encapsulated cation. { 'sfir·ənd }

spherical powder [MATER] A powder consisting of globular-shaped particles. { 'sfir·ə·kəl 'paüd·ər }

spheroidal graphite cast iron See nodular cast iron. { sfir'óid·əl ¦gra,fīt 'kast 'ī·ərn }

spheroidal group [CRYSTAL] A group in the tetragonal symmetry system; the sphenoid is the typical form. { sfir'óid·əl 'grüp }

spheroidized carbides [MET] Globular forms of carbide, as formed in spheroidized steel. { 'sfir·ə,dīzd 'kär,bīdz }

spheroidized steel [MET] Steel that has been

heat-treated to produce a spheroidized carbide structure. { 'sfir·ə,dīzd 'stēl }

spheroidizing [MET] Heating steels just below Ae₁ until the shape of cementite particles becomes relatively spherical. { 'sfir·ə,dīz·iŋ }

spider [MET] In founding, a device that consists of a frame with radiating arms or members and is used for strengthening a core or mold. { 'spīd·ər }

spiegeleisen [MET] An iron alloy containing 15–30% manganese and 5% carbon used in steelmaking. { 'spē·gə,līz·ən }

spike [SOLID STATE] A sputtering event in which the process from impact of a bombarding projectile to the ejection of target atoms involves motion of a large number of particles in the target, so that collisions between particles become significant. { spīk }

spincasting [ENG] A technique for manufacturing telescope mirrors in which molten glass is poured into a rotating mold and, as the glass cools and solidifies, the surface of the relatively thin mirror takes on a shape that is relatively close to the desired one, reducing substantially the need for grinding away excess glass. { 'spin ,kast·iŋ }

spin-density wave [SOLID STATE] The ground state of a metal in which the conduction-electron spin density has a sinusoidal variation in space. { 'spin ¦den·səd·ē 'wāv }

spin electronics See magnetoelectronics. { 'spin ,i·lek,trän·iks }

spin glass [SOLID STATE] A substance in which the atomic spins are oriented in random but fixed directions. { 'spin ,glas }

spin label [PHYS CHEM] A molecule which contains an atom or group of atoms exhibiting an unpaired electron spin that can be detected by electron spin resonance (ESR) spectroscopy and can be bonded to another molecule. { 'spin ,lā·bəl }

spin-lattice interaction [SOLID STATE] The state of a solid when the energy of electron spins is being shared with the thermal-vibration energy of the solid as a whole. { 'spin ¦lad·əs ,in·tə ,rak·shən }

spin-lattice relaxation [SOLID STATE] Magnetic relaxation in which the excess potential energy associated with electron spins in a magnetic field is transferred to the lattice. { 'spin ¦lad·əs ,rē,lak ,sā·shən }

spin magnetism [SOLID STATE] Paramagnetism or ferromagnetism that arises from polarization of electron spins in a substance. { 'spin ,mag·nə ,tiz·əm }

spin paramagnetism [SOLID STATE] Paramagnetism that arises from the electron spins in a substance. { 'spin ,par·ə'mag·nə,tiz·əm }

spin-polarized low-energy electron diffraction [SOLID STATE] A version of low-energy electron diffraction in which electrons in the incident beam have their spins aligned in one direction; used in studies of the magnetic properties of atoms near the surface of a material. { 'spin ¦pō·lə,rīzd ¦lō 'en·ər·jē i'lek,trän di,frak·shən }

spin resonance See magnetic resonance. { 'spin ,rez·ən·əns }

spin-spin relaxation [SOLID STATE] Magnetic relaxation, observed after application of weak magnetic fields, in which the excess potential energy associated with electron spins in a magnetic field is redistributed among the spins, resulting in heating of the spin system. { 'spin 'spin ,rē,lak,sā·shən }

spin temperature [SOLID STATE] For a system of electron spins in a lattice, a temperature such that the population of the energy levels of the spin system is given by the Boltzmann distribution with this temperature. { 'spin ,tem·prə·chər }

spintronics See magnetoelectronics. { spin 'trän·iks }

spin wave [SOLID STATE] A sinusoidal variation, propagating through a crystal lattice, of that angular momentum which is associated with magnetism (mostly spin angular momentum of the electrons). { 'spin ,wāv }

spirit stain [MATER] A dye dissolved in methylated spirits; used to stain wood surfaces. { 'spir·ət ¦stān }

spirit varnish [MATER] An artificial varnish consisting of resin, asphalt, or a cellulose ester dissolved in a volatile solvent. { 'spir·ət ,vär·nish }

splicing tape [MATER] A pressure-sensitive nonmagnetic tape used for splicing magnetic tape and motion picture film; it has a hard adhesive that will not ooze and gum up the equipment or cause adjacent layers of tape or film on the reel to stick together. { 'splīs·iŋ ,tāp }

split die [MET] **1.** A screw-thread die made in one piece with a longitudinal slit connecting the outside to the central hole which allows size adjustment. **2.** See segment die. { 'split 'dī }

split interstitial [CRYSTAL] A crystal defect in which a displaced atom forms a bond with a normal atom in such a way that neither atom is on the normal site but the two are symmetrically displaced from it. { 'split ,in·tər'stish·əl }

sponge gold See cake of gold. { 'spənj ¦gōld }

sponge iron [MET] Iron in porous or powder form made without fusion by heating iron ore in a reducing gas or with charcoal. { 'spənj ¦ī·ərn }

sponge metal [MET] Any porous metal made by decomposition or reduction of a compound without melting. { 'spənj ¦med·əl }

sponge rubber See rubber sponge. { 'spənj ¦rəb·ər }

spontaneous magnetization [ELECTROMAG] Magnetization which a substance possesses in the absence of an applied magnetic field. { spän'tā·nē·əs ,mag·nə·də'zā·shən }

spontaneous process [THERMO] A thermodynamic process which takes place without the application of an external agency, because of the inherent properties of a system. { spän'tā·nē·əs 'prä·səs }

spot welding [MET] Resistance welding in which fusion is localized in small circular areas; sometimes also accomplished by various arc-welding processes. { 'spät ,weld·iŋ }

spray quenching |MET| Rapid cooling in a spray of water or oil. { 'sprā ,kwench·iŋ }

spray transfer |MET| In arc welding, transfer of filler metal across the arc to the workpiece in the form of droplets. { 'sprā ,tranz·fər }

spreadable life *See* pot life. { 'spred·ə·bəl ,līf }

spreading coefficient |THERMO| The work done in spreading one liquid over a unit area of another, equal to the surface tension of the stationary liquid, minus the surface tension of the spreading liquid, minus the interfacial tension between the liquids. { 'spred·iŋ ,kō·i,fish·ənt }

springback |MET| **1.** Return of a metal part to its original shape after release of stress. **2.** The degree to which a metal returns to its original shape after forming operations. **3.** In flash, upset or pressure welding, the deflection in the welding machine caused by the upset pressure. { 'spriŋ,bak }

spring brass |MET| Common brass containing 70–72% copper which has been cold-worked to make it stiffer. { 'spriŋ ,bras }

spring modulus |MECH| The additional force necessary to deflect a spring an additional unit distance; if a certain spring has a modulus of 100 newtons per centimeter, a 100-newton weight will compress it 1 centimeter, a 200-newton weight 2 centimeters, and so on. { 'spriŋ 'mäj·ə·ləs }

spring steel |MET| Carbon or low-alloy steel which can be processed to give it the hardness and yield strength needed in springs. { 'spriŋ ¦stēl }

spring temper |MET| **1.** A steel temper characterized by an increased upper limit of elasticity; obtained by hardening and tempering in the usual way, then reheating until the steel turns blue. **2.** A similar temper in brass obtained by cold rolling. { 'spriŋ ,tem·pər }

spun glass |MATER| Blown glass made of fine threads of glass. { 'spən 'glas }

sputtering |ELECTR| **1.** Also known as cathode sputtering. **2.** The ejection of atoms or groups of atoms from the surface of the cathode of a vacuum tube as the result of heavy-ion impact. **3.** The use of this process to deposit a thin layer of metal on a glass, plastic, metal, or other surface in vacuum. { 'spəd·ə·riŋ }

sp vol *See* specific volume.

square groove weld |MET| A groove weld in which the joint edges are square. { 'skwer ¦grüv ,weld }

squeeze time |MET| In resistance welding, the time between the initial application of the current and of the pressure. { 'skwēz ,tīm }

Sr *See* strontium.

SS *See* stainless steel.

stability |CHEM| The property of a chemical compound which is not readily decomposed and does not react with other compounds. |PHYS| **1.** The property of a system which does not undergo any change without the application of an external agency. **2.** The property of a system in which any departure from an equilibrium state gives rise to forces or influences which tend to return the system to equilibrium. Also known as static stability. { stə'bil·əd·ē }

stabilizer |MATER| Any substance that tends to maintain the physical and chemical properties of a material. { 'stā·bə,līz·ər }

stabilizing treatment |MET| Any of various treatments intended to promote dimensional stability of a metal part or stabilize the structure of an alloy. { 'stā·bə,līz·iŋ ,trēt·mənt }

stable |PHYS| Not subject to any change without the application of an external agency, such as radiation; said of a molecule, atom, nucleus, or elementary particle. { 'stā·bəl }

stable-base film |MATER| A particular type of film having high stability in regard to shrinkage and stretching. { 'stā·bəl ¦bās 'film }

stable system |PHYS| A system that returns to a stationary state following sufficiently small perturbations. { ¦stā·bəl ¦sis·təm }

stack |MET| The cone-shaped section of a blast furnace or cupola above the hearth and melting zone and extending to the throat. { stak }

stack cutting |MET| Cutting a stack of metal plates with a single cut using a stream of oxygen. { 'stak ,kəd·iŋ }

stacking fault |CRYSTAL| A defect in a face-centered cubic or hexagonal close-packed crystal in which there is a change from the regular sequence of positions of atomic planes. { 'stak·iŋ ,fólt }

stack welding |MET| Simultaneous spot welding of stacked plates. { 'stak ,weld·iŋ }

Staebler-Wronski effect |SOLID STATE| A reversible change (usually a reduction) in the dark conductivity and photoconductivity of hydrogenated amorphous silicon during and following illumination by light with sufficient energy to produce electron-hole pairs. { ¦stāb·lər 'rän·skē i,fekt }

staggered-intermittent fillet welding |MET| Making a line of intermittent fillet welds on each side of a joint in a manner such that the increments on one side are not opposite to those on the other side. { 'stag·ərd ,in·tər'mit·ənt 'fil·ət ,weld·iŋ }

stain |MATER| **1.** A nonprotective coloring matter used on wood surfaces; imparts color without obscuring the wood grains. **2.** Any colored, organic compound used to stain tissues, cells, cell components, cell contents, or other biological substrates for microscopic examination. { stān }

stainless alloy |MET| Any of a large and complex group of corrosion-resistant iron-chromium alloys (containing 10% or more chromium), sometimes containing other elements, such as nickel, silicon, molybdenum, tungsten, and niobium. Commonly known as stainless steel (SS). { 'stān·ləs 'al,ói }

stainless-clad steel |MET| Steel clad on one or two sides with a stainless steel to provide a surface that is corrosion-resistant and attractive. { 'stān·ləs ¦klad 'stēl }

stainless iron *See* ferritic stainless steel. { 'stān·ləs 'ī·ərn }

stainless steel See stainless alloy. { 'stān·ləs 'stēl }

stand [MET] A set of rolls used in a metal-rolling process. { stand }

standard atmosphere See atmosphere. { 'stan·dərd 'at·mə,sfir }

standard conditions [PHYS] **1.** A temperature of 0°C and a pressure of 1 atmosphere (760 torr). Also known as normal temperature and pressure (NTP); standard temperature and pressure (STP). **2.** According to the American Gas Association, a temperature of 60°F (15⅗°C) and a pressure of 762 millimeters (30 inches) of mercury. **3.** According to the Compressed Gas Institute, a temperature of 20°C (68°F) and a pressure of 1 atmosphere. [SOLID STATE] The allotropic form in which a substance most commonly occurs. { 'stan·dərd kən'dish·ənz }

standard free-energy increase [THERMO] The increase in Gibbs free energy in a chemical reaction, when both the reactants and the products of the reaction are in their standard states. { 'stan·dərd 'frē ¦en·ər·jē 'in,krēs }

standard gold [MET] A gold alloy containing 10% copper; at one time used for legal coinage in the United States. { 'stan·dərd 'gōld }

standard heat of formation [THERMO] The heat needed to produce one mole of a compound from its elements in their standard state. { 'stan·dərd 'hēt əv fór'mā·shən }

standard plane [CRYSTAL] The crystal plane whose Miller indices are (111), that is, whose intercepts on the crystal axes are proportional to the corresponding sides of a unit cell. { 'stan·dərd 'plān }

standard sand [MATER] A silica sand that is free of organic matter and is used in making test samples of concrete and cement. { 'stan·dərd 'sand }

standard temperature and pressure See standard conditions. { 'stan·dərd 'tem·prə·chər ən 'presh·ər }

stannane See tin hydride. { 'sta,nān }

stannic acid See stannic oxide. { 'stan·ik 'as·əd }

stannic anhydride See stannic oxide. { 'stan·ik an'hī,drīd }

stannic bromide [INORG CHEM] SnBr₄ Water- and alcohol-soluble, white crystals that fume when exposed to air, and melt at 31°C; used in mineral separations. Also known as tin bromide; tin tetrabromide. { 'stan·ik 'brō,mīd }

stannic chloride [INORG CHEM] SnCl₄ A colorless, fuming liquid; soluble in cold water, alcohol, carbon disulfide, and oil of turpentine; decomposed by hot water; boils at 114°C; used as a conductive coating and a sugar bleach, and in drugs, ceramics, soaps, and blueprinting. Also known as tin chloride; tin tetrachloride. { 'stan·ik 'klòr,īd }

stannic chromate [INORG CHEM] Sn(CrO₄)₂ Toxic, brownish-yellow crystals, slightly soluble in water; used to decorate porcelain and china. Also known as tin chromate. { 'stan·ik 'krō ,māt }

stannic iodide [INORG CHEM] SnI₄ Yellow-reddish crystals; insoluble in water, soluble in alcohol, ether, chloroform, carbon disulfide, and benzene; decomposed by water, melt at 144°C, sublime at 180°C. Also known as tin iodide; tin tetraiodide. { 'stan·ik 'ī·ə,dīd }

stannic oxide [INORG CHEM] SnO₂ A white powder; insoluble in water, soluble in concentrated sulfuric acid; melts at 1127°C; used in ceramic glazes and colors, special glasses, putty, and cosmetics, and as a catalyst. Also known as flowers of tin; stannic acid; stannic anhydride; tin dioxide; tin oxide; tin peroxide. { 'stan·ik 'äk ,sīd }

stannic sulfide [INORG CHEM] SnS₂ A yellow-brown powder; insoluble in water, soluble in alkaline sulfides; decomposes at red heat; used as a pigment and for imitation gilding. Also known as artificial gold; mosaic gold; tin bisulfide. { 'stan·ik 'səl,fīd }

stannous bromide [INORG CHEM] SnBr₂ A yellow powder; soluble in water, alcohol, acetone, ether, and dilute hydrochloric acid; browns in air; melts at 215°C. Also known as tin bromide. { 'stan·əs 'brō,mīd }

stannous chloride [INORG CHEM] SnCl₂ White crystals; soluble in water, alcohol, and alkalies; oxidized in air to the oxychloride; melt at 247°C; used as a chemical intermediate, reducing agent, and ink-stain remover, and for silvering mirrors. Also known as tin chloride; tin crystals; tin dichloride; tin salts. { 'stan·əs 'klòr,īd }

stannous chromate [INORG CHEM] SnCrO₄ A brown powder; very slightly soluble in water; used to decorate porcelain. Also known as tin chromate. { 'stan·əs 'krō,māt }

stannous fluoride [INORG CHEM] SnF₂ A white, lustrous powder; slightly soluble in water; used to fluoridate toothpaste and as a medicine. { 'stan·əs 'flùr,īd }

stannous oxide [INORG CHEM] SnO An air-unstable, brown to black powder; insoluble in water, soluble in acids and strong bases; decomposes when heated; used as a reducing agent and chemical intermediate, and for glass plating. Also known as tin oxide; tin protoxide. { 'stan·əs 'äk,sīd }

stannous sulfate [INORG CHEM] SnSO₄ Heavy light-colored crystals; decomposes rapidly in water, loses SO₂ at 360°C; used for dyeing and tin plating. Also known as tin sulfate. { 'stan·əs 'səl,fāt }

stannous sulfide [INORG CHEM] SnS Dark crystals; insoluble in water, soluble (with decomposition) in concentrated hydrochloric acid; melts at 880°C; used as an analytical reagent and catalyst, and in bearing material. Also known as tin monosulfide; tin protosulfide; tin sulfide. { 'stan·əs 'səl,fīd }

stannum [CHEM] The Latin name for tin, thus the symbol Sn for the element. { 'stan·əm }

Stanton number [THERMO] A dimensionless number used in the study of forced convection, equal to the heat-transfer coefficient of a fluid divided by the product of the specific heat at constant pressure, the fluid density, and the fluid velocity. Symbolized N_{St}. Also known as Margoulis number (M). { 'stant·ən ,nəm·bər }

starburst polymer See dendrimer. { ,stär,bərst 'päl·ə·mər }

star grain [MATER] A rocket propellant grain with a cavity of star-shaped cross section. { 'stär ,grān }

Stark-Einstein law See Einstein photochemical equivalence law. { 'stärk 'īn,stīn ,lȯ }

Stark number See Stefan number. { 'stärk ,nəm·bər }

star polymer [POLYM CHEM] A macromolecule having a small core of molecules (branch point) with branches radiating from the core. { 'stär ¦päl·ə·mər }

starting mix [MATER] In pyrotechnic devices, an easily ignited mixture which transmits flame from an initiating device to a less readily ignitable composition. { 'stärd·iŋ ,miks }

starting sheet [MET] A thin sheet of metal used as the initial cathode in electrowinning or electrorefining. { 'stärd·iŋ ,shēt }

state parameter See thermodynamic function of state. { 'stāt pə,ram·əd·ər }

state variable See thermodynamic function of state. { 'stāt ,ver·ē·ə·bəl }

static charge [ELEC] An electric charge accumulated on an object. { 'stad·ik 'chärj }

static discharger [ELEC] A rubber-covered cloth wick about 6 inches (15 centimeters) long, sometimes attached to the trailing edges of the surfaces of an aircraft to discharge static electricity in flight. { 'stad·ik 'dis,chär·jər }

static electricity [ELEC] **1.** The study of the effects of macroscopic charges, including the transfer of a static charge from one object to another by actual contact or by means of a spark that bridges an air gap between the objects. **2.** See electrostatics. { 'stad·ik ,i,lek'tris·əd·ē }

static equilibrium See equilibrium. { 'stad·ik ,ē·kwə'lib·rē·əm }

static load [MECH] A nonvarying load; the basal pressure exerted by the weight of a mass at rest, such as the load imposed on a drill bit by the weight of the drill-stem equipment or the pressure exerted on the rocks around an underground opening by the weight of the superimposed rocks. Also known as dead load. { 'stad·ik 'lōd }

static magnetism See magnetostatics. { 'stad·ik 'mag·nə,tiz·əm }

static stability See stability. { 'stad·ik stə'bil·əd·ē }

statuary bronze [MET] Special bronze alloys used for casting statues and other ornamental objects; a typical bronze for statuary work contains 90% copper, 6% tin, 3% zinc, and 1% lead. { 'stach·ə,wer·ē ¦bränz }

steadite [MET] A hard structural constituent of cast iron consisting of the eutectic of ferrite and iron phosphide (Fe_3P); composition of the eutectic is 10.2% phosphorus and 89.8% iron; melts at 1049°C (1920°F). { 'ste,dīt }

steady-state conduction [THERMO] Heat conduction in which the temperature and heat flow at each point does not change with time. { 'sted·ē ¦stāt kən'dək·shən }

steady-state creep See secondary creep. { 'sted·ē ¦stāt 'krēp }

steady-state vibration [MECH] Vibration in which the velocity of each particle in the system is a continuous periodic quantity. { 'sted·ē ¦stāt vī'brā·shən }

steam blow See blister. { 'stēm ,blō }

steam bronze [MET] A leaded tin-bronze containing 88% copper, 6% tin, 4.5% zinc, 1.5% lead; used for steam valve bodies, gears, and bearings. { 'stēm ,bränz }

steam cycle See Rankine cycle. { 'stēm ,sī·kəl }

steam-hammer oil See tempering oil. { stēm ham·ər ,ȯil }

steam line [THERMO] A graph of the boiling point of water as a function of pressure. { 'stēm ,līn }

steam point [THERMO] The boiling point of pure water whose isotopic composition is the same as that of sea water at standard atmospheric pressure; it is assigned a value of 100°C on the International Practical Temperature Scale of 1968. { 'stēm ,pȯint }

steam-refined asphalt [MATER] Petroleum-derived asphalt that has been refined in the presence of steam during the distillation of crude oil. { 'stēm ri¦fīnd 'as,fȯlt }

Steckel rolling [MET] Cold metal rolling in which the strip is pulled through idler rolls by front tension only; the direction is reversed until the desired thickness is attained. { 'stek·əl ,rōl·iŋ }

steel [MET] An iron base alloy, malleable under proper conditions, containing up to about 2% carbon. { stēl }

steel bronze [MET] A hardened bronze consisting of 92% copper and 8% tin; used as a substitute for steel in guns. { 'stēl 'bränz }

steel converter [MET] A retort in which cast iron is converted to steel; an example is the Bessemer converter. { 'stēl kən,vərd·ər }

steel emery [MET] An abrasive material composed of chilled iron produced by forcing iron through a steam jet; used in tumbling barrels and for grinding stones. { 'stēl 'em·ə·rē }

steel foil [MET] A very thin sheet of steel, the thickness of which is measured in thousandths of an inch. { 'stēl 'fȯil }

steelmaking [MET] Any of various processes for making steel from pig iron. { 'stēl,māk·iŋ }

steel wool [MET] Fine steel threads matted into a mass. { 'stēl 'wùl }

Stefan number [THERMO] A dimensionless number used in the study of radiant heat transfer, equal to the Stefan-Boltzmann constant times the cube of the temperature times the thickness of a layer divided by the layer's thermal conductivity. Symbolized St. Also known as Stark number (Sk). { 'shte,fän ,nəm·bər }

Steinmetz coefficient [ELECTROMAG] The constant of proportionality in Steinmetz's law. { 'stīn,mets ,kō·i,fish·ənt }

Steinmetz's law [ELECTROMAG] The energy converted into heat per unit volume per cycle during a cyclic change of magnetization is proportional to the maximum magnetic induction raised to the 1.6 power, the constant of proportionality depending only on the material. { 'stīn,mets·əz ,lō }

stellar crystal See plane-dendritic crystal. { 'stel·ər 'krist·əl }

stellite [MET] A hard, wear- and corrosion-resistant family of nonferrous alloys of cobalt (20–65), chromium (11–32), and tungsten (2–5); resistance to softening is exceptionally high at high temperature. { 'ste,līt }

STEM See scanning transmission electron microscope. { stem }

stem correction [THERMO] A correction which must be made in reading a thermometer in which part of the stem, and the thermometric fluid within it, is at a temperature which differs from the temperature being measured. { 'stem kə ,rek·shən }

stench See odorant. { stench }

step brazing [MET] Brazing consecutive joints at sequentially lower temperatures to maintain the integrity of preceding joints. { 'step ,brāz·iŋ }

step soldering [MET] Soldering consecutive joints at sequentially lower temperatures to maintain the integrity of preceding joints. { 'step ,säd·ə·riŋ }

stereocenter [ORG CHEM] A (chiral) carbon atom that has four different substituents bonded to it. Also known as stereogenic atom. { 'ster·ē·ō,sen·tər }

stereogenic atom See stereocenter. { ,ster·ē·ə ,jen·ik 'ad·əm }

stereogenic center See asymmetric carbon atom. { ,ster·ē·ə¦jen·ik 'sen·tər }

stereognomogram [CRYSTAL] The projection resulting from the superposition of the projection planes of a stereogram and a gnomogram. { ¦ster·ē·ə'nō·mə,gram }

stereographic projection [CRYSTAL] A method of displaying the positions of the poles of a crystal in which poles are projected through the equatorial plane of the reference sphere by lines joining them with the south pole for poles in the upper hemisphere, and with the north pole for poles in the lower hemisphere. { ¦ster·ē·ə ¦graf·ik prə'jek·shən }

stereoisomers [ORG CHEM] Compounds whose molecules have the same number and kind of atoms and the same atomic arrangement, but differ in their spatial relationship. { ¦ster·ē·ō'ī·sə·mərz }

stereoregular polymer [POLYM CHEM] A polymer with a regularity or symmetry in the structural arrangement of its molecules. See tactic polymer. { ¦ster·ē·ə'reg·yə·lər 'päl·i·mər }

stereorubber [POLYM CHEM] Synthetic rubber, cis-polyisoprene, a polymer with stereospecificity. { 'ster·ō·ə,rəb·ər }

stereoselective reaction [ORG CHEM] A chemical reaction in which one stereoisomer is produced or decomposed more rapidly than another. Also known as enantioselective reaction. { ¦ster·ē·ə·si'lek·tiv rē'ak·shən }

stereospecificity [ORG CHEM] The condition of a polymer whose molecular structure has a fixed spatial (geometric) arrangement of its constituent atoms, thus having crystalline properties; for example, synthetic natural rubber, cis-polyisoprene. { ¦ster·ē·ō·spə·sə'fis·ə d·ē }

stereospecific synthesis [ORG CHEM] Catalytic polymerization of monomer molecules to produce stereospecific polymers, as with Ziegler or Natta catalysts (derived from a transition metal halide and a metal alkyl). { ¦ster·ē·ō·spə'sif·ik 'sin·thə·səs }

steric effect [PHYS CHEM] The influence of the spatial configuration of reacting substances upon the rate, nature, and extent of reaction. { 'ster·ik i,fekt }

steric hindrance [ORG CHEM] The prevention or retardation of chemical reaction because of neighboring groups on the same molecule; for example, ortho-substituted aromatic acids are more difficult to esterify than are the meta and para substitutions. { 'ster·ik 'hin·drəns }

sterling silver [MET] A silver alloy having a defined standard of purity of 92.5% silver and the remaining 7.5% usually of copper. { 'stər·liŋ 'sil·vər }

sterrometal [MET] Hard brass containing a small amount of iron and manganese; used for hydraulic cylinders and marine castings. { 'ste·rō ,med·əl }

Stetefeldt furnace [MET] A furnace for desulfurizing and chloridizing silver ores; the ores are powdered, mixed with salt, and dropped through a hot atmosphere. { 'sted·ə,felt ,fər·nəs }

stibide See antimonide. { 'sti,bīd }

stibium [CHEM] The Latin name for antimony, thus the symbol Sb for the element. { 'stib·ē·əm }

stibnate See potassium antimonate. { 'stib ,nāt }

sticking coefficient [PHYS CHEM] The fraction of all atoms incident on a surface that are adsorbed on the surface. { 'stik·iŋ ,kō·i,fish·ənt }

stick shellac [MATER] Shellac in the form of a solid stick and in a variety of colors; used for filling imperfections in wood. { 'stik shə,lak }

stiffness [MECH] The ratio of a steady force acting on a deformable elastic medium to the resulting displacement. { 'stif·nəs }

stiffness constant [MECH] Any one of the coefficients of the relations in the generalized Hooke's law used to express stress components as linear functions of the strain components. Also known as elastic constant. { 'stif·nəs ,kän·stənt }

stir-in resin [MATER] A vinyl resin that does not require grinding in order to disperse in a plastisol or organisol. { 'stir¦in 'rez·ən }

Stirling cycle [THERMO] A regenerative thermodynamic power cycle using two isothermal and two constant volume phases. { 'stir·liŋ ,sī·kəl }

stirring effect [ELECTROMAG] The circulation in a molten metal carrying electric current as a result of the combined forces of the pinch and motor effects. { 'stər·iŋ i,fekt }

stitch welding |MET| A series of spaced spot resistance welds. { 'stich ,weld·iŋ }

Stoddard solvent |MATER| A petroleum naphtha product with a comparatively narrow boiling range; used mostly for dry cleaning. { 'städ·ərd ,säl·vənt }

stoichiometry |PHYS CHEM| The numerical relationship of elements and compounds as reactants and products in chemical reactions. { ,stói·kē'äm·ə·trē }

stoking |MET| Presintering or sintering a metal powder in such a way as to advance the compacts through the furnace at a fixed rate. Also known as continuous sintering. { 'stōk·iŋ }

stoneware |MATER| Vitrified ware with impermeable surface; used for corrosive materials in the laboratory and for some industrial operations. { 'stōn,wer }

stopoff See resist. { 'stäp,óf }

stopping off |MET| **1.** Local deposition of a protective coating, such as copper or fireclay, to prevent carburization, decarburization, or nitriding during heat treatment. **2.** Filling up a portion of the mold cavity to keep out molten metal. **3.** Applying a nonconducting layer to avoid electrodeposition in certain areas. { 'stäp·iŋ 'óf }

storage battery |ELEC| A connected group of two or more storage cells or a single storage cell. Also known as accumulator; accumulator battery; rechargeable battery; secondary battery. { 'stòr·ij ,bad·ə·rē }

storage cell |ELEC| An electrolytic cell for generating electric energy, in which the cell after being discharged may be restored to a charged condition by sending a current through it in a direction opposite to that of the discharging current. Also known as secondary cell. { 'stòr·ij ,sel }

storage factor See Q. { 'stór·ij ,fak·tər }

stored-energy welding |MET| Welding by means of energy accumulated electrostatically, electrochemically, or electromagnetically at a low rate. { 'stòrd ¦en·ər·jē 'weld·iŋ }

STPP See sodium tripolyphosphate.

straight polarity |MET| Arrangement of an arc welding circuit in which the electrode is connected to the negative terminal. { 'strät pə'lar·əd·ē }

strain |MECH| Change in length of an object in some direction per unit undistorted length in some direction, not necessarily the same; the nine possible strains form a second-rank tensor. { strān }

strain aging |MET| Change of mechanical properties of a metal by aging induced by plastic deformation. { 'strān ,āj·iŋ }

strain energy |MECH| The potential energy stored in a body by virtue of an elastic deformation, equal to the work that must be done to produce this deformation. { 'strān ,en·ər·jē }

strain gage |ENG| A device which uses the change of electrical resistance of a wire under strain to measure pressure. { 'strān ,gāj }

strain hardening |MET| Increasing the hardness and tensile strength of a metal by cold plastic deformation. { 'strān ,härd·ən·iŋ }

strain rate |MECH| The time rate for the usual tensile test. { 'strān ,rāt }

strain rosette |MECH| A pattern of intersecting lines on a surface along which linear strains are measured to find stresses at a point. { 'strān rō ,zet }

Strasbourg turpentine |MATER| A balsam from the European white fir; a heavy, thick material, it is sometimes used in painting mediums and glazes, but an excessive amount causes smearing and slow drying of the paint. { 'stras,bərg 'tər·pən,tīn }

stratified film |PHYS CHEM| A film in which two thicknesses are present in a fixed configuration for a significant period of time. { 'strad·ə,fīd 'film }

strawboard |MATER| A type of pasteboard in which straw and a bonding material are incorporated with other paper materials having a low cellulose content. Also known as compressed straw slab. { 'strò,bórd }

stray current corrosion |MET| Corrosion of metals caused by a stray current. { 'strā ¦kə·rənt kə'rō·zhən }

strength |MECH| The stress at which material ruptures or fails. { streŋkth }

stress |MECH| The force acting across a unit area in a solid material resisting the separation, compacting, or sliding that tends to be induced by external forces. { stres }

stress analysis |PHYS| The determination of the stresses produced in a solid body when subjected to various external forces. { 'stres ə,nal·ə·səs }

stress concentration |MECH| A condition in which a stress distribution has high localized stresses; usually induced by an abrupt change in the shape of a member; in the vicinity of notches, holes, changes in diameter of a shaft, and so forth, maximum stress is several times greater than where there is no geometrical discontinuity. { 'stres ,kän·sən,trā·shən }

stress concentration factor |MECH| A theoretical factor K_t expressing the ratio of the greatest stress in the region of stress concentration to the corresponding nominal stress. { 'stres ,kän·sən ¦trā·shən ,fak·tər }

stress corrosion |MET| Corrosion that is accelerated by stress, applied or residual, in a metal. { 'stres kə,rō·zhən }

stress-corrosion cracking |MET| Failure by cracking under the conjoint action of a constant tensile stress, which is applied to residual, in certain chemical environments specific to the metal. { 'stres kə¦rō·zhən ,krak·iŋ }

stress crack |MECH| An external or internal crack in a solid body (metal or plastic) caused by tensile, compressive, or shear forces. { 'stres ,krak }

stress intensity |MECH| Stress at a point in a structure due to pressure resulting from combined tension (positive) stresses and compression (negative) stresses. { 'stres in,ten·səd·ē }

stress lines See isostatics. { 'stres ,līnz }

stress-optic law [OPTICS] In a transparent, isotropic plate subjected to a biaxial stress field, the relative retardation R_t between the two components produced by temporary double refraction is equal to $Ct(p - q)$, which in turn is equal to $n\lambda$; C is the stress-optic coefficient, t the plate thickness, p and q the principal stresses, n the number of fringes which have passed the point during application of the load, and λ the wavelength of the light. { 'stres 'äp·tik ,lò }

stress raiser [MET] A notch, hole, or other discontinuity in contour or structure which causes localized stress concentration. { 'stres ,rā·zər }

stress range [MECH] The algebraic difference between the maximum and minimum stress in one fatigue test cycle. { 'stres ,rānj }

stress ratio [MECH] The ratio of minimum to maximum stress in fatigue testing, considering tensile stresses as positive and compressive stresses as negative. { 'stres ,rā·shō }

stress relief cracking [MET] Cracking between metal grains in the heat-affected zone of a weldment during exposure to high temperatures. { 'stres ri¦lēf ,krak·iŋ }

stress relieving [MET] Low-temperature heating to reduce residual stress. { 'stres ri,lēv·iŋ }

stress-strain curve See deformation curve. { 'stres 'strān ,kərv }

stress trajectories See isostatics. { 'stres trə ,jek·trēz }

stretcher leveling [MET] Removing warp and distortion from a piece of metal by gripping it at each end and subjecting it to stress beyond the yield strength. Also known as patent leveling. { 'strech·ər 'lev·ə·liŋ }

stretcher strains See Lüders' lines. { 'strech·ər ,strānz }

strike [MET] **1.** A very thin, initially electroplated film or the plating solution with which to deposit such a film. **2.** A local crater in a metal surface due to accidental contact with the welding electrode. { strīk }

strike plating [MET] Applying a thin electroplated film prior to depositing the principal electroplate. { 'strīk ,plād·iŋ }

string [MECH] A solid body whose length is many times as large as any of its cross-sectional dimensions, and which has no stiffness. { striŋ }

string bead [MET] A continuous weld bead made without appreciable transverse oscillation. { 'striŋ ,bēd }

string electrometer [ENG] An electrometer in which a conducting fiber is stretched midway between two oppositely charged metal plates; the electrostatic field between the plates displaces the fiber laterally in proportion to the voltage between the plates. { 'striŋ ,i,lek'träm·əd·ər }

stringer [MET] An elongated mass of microconstituents or foreign material in wrought metal oriented in the direction of working. { 'striŋ·ər }

string galvanometer [ENG] A galvanometer consisting of a silver-plated quartz fiber under tension in a magnetic field, used to measure oscillating currents. Also known as Einthoven galvanometer. { 'striŋ ,gal·və'näm·əd·ər }

strip [MATER] A long, narrow piece of rigid material of uniform width. { strip }

stripe See stripe phase. { strīp }

stripe phase [SOLID STATE] A phase exhibited by the electrons in certain solid systems, such as doped antiferomagnets and quantum Hall systems, in which the behavior of the electrons is similar to that of the molecules in a liquid crystal, exhibiting properties of both a crystal (orientational order and anisotropy) and a liquid (absence of space periodicity). Also known as stripe. { strīp ,fāz }

stripper punch [MET] In powder metallurgy, a device used as the bottom or top of the die cavity which can be pushed into the die to eject the formed compact. { 'strip·ər ,pənch }

stripping [MET] Removing a coating from the surface of a metal. { 'strip·iŋ }

strong acid [CHEM] An acid with a high degree of dissociation in solution, for example, mineral acids, such as hydrochloric acid, HCl, sulfuric acid, H_2SO_4, or nitric acid, HNO_3. { 'stròŋ 'as·əd }

strong base [CHEM] A base with a high degree of dissociation in solution, for example, sodium hydroxide, NaOH, potassium hydroxide, KOH. { 'stròŋ 'bās }

strontia See strontium oxide. { 'strän·chə }

strontium [CHEM] A metallic element in group II, symbol Sr, with atomic number 38, atomic weight 87.62; flammable, soft, pale-yellow solid; soluble in alcohol and acids, decomposes in water; melts at 770°C, boils at 1380°C; chemistry is similar to that of calcium; used as electron-tube getter. { 'strän·tē·əm }

strontium bromide [INORG CHEM] $SrBr_2·6H_2O$ A white, hygroscopic powder soluble in water and alcohol; loses water at 180°C, melts at 643°C; used in medicine and as an analytical reagent. { 'strän·tē·əm 'brō,mīd }

strontium carbonate [INORG CHEM] $SrCO_3$ A white powder slightly soluble in water, decomposes at 1340°C; used to make TV-tube glass, strontium salts, and ceramic ferrites, and in pyrotechnics. { 'strän·tē·əm 'kär·bə,nāt }

strontium chlorate [INORG CHEM] $Sr(ClO_3)_2$ Shock-sensitive, highly combustible, white, water-soluble crystals that decompose at 120°C; used in pyrotechnics and tracer bullets. { 'strän·tē·əm 'klòr,āt }

strontium chloride [INORG CHEM] $SrCl_2$ Water- and alcohol-soluble white crystals, melts at 872°C; used in medicine and pyrotechnics and to make strontium salts. { 'strän·tē·əm 'klòr ,īd }

strontium chromate [INORG CHEM] $SrCrO_4$ A light yellow, rust- and corrosion-resistant pigment used in metal coatings and for pyrotechnics. { 'strän·tē·əm 'krō,māt }

strontium dioxide See strontium peroxide. { 'strän·tē·əm dī'äk,sīd }

strontium fluoride [INORG CHEM] SrF_2 A white powder, soluble in hydrochloric acid and hydrofluoric acid; used in medicine and for single crystals for lasers. { 'strän·tē·əm 'flùr,īd }

strontium hydrate See strontium hydroxide.
{ 'strän·tē·əm 'hī,drāt }

strontium hydroxide [INORG CHEM] $Sr(OH)_2$ Colorless deliquescent crystals that absorb carbon dioxide from air, soluble in hot water and acids, melts at $375°C$; used by the sugar industry, in lubricants and soaps, and as a plastic stabilizer. Also known as strontium hydrate. { 'strän·tē·əm hī'dräk,sīd }

strontium iodide [INORG CHEM] SrI_2 Air-yellowing, white crystals that decompose in moist air, melts at $515°C$; used in medicine and as a chemicals intermediate. { 'strän·tē·əm 'ī·ə ,dīd }

strontium monosulfide See strontium sulfide. { 'strän·tē·əm ,män·ə'səl,fīd }

strontium nitrate [INORG CHEM] $Sr(NO_3)_2$ A white, water-soluble powder melting at $570°C$; used in pyrotechnics, signals and flares, medicine, and matches, and as a chemicals intermediate. { 'strän·tē·əm 'nī,trāt }

strontium oxalate [INORG CHEM] $SrC_2O_4·H_2O$ A white powder that loses water at $150°C$; used in pyrotechnics and tanning. { 'strän·tē·əm 'äk·sə ,lāt }

strontium oxide [INORG CHEM] SrO A grayish powder, melts at $2430°C$, becomes the hydroxide in water; used in medicine, pyrotechnics, pigments, greases, soaps, and as a chemicals intermediate. Also known as strontia. { 'strän·tē·əm 'äk,sīd }

strontium peroxide [INORG CHEM] SrO_2 A strongly oxidizing, fire-hazardous, white, alcohol-soluble powder that decomposes in hot water; used in medicine, bleaching, and fireworks. Also known as strontium dioxide. { 'strän·tē·əm pər'äk,sīd }

strontium sulfate [INORG CHEM] $SrSO_4$ White crystals insoluble in alcohol, slightly soluble in water and concentrated acids, melts at $1605°C$; used in paper manufacture, pyrotechnics, ceramics, and glass. { 'strän·tē·əm 'səl,fāt }

strontium sulfide [INORG CHEM] SrS A gray powder with a hydrogen sulfide aroma in moist air, slightly soluble in water, soluble (with decomposition) in acids, melts above $2000°C$; used in depilatories and luminous paints and as a chemicals intermediate. Also known as strontium monosulfide. { 'strän·tē·əm 'səl,fīd }

strontium titanate [INORG CHEM] $SrTiO_3$ A solid material, insoluble in water and melting at $2060°C$; used in electronics and electrical insulation. { 'strän·tē·əm 'tīt·ən,āt }

structural adhesive [MATER] An adhesive capable of bearing loads of considerable magnitude; a structural adhesive will not fail when a bonded joint prepared from the thickness of metal, or other material typical for that industry, is stressed to its yield point. { 'strək·chə·rəl ad'hē·siv }

structural clay tile [MATER] Hollow burned-clay masonry unit with parallel cells used as facing tile, load-bearing tile, partition tile, fireproofing tile, furring tile, floor tile, and header tile. { 'strək·chə·rəl 'klā 'tīl }

structural concrete [MATER] A special type of concrete that is capable of carrying a structural load or forming an integral part of a structure. { 'strək·chə·rəl ,kän'krēt }

structural foam [MATER] A type of cellular plastic with a dense outer skin surrounding a foamed core. { 'strək·chə·rəl ,fōm }

structural formula [CHEM] A system of notation used for organic compounds in which the exact structure, if it is known, is given in schematic representation. { 'strək·chə·rəl 'fȯr·myə·lə }

structural glass [MATER] A vitreous material cast as cubes, rectangular blocks, and rectangular plates to be used as a decorative covering for masonry walls. { 'strək·chə·rəl 'glas }

structural shape [MET] A piece of metal of a standard design used in construction. { 'strək·chə·rəl 'shāp }

structural steel [MET] Steel used in engineering structures, usually manufactured by either open-hearth or the electric-furnace process. { 'strək·chə·rəl 'stēl }

structural tile [MATER] A hollow clay product which may be load-bearing or non-load-bearing; used for facing, flooring, or partitions. { 'strək·chə·rəl 'tīl }

structure amplitude [SOLID STATE] The absolute value of a structure factor. { 'strək·chər ,am·plə,tüd }

structure cell See unit cell. { 'strək·chər ,sel }

structure factor [SOLID STATE] A factor which determines the amplitude of the beam reflected from a given atomic plane in the diffraction of an x-ray beam by a crystal, and is equal to the sum of the atomic scattering factors of the atoms in a unit cell, each multiplied by an appropriate phase factor. { 'strək·chər ,fak·tər }

structure-sensitive property [SOLID STATE] A property of a substance that depends on impurities and the imperfections of the crystal structure. { ¦strək·chər ¦sen·sə·tiv 'präp·ərd·ē }

structure type [CRYSTAL] The structural arrangement of a crystal, regardless of the atomic elements present; it corresponds to the crystal's space group. { 'strək·chər ,tīp }

stucco [MATER] A smooth plasterlike material applied to the outside wall or other exterior surface of a building or structure. { 'stək·ō }

stud welding [MET] Arc-welding using the heat of an electric arc produced between a metal stud and another part, and then bringing the parts together under pressure. { 'stəd ,weld·iŋ }

styrene [ORG CHEM] $C_6H_5CH:CH_2$ A colorless, toxic liquid with a strong aroma; insoluble in water, soluble in alcohol and ether; polymerizes rapidly, can become explosive; boils at $145°C$; used to make polymers and copolymers, polystyrene plastics, and rubbers. Also known as phenylethylene; styrene monomer; vinylbenzene. { 'stī,rēn }

styrene-acrylonitrile resin [POLYM CHEM] A thermoplastic copolymer of styrene and acrylonitrile with good stiffness and resistance to scratching, chemicals, and stress. Also known as SAN. { 'stī,rēn ,ak·rə'län·ə·trəl 'rez·ən }

styrene-butadiene rubber [POLYM CHEM] The most common type of synthetic rubber, made by the copolymerization of styrene and butadiene monomers; used in tires, footwear, adhesives, and sealants. Also known as SBR. { 'stī,rēn ,byüd·ə'dī,ēn 'rəb·ər }

styrene monomer *See* styrene. { 'stī,rēn 'män·ə·mər }

styrene plastic [POLYM CHEM] A plastic made by the polymerization of styrene or the copolymerization of styrene with other unsaturated compounds. { 'stī,rēn 'plas·tik }

styrene-rubber plastic [POLYM CHEM] A plastic-rubber mixture consisting of at least 50% of a styrene plastic combined with rubber and various compounding ingredients. { 'stī,rēn ¦rəb·ər 'plas·tik }

subboundary structure [MET] A network of low-angle grain boundaries of less than one degree within the main crystals of a metal. { ¦səb'bau̇n·drē 'strək·chər }

subgrain [MET] The portion of a metal crystal or grain with an orientation that differs slightly from the orientation of neighboring portions of the same crystal. { 'səb,grān }

sublimation [THERMO] The process by which solids are transformed directly to the vapor state or vice versa without passing through the liquid phase. { ,səb·lə'mā·shən }

sublimation cooling [THERMO] Cooling caused by the extraction of energy to produce sublimation. { ,səb·lə'mā·shən ¦kül·iŋ }

sublimation curve [THERMO] A graph of the vapor pressure of a solid as a function of temperature. { ,səb·lə'mā·shən ¦kərv }

sublimation energy [THERMO] The increase in internal energy when a unit mass, or 1 mole, of a solid is converted into a gas, at constant pressure and temperature. { ,səb·lə'mā·shən ¦en·ər·jē }

sublimation point [THERMO] The temperature at which the vapor pressure of the solid phase of a compound is equal to the total pressure of the gas phase in contact with it; analogous to the boiling point of a liquid. { ,səb·lə'mā·shən ¦pȯint }

sublimation pressure [THERMO] The vapor pressure of a solid. { ,səb·lə'mā·shən ¦presh·ər }

sublime [THERMO] To change from the solid to the gaseous state without passing through the liquid phase. { sə'blīm }

submerged-arc furnace [MET] An arc-heating furnace in which the arcs may be completely submerged under the charge or in the molten bath under the charge. { səb'mərjd ¦ärk 'fər·nəs }

submerged-arc welding [MET] Arc welding with a bare metal electrode, the arc and tip of the electrode being shielded by a blanket of granular, fusible material. { səb'mərjd ¦ärk 'weld·iŋ }

submetallic [OPTICS] Referring to a luster intermediate between metallic and nonmetallic, such as exhibited by the mineral chromite. { ¦səb·mə'tal·ik }

subscale [MET] Oxidation occurring within a metal instead of on the surface. { 'səb,skāl }

subsieve fraction [MET] The fraction of particles of a metal powder which pass through a standard 44-micrometer sieve. { 'səb,siv 'frak·shən }

substance [PHYS] Tangible material, occurring in macroscopic amounts. { 'səb·stəns }

substantive dye *See* direct dye. { 'səb·stən·tiv 'dī }

substituent [ORG CHEM] An atom or functional group substituted for another in a chemical structure. { səb'stich·ə·wənt }

substitutional impurity [SOLID STATE] An atom or ion which is not normally found in a solid, but which resides at the position where an atom or ion would ordinarily be located in the lattice structure, and replaces it. { ,səb·stə'tü·shən·əl im'pyu̇r·əd·ē }

substitution solid solution [MET] A solid alloy having the atoms of the solute located at some lattice of points of the solvent. { ,səb·stə'tü·shən 'säl·əd sə'lü·shən }

substrate [ELECTR] The physical material on which a microcircuit is fabricated; used primarily for mechanical support and insulating purposes, as with ceramic, plastic, and glass substrates; however, semiconductor and ferrite substrates may also provide useful electrical functions. [ENG] Basic surface on which a material adheres, for example, paint or laminate. [ORG CHEM] A compound with which a reagent reacts. { 'səb ,strāt }

subtractive primaries [OPTICS] The three colors, usually yellow, magenta, and cyan (greenish-blue), which are mixed together in a subtractive process. { səb'trak·tiv 'prī,mer·ēz }

subtractive process [OPTICS] The process of producing colors by mixing absorbing media or filters of subtractive primary colors. { səb'trak·tiv 'prä·səs }

Suhl amplifier [SOLID STATE] A parametric microwave amplifier which utilizes the instability of certain spin waves in a ferromagnetic material subjected to intense microwave fields. { 'su̇l ,am·plə,fī·ər }

sulfamic acid [INORG CHEM] HSO_3NH_2 White, nonvolatile crystals slightly soluble in water and organic solvents, decomposes at 205°C; used to clean metals and ceramics, and as a plasticizer, fire retardant, chemical intermediate, and textile and paper bleach. { ¦səl¦fam·ik 'as·əd }

sulfate [CHEM] **1.** A compound containing the $-SO_4$ group, as in sodium sulfate, Na_2SO_4. **2.** A salt of sulfuric acid. { 'səl,fāt }

sulfate paper [MATER] Paper made by the sulfate process, and which cannot be bleached as white as soda or sulfite paper; strong papers, such as kraft paper, are made from unbleached sulfate materials. { 'səl,fāt ,pā·pər }

sulfhydryl compound [CHEM] A compound with a −SH group. Also known as a mercapto compound. { ¦səlf'hī·drəl 'käm,pau̇nd }

sulfide [CHEM] Any compound with one or more sulfur atoms in which the sulfur is connected directly to a carbon, metal, or other nonoxygen atom; for example, sodium sulfide, Na_2S. { 'səl ,fīd }

sulfite |INORG CHEM| M_2SO_3 A salt of sulfurous acid, for example, sodium sulfite, Na_2SO_3. { 'səl ‚fīt }

sulfite paper |MATER| Paper made from sulfite pulp. { 'səl‚fīt ‚pā·pər }

sulfite pulp |MATER| Wood chips digested with a solution of magnesium, ammonium, or calcium disulfite, with free sulfur dioxide present; used to make paper and paper products from spruce and other coniferous woods. { 'səl‚fīt ‚pəlp }

sulfo- |CHEM| Prefix for a compound with either a divalent sulfur atom, or the presence of $-SO_3H$, the sulfo group in a compound. Also spelled sulpho-. { 'səl·fō or 'səl·fə }

sulfocyanate See thiocyanate. { ‚səl·fō'sī·ə ‚nāt }

sulfocyanic acid See thiocyanic acid. { ‚səl· fō·sī¦an·ik 'as·əd }

sulfocyanide See thiocyanate. { ‚səl·fō'sī·ə ‚nīd }

sulfonate [CHEM] A sulfuric acid derivative or a sulfonic acid ester containing a $-SO_3-$ group. |ORG CHEM| Any of a group of petroleum hydrocarbons derived from sulfuric-acid treatment of oils, used as synthetic detergents, emulsifying and wetting agents, and chemical intermediates. { 'səl·fə‚nāt }

sulfonated castor oil See Turkey red oil. { ‚səl·fə ‚nād·əd 'kas·tər ‚oil }

sulfonyl [CHEM] Also known as sulfuryl. **1.** A compound containing the radical $-SO_2-$. **2.** A prefix denoting the presence of a sulfone group. { 'səl·fə‚nil }

sulfonyl chloride See sulfuryl chloride. { 'səl·fə ‚nil 'klȯr‚īd }

sulfur |CHEM| A nonmetallic element in group 16, symbol S, atomic number 16, atomic weight 32.06, existing in a crystalline or amorphous form and in four stable isotopes; used as a chemical intermediate and fungicide, and in rubber vulcanization. { 'səl·fər }

sulfurated lime See calcium sulfide. { ‚səl· fə'rād·əd 'līm }

sulfur bichloride See sulfur dichloride. { 'səl· fər bī'klȯr‚īd }

sulfur bromide |INORG CHEM| S_2Br_2 A toxic, irritating, yellow liquid that reddens in air, soluble in carbon disulfide, decomposes in water, boils at 54°C. Also known as sulfur monobromide. { 'səl·fər 'brō‚mīd }

sulfur cement |MATER| Cement used for connecting iron parts; made of equal parts of sulfur and pitch. { 'səl·fər si‚ment }

sulfur chloride |INORG CHEM| S_2Cl_2 A combustible, water-soluble, oily, fuming, amber to yellow-red liquid with an irritating effect on the eyes and lungs, boils at 138°C; used to make military gas and insecticides, in rubber substitutes and cements, to purify sugar juices, and as a chemical intermediate. Also known as sulfur subchloride. { 'səl·fər 'klȯr‚īd }

sulfur dichloride |INORG CHEM| SCl_2 A red-brown liquid boiling (when heated rapidly) at 60°C, decomposes in water; used to make insecticides, for rubber vulcanization, and as a chemical

intermediate and a solvent. Also known as sulfur bichloride. { 'səl·fər dī'klȯr‚īd }

sulfur dioxide |INORG CHEM| SO_2 A toxic, irritating, colorless gas soluble in water, alcohol, and ether; boils at $-10°C$; used as a chemical intermediate, in artificial ice, paper pulping, and ore refining, and as a solvent. Also known as sulfurous acid anhydride. { 'səl·fər dī'äk‚sīd }

sulfur dome |MET| An inverted container containing a high concentration of sulfur dioxide gas, used in die casting to cover a pot of molten magnesium to prevent burning. { 'səl·fər ‚dōm }

sulfur hexafluoride |INORG CHEM| SF_6 A colorless gas soluble in alcohol and ether, slightly soluble in water, sublimes at $-64°C$; used as a dielectric in electronics. { 'səl·fər ¦hek·sə'flúr ‚īd }

sulfuric acid |INORG CHEM| H_2SO_4 A toxic, corrosive, strongly acid, colorless liquid that is miscible with water and dissolves most metals, and melts at 10°C; used in industry in the manufacture of chemicals, fertilizers, and explosives, and in petroleum refining. Also known as dipping acid; oil of vitriol; vitriolic acid. { ‚səl¦fyúr·ik 'as·əd }

sulfuric chloride See sulfuryl chloride. { ¦səl ¦fyúr·ik 'klȯr‚īd }

sulfur iodide See sulfur iodine. { ¦səl¦fyúr·ik 'ī·ə ‚dīd }

sulfur iodine |INORG CHEM| I_2S_2 A gray-black brittle mass with an iodine aroma and a metallic luster, insoluble in water, soluble in carbon disulfide; used in medicine. Also known as iodine bisulfide; iodine disulfide; sulfur iodide. { ¦səl ¦fyúr·ik 'ī·ə‚dīn }

sulfur monobromide See sulfur bromide. { 'səl· fər ¦män·ə'brō‚mīd }

sulfur monoxide |INORG CHEM| SO A gas at ordinary temperatures; produces an orange-red deposit when cooled to temperatures of liquid air; prepared by passing an electric discharge through a mixture of sulfur vapor and sulfur dioxide at low temperature. { 'səl·fər mə'näk ‚sīd }

sulfurous acid |INORG CHEM| H_2SO_3 An unstable, water-soluble, colorless liquid with a strong sulfur aroma; derived from absorption of sulfur dioxide in water; used in the synthesis of medicine and chemicals, manufacture of paper and wine, brewing, metallurgy, and ore flotation, as a bleach and analytic reagent, and to refine petroleum products. { 'səl·fə·rəs 'as·əd }

sulfurous acid anhydride See sulfur dioxide. { 'səl·fə·rəs 'as·əd an'hī‚drīd }

sulfurous oxychloride See thionyl chloride. { 'səl·fə·rəs ¦äk·sē'klȯr‚īd }

sulfur oxide |INORG CHEM| An oxide of sulfur, such as sulfur dioxide, SO_2, and sulfur trioxide, SO_3. { 'səl·fər 'äk‚sīd }

sulfur oxychloride See thionyl chloride. { 'səl·fər ¦äk·sē'klȯr‚īd }

sulfur subchloride See sulfur chloride. { 'səl· fər ¦səb'klȯr‚īd }

sulfur trioxide |INORG CHEM| SO_3 A toxic, irritating liquid in three forms, α, β, γ, with respective melting points of 62°C, 33°C, and 17°C; a strong oxidizing agent and fire hazard; used for sulfonation of organic chemicals. { 'səl·fər trī'äk,sīd }

sulfuryl See sulfonyl. { 'səl·fə,ril }

sulfuryl chloride [INORG CHEM] SO₂Cl₂ A colorless liquid with a pungent aroma, boils at 69°C, decomposed by hot water and alkalies; used as a chlorinating agent and solvent and for pharmaceuticals, dyestuffs, rayon, and poison gas. Also known as sulfonyl chloride; sulfuric chloride. { 'səl·fə,ril 'klȯr,īd }

sulfuryl fluoride [INORG CHEM] SO₂F₂ A colorless gas with a melting point of −136.7°C and a boiling point of 55.4°C; used as an insecticide and fumigant. { 'səl·fə,ril 'flu̇r,īd }

sullage [MET] Scoria flowing on molten metal carried in a ladle. { 'səl·ij }

sumac wax See Japan wax. { 'sü,mak ,waks }

sumi ink See india ink. { 'sü·mē ,iŋk }

summer black oil See tempering oil. { 'səm·ər 'blak ,ȯil }

superacid [CHEM] **1.** An acidic medium that has a proton-donating ability equal to or greater than 100% sulfuric acid. **2.** A solution of acetic or phosphoric acid. { ¦sü·pər'as·əd }

superalloy [MET] A thermally resistant alloy for use at elevated temperatures where high stresses and oxidation are encountered. { ¦sü·pər'al,ȯi }

supercalendered finish [MATER] A shiny, smooth-surface paper obtained by passing the paper between alternating fiber-filled and steel rolls with the application of steam and pressure. { ¦sü·pər'kal·ən·dərd 'fin·ish }

supercompressibility factor See compressibility factor. { ¦sü·pər·kəm,pres·ə'bil·əd·ē ,fak·tər }

superconducting alloy [MET] An alloy capable of exhibiting superconductivity, such as an alloy of niobium and zirconium or an alloy of lead and bismuth. { ¦sü·pər·kən'dəkt·iŋ 'al,ȯi }

superconducting ball [CRYO] A ball, typically with a radius of about 0.25 millimeter (0.01 inch), that is formed from the aggregation of several million microscopic superconducting particles in a strong electric field. { ¦sü·pər·kən,duk·tiŋ 'bȯl }

superconducting circuit [CRYO] An electric circuit having elements which are in a superconducting state at least part of the time, such as a cryotron. { ¦sü·pər·kən'dəkt·iŋ 'sər·kət }

superconducting device See cryogenic device. { ¦sü·pər·kən'dəkt·iŋ di'vīs }

superconducting magnet [CRYO] An electromagnet whose coils are made of a type II superconductor with a high transition temperature and extremely high critical field, such as niobium tin, Nb₃Sn; it is capable of generating magnetic fields of 100,000 oersteds (8,000,000 amperes per meter) and more with no steady power dissipation. { ¦sü·pər·kən'dəkt·iŋ 'mag·nət }

superconducting material See superconductor. { ¦sü·pər·kən'dəkt·iŋ mə'tir·ē·əl }

superconducting memory [CRYO] A computer memory made up of a number of cryotrons,

thin-film cryotrons, superconducting thin films, or other superconducting storage devices; these operate only under cryogenic conditions and dissipate power only during the read or write operation, which permits construction of large, dense memories. { ¦sü·pər·kən'dəkt·iŋ 'mem·rē }

superconducting metal [MET] A metal capable of exhibiting superconductivity. { ¦sü·pər·kən'dəkt·iŋ 'med·əl }

superconducting thin film [CRYO] A thin film of indium, tin, or other superconducting element, used as a cryogenic switching or storage device, as in a thin-film cryotron. { ¦sü·pər·kən'dəkt·iŋ 'thin 'film }

superconductivity [SOLID STATE] A property of many metals, alloys, and chemical compounds at temperatures near absolute zero by virtue of which their electrical resistivity vanishes and they become strongly diamagnetic. { ¦sü·pər ,kän,dək'tiv·əd·ē }

superconductor [SOLID STATE] Any material capable of exhibiting superconductivity; examples include iridium, lead, mercury, niobium, tin, tantalum, vanadium, and many alloys. Also known as cryogenic conductor; superconducting material. { ¦sü·pər·kən'dək·tər }

supercooling [THERMO] Cooling of a substance below the temperature at which a change of state would ordinarily take place without such a change of state occurring, for example, the cooling of a liquid below its freezing point without freezing taking place; this results in a metastable state. { ¦sü·pər'kül·iŋ }

supercritical [THERMO] Property of a gas which is above its critical pressure and temperature. { ¦sü·pər'krid·ə·kəl }

supercritical fluid [THERMO] A fluid at a temperature and pressure above its critical point; also, a fluid above its critical temperature regardless of pressure. { ¦sü·pər¦krid·ə·kəl 'flü·əd }

supercurrent [SOLID STATE] In the two-fluid model of superconductivity, the current arising from motion of superconducting electrons, in contrast to the normal current. { ¦sü·pər'kə·rənt }

superexchange [SOLID STATE] A phenomenon in which two electrons from a double negative ion (such as oxygen) in a solid go to different positive ions and couple with their spins, giving rise to a strong antiferromagnetic coupling between the positive ions, which are too far apart to have a direct exchange interaction. { ¦sü·pər·iks'chānj }

superficial expansivity See coefficient of superficial expansion. { ¦sü·pər¦fish·əl ,ik,span'siv· əd·ē }

superficial Rockwell hardness test [MET] A test to determine surface hardness of thin sheet material which applies relatively light loads producing minimal penetration and damage. { ¦sü·pər¦fish·əl 'räk,wel 'härd·nəs ,test }

superfines [MET] The portion of a metal powder composed of particles smaller than 10 micrometers. { 'sü·pər,fīnz }

superfinishing [MET] Fine honing of a metal surface with abrasive stones. { ¦sü·pər'fin·ə·shiŋ }

superfluidity |CRYO| The frictionless flow of liquid helium at temperatures very close to absolute zero through holes as small as 10^{-7} centimeter in diameter, and for particle velocities below a few centimeters per second. { ¦sü·pər·flü'id·əd·ē }

superheat |THERMO| Sensible heat in a gas above the amount needed to maintain the gas phase. { 'sü·pər¸hēt }

superheated vapor |THERMO| A vapor that has been heated above its boiling point. { ¦sü·pər'hēd·əd 'vā·pər }

superheating |THERMO| Heating of a substance above the temperature at which a change of state would ordinarily take place without such a change of state occurring, for example, the heating of a liquid above its boiling point without boiling taking place; this results in a metastable state. { ¦sü·pər'hēd·iŋ }

superheavy element |INORG CHEM| A chemical element with an atomic number of 110 or greater. { ¦sü·pər'hev·ē 'el·ə·mənt }

superionic conduction |SOLID STATE| Extremely fast conduction of ions in certain inorganic crystalline solids, approaching the ionic conductivity of aqueous sodium chloride. { ¦sü·pər¸ī'än·ik kən'dək·shən }

superionic conductor |SOLID STATE| An ionic solid whose ionic conductivity is extremely high, on the order of 100 times that normally observed. { ¦sü·pər¸ī¦än·ik kən'dək·tər }

superlattice |ELECTR| A structure consisting of alternating layers of two different semiconductor materials, each several nanometers thick. |SOLID STATE| An ordered arrangement of atoms in a solid solution which forms a lattice superimposed on the normal solid solution lattice. Also known as artificial crystal; artificially layered structure; superstructure. { ¦sü·pər'lad·əs }

superleak See lambda leak. { 'sü·pər¸lēk }

supermolecule |PHYS CHEM| A single quantum-mechanical entity presumably formed by two reacting molecules and in existence only during the collision process; a concept in the hard-sphere collision theory of chemical kinetics. { ¦sü·pər'mäl·ə¸kyül }

superparamagnetic particle |SOLID STATE| A crystalline grain in a magnetic medium that is so small that its magnetic properties decrease with time due to thermal fluctuations. { ¦sü·pər¸par·ə·mag¸ned·ik 'pärd·i·kəl }

superparamagnetism See collective paramagnetism. { ¦sü·pər¸par·ə'mag·nə¸tiz·əm }

superphosphate |MATER| The most important phosphorus fertilizer, derived by action of sulfuric acid on phosphate rock (mostly tribasic calcium phosphate) to produce a mix of gypsum and monobasic calcium phosphate. { ¦sü·pər'fä¸sfāt }

superplasticity |MET| The unusual ability of some metals and alloys to elongate uniformly by thousands of percent at elevated temperatures, much like hot polymers or glasses. { ¦sü·pər·pla'stis·əd·ē }

superposition principle See principle of superposition. { ¸sü·pər·pə'zish·ən 'prin·sə·pəl }

supersaturation |PHYS CHEM| The condition existing in a solution when it contains more solute than is needed to cause saturation. { ¦sü·pər¸sach·ə'rā·shən }

superstructure See superlattice. { 'sü·pər ¸strək·chər }

supramolecular chemistry |CHEM| A highly interdisciplinary field covering the chemical, physical, and biological features of complex chemical species held together and organized by means of intermolecular (noncovalent) bonding interactions such as hydrogen bonds, van der Waals forces, and hydrophobic interactions. { ¸sü·prə·mə¸lek·yə·lər 'kem·ə·strē }

supramolecule |PHYS CHEM| A stable system formed by two or more molecules held together and organized by intermolecular (noncovalent bonding) interactions. { ¸sü·prə'mäl·ə¸kyül }

surface |ENG| The outer part (skin with a thickness of zero) of a body; can apply to structures, to micrometer-sized particles, or to extended-surface zeolites. { 'sər·fəs }

surface-active agent |MATER| A soluble compound that reduces the surface tension of liquids, or reduces interfacial tension between two liquids or a liquid and a solid. Also known as surfactant. { 'sər·fəs ¦ak·tiv 'ā·jənt }

surface alloying |MET| Deposition and metallurgical bonding of additional metals or alloys on the surfaces of ferrous or nonferrous metals; such additional materials become an integral part of the total mass, as distinguished from coatings which are bonded mechanically. { 'sər·fəs 'al ¸ȯi·iŋ }

surface analyzer |ENG| An instrument that measures or records irregularities in a surface by moving the stylus of a crystal pickup or similar device over the surface, amplifying the resulting voltage, and feeding the output voltage to an indicator or recorder that shows the surface irregularities magnified as much as 50,000 times. { 'sər·fəs ¸an·ə¸līz·ər }

surface area |ENG| Measurement of the extent of the area (without allowance for thickness) covered by a surface. { 'sər·fəs ¸er·ē·ə }

surface color |OPTICS| The color of light reflected from the surface of a body; in contrast to the color of light that is reflected after penetrating some distance into the body. { 'sər·fəs ¸kəl·ər }

surface density |PHYS| The quantity of anything distributed over a surface per unit area of surface. { 'sər·fəs ¸den·səd·ē }

surface energy |FL MECH| The energy per unit area of an exposed surface of a liquid; generally greater than the surface tension, which equals the free energy per unit surface. { 'sər·fəs ¸en·ər¸jē }

surface finish |ENG| The surface roughness of a component after final treatment, measured by a surface profile. { 'sər·fəs ¸fin·ish }

surface hardening |MET| Hardening the surface of steel by one of several processes, such as carburizing, carbonitriding, nitriding, flame or induction hardening, and surface working. { 'sər·fəs ¸härd·ən·iŋ }

surface leakage |ELEC| The passage of current over the surface of an insulator. { 'sər·fəs ˌlē·kij }

surface machining |MET| The cutting of three-dimensional shapes in a piece of work, by using numerical control equipment to transmit predetermined paths and designs. { 'sər·fəs mə ˌshēn·iŋ }

surface physics |SOLID STATE| The study of the structure and dynamics of atoms and their associated electron clouds in the vicinity of a surface, usually at the boundary between a solid and a low-density gas. { 'sər·fəs 'fiz·iks }

surface plasmon |SOLID STATE| A quantum of a collective oscillation of charges on the surface of a solid induced by a time-varying electric field. { 'sər·fəs 'plaz·ˌmän }

surface pressure *See* film pressure. { 'sər·fəs ˌpresh·ər }

surface reaction |CHEM| A chemical reaction carried out on a surface as on an adsorbent or solid catalyst. { 'sər·fəs rē·ˌak·shən }

surface recombination rate |SOLID STATE| The rate at which free electrons and holes at the surface of a semiconductor recombine, thus neutralizing each other. { ¦sər·fəs rē·ˌkäm·bə'nā·shən ˌrāt }

surface recombination velocity |SOLID STATE| A measure of the rate of recombination between electrons and holes at the surface of a semiconductor, equal to the component of the electron or hole current density normal to the surface divided by the excess electron or hole volume charge density close to the surface. { 'sər·fəs rē ˌkäm·bə'nā·shən və‚läs·əd·ē }

surface resistivity |ELEC| The electric resistance of the surface of an insulator, measured between the opposite sides of a square on the surface; the value in ohms is independent of the size of the square and the thickness of the surface film. { 'sər·fəs ˌrē‚zis'tiv·əd·ē }

surface rolling |MET| A cold-rolling process for hardening the surface of a metal. { 'sər·fəs ˌrōl·iŋ }

surface roughness |ENG| The closely spaced unevenness of a solid surface (pits and projections) that results in friction for solid-solid movement or for fluid flow across the solid surface. { 'sər·fəs ˌrəf·nəs }

surface state |SOLID STATE| An electron state in a semiconductor whose wave function is restricted to a layer near the surface. { 'sər·fəs ˌstāt }

surface tension |FL MECH| The force acting on the surface of a liquid, tending to minimize the area of the surface; quantitatively, the force that appears to act across a line of unit length on the surface. Also known as interfacial force; interfacial tension; surface tensity. { 'sər·fəs ˌten·chən }

surface tension number |FL MECH| A dimensionless number used in studying mass transfer in packed columns equal to the square of the dynamic viscosity of a fluid times the length of the perimeter of a packing element, divided by the product of the surface area of the packing

element, the surface tension, and the density of the liquid. Symbol T_s. { 'sər·fəs ˌten·chən ˌnəm·bər }

surface tensity *See* surface tension. { 'sər·fəs ˌten·səd·ē }

surface treating |ENG| Any method of treating a material (metal, polymer, or wood) so as to alter the surface, rendering it receptive to inks, paints, lacquers, adhesives, and various other treatments, or resistant to weather or chemical attack. { 'sər·fəs ˌtrēd·iŋ }

surfacing |MET| Depositing filler metal on a metal surface by welding or spraying. { 'sər·fə·siŋ }

surfactant *See* surface-active agent. { sər'fek·tənt }

susceptance |ELEC| The imaginary component of admittance. { sə'sep·təns }

susceptance standard |ELEC| Standard that introduces calibrated small values of shunt capacitance into 50-ohm coaxial transmission arrays. { sə'sep·təns ˌstan·dərd }

susceptibility *See* electric susceptibility; magnetic susceptibility. { sə‚sep·tə'bil·əd·ē }

suspended solids *See* suspension. { sə'spen·dəd 'säl·ədz }

suspended transformation |THERMO| The cessation of change before true equilibrium is reached, or the failure of a system to change immediately after a change in conditions, such as in supercooling and other forms of metastable equilibrium. { sə'spen·dəd ˌtranz·fər'mā·shən }

suspension |CHEM| A mixture of fine, nonsettling particles of any solid within a liquid or gas, the particles being the dispersed phase, while the suspending medium is the continuous phase. Also known as suspended solids. { sə'spen·shən }

swaging |MET| Tapering a rod or tube or reducing its diameter by any of several methods, such as forging, squeezing, or hammering. Also known as cressing. { 'swāj·iŋ }

swarf cut |MET| The removal and cutting of metal in which the axis of the cutting tool is varied with respect to the part being machined. { 'swórf ˌkət }

sweat |MET| Exudate of low-melting-point constituents from a metal on solidification. { swet }

sweated wax |MATER| A white, moisture-free petroleum wax with the oil removed by a sweating process, in which the unrefined wax is heated in shallow pans. { 'swed·əd 'waks }

sweating out |MET| Bringing small globules of low-melting constituents to the surface of an alloy during heat treatment (as lead out of bronze). { 'swed·iŋ ˌaút }

sweat soldering |MET| Soldering two parts by precoating with solder and merging them by application of heat. { 'swet ˌsäd·ə·riŋ }

sweep |MET| A profile pattern used to form molds for symmetrical articles made by sweep casting. { swēp }

symbol |CHEM| Letter or combination of letters and numbers that represent various conditions or properties of an element, for example, a normal atom, O (oxygen); with its atomic weight, ^{16}O; its atomic number, $_8^{16}O$; as a molecule, O_2; as an ion, O^{2+}; in excited state, O^*; or as an isotope, ^{18}O. { 'sim·bəl }

symmetry axis See rotation axis. { 'sim·ə,trē ,ak·səs }

symmetry axis of rotary inversion See rotoinversion axis. { 'sim·ə,trē ,ak·səs əv 'rōd·ə·rē in'vər·zhən }

symmetry axis of rotoinversion See rotoinversion axis. { 'sim·ə,trē ,ak·səs əv ¦rōd·ō·in 'vər·zhən }

symmetry class See crystal class. { 'sim·ə,trē ,klas }

symmetry element [CRYSTAL] 1. Some combination of rotations and reflections and translations which brings a crystal into a position that cannot be distinguished from its original position. Also known as symmetry operation; symmetry transformation. 2. The rotational axes, mirror planes, and center of symmetry characteristic of a given crystal. { 'sim·ə,trē ,el·ə·mənt }

symmetry function See symmetry transformation. { 'sim·ə,trē ,fəŋk·shən }

symmetry operation See symmetry element. { 'sim·ə,trē ,äp·ə,rā·shən }

symmetry operation of the second kind [CRYSTAL] A combination of rotations, reflections, and translations that brings a crystal into a position that is a mirror image of its original position. { ¦sim·ə·trē ,äp·ə,rā·shən əv tha 'sek·ənd ,kīnd }

symmetry plane See plane of mirror symmetry. { 'sim·ə,trē ,plān }

symmetry transformation See symmetry element. { 'sim·ə,trē ,tranz·fər'mā·shən }

synchronous timing [MET] Regulating the welding-transformer primary current in spot, seam, or projection welding so that the following conditions prevail: the first half-cycle is initiated at the proper time in relation to the voltage to ensure a balanced current wave, each succeeding half-cycle is essentially the same as the first, and the last half-cycle is of the opposite polarity to the first. { 'siŋ·krə·nəs 'tīm·iŋ }

syndiotactic polymer [POLYM CHEM] A vinyl polymer in which the side chains alternate regularly above and below the plane of the backbone. { ¦sin·dē·ə¦tak·tik 'päl·i·mər }

syneresis [CHEM] Spontaneous separation of a liquid from a gel or colloidal suspension due to contraction of the gel. { sə'ner·ə·səs }

synergism [MATER] An action where the total effect of two active components in a mixture is greater than the sum of their individual effects, for example, a mixture volume that is greater

than the sum of the individual volumes, or in resin formulation, the use of two or more stabilizers, where the combination improves polymer stability more than expected from the additive effect of the stabilizers. { 'sin·ər ,jiz·əm }

synergist [MATER] A material that enhances the effect of another material so that when they are combined the total effect is greater than the sum of their individual effects. { 'sin·ər,jist }

syntactic foam [MATER] A cellular polymer made by dispersing rigid, microscopic particles in a fluid polymer and then curing it. { sin'tak·tik 'fōm }

syntectic [MET] Isothermal, reversible conversion of a solid phase into two conjugate liquid phases on applying heat. { sin'tek·tik }

synthesis [CHEM] Any process or reaction for building up a complex compound by the union of simpler compounds or elements. { 'sin·thə·səs }

synthetic fiber See artificial fiber. { sin'thed·ik 'fī·bər }

synthetic gem [MATER] A precious or semiprecious stone made by artificial processes, for example, synthetic diamonds made by extreme heat and pressure on carbon, used industrially; and synthetic rubies made by high-temperature crystallization of aluminum oxide, used in laser equipment. { sin'thed·ik 'jem }

synthetic graphite [MATER] Graphitic crystalline material made by the high-temperature and pressure processing of carbon. { sin'thed·ik 'gra ,fīt }

synthetic mica [MATER] A fluorphlogopite mica made artificially by heating a large batch of raw material in an electric resistance furnace and allowing the mica to crystallize from the melt during controlled slow cooling. { sin'thed·ik 'mī·kə }

synthetic quartz [MATER] A quartz crystal grown commercially at high temperature and pressure around a seed of quartz suspended in a solution which contains scraps of natural quartz crystals. { sin'thed·ik 'kwórts }

synthetic resin [POLYM CHEM] Amorphous, organic, semisolid, or solid material derived from the polymerization of unsaturated monomers such as ethylene, butylene, propylene, and styrene. { sin'thed·ik 'rez·ən }

synthetic rubber [POLYM CHEM] Synthetic products whose properties are similar to those of natural rubber, including elasticity and ability to be vulcanized; usually produced by the polymerization or copolymerization of petroleum-derived olefinic or other unsaturated compounds. { sin'thed·ik 'rəb·ər }

T

Ta *See* tantalum.

table salt *See* sodium chloride. { 'tā·bəl ,sȯlt }

tabular crystal |CRYSTAL| A crystal that appears broad and flat due to two prominent parallel faces. { 'tab·yə·lər 'krist·əl }

tachiol *See* silver fluoride. { 'tak·ē,ȯl }

tack |MATER| Adhesive stickiness, such as occurs on the surface of a varnish or ink that has not completely dried. { tak }

tackifier |MATER| A type of resinous material added to an elastomer to produce adhesives that will adhere on contact. { 'tak·ə,fī·ər }

tackiness agent |MATER| An additive used to impart adhesive properties to otherwise non-adhesive materials, such as oils and greases. { 'tak·ē·nəs ,ā·jənt }

tacking |MET| Making small, isolated tack welds. { 'tak·iŋ }

tack weld |MET| A joint between two pieces of metal made by welding at isolated points. { 'tak ,weld }

tacky *See* tacky dry. { 'tak·ē }

tacky dry |MATER| **1.** Also known as tacky. **2.** In drying of adhesive, that stage at which the volatile constituents have evaporated or have been absorbed sufficiently to give the adhesive the desired degree of tack. **3.** In drying of paint, that stage at which the paint surface feels sticky when lightly touched. { ¦tak·ē 'drī }

tactic polymer *See* stereoregular polymer. { 'tak·tə·kəl 'päl·i·mər }

tall oil |MATER| A yellow-black, malodorous, resinous admixture of rosin, fatty acids, sterols, high-molecular-weight alcohols, and other materials, derived from wood-pulping waste liquors; used in paint drying oils, alkyd resins, linoleum, soaps, lubricants, and greases. Also known as liquid rosin; tallol. { 'täl ,ȯil }

tallol *See* tall oil. { 'tä,lȯl }

tandem mill |MET| A rolling mill consisting of two or more stands in succession, synchronized so that the metal passes directly from one to another. { 'tan·dəm 'mil }

tangent bending |MET| In a single piece of metal, forming a series of identical bends with parallel axes. { 'tan·jənt ,bend·iŋ }

tangential stress *See* shearing stress. { tan 'jen·chəl 'stres }

tangent welding |MET| Arc welding in which two or more electrodes are in a plane parallel to the line of travel. { 'tan·jənt ¦weld·iŋ }

tangling |SOLID STATE| The reduction of motion of dislocations in a substance by increasing the total number of dislocations, so that they tangle and interfere with each other's motions. { 'taŋ·gliŋ }

tantalic acid anhydride *See* tantalum oxide. { tan'tal·ik 'as·əd an'hī,drīd }

tantalic chloride *See* tantalum chloride. { tan'tal·ik 'klȯr,īd }

tantalum |CHEM| A metallic transition element, symbol Ta, atomic number 73, atomic weight 180.9479; black powder or steel-blue solid soluble in fused alkalies, insoluble in acids (except hydrofluoric and fuming sulfuric); melts about 3000°C. |MET| A lustrous, platinum-gray ductile metal used in making dental and surgical tools, pen points, and electronic equipment. { 'tant·əl·əm }

tantalum carbide |INORG CHEM| TaC Hard, chemical-resistant crystals melting at 3875°C; used in cutting tools and dies. { 'tant·əl·əm 'kär,bīd }

tantalum chloride |INORG CHEM| TaCl₅ A highly reactive, pale-yellow powder decomposing in moist air; soluble in alcohol and potassium hydroxide; melts at 221°C; used to produce tantalum and as a chemical intermediate. Also known as tantalic chloride; tantalum pentachloride. { 'tant·əl·əm 'klȯr,īd }

tantalum nitride |INORG CHEM| TaN A very hard, black, water-insoluble solid, melting at 3360°C. { 'tant·əl·əm 'nī,trīd }

tantalum oxide |INORG CHEM| Ta₂O₅ Prisms insoluble in water and acids (except for hydrofluoric); melts at 1800°C; used to make tantalum, in optical glass and electronic equipment, and as a chemical intermediate. Also known as tantalic acid anhydride; tantalum pentoxide. { 'tant·əl·əm 'äk,sīd }

tantalum pentachloride *See* tantalum chloride. { 'tant·əl·əm ¦pen·tə'klȯr,īd }

tantalum pentoxide *See* tantalum oxide. { 'tant·əl·əm pen'täk,sīd }

tantiron |MET| An iron alloy containing silicon, carbon, manganese, phosphorus, and sulfur; used for chemical equipment where resistance to acids is needed. { 'tan,tī·ərn }

tap |MET| **1.** A quantity of molten metal run out

from a furnace at one time. **2.** To remove excess slag from the floor of a pot furnace. { tap }

tap crystal |ELECTR| Compound semiconductor that stores current when stimulated by light and then gives up energy as flashes of light when it is physically tapped. { 'tap ‚krist·əl }

tap density |MET| The apparent density of a volume of metal powder obtained when its receptacle is tapped or vibrated. { 'tap ‚den·səd·ē }

taphole |MET| A hole in a furnace or ladle through which molten metal is tapped. { 'tap ‚hōl }

tapping |MET| Opening the pouring hole of a melting furnace to remove molten metal. { 'tap·iŋ }

tar |MATER| A viscous material composed of complex, high-molecular-weight compounds derived from the distillation of petroleum or the destructive distillation of wood or coal. { tär }

tar acid |MATER| A mixture of phenols (phenols, cresols, and xylenols) found in tars and tar distillates; toxic, combustible, and soluble in alcohol and coal-tar hydrocarbons; used as a wood preservative and an insecticide for farm animals and also to make disinfectants. { 'tär ‚as·əd }

target compound |ORG CHEM| In chemical synthesis, the molecule of interest. { 'tär·gət ‚käm ‚paúnd }

tarnish |MET| Discoloration of a metal surface due to the formation of a thin film of oxide, sulfide, or some other corrosion product. { 'tär·nish }

tar paper |MATER| Heavy construction paper coated or impregnated with tar. { 'tär ‚pā·pər }

tarpaulin |MATER| A sheet of waterproof canvas or other material; used to cover and protect construction materials and equipment, athletic fields, vehicles, or other exposed objects. { 'tär·pə·lən }

Taylor-Orowan dislocation *See* edge dislocation. { 'tā·lər ō'rō·wən ‚dis·lō‚kā·shən }

Taylor process |MET| A process for making extremely fine wire by stretching wire in a glass tube at elevated temperatures, or drawing wire through a bead of molten glass and then through dies. { 'tā·lər ‚prä·səs }

Tb *See* terbium.

Tc *See* technetium.

Te *See* tellurium.

teakwood |MATER| The strong, durable, yellowish-brown wood that is obtained from the teak tree, *Tectona grandis*. { 'tēk‚wúd }

tear strength |MECH| The force needed to initiate or to continue tearing a sheet or fabric. { 'ter ‚streŋkth }

technetium |CHEM| A transition element, symbol Tc, atomic number 43; derived from uranium and plutonium fission products; chemically similar to rhenium and manganese; isotope ^{99}Tc has a half-life of 200,000 years; used to absorb slow neutrons in reactor technology. |MET| Silvergray metal with a high melting point, slightly magnetic. { tek'nē·shē·əm }

technical cohesive strength |MET| Fracture

stress in a notched tensile test. { 'tek·nə·kəl kō'he·siv 'streŋkth }

teeming |MET| Pouring molten metal, usually a ferrous metal, into an ingot mold from a furnace or ladle. { 'tēm·iŋ }

telluric acid |INORG CHEM| H_6TeO_6 Toxic white crystals, slightly soluble in cold water, soluble in hot water and alkalies; melts at 136°C; used as an analytical reagent. Also known as hydrogen tellurate. { tə'lúr·ik 'as·əd }

tellurium |CHEM| A member of group 16, symbol Te, atomic number 52, atomic weight 127.60; dark-gray crystals, insoluble in water, soluble in nitric and sulfuric acids and potassium hydroxide; melts at 452°C, boils at 1390°C; used in alloys (with lead or steel), glass, and ceramics. { tə'lúr·ē·əm }

tellurium dibromide |INORG CHEM| $TeBr_2$ Toxic, hygroscopic, green- or gray-black crystals with violet vapor, soluble in ether, decomposes in water, and melts at 210°C. { tə'lúr·ē·əm dī'brō ‚mīd }

tellurium dichloride |INORG CHEM| $TeCl_2$ A toxic, amorphous, black or green-yellow powder decomposing in water, melting at 209°C. { tə'lúr·ē·əm dī'klór‚īd }

tellurium dioxide |INORG CHEM| TeO_2 The most stable oxide of tellurium, formed when tellurium is burned in oxygen or air or by oxidation of tellurium with cold nitric acid; crystallizes as colorless, tetragonal, hexagonlike crystals that melt at 452°C. { tə'lúr·ē·əm dī'äk‚sīd }

tellurium disulfide |INORG CHEM| TeS_2 A toxic, red powder, insoluble in water and acids. Also known as tellurium sulfide. { tə'lúr·ē·əm dī'səl ‚fīd }

tellurium hexafluoride |INORG CHEM| TeF_6 A colorless gas which is formed from the elements tellurium and fluorine; it is slowly hydrolyzed by water. { tə'lúr·ē·əm ‚hek·sə'flúr‚īd }

tellurium monoxide |INORG CHEM| TeO A black, amorphous powder, stable in cold dry air; formed by heating the mixed oxide $TeSO_3$. { tə'lúr·ē·əm mə'näk‚sīd }

tellurium sulfide *See* tellurium disulfide. { tə 'lúr·ē·əm 'səl‚fīd }

tellurous acid |INORG CHEM| H_2TeO_3 Toxic, white crystals, soluble in alkalies and acids, slightly soluble in water and alcohol; decomposes at 40°C. { 'tel·yə·rəs 'as·əd }

temper |ENG| **1.** To moisten and mix clay, plaster, or mortar to the proper consistency for use. **2.** *See* anneal. |MET| **1.** The hardness and strength of a rolled metal. **2.** The nominal carbon content of steel. **3.** To soften hardened steel or cast iron by reheating to some temperature below the eutectoid temperature. **4.** An alloy added to pure tin to make the finest pewter. { 'tem·pər }

tempera |MATER| **1.** An opaque watercolor paint consisting of pigment ground in water and mixed with egg yolk. **2.** A poster paint that uses glue or gum as a binder. { 'tem·pə·rə }

temperature |THERMO| A property of an object which determines the direction of heat flow when

340

the object is placed in thermal contact with another object: heat flows from a region of higher temperature to one of lower temperature; it is measured either by an empirical temperature scale, based on some convenient property of a material or instrument, or by a scale of absolute temperature, for example, the Kelvin scale. { 'tem·prə·chər }

temperature bath [THERMO] A relatively large volume of a homogeneous substance held at constant temperature, so that an object placed in thermal contact with it is maintained at the same temperature. { 'tem·prə·chər ˌbath }

temperature coefficient [PHYS] The rate of change of some physical quantity (such as resistance of a conductor or voltage drop across a vacuum tube) with respect to temperature. { 'tem·prə·chər ˌkō·i,fish·ənt }

temperature color scale [THERMO] The relation between an incandescent substance's temperature and the color of the light it emits. { 'tem·prə·chər 'kəl·ər ˌskāl }

temperature gradient [THERMO] For a given point, a vector whose direction is perpendicular to an isothermal surface at the point, and whose magnitude equals the rate of change of temperature in this direction. { 'tem·prə·chər ˌgrād·ē·ənt }

temperature-indicating compound [MATER] A temperature-sensitive material with a predetermined melting point; used to indicate when a predesignated temperature is reached in such processes as heat treating, welding, and molding. { 'tem·prə·chər ¦in·də,kād·iŋ 'käm,pau̇nd }

temperature scale [THERMO] An assignment of numbers to temperatures in a continuous manner, such that the resulting function is single valued; it is either an empirical temperature scale, based on some convenient property of a substance or object, or it measures the absolute temperature. { 'tem·prə·chər ˌskāl }

temperature wave [CRYO] A disturbance in which a variation in temperature propagates through a medium; the chief example of this is second sound. Also known as thermal wave. { 'tem·prə·chər ˌwāv }

temper brittleness [MET] A brittle state resulting when certain low-alloy steels are slowly cooled in a range of 600–300°C or reheated in this range after quenching from 600°C. { 'tem·pər ˌbrid·əl·nəs }

tempered glass [MATER] Glass that has been prestressed by heating followed by sudden quenching to give it two to four times the strength of ordinary glass. Also known as toughened glass. { ¦tem·pərd 'glas }

tempering [MATER] Impregnating wood fibers or composition board with an oxidizing resin or drying oil followed by heat treatment, to improve strength, durability, and water resistance. [MET] Heat treatment of hardened steels to temperatures below the transformation temperature range, usually to improve toughness. { 'tem·pə·riŋ }

tempering oil [MATER] A high-viscosity neutral petroleum oil, such as a steam cylinder stock, used for the drawing or tempering of steel. Also known as steam-hammer oil; summer black oil. { 'tem·pə·riŋ ˌȯil }

temper time [MET] In resistance welding, that part of the postweld interval during which the current is suitable for tempering or heat treatment. { 'tem·pər ˌtīm }

tempilstick [MATER] A crayon that, when applied to a surface, indicates when the surface temperature exceeds a given value by changing color. { 'tem·pəl,stik }

tensile bar [ENG] A molded, cast, or machined specimen of specified cross-sectional dimensions used to determine the tensile properties of a material by use of a calibrated pull test. Also known as tensile specimen; test specimen. { 'ten·səl ˌbär }

tensile modulus [MECH] The tangent or secant modulus of elasticity of a material in tension. { 'ten·səl ˌmäj·ə·ləs }

tensile specimen See tensile bar. { 'ten·səl ˌspes·ə·mən }

tensile strength [MECH] The maximum stress a material subjected to a stretching load can withstand without tearing. Also known as hot strength. { 'ten·səl ˌstreŋkth }

tensile stress [MECH] Stress developed by a material bearing a tensile load. { 'ten·səl ˌstres }

tensiometer method [FL MECH] A method of determining the surface tension of a liquid that involves measuring the force necessary to remove a ring of known radius from the liquid surface, usually by means of a torsion balance. { ˌten·sē'äm·əd·ər ˌmeth·əd }

tensiometry [ENG] A discipline concerned with the measurement of tension or tensile strength. { ˌten·sē'äm·ə·trē }

tension [MECH] **1.** The condition of a string, wire, or rod that is stretched between two points. **2.** The force exerted by the stretched object on a support. { 'ten·chən }

tensometer [ENG] A portable machine that is used to measure the tensile strength and other mechanical properties of materials. { ten'säm·əd·ər }

teraohmmeter [ENG] An ohmmeter having a teraohm range for measuring extremely high insulation resistance values. { ¦ter·ə'ōm,mēd·ər }

terbia See terbium oxide. { 'tər·bē·ə }

terbium [CHEM] A rare-earth element, symbol Tb, in the yttrium subgroup of the transition elements, atomic number 65, atomic weight 158.9254. { 'tər·bē·əm }

terbium chloride [INORG CHEM] $TbCl_3 \cdot 6H_2O$ Water- and alcohol-soluble, hygroscopic, colorless, transparent prisms; anhydrous form melts at 588°C. { 'tər·bē·əm 'klȯr,īd }

terbium nitrate [INORG CHEM] $Tb(NO_3)_3 \cdot 6H_2O$ A colorless, fire-hazardous (strong oxidant) powder, soluble in water; melts at 89°C. { 'tər·bē·əm 'nī,trāt }

terbium oxide [INORG CHEM] Tb_2O_3 A slightly hygroscopic, dark-brown powder soluble in dilute acids, absorbs carbon dioxide from air. Also known as terbia. { 'tər·bē·əm 'äk‚sĭd }

terephthalic acid [ORG CHEM] $C_6H_4(COOH)_2$ A combustible white powder, insoluble in water, soluble in alkalies, sublimes above $300°C$; used to make polyester resins for fibers and films and as an analytical reagent and poultry-feed additive. { ‚ter·əf‚thal·ik 'as·əd }

terminal area [ELECTR] The enlarged portion of conductor material surrounding a hole for a lead on a printed circuit. Also known as land; pad. { 'tər·mən·əl ‚er·ē·ə }

termination step [CHEM] In a chain reaction, the mechanism that halts the reaction. { ‚tər·mə'nā·shən ‚steps }

ternary alloy [MET] An alloy composed of three principal elements. { 'tər·nə·rē 'al‚ói }

ternary compound [CHEM] A molecule consisting of three different types of atoms; for example, sulfuric acid, H_2SO_4. { 'tər·nə·rē 'käm‚paúnd }

terne [MET] A lead alloy having a composition of 10–20% tin and 80–90% lead; used to coat iron or steel surfaces. { tərn }

terneplate [MATER] A sheet of iron or steel coated with a lead-tin alloy; used chiefly for roofing. { 'tərn‚plāt }

terpene [ORG CHEM] **1.** $C_{10}H_{16}$ A moderately toxic, flammable, unsaturated hydrocarbon liquid found in essential oils and plant oleoresins; used as an intermediate for camphor, menthol, and terpineol. **2.** A class of naturally occurring compounds whose carbon skeletons are composed exclusively of isopentyl (isoprene) C_5 units. Also known as isoprenoid. { 'tər‚pēn }

terpenoid [ORG CHEM] Any compound with an isoprenoid structure similar to that of the terpene hydrocarbons. { 'tər·pə‚nóid }

terpolymer [POLYM CHEM] A polymer that contains three distinct monomers; for example, acrylonitrile-butadiene-styrene terpolymer, ABS. { ‚tər'päl·i·mər }

terra alba [MATER] Pure white uncalcined gypsum, used as a filler in paper and paints and as a nutrient in growing yeast. { 'ter·ə 'al·bə }

terra-cotta [MATER] A brownish-orange clay used in the production of high-quality earthenware, vases, and statuettes, and for tile floors and roofing. { ‚ter·ə‚käd·ə }

terrazzo [MATER] A mosaic flooring surface made by embedding marble or granite chips in mortar, allowing the mortar to harden, and then grinding and polishing the surface. { tə'rät‚sō }

tertiary creep [MET] Creep strain occurring at an accelerating rate leading to fracture. { 'tər·shē ‚er·ē 'krēp }

tertiary pyroelectricity [SOLID STATE] The polarization due to temperature and gradients and corresponding nonuniform stresses and strains when the crystal is heated nonuniformly; found in pyroelectric and nonpyroelectric crystals, that is, crystals which have no polar directions. Also known as false pyroelectricity. { 'tər·shē‚er·ē ‚pī·rō‚i‚lek'tris·əd·ē }

tertiary sodium phosphate See trisodium phosphate. { 'tər·shē‚er·ē 'sōd·ē·əm 'fä‚sfāt }

tesla [ELECTROMAG] The International System unit of magnetic flux density, equal to one weber per square meter. Symbolized T. { 'tes·lə }

tessera [MATER] A small rectangular piece of ceramic tile, stone, or other material used in a mosaic design. { 'tes·ə·rə }

test chamber [ENG] A place, section, or room having special characteristics where a person or object is subjected to experimental procedures, as an altitude chamber. { 'test ‚chām·bər }

test specimen See tensile bar. { 'test ‚spes·ə·mən }

tetrachlorosilane See silicon tetrachloride. { ‚te·trə‚klór·ə'sī‚lān }

tetrad axis [CRYSTAL] A rotation axis whose multiplicity is equal to 4. { 'te‚trad ‚ak·səs }

tetradentate ligand [INORG CHEM] A chelating agent which has four groups capable of attachment to a metal ion. Also known as quadridentate ligand. { ‚te·trə'den‚tāt 'līg·ənd }

tetrafluorohydrazine [INORG CHEM] F_2NNF_2 A colorless liquid or gas with a calculated boiling point of $-73°C$; used as an oxidizer in rocket fuels. { ‚te·trə‚flúr·ō'hī·drə‚zēn }

tetragonal lattice [CRYSTAL] A crystal lattice in which the axes of a unit cell are perpendicular, and two of them are equal in length to each other, but not to the third axis. { te'trag·ən·əl 'lad·əs }

tetragonal trisoctahedron See trapezohedron. { te'trag·ən·əl ‚tris‚äk·tə'hē·drən }

tetragonal tristetrahedron See deltohedron. { te'trag·ən·əl ‚tris‚te·trə'hē·drən }

tetrahedron [CRYSTAL] An isometric crystal form in cubic crystals, in the shape of a four-faced polyhedron, each face of which is a triangle. { ‚te·trə'hē·drən }

tetrahexahedron [CRYSTAL] A form of regular crystal system with four triangular isosceles faces on each side of a cube; there are altogether 24 congruent faces. { ‚te·trə'hek·sə‚drän }

tetramer [POLYM CHEM] A polymer that results from the union of four identical monomers; for example, the tetramer C_8H_8 forms from union of four molecules of C_2H_2. { 'te·trə·mər }

tetraphosphorus trisulfide See phosphorus sesquisulfide. { ‚te·trə'fä·sfə·rəs trī'səl‚fīd }

tetrapotassium pyrophosphate [INORG CHEM] Potassium pyrophosphate. { ‚te·trə·pə'tas·ē·əm ‚pī·rō'fä ‚sfāt }

tetrasodium pyrophosphate See sodium pyrophosphate. { ‚te·trə‚sōd·ē·əm ‚pī·rō'fä‚sfāt }

tetratohedral crystal [CRYSTAL] A crystal which has one quarter of the maximum number of faces allowed by the crystal system to which the crystal belongs. { 'te·trəd‚ō'hē·drəl 'krist·əl }

textile [MATER] A material made of natural or artificial fibers and used for the manufacture of items such as clothing and furniture fittings. { 'tek·stəl }

textile oil [MATER] A specially compounded oil used to condition raw textile fibers, yarn, and fabric for manufacturing and finishing operations. { 'teks·təl ‚óil }

textile softener [MATER] A chemical that attaches molecularly to textile fibers with the polar (charged) end of the cation oriented toward the fiber and the fatty tail exposed to give a feeling of softness to the fabric. { 'tek·stəl ˌsȯf·nər }

texture |CRYSTAL| The nature of the orientation, shape, and size of the small crystals in a polycrystalline solid. { 'teks·chər }

Th See thorium.

thallium |CHEM| A metallic element in group 13, symbol Tl, atomic number 81, atomic weight 204.383; insoluble in water, soluble in nitric and sulfuric acids, melts at 302°C, boils at 1457°C. |MET| Bluish-white metal with tinlike malleability, but a little softer; used in alloys. { 'thal·ē·əm }

thallium bromide |INORG CHEM| TlBr A toxic, yellowish powder soluble in alcohol, slightly soluble in water, melts at 460°C; used in infrared radiation transmitters and detectors. Also known as thallous bromide. { 'thal·ē·əm 'brō,mīd }

thallium carbonate |INORG CHEM| Tl₂CO₃ Toxic, shiny, colorless needles soluble in water, insoluble in alcohol, melts at 272°C; used as an analytical reagent and in artificial gems. Also known as thallous carbonate. { 'thal·ē·əm 'kär·bə,nāt }

thallium chloride |INORG CHEM| TlCl A white, toxic, light-sensitive powder, slightly soluble in water, insoluble in alcohol, melts at 430°C; used as a chlorination catalyst and in medicine and suntan lamps. Also known as thallous chloride. { 'thal·ē·əm 'klȯr,īd }

thallium hydroxide |INORG CHEM| TlOH·H₂O Toxic yellow, water- and alcohol-soluble needles, decomposes at 139°C; used as an analytical reagent. Also known as thallous hydroxide. { 'thal·ē·əm hī'dräk,sīd }

thallium iodide |INORG CHEM| TlI A toxic, yellow powder, insoluble in alcohol, slightly soluble in water, melts at 440°C; used in infrared radiation transmitters and in medicine. Also known as thallous iodide. { 'thal·ē·əm 'ī·ə,dīd }

thallium monoxide |INORG CHEM| Tl₂O Black, toxic, water- and alcohol-soluble powder, melts at 300°C; used as an analytical reagent and in artificial gems and optical glass. Also known as thallium oxide; thallous oxide. { 'thal·ē·əm mə'näk,sīd }

thallium nitrate |INORG CHEM| TlNO₃ Colorless, toxic, fire-hazardous crystals soluble in hot water, insoluble in alcohol, melts at 206°C, decomposes at 450°C; used as an analytical reagent and in pyrotechnics. Also known as thallous nitrate. { 'thal·ē·əm 'nī,trāt }

thallium oxide See thallium monoxide. { 'thal·ē·əm 'äk,sīd }

thallium sulfate |INORG CHEM| Tl₂SO₄ Toxic, water-soluble, colorless crystals melting at 632°C; used as an analytical reagent and in medicine, rodenticides, and pesticides. Also known as thallous sulfate. { 'thal·ē·əm 'səl,fāt }

thallium sulfide |INORG CHEM| Tl₂S Lustrous, toxic, blue-black crystals insoluble in water, alcohol, and ether, soluble in mineral acids, melts at 448°C; used in infrared-sensitive devices. Also known as thallous sulfide. { 'thal·ē·əm 'səl,fīd }

thallous bromide See thallium bromide. { 'thal·əs 'brō,mīd }

thallous carbonate See thallium carbonate. { 'thal·əs 'kär·bə,nāt }

thallous chloride See thallium chloride. { 'thal·əs 'klȯr,īd }

thallous hydroxide See thallium hydroxide. { 'thal·əs hī'dräk,sīd }

thallous iodide See thallium iodide. { 'thal·əs 'ī·ə,dīd }

thallous nitrate See thallium nitrate. { 'thal·əs 'nī,trāt }

thallous oxide See thallium monoxide. { 'thal·əs 'äk,sīd }

thallous sulfate See thallium sulfate. { 'thal·əs 'səl,fāt }

thallous sulfide See thallium sulfide. { 'thal·əs 'səl,fīd }

therm |THERMO| A unit of heat energy, equal to 100,000 international table British thermal units, or approximately 1.055×10^8 joules. { thərm }

thermal |THERMO| Of or concerning heat. { 'thər·məl }

thermal agitation |SOLID STATE| Random movements of the free electrons in a conductor, producing noise signals that may become noticeable when they occur at the input of a high-gain amplifier. Also known as thermal effect. { 'thər·məl ˌaj·ə'tā·shən }

thermal analysis |MET| Determining transformations in a metal by observing the temperature-time relationship during uniform cooling or heating; phase tranformations are indicated by irregularities in a smooth curve. { 'thər·məl ə'nal·ə·səs }

thermal capacitance |THERMO| The ratio of the entropy added to a body to the resulting rise in temperature. { 'thər·məl kə'pas·əd·əns }

thermal capacity See heat capacity. { 'thər·məl kə'pas·əd·ē }

thermal charge See entropy. { 'thər·məl ¦chärj }

thermal conductance |THERMO| The amount of heat transmitted by a material divided by the difference in temperature of the surfaces of the material. Also known as conductance. { 'thər·məl kən'dək·təns }

thermal conductimetry |THERMO| Measurement of thermal conductivities. { 'thər·məl ˌkän ˌdək'tim·ə·trē }

thermal conductivity |THERMO| The heat flow across a surface per unit area per unit time, divided by the negative of the rate of change of temperature with distance in a direction perpendicular to the surface. Also known as coefficient of conductivity; heat conductivity. { 'thər·məl ˌkan,dək'tiv·əd·ē }

thermal conductor |THERMO| A substance with a relatively high thermal conductivity. { 'thər·məl kən'dək·tər }

thermal convection See heat convection. { 'thər·məl kən'vek·shən }

thermal coulomb |THERMO| A unit of entropy equal to 1 joule per kelvin. { 'thər·məl 'kü,läm }

thermal cutting |MET| A group of processes to sever metals by melting or by chemical reaction of oxygen with the metal at elevated temperatures. { 'thər·məl 'kəd·iŋ }

thermal diffusivity See diffusivity. { 'thər·məl ˌdi·fyü'siv·əd·ē }

thermal effect See thermal agitation. { 'thər·məl i,fekt }

thermal efficiency See efficiency. { 'thər·məl i'fish·ən·sē }

thermal effusion See thermal transpiration. { 'thər·məl e'fyü·zhən }

thermal emissivity See emissivity. { 'thər·məl ˌē·mi'siv·əd·ē }

thermal equilibrium [THERMO] Property of a system all parts of which have attained a uniform temperature which is the same as that of the system's surroundings. { 'thər·məl ˌē·kwə'lib·rē·əm }

thermal expansion [PHYS] The dimensional changes exhibited by solids, liquids, and gases for changes in temperature while pressure is held constant. { 'thər·məl ik'span·chən }

thermal expansion coefficient [PHYS] The fractional change in length or volume of a material for a unit change in temperature. { 'thər·məl ik¦span·chən ˌkō·i,fish·ənt }

thermal farad [THERMO] A unit of thermal capacitance equal to the thermal capacitance of a body for which an increase in entropy of 1 joule per kelvin results in a temperature rise of 1 kelvin. { 'thər·məl 'far,ad }

thermal flux See heat flux. { 'thər·məl 'fləks }

thermal henry [THERMO] A unit of thermal inductance equal to the product of a temperature difference of 1 kelvin and a time of 1 second divided by a rate of flow of entropy of 1 watt per kelvin. { 'thər·məl 'hen·rē }

thermal hysteresis [THERMO] A phenomenon sometimes observed in the behavior of a temperature-dependent property of a body; it is said to occur if the behavior of such a property is different when the body is heated through a given temperature range from when it is cooled through the same temperature range. { 'thər·məl ˌhis·tə'rē·səs }

thermal inductance [THERMO] The product of temperature difference and time divided by entropy flow. { 'thər·məl in'dək·təns }

thermal magnon [SOLID STATE] A magnon with a relatively short wavelength, on the order of 10^{-6} centimeter. { 'thər·məl 'mag,nän }

thermal ohm [THERMO] A unit of thermal resistance equal to the thermal resistance for which a temperature difference of 1 kelvin produces a flow of entropy of 1 watt per kelvin. Also known as fourier. { 'thər·məl 'ōm }

thermal potential difference [THERMO] The difference between the thermodynamic temperatures of two points. { 'thər·məl pə¦ten·chəl 'dif·rəns }

thermal pulse method [SOLID STATE] A method of measuring properties of insulating and conducting crystals, in which a heat pulse of known duration is measured after propagating through a crystal; the pulse can be generated by directing a laser pulse at an absorbing film evaporated onto one face of the crystal, and detected by a thin-film circuit on the other face. { 'thər·məl ¦pəls ˌmeth·əd }

thermal radiation See heat radiation. { 'thər·məl ˌrād·ē'ā·shən }

thermal resistance [THERMO] A measure of a body's ability to prevent heat from flowing through it, equal to the difference between the temperatures of opposite faces of the body divided by the rate of heat flow. Also known as heat resistance. { 'thər·məl ri'zis·təns }

thermal resistivity [THERMO] The reciprocal of the thermal conductivity. { 'thər·məl rē ˌzis'tiv·əd·ē }

thermal scattering [SOLID STATE] Scattering of electrons, neutrons, or x-rays passing through a solid due to thermal motion of the atoms in the crystal lattice. { 'thər·məl 'skad·ə·riŋ }

thermal shock [MECH] Stress produced in a body or in a material as a result of undergoing a sudden change in temperature. { 'thər·məl 'shäk }

thermal spraying [MET] Spraying finely divided particles of powder or droplets of atomized metal wire or rod for coating a substrate. { 'thər·məl 'sprā·iŋ }

thermal stress [MECH] Mechanical stress induced in a body when some or all of its parts are not free to expand or contract in response to changes in temperature. { 'thər·məl 'stres }

thermal stress cracking [MECH] Crazing or cracking of materials (plastics or metals) by overexposure to elevated temperatures and sudden temperature changes or large temperature differentials. { 'thər·məl ¦stres 'krak·iŋ }

thermal transpiration [THERMO] The formation of a pressure gradient in gas inside a tube when there is a temperature gradient in the gas and when the mean free path of molecules in the gas is a significant fraction of the tube diameter. Also known as thermal effusion. { 'thər·məl ˌtranz·pə'rā·shən }

thermal value [THERMO] Heat produced by combustion, usually expressed in calories per gram or British thermal units per pound. { 'thər·məl ˌval·yü }

thermal volt See kelvin. { 'thər·məl 'vōlt }

thermal wave See temperature wave. { 'thər·məl 'wāv }

thermie [THERMO] A unit of heat energy equal to the heat energy needed to raise 1 tonne of water from 14.5°C to 15.5°C at a constant pressure of 1 standard atmosphere; equal to 10^6 fifteen-degrees calories or $(4.1855 \pm 0.0005) \times 10^6$ joules. Abbreviated th. { 'thər·mē }

thermionic [ELECTR] Pertaining to the emission of electrons as a result of heat. { ˌthər·mē'än·ik }

thermit See thermite. { 'thər·mət }

thermite [MATER] A fire-hazardous mixture of ferric oxide and powdered aluminum; upon ignition by a magnesium ribbon, it reaches a temperature of 4000°F (2200°C), sufficient to soften steel; used for industrial purposes or as an incendiary bomb. Also spelled thermit. { 'thər ˌmīt }

thermit process [MET] An exothermic reaction when heating finely divided aluminum on a metal oxide causing reduction of the oxide. { 'thər·mət ˌprä·səs }

thermit welding [MET] Welding with molten iron which is obtained by igniting aluminum and an iron oxide in a crucible, whereby the aluminum floats to the top of the molten metal and is poured off. { 'thər·mət ˌweld·iŋ }

thermochemical calorie See calorie. { ¦thər·mō'kem·ə·kəl 'kal·ə·rē }

thermochromism [PHYS] A reversible change in the color of a substance as its temperature is varied. { ˌthər·mə'krō,miz·əm }

thermocompression bonding [ENG] Use of a combination of heat and pressure to make connections, as when attaching beads to integrated-circuit chips; examples include wedge bonding and ball bonding. { ¦thər·mō·kəm'presh·ən 'bänd·iŋ }

thermocouple [ENG] A device consisting basically of two dissimilar conductors joined together at their ends; the thermoelectric voltage developed between the two junctions is proportional to the temperature difference between the junctions, so the device can be used to measure the temperature of one of the junctions when the other is held at a fixed, known temperature, or to convert radiant energy into electric energy. { 'thər·mə,kəp·əl }

thermodynamic cycle [THERMO] A procedure or arrangement in which some material goes through a cyclic process and one form of energy, such as heat at an elevated temperature from combustion of a fuel, is in part converted to another form, such as mechanical energy of a shaft, the remainder being rejected to a lower temperature sink. Also known as heat cycle. { ¦thər·mō·dī'nam·ik 'sī·kəl }

thermodynamic equation of state [THERMO] An equation that relates the reversible change in energy of a thermodynamic system to the pressure, volume, and temperature. { ¦thər·mō·dī'nam·ik i'kwā·zhən əv 'stāt }

thermodynamic equilibrium [THERMO] Property of a system which is in mechanical, chemical, and thermal equilibrium. { ¦thər·mō·dī'nam·ik ˌē·kwə'lib·rē·əm }

thermodynamic function of state [THERMO] Any of the quantities defining the thermodynamic state of a substance in thermodynamic equilibrium; for a perfect gas, the pressure, temperature, and density are the fundamental thermodynamic variables, any two of which are, by the equation of state, sufficient to specify the state. Also known as state parameter; state variable; thermodynamic variable. { ¦thər·mō·dī'nam·ik 'fəŋk·shən əv 'stāt }

thermodynamic potential [THERMO] One of several extensive quantities which are determined by the instantaneous state of a thermodynamic system, independent of its previous history, and which are at a minimum when the system is in thermodynamic equilibrium under specified conditions. { ¦thər·mō·dī'nam·ik pə'ten·chəl }

thermodynamic potential at constant volume See free energy. { ¦thər·mō·dī'nam·ik pe¦ten·chəl at 'kän·stənt 'väl·yəm }

thermodynamic principles [THERMO] Laws governing the conversion of energy from one form to another. { ¦thər·mō·dī'nam·ik 'prin·sə·pəlz }

thermodynamic probability [THERMO] Under specified conditions, the number of equally likely states in which a substance may exist; the thermodynamic probability Ω is related to the entropy S by $S = k \ln \Omega$, where k is Boltzmann's constant. { ¦thər·mō·dī'nam·ik ˌpräb·ə'bil·əd·ē }

thermodynamic process [THERMO] A change of any property of an aggregation of matter and energy, accompanied by thermal effects. { ¦thər·mō·dī'nam·ik 'prä·səs }

thermodynamic property [THERMO] A quantity which is either an attribute of an entire system or is a function of position which is continuous and does not vary rapidly over microscopic distances, except possibly for abrupt changes at boundaries between phases of the system; examples are temperature, pressure, volume, concentration, surface tension, and viscosity. Also known as macroscopic property. { ¦thər·mō·dī'nam·ik 'präp·ərd·ē }

thermodynamic system [THERMO] A part of the physical world as described by its thermodynamic properties. { ¦thər·mō·dī'nam·ik 'sis·təm }

thermodynamic temperature scale [THERMO] Any temperature scale in which the ratio of the temperatures of two reservoirs is equal to the ratio of the amount of heat absorbed from one of them by a heat engine operating in a Carnot cycle to the amount of heat rejected by this engine to the other reservoir; the Kelvin scale and the Rankine scale are examples of this type. { ¦thər·mō·dī'nam·ik 'tem·prə·chər ˌskāl }

thermodynamic variable See thermodynamic function of state. { ¦thər·mō·dī'nam·ik 'ver·ē·ə·bəl }

thermoelasticity [PHYS] Dependence of the stress distribution of an elastic solid on its thermal state, or of its thermal conductivity on the stress distribution. { ¦thər·mō·i,las 'tis·əd·ē }

thermoelectric effect See thermoelectricity. { ¦thər·mō·i'lek·trik i¦fekt }

thermoelectricity [PHYS] The direct conversion of heat into electrical energy, or the reverse; it encompasses the Seebeck, Peltier, and Thomson effects but, by convention, excludes other electrothermal phenomena, such as thermionic emission. Also known as thermoelectric effect. { ¦thər·mō,i,lek'tris·əd·ē }

thermoelectric material [ELECTR] A material that can be used to convert thermal energy into electric energy or provide refrigeration directly from electric energy; good thermoelectric materials include lead telluride, germanium telluride, bismuth telluride, and cesium sulfide. { ¦thər·mō·i'lek·trik mə'tir·ē·əl }

thermoelectric properties |PHYS| Properties of materials associated with thermoelectricity, namely, the electromotive force generated in the Seebeck effect, the heat generated or absorbed in the Peltier and Thomson effects, and the influence of magnetic fields upon these quantities. { ¦thər·mō·i'lek·trik 'präp·ərd·ēz }

thermoelectric series |MET| A series of metals arranged in order of their thermoelectric voltage-generating ratings with respect to some reference metal, such as lead. { ¦thər·mō·i'lek·trik ˌsir·ēz }

thermogalvanic corrosion |MET| Corrosion associated with the passage of an electric current in which the anode and cathode are at different temperatures, the anode usually being the colder of the two. { ¦thər·mō·gal'van·ik kə'rō·zhən }

thermolysis See pyrolysis. { 'thər·mäl·ə·səs }

thermomagnetic effect |PHYS| An electrical or thermal phenomenon occurring when a conductor or semiconductor is placed simultaneously in a temperature gradient and a magnetic field; examples are the Ettingshausen-Nernst effect and the Righi-Leduc effect. { ¦thər·mō·mag'ned·ik i¦fekt }

thermometal |MET| A bimetallic strip which, on temperature change, deflects because of differences in the coefficients of expansion of the two bonded metals. { 'thər·mō,med·əl }

thermometric analysis |PHYS CHEM| A method for determination of the transformations a substance undergoes while being heated or cooled at an essentially constant rate, for example, freezing-point determinations. { ¦thər·mə ¦me·trik 'nal·ə·səs }

thermometric conductivity See diffusivity. { ¦thər·mə¦me·trik ˌkän,dək,tiv·əd·ē }

thermometric fluid |THERMO| A fluid that has properties, such as a large and uniform thermal expansion coefficient, good thermal conductivity, and chemical stability, that make it suitable for use in a thermometer. { ˌthər·mə¦me·trik 'flü·əd }

thermometric property |THERMO| A physical property that changes in a known way with temperature, and can therefore be used to measure temperature. { ¦thər·mə¦me·trik 'präp·ərd·ē }

thermometry |THERMO| The science and technology of measuring temperature, and the establishment of standards of temperature measurement. { thər'mäm·ə·trē }

thermooptic effect |PHYS| The change in optical properties of a material because of heat radiation. { ¦thər·mō'äp·tik ə'fekt }

thermophoresis |THERMO| The movement of particles in a thermal gradient from high to low temperatures. { ˌthər·mə·fə'rē·səs }

thermoplastic |POLYM CHEM| A polymeric material with a linear macromolecular structure that will repeatedly soften when heated and harden when cooled; for example, styrene, acrylics, polyethylenes, vinyls, nylons, and fluorocarbons. Also known as thermoplastic resin. { 'thər·mə ˌplas·tik }

thermoplastic insulation |MATER| Electrical insulation made of a thermoplastic material. { ¦thər·mə¦plas·tik ˌin·sə'lā·shən }

thermoplastic resin See thermoplastic. { ¦thər·mə¦plas·tik 'rez·ən }

thermosets |POLYM CHEM| Polymeric materials that usually have a crosslinked network. They are formed into a permanent shape and are cured (set) by a chemical reaction that may require heat and pressure. Thermoset polymers are typically insoluble, and cannot be remelted or reformed into another shape after curing. Also known as thermosetting resin. { 'thər·mə,sets }

thermosetting resin See thermosets. { 'thər·mə,sed·iŋ 'rez·ən }

thermotropic liquid crystal |PHYS CHEM| A liquid crystal prepared by heating the substance. { ¦thər·mō¦träp·ik 'lik·wəd 'krist·əl }

thetagram |THERMO| A thermodynamic diagram with coordinates of pressure and temperature, both on a linear scale. { 'thād·ə,gram }

Thévenin's theorem |ELEC| A theorem in network problems which allows calculation of the performance of a device from its terminal properties only: the theorem states that at any given frequency the current flowing in any impedance, connected to two terminals of a linear bilateral network containing generators of the same frequency, is equal to the current flowing in the same impedance when it is connected to a voltage generator whose generated voltage is the voltage at the terminals in question with the impedance removed, and whose series impedance is the impedance of the network looking back from the terminals into the network with all generators replaced by their internal impedances. Also known as Helmholtz's theorem. { tā·vò'naz ,thir·əm }

thick-film circuit |ELECTR| A microcircuit in which passive components, of a ceramic-metal composition, are formed on a ceramic substrate by successive screen-printing and firing processes, and discrete active elements are attached separately. { 'thik ¦film 'sər'kət }

thick-film hybrid |ELECTR| An assembly consisting of a thick-film circuit pattern with mounting positions for the insertion of conventional silicon devices. { ,thik ,film 'hī·brəd }

thickness gage |ENG| A gage for measuring the thickness of a sheet of material, the thickness of an object, or the thickness of a coating; examples include penetration-type and backscattering radioactive thickness gages and ultrasonic thickness gages. { 'thik·nəs ,gāj }

thin film |ELECTR| A film a few molecules thick deposited on a glass, ceramic, or semiconductor substrate to form a capacitor, resistor, coil, cryotron, or other circuit component. |MATER| A film of a material from one to several hundred molecules thick deposited on a solid substrate such as glass or ceramic or as a layer on a supporting liquid. { 'thin 'film }

thin-film circuit |ELECTR| A circuit in which the passive components and conductors are produced as films on a substrate by evaporation or sputtering; active components may be similarly produced or mounted separately. { 'thin ¦film 'sər·kət }

thin-film ferrite coil [ELECTROMAG] An inductor made by depositing a thin flat spiral of gold or other conducting metal on a ferrite substrate. { 'thin ¦film 'fe,rīt 'kȯil }

thin-film integrated circuit [ELECTR] An integrated circuit consisting entirely of thin films deposited in a patterned relationship on a substrate. { 'thin ¦film 'int·ə,grād·əd 'sər·kət }

thin-film material [ELECTR] A material that can be deposited as a thin film in a desired pattern by a variety of chemical, mechanical, or high-vacuum evaporation techniques. { 'thin ¦film mə'tir·ē·əl }

thin-film semiconductor [ELECTR] Semiconductor produced by the deposition of an appropriate single-crystal layer on a suitable insulator. { 'thin ¦film 'sem·i·kən,dək·tər }

thin-film transducer [SOLID STATE] A film a few molecules thick, usually consisting of cadmium sulfide, evaporated on a crystal substrate, used to convert microwave radiation into hypersonic sound waves in the crystal. { 'thin ¦film tranz'dü·sər }

thinner [MATER] A liquid used to thin paint, varnish, cement, or other material to a desired consistency. { 'thin·ər }

thio- [CHEM] A chemical prefix derived from the Greek *theion*, meaning sulfur; indicates the replacement of an oxygen in an acid radical by sulfur with a negative valence of 2. { 'thī·ō }

thiocyanate [INORG CHEM] A salt of thiocyanic acid that contains the —SCN radical; for example, sodium thiocyanate, NaSCN. Also known as sulfocyanate; sulfocyanide; thiocyanide. { ¦thī·ō'sī·ə,nāt }

thiocyanic acid [INORG CHEM] HSC:N A colorless, water-soluble liquid decomposing at 200°C; used to inhibit paper deterioration due to the action of light, and (in the form of organic esters) as an insecticide. Also known as rhodanic acid; sulfocyanic acid. { ¦thī·ō·sī¦an·ik 'as·əd }

thiocyanide See thiocyanate. { ¦thī·ō'sī·ə,nīd }

thiocyanogen [INORG CHEM] NCSSCN White, light-unstable rhombic crystals melting at −2°C. { ¦thī·ō·sī'an·ə·jən }

thioindigo [MATER] A group of sulfur dyes made by treating the appropriate organic compound with sodium sulfide; colors are fast to washing and light. { ¦thī·ō'in·də,gō }

thionic acid [INORG CHEM] $H_2S_xO_6$, where x varies from 2 to 6. [ORG CHEM] An organic acid with the radical —CSOH. { thī'än·ik 'as·əd }

thionyl chloride [INORG CHEM] $SOCl_2$ A toxic, yellowish to red liquid with a pungent aroma, soluble in benzene, decomposes in water and at 140°C; boils at 79°C; used as a chemical intermediate and catalyst. Also known as sulfur oxychloride; sulfurous oxychloride. { 'thī·ən·əl 'klȯr,īd }

thiosulfate [INORG CHEM] $M_2S_2O_3$ A salt of thiosulfuric acid and a base; for example, reaction of sodium hydroxide and thiosulfuric acid to produce sodium thiosulfate. { ¦thī·ə'səl,fāt }

thiosulfuric acid [INORG CHEM] $H_2S_2O_3$ An unstable acid that decomposes readily to form sulfur and sulfurous acid. { ¦thī·ə,səl'fyúr·ik 'as·əd }

third law of thermodynamics [THERMO] The entropy of all perfect crystalline solids is zero at absolute zero temperature. { 'thərd 'lȯ əv ¦thər·mō·də'nam·iks }

third sound [CRYO] A type of wave propagated in thin films of superfluid helium (helium II), consisting of variations in film thickness and temperature. { 'thərd ,saúnd }

13.0 temperature See annealing point. { ¦thər ,tēn 'tem·prə·chər }

thixotropy [PHYS CHEM] The reversible property of certain gels and pastes where the viscosity will decrease (become fluid) on shearing and increase (thicken) on standing. { thik'sä·trə·pē }

Thomas converter [MET] A basic Bessemer converter; that is, one in which air is forced upward through holes in the bottom of the steel container which has a basic lining, usually dolomite, and which employs a basic slag. { 'täm·əs kən ,vərd·ər }

thoria See thorium dioxide. { 'thȯr·ē·ə }

thorium [CHEM] An element of the actinium series, symbol Th, atomic number 90, atomic weight 232; soft, radioactive, insoluble in water and alkalies, soluble in acids, melts at 1750°C, boils at 4500°C. [MET] A heavy malleable metal that changes from silvery-white to dark gray or black in air; potential source of nuclear energy; used in manufacture of sunlamps. { 'thȯr·ē·əm }

thorium anhydride See thorium dioxide. { 'thȯr·ē·əm an'hī,drīd }

thorium carbide [INORG CHEM] ThC_2 A yellow solid melting at above 2630°C, decomposes in water; used in nuclear fuel. { 'thȯr·ē·əm 'kär ,bīd }

thorium chloride [INORG CHEM] $ThCl_4$ Hygroscopic, toxic colorless crystal needles soluble in alcohol, melts at 820°C, decomposes at 928°C; used in incandescent lighting. Also known as thorium tetrachloride. { 'thȯr·ē·əm 'klȯr,īd }

thorium dioxide [INORG CHEM] ThO_2 A heavy, white powder soluble in sulfuric acid, insoluble in water, melts at 3300°C; used in medicine, ceramics, flame spraying, and electrodes. Also known as thoria; thorium anhydride; thorium oxide. { 'thȯr·ē·əm dī'äk,sīd }

thorium fluoride [INORG CHEM] ThF_4 A white, toxic powder, melts at 1111°C; used to make thorium metal and magnesium-thorium alloys and in high-temperature ceramics. { 'thȯr·ē·əm 'flúr,īd }

thorium nitrate [INORG CHEM] $Th(NO_3)_4·4H_2O$ Explosive white crystals soluble in water and alcohol, strong oxidizer; the anhydrous form decomposes at 500°C; used in medicine and as an analytical reagent. { 'thȯr·ē·əm 'nī,trāt }

thorium oxide See thorium dioxide. { 'thȯr·ē·əm 'äk,sīd }

thorium sulfate [INORG CHEM] $Th(SO_4)_2·8H_2O$ A white powder soluble in ice water, loses water at 42° and 400°C. Also known as normal thorium sulfate. { 'thór·ē·əm 'səl,fāt }

thorium tetrachloride See thorium chloride. { 'thór·ē·əm ¦te·trə'klór,īd }

threeling See trilling. { 'thrēl·iŋ }

three-point bending [MET] Bending a piece of metal by placing the specimen on two supports and then applying a load on it between the supported ends. { 'thrē ¦póint 'bend·iŋ }

three-quarters hard [MET] A temper designation for various nonferrous metals, such as aluminum, copper, and magnesium alloys, expressing degree of hardness achieved by mechanical working. { 'thrē ¦kwórd·ərz 'härd }

threshold contrast [OPTICS] The smallest contrast of luminance (or brightness) that is perceptible to the human eye under specified conditions of adaptation luminance and target visual angle. Also known as contrast sensitivity; contrast threshold; liminal contrast. { 'thresh,hōld ¦kän ,trast }

threshold illuminance [OPTICS] The lowest value of illuminance which the eye is capable of detecting under specified conditions of background luminance and degree of dark adaptation of the eye. Also known as flux density threshold. { 'thresh,hōld i'lü·mə·nəns }

throat depth [MET] The distance from the center line of the electrodes or platens of a resistance welding machine to the nearest point of interference for flat work. { 'thrōt ,depth }

throat of fillet weld [MET] The thinnest part of a fillet weld, or the shortest distance from the root of a fillet weld to its face. { 'thrōt əv 'fil·ət 'weld }

throttling [THERMO] An adiabatic, irreversible process in which a gas expands by passing from one chamber to another chamber which is at a lower pressure than the first chamber. { 'thräd·əl·iŋ }

through weld [MET] A long weld made through the unbroken surface of one member to the other member in a lap or tree joint. { 'thrü ,weld }

throwing power [MET] The ability of an electroplating solution to deposit metal uniformly on an irregularly shaped cathode. { 'thrō·iŋ ,paù·ər }

thulia See thulium oxide. { 'thü·lē·ə }

thulium [CHEM] A rare-earth element, symbol Tm, of the lanthanide group, atomic number 69, atomic weight 168.9342; reacts slowly with water, soluble in dilute acids, melts at 1550°C, boils at 1727°C; the dust is a fire hazard; used as x-ray source and to make ferrites. { 'thü·lē·əm }

thulium chloride [INORG CHEM] $TmCl_3·7H_2O$ Green, deliquescent crystals soluble in water and alcohol; melts at 824°C. { 'thü·lē·əm 'klór ,īd }

thulium oxide [INORG CHEM] Tm_2O_3 A white, slightly hygroscopic powder that absorbs water and carbon dioxide from the air, and is slowly soluble in strong acids; used to make thulium metal. Also known as thulia. { 'thü·lē·əm 'äk ,sīd }

Ti See titanium.

tight binding approximation [SOLID STATE] A method of calculating energy states and wave functions of electrons in a solid in which the wave function is assumed to be a sum of pure atomic wave functions centered about each of the atoms in the lattice, each multiplied by a phase factor; it is suitable for deep-lying energy levels. { 'tīt ¦bīnd·iŋ ə,präk·sə'mā·shən }

TIG welding See tungsten-inert gas welding. { ¦tē¦ī¦jē ,weld·iŋ }

tile [MATER] **1.** A piece of fired clay, stone, concrete, or other material used ornamentally to cover roofs, floors, or walls. **2.** A hollow building unit made of burned clay or other material. { tīl }

tileboard [MATER] A type of wallboard used for interior finishing in which the outer surface is a layer of hard glossy material, usually simulating tile. { 'tīl,bórd }

tilt boundary [SOLID STATE] A boundary between two crystals that differ in orientation by only a few degrees, consisting of a series of edge dislocations; it is formed during polygonization. Also known as bend plane; polygon wall. { 'tilt ,baùn·dre }

tilt mold [MET] A mold that rotates from a horizontal to a vertical position during filling to reduce agitation and risk of dross entrapment. { 'tilt ,mōld }

time of set [MATER] The time required for freshly mixed concrete to stiffen (initial set, about 1 hour) or to attain a minimum specified hardness (final set, about 10 hours); actual times vary with the type of cement used. { 'tīm əv 'set }

time quenching [MET] Interrupted quenching in which the time in the quenching medium is controlled. { 'tīm ,kwench·iŋ }

tin [CHEM] Metallic element in group IV, symbol Sn, atomic number 50, atomic weight 118.69; insoluble in water, soluble in acids and hot potassium hydroxide solution; melts at 232°C, boils at 2260°C. [MET] A lustrous silver-white ductile, malleable metal used in alloys, for solder, terneplate, and tinplate. { tin }

tin bisulfide See stannic sulfide. { 'tin bī'səl ,fīd }

tin bromide See stannic bromide;stannous bromide. { tin 'brō,mīd }

tin bronze [MET] A tin-copper alloy. { 'tin 'bränz }

tin chloride See stannic chloride;stannous chloride. { 'tin 'klór,īd }

tin chromate See stannic chromate;stannous chromate. { 'tin 'krō,māt }

tin crystals See stannous chloride. { 'tin ,krist·əlz }

tin dichloride See stannous chloride. { 'tin dī'klór,īd }

tin difluoride See stannous fluoride. { 'tin dī'flùr,īd }

tin dioxide See stannic oxide. { 'tin dī 'äk,sīd }

tin fluoride See stannous fluoride. { 'tin 'flùr ,īd }

tinfoil [MET] Foil made of tin or a tin alloy. { 'tin ,fóil }

tin hydride [INORG CHEM] SnH_4 A gas boiling at −52°C. Also known as stannane. { 'tin 'hī,drīd }

tin iodide See stannic iodide. { 'tin 'ī·ə,dīd }
tin monosulfide See stannous sulfide. { 'tin ¦män·ə'səl,fīd }
tinned wire [MET] Copper wire that has been coated during manufacture with a layer of tin or solder to prevent corrosion and simplify soldering of connections. { 'tind 'wīr }
tinning [MET] **1.** Covering or preserving with tin. **2.** A protective coating of tin. { 'tin·iŋ }
tin oxide See stannic oxide; stannous oxide. { 'tin 'äk,sīd }
tin peroxide See stannic oxide. { 'tin pə'räk,sīd }
tin pest [MET] Transformation of tin to a brittle, gray variety occurring spontaneously at temperatures below 0°C. { 'tin 'pest }
tinplate [MET] Thin sheet iron or steel coated with tin. { 'tin,plāt }
tin protosulfide See stannous sulfide. { 'tin ¦prōd·ō'səl,fīd }
tin protoxide See stannous oxide. { 'tin prə'täk ,sīd }
tin salts See stannous chloride. { 'tin ,sȯlts }
tin sulfate See stannous sulfate. { 'tin 'səl,fāt }
tin sulfide See stannous sulfide. { 'tin 'səl,fīd }
tin sweat [MET] Exudation of tin-rich low-melting-point material from a tin-bronze surface as a result of inverse segregation in bronze casting, or overheating of the alloy. { 'tin ,swet }
tint [OPTICS] The mixture of a pure color with white. { 'tint }
tin tetrabromide See stannic bromide. { 'tin ¦te·trə'brō,mīd }
tin tetrachloride See stannic chloride. { 'tin ¦te·trə'klȯr,īd }
tin tetraiodide See stannic iodide. { 'tin ¦te·trə'ī·ə,dīd }
tintometer [OPTICS] A device used to estimate the intensity of a colored solution by comparing it with standard solutions or colored glass slides, as with the Lovibond tintometer. { tin 'täm·əd·ər }
tissue culture [BIOMATER] Growth of tissue cells in artificial media. { 'tish·ü ,kəl·chər }
tissue engineering [BIOMATER] The creation of tissues or organs to replace lost form or function. { 'tish·ü ,en·jə,nir·iŋ }
tissue paper [MATER] Extremely lightweight paper, available in many colors and used in craft projects and in collage painting. { 'tish·ü ,pā·pər }
titanate [INORG CHEM] A salt of titanic acid; titanates of the M_2TiO_3 type are called metatitanates, those of the M_4TiO_4 type are called orthotitanates; an example is sodium titanate, $(Na_2O)_2Ti_2O_5$. { 'tīt·ən,āt }
titanellow See titanium trioxide. { ¦tīt·ən'el·ō }
titania See titanium dioxide. { tī'tā·nē·ə }
titanic acid [INORG CHEM] H_2TiO_3 A white, water-insoluble powder; used as a dyeing mordant. Also known as metatitanic acid; titanic hydroxide. { tī'tan·ik 'as·əd }
titanic anhydride See titanium dioxide. { tī'tan·ik an'hī,drīd }
titanic chloride See titanium tetrachloride. { tī'tan·ik 'klȯr,īd }

titanic hydroxide See titanic acid. { tī'tan·ik hī'dräk,sīd }
titanic sulfate See titanium sulfate. { tī'tan·ik 'səl,fāt }
titanium [CHEM] A metallic transition element, symbol Ti, atomic number 22, atomic weight 47.90; ninth most abundant element in the earth's crust; insoluble in water, melts at 1660°C, boils above 3000°C. [MET] A lustrous, silvery-gray, strong, light metal that is hard and brittle when cold, malleable when heated, and ductile when pure; used in the pure state or in alloys for aircraft and chemical-plate metals, for surgical instruments, and in cermets, and metal-ceramic brazing. { tī'tā·nē·əm }
titanium boride [INORG CHEM] TiB_2 A hard solid that resists oxidation at elevated temperatures and melts at 2980°C; used as a refractory and in alloys, high-temperature electrical conductors, and cermets. { tī'tā·nē·əm 'bȯr,īd }
titanium carbide [INORG CHEM] TiC Very hard gray crystals insoluble in water, soluble in nitric acid and aqua regia, melts at about 3140°C; used in cermets, arc-melting electrodes, and tungsten-carbide tools. { tī'tā·nē·əm 'kär,bīd }
titanium chloride See titanium dichloride. { tī'tā·nē·əm 'klȯr,īd }
titanium dichloride [INORG CHEM] $TiCl_2$ A flammable, alcohol-soluble, black powder that decomposes in water, and in vacuum at 475°C, and burns in air. Also known as titanium chloride. { tī'tā·nē·əm dī'klȯr,īd }
titanium dioxide [INORG CHEM] TiO_2 A white, water-insoluble powder that melts at 1560°C, and which is produced commercially from the titanium dioxide minerals ilmenite and rutile; used in paints and cosmetics. Also known as titania; titanic anhydride; titanium oxide; titanium white. { tī'tā·nē·əm dī'äk,sīd }
titanium hydride [INORG CHEM] TiH_2 A black metallic powder whose dust is an explosion hazard and which dissociates above 288°C; used in powder metallurgy, hydrogen production, foamed metals, glass solder, and refractories, and as an electronic gas getter. { tī'tā·nē·əm 'hī,drīd }
titanium nitride [INORG CHEM] TiN Golden-brown brittle crystals melting at 2927°C; used in refractories, alloys, cermets, and semiconductors. { tī'tā·nē·əm 'nī,trīd }
titanium oxide See titanium dioxide; titanium trioxide { tī'tā·nē·əm 'äk,sīd }
titanium peroxide See titanium trioxide. { tī'tā·nē·əm pə'räk,sīd }
titanium sesquisulfate See titanous sulfate. { tī'tā·nē·əm ¦ses·kwə'səl,fāt }
titanium sulfate [INORG CHEM] $Ti(SO_4)_2 \cdot 9H_2O$ Caked solid, soluble in water, toxic, highly acidic; used as a dye stripper, reducing agent, laundry chemical, and in treatment of chrome yellow colors. Also known as titanic sulfate; titanyl sulfate. { tī'tā·nē·əm 'səl,fāt }
titanium tetrachloride [INORG CHEM] $TiCl_4$ A colorless, toxic liquid soluble in water, fumes when exposed to moist air, boils at 136°C; used

349

to make titanium and titanium salts, as a dye mordant and polymerization catalyst, and in smoke screens and pigments. Also known as titanic chloride. { tī'tā·nē·əm ¦te·trə'klȯr,īd }

titanium trichloride [INORG CHEM] $TiCl_3$ Toxic, dark-violet, deliquescent crystals soluble in alcohol and some amines, decomposes in water with heat evolution, decomposes above 440°C; used as a reducing agent, chemical intermediate, polymerization catalyst, and laundry stripping agent. Also known as titanous chloride. { tī'tā·nē·əm trī'klȯr,īd }

titanium trioxide [INORG CHEM] TiO_3 Yellow titanium oxide used to make ivory shades in ceramics. Also known as titanellow; titanium oxide; titanium peroxide. { tī'tā·nē·əm trī'äk ,sīd }

titanium white *See* titanium dioxide. { tī'tā· nē·əm 'wīt }

titanous chloride *See* titanium trichloride. { tī'tan·əs 'klȯr,īd }

titanous sulfate [INORG CHEM] $Ti_2(SO_4)_3$ Green crystals soluble in dilute hydrochloric and sulfuric acids, insoluble in water and alcohol; used as a textile reducing agent. Also known as titanium sesquisulfate. { tī'tan·əs 'səl,fāt }

titanyl sulfate *See* titanium sulfate. { 'tīt·ən·əl 'səl,fāt }

Tl *See* thallium.

Tm *See* thulium.

toe [MET] The junction between the face of a weld and the base metal. { tō }

toe crack [MET] A crack in the base metal at the toe of a weld. { 'tō ,krak }

tong hold [MET] The end of a forging billet that is gripped by the operator's tongs; it is removed at the end of the forging operation. { 'täŋ ,hōld }

tool steel [MET] Any of various steels capable of being hardened sufficiently so as to be a suitable material for making cutting tools. { 'tül ¦stēl }

top and bottom process [MET] A process in which sodium sulfide is added to molten copper-nickel sulfide to form a two-layer melt, with the bulk of the nickel in the bottom layer. { ¦täp ən ¦bäd·əm ,prä·səs }

top-blown rotary converter [MET] A rotary converter used for making nickel and steel; oxygen and other gases are fed to the furnace by a lance at the elevated end of the converter to permit formation of a metal product without oxidation. { ¦täp ¦blōn 'rōd·ə·rē kən'vərd·ər }

topological order [CRYO] An internal order in a quantum Hall state which describes the quantum motions of the electrons with respect to one another; it is different from any other known order in not being associated with any symmetries or breaking of symmetries. { ,täp·ə,läj·ə·kəl 'ȯrd·ər }

torsion [MECH] A twisting deformation of a solid body about an axis in which lines that were initially parallel to the axis become helices. { 'tȯr·shən }

torsional compliance [MECH] The reciprocal of the torsional rigidity. { ¦tȯr·shə·nəl kəm'plī·əns }

torsional hysteresis [MECH] Dependence of the torques in a twisted wire or rod not only on the present torsion of the object but on its previous history of torsion. { ¦tȯr·shə·nəl ,his·tə'rē·səs }

torsional modulus [MECH] The ratio of the torsional rigidity of a bar to its length. Also known as modulus of torsion. { 'tȯr·shən·əl 'mäj·ə·ləs }

torsional rigidity [MECH] The ratio of the torque applied about the centroidal axis of a bar at one end of the bar to the resulting torsional angle, when the other end is held fixed. { 'tȯr·shən·əl ri'jid·əd·ē }

total carbon [MET] The sum of free and combined carbon in a ferrous alloy, especially steel. { 'tōd·əl 'kär·bən }

total cyanide [MET] Total amount of cyanide contained in an electroplating bath, including both simple and complex ions. { 'tōd·əl 'sī·ə ,nīd }

total heat *See* enthalpy. { 'tōd·əl 'hēt }

total wetting [FL MECH] The situation in which a liquid surface meets a solid surface with zero contact angle. { ,tōd·əl 'wed·iŋ }

toughened glass *See* tempered glass. { 'təf· ənd ,glas }

toughness [MECH] A property of a material capable of absorbing energy by plastic deformation; intermediate between softness and brittleness. { 'təf·nəs }

tough pitch copper [MET] Copper refined in a reverberatory furnace to adjust the oxygen content to 0.2–0.5%. { 'təf 'pich ,käp·ər }

tracer [CHEM] A foreign substance, usually radioactive, that is mixed with or attached to a given substance so the distribution or location of the latter can later be determined; used to trace chemical behavior of a natural element in an organism. Also known as tracer element. [ENG] A thread of contrasting color woven into the insulation of a wire for identification purposes. { 'trā·sər }

tracer element *See* tracer. { 'trā·sər ,el·ə·mənt }

tracing paper [MATER] Thin paper used both for tracing and original drawings, and of various types and surfaces. { 'trās·iŋ ,pā·pər }

transactinide elements [CHEM] In the periodic table, elements with atomic numbers higher than 103. { tranz'ak·tə,nīd ,el·ə·məns }

transcrystalline [MET] Across the crystals of a metal; used of cracks in metals. Also known as intracrystalline; transgranular. { ¦tranz'krist· əl·ən }

transferred arc [MET] In plasma arc welding, an arc established between the electrode and the workpiece. { 'tranz'fərd 'ärk }

transferred-electron effect [SOLID STATE] The variation in the effective drift mobility of charge carriers in a semiconductor when significant numbers of electrons are transferred from a low-mobility valley of the conduction band in a zone to a high-mobility valley, or vice versa. { 'tranz'fərd i¦lek,trän i,fekt }

transformation *See* inversion. { ,tranz·fər'mā· shən }

transformation temperature [MET] 1. The temperature at which a change in phase occurs in a metal during heating or cooling. 2. The maximum or minimum temperature of a transformation temperature range. { ,tranz·fər'mā·shən ,tem·prə·chər }

transformation-temperature ranges [MET] The ranges of temperatures within which austenite forms during heating and transforms during cooling. { ,tranz·fər'mā·shən 'tem·prə·chər ,rān·jəz }

transformation twin [CRYSTAL] A crystal twin developed by a growth transformation from a higher to a lower symmetry. { ,tranz·fər'mā·shən ,twin }

transgranular See transcrystalline. { ¦tranz 'gran·yə·lər }

transient grating photoacoustics See impulsive stimulated thermal scattering. { ,tranch·ənt ¦grād·iŋ ,fōd·ō·ə'kü·stiks }

transient liquid phase bonding See diffusion brazing. { ¦tranch·ənt ,lik·wəd,fāz 'bän·diŋ }

transistor [ELECTR] An active component of an electronic circuit consisting of a small block of semiconducting material to which at least three electrical contacts are made, usually two closely spaced rectifying contacts and one ohmic (nonrectifying) contact; it may be used as an amplifier, detector, or switch. { tran'zis·tər }

transition [THERMO] A change of a substance from one of the three states of matter to another. { tran'zish·ən }

transition element [CHEM] One of a group of metallic elements in which the members have the filling of the outermost shell to 8 electrons interrupted to bring the penultimate shell from 8 to 18 or 32 electrons; includes elements 21 through 29 (scandium through copper), 39 through 47 (yttrium through silver), 57 through 79 (lanthanum through gold), and all known elements from 89 (actinium) on. Also known as transition metal. { tran'zish·ən ,el·ə·mənt }

transition lattice [MET] An unstable, intermediate configuration formed in a metal lattice during solid-state reactions such as precipitation or transformation. { tran'zish·ən ,lad·əs }

transition metal See transition element. { tran 'zish·ən ,med·əl }

transition point [THERMO] Either the temperature at which a substance changes from one state of aggregation to another (a first-order transition), or the temperature of culmination of a gradual change, such as the lambda point, or Curie point (a second-order transition). Also known as transition temperature. { tran'zish·ən ,pòint }

transition region [SOLID STATE] The region between two homogeneous semiconductors in which the impurity concentration changes. { tran'zish·ən ,rē·jən }

transition temperature [CHEM] The temperature at which an enantiotropic polymorph is converted into a different form. [MET] The temperature at which a fracture changes from tough to brittle in various tests, such as notched-bar impact test. [THERMO] See transition point. { tran'zish·ən ,tem·prə·chər }

translation gliding See crystal gliding. { tran 'slā·shən ,glīd·iŋ }

translation group [CRYSTAL] The collection of all translation operations which carry a crystal lattice into itself. { tran'slā·shən ,grüp }

translucent medium [OPTICS] A medium which transmits rays of light so diffused that objects cannot be seen distinctly; examples are various forms of glass which admit considerable light but impede vision. { tran'slüs·əns 'mēd·ē·əm }

transmission electron microscope [ELECTR] A type of electron microscope in which the specimen transmits an electron beam focused on it, image contrasts are formed by the scattering of electrons out of the beam, and various magnetic lenses perform functions analogous to those of ordinary lenses in a light microscope. { tranz'mish·ən i'lek,trän 'mī·krə,skōp }

transparency [OPTICS] The ability of a substance to transmit light of different wavelengths, sometimes measured in percent of radiation which penetrates a distance of 1 meter. { tranz'par·ən·sē }

transparent medium [OPTICS] 1. A medium which has the property of transmitting rays of light in such a way that the human eye may see through the medium distinctly. 2. A medium transparent to other regions of the electromagnetic spectrum, such as x-rays and microwaves. { tranz'par·ənt 'mēd·ē·əm }

transplutonium element [INORG CHEM] An element having an atomic number greater than that of plutonium (94). { ¦tranz·plə'tō·nē·əm 'el·ə·mənt }

transuranic elements [CHEM] Elements that have atomic numbers greater than 92; all are radioactive and are products of artificial nuclear changes. Also known as transuranium elements. { ¦tranz·yù'ran·ik 'el·ə·məns }

transuranium elements See transuranic elements. { ¦tranz·yù'rā·nē·əm 'el·ə·məns }

transverse piezoelectric effect [SOLID STATE] The manifestation of the piezoelectric effect in which the applied stress is perpendicular to the direction of the resultant electric field, or in which the applied electric field is perpendicular to the direction of the resultant stress. { ¦tranz,vərs pē ,āt·sō·i'lek·trik i,fekt }

trap [SOLID STATE] Any irregularity, such as a vacancy, in a semiconductor at which an electron or hole in the conduction band can be caught and trapped until released by thermal agitation. Also known as semiconductor trap. { trap }

trapezohedron [CRYSTAL] An isometric crystal form of 24 faces, each face of which is an irregular four-sided figure. Also known as icositetrahedron; leucitohedron; tetragonal trisoctahedron. { trə¦pē·zō¦hē·drən }

tree [MET] A projecting treelike aggregate of crystals formed at areas of high local current density in electroplating. { trē }

Tresca criterion [MECH] The assumption that plastic deformation of a material begins when the difference between the maximum and minimum principal stresses equals twice the yield stress in shear. { 'tres·kə krī,tir·ē·ən }

triad axis |CRYSTAL| A rotation axis whose multiplicity is equal to 3. { 'trī,ad ,ak·səs }

tribasic calcium phosphate See calcium phosphate. { trī'bā·sik 'kal·sē·əm 'fä,sfāt }

tribasic zinc phosphate See zinc phosphate. { trī'bā·sik 'ziŋk 'fä,sfāt }

triboelectricity See frictional electricity. { ¦trī-bō,i,lek'tris·əd·ē }

tricalcium phosphate See calcium phosphate. { trī'kal·sē·əm 'fä,sfāt }

trichloronitromethane See chloropicrin. { trī ¦klȯr·ō¦nī·trō'meth,ān }

trichroism |OPTICS| Phenomenon exhibited by certain optically anisotropic transparent crystals when subjected to white light, in which a cube of the material is found to transmit a different color through each of the three pairs of parallel faces. { 'trī,krō,iz·əm }

trichromatic theory |OPTICS| A theory of color vision which states that three primary colors may be chosen in such a way that, combined in various proportions, they can match any color. { ¦trī·krō'mad·ik 'thē·ə·rē }

triclinic crystal |CRYSTAL| A crystal whose unit cell has axes which are not at right angles, and are unequal. Also known as anorthic crystal. { trī'klin·ik 'krist·əl }

triclinic system |CRYSTAL| The most general and least symmetric crystal system, referred to by three axes of different length which are not at right angles to one another. { trī'klin·ik 'sis,təm }

tridentate ligand |INORG CHEM| A chelating agent having three groups capable of attachment to a metal ion. { trī'den,tāt 'līg·ənd }

trigonal lattice See rhombohedral lattice. { trī'gōn·əl 'lad·əs }

trigonal system |CRYSTAL| A crystal system which is characterized by threefold symmetry, and which is usually considered as part of the hexagonal system since the lattice may be either hexagonal or rhombohedral. { trī'gōn·əl 'sis,təm }

trill See trilling. { tril }

trilling |CRYSTAL| A cyclic crystal twin consisting of three individual crystals. Also known as threeling; trill. { 'tril·iŋ }

trimercuric orthophosphate See mercuric phosphate. { ,trī·mər'kyùr·ik ¦ȯr·thō'fä,sfāt }

trimercurous orthophosphate See mercurous phosphate. { ,trə·mər'kyùr·əs ¦ȯr·thō'fä,sfāt }

trimming |MET| Removing irregular edges from a drawn part, or parting-line flash from a forging, or gates, risers, and fins from a casting. { 'trim·iŋ }

Trinidad asphalt |MATER| Natural asphaltic material found in Trinidad; contains about 47% bitumen and 28% clay, and the remainder is water. { 'trin·ə,dad 'as,fȯlt }

trioxygen See ozone. { trī'äk·sə·jən }

triple point |PHYS CHEM| A particular temperature and pressure at which three different phases of one substance can coexist in equilibrium. { 'trip·əl 'pȯint }

tripotassium orthophosphate See potassium phosphate. { ,trī·pə'tas·ē·əm ¦ȯr·thō'fä,sfāt }

trisodium orthophosphate See trisodium phosphate. { trī'sōd·ē·əm ¦ȯr·thō'fä,sfāt }

trisodium phosphate |INORG CHEM| Na_3PO_4 A water-soluble crystalline compound; used as a cleaning compound and as a water softener. Abbreviated TSP. Also known as tertiary sodium phosphate; trisodium orthophosphate. { trī'sōd·ē·əm 'fä,sfāt }

triuranium octoxide |INORG CHEM| U_3O_8 Olive green to black crystals or granules, soluble in nitric acid and sulfuric acid; decomposes at 1300°C; used in nuclear technology and in the preparation of other uranium compounds. Also known as uranous-uranic oxide; uranyl uranate. { ,trī·yù'rā·nē·əm ak'täk,sīd }

Trouton's rule |THERMO| The rule that, for a nonassociated liquid, the latent heat of vaporization in calories is equal to approximately 22 times the normal boiling point on the Kelvin scale. { 'traùt·ənz ,rül }

true porcelain See porcelain. { 'trü 'pȯr·slən }

trypan blue |MATER| An acid diazo dye of the benzopurpurine series used as a vital stain. { 'tri,pan 'blü }

TSP See trisodium phosphate.

TSPP See sodium pyrophosphate.

tubercle |MET| A mound of corrosive products on the surface of a metal that is subjected to local corrosive attack. { 'tü·bər·kəl }

tuberculation |MET| Corrosive attack with formation of tubercles. { tə,bər·kyə'lā·shən }

tube reducing |MET| Reducing the diameter and wall thickness of tubing by means of a mandrel and rolls. { 'tüb ri,düs·iŋ }

tube stock |MET| Semifinished metal tubing. { 'tüb ,stäk }

tumble-plating process |MET| A method of zinc-coating small metal parts by first applying zinc powder with an adhesive, then tumbling with glass beads to roll out the powder into a continuous coat. { 'təm·bəl ¦plād·iŋ ,prä·səs }

tunable laser |OPTICS| A laser in which the frequency of the output radiation can be tuned over part or all of the ultraviolet, visible, and infrared regions of the spectrum. { 'tü·nə·bəl 'lā·zər }

tundish |MET| A funnel or pouring basin used for transferring a stream of molten metal. { 'tən ,dish }

tung oil |MATER| A yellow, combustible drying oil extracted from the seed of the tung tree; soluble in ether, chloroform, carbon disulfide, and oils; used in formulations for paints, varnishes, varnish driers, paper waterproofing, and linoleum. Also known as China wood oil. { 'təŋ ,ȯil }

tungstate |INORG CHEM| M_2WO_4 A salt of tungstic acid; for example, sodium tungstate, Na_2WO_4. { 'təŋ,stāt }

tungstate white See barium tungstate. { 'təŋ ,stāt 'wīt }

tungsten |CHEM| Also known as wolfram. A metallic transition element, symbol W, atomic number 74, atomic weight 183.85; soluble in mixed nitric and hydrofluoric acids; melts at 3400°C. |MET| A hard, brittle, ductile, heavy gray-white metal used in the pure form chiefly for electrical purposes and with other substances in dentistry, pen points, x-ray-tube targets, phonograph needles, and high-speed tool metal, and as a radioactive shield. { 'təŋ·stən }

tungsten boride [INORG CHEM] WB₂ A silvery solid; insoluble in water, soluble in aqua regia and concentrated acids; melts at 2900°C; used as a refractory for furnaces and chemical process equipment. { 'təŋ·stən 'bȯr,īd }

tungsten carbide [INORG CHEM] WC A hard, gray powder; insoluble in water; readily attacked by nitric-hydrofluoric acid mixture; melts at 2780°C; used in tools, dies, ceramics, cermets, and wear-resistant mechanical parts, and as an abrasive. { 'təŋ·stən 'kär,bīd }

tungsten carbonyl See tungsten hexacarbonyl. { 'təŋ·stən 'kär·bə,nil }

tungsten disulfide [INORG CHEM] WS₂ A grayish-black solid with a melting point above 1480°C; used as a lubricant and aerosol. { 'təŋ·stən dī'səl,fīd }

tungsten hexacarbonyl [INORG CHEM] W(CO)₆ A white, refractive, crystalline solid which decomposes at 150°C; used for tungsten coatings on base metals. Also known as tungsten carbonyl. { 'təŋ·stən ¦hek·sə'kär·bə,nil }

tungsten hexachloride [INORG CHEM] WCl₆ Dark blue or violet crystals with a melting point of 275°C; soluble in organic solvents; used for tungsten coatings on base metals and as a catalyst for olefin polymers. { 'təŋ·stən ¦hek·sə'klȯr,īd }

tungsten-inert gas welding [MET] Welding in which an arc plasma from a nonconsumable tungsten electrode radiates heat onto the work surface, to create a weld puddle in a protective atmosphere provided by a flow of inert shielding gas; heat must then travel by conduction from this puddle to melt the desired depth of weld. Abbreviated TIG welding. { 'təŋ·stən i'nərt ¦gas ,weld·iŋ }

tungsten oxychloride [INORG CHEM] WOCl₄ Dark red crystals with a melting point of approximately 211°C; soluble in carbon disulfide; used for incandescent lamps. { 'təŋ·stən ¦äk·sə'klȯr,īd }

tungsten steel [MET] Steel containing tungsten with other alloys; formerly used for cutting and forging tools but replaced by high-speed steel. { 'təŋ·stən 'stēl }

tungstic acid [INORG CHEM] H₂WO₄ A yellow powder; insoluble in water, soluble in alkalies; used as a color-resist mordant for textiles, as an ingredient in plastics, and for the manufacture of tungsten metal products. Also known as orthotungstic acid; wolframic acid. { 'təŋ·stik 'as·əd }

tungstic acid anhydride See tungstic oxide. { 'təŋ·stik 'as·əd an'hī,drīd }

tungstic anhydride See tungstic oxide. { 'təŋ·stik an'hī,drīd }

tungstic oxide [INORG CHEM] WO₃ A heavy, canary-yellow powder; soluble in caustic, insoluble in water; melts at 1473°C; used in alloys, in fabric fireproofing, for ceramic pigments, and for the manufacture of tungsten metal. Also known as wolframic acid; tungstic acid anhydride; tungstic anhydride; tungstic trioxide. { 'təŋ·stik 'äk,sīd }

tungstic trioxide See tungstic oxide. { 'təŋ·stik trī'äk,sīd }

tunnel-bearing grease [MATER] Lubricating grease for the main engine and propeller shaft (in the shaft tunnel) of ships. { 'tən·əl ¦ber·iŋ ,grēs }

tunneling magnetoresistance [SOLID STATE] A type of magnetoresistance displayed by a trilayer thin-film structure consisting of two metallic ferromagnetic thin films sandwiching an insulating film that is thin enough (less than about 2 nanometers) that electrons can pass through it via quantum-mechanical tunneling. Also known as junction magnetoresistance. { ,tən·əl·iŋ mag,ned·ō·ri'zis·təns }

tunneling microscope See scanning tunneling microscope. { 'tən·əl·iŋ 'mī·krə,skōp }

Turkey red oil [MATER] Sulfonated castor oil, soluble in water; autoignites at 833°F (445°C); used in textiles, leather, and paper coatings, for manufacture of soaps, and as an alizarin dye assistant. Also known as soluble castor oil; sulfonated castor oil. { 'tər·kē ¦red ,ȯil }

Turk's head rolls [MET] A group of four idler rolls, arranged in a square or rectangular pattern, through which strip metal can be drawn to form angled sections. { 'tərks ¦hed ,rōlz }

Turnbull's blue [INORG CHEM] A blue pigment that precipitates from the reaction of potassium ferricyanide with a ferrous salt. { 'tərn,bülz'blü }

turpentine [MATER] An essential oil produced by steam distillation of pine woods and from gum turpentine; used as a solvent and a thinner for paints and varnishes. { 'tər·pən,tīn }

tusche [MATER] A liquid lithographic ink that can be used with pen or brush on lithographic stones or metal plates; it is also used in the silk-screen process in a glue-resist system of pattern making; an extremely greasy material. { 'tüsh·ə }

tuyere [MET] An opening in the shell and refractory lining of a furnace through which air is forced. { twē'yer }

twin See twin crystal. { 'twin }

twin axis [CRYSTAL] The crystal axis about which one individual of a twin crystal may be rotated (usually 180°) to bring it into coincidence with the other individual. { 'twin 'ak·səs }

twin band [MET] A line on a polished or etched surface representing the section through crystal twins. { 'twin 'band }

twin boundary [CRYSTAL] A grain boundary whose lattice structures are mirror images of each other in the plane of the boundary. { 'twin 'baůn·drē }

twin crystal [CRYSTAL] A compound crystal which has one or more parts whose lattice structure is the mirror image of that in the other parts of the crystal. Also known as crystal twin; twin. { 'twin 'krist·əl }

twine [MATER] A strong string made up of two or more strands twisted together. { twīn }

twin law [CRYSTAL] A statement relating two or more individuals of a twin to one another in terms of their crystallography (twin plane, twin axis, and so on). { 'twin ˌlȯ }

twinning [CRYSTAL] The development of a twin crystal by growth, translation, or gliding. { 'twin·iŋ }

twinning plane See twin plane. { 'twin·iŋ ˌplān }

twin plane [CRYSTAL] The plane common to and across which the individual crystals or components of a crystal twin are symmetrically arranged or reflected. Also known as twinning plane. { 'twin ˌplān }

twist boundary [SOLID STATE] A boundary between two crystals that differ in orientation by only a few degrees, consisting of a series of screw dislocations. { 'twist ˌbaún·drē }

twister [SOLID STATE] A piezoelectric crystal that generates a voltage when twisted. { 'twis·tər }

two-carrier theory [SOLID STATE] A theory of the conduction properties of a material in bulk or in a rectifying barrier which takes into account the motion of both electrons and holes. { 'tü ˌkar·ē·ər 'thē·ə·rē }

two-dimensional electron gas [SOLID STATE] A system of electrons that are confined by opposing forces to a thin planar region adjacent to an interface or within a thin layer of material, but are free to move along the plane scattering off each other. { ˌtü di'men·shən·əl i¦lek¦trän 'gas }

two-fluid model [CRYO] A theoretical model of helium II which assumes that it consists of two interpenetrating components, a normal fluid and a superfluid with zero entropy, viscosity, and thermal conductivity. { 'tü,flü·əd ˌmäd·əl }

two-part adhesive [MATER] A glue supplied in two parts, a resin and an accelerator, which are mixed only just before application. { 'tü 'pärt ad'hē·siv }

two-phase flow [CRYO] Flow of helium II, or of electrons in a superconductor, thought of as consisting of two interpenetrating, noninteracting fluids, a superfluid component which exhibits no resistance to flow and is responsible for superconducting properties, and a normal component, which behaves as does an ordinary fluid or as conduction electrons in a nonsuperconducting metal. [FL MECH] Cocurrent movement of two phases (for example, gas and liquid) through a closed conduit or duct (for example, a pipe). { 'tü ¦fāz 'flō }

type II superconductor [CRYO] A superconductor for which there are two critical magnetic fields; magnetic flux is completely excluded from the interior of the material only at field strengths below the smaller critical field, and at field strengths between the two critical fields the magnetic flux consists of flux vortices in the form of filaments embedded in the superconducting material. Also known as high-field superconductor (HFS). { 'tīp ¦tü ¦sü·pər·kən'dək·tər }

type I superconductor [CRYO] A superconductor for which there is a single critical magnetic field; magnetic flux is completely excluded from the interior of the material at field strengths below this critical field, while at field strengths above the critical field, magnetic flux penetrates the superconductor completely and it reverts to the normal state. { 'tīp ¦wən ¦sü·pər·kən'dək·tər }

type metal [MET] Any of various low-melting-point alloys, composed mainly of lead (50–90%), antimony (2–30%), and tin (2–20%), used for casting printers' type. { 'tīp ˌmed·əl }

U

U *See* uranium.

U center [CRYSTAL] The color-center type of point lattice defect in ionic crystals created by the incorporation of an impurity such as hydrogen into alkali halides. { 'yü ,sen·tər }

Ugine Sejournet process *See* Sejournet process. { ü,zhēn se·zhər'nä ,prä·səs }

ultimate elongation [MET] The percentage of permanent deformation remaining after tensile rupture. { 'əl·tə·mət ,ē·lȯŋ'gā·shən }

ultimate load *See* breaking load. { 'əl·tə·mət ,lȯd }

ultimate set [ENG] The ratio of the length of a specimen plate or bar before testing to the length at the moment of fracture; usually expressed as a percentage. [MATER] The final degree of firmness achieved by a plastic compound as a result of curing, evaporation of the volatile materials, and polymerization at the surface. { 'əl·tə·mət 'set }

ultimate strength [MECH] The tensile stress, per unit of the original surface area, at which a body will fracture, or continue to deform under a decreasing load. { 'əl·tə·mət 'streŋkth }

ultramarine blue [INORG CHEM] A blue pigment; a powder with heat resistance, used for enamels on toys and machinery, white baking enamels, printing inks, and cosmetics, and in textile printing. { ¦əl'trə·mə'rēn 'blü }

ultraperformance composite [MATER] A material that is made from a combination of polymers and can perform under extreme atmospheric conditions for extended periods. { ,əl·trə·pər ¦fȯr·məns kəm'päz·ət }

ultrasonic bonding [MET] Bonding of two identical or dissimilar metals by mechanical pressure combined with a wiping motion produced by ultrasonic vibration. { ¦əl·trə'sän·ik 'bänd·iŋ }

ultrasonic metal inspection [MET] The application of ultrasonic vibrations to materials for detection of flaws, pits, or wall thickness. { ¦əl· trə'sän·ik 'med·əl in,spek·shən }

ultrasonic soldering [MET] A method used for aluminum soldering by ultrasonically vibrating the soldering iron to disrupt the oxide film on the metal. { ¦əl·trə'sän·ik 'säd·ə·riŋ }

ultrasonic testing [ENG] A nondestructive test method that employs high-frequency mechanical vibration energy to detect and locate structural discontinuities or differences and to measure thickness of a variety of materials. { ¦əl·trə'sän·ik 'test·iŋ }

ultrasonic thickness gage [ENG] A thickness gage in which the time of travel of an ultrasonic beam through a sheet of material is used as a measure of the thickness of the material. { ¦əl·trə'sän·ik 'thik·nəs ,gāj }

ultrasonic welding [MET] A nonfusion welding process in which the atomic movement required for coalescence is stimulated by ultrasonic vibrations. { ¦əl·trə'sän·ik 'weld·iŋ }

ultraspeed welding *See* commutator-controlled welding. { 'əl·trə,spēd 'weld·iŋ }

ultrastable material [MATER] A material having extremely low thermal expansivity, extremely high temporal stability, and high stiffness-to-density ratio. { ¦əl·trə¦stā·bəl mə'tir·ē·əl }

ultraviolet [PHYS] Pertaining to ultraviolet radiation. Abbreviated UV. { ¦əl·trə'vī·lət }

ultraviolet absorber [OPTICS] Any substance that absorbs ultraviolet radiant energy, then dissipates the energy in a harmless form; used in plastics and rubbers to decrease light sensitivity. { ¦əl·trə'vī·lət əb'sȯr·bər }

ultraviolet absorber fixative [MATER] A protective fixative that includes a material to filter ultraviolet light from the sun and from artificial light; it helps to keep colors from fading. { ¦əl·trə'vī·lət əb¦sȯr·bər ,fik·səd·iv }

ultraviolet absorption [OPTICS] Absorption of specific ultraviolet radiation wavelengths by a material; for example, by a sample solution during spectroscopic analysis. { ¦əl·trə'vī·lət əb'sȯrp·shən }

ultraviolet light *See* ultraviolet radiation. { ¦əl· trə'vī·lət 'līt }

ultraviolet radiation [ELECTROMAG] Electromagnetic radiation in the wavelength range 4–400 nanometers; this range begins at the short-wavelength limit of visible light and overlaps the wavelengths of long x-rays (some scientists place the lower limit at higher values, up to 40 nanometers). Also known as ultraviolet light. { ¦əl·trə'vī·lət ,rād·ē'ā·shən }

ultraviolet spectrum [ELECTROMAG] **1.** The range of wavelengths of ultraviolet radiation, covering 4–400 nanometers. **2.** A display or graph of the intensity of ultraviolet radiation emitted or absorbed by a material as a function of wavelength or some related parameter. { ¦əl·trə'vī·lət 'spek·trəm }

ultraviolet stabilizer See UV stabilizer. { ¦əl·trə'vī·lət 'stā·bə‚līz·ər }

umber [MATER] A naturally occurring brown siliceous earth, deriving its color from iron oxides and manganese oxide; used as a paint pigment. { 'əm·bər }

Umklapp process [SOLID STATE] The interaction of three or more waves in a solid, such as lattice waves or electron waves, in which the sum of the wave vectors is not equal to zero but, rather, is equal to a vector in the reciprocal lattice. Also known as flip-over process. { 'úm‚kläp ‚prä·səs }

unavailable energy [THERMO] That part of the energy which, when an irreversible process takes place, is initially in a form completely available for work and is converted to a form completely unavailable for work. { ¦ən·ə¦vāl·ə·bəl 'en·ər·jē }

unbonded strain gage [ENG] A type of strain gage that consists of a grid of fine wires strung under slight tension between a stationary frame and a movable armature; pressure applied to the bellows or to the diaphragm sensing element moves the armature with respect to the frame, increasing tension in one half of the filaments and decreasing tension in the rest. { ¦ən'bän·dəd 'strān ‚gāj }

unctuous [MATER] Greasy, oily, or soapy to the touch. { 'əŋk·shəs }

underbead crack [MET] A crack in the heat-affected zone of a weldment which usually does not reach the base-metal surface. { 'ən·dər‚bēd ‚krak }

undercoat See undercoater;undercoating. { 'ən·dər‚kōt }

undercoater [MATER] A type of solvent-thinned paint that is formulated to have good adhesion to a substrate and furnish a good base for additional coats of paint, and having a relatively small amount of pigment. Also known as undercoat. { ¦ən·dər¦kōd·ər }

undercoating [MATER] **1.** A waterproof protective coating applied to the underside of a vehicle to resist corrosion. **2.** A coat of paint over which another coat is applied. Also known as undercoat. { 'ən·dər‚kōd·iŋ }

undercooling [MET] Cooling a metal below the transformation temperature without obtaining the transformation. { 'ən·dər‚kül·iŋ }

undercooling effect [SOLID STATE] The effect whereby a superconductor can be cooled below its critical temperature without the onset of superconductivity. { ¦ən·dər¦kül·iŋ i‚fekt }

undercut [ELECTR] Undesirable lateral etching by chemicals in the fabrication of semiconductor devices. [ENG] Underside recess either cut or molded into an object so as to leave a topside lip or protuberance. [MET] An unfilled groove melted into the base metal at the toe of a weld. To fail to machine a part to a sufficient extent. { 'ən·dər‚kət }

underdraft [MET] Downward curving of a metal part on leaving the rolls because of higher speed of the upper roll. { 'ən·dər‚draft }

underfill [MET] A depression on the face of the weld which falls below the surface of the adjacent base metal. { 'ən·dər‚fil }

underhead crack [MET] A subsurface crack in the heat-affected zone of the base metal near a weld. { 'ən·dər‚hed 'krak }

understressing [MET] Repeated stressing below the fatigue limit or below the final applied stress; can improve fatigue properties as a result of strain-aging effects. { ¦ən·dər¦stres·iŋ }

unglazed tile [MATER] A hard, dense tile of homogeneous composition that derives its texture and color from the materials and method of manufacture. { ¦ən‚glāzd 'tīl }

uniaxial crystal [OPTICS] A doubly refracting crystal which has a single axis along which light can propagate without exhibiting double refraction. { ¦yü·nē'ak·sē·əl 'krist·əl }

uniform corrosion [MET] Corrosion which takes place uniformly over the entire exposed surfaces. { 'yü·nə‚fórm kə'rō·zhən }

uniform luminance [OPTICS] Property of a surface for which the luminous intensity of any area of the surface is proportional to the area. { 'yü·nə‚fórm 'lü·mə·nəns }

unit cell [CRYSTAL] A parallelepiped which will fill all space under the action of translations which leave the crystal lattice unchanged. Also known as structure cell. { 'yü·nət 'sel }

unit die [MET] A die block having more than one cavity insert and allowing several different castings to be made. { 'yü·nət 'dī }

unit magnetic pole [ELECTROMAG] Two equal magnetic poles of the same sign have unit value when they repel each other with a force of 1 dyne if placed 1 centimeter apart in a vacuum. { 'yü·nət mag'ned·ik 'pōl }

unit power [MET] A unit describing machinability of a metal; the power needed to remove a unit volume in unit time, usually expressed as horsepower per cubic inch per minute. { 'yü·nət 'pau̇·ər }

unit strain [MECH] **1.** For tensile strain, the elongation per unit length. **2.** For compressive strain, the shortening per unit length. **3.** For shear strain, the change in angle between two lines originally perpendicular to each other. { 'yü·nət 'strān }

unit stress [MECH] The load per unit of area. { 'yü·nət 'stres }

univariant system [THERMO] A system which has only one degree of freedom according to the phase rule. { ¦yü·nə¦ver·ē·ənt 'sis·təm }

universal constants See fundamental constants. { ¦yü·nə¦vər·səl 'kän·stəns }

universal gas constant See gas constant. { ¦yü·nə¦vər·səl 'gas ‚kän·stənt }

universal mill [MET] A rolling mill having both horizontal and vertical sets of rolls. { ¦yü·nə¦vər·səl 'mil }

unpolarized light [OPTICS] Light in which the electric vector is oriented in a random, unpredictable fashion. { ¦ən'pō·lə‚rīzd 'līt }

unsaturated compound [CHEM] Any chemical compound with more than one bond between adjacent atoms, usually carbon, and thus reactive toward the addition of other atoms at that point; for example, olefins, diolefins, and unsaturated fatty acids. { ¦ən'sach·ə,räd·əd 'käm,paúnd }

unsaturated hydrocarbon |ORG CHEM| One of a class of hydrocarbons that have at least one double or triple carbon-to-carbon bond that is not in an aromatic ring; examples are ethylene, propadiene, and acetylene. { ¦ən'sach·ə,räd·əd ¦hī-drə'kär·bən }

unsaturation |ORG CHEM| A state in which the atomic bonds of an organic compound's chain or ring are not completely satisfied (that is, not saturated); usually applies to carbon, but can include other ring or chain atoms; unsaturation usually results in a double bond (as for olefins) or a triple bond (as for the acetylenes). { ¦ən ,sach·ə'rā-shən }

upper consolute temperature See consolute temperature. { 'əp·ər 'kän·sə,lüt 'tem·prə·chər }

upper critical field |SOLID STATE| The magnetic field strength above which a type II superconductor is completely normal. Symbolized H_{c2}. { ¦əp·ər ¦krid·i·kəl 'fēld }

upper critical solution temperature See consolute temperature. { 'əp·ər ¦krid·ə·kəl sə'lü·shən 'tem·prə·chər }

upper punch |MET| In powder metallurgy, the member of the die assembly that moves downward into the die body to transmit pressure to the metal powder in the cavity. { 'əp·ər ,pənch }

upset |MATER| A defect occurring in timber as a result of a severe blow that breaks the fibers across the grain of the wood. |MET| A localized increase in the cross-sectional area of a metal during working, caused by the application of pressure; enables a head to be formed on fasteners such as bolts. { 'əp,set (noun); əp'set (verb) }

upsetting test |MET| A test used to identify the role of variables in forging, which demonstrates that the force to forge is a function of the strength of the material, coefficient of friction, and ratio of the lateral to thickness dimensions of the workpiece. { ¦əp¦sed·iŋ ,test }

upset welding |MET| A resistance welding process in which coalescence is produced simultaneously over the entire area of abutting surfaces or progressively along a joint; pressure is applied before heating is started and is maintained throughout the heating period. { 'əp ,set ,weld·iŋ }

upslope time |MET| In resistance welding, the time associated with an increase in current when slope control is used. { 'əp,slōp 'tīm }

uranate |INORG CHEM| A salt of uranic acid; for example, sodium uranate, Na_2UO_4. { 'yür·ə ,nāt }

urania See uranium dioxide. { yə'rā-nē·ə }

uranic chloride See uranium tetrachloride. { yü 'ran·ik 'klȯr,īd }

uranic oxide See uranium dioxide. { yü'ran·ik 'äk,sīd }

uranium |CHEM| A metallic element in the actinide series, symbol U, atomic number 92,

atomic weight 238.03; highly toxic and radioactive; ignites spontaneously in air and reacts with nearly all nonmetals; melts at 1132°C, boils at 3818°C; used in nuclear fuel and as the source of ^{235}U and plutonium. |MET| A dense, silvery, ductile, strongly electropositive metal. { yə'rā-nē·əm }

uranium acetate See uranyl acetate. { yə'rā-nē·əm 'as·ə,tāt }

uranium carbide |INORG CHEM| One of the carbides of uranium, such as uranium monocarbide; used chiefly as a nuclear fuel. { yə'rā-nē·əm 'kär ,bīd }

uranium dioxide |INORG CHEM| UO_2 Black, highly toxic, spontaneously flammable, radioactive crystals; insoluble in water, soluble in nitric and sulfuric acids; melts at approximately 3000°C; used to pack nuclear fuel rods and in ceramics, pigments, and photographic chemicals. Also known as urania; uranic oxide; uranium oxide. { yə'rā-nē·əm dī'äk,sīd }

uranium hexafluoride |INORG CHEM| UF_6 Highly toxic, radioactive, corrosive, colorless crystals; soluble in carbon tetrachloride, fluorocarbons, and liquid halogens; it reacts vigorously with alcohol, water, ether, and most metals, and it sublimes; used to separate uranium isotopes in the gaseous-diffusion process. { yə'rā-nē·əm ¦hek·sə'flùr,īd }

uranium hydride |INORG CHEM| UH_3 A highly toxic, gray to black powder that ignites spontaneously in air, and that conducts electricity; used for making powdered uranium metal, for hydrogen-isotope separation, and as a reducing agent. { yə'rā-nē·əm 'hī,drīd }

uranium nitrate See uranyl nitrate. { yə'rā-nē·əm 'nī,trāt }

uranium oxide See uranium dioxide; uranium trioxide. { yə'rā-nē·əm 'äk,sīd }

uranium sulfate See uranyl sulfate. { yə'rā-nē·əm 'səl,fāt }

uranium tetrachloride |INORG CHEM| UCl_4 Poisonous, radioactive, hygroscopic, dark-green crystals; soluble in alcohol and water; melts at 590°C, boils at 792°C. Also known as uranic chloride. { yə'rā-nē·əm ¦te·trə'klȯr,īd }

uranium tetrafluoride |INORG CHEM| UF_4 Toxic, radioactive, corrosive green crystals; insoluble in water; melts at 1036°C; used in the manufacture of uranium metal. Also known as green salt. { yə'rā-nē·əm ¦te·trə'flùr,īd }

uranium trioxide |INORG CHEM| UO_3 A poisonous, radioactive, red to yellow powder; soluble in nitric acid, insoluble in water; decomposes when heated; used in ceramics and pigments and for uranium refining. Also known as orange oxide; uranium oxide. { yə'rā-nē·əm trī'äk,sīd }

uranous-uranic oxide See triuranium octoxide. { 'yür·ə·nəs yə'ran·ik 'äk,sīd }

uranyl acetate |INORG CHEM| $UO_2(C_2H_3O_2)_2 \cdot 2H_2O$ Poisonous, radioactive yellow crystals, decomposed by light; soluble in cold water, decomposes in hot water; loses water of crystallization at 110°C, decomposes at 275°C; used in medicine and as an analytical reagent and bacterial oxidant. Also known as uranium acetate. { 'yùr·ə,nil 'as·ə,tāt }

uranyl nitrate [INORG CHEM] $UO_2(NO_3)_2·6H_2O$ Toxic, explosive, unstable yellow crystals; soluble in water, alcohol, and ether; melts at 60°C and boils at 118°C; used in photography, in medicine, and for uranium extraction and uranium glaze. Also known as uranium nitrate; yellow salt. { 'yùr·ə,nil 'nī,trāt }

uranyl salts [INORG CHEM] Salts of UO_3 that ionize to form UO_2^{2+} and that are yellow in solution; for example, uranyl chloride, UO_2Cl_2. { 'yùr·ə,nil ,sòls }

uranyl sulfate [INORG CHEM] $UO_2SO_4·3\frac{1}{2}H_2O$ and $UO_2SO_4·3H_2O$ Poisonous, radioactive yellow crystals; soluble in water and concentrated hydrochloric acid; used as an analytical reagent. Also known as uranium sulfate. { 'yùr·ə,nil 'səl ,fāt }

uranyl uranate See triuranium octoxide. { 'yùr·ə,nil 'yùr·ə,nāt }

urea [ORG CHEM] $CO(HN_2)_2$ A natural product of protein metabolism found in urine; synthesized as white crystals or powder with a melting point of 132.7°C; soluble in water, alcohol, and benzene; used as a fertilizer, in plastics, adhesives, and flameproofing agents, and in medicine. Also known as carbamide. { yù 'rē·ə }

urea anhydride See cyanamide. { yù'rē·ə an'hī ,drīd }

urea-formaldehyde resin [POLYM CHEM] A synthetic thermoset resin derived by the reaction of urea (carbamide) with formaldehyde or its polymers. Also known as urea resin. { yù'rē·ə fòr'mal·də,hīd 'rez·ən }

urea resin See urea-formaldehyde resin. { yù 'rē·ə 'rez·ən }

urethane [ORG CHEM] $CO(NH_2)OC_2H_5$ A combustible, toxic, colorless powder; soluble in water and alcohol; melts at 49°C; used as a solvent and chemical intermediate and in biochemical research and veterinary medicine. Also known as ethyl carbamate; ethyl urethane. { 'yùr·ə,thān }

usable iron ore [MET] A steel industry term for high-grade iron ore, concentrates, or agglomerates which can be used in blast furnaces or other processing plants. { 'yüz·ə·bəl 'ī·ərn ,òr }

usable life See pot life. { 'yüz·ə·bəl 'līf }

UV See ultraviolet.

uviol glass [MATER] A type of glass that is highly transparent to ultraviolet radiation. { 'yü·vē,òl ,glas }

UV stabilizer [CHEM] Any chemical compound that, admixed with a thermoplastic resin, selectively absorbs ultraviolet rays; used to prevent ultraviolet degradation of polymers. Also known as ultraviolet stabilizer. { ¦yü¦vē 'stā·bə,līz·ər }

V

V See electric potential; vanadium; volt.

VA See volt-ampere.

vacancy |SOLID STATE| A defect in the form of an unoccupied lattice position in a crystal. { 'vā·kən·sē }

vacuum-arc casting |MET| A process for producing metal ingots that are low in oxygen and need no further treatment to remove it, whereby an ultrapure metal or metal alloy powder is pressed, sintered, and melted (all under vacuum). { ,vak·yəm ,ärk 'kast·iŋ }

vacuum brazing |MET| Brazing utilizing a chamber at subatmospheric pressure. { 'vak·yəm 'brāz·iŋ }

vacuum casting |MET| Metal casting in a vacuum. { 'vak·yəm 'kast·iŋ }

vacuum degassing |MET| A process for removing gases from a metal either by melting or heating the solid metal in a vacuum. { 'vak·yəm dē'gas·iŋ }

vacuum deposition |MET| Deposition of a thin coating of metal by condensation on a cool work surface in vacuum. { 'vak·yəm ,dep·ə'zish·ən }

vacuum fusion |MET| A technique for determining the oxygen, hydrogen, and sometimes nitrogen content of metals; can be applied to a wide variety of metals with the exception of alkali and alkaline earth metals. { 'vak·yəm 'fyü·zhən }

vacuum metallizing |MET| A special form of vapor deposition in which a metal coating is applied to the surface of a polymer. { ¦vak·yüm 'med·əl,īz·iŋ }

vacuum metallurgy |MET| The melting, shaping, and treating of metals and alloys under reduced pressure that ranges from subatmospheric pressure to ultra-high vacuum. { 'vak·yəm 'med·əl ,ər·jē }

vacuum plating See vapor deposition. { 'vak·yəm 'plād·iŋ }

valence |CHEM| A positive number that characterizes the combining power of an element for other elements, as measured by the number of bonds to other atoms which one atom of the given element forms upon chemical combination; hydrogen is assigned valence 1, and the valence is the number of hydrogen atoms, or their equivalent, with which an atom of the given element combines. { 'vā·ləns }

valence angle See bond angle. { 'vā·ləns ,aŋ·gəl }

valence band |SOLID STATE| The highest electronic energy band in a semiconductor or insulator which can be filled with electrons. { 'vā·ləns ,band }

valence-bond theory |CHEM| A theory of the structure of chemical compounds according to which the principal requirements for the formation of a covalent bond are a pair of electrons and suitably oriented electron orbitals on each of the atoms being bonded; the geometry of the atoms in the resulting coordination polyhedron is coordinated with the orientation of the orbitals on the central atom. { 'vā·ləns ¦bänd ,thē·ə·rē }

valence crystal See covalent crystal. { 'vā·ləns ,krist·əl }

valence electron See conduction electron. { 'vā·ləns i,lek,trän }

valence number |CHEM| A number that is equal to the valence of an atom or ion multiplied by +1 or −1, depending on whether the ion is positive or negative, or equivalently on whether the atom in the molecule under consideration has lost or gained electrons from its free state. { 'vā·ləns ,nəm·bər }

valency effect |SOLID STATE| The dependence of the transition temperature of a superconductor on the concentration of doping atoms. { 'vā·lən·sē i,fekt }

vanadic acid |INORG CHEM| Any of various acids that do not exist in a pure state and are found in various alkali and other metal vanadates; forms are meta- (HVO_3), ortho- (H_3VO_4), and pyro- ($H_4V_2O_7$). { və'nād·ik 'as·əd }

vanadic acid anhydride See vanadium pentoxide. { və'nād·ik 'as·əd an'hī,drīd }

vanadic sulfate See vanadyl sulfate. { və'nād·ik 'səl,fāt }

vanadic sulfide See vanadium sulfide. { və 'nād·ik 'səl,fīd }

vanadium |CHEM| A metallic transition element, symbol V, atomic number 23; soluble in strong acids and alkalies; melts at 1900°C; boils about 3000°C; used as a catalyst. |MET| A silvery-white, ductile metal resistant to corrosion; used in alloy steels and as an x-ray target. { və'nād·ē·əm }

vanadium carbide |INORG CHEM| VC Hard, black crystals, melting at 2800°C, boiling at

3900°C; insoluble in acids, except nitric acid; used in cutting-tool alloys and as a steel additive. { və'nād·ē·əm 'kär,bīd }

vanadium dichloride [INORG CHEM] VCl₂ Toxic, green crystals, soluble in alcohol and ether; decomposes in hot water; used as a reducing agent. Also known as vanadous chloride. { və'nād·ē·əm dī'klȯr,īd }

vanadium oxide [INORG CHEM] A compound of vanadium with oxygen, for example, vanadium tetroxide (V_2O_4), vanadium trioxide or sesquioxide (V_2O_3), vanadium oxide (VO), and vanadium pentoxide (V_2O_5). { və'nād·ē·əm 'äk,sīd }

vanadium oxytrichloride [INORG CHEM] VOCl₃ A toxic, yellow liquid that dissolves or reacts with many organic substances; hydrolyzes in moisture; boils at 126°C; used as an olefin-polymerization catalyst and in organovanadium synthesis. { və'nād·ē·əm ,äk·sē·trī'klȯr,īd }

vanadium pentasulfide See vanadium sulfide. { və'nād·ē·əm ¦pen·tə'səl,fīd }

vanadium pentoxide [INORG CHEM] V_2O_5 A toxic, yellow to red powder, soluble in alkalies and acids, slightly soluble in water; melts at 690°C; used in medicine, as a catalyst, as a ceramics coloring, for ultraviolet-resistant glass, photographic developers, textiles dyeing, and nuclear reactors. Also known as vanadic acid anhydride. { və'nād·ē·əm 'pen'täk,sīd }

vanadium sesquioxide See vanadium trioxide. { və'nād·ē·əm ¦ses·kwē'äk,sīd }

vanadium steel [MET] A low-alloy steel containing 0.10–0.15% vanadium. { və'nād·ē·əm 'stēl }

vanadium sulfate See vanadyl sulfate. { və'nād·ē·əm 'səl,fāt }

vanadium sulfide [INORG CHEM] V_2S_5 A toxic, black-green powder; insoluble in water, soluble in alkalies and acids; decomposes when heated; used to make vanadium compounds. Also known as vanadic sulfide; vanadium pentasulfide. { və'nād·ē·əm 'səl,fīd }

vanadium tetrachloride [INORG CHEM] VCl₄ A toxic, red liquid; soluble in ether and absolute alcohol; boils at 154°C; used in medicine and to manufacture vanadium and organovanadium compounds. { və'nād·ē·əm ¦te·trə'klȯr,īd }

vanadium tetraoxide [INORG CHEM] V_2O_4 A toxic blue-black powder; insoluble in water, soluble in alkalies and acids; melts at 1967°C; used as a catalyst. { və'nād·ē·əm ¦te·trə'äk,sīd }

vanadium trichloride [INORG CHEM] VCl₃ Toxic, deliquescent, pink crystals; soluble in ether and absolute alcohol; decomposes in water and when heated; used to prepare vanadium and organovanadium compounds. { və'nād·ē·əm ¦trī'klȯr,īd }

vanadium trioxide [INORG CHEM] V_2O_3 Toxic, black crystals; soluble in alkalies and hydrofluoric acid, slightly soluble in water; melts at 1970°C; used as a catalyst. Also known as vanadium sesquioxide. { və'nād·ē·əm trī'äk,sīd }

vanadous chloride See vanadium dichloride. { və'nād·əs 'klȯr,īd }

vanadyl chloride [INORG CHEM] $V_2O_2Cl_4·5H_2O$ Toxic, deliquescent, water- and alcohol-soluble

green crystals; used to mordant textiles. { və'nād·əl 'klȯr,īd }

vanadyl sulfate [INORG CHEM] $VOSO_4·2H_2O$ Blue, toxic, water-soluble crystals; used as a reducing agent, catalyst, glass and ceramics colorant, and mordant. Also known as vanadic sulfate; vanadium sulfate. { və'nād·əl 'səl,fāt }

van der Waals adsorption [PHYS CHEM] Adsorption in which the cohesion between gas and solid arises from van der Waals forces. { 'van dər ,wȯlz ad,sȯrp·shən }

van der Waals attraction See van der Waals force. { 'van dər ,wȯlz ə,trak·shən }

van der Waals equation [PHYS CHEM] An empirical equation of state which takes into account the finite size of the molecules and the attractive forces between them: $p = [RT/(v − b)] − (a/v^2)$, where p is the pressure, v is the volume per mole, T is the absolute temperature, R is the gas constant, and a and b are constants. { 'van dər ,wȯlz i,kwā·zhən }

van der Waals force [PHYS CHEM] An attractive force between two atoms or nonpolar molecules, which arises because a fluctuating dipole moment in one molecule induces a dipole moment in the other, and the two dipole moments then interact. Also known as dispersion force; van der Waals attraction. { 'van dər ,wȯlz ,fȯrs }

van der Waals structure [CRYSTAL] The structure of a molecular crystal. { 'van dər ,wȯlz ,strək·chər }

van der Waals surface tension formula [THERMO] An empirical formula for the dependence of the surface tension on temperature: $\gamma = Kp_c^{2/3} T_c^{1/3} (1 − T/T_c)^n$, where γ is the surface tension, T is the temperature, T_c and p_c are the critical temperature and pressure, K is a constant, and n is a constant equal to approximately 1.23. { 'van dər ,wȯlz 'sər·fəs ,ten·chən ,fȯr·myə·lə }

van't Hoff equation [PHYS CHEM] An equation for the variation with temperature T of the equilibrium constant K of a gaseous reaction in terms of the heat of reaction at constant pressure, ΔH: $d(\ln K)/dT = \Delta H/RT^2$, where R is the gas constant. Also known as van't Hoff isochore. { van'tȯf i,kwā·zhən }

van't Hoff isochore See van't Hoff equation. { van'tȯf 'ī·sə,kȯr }

van't Hoff isotherm [PHYS CHEM] An equation for the change in free energy during a chemical reaction in terms of the reaction, the temperature, and the concentration and number of molecules of the reactants. { van'tȯf 'ī·sə,thərm }

vapor [THERMO] A gas at a temperature below the critical temperature, so that it can be liquefied by compression, without lowering the temperature. { 'vā·pər }

vapor blasting [MET] Cleaning the surface of a metal with a fine abrasive suspended in water and propelled at high speed by air or steam. Also known as liquid honing; vapor honing. { 'vā·pər 'blast·iŋ }

vapor cycle [THERMO] A thermodynamic cycle, operating as a heat engine or a heat pump, during which the working substance is in, or passes through, the vapor state. { 'vā·pər ˌsī·kəl }

vapor deposition [MET] Producing a film of metal on a heated surface, often in a vacuum, either by decomposition of the vapor of a compound at the work surface or by direct reaction between the work surface and the vapor. Also known as vacuum plating. { 'vā·pər ˌdep·ə'zish·ən }

vapor honing See vapor blasting. { 'vā·pər 'hōn·iŋ }

vaporization See volatilization. { ˌvā·pə·rə'zā·shən }

vaporization coefficient [THERMO] The ratio of the rate of vaporization of a solid or liquid at a given temperature and corresponding vapor pressure to the rate of vaporization that would be necessary to produce the same vapor pressure at this temperature if every vapor molecule striking the solid or liquid were absorbed there. { ˌvā·pə·rə'zā·shən ˌkō·ə·fish·ənt }

vapor-phase epitaxy [SOLID STATE] The use of chemical vapor deposition to grow epitaxial layers. Abbreviated VPE. { 'vā·pərˌfāz 'ep·ə,tak·sē }

vapor pressure [THERMO] For a liquid or solid, the pressure of the vapor in equilibrium with the liquid or solid. { 'vā·pər ˌpresh·ər }

variable-thickness microbridge [CRYO] A Josephson junction formed by a short, narrow constriction in a thin superconducting film, which is thinner than the rest of the film. { 'ver·ē·ə·bəl ˌthik·nəs 'mī·krō,brij }

varnish [MATER] A transparent surface coating which is applied as a liquid and then changes to a hard solid; all varnishes are solutions of resinous materials in a solvent. { 'vär·nish }

varnish makers' and painters' naphtha [MATER] A petroleum naphtha that has a narrow boiling range and is used mainly as a thinner in paint and varnish. Abbreviated VM & P naphtha. { 'vär·nish ˌmāk·ərz ən ˌpān·tərz 'naf·thə }

vat dye [MATER] One of the dyes that are easily reduced to a soluble and colorless form in which they easily impregnate fibers; subsequent oxidation produces the final color; examples are indigo and indanthrene blue. { 'vat ˌdī }

vat printing assistant [MATER] The carrier for the dye in the printing of fabrics with vat dyes; a mixture of gums and reducing and wetting agents to assist in penetrating the fabric. { 'vat ˌprint·iŋ ə,sis·tənt }

Vegard's Law [MET] Linear relation between lattice parameters and composition of solid solution alloys expressed as atomic percentage. { ve'gärz ˌlò }

vehicle [MATER] The fluid component of a paint or printing ink; acts as a carrier for the pigment. { 've·ə·kəl }

veining [MET] Lines in a polished and etched metal surface marking slight imperfections in structure of an otherwise single grain. { 'vān·iŋ }

vellum [MATER] A high-grade paper made to resemble genuine parchment. { 'vel·əm }

velour paper [MATER] A paper with a velvetlike finish, produced by flocking the surface with fine bits of rayon, nylon, cotton, or wool; it is

sometimes embossed in various patterns. { və'lúr ˌpā·pər }

veneer [MATER] **1.** A thin sheet of wood of uniform thickness used for facing furniture or, when bonded, used to make plywood. **2.** A facing, as of brick or marble, on the outside of a wall. { və'nir }

Venetian red [INORG CHEM] A pigment with a true red hue; contains 15–40% ferric oxide and 60–80% calcium sulfate. { və'nēsh·ən 'red }

vent [MET] A small opening in a casting mold to allow for the escape of gases. { vent }

vermilion See mercuric sulfide. { vər'mil·yən }

vernier caliper [ENG] A caliper rule with an attached vernier scale. { 'vər·nē·ər 'kal·ə·pər }

vertical-position welding [MET] Welding in which the weld axis is essentially vertical. { 'vərd·ə·kəl pə¦zish·ən 'weld·iŋ }

vertical retort process [MET] A zinc-smelting method using a vertical retort of silicon carbide brick. Also known as New Jersey retort process. { 'vərd·ə·kəl ri¦tórt ˌprä·səs }

vibrational energy [PHYS CHEM] For a diatomic molecule, the difference between the energy of the molecule idealized by setting the rotational energy equal to zero, and that of a further idealized molecule which is obtained by gradually stopping the vibration of the nuclei without placing any new constraint on the motions of electrons. { vī'brā·shən·əl 'en·ər·jē }

vicinal [ORG CHEM] Referring to neighboring or adjoining positions on a carbon structure (ring or chain). { 'vis·ən·əl }

vicinal faces [CRYSTAL] Macroscopic crystal faces which are inclined only a few minutes of arc to crystal faces with low Miller indices, and which therefore must have high Miller indices themselves. { 'vis·ən·əl ˌfās·əz }

Vickers hardness test See diamond-pyramid hardness test. { 'vik·ərz 'härd·nəs ˌtest }

vinyl acetal resin [POLYM CHEM] $[CH_2CH(OC_2H_5)]_x$ A colorless, odorless, light-stable thermoplastic that is unaffected by water, gasoline, or oils; soluble in lower alcohols, benzene, and chlorinated hydrocarbons; used in lacquers, coatings, and molded objects. Also known as polyvinyl acetal resin. { 'vīn·əl 'as·ə ˌtal 'rez·ən }

vinyl acetate [ORG CHEM] $CH_3COOCH:CH_2$ A colorless, water-insoluble, flammable liquid that boils at 73°C; used as a chemical intermediate and in the production of polymers and copolymers (for example, the polyvinyl resins). { 'vīn·əl 'as·ə,tāt }

vinyl acetate resin [POLYM CHEM] $(CH_2:CHOOCCH_3)_x$ An odorless thermoplastic formed by the polymerization of vinyl acetate; resists attack by water, gasoline, and oils; soluble in lower alcohols, benzene, and chlorinated hydrocarbons; used in lacquers, coatings, and molded products. { 'vīn·əl 'as·ə,tāt 'rez·ən }

vinyl alcohol [ORG CHEM] $CH_2:CHOH$ A flammable, unstable liquid found only in ester or polymer form. Also known as ethenol. { 'vīn·əl 'al·kə,hòl }

vinylbenzene See styrene. { ¦vīn·əl'ben,zēn }
vinyl chloride |ORG CHEM| CH_2:CHCl A flammable, explosive gas with an ethereal aroma; soluble in alcohol and ether, slightly soluble in water; boils at $-14°C$; an important monomer for polyvinyl chloride and its copolymers; used in organic synthesis and in adhesives. Also known as chloroethene { 'vīn·əl 'klȯr,īd }
vinyl chloride resin |POLYM CHEM| $(CH_2CHCl)_x$ A white-power polymer made by the polymerization of vinyl chloride; used to make chemical-resistant pipe (when unplasticized) or bottles and parts (when plasticized). { 'vīn·əl 'klȯr,īd ,rez·ən }
vinylcyanide See acrylonitrile. { ¦vīn·əl'sī·ə ,nīd }
vinyl ether |ORG CHEM| CH_2:CHOCH:CH_2 A colorless, light-sensitive, flammable, explosive liquid; soluble in alcohol, acetone, ether, and chloroform, slightly soluble in water; boils at $39°C$; used as an anesthetic and a comonomer in polyvinyl chloride polymers. Also known as divinyl ether; divinyl oxide. { 'vīn·əl 'ē·thər }
vinyl ether resin |POLYM CHEM| Any of a group of vinyl ether polymers; for example, polyvinyl methyl ether, polyvinyl ethyl ether, and polyvinyl butyl ether. { 'vīn·əl 'ē·thər 'rez·ən }
vinyl group |ORG CHEM| CH_2=CH– A group of atoms derived when one hydrogen atom is removed from ethylene. { 'vīn·əl ,grüp }
vinylidene chloride |ORG CHEM| CH_2:CCl_2 A colorless, flammable, explosive liquid, insoluble in water; boils at $37°C$; used to make polymers copolymerized with vinyl chloride or acrylonitrile (Saran). { vī'nil·ə,dēn 'klȯr,īd }
vinylidene resin |POLYM CHEM| A polymer made up of the $(-H_2CCX_2-)$ unit, with X usually a chloride, fluoride, or cyanide radical. Also known as polyvinylidene resin. { vī'nil·ə,dēn 'rez·ən }
vinyl plastic See polyvinyl resin. { 'vīn·əl 'plas·tik }
vinyl polymerization |POLYM CHEM| Addition polymerization where the unsaturated monomer contains a CH_2=C– group. { 'vīn·əl pə,lim·ə·rə'zā·shən }
violet |OPTICS| The hue evoked in an average observer by monochromatic radiation having a wavelength in the approximate range from 390 to 455 nanometers; however, the same sensation can be produced in a variety of other ways. { 'vī·ə·lət }
virial coefficients |THERMO| For a given temperature T, one of the coefficients in the expansion of P/RT in inverse powers of the molar volume, where P is the pressure and R is the gas constant. { 'vir·ē·əl ,kō·i'fish·əns }
virtual entropy |THERMO| The entropy of a system, excluding that due to nuclear spin. Also known as practical entropy. { 'vər·chə·wəl 'en·trə·pē }
viscoelastic fluid |FL MECH| A fluid that displays viscoelasticity. { ¦vis·kō·i¦las·tik 'flü·əd }
viscoelasticity |MECH| Property of a material which is viscous but which also exhibits cer-

tain elastic properties such as the ability to store energy of deformation, and in which the application of a stress gives rise to a strain that approaches its equilibrium value slowly. { ¦vis·kō·i¦las·tis·əd·ē }
viscoelastic theory |MECH| The theory which attempts to specify the relationship between stress and strain in a material displaying viscoelasticity. { ¦vis·kō·i¦las·tik 'thē·ə·rē }
viscometer |ENG| An instrument designed to measure the viscosity of a fluid. { vi'skäm· əd·ər }
viscometric analysis |FL MECH| Measurement of the flow properties of substances by viscometry. { ¦vis·kə¦me·trik ə'nal·ə·səs }
viscometry |ENG| A branch of rheology; the study of the behavior of fluids under conditions of internal shear; the technology of measuring viscosities of fluids. { vi'skäm·ə·trē }
viscosity |FL MECH| The resistance that a gaseous or liquid system offers to flow when it is subjected to a shear stress. Also known as flow resistance; internal friction. { vi'skäs· əd·ē }
viscosity coefficient |FL MECH| An empirical number used in equations of fluid mechanics to account for the effects of viscosity. { vi'skäs·əd·ē ,kō·i,fish·ənt }
viscosity curve |FL MECH| A graph showing the viscosity of a liquid or gaseous material as a function of temperature. { vi'skäs·əd·ē ,kərv }
viscous flow |FL MECH| **1.** The flow of a viscous fluid. **2.** The flow of a fluid through a duct under conditions such that the mean free path is small in comparison with the smallest, transverse section of the duct. { 'vis·kəs 'flō }
viscous fluid |FL MECH| A fluid whose viscosity is sufficiently large to make the viscous forces a significant part of the total force field in the fluid. { 'vis·kəs 'flü·əd }
viscous force |FL MECH| The force per unit volume or per unit mass arising from viscous effects in fluid flow. { 'vis·kəs 'fȯrs }
visible radiation See light. { 'viz·ə·bəl ,rād·ē'ā·shən }
vitreous enamel |MATER| A glass coating applied to a metal by covering the surface with a powdered glass frit and heating until fusion occurs. Also known as porcelain enamel. { 'vi·trē·əs i'nam·əl }
vitreous silica See silica glass. { 'vi·trē·əs 'sil·ə· kə }
vitreous state |SOLID STATE| A solid state in which the atoms or molecules are not arranged in any regular order, as in a crystal, and which crystallizes only after an extremely long time. Also known as glassy state. { 'vi·trē·əs ,stāt }
vitrification |ENG| Heat treatment of a material such as a ceramic to produce a glazed surface. { ,vi·trə·fə'kā·shən }
vitrified brick |MATER| A type of brick that has been glazed to render it impervious to water and highly resistant to corrosion. { ¦vi·trə,fīd 'brik }

vitrified-clay pipe [MATER] A pipe, made of clay treated in a kiln to induce vitrification, with the surface glazed for watertightness; used for drainage. { 'vi·trə‚fīd ¦klā 'pīp }

vitriolic acid See sulfuric acid. { 'vi·trē‚äl·ik 'əs·ed }

VM & P naphtha See varnish makers' and painters' naphtha. { ¦vē¦em ən ¦pē 'naf·thə }

voidage [MATER] The volume of the voids in a sample of powder divided by its overall volume (that is, the total volume occupied by the voids and the solid material). { 'void·ij }

volatile [CHEM] Readily passing off by evaporation. { 'väl·əd·əl }

volatility [THERMO] The quality of having a low boiling point or subliming temperature at ordinary pressure or, equivalently, of having a high vapor pressure at ordinary temperatures. { ‚väl·ə'til·əd·ē }

volatilization [THERMO] The conversion of a chemical substance from a liquid or solid state to a gaseous or vapor state by the application of heat, by reducing pressure, or by a combination of these processes. Also known as vaporization. { ‚väl·əd·əl·ə'zā·shən }

volt [ELEC] The unit of potential difference or electromotive force in the meter-kilogram-second system, equal to the potential difference between two points for which 1 coulomb of electricity will do 1 joule of work in going from one point to the other. Symbolized V. { vōlt }

Volta effect See contact potential difference. { 'vōl·tə i‚fekt }

voltage [ELEC] Potential difference or electromotive force measured in volts. { 'vōl·tij }

voltage drop [ELEC] The voltage developed across a component or conductor by the flow of current through the resistance or impedance of that component or conductor. { 'vōl·tij ‚dräp }

voltage gradient [ELEC] The voltage per unit length along a resistor or other conductive path. { 'vōl·tij ‚grād·ē·ənt }

voltaic cell [ELEC] A primary cell consisting of two dissimilar metal electrodes in a solution that acts chemically on one or both of them to produce a voltage. { vōl'tā·ik 'sel }

voltameter See coulometer. { väl'tam·əd·ər }

voltammetry [PHYS CHEM] Any electrochemical technique in which a faradaic current passing through the electrolysis solution is measured while an appropriate potential is applied to the polarizable or indicator electrode; for example, polarography. { vōlt'äm·ə·trē }

volt-ampere [ELEC] The unit of apparent power in the International System; it is equal to the apparent power in a circuit when the product of the root-mean-square value of the voltage, expressed in volts, and the root-mean-square value of the current, expressed in amperes, equals 1. Abbreviated VA. { 'vōlt 'am ‚pir }

volt-ampere hour [ELEC] A unit for expressing the integral of apparent power over time, equal to the product of 1 volt-ampere and 1 hour, or to 3600 joules. { 'vōlt 'am‚pir 'aúr }

Volterra dislocation [SOLID STATE] A model of a dislocation which is formed in a ring of crystalline material by cutting the ring, moving the cut surfaces over each other, and then rejoining them. { vol'ter·ə ‚dis·lō'kā·shən }

voltmeter [ENG] An instrument for the measurement of potential difference between two points, in volts or in related smaller or larger units. { 'vōlt‚mēd·ər }

voltmeter-ammeter [ENG] A voltmeter and an ammeter combined in a single case but having separate terminals. { 'vōlt‚mēd·ər 'am‚ēd·ər }

volt-ohm-milliammeter [ENG] A test instrument having a number of different ranges for measuring voltage, current, and resistance. Also known as circuit analyzer; multimeter; multiple-purpose tester. { 'vōlt 'ōm ¦mil·ē'am‚ēd·ər }

volume lifetime [SOLID STATE] Average time interval between the generation and recombination of minority carriers in a homogeneous semiconductor. { 'väl·yəm 'līf‚tīm }

volumenometer [ENG] An instrument for determining the volume of a body by measuring the pressure in a closed air space when the specimen is present and when it is absent. { väl ‚yü·mə'näm·əd·ər }

volume recombination rate [SOLID STATE] The rate at which free electrons and holes within the volume of a semiconductor recombine and thus neutralize each other. { 'väl·yəm 'rē ‚käm·bə'nā·shən ‚rāt }

volume resistivity [ELEC] Electrical resistance between opposite faces of a 1-centimeter cube of insulating material, commonly expressed in ohm-centimeters. Also known as specific insulation resistance. { 'väl·yəm ‚rē‚zis'tiv·əd·ē }

volume susceptibility [PHYS CHEM] The magnetic susceptibility of a specified volume (for example, 1 cubic centimeter) of a magnetically susceptible material. { 'väl·yəm sə‚sep·tə 'bil·əd·ē }

volumeter [ENG] Any instrument for measuring volumes of gases, liquids, or solids. { 'väl·yə ‚mēd·ər }

volumetric strain [MECH] One measure of deformation; the change of volume per unit of volume. { ¦väl·yə¦me·trik 'strān }

von Mises yield criterion [MECH] The assumption that plastic deformation of a material begins when the sum of the squares of the principal components of the deviatoric stress reaches a certain critical value. { fön ¦mēz·əz 'yēld ‚krī ‚tir·ē·ən }

vortex [FL MECH] **1.** Any flow possessing vorticity; for example, an eddy, whirlpool, or other rotary motion. **2.** A flow with closed streamlines, such as a free vortex or line vortex. [SOLID STATE] See fluxoid. { 'vòr‚teks }

VPE See vapor-phase epitaxy.

vulcanization [POLYM CHEM] A chemical reaction of sulfur (or other vulcanizing agent) with rubber or plastic to cause cross-linking of the polymer chains; it increases strength and resiliency of the polymer. Also known as cure. { ˌvəl·kə·nə'zā·shən }

vulcanized fiber [MATER] A laminated plastic made by chemically treating layers of 100% rag-content paper to gelatinize the paper and fuse the layers into a solid mass; when dried under pressure, it forms a hard, tough material having good electrical properties along with mechanical strength and dimensional stability. { 'vəl·kə ˌnīzd 'fī·bər }

W

W *See* tungsten; watt.

wafer |ELECTR| A thin semiconductor slice on which matrices of microcircuits can be fabricated, or which can be cut into individual dice for fabricating single transistors and diodes. { 'wā·fər }

wallboard |MATER| Panels of various materials for surfacing ceilings and walls, including asbestos cement sheet, plywood, gypsum plasterboard, and laminated plastics. { 'wȯl,bȯrd }

wall energy |SOLID STATE| The energy per unit area of the boundary between two ferromagnetic domains which are oriented in different directions. { 'wȯl ¦en·ər·jē }

wall superheat |THERMO| The difference between the temperature of a surface and the saturation temperature (boiling point at the ambient pressure) of an adjacent liquid that is heated by the surface. { ¦wȯl 'sü·pər,hēt }

wandering sequence |MET| A welding sequence in which the increments of weld bead are longitudinally deposited in a random fashion. { 'wän·də·riŋ ,sē·kwəns }

wane |MATER| A rounded edge of bark along an edge or at a corner of a section of lumber. { wān }

Wannier function |SOLID STATE| The Fourier transform of a Bloch function defined for an entire band, regarded as a function of the wave vector. { vän'yā ,fəŋk·shən }

warning agent *See* odorant. { 'wȯrn·iŋ ,ā·jənt }

warpage |MECH| The action, process, or result of twisting or turning out of shape. { 'wȯr·pij }

wash |MET| **1.** A coating applied to the face of a mold prior to casting. **2.** A sand expansion defect on the surface finish of a casting due to radiation from the metal rising in the mold and causing increased volume and shear of the interface sand on the upper layers. { wäsh }

washing soda *See* sal soda. { 'wäsh·iŋ ,sōd·ə }

Washita stone |MATER| A relatively porous and not very dense oilstone, used chiefly for whetstones and for sharpening coarse tools. { 'wäsh·əd·ə ,stōn }

wash metal |MET| Molten metal used to clean out a furnace, ladle, or other container. { 'wäsh ,med·əl }

wash primer |MET| A synthetic vehicle primer containing phosphoric acid and zinc chromate; used as a corrosion-inhibiting first paint coat on metals. { 'wäsh ,prī·mər }

water |CHEM| H_2O Clear, odorless, tasteless liquid that is essential for most animal and plant life and is an excellent solvent for many substances; melting point $0°C$ $(32°F)$, boiling point $100°C$ $(212°F)$; the chemical compound may be termed hydrogen oxide. { 'wȯd·ər }

water-base paint |MATER| Paint in which the vehicle or binder is dissolved in water or in which the vehicle or binder is dispersed as an emulsion; an example of the dispersion type is latex paint. Also known as water-thinned paint. { 'wȯd·ər ¦bās 'pānt }

water break |MET| A break in the continuity of the film of water on the surface of a metal withdrawn from an aqueous bath. { 'wȯd·ər ,brāk }

watercolor |MATER| A pigment ground in a solution of gum arabic, water, and plasticizer, such as glycerin; the glycerin film retards drying in the tube and prevents brittleness in the paint film. { 'wȯd·ər,kəl·ər }

watercolor paper |MATER| A special drawing paper with a surface texture suitable to accept watercolors; the better grades can withstand the harsh scraping that is sometimes necessary to produce highlights; for permanent painting, the paper should be 100% rag, not wood pulp. { 'wȯd·ər,kəl·ər ,pā·pər }

watercolor pigment |INORG CHEM| A permanent pigment used in watercolor painting, for example, titanium oxide (white). { 'wȯd·ər,kəl·ər ,pig·mənt }

water glass *See* sodium silicate. { 'wȯd·ər ,glas }

water of crystallization *See* water of hydration. { 'wȯd·ər əv ,krist·əl·ə'zā·shən }

water of hydration |CHEM| Water present in a definite amount and attached to a compound to form a hydrate; can be removed, as by heating, without altering the composition of the compound. Also known as water of crystallization. { 'wȯd·ər əv hī'drā·shən }

waterproof grease |MATER| A viscous lubricating material that does not dissolve in water and that resists being washed out of bearings or gears; it usually has a low content of oil and metallic soaps of aluminum, barium, calcium, or strontium. { 'wȯd·ər,prüf 'grēs }

waterproofing agent |MATER| A substance used to make textiles, paper, wood, and other porous or absorbent materials impervious to penetration by water. { ¦wȯd·ər¦prüf·iŋ ‚ā·jənt }

water-pump lubricant |MATER| A lubricating grease suitable for the types of automotive water pumps that require grease lubrication. { 'wȯd·ər ‚pəmp ‚lü·brə·kənt }

water putty |MATER| A powder that forms a puttylike paste when mixed with water and is used to fill small holes or cracks in wood. { 'wȯd·ər ‚pəd·ē }

water-reducing agent |MATER| An additive for freshly mixed mortar or concrete for increasing workability without increasing water content, or for maintaining workability with a reduced amount of water. { 'wȯd·ər ri‚düs·iŋ ‚ā·jənt }

water repellent |MATER| Chemicals used to treat textiles, leather, paper, or wood to make them resistant (but not proof) to wetting by water; includes various types of resins, aluminum of zirconium acetates, or latexes. { 'wȯd·ər ri‚pel·ənt }

water-thinned paint See water-base paint. { 'wȯd·ər ¦thind ¦pānt }

water white |CHEM| A grade of color for liquids that has the appearance of clear water; for petroleum products, a plus 21 in the scale of the Saybolt chromometer. { 'wȯd·ər 'wīt }

watt |PHYS| The unit of power in the meter-kilogram-second system of units, equal to 1 joule per second. Symbolized W. { wät }

watt-hour |ELEC| A unit of energy used in electrical measurements, equal to the energy converted or consumed at a rate of 1 watt during a period of 1 hour, or to 3600 joules. Abbreviated Wh. { 'wät ‚aủr }

wattle gum |MATER| A gum arabic extracted from the Australian and East African trees of the genus *Acacia*, as A. *dealbata*, and other species (called mimosa in Kenya); contains 65% tannin. { 'wad·əl ‚gəm }

watt-second |PHYS| Amount of energy corresponding to 1 watt acting for 1 second; 1 watt-second is equal to 1 joule. { 'wät ¦sek·ənd }

Watt's law |THERMO| A law which states that the sum of the latent heat of steam at any temperature of generation and the heat required to raise water from 0°C to that temperature is constant; it has been shown to be substantially in error. { 'wäts ‚lȯ }

wavelength |PHYS| The distance between two points having the same phase in two consecutive cycles of a periodic wave, along a line in the direction of propagation. { 'wāv‚leŋkth }

wave optics |OPTICS| The branch of optics which treats of light (or electromagnetic radiation in general) with explicit recognition of its wave nature. { 'wāv ‚äp·tiks }

wave polarization See polarization. { 'wāv ‚pō·lə·rə‚zā·shən }

wave theory of light |OPTICS| A theory which assumes that light is a wave motion, rather than a stream of particles. { 'wāv 'thē·ə·rē əv 'līt }

wave-vector space |SOLID STATE| The space of the wave vectors of the state functions of some system; this would be used, for example, for electron wave functions in a crystal and thermal vibrations of a lattice. Also known as **k**-space; reciprocal space. { 'wāv ¦vek·tər ‚spās }

wax |MATER| Any of a group of substances resembling beeswax in appearance and character, and in general distinguished by their composition of esters and higher alcohols, and by their freedom from fatty acids. { waks }

wax distillate |MATER| A neutral distillate from distillation of crude oil that contains a high percentage of crystallizable paraffin wax; used as a primary base for paraffin wax and neutral lubricating oils. { 'waks 'dis·tə·lət }

waxed paper |MATER| Paper that is treated or coated with wax to make it waterproof and grease-proof; used for wrapping. { 'wakst 'pā·pər }

wax stain |MATER| A semitransparent pigment mixed with beeswax, and thinned with turpentine. { 'waks 'stān }

wear |ENG| Deterioration of a surface due to material removal caused by relative motion between it and another part. { wer }

web |GRAPHICS| The continuous length of paper formed when paper pulp moves through a papermaking machine; the web is then cut into sheets or wrapped onto rolls. |MET| In forging, the thin section of metal remaining at the bottom of a depression or at the location of the punches. { web }

Weber number 1 |FL MECH| A dimensionless number used in the study of surface tension waves and bubble formation, equal to the product of the square of the velocity of the wave or the fluid velocity, the density of the fluid, and a characteristic length, divided by the surface tension. Symbolized N_{We1}, We. { 'vā·bər ¦nəm·bər 'wən }

Weber number 2 |FL MECH| A dimensionless number, equal to the square root of Weber number 1. Symbolized N_{We2}. { 'vā·bər ¦nəm·bər 'tü }

Weissenberg method |SOLID STATE| A method of studying crystal structure by x-ray diffraction in which the crystal is rotated in a beam of x-rays, and a photographic film is moved parallel to the axis of rotation; the crystal is surrounded by a sleeve which has a slot that passes only diffraction spots from a single layer of the reciprocal lattice, permitting positive identification of each spot in the pattern. { 'vīs·ən‚berk ‚meth·əd }

Weiss molecular field |SOLID STATE| The effective magnetic field postulated in the Weiss theory of ferromagnetism, which acts on atomic magnetic moments within a domain, tending to align them, and is in turn generated by these magnetic moments. { 'ves mə'lek·yə·lər 'fēld }

Weiss theory |SOLID STATE| A theory of ferromagnetism based on the hypotheses that below the Curie point a ferromagnetic substance is composed of small, spontaneously magnetized regions called domains, and that each domain is spontaneously magnetized because a strong molecular magnetic field tends to align the individual atomic magnetic moments within the domain. Also known as molecular field theory. { 'ves ,thē·ə·rē }

weld |MET| A union made between two metals by welding. { weld }

weldability |MET| Suitability of a metal to be welded under specified conditions. { ,wel·də'bil·əd·ē }

weld bead |MET| A deposit of filler metal from a single welding pass. Also known as bead. { 'weld ,bēd }

weldbonding |MET| A process for joining metals in which adhesive, typically an epoxy paste, is applied to the parts, which are then clamped together, spot-welded, and put into an oven (250°F, or 121°C, for 1 hour) to cure the adhesive. { 'weld,bänd·iŋ }

weld decay |MET| Intercrystalline corrosion of austenitic stainless steels near welded areas; caused by chromium carbide precipitation along grain boundaries of alloy subject to prolonged heating in the temperature range 400–850°C. { 'weld di,kā }

weld delay time |MET| Delay of the time current in spot, seam, or projection welding with respect to starting the forge delay timer used to synchronize pressure and heat. { 'weld di'lā ,tīm }

welder |MET| 1. A machine used in welding. Also known as welding machine. 2. A person who performs a welding operation. { 'wel·dər }

weld gage |ENG| A device used to check the shape and size of welds. { 'weld ,gāj }

welding |MET| Joining two metals by applying heat to melt and fuse them, with or without filler metal. { 'weld·iŋ }

welding cycle |MET| The complete sequence of events involved in making a resistance weld. { 'weld·iŋ ,sī·kəl }

welding electrode |MET| 1. In arc welding, the current-carrying rod or rods used to strike an arc between rod and work. 2. In resistance welding, the component of a machine through which current and pressure are applied to the work. { 'weld·iŋ i,lek,trōd }

welding force See electrode force. { 'weld·iŋ ,fòrs }

welding ground See work lead. { 'weld·iŋ ,graùnd }

welding machine See welder. { 'weld·iŋ mə ,shēn }

welding rod |MET| Filler metal in the form of a rod or heavy wire. { 'weld·iŋ ,räd }

welding schedule |MET| A record of all welding machine settings plus identification of the machine needed to produce a weld for a given material of a given size and finish. { 'weld·iŋ ,skej·əl }

welding sequence |MET| The order for welding component parts of a weldment or structure. { 'weld·iŋ ,sē·kwəns }

welding stress |MET| Residual stress resulting from localized heating and cooling during welding. { 'weld·iŋ ,stress }

welding tip |ENG| A replaceable nozzle for a gas torch used in welding. |MET| An electrode used in spot or projection welding. { 'weld·iŋ ,tip }

welding torch |ENG| A gas-mixing and burning tool for the welding of metal. { 'weld·iŋ ,tòrch }

weld interval |MET| The total heat and cool times for making one multiple-impulse weld. { 'weld ,in·tər·vəl }

weld-interval timer |ENG| A device used to control weld interval. { 'weld |in·tər·vəl ,tīm·ər }

weld line |MET| The junction of the weld metal and base metal, or the junction of base-metal parts when filler metal is not used. { 'weld ,līn }

weldment |ENG| An assembly or structure whose component parts are joined by welding. { 'weld·mənt }

weld metal |MET| The metal constituting the fused zone in spot, seam, or projection welding. { 'weld ,med·əl }

weld time |MET| The time that the welding current is applied to the work in single-impulse and flash welding. { 'weld ,tīm }

weld zone |MET| The region of a weld that includes both the weld metal and the heat-affected zone. { 'weld ,zōn }

Werner complex See coordination compound. { 'ver·nər ,käm,pleks }

West African cedar See Sapele mahogany. { 'west |af·rə·kən 'sēd·ər }

wet |PHYS| A liquid is said to wet a solid if the contact angle between the solid and the liquid, measured through the liquid, lies between 0 and 90°, and not to wet the solid if the contact angle lies between 90 and 180°. { wet }

wet blasting |MET| Liquid honing in which an impeller wheel drives the liquid suspension. { 'wet 'blast·iŋ }

wet corrosion |MET| Corrosion caused by exposure to aqueous solutions. { 'wet kə'rō·zhən }

wet drawing |MET| Drawing in which the dies and blocks are completely immersed in the lubricant. { 'wet |drò·iŋ }

wet rot |MATER| Fungal decay of wood with a high moisture content. { 'wet ,rät }

wet strength |MATER| 1. The strength of a material saturated with water. 2. The ability to withstand water (as for paper products) with a wet-strength additive or resin finish. { 'wet ,streŋkth }

wet-strength paper |MATER| Paper with increased water resistance due to processing and interlocking of fibers, as well as impregnation with small amounts of resins (for example, melamine or urea formaldehyde). Also known as wet-strong paper. { 'wet |streŋkth 'pā·pər }

wet-strong paper See wet-strength paper. { 'wet |stroŋ 'pā·pər }

wettability |CHEM| The ability of any solid surface to be wetted when in contact with a liquid; that is, the surface tension of the liquid is reduced so that the liquid spreads over the surface. { ,wed·ə'bil·əd·ē }

wetted |CHEM| Pertaining to material that has accepted water or other liquid, either on its surface or within its pore structure. { 'wed·əd }

wetting |MET| Spreading liquid filler metal or flux on a solid base metal. { 'wed·iŋ }

wetting agent |CHEM ENG| A substance that

reduces the surface tension of a liquid, increasing the rate at which it spreads across a surface when it is added to the liquid in small amounts. { 'wed·iŋ ,ā·jənt }

wetting angle [FL MECH] A contact angle which lies between 0 and 90°. { 'wed·iŋ ,aŋ·gəl }

whetstone [MATER] Any hard, fine-grained, naturally occurring, usually siliceous rock suitable for sharpening cutting instruments. { ,wet,stōn }

whisker See crystal whisker. { 'wis·kər }

white carbon black [MATER] A white silica powder made from silicon tetrachloride; used as a replacement for carbon black in rubber compounding. { 'wīt 'kär·bən 'blak }

white cast iron [MET] An extremely hard cast iron, rapidly cooled from the melt; contains about 3% carbon in the form of cementite and fine pearlite. { 'wīt 'kast 'ī·ərn }

white cement [MATER] Pure white portland cement, made from raw materials with a low iron content, or by using a reducing flame to fire the clinker. { 'wīt si'ment }

white copperas See zinc sulfate. { 'wīt 'käp·rəs }

white graphite See hexagonal boron nitride. { ,wīt 'graf,īt }

whiteheart malleable iron [MET] White cast iron malleableized and decarburized by heat treatment in an oxidizing material at 900°C for 100–150 hours; decarburization produces a light-colored fracture, in contrast to blackheart malleable iron, which is not decarburized. Also known as blackheart malleable iron. { 'wīt,härt 'mal·yə·bəl 'ī·ərn }

white iron [MET] A brittle cast iron whose total carbon content is in the combined forms, and containing little or no graphite; a fresh fracture is white. { 'wīt 'ī·ərn }

white lead [INORG CHEM] Basic lead carbonate of variable composition, the oldest and most important lead paint pigment; also used in putty and ceramics. { 'wīt 'led }

white light [OPTICS] Any radiation producing the same color sensation as average noon sunlight. { 'wīt 'līt }

white metal [MET] **1.** Any of several white-colored metals and their alloys of relatively low melting points, such as lead, tin, antimony, and zinc. **2.** A copper matte of about 77% copper, obtained from the smelting of sulfide copper ores. { 'wīt ,med·əl }

white-metal bearing alloy See lead-base babbitt. { 'wīt ,med·əl ¦ber·iŋ 'al,ȯi }

white phosphorus [CHEM] The element phosphorus in its allotropic form, a soft, waxy, poisonous solid melting at 44.5°C; soluble in carbon disulfide, insoluble in water and alcohol; self-igniting in air. Also known as yellow phosphorus. { 'wīt 'fä·sfə·rəs }

white portland cement [MATER] Finely ground portland cement made from pure calcite limestone and white clay. { 'wīt 'pȯrt·lənd si'mənt }

white vitriol See zinc sulfate. { 'wīt 'vi·trē,ȯl }

whitewash [MATER] A simple mixture of hy-drated lime and water, used mostly for painting fences and outbuildings; common whitewash is not water-resistant and rubs off easily. { 'wīt ,wäsh }

whr See watt-hour.

Wiedemann-Franz law [SOLID STATE] The law that the ratio of the thermal conductivity of a metal to its electrical conductivity is a constant, independent of the metal, times the absolute temperature. Also known as Lorentz relation. { 'vēd·ə,män 'fränts ,lȯ }

Wigner-Seitz cell [CRYSTAL] A polyhedron about an atom in a face-centered cubic structure, made by drawing planes which perpendicularly bisect the lines to the nearest neighbors; in a body-centered cubic structure, bisecting planes of lines to nearest neighbors and next-nearest neighbors are used; such polyhedra fill space. { 'wig·nər 'zīts ,sel }

Wigner-Seitz method [SOLID STATE] A method of approximating the band structure of a solid: Wigner-Seitz cells surrounding atoms in the solid are approximated by spheres, and band solutions of the Schrödinger equation for one electron are estimated by using the assumption that an electronic wave function is the product of a plane wave function and a function whose gradient has a vanishing radial component at the sphere's surface. { 'wig·nər 'zīts ,meth·əd }

wildness [MET] A condition that exists when molten metal, during cooling, evolves so much gas that it becomes violently agitated, forcibly ejecting metal from its container. { 'wīl·nəs }

winding [ELEC] **1.** One or more turns of wire forming a continuous coil for a transformer, relay, rotating machine, or other electric device. **2.** A conductive path, usually of wire, that is inductively coupled to a magnetic storage core or cell. [MATER] A material that is wound or coiled around a cylindrical object such as a mandrel. { 'wīnd·iŋ }

window [ELECTR] A material having minimum absorption and minimum reflection of radiant energy, sealed into the vacuum envelope of a microwave or other electron tube to permit passage of the desired radiation through the envelope to the output device. [MATER] A globular defect in a thermoplastic sheet or film caused by incomplete plasticization; similar to a fisheye. { 'win·dō }

wiped joint [MET] A joint wherein filler metal is applied in liquid form, and the joint is wiped mechanically to distribute the metal. { 'wīpt 'jȯint }

wiping effect [MET] Activation of a metal surface by mechanically rubbing or wiping to enhance the formation of a conversion coating. { 'wīp·iŋ i,fekt }

wire [ELEC] A single bare or insulated metallic conductor having solid, stranded, or tinsel construction, designed to carry current in an electric circuit. Also known as electric wire. [MET] A thin, flexible, continuous length of metal, usually of circular cross section. { wīr }

wire bonding |ELECTR| **1.** A method of connecting integrated-circuit chips to their substrate, using ultrasonic energy to weld very fine wires mechanically from metallized terminal pads along the periphery of the chip to corresponding bonding pads on the substrate. **2.** The attachment of very fine aluminum or gold wire (by thermal compression or ultrasonic welding) from metallized terminal pads along the periphery of an integrated circuit chip to corresponding bonding pads on the surface of the package leads. { 'wīr ,bänd·iŋ }

wire-cut brick |MATER| A brick cut from clay shaped by extrusion before burning; the long bar of extruded clay is cut into bricks by a set of wires 9 inches (23 centimeters) apart. { 'wīr ¦kət 'brik }

wire drawing |MET| The reduction of the diameter of a metal rod or wire by pulling it through a die or a series of dies. { 'wī·ər ,drȯ·iŋ }

wire-fabric reinforcing |CIV ENG| Reinforcing concrete or mortar with a welded wire fabric. { 'wīr ¦fab·rik ,rē·ən'fȯrs·iŋ }

wire fusing current |ELEC| The electric current which will cause a wire to melt. { 'wīr ¦fyüz·iŋ ,kə·rənt }

wire glass |MATER| Sheet glass with woven wire mesh embedded in the center of the sheet; used in building construction for windows, doors, floors, and skylights. { 'wīr ,glas }

wire insulation |MATER| A flexible insulation used to cover an electric wire. { 'wīr ,in·sə'lā·shən }

wire rod |MET| A metal rod used in wiredrawing. { 'wīr ,räd }

wire-wound cryotron |CRYO| A cryotron that consists of a central insulated wire surrounded by a control coil; it is designed so that a relatively small current passed through the control coil produces a magnetic field which makes the gate resistive. { 'wīr ¦wau̇nd 'krī·ə,trän }

Wobbe index |THERMO| A measure of the amount of heat released by a gas burner with a constant orifice, equal to the gross calorific value of the gas in British thermal units per cubic foot at standard temperature and pressure divided by the square root of the specific gravity of the gas. { 'wä·bə ,in,deks }

wolfram *See* tungsten. { 'wu̇l·frəm }

wolframic acid *See* tungstic acid. { wu̇l'fram·ik 'as·əd }

wolfram white *See* barium tungstate. { 'wu̇l·frəm wīt }

wood |MATER| Lumber or timber obtained from trees. { wu̇d }

wood filler |MATER| A paste designed to fill the pores of open-grained woods such as ash, chestnut, mahogany, and oak. { 'wu̇d 'fil·ər }

wood flour |MATER| Dried wood ground to a very fine powder and used in plastic wood, in molding of certain plastics, as an extender in some glues, and in metal-casting operations. { 'wu̇d 'flau̇r }

wood physics |MATER| The area of wood science concerned with the physical and mechanical properties of wood and the factors which affect them. { 'wu̇d 'fiz·iks }

wood preservative |MATER| A material used to coat wood to kill insects and fungi, but not usually classed as an insecticide; coal tar creosote and its derivatives are the most widely used wood preservatives. { 'wu̇d pri,zər·vəd·iv }

wood pulp *See* pulp. { 'wu̇d ,pəlp }

Wood's glass |MATER| A type of glass that has a high transmission factor for ultraviolet radiation but is relatively opaque to visible radiation. { 'wu̇dz ,glas }

Wood's metal |MET| A fusible alloy of the Cerro Corporation that contains 50% bismuth, 25% lead, 12.5% tin, and 12.5% cadmium, and melts at 158°F (70–72°C); used for automatic sprinkler plugs. { 'wu̇dz ,med·əl }

woody structure |MET| A fibrous appearance in a fracture, particularly found in wrought iron and extruded aluminum alloys, usually associated with elongated inclusions or grains. { 'wu̇d·ē 'strək·chər }

wool fat *See* wool grease. { 'wu̇l ,fat }

wool grease |MATER| A highly complex mixture of wax ester, alcohols, and fatty acids coating the surface of sheep wool fibers and obtained by scouring the wool with soap or synthetic detergent; used in the manufacture of lanolin and its derivatives, for dressing leather, and in lubricating and slushing oils, soaps, and ointments. Formerly known as degras. Also known as wool fat; wool wax. { 'wu̇l ,grēs }

wool wax *See* wool grease. { 'wu̇l ,waks }

workability |MATER| The ease with which concrete can be placed. { ,wər·kə'bil·əd·ē }

work angle |MET| In arc welding, the angle in a plane normal to the weld axis between the electrode and one member of the joint. { 'wərk ,aŋ·gəl }

work function |SOLID STATE| The minimum energy needed to remove an electron from the Fermi level of a metal to infinity; usually expressed in electronvolts. |THERMO| *See* free energy. { 'wərk ,fəŋk·shən }

work hardening |MET| Increased hardness accompanying plastic deformation of a metal below the recrystallization temperature range. { 'wərk ¦härd·ən·iŋ }

working life *See* work life. { ,wərk·iŋ ,līf }

work lead |MET| The electrical conductor connecting the source of current to the work in arc welding. Also known as ground lead; welding ground. { 'wərk ,lēd }

work life |POLYM CHEM| The period of time a resin or an adhesive will remain usable after it is mixed with a catalyst and other ingredients. Also known as pot life; working life. { 'wərk ,līf }

work of adhesion *See* adhesional work. { 'wərk əv ad'hē·zhən }

worm |MET| Sweat of molten metal which exudes through the crust of solidifying metal in a casting, and is caused by gas evolution. { wərm }

wove paper |MATER| Paper characterized by a uniform, unlined surface and a soft, smooth finish. { 'wōv ,pā·pər }

wrinkling |MATER| **1.** Distortion occurring in a film of paint that gives the appearance of ripples. **2.** Development of a roughened surface on a film of sealant. Also known as crinkling; riveling. |MET| Waviness around the edges of a drawn metal product. { 'riŋk·liŋ }

wrought alloy |MET| An alloy that has been mechanically worked after casting. { 'rȯt 'al ‚ȯī }

wrought iron |MET| A commercial iron consisting of slag fibers, primarily iron silicate, embedded in a ferrite matrix. { 'rȯt 'ī·ərn }

Wurster process *See* air-suspension encapsulation. { 'wər·stər ‚prä·səs }

X

xanthan gum |ORG CHEM| A high-molecular-weight (5–10 million) water-soluble natural gum; a heteropolysaccharide made up of building blocks of D-glucose, D-mannose, and D-glucuronic acid residues; produced by pure culture fermentation of glucose with *Xanthomonas campestris*. { 'zan·thən ‚gəm }

x axis |CRYSTAL| A reference axis within a quartz crystal. { 'eks ‚ak·səs }

X cut |CRYSTAL| A quartz-crystal cut made in such a manner that the *x* axis is perpendicular to the faces of the resulting slab. { 'eks ‚kət }

Xe *See* xenon.

xenocryst |CRYSTAL| A crystal in igneous rock that resembles a phenocryst and is foreign to the enclosing body of rock. Also known as chadacryst. { 'zēn·ə‚krist }

xenon |CHEM| An element, symbol Xe, member of the noble gas family, group 0, atomic number 54, atomic weight 131.291; colorless, boiling point $-108°C$ (1 atmosphere, or 101,325 pascals), noncombustible, nontoxic, and nonreactive; used in photographic flash lamps, luminescent tubes, and lasers, and as an anesthetic. { 'zē‚nän }

xerogel |CHEM| **1.** A gel whose final form contains little or none of the dispersion medium used. **2.** An organic polymer capable of swelling in suitable solvents to yield particles possessing a three-dimensional network of polymer chains. { 'zer·ə‚jel }

x-ray crystallography |CRYSTAL| The study of crystal structure by x-ray diffraction techniques. Also known as roentgen diffractometry. { 'eks ‚rā ‚krist·əl'äg·rə·fē }

x-ray diffraction analysis |CRYSTAL| Analysis of the crystal structure of materials by passing x-rays through them and registering the diffraction (scattering) image of the rays. { 'eks ‚rā di'frak·shən ə‚nal·ə·səs }

x-ray lithography |ELECTR| Lithography in which the resist is exposed to a well-collimated, high-intensity x-ray beam projected through a special mask in close proximity to the silicon slice. { 'eks ‚rā li'thäg·rə·fē }

x-ray microscope |ENG| **1.** A device in which an ultra-fine-focus x-ray tube or electron gun produces an electron beam focused to an extremely small image on a transmission-type x-ray target that serves as a vacuum seal; the magnification is by projection; specimens being examined can thus be in air, as also can the photographic film that records the magnified image. **2.** Any of several instruments which utilize x-radiation for chemical analysis and for magnification of 100–1000 diameters; it is based on contact or projection microradiography, reflection x-ray microscopy, or x-ray image spectrography. { 'eks ‚rā 'mī·krə‚skōp }

x-ray powder diffractometer *See* powder diffraction camera. { 'eks ‚rā 'paůd·ər ‚di‚frak'täm·əd·ər }

x-ray powder method *See* powder method. { 'eks ‚ā 'paůd·ər ‚meth·əd }

x-ray thickness gage |ENG| A thickness gage used for measuring and indicating the thickness of moving cold-rolled sheet steel during the rolling process without making contact with the sheet; an x-ray beam directed through the sheet is absorbed in proportion to the thickness of the material and its atomic number. { 'eks ‚rā 'thik·nəs ‚gāj }

Y

Y *See* yttrium.

yacca gum *See* acaroid resin. { 'yak·ə ‚gəm }

yag laser *See* yttrium-aluminum-garnet laser. { 'yag ‚lā·zər }

y axis [CRYSTAL] A line perpendicular to two opposite parallel faces of a quartz crystal. { 'wī ‚ak·səs }

Yb *See* ytterbium.

Y block [MET] A Y-shaped test casting used to appraise low-shrinkage alloys. { 'wī ‚bläk }

Y cut [CRYSTAL] A quartz-crystal cut such that the y axis is perpendicular to the faces of the resulting slab. { 'wī ‚kət }

yellow [OPTICS] The hue evoked in an average observer by monochromatic radiation having a wavelength in the approximate range from 577 to 597 nanometers; however, the same sensation can be produced in a variety of other ways. { yel·ō }

yellow cedar *See* Alaska cedar. { 'yel·ō 'sēd·ər }

yellow cypress *See* Alaska cedar. { 'yel·ō 'sī·prəs }

yellow metal *See* Muntz metal. { yel·ō ‚med·əl }

yellow ocher [MATER] A form of limonite used as a yellow pigment. { ¦yel·ō 'ō·kər }

yellow phosphorus *See* white phosphorus. { 'yel·ō 'fä·sfə·rəs }

yellow precipitate *See* mercuric oxide. { 'yel·ō pri'sip·ə‚tāt }

yellow prussiate of potash *See* potassium ferrocyanide. { 'yel·ō 'prəs·ē‚āt əv 'päd‚ash }

yellow prussiate of soda *See* sodium ferrocyanide. { 'yel·ō 'prəs·ē‚āt əv 'sōd·ə }

yellow salt *See* uranyl nitrate. { 'yel·ō 'sòlt }

yellow scale [MATER] The commercial name for low-grade paraffin wax. { 'yel·ō 'skāl }

yield [MECH] That stress in a material at which plastic deformation occurs. { yēld }

yield point [MECH] The lowest stress at which strain increases without increase in stress. { 'yēld ‚pòint }

yield strength [MECH] The stress at which a material exhibits a specified deviation from proportionality of stress and strain. { 'yēld ‚streŋkth }

yield stress [FL MECH] The minimum stress needed to cause a Bingham plastic to flow. [MECH] The lowest stress at which extension of the tensile test piece increases without increase in load. { yēld ‚stres }

yield temperature [ENG] The temperature at which a fusible plug device melts and is dislodged by its holder and thus relieves pressure in a pressure vessel; it is caused by the melting of the fusible material, which is then forced from its holder. { 'yēld ‚tem·prə·chər }

yig *See* yttrium iron garnet. { yig }

Young-Helmholtz laws [MECH] Two laws describing the motion of bowed strings; the first states that no overtone with a node at the point of excitation can be present; the second states that when the string is bowed at a distance of $1/n$ times the string's length from one of the ends, where n is an integer, the string moves back and forth with two constant velocities, one of which has the same direction as that of the bow and is equal to it, while the other has the opposite direction and is $n - 1$ times as large. { ¦yəŋ 'helm‚hōlts ‚lóz }

Young's modulus [MECH] The ratio of a simple tension stress applied to a material to the resulting strain parallel to the tension. Also known as modulus of elasticity. { 'yəŋz ‚mäj·ə·ləs }

ytterbia *See* ytterbium oxide. { i'tər·bē·ə }

ytterbium [CHEM] A rare-earth metal of the yttrium subgroup, symbol Yb, atomic number 70, atomic weight 173.04; lustrous, malleable, soluble in dilute acids and liquid ammonia, reacts slowly with water; melts at 824°C, boils at 1427°C; used in chemical research, lasers, garnet doping, and x-ray tubes. { i'tər·bē·əm }

ytterbium oxide [INORG CHEM] Yb_2O_3 A colorless compound, melts at 2346°C, dissolves in hot dilute acids; used to prepare alloys, ceramics, and special glasses. Also known as ytterbia. { i'tər·bē·əm 'äk‚sīd }

yttria *See* yttrium oxide. { 'i·trē·ə }

yttrium [CHEM] A rare-earth metal, symbol Y, atomic number 39, atomic weight 88.9059; dark-gray, flammable (as powder), soluble in dilute acids and potassium hydroxide solution, and decomposes in water; melts at 1500°C, boils at 2927°C; used in alloys and nuclear technology and as a metal deoxidizer. { 'i·trē·əm }

yttrium-aluminum-garnet laser [OPTICS] A four-level infrared laser in which the active material is neodymium ions in an yttrium-aluminum-garnet crystal; it can provide a continuous output power of several watts. Abbreviated yag laser. { 'i·trē·əm ə¦lü·mə·nəm ¦gär·nət 'lā·zər }

yttrium chloride [INORG CHEM] $YCl_3 \cdot 6H_2O$ Reddish, transparent, water- and alcohol-soluble prisms; decomposes at 100°C; used as an analytical reagent. { 'i·trē·əm 'klȯr,īd }

yttrium iron garnet [MATER] $Y_3Fe_5O_{12}$ A synthetic ferrimagnetic material with the garnet crystal structure; used in microwave ferrite devices because of its very narrow ferromagnetic resonance absorption line. Abbreviated yig. { 'i·trē·əm 'ī·ərn 'gär ·nət }

yttrium oxide [INORG CHEM] Y_2O_3 A yellowish powder, insoluble in water, soluble in dilute acids; used as television tube phosphor and microwave filters. Also known as yttria. { 'i·trē·əm 'äk,sīd }

yttrium sulfate [INORG CHEM] $Y_2(SO_4)_3 \cdot 8H_2O$ Reddish crystals that are soluble in concentrated sulfuric acid, slightly soluble in water; decomposes at 700°C; used as an analytical reagent. { 'i·trē·əm 'səl,fāt }

Z

Zanzibar gum |MATER| A combustible, hard, fossil-type copal, which is insoluble in most solvents and melts at about 245°C; used in varnishes. { 'zan·zə,bär ,gəm }

z axis |CRYSTAL| The optical axis of a quartz crystal, perpendicular to both the x and y axes. { 'zē ,ak·səs }

zein |MATER| A combustible, white to yellowish protein powder derived from corn; insoluble in water, soluble in dilute alcohol; used in inks, fibers, microencapsulation, and coatings for paper and food. { 'zē·ən }

zeolite catalyst |INORG CHEM| Hydrated aluminum and calcium (or sodium) silicates (for example, $CaO·2Al_2O_3·5SiO_2$ or $Na_2O·2Al_2O_3·5SiO_2$) made with controlled porosity; used as a catalytic cracking catalyst in petroleum refineries, or loaded with catalyst for other chemical reactions. { 'zē·ə,līt 'kad·əl·əst }

zerogel |CHEM| A gel which has dried until apparently solid; sometimes it will swell or redisperse to form a sol when treated with a suitable solvent. { 'zir·ō,jel }

zeroth law of thermodynamics |THERMO| A law that if two systems are separately found to be in thermal equilibrium with a third system, the first two systems are in thermal equilibrium with each other, that is, all three systems are at the same temperature. { ¦zir,ōth ¦lȯ əv ,thər·mō·dī'nam·iks }

Ziegler catalyst |MATER| A special catalyst developed to produce stereospecific polymers, and derived from a transition-metal halide and a metal hydride or metal alkyl. { 'zē·glər ,kad·əl·əst }

zigzag nanotube |PHYS CHEM| A carbon nanotube formed from a graphite sheet that is rolled up so that it has a zigzag edge. { ,zig,zag 'nan·ō ,tüb }

zinc |CHEM| A metallic transition element, symbol Zn, atomic number 30, atomic weight 65.38; explosive as powder; soluble in acids and alkalies, insoluble in water; strongly electropositive; melts at 419°C; boils at 907°C. |MET| A shiny, bluish-white, lustrous metal that is ductile when pure; used in alloys, metal coatings, electrical fuses, anodes, and dry cells. { ziŋk }

zinc arsenite |INORG CHEM| $Zn(AsO_2)_2$ A toxic white powder that is insoluble in water, soluble in alkalies; used as an insecticide and timber preservative. Also known as zinc metaarsenite. { 'ziŋk 'ärs·ən,īt }

zincate |INORG CHEM| A reaction product of zinc with an alkali metal or with ammonia; for example, sodium zincate, Na_2ZnO_2. { 'ziŋ,kāt }

zinc baryta white *See* lithopone. { 'ziŋk bə¦rīd·ə 'wīt }

zinc borate |INORG CHEM| $3ZnO·2B_2O_3$ A white, amorphous powder that is soluble in dilute acids, slightly soluble in water; melts at 980°C; used in medicine, as a ceramics flux, as an inhibitor for mildew, and to fireproof textiles. { 'ziŋk 'bȯr,āt }

zinc bromide |INORG CHEM| $ZnBr_2$ Water- and alcohol-soluble, white crystals that melt at 294°C; used in medicine, manufacture of rayon, and photography, and in a radiation viewing screen. { 'ziŋk 'brō,mīd }

zinc carbonate |INORG CHEM| $ZnCO_3$ White crystals that are insoluble in water, soluble in alkalies and acids; used in ceramics and ointments, and as a fireproofing agent and feed additive. { 'ziŋk 'kär·bə,nāt }

zinc chloride |INORG CHEM| $ZnCl_2$ Water- and alcohol-soluble, white, fire-hazardous crystals that melt at 290°C, and are irritating to the skin; used as a catalyst and in electroplating, wood preservation, textile processing, petroleum refining, medicine, and feed additives. { 'ziŋk 'klȯr,īd }

zinc chromate |INORG CHEM| $ZnCrO_4$ A toxic, yellow powder that is insoluble in water, soluble in acids; used as a pigment in paints (artists', automotive, primer), varnishes, linoleum, and epoxy laminates. { 'ziŋk 'krō,māt }

zinc cyanide |INORG CHEM| $Zn(CN)_2$ A toxic, white powder that is insoluble in water and alcohol, soluble in alkalies and dilute acids; melts at 800°C; used as an analytical reagent and insecticide, and in medicine and metal plating. { 'ziŋk 'sī·ə,nīd }

zinc dust |MATER| Finely divided zinc metal, at least 97% pure, that is used as a pigment in paint primer for galvanized iron and other metal substrates. { 'ziŋk 'dəst }

zinc fluoride |INORG CHEM| ZnF_2 A toxic white powder that is slightly soluble in water and melts at 872°C; used in enamels, ceramic glazes, and galvanizing. { 'ziŋk 'flu̇r,īd }

zinc halide |INORG CHEM| A binary compound of zinc and a halogen; for example, $ZnBr_2$, $ZnCl_2$, ZnF_2, and ZnI_2. { 'ziŋk 'ha,līd }

zinc hydroxide

zinc hydroxide [INORG CHEM] $Zn(OH)_2$ Colorless, water-soluble crystals that decompose at $125°C$; used as a chemical intermediate and in rubber compounding and surgical dressings. { 'ziŋk hī'dräk,sīd }

zinc metaarsenite See zinc arsenite. { 'ziŋk ¦med·ə'ärs·ən,āt }

zinc orthoarsenate [INORG CHEM] $Zn_3(AsO_4)_2$ A toxic white powder that is insoluble in water, soluble in alkalies; used as an insecticide and wood preservative. { 'ziŋk ¦ör·thō'ärs·ən,āt }

zinc orthophosphate See zinc phosphate. { 'ziŋk ¦ör·thō'fä,sfāt }

zinc oxide [INORG CHEM] ZnO A bitter-tasting, white to gray powder that is insoluble in water, soluble in alkalies and acids; melts at $1978°C$; used as a pigment, mold-growth inhibitor, and dietary supplement, and in cosmetics, electronics, and color photography. { 'ziŋk 'äk,sīd }

zinc phosphate [INORG CHEM] $Zn_3(PO_4)_2$ A white powder that is insoluble in water, soluble in acids and ammonium hydroxide; melts at $900°C$; used in coatings for steel, aluminum, and other metals, and in dental cements and phosphors. Also known as tribasic zinc phosphate; zinc orthophosphate. { 'ziŋk 'fä,sfāt }

zinc phosphide [INORG CHEM] Zn_3P_2 A toxic, alcohol-insoluble, gray gritty powder that reacts violently with oxidizing agents; melts at over $420°C$, decomposes in water; used as a rat poison and in medicine. { 'ziŋk 'fä,sfīd }

zinc selenide [INORG CHEM] $ZnSe$ A water-insoluble, moderately toxic, yellow to reddish solid that is a fire hazard when in contact with water and acids; melts above $1100°C$; used as infrared optical windows. { 'ziŋk 'sel·ə,nīd }

zinc sulfate [INORG CHEM] $ZnSO_4·7H_2O$ Efflorescent, water-soluble, colorless crystals with an astringent taste; used to preserve skins and wood and as a paper bleach, analytical reagent, feed additive, and fungicide. Also known as white copperas; white vitriol; zinc vitriol. { 'ziŋk 'səl,fāt }

zinc sulfide [INORG CHEM] ZnS A yellowish powder that is insoluble in water, soluble in acids; exists in two crystalline forms (alpha, or wurtzite, and beta, or sphalerite); beta becomes alpha at $1020°C$, and sublimes at $1180°C$; used as a pigment for paints and linoleum, in opaque glass, rubber, and plastics, for hydrosulfite dyeing process, as x-ray and television screen phosphor, and as a fungicide. { 'ziŋk 'səl,fīd }

zinc sulfide white See lithopone. { 'ziŋk 'səl,fīd 'wīt }

zinc telluride [INORG CHEM] $ZnTe$ Moderately toxic, reddish crystals that melt at $1238°C$ and decompose in water. { 'ziŋk 'tel·yə,rīd }

zinc vitriol See zinc sulfate. { 'ziŋk 'vi·trē,ól }

zinc white See Chinese white. { 'ziŋk 'wīt }

zircaloy [MET] Any member of a group of alloys containing mainly zirconium that possess resistance to corrosion and stability over a wide range of temperatures and types of radiation. { 'zərk·ə ,lòi }

zirconia See zirconium oxide. { ,zər'kō·nē·ə }

zirconia brick [MATER] A type of brick containing zirconium oxide, used to line metallurgical furnaces. { ,zər'kō·nē·ə 'brik }

zirconic anhydride See zirconium oxide. { ,zər 'kän·ik an'hī,drīd }

zirconium [CHEM] A metallic transition element, symbol Zr, atomic number 40, atomic weight 91.22; occurs as crystals, flammable as powder; insoluble in water, soluble in hot, concentrated acids; melts at $1850°C$, boils at $4377°C$. [MET] A hard, lustrous, grayish metal that is strong and ductile; used in alloys, pyrotechnics, welding fluxes, and explosives. { ,zər'kō·nē·əm }

zirconium boride [INORG CHEM] ZrB_2 A hard, toxic, gray powder that melts at $3000°C$; used as an aerospace refractory, in cutting tools, and to protect thermocouple tubes. Also known as zirconium diboride. { ,zər'kō·nē·əm 'bòr,īd }

zirconium carbide [INORG CHEM] ZrC Hard, gray crystals that are soluble in water, soluble in acids; as powder, it ignites spontaneously in air; melts at $3400°C$, boils at $5100°C$; used as an abrasive, refractory, and metal cladding, and in cermets, incandescent filaments, and cutting tools. { ,zər'kō·nē·əm 'kär,bīd }

zirconium chloride See zirconium tetrachloride. { ,zər'kō·nē·əm 'klòr,īd }

zirconium diboride See zirconium boride. { ,zər 'kō·nē·əm dī'bòr,īd }

zirconium dioxide See zirconium oxide. { ,zər 'kō·nē·əm dī'äk,sīd }

zirconium halide [INORG CHEM] A compound of zirconium with a halogen; for example, $ZrBr_2$, $ZrCl_2$, $ZrCl_3$, $ZrCl_4$, $ZrBr_2$, $ZrBr_3$, ZrF_4, and ZrI_4. { ,zər'kō·nē·əm 'ha,līd }

zirconium hydride [INORG CHEM] ZrH_2 A flammable, gray-black powder; used in powder metallurgy and nuclear moderators, and as a reducing agent, vacuum-tube getter, and metal-foaming agent. { ,zər'kō·nē·əm 'hī,drīd }

zirconium hydroxide [INORG CHEM] $Zr(OH)_4$ A toxic, amorphous white powder; insoluble in water, soluble in dilute mineral acids; decomposes at $550°C$; used in pigments, glass, and dyes, and to make zirconium compounds. { ,zər'kō·nē·əm hī'dräk,sīd }

zirconium nitride [INORG CHEM] ZrN A hard, brassy powder that is soluble in concentrated acids; melts at $2930°C$; used in refractories, cermets, and laboratory crucibles. { ,zər'kō·nē·əm 'nī,trīd }

zirconium orthophosphate See zirconium phosphate. { ,zər'kō·nē·əm ¦ór·thō'fä,sfāt }

zirconium oxide [INORG CHEM] ZrO_2 A toxic, heavy white powder that is insoluble in water, soluble in mineral acids; melts at $2700°C$; used in ceramic glazes, special glasses, and medicine, and to make piezoelectric crystals. Also known as zirconia; zirconic anhydride; zirconium dioxide. { ,zər'kō·nē·əm 'äk,sīd }

zirconium oxychloride [INORG CHEM] $ZrOCl_2·8H_2O$ White crystals that are soluble in water, insoluble in organic solvents, and acidic in aqueous solution; used for textile dyeing and oilfield acidizing, in cosmetics and greases, and for antiperspirants and water repellents. Also known as zirconyl chloride. { ˌzər'kō·nē·əm ¦äk·sē'klȯr ˌīd }

zirconium phosphate |INORG CHEM| $ZrO(H_2PO_4)_2 \cdot 3H_2O$ A toxic, dense white powder that is insoluble in water, soluble in acids and organic solvents; decomposes on heating; used as an analytical reagent, coagulant, and radioactive-phosphor carrier. Also known as zirconium orthophosphate. { ˌzər'kō·nē·əm 'fä ˌsfāt }

zirconium tetrachloride |INORG CHEM| $ZrCl_4$ Toxic, alcohol-soluble, white lustrous crystals; sublimes above 300°C and decomposes in water; used to make pure zirconium and for water-repellent textiles and as an analytical reagent. Also known as zirconium chloride. { ˌzər'kō·nē·əm ¦te·trə'klȯr,īd }

zircon sand |MATER| A refractory sand consisting principally of zirconium silicate and characterized by low thermal expansion and high thermal conductivity. { ¦zər,kän 'sand }

zirconyl chloride See zirconium oxychloride. { 'zər·kən·əl 'klȯr,īd }

Zn See zinc.

zone |CRYSTAL| A set of crystal faces which intersect (or would intersect, if extended) along edges which are all parallel. { zōn }

zone axis |CRYSTAL| A line through the center of a crystal which is parallel to all the faces of a zone. { 'zōn 'ak·səs }

zone indices |CRYSTAL| Three integers identifying a zone of a crystal; they are the crystallographic coordinates of a point joined to the origin by a line parallel to the zone axis. { 'zōn 'in·də ˌsēz }

zone law |CRYSTAL| A law which states that the Miller indices (h, k, l) of any crystal plane lying in a zone with zone indices (u, v, w) satisfy the equation $hu + lv + kw = 0$. { 'zōn ˌlȯ }

zone purification See zone refining. { 'zōn ˌpyúr·ə·fə'kā·shən }

zone refining |MET| A technique to purify materials in which a narrow molten zone is moved slowly along the complete length of the specimen to bring about impurity segregation, and which depends on differences in composition of the liquid and solid in equilibrium. Also known as zone purification. { 'zōn ri'fīn·iŋ }

zoning |CRYSTAL| A variation in the composition of a crystal from core to margin due to a separation of the crystal phases during its growth by loss of equilibrium in a continuous reaction series. { 'zōn·iŋ }

Zr See zirconium.

zwitterion See dipolar ion. { 'tsfid·ər,ī,än }

Appendix

Equivalents of commonly used units for the U.S. Customary System and the metric system

1 inch = 2.5 centimeters (25 millimeters)	1 centimeter = 0.4 inch	1 inch = 0.083 foot
1 foot = 0.3 meter (30 centimeters)	1 meter = 3.3 feet	1 foot = 0.33 yard (12 inches)
1 yard = 0.9 meter	1 meter = 1.1 yards	1 yard = 3 feet (36 inches)
1 mile = 1.6 kilometers	1 kilometer = 0.62 mile	1 mile = 5280 feet (1760 yards)

1 acre = 0.4 hectare	1 hectare = 2.47 acres	
1 acre = 4047 square meters	1 square meter = 0.00025 acre	

1 gallon = 3.8 liters	1 liter = 1.06 quarts = 0.26 gallon	1 quart = 0.25 gallon (32 ounces; 2 pints)
1 fluid ounce = 29.6 milliliters	1 milliliter = 0.034 fluid ounce	1 pint = 0.125 gallon (16 ounces)
32 fluid ounces = 946.4 milliliters		1 gallon = 4 quarts (8 pints)

1 quart = 0.95 liter	1 gram = 0.035 ounce	1 ounce = 0.0625 pound
1 ounce = 28.35 grams	1 kilogram = 2.2 pounds	1 pound = 16 ounces
1 pound = 0.45 kilogram	1 kilogram = 1.1×10^{-3} ton	1 ton = 2000 pounds
1 ton = 907.18 kilograms		

$°F = (1.8 \times °C) + 32$ $°C = (°F - 32) \div 1.8$

Conversion factors for the U.S. Customary System, metric system, and International System

A. Units of length

Units	cm	m	in.	ft	yd	mi
1 cm	= 1	0.01	0.3937008	0.03280840	0.01093613	6.213712×10^{-6}
1 m	= 100.	1	39.37008	3.280840	1.093613	6.213712×10^{-4}
1 in.	= 2.54	0.0254	1	0.08333333...	0.02777777...	1.578283×10^{-5}
1 ft	= 30.48	0.3048	12.	1	0.3333333...	$1.893939... \times 10^{-4}$
1 yd	= 91.44	0.9144	36.	3.	1	$5.681818... \times 10^{-4}$
1 mi	= 1.609344×10^{5}	1.609344×10^{3}	6.336×10^{4}	5280.	1760.	1

B. Units of area

Units	cm^{2}	m^{2}	$in.^{2}$	ft^{2}	yd^{2}	mi^{2}
1 cm^{2}	= 1	10^{-4}	0.1550003	1.076391×10^{-3}	1.195990×10^{-4}	3.861022×10^{-11}
1 m^{2}	= 10^{4}	1	1550.003	10.76391	1.195990	3.861022×10^{-7}
1 $in.^{2}$	= 6.4516	6.4516×10^{-4}	1	$6.944444... \times 10^{-3}$	7.716049×10^{-4}	2.490977×10^{-10}
1 ft^{2}	= 929.0304	0.09290304	144.	1	0.1111111...	3.587007×10^{-8}
1 yd^{2}	= 8361.273	0.8361273	1296.	9.	1	3.228306×10^{-7}
1 mi^{2}	= 2.589988×10^{10}	2.589988×10^{6}	4.014490×10^{9}	2.78784×10^{7}	3.0976×10^{6}	1

C. Units of volume

Units	m^3	cm^3	liter	$in.^3$	ft^3	qt	gal
1 m³	= 1	10^6	10^3	6.102374×10^4	35.31467×10^{-3}	1.056688	264.1721
1 cm³	= 10^{-6}	1	10^{-3}	0.06102374	3.531467×10^{-5}	1.056688×10^{-3}	2.641721×10^{-4}
1 liter	= 10^{-3}	1000.	1	61.02374	0.03531467	1.056688	0.2641721
1 in.³	= 1.638706×10^{-5}	16.38706	0.01638706	1	5.787037×10^{-4}	0.01731602	4.329004×10^{-3}
1 ft³	= 2.831685×10^{-2}	28316.85	28.31685	1728.	1	2.992208	7.480520
1 qt	= 9.46352×10^{-4}	946.359	0.946351	57.75	0.03342014	1	0.25
1 gal (U.S.)	= 3.785412×10^{-3}	3785.412	3.785412	231.	0.1336806	4.	1

D. Units of mass

Units	g	kg	oz	lb	metric ton	ton
1 g	= 1	10^{-3}	0.03527396	2.204623×10^{-3}	10^{-6}	1.102311×10^{-6}
1 kg	= 1000.	1	35.27396	2.204623	10^{-3}	1.102311×10^{-3}
1 oz (avdp)	= 28.34952	0.02834952	1	0.0625	2.834952×10^{-5}	3.125×10^{-5}
1 lb (avdp)	= 453.5924	0.4535924	16.	1	4.535924×10^{-4}	$5. \times 10^{-4}$
1 metric ton	= 10^8	1000.	35273.96	2204.623	1	1.102311
1 ton	= 907184.7	907.1847	32000.	2000.	0.9071847	1

Conversion factors for the U.S. Customary System, metric system, and International System (*cont.*)

E. Units of density

Units	$g \cdot cm^{-3}$	$g \cdot L^{-1}$, $kg \cdot m^{-3}$	$oz \cdot in.^{-3}$	$lb \cdot in.^{-3}$	$lb \cdot ft^{-3}$	$lb \cdot gal^{-1}$
1 $g \cdot cm^{-3}$	= 1	1000.	0.5780365	0.03612728	62.42795	8.345403
1 $g \cdot L^{-1}$, $kg \cdot m^{-3}$	= 10^{-3}	1	5.780365×10^{-4}	3.612728×10^{-5}	0.06242795	8.345403×10^{-3}
1 $oz \cdot in.^{-3}$	= 1.729994	1729.994	1	0.0625	108.	14.4375
1 $lb \cdot in.^{-3}$	= 27.67991	27679.91	16.	1	1728.	231.
1 $lb \cdot ft^{-3}$	= 0.01601847	16.01847	9.259259×10^{-3}	5.787037×10^{-4}	1	0.1336806
1 $lb \cdot gal^{-1}$	= 0.1198264	119.8264	4.749536×10^{-3}	4.329004×10^{-3}	7.480519	1

F. Units of pressure

Units	Pa, $N \cdot m^{-2}$	$dyn \cdot cm^{-2}$	bar	atm	$kgf \cdot cm^{-2}$	mmHg (torr)	in. Hg	$lbf \cdot in.^{-2}$
1 Pa, 1 $N \cdot m^{-2}$	= 1	10	10^{-5}	9.869233×10^{-6}	1.019716×10^{-5}	7.500617×10^{-3}	2.952999×10^{-4}	1.450377×10^{-4}
1 $dyn \cdot cm^{-2}$	= 0.1	1	10^{-6}	9.869233×10^{-7}	1.019716×10^{-6}	7.500617×10^{-4}	2.952999×10^{-5}	1.450377×10^{-5}
1 bar	= 10^5	10^6	1	0.9869233	1.019716	750.0617	29.52999	14.50377
1 atm	= 101325	101325.0	1.01325	1	1.033227	760.	29.92126	14.69595
1 $kgf \cdot cm^{-2}$	= 98066.5	980665	0.980665	0.9678411	1	735.5592	28.95903	14.22334
1 mmHg (torr)	= 133.3224	1333.224	1.333224×10^3	1.315789×10^{-3}	1.359510×10^{-3}	1	0.03937008	0.01933678
1 in. Hg	= 3386.388	33863.88	0.03386388	0.03342105	0.03453155	25.4	1	0.4911541
1 $lbf \cdot in.^{-2}$	= 6894.757	68947.57	0.06894757	0.06804596	0.07030696	51.71493	2.036021	1

G. Units of energy

Units	g mass (energy equiv.)	J	eV	cal	cal_{IT}	Btu_{IT}	kWh	hp-h	ft-lbf	$ft^3 \cdot lbf \cdot in.^{-2}$	liter-atm
1 g mass (energy equiv.)	1	8.987552×10^{13}	5.609589×10^{32}	2.148076×10^{13}	2.146640×10^{13}	8.518555×10^{10}	2.496542×10^{7}	3.347918×10^{7}	6.628878×10^{13}	4.603388×10^{11}	8.870024×10^{11}
1 J	1.112650×10^{-14}	1	6.241510×10^{18}	0.2390057	0.2388459	9.478172×10^{-4}	$2.777777\ldots \times 10^{-7}$	3.725062×10^{-7}	0.7375622	5.121960×10^{-3}	9.869233×10^{-3}
1 eV	1.782662×10^{-33}	1.602176×10^{-19}	1	3.829293×10^{-20}	3.826733×10^{-20}	1.518570×10^{-22}	4.450490×10^{-26}	5.96206×10^{-26}	1.181705×10^{-19}	8.206283×10^{-22}	1.581225×10^{-21}
1 cal	4.655328×10^{-14}	4.184	2.611448×10^{19}	1	0.9993312	3.965667×10^{-3}	$1.1622222\ldots \times 10^{-6}$	1.558562×10^{-6}	3.085960	2.143028×10^{-2}	0.04129287
1 cal_{IT}	4.658443×10^{-14}	4.1868	2.613195×10^{19}	1.000669	1	3.968321×10^{-3}	1.163×10^{-6}	1.559609×10^{-6}	3.088025	2.144462×10^{-2}	0.04132050
1 Btu_{IT}	1.173908×10^{-11}	1055.056	6.585141×10^{21}	252.1644	251.9958	1	2.930711×10^{-4}	3.930148×10^{-4}	778.1693	5.403953	10.41259
1 kWh	4.005540×10^{-8}	3600000.	2.246944×10^{25}	860420.7	859845.2	3412.142	1	1.341022	2655224.	18349.06	35529.24
1 hp-h	2.986931×10^{-8}	2384519.	1.675545×10^{25}	641615.6	641186.5	2544.33	0.7456998	1	1980000.	13750.	26494.15
1 ft-lbf	1.508551×10^{-14}	1.355818	8.462351×10^{18}	0.3240483	0.3238315	1.285067×10^{-3}	3.766161×10^{-7}	$5.050505\ldots \times 10^{-7}$	1	$6.944444\ldots \times 10^{-3}$	0.01338088
1 $ft^3\,lbf \cdot in.^{-2}$	2.172313×10^{-12}	195.2378	1.218579×10^{21}	46.66295.	46.63174	0.1850497	5.423272×10^{-5}	$7.272727\ldots \times 10^{-5}$	144.	1	1.926847
1 liter-atm	1.127393×10^{-12}	101.325	6.324210×10^{20}	24.21726	24.20106	0.09603757	2.814583×10^{-5}	3.774419×10^{-5}	74.73349	0.5189825	1

Periodic table

(The atomic numbers are listed above the
symbols identifying the elements. The heavy
line separates metals from nonmetals.)

1s	1 H Hydrogen	2 He Helium

s	1	2
	3 Li Lithium	4 Be Beryllium
	11 Na Sodium	12 Mg Magnesium
	19 K Potassium	20 Ca Calcium
	37 Rb Rubidium	38 Sr Strontium
	55 Cs Cesium	56 Ba Barium
	87 Fr Francium	88 Ra Radium

d	3	4	5	6	7	8	9	10	11	12
	21 Sc Scandium	22 Ti Titanium	23 V Vanadium	24 Cr Chromium	25 Mn Manganese	26 Fe Iron	27 Co Cobalt	28 Ni Nickel	29 Cu Copper	30 Zn Zinc
	39 Y Yttrium	40 Zr Zirconium	41 Nb Niobium	42 Mo Molybdenum	43 Tc Technetium	44 Ru Ruthenium	45 Rh Rhodium	46 Pd Palladium	47 Ag Silver	48 Cd Cadmium
	71 Lu Lutetium	72 Hf Hafnium	73 Ta Tantalum	74 W Tungsten	75 Re Rhenium	76 Os Osmium	77 Ir Iridium	78 Pt Platinum	79 Au Gold	80 Hg Mercury
	103 Lr Lawrencium	104 Rf Rutherfor- dium	105 Db Dubnium	106 Sg Seaborgium	107 Bh Bohrium	108 Hs Hassium	109 Mt Meitnerium	110 Ds Darmstadtium	111	112

p	13	14	15	16	17	18
	5 B Boron	6 C Carbon	7 N Nitrogen	8 O Oxygen	9 F Fluorine	10 Ne Neon
	13 Al Aluminum	14 Si Silicon	15 P Phosphorus	16 S Sulfur	17 Cl Chlorine	18 Ar Argon
	31 Ga Gallium	32 Ge Germanium	33 As Arsenic	34 Se Selenium	35 Br Bromine	36 Kr Krypton
	49 In Indium	50 Sn Tin	51 Sb Antimony	52 Te Tellurium	53 I Iodine	54 Xe Xenon
	81 Tl Thallium	82 Pb Lead	83 Bi Bismuth	84 Po Polonium	85 At Astatine	86 Rn Radon
	113	114	115	116	117	118

f														
	57 La Lanthanum	58 Ce Cerium	59 Pr Praseodymium	60 Nd Neodymium	61 Pm Promethium	62 Sm Samarium	63 Eu Europium	64 Gd Gadolinium	65 Tb Terbium	66 Dy Dysprosium	67 Ho Holmium	68 Er Erbium	69 Tm Thulium	70 Yb Ytterbium
	89 Ac Actinium	90 Th Thorium	91 Pa Protactinium	92 U Uranium	93 Np Neptunium	94 Pu Plutonium	95 Am Americium	96 Cm Curium	97 Bk Berkelium	98 Cf Californium	99 Es Einsteinium	100 Fm Fermium	101 Md Mendelevium	102 No Nobelium

Physical properties of crystalline optical materials

Material	Symbol	Refractive index (wavelength = 500 nm)	Density, g/cm³	Melting or softening temperature, K
Germanium	Ge	4	5.33	1210
Lithium fluoride	LiF	1.394	3.5	1140
Magnesium fluoride	MgF₂	1.39	3.18	1528
Sodium chloride	NaCl	1.53	2.17	1070
Zinc sulfide	ZnS	2.42	4.08	2100
Zinc selenide	ZnSe	2.43	5.42	1790
Barium titanate	BaTiO₃	—	5.9	1870
Cesium iodide	CsI	1.75	4.51	894
Diamond	C	2.4	3.51	3770
Lanthanum fluoride	LaF₃	1.6	5.94	1766
Magnesia	MgO	1.74	3.585	3053
Potassium chloride	KCl	1.49	1.98	1050
Sapphire	Al₂O₃	1.77	3.98	2300
Crystal quartz	SiO₂	1.55	2.65	1740
Barium fluoride	BaF₂	1.47	4.89	1550
Calcium fluoride	CaF₂	1.43	3.18	1630
Cadmium telluride	CdTe	2.69	6.2	1320
Calcite	CaCO₃	n_o = 1.665, n_e = 1.490	2.710	1612
Cuprous chloride	CuCl	2.0	4.14	695
Gallium phosphide	GaP	3.65	4.13	1623
Indium arsenide	InAs	4.5	—	1215
Lead fluoride	PbF₂	1.78	8.24	1100
Lead sulfide	PbS	4.3 at 3 μm	7.5	1387
Silicon carbide	SiC	2.68	3.217	3000
Selenium	Se	2.83	4.82	490
Silicon	Si	3.45	2.329	1690

Appendix

The 14 Bravais lattices, derived by centering of the seven crystal classes (P and R) defined by symmetry operators

Bravais lattice cells	Examples
 Cubic P Cubic I Cubic F	Copper (Cu), silver (Ag), sodium chloride (NaCl)
 Tetragonal P Tetragonal I	While tin (Sn), rutile (TiO_2), β-spodumene ($LiAlSi_2O_6$)
 P C I F Orthorhombic	Gallium (Ga), perovskite ($CaTiO_3$)
 Monoclinic P Monoclinc C	Gypsum ($CaSO_4 \cdot 2H_2O$)
 Triclinic P	Potassium chromate (K_2CrO_7)
 Trigonal R (rhombohedral)	Calcite ($CaCO_3$), arsenic (As), bismuth (Bi)
 Trigonal and hexagonal C (or P)	Zinc (Zn), cadmium (Cd), quartz (SiO_2) |P|

Industrially important synthetic fibers

Fiber	Definition	Uses
Acrylic	85% or more acrylonitrile units	Carpeting, skirts, socks, slacks, sweaters, blankets, draperies
Modacrylic	35–85% acrylonitrile units	Simulated fur, scatter rugs, stuffed toys, paint rollers, carpets, hairpieces
Polyester	85% or greater ester units	Permanent press wear, skirts, shirts, slacks, underwear, blouses, rope, fish nets, tire cord, sails, thread
Nylon	Polyamide	Carpeting, upholstery, tents, blouses, sails, suits, stretch fabrics, tire cord, curtains, rope, nets, parachutes
Polyurethane	85% or more urethane units	Girdles, bras, slacks, bathing suits, pillows
Rayon	Regenerated cellulose with substituents replacing not more than 15% of the hydrogen's of the hydroxyl groups	Dresses, suits, slacks, tire cord, ties, curtains, blankets, blouses
(Rayon) acetate	More than 92% OH substituted	Dresses, shirts, slacks, cigarette filters, upholstery, draperies
Triacetate	Reaction of cellulose with acetic acid or acetic anhydride	Skirts, dresses, sportswear
Fibrous glass		Composites, insulation
Olefins	Polypropylene, polyethylene	Synthetic grass, rugs

Synthetic elastomers

Elastomer	Use
Acrylonitrile-butadiene-styrene, ABS	Oil hoses, fuel tanks, pipe, appliance and automotive housings
Butadiene rubber, BR	Tire tread, hose, belts
Butyl rubber, IIR	Inner tubes, cable sheathing, roofing, seals, tank liners, coated fabrics
Chloroprene rubber, polychloroprene, CR	Tire and cable insulation, hoses, footwear, mechanical automotive products
Epichlorohydrine (epoxy copolymers)	Seals, gaskets, wire and cable insulation
Ethylene-propylene rubbers, EP, EPDM	Cable insulation, window strips
Fluoroelastomers	Wire and cable insulation, aerospace applications
Ionomers, mostly copolymers of ethylene and acid-containing monomers reacted with metal ions	Golf ball covers, shoe soles, weather stripping
Natural rubber, polyisoprene, NR	Tires, bushings, couplings, seals, footwear, belting
Nitrile rubber (random copolymer of butadiene and acrylonitrile), NBR	Seals, automotive parts that are in contact with oils and gas, footwear, hoses
Polysulfide	Adhesives, sealants, hose binders
Polyurethanes	Sealing and joints, printing rollers, tires, footwear, wire and cable covering
Silicons, mostly polydimethylsiloxanes	Medical-application body parts, outer space applications, flexible molds, gaskets, seals
Styrene-butadiene rubber, SBR	Tire tread, footwear, wire and cable covering

Important plastics

Plastic	Use
Epoxies	Coatings, laminates, composites, molding, flooring
Urea-formaldehyde resins	Molding compounds, dinnerware
Phenol-formaldehyde resins	Molding compounds
Melamine-formaldehyde resins	Dinnerware, table tops
Polytetrafluoroethylene	Electrical components, nonsticking surfaces, gaskets
Polypropylene	Automotive parts, toys, housewares, appliance parts
Polystyrene	Containers, recreational equipment, housewares, appliance parts
Poly(vinyl chloride) and copolymers	Pipes, fittings, sheets, flooring materials, automotive parts
Polycarbonates	Tail light lens, bullet-resistant vests, appliance housings, signs, bottles
Polysulfones	Mechanical parts, small appliances, electrical connectors
Poly(phenylene sulfide)	Electrical and mechanical parts
Polyesters	Apparel, home furnishings
Polyethylenes	Containers, bottles, housewares, pipe and fittings
Styrene-acrylonitriles	Appliance housings, housewares
Poly(methyl methacrylate)	Signs, glazing, lighting, fixtures, automotive lenses, solar panels
Poly(phenylene oxide)	Business machine housings, electrical parts, automotive parts

Compositions and melting temperatures of eutectic fusible alloys

T_E, °F (°C)	Composition, percentage by weight				
	Bi	Pb	Sn	Cd	Other
116.2 (46.8)	44.70	22.60	8.30	5.30	19.10 In
136 (58)	49.00	18.00	12.00	—	21.00 In
158 (70)	50.00	26.70	13.30	10.00	—
196.7 (91.5)	51.60	40.20	—	8.20	—
203 (95)	52.50	32.00	15.50	—	—
216.5 (102.5)	54.00	—	26.00	20.00	—
255 (124)	55.50	44.50	—	—	—
281.3 (138.5)	58.00	—	42.00	—	—
288 (142)	—	30.60	51.20	18.20	—
291 (144)	60.00	—	—	40.00	—
351 (177)	—	—	67.75	32.25	—
361 (183)	—	38.14	61.86	—	—
390 (199)	—	—	91.00	—	9.00 Zn
430.3 (221.3)	—	—	96.50	—	3.50 Ag
457 (236)	—	79.7	—	17.7	2.60 Sb
477 (247)	—	87.0	—	—	13.00 Sb

Appendix

Phase diagram of the eutectic silver-copper (Ag-Cu) system

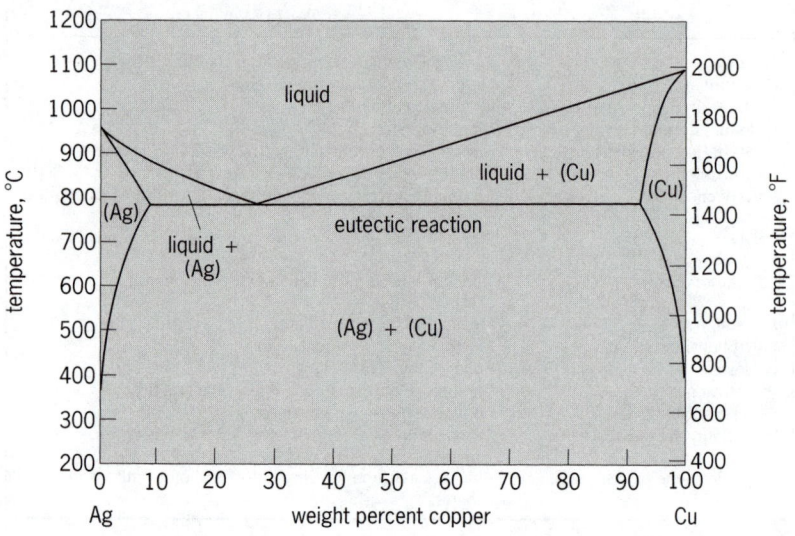

Characteristics of smart materials

Class	Primary transduction characteristic	Example	Primary uses in smart structures	Performance characteristics
Piezoelectric materials	Electrical field to mechanical strain or mechanical strain to electrical charge	Lead zirconate titanate (a ceramic)	As sensors, actuators, or both, primarily used for vibration suppression applications	Reasonably linear sensor/actuator capable of operating at high frequencies but low strain levels; bidirectional, but cannot operate statically
Electrostrictive materials	Electrical field to mechanical strain	Lead magnesium niobate (a ceramic)	As actuators for vibration suppression and static precision alignment	More strain capability than piezoelectric materials; can operate statically but is highly nonlinear, is unidirectional, and requires a mechanical preload
Magnetostrictive materials	Magnetic field to strain	Terbium-iron alloy (Terfenol)	Similar to electrostrictive materials	Similar to electrostrictive materials, requires a magnetic bias
Shape memory alloy	Temperature to mechanical strain and stiffness	Alloys of nickel and titanium (nitinol)	As actuators that provide large static shape change	Large strain capability limited to static or very low frequency
Electrorheological fluids	Electric field to fluid viscosity	Colloidal suspension of electrically charged particles	As adaptive shock absorbers for actively controlled viscous damping	Significant change in viscosity is provided, but usually requires relatively high-voltage electric fields
Magnetorheological fluids	Magnetic field to fluid viscosity	Colloidal suspension of magnetic particles	Similar to electrorheological fluids	High voltages are not required, but heavier systems are needed to generate magnetic fields

Dimensional range (diameters) of microstructural features in composite and conventional materials (1 cm = 0.39 in.)

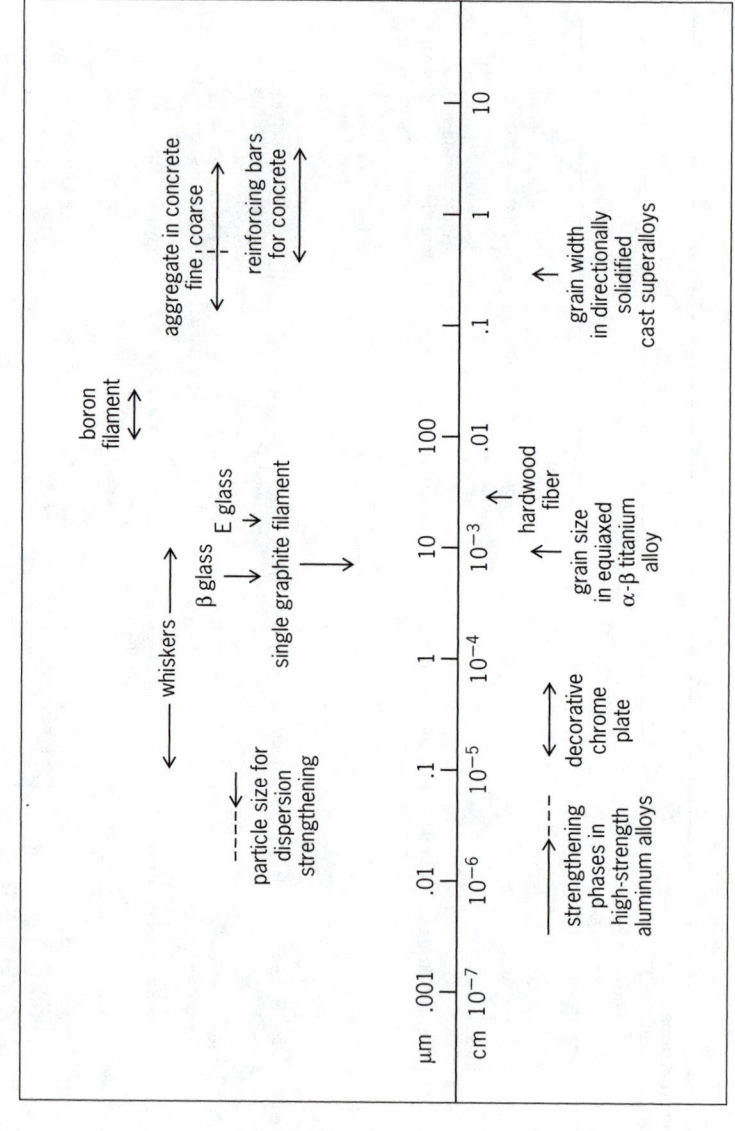

Typical creep curves for materials

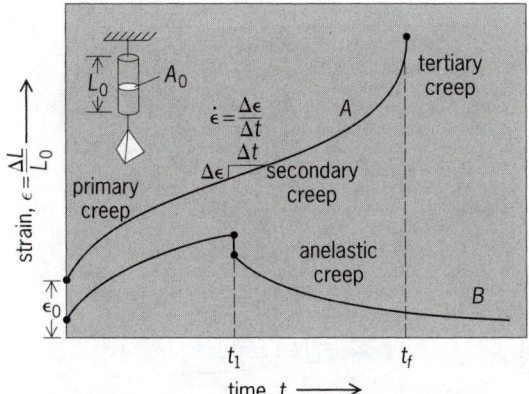

L_0, initial length of a body; ΔL, its increase in length
time, t
loading strain, ϵ_0
secondary creep rate, $\dot{\epsilon}$
time to failure, t_f

Temperature coefficients of linear thermal expansion for typical substances at room temperature

Substance	Coefficient of linear expansion per °F (°C) × 10^6
Aluminum, commercial	13 (24)
Copper	9 (17)
Diamond	0.6 (1)
Glass, commercial	6 (11)
Glass, Pyrex	2 (3)
Granite	4.6 (8.3)
Ice	28 (50)
Iron	7 (12)
Invar alloy	0.5 (0.9)
Quartz, crystalline	3 (5)
Quartz, fused	0.3 (0.5)
Oak, along fiber	3 (5)
Oak, across fiber	30 (54)
Rubber, hard	44 (80)

Appendix

Thermal conductivities of some typical gases, liquids, and solids [°C = (°F − 32)/1.8]

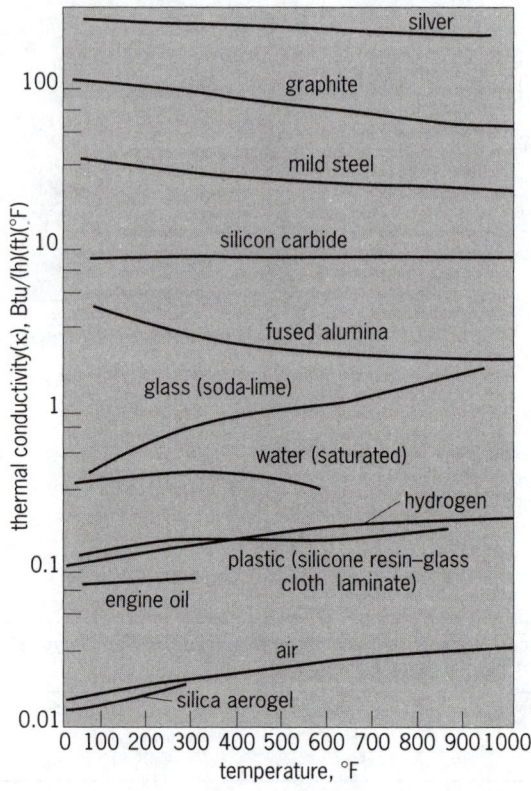

Resistivities of some materials at room temperature

Material	ohm · m
Poly(tetrafluoroethylene)	10^{18}
Quartz	10^{16}
Diamond	10^{12}
Glass	10^{11}
Silicon, pure	10^{3}
Germanium, pure	5
Metglass 2204 ($Ti_{50}Be_{40}Zr_{10}$)	3
Bismuth	1.2×10^{-6}
Antimony	4.2×10^{-7}
Nichrome	10^{-6}
Iron	10×10^{-8}
Potassium	7×10^{-8}
Copper	1.7×10^{-8}

Coefficients of viscosity of selected gases and liquids

Fluid	Temperature, °C (°F)	μ, centipoise
Air (gas)	−104 (−155)	0.01130
	0 (32)	0.01708
	74 (165)	0.02102
Hydrogen (gas)	0 (32)	0.00835
Ethane (gas)	0 (32)	0.00848
Water (liquid)	5 (41)	1.519
	20 (68)	1.002
	95 (203)	0.2975
Mercury (liquid)	0 (32)	1.685
Ethyl alcohol (liquid)	0 (32)	1.773
Olive oil (liquid)	10 (50)	138.0
Glycerin	0 (32)	12,110

Viscosity conversions

Unit	poise	cp	Pa · s	$lb_m/(ft \cdot s)$	$lb_f \cdot s/ft^2$
1 poise* =	1	100	0.1	6.72×10^{-2}	2.089×10^{-3}
1 centipoise =	0.01	1	0.001	6.72×10^{-4}	2.089×10^{-5}
1 pascal-second† =	10	1000	1	0.672	2.089×10^{-2}
1 $lb_m/(ft \cdot s)$ =	14.88	1488	1.488	1	3.108×10^{-2}
1 $lb_f \cdot s/ft^2$ =	478.8	4.788×10^4	47.88	32.17	1

*1 poise = 1 dyne · s/cm² = 1 g/(cm · s).
†1 Pa · s = 1 kg/(m · s).